THE UNDERWATER HANDBOOK

A Guide to Physiology and
Performance for the Engineer

THE UNDERWATER HANDBOOK

A Guide to Physiology and Performance for the Engineer

Edited by

Charles W. Shilling
Margaret F. Werts
and
Nancy R. Schandelmeier

*Science Communication Division
Department of Medical and Public Affairs
The Medical Center
The George Washington University
Washington, D.C.*

PLENUM PRESS · NEW YORK AND LONDON

Library of Congress Cataloging in Publication Data

Main entry under title:

The Underwater handbook.

Includes bibliographies and index.
1. Underwater physiology. 2. Submarine medicine. 3. Diving, Submarine. 4.
Oceanography. I. Shilling, Charles Wesley, 1901-   II. Werts, Margaret F.
III. Schandelmeier, Nancy R.
RC1015.U5   616.9'8022   76-7433
ISBN 0-306-30843-6

The preparation of this handbook was supported jointly by the Research and Devel-
opment Command of the Bureau of Medicine and Surgery, and the Office of Naval
Research, United States Department of the Navy, under ONR Contract N00014-67-
A-0214-0013 with the Science Communication Division, Department of Medical and
Public Affairs, Medical Center, The George Washington University, Washington, D.C.

Contributors and Reviewers

Because of the extensive and almost continual reorganization of the various sections of this handbook, it is exceedingly difficult to match authors and reviewers with the final version of each chapter. For this reason, all individuals who contributed or reviewed material are listed alphabetically as follows:

Neil R. Anderson
Arthur J. Bachrach
Albert R. Behnke
Peter B. Bennett
Mildred Benton
Thomas E. Berghage
Robert C. Bornmann
Hugh M. Bowen
Richard G. Buckles
Ermine Christian
Joseph C. Farmer
Robert W. Hamilton
John Hardy
J. Donald Harris
Suk Ki Hong
Marlene Jorris

Jo Ann S. Kinney
Theodore D. Langley
Leonard M. Libber
Joseph B. MacInnis
Gerald S. Malecki
Dennis C. Pauli
Lawrence W. Raymond
Nancy R. Schandelmeier
Thomas C. Schmidt
Charles W. Shilling
Janice Smith
Gilbert C. Tolhurst
Willard S. Vaughan, Jr.
Paul Webb
Margaret F. Werts
J. Murray Young

Advisory Committee

Robert C. Bornmann
E. Fisher Coil
Suzanne Kronheim

Leonard M. Libber
Gerald S. Malecki
Martin A. Tolcott

Foreword

This handbook attempts to translate data on various parameters of man's capability in underwater and hyperbaric environments for those without a background in the life sciences.

Accomplishing any multifaceted task requires team work, and effective team work depends on facile communication among all participants. To communicate properly, all parties must understand each other's problems and be able to speak a similar language. To this end we believe that this publication will go a long way in furthering the understanding and communication necessary for maximum achievement.

The U. S. Navy has a fundamental interest in all types of activities connected with the ocean and is especially interested in the growing field of manned underwater and hyperbaric activities. Thus, the manuscript for this comprehensive book was developed under Office of Naval Research contract N00014-67-A-0214-0013 with The George Washington University. We acknowledge with appreciation the financial support and technical guidance for this undertaking by the Naval Medical Research and Development Command of the Bureau of Medicine and Surgery as well as by the Engineering Psychology Program and the Physiology Program of the Office of Naval Research.

JOSEPH P. POLLARD
Director
Biological and Medical
Sciences Division
Office of Naval Research

Preface

A need was felt for a book that would document the relationship of the human being to the underwater hyperbaric environment in such a way that the individual unfamiliar with the psychological or biomedical jargon could still understand and appreciate the information.

This book is not meant to be a medical text. It was designed to meet the technical and scientific needs of the engineer, the manager of underwater activity, and the interested layman.

This book was prepared under support by the Department of the Navy, Contract No. N00014-67-A-0214-0013, issued by the Office of Naval Research. However, the content does not necessarily reflect the position or the policy of the Department of the Navy or the Government, and no official endorsement should be inferred.

The United States Government has a royalty-free nonexclusive and irrevocable license throughout the world for Government purposes to publish, translate, reproduce, deliver, perform, dispose of, and to authorize others so to do, all or any portion of this work.

C. W. SHILLING

Contents

III
Man in the Ocean Environment: Physical Factors

IV
Man in the Ocean Environment: Physiological Factors

V
Man in the Ocean Environment: Psychophysiological Factors

VI
Man in the Ocean Environment: Performance

VII
Decompression Sickness

VIII
Operational Safety Considerations

X
Underwater Communications

XI
Selection and Training of Divers

The Human Machine

A. *Introduction*

Throughout this handbook man will be considered in his relationship to the surrounding environment. He may be under water wearing a scuba outfit or living in an
underwater habitat, but, because of his respiratory needs, the environment will always
involve air or other respirable gas either at atmospheric pressure or under increased
pressure. Man will also be considered in relationship to the various pieces of equipment he uses, works with, or lives in. In many situations he will be part of a closed
loop, but always with remarkable feedback potential. The cybernetic man–machine
relationship will be evident in most of the underwater work man engages in.

In order to understand management problems involving an underwater environment and in order to develop underwater equipment and habitats, it is important to
understand man—his ability, his adaptability, and his strengths and limitations. To
do this we will consider man as a human machine. Hempleman (1972) says: "Viewing
men as machines it is quickly realized that they represent a very versatile and adaptable
proposition, and, in fact, are a well used machine for all underwater work to depths
not exceeding those of the Continental Shelf, i.e., about 300 meters. Like all machines,
they have their working conditions and limitations . . .." It is the human machine and
its relationship to the underwater environment that will be considered in this book.

Man differs from the ordinary, inanimate machine in that he is alive. In engineering terms it is possible to say that "life" is a system attribute and not an attribute of
molecular components. Man is an "animate" system since the system has the "capacity
to sustain itself in a state in which observable processes occur without causing the
system itself to follow a path toward the most general mechanical, chemical, thermal,
and electrical equilibrium for all the processes observed" (Brown *et al.* 1971). To
understand more fully what is meant by living, note the following essential characteristics of living matter.

Organization. Organization is an essential characteristic of living things. For
example, although various chemical substances, such as carbon, nitrogen, oxygen,
and various minerals, are known to exist in nature abundantly, it is their organization
in the protoplasm of the cell that makes the difference between living and nonliving

1

matter. In fact the single cell is the smallest system that unambiguously has all the attributes that define a system as "living."

Growth. Another indication of life is growth. This occurs, not as inorganic objects grow, by addition to the outside (accretion), but by growth from within (cell replication). Each cell reproduces its own type through information stored in nucleic acid polymers in each cell. Thus the human machine is continuously self-constructed. This is possible through the ingestion of food, the breathing of oxygen, and the resulting chemical transformation of the food into chemical building blocks.

Respiration. Respiration in some form is also characteristic of life. In the animal, oxygen is breathed in, and is used to oxidize the food substances; energy is thus released for movement or for forming new protoplasm, or else it is stored as reserve.

Waste elimination. Plants and animals get rid of waste materials, especially if they are harmful. Animals eliminate waste through the lungs, the skin, the kidneys, and the intestines.

Irritability. The ability to respond to stimuli is important for the continuance of life. The amoeba and the human being both respond to the prick of a pin! It takes little imagination to realize how important this characteristic is.

Adaptation. Ability to adapt to environment is another characteristic of life that is essential to the preservation of the species.

Movement. Movement is demonstrated alike by the flowing of the one-cell amoeba and the fingers of the pianist. Of course, an automobile exhibits movement, but it cannot sense what is going on around it and respond to these stimuli. Its movement is not purposeful and never develops into behavior.

Reproduction. The ability to reproduce in kind is the greatest distinguishing feature of living organisms, and without it life would cease to exist. "Living systems are . . . self-reproducing hereditary systems whose characteristics may change in time through adaptation of the phenotype to the environment, and through the play of chance, the environment, and mating habits upon the genotype and the gene pool" (Brown *et al.* 1971).

In the ways noted above the human machine shows its biological nature and in these ways it differs from an inanimate, man-made machine.

It should be noted that biological systems obey the known laws of thermodynamics, and the laws of physics and chemistry in general.

B. *The Mechanics of Movement*

All operational machines have moving parts. Some machines are designed to move from place to place, while others are stationary but do their work by executing the various movements for which they are designed. The human machine, however, is the most remarkable of all because it is self-motivated and self-directed and can develop an enormous number of skills as a consequence of the wide range of movement possible for various parts of the body. We now consider how this movement is possible.

1. *Bones*

The bones form the framework which supports and gives shape to the body; they afford attachment for the tendons, muscles, and ligaments which move the body; and they protect certain vital organs (e.g., the brain). Since the bones are linked to each other and act together, body movement is made possible. The bones are hard structures made up of calcium, phosphorus, other mineral salts, and some organic substances. Figure I-1 illustrates the construction. Even though bone is hard, it is also elastic, allowing a certain amount of "give," and, since it is living tissue, it will repair a fracture. There are 206 bones in the human skeleton. The femur or long leg bone shown in Figure I-1 is the bone most often involved in dysbaric osteonecrosis (known formerly as aseptic bone necrosis), which occurs in caisson workers and divers. The blood supply to the bones is not extensive and the cause of the condition is considered to be mechanical action of gas bubbles blocking the blood flow, thus causing death of the bone tissue.

2. *Joints*

Obviously no motion would be possible for the human machine if it were not for the joints. Wherever two bones are attached together, a joint is formed. There are three types of joints in the body, as shown in Figure I-2.

The immovable joints are mostly in the skull, the slightly movable joints are between the vertebrae of the spinal column, and the freely movable joints are in such locations as the knee, shoulder, hip, and elbow. The freely movable joints may be further classified as hinge joints, socket joints, gliding joints, and pivot joints.

Joint movements allow for such actions as flexion, extension, abduction, adduction, rotation, pronation, supination, eversion, and inversion. As can be seen in Figure I-2, the joints are closed in a sort of watertight sac containing a small amount of lubricating fluid, which enables the joint to work with very little friction. The ligaments join the two bones together and keep them from getting out of place.

3. *Muscles*

There are more than 500 muscles, which ensure life and movement for the human machine, and they account for one-half of the body weight. The voluntary or skeletal muscles are controlled through the central nervous system. They are located chiefly in the face, neck, limbs, and other parts of the trunk, and since they are attached to the skeleton, provide for movement of the body. The involuntary muscles, over which there is no voluntary control, are the heart muscles and the muscles of the intestines, the stomach, the blood vessels, and the other organs. These are constantly in motion as long as life continues.

Thus the work of the muscle is: to provide involuntary movement, as in peristalsis in the intestines; to maintain posture through muscle tone, as in the muscles of the head, neck, and shoulders; to perform work as called upon; and to produce heat

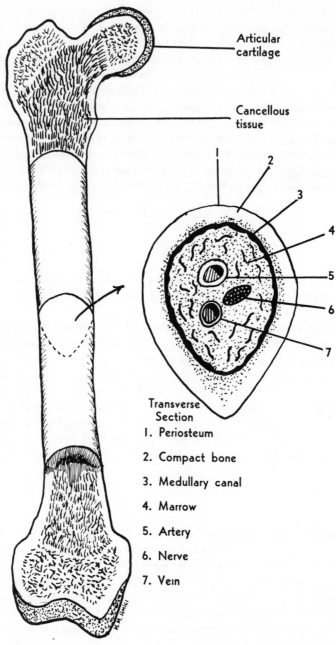

Articular
cartilage

Cancellous
tissue

Transverse
Section

1. Periosteum

2. Compact bone

3. Medullary canal

4. Marrow

5. Artery

6. Nerve

7. Vein

Figure I-1. Structure of a typical long bone (femur) (Shilling 1965).

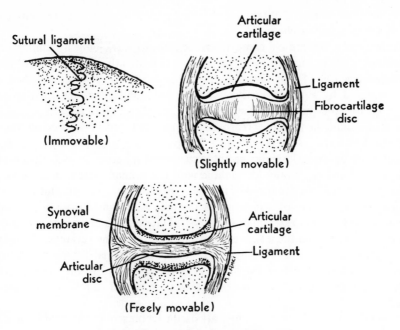

Figure I-2. Typical joints (Shilling 1965).

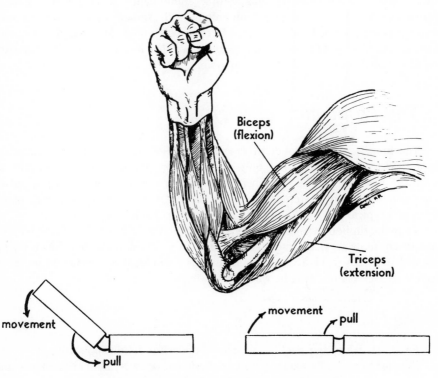

Figure I-3. Mechanics of muscle action (Shilling 1965).

through chemical changes that take place during muscle activity, thus becoming a part of the heat-producing mechanism.

Mechanically, the muscle can exert force and do work only by pulling or contraction, and not by pushing or expansion. Consequently, muscles must work in opposing groups, one of which reverses the action of the other. When attached to two bones connected by a joint, muscles work in antagonistic groups so that one set produces flexion and the other produces extension. (See Figure I-3.)

Let us examine the human forearm in engineering terms under a condition where it is subjected to a static load (Figure I-4). The problem is simplified by assuming only the relationship between the static weight load w and the vertical muscle (biceps) force. The physical model is visually identifiable with the human system it represents and describes the behavior of the system with the same variables. The graphical model is a way of describing the plot of two variables and may be obtained from direct measurements of either the human system or the physical model. The mathematical model is a way of describing the mathematical expressions that relate the variables of the system under investigation. The block-diagram model is a pictorial representation of the pertinent mathematical relationships that govern the system. In this simple system it describes in diagrammatic form how an excitation signal is related to a resultant response.

Also in this figure electrical analogs are used to represent the physiological system. In the upper analog a voltage e represents the applied weight force w, and the current flow i represents the resultant muscle-force component f. The electrical circuit is arranged so that the current varies with voltage in the same manner as vertical muscle force varies with applied weight. A plot of e versus i would thus have the same shape as the graphical model of the physiological system. The lower two circuits may also be considered analogs of the physiological system.

C. *The Production of Energy*

The human machine takes in raw materials in the form of food, water, and oxygen (air), and produces not only materials for building and repairing its own structure, power to run its machinery, and fuel or energy to do its work, but also stores as a reserve any materials not immediately utilized. The fuel or energy requirements of a typical 154-lb (70-kg) man are shown in Tables I-1 and I-2.

To maintain body weight without loss or gain, the individual in Table I-1 would have to consume food in amounts and kind to yield 3450 calories. Since we have assumed this man to be a normal individual, he would, if he consumed more, gain weight; if he consumed less, he would lose weight.

1. *The Fuel*

Food is the fuel and may be considered under the following headings.

Proteins are usually eaten as animal flesh and glands; animal products (milk, cheese, eggs); and vegetables (peas, beans, nuts).

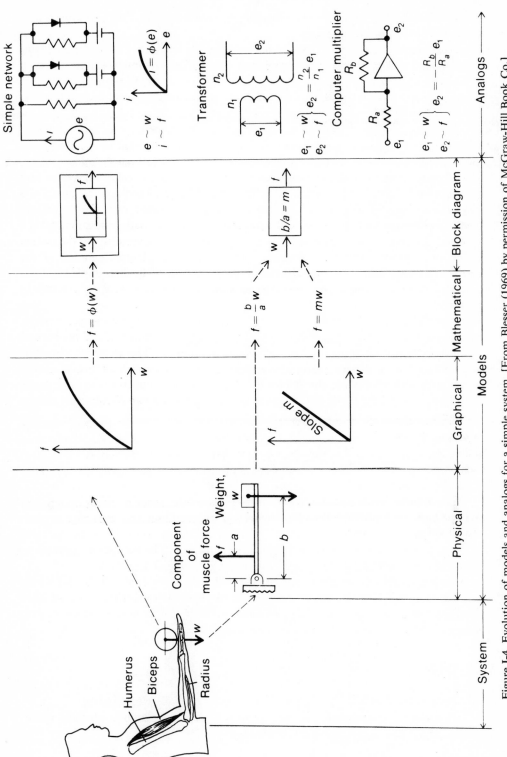

Figure I-4. Evolution of models and analogs for a simple system. [From Blesser (1969) by permission of McGraw-Hill Book Co.]

Table I-1

Relationship of Activities and Energy Expenditures
(Shilling 1965)

Form of activity	Expenditure, cal/hr
Sleeping	65
Awake, lying still	77
Standing at attention	115
Typewriting rapidly	140
Sweeping bare floors (38 strokes/min)	169
Walking slowly (2.6 mph)	200
Carpentry, metal working, industrial painting	240
Walking moderately fast	300
Sawing wood	480
Swimming	500
Running (5.3 mph)	570
Walking upstairs	1100

The probable energy requirements for one day of a 154-lb individual doing physical work may be estimated as follows:

	cal
8 hr of sleep (65 cal/hr)	520
3 hr of light exercise, such as going to and from work (170 cal/hr)	510
8 hr of shipside painting (240 cal/hr)	1920
5 hr of sitting at rest (100 cal/hr)	500
Total for the day	3450

Carbohydrates are the quick-energy foods usually eaten in the form of starches or sugars. Starches are usually eaten as grains, breakfast foods, and breads; as tubers and roots, such as potatoes, turnips, carrots, beets, and so forth; and as vegetables, such as pumpkin and winter squash. Sugars are eaten as sugar, syrup, honey, jelly, jam, candy, and fruits.

Fats are eaten as animal fat in the form of meat, fish, lard, butter, cream, cheese, and eggs; and as vegetable fat, such as olive oil, corn, cottonseed oil, coconut oil, nuts, and chocolate.

Table I-2

Energy Expenditures in Swimming for 70-kg
Man[a]

Swimming activity	Energy expended, kcal/hr
Breast stroke 1 mph	410
Crawl stroke 1 mph	420
Breast stroke 1.6 mph	490
Crawl stroke 1.6 mph	700
Breast stroke 2.2 mph	1600
Crawl stroke 2.2 mph	1850

[a] From data of Morehouse and Miller, quoted in Beckman (1964).

Table I-3

Body Utilization of Food

(Shilling 1965)

Purpose	Food	Specific uses
Energy	Carbohydrates	Used mainly by muscles; need the presence of insulin from the pancreas for their use
	Fats	Can be burned completely only when some carbohydrate is present
	Protein	Can be burned for energy if amounts eaten are above needs of the body cells
	Vitamins, minerals	Regulate cell metabolism
Structure	Protein	For structure of every cell in body
	Water	Makes up $\frac{2}{3}$ of the body weight
	Minerals	Used in structure of bones, teeth, blood, and gland secretions
	Fats	Used mainly in structure of fatty tissue
	Vitamins	Regulate growth
Storage	Fats	A stable form of stored food
	Carbohydrates	As glycogen in the liver and muscles, otherwise changed to fat for storage
	Protein, minerals and vitamins	Stored in small amounts, especially in the liver
Regulation	Vitamins, minerals, water, cellulose	Regulate body processes and ensure normal growth and health

Minerals are extremely important for the body's growth and maintenance; and vitamins are also essential to life. Table I-3 lists the classes of foods from a chemical standpoint, by the purpose for which they are used, and by specific uses in the body. (See also section on Diet.)

2. *The Chemical Laboratory*

Body cells cannot use food as it occurs in nature or even after it is cooked, but by the process of digestion it is broken down both mechanically and chemically, and thus changed into useable form—for example, from carbohydrates into glucose; fats into fatty acids; and proteins into amino acids. The various parts of the digestive system are shown in Figure I-5.

D. *The Transportation System*

Even after the food is broken down into a fluid form usable by the body cells it must be transported to every cell. This is achieved through two systems: the blood and blood vascular system, and the lymph and lymphatic system.

The blood vascular system is a hydraulic system consisting of the heart (the pump) and blood vessel system (the pipes). The lymph and lymph vascular system is another

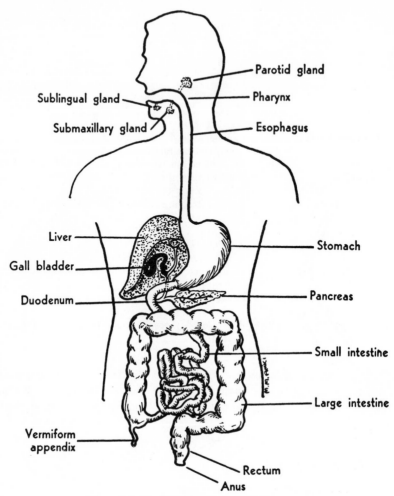

Figure I-5. The digestive system (Shilling 1965).

separate circulatory system in the body. It is so extensive that all body cells are bathed in its clear lymph, which is actually blood plasma filtered through the walls of the capillaries. By these systems oxygen and food are brought to the cells and carbon dioxide and waste taken away; thus body action is made possible as the "food" is metabolized.

Because the circulatory system is so important in considering tissue saturation with various gas mixtures, and particularly in relation to the production of compressed-air illness, we will consider it in more detail.

1. The Heart

The heart (Figure I-6) is a hollow muscular organ which is located in the front and center of the chest between the lungs. It is about the size of the individual's closed

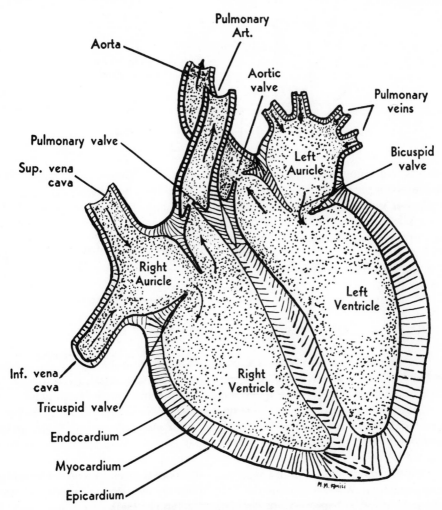

Figure I-6. Diagram of the heart (Shilling 1965).

fist and looks somewhat like a strawberry in shape, with the point downward and toward the left. The special heart muscle is so constructed that the heart contracts with a wringing type of motion that literally squeezes the blood into the blood vessel system. The heart operates without any conscious direction; it beats about 100,800 times in 24 h, while pumping some ten thousand quarts of blood through its valves and chambers. The heart has its own system of blood vessels, called coronary arteries, and although it weighs only one two-hundredths of the total body weight, it requires one-twentieth of the blood for its own circulation.

Almost every anatomical part of the body and its function can be considered in mechanical engineering, mathematical, and electrical terms. Let us, as an example, consider the mitral valve located between the left auricle (or atrium) and the left ventricle. This valve is essentially a mechanical check valve (see Figure I-7), which permits flow when the atrial pressure P_a exceeds ventricular pressure P_v and which

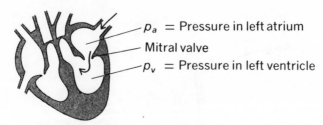

Figure I-7. Diagrammatic sketch of the heart. [From Blesser
(1969) by permission of McGraw-Hill Book Co.]

prevents return flow when the ventricular pressure exceeds the atrial pressure. "This
unidirectional property of the mitral valve is found in many other devices, the most
familiar of which are the mechanical check valve and the electrical diode. A diagrammatic representation of the equivalency of these devices is shown in Figure I-8;
the mitral valve and the check valve permit flow in the direction shown when $P_a > P_v$
but flow ceases when $P_a < P_v$. In a similar manner, the diode permits current flow
when $e_a > e_v$; flow is reduced practically to zero when the voltage is reversed"
(Blesser 1969).

2. Arteries, Capillaries, and Veins

The arteries are elastic tubes that carry the blood from the heart to all parts of
the body. They somewhat resemble a tree with the largest vessel, the aorta, as the
trunk, and with the arteries as large branches and the arterioles as small twigs.
The capillaries are the final, minute subdivisions of the arteries where the exchange
of materials between the cells and the blood takes place. They form a very dense,
interlocking network in all parts of the body. It is thus obvious that the gas being
breathed is carried to all parts of the body.

Another important point in tissue saturation with a given gas concerns capillary
accommodation. When a part of the body is not in use, only a few of the many
capillaries in the bed are open to circulation of blood. But just as soon as activity
begins for that particular portion of the body, the other capillaries in the bed open to

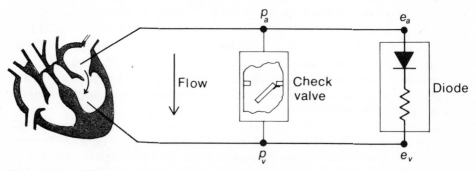

Figure I-8. Mechanical and electrical analog of mitral valve. [From Blesser (1969) by permission of
McGraw-Hill Book Co.]

circulation; and when the tissue is very active, all of them will open, allowing a marked increase in blood supply to the active area. It is thus obvious why the "bends" often occurs in the arm that has been doing the hard work under pressure, and that heavy physical exertion under pressure leads to more general saturation of the tissues and the increased likelihood of developing decompression sickness. It is also obvious why exercise is recommended during decompression to increase the speed of elimination of nitrogen or other inert gas stored in the tissues.

The veins are hollow elastic tubes that pick up the blood from the capillaries and carry it back to the heart. They are similar to the arteries, although their walls are thinner, with less muscular tissue.

3. *The Blood*

The blood is a fluid tissue that circulates through the blood vessels of the body. It consists of the fluid plasma containing freely flowing cells, called red cells, white cells, and platelets. The blood serves first as a fluid conveyer belt to carry all the food, hormones, oxygen, and other essential nutrients of life to each of the body cells. As it is being returned from the cells to the heart, the blood serves as a type of sewage system for carrying away body waste, which eventually goes to the organs of excretion—the kidneys, intestines, lungs, and skin.

Red cells. The red blood cells, circular disks or saucers, are formed in the bone marrow. They are about 1/3200 in. in diameter. The adult male has about five million red blood cells per cubic millimeter of blood. Each of these red blood cells carries a chemical called hemoglobin that gives the cells the chemical ability to pick up oxygen, carry it to the body cells, give it up, and then pick up carbon dioxide and bring it back to the lungs for elimination. The red blood cells have a life span of about one month and are constantly being replaced.

White cells. The white blood cells or leucocytes are of several types and in a healthy adult number only about 6000–8000/mm^3 of blood. They are disease fighters with the ability to move through minute openings and to attack and engulf solid particles, bacteria, and other foreign material. Their life span is less than that of the red cells.

Platelets. Blood platelets are round bodies, smaller than the red blood cells, and they number about 300,000–800,000/mm^3 of blood. They are essential for emergency repair, for they are involved in clotting of the blood and thus in stopping bleeding or hemorrhage associated with wounds. There is additional interest in platelets in hyperbaric medicine because of their possible relationship to sludging of the blood in decompression sickness.

4. *Blood Circulation*

To better understand tissue saturation during compression and the elimination of inert gas during decompression, it is well to trace the circulation of blood throughout the body. To follow the flow, note Figure I-9 and also Figure I-6. We will start with the impure venous blood, which is returning from the body tissues where it has

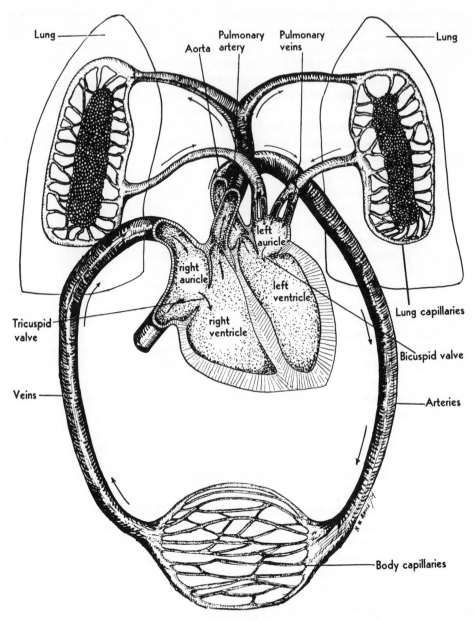

Figure I-9. Diagram of body blood circulation (Shilling 1965).

come in equilibrium with the tissue gas tension and where it has picked up CO_2 and other impurities. The capillaries empty into the small veins and thus eventually through the superior or inferior vena cava into the right auricle, and through the tricuspid valve into the right ventricle. As the right ventricle contracts, the tricuspid valve snaps shut and the blood is forced through the pulmonary valve into the pulmonary artery. The pulmonary artery itself divides, one branch going to each lung, where again it divides into smaller arteries, arterioles, and capillaries. Thus the blood flowing through this system finally reaches the tiny air sacs of the lung where it is in contact, through a permeable membrane, with the gas being breathed by the individual. Here the blood gives up its load of carbon dioxide and takes up oxygen to be carried back to the tissues. But also the fluid element (plasma) comes into equilibrium with the gas phase in the air sacs of the lung. If it is air under pressure, the nitrogen and additional oxygen are taken up by the plasma; if it is helium, then the helium and oxygen are taken up in accordance with the gas laws and the pressure of the environmental gas mixture being breathed. The blood then flows into the pulmonary veins, which empty into the left auricle of the heart. The circulation from the right side of the heart to the left side of the heart is called the pulmonary circulation, and takes only 10 sec.

From the left auricle, which acts as a receiving chamber, the blood passes through the bicuspid valve into the left ventricle. As the left ventricle contracts, this valve snaps shut and the blood is forced through the aortic valve out into the body's largest blood vessel, the aorta, and from there it flows into the smaller subdivision, the arterial branches, to smaller arteries, to arterioles, to capillaries, and thus to every cell of the body.

Although it is well to think of this circulatory process in engineering terms, it is most difficult to model. As pointed out by Brown *et al.* (1971): "Unlike in most engineering fluid systems, mathematical modeling problems in biological fluid systems are extremely complex. The complexity in these systems arises from four factors: (a) distensibility of the fluid vessels, (b) non-Newtonian nature of biological fluids, the blood in particular, (c) numerous branchings of the vessels, and (d) pulsating nature of pressures. Various neural controls which the central nervous system (CNS) exerts to regulate flow and distribution make the task of mathematical modeling even more difficult."

E. *Temperature Control*

Man, as a warm-blooded, thermostable animal, is able to maintain his body temperature within 1°F variation under normal environmental conditions. The normal body temperature for the healthy adult is shown in Figure I-10.

In order to logically present the entire mechanism of temperature control, the material of this section will be presented under the following headings: heat production, heat distribution, heat loss, the thermostat, and the insulating walls. Even though they are an essential part of temperature control, the air ducts (or the respiratory system) are so important to underwater activity that the essential anatomy and physiology are treated separately in the following section of this chapter under the heading, gas exchange.

Heat production is accomplished by metabolism (burning the fuel) in the cells of the body and, under normal conditions, is all that is needed. Man has developed clothing to help maintain body heat, and modern man has developed elaborate methods of both heating and cooling the environment so as to provide a comfortable temperature for the body.

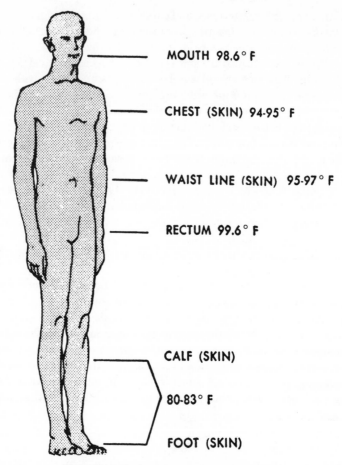

MOUTH 98.6° F

CHEST (SKIN) 94-95° F

WAIST LINE (SKIN) 95-97° F

RECTUM 99.6° F

CALF (SKIN)

80-83° F

FOOT (SKIN)

Figure I-10. Body temperature (Shilling 1965).

The oxygen-consumption rate of nude subjects in 10°C (50°F) water may be as much as nine times the resting rate. Obviously the body cannot keep up with the cooling effect of immersion in cold water, so the duration of exposure must be short or special heated suits must be worn.

Heat distribution is accomplished by the blood circulatory system. The many capillaries in the skin and superficial layers of tissue with their vasomotor dilating-contracting mechanism are an important part of the air-conditioning system.

Heat loss comes about through respiration (one-third of the loss in a cold environment), the evaporation of sweat, and by radiation and conduction from the skin. Heat loss can be partially blocked by heated suits and by warming the inspired air.

The *thermostat* is a nerve center in the brain which sets in motion the blood-cooling mechanism, or conversely the warming by muscle action. Since this thermostat operates on the basis of changes in blood chemistry, it is called a "respiratory chemostat" by Brown *et al.* (1971). They have diagrammed this function cleverly, as shown in Figure I-11. The controller (chemostat) functions to maintain the blood

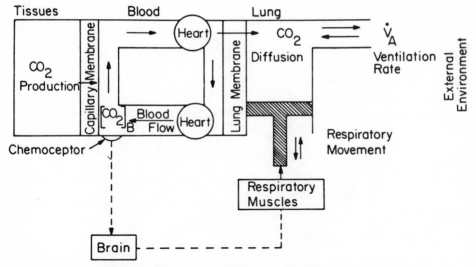

Figure I-11. Schematic diagram of the respiratory chemostat (Brown *et al.* 1971).

concentrations of the gases, oxygen and carbon dioxide, and of the hydrogen ion within limits compatible with life. Variations from normal in the concentrations of these chemical species in the blood caused by metabolic changes, as in exercise or by breathing abnormal gas mixtures, are detected by the chemoreceptor. The chemoreceptor responds to changes in the blood chemistry by modifying the rate of transmitted nerve impulses to the brain, which in turn controls the muscles involved in changing the rate and depth of breathing. The ventilation rate is the final control signal. The *insulating wall* of the human machine is the skin, which is a protective covering for the body with the following functions: holding the underlying muscles and tissues in place; preventing undue loss of moisture from these tissues; serving as defense against the entrance of infection; constituting a large-area organ of sensation; making the body more beautiful; and acting as an insulating wall in the human machine's temperature-control mechanism. There are about 20 ft^2 of body skin, all of which is well supplied with blood capillaries. Thus, we have a very large surface area which can dissipate a great amount of heat when exposed to cooler air or water.

F. Gas Exchange

The primary role of the respiratory system is to supply oxygen for the combustion of foods (metabolism) and to eliminate carbon dioxide. In order to understand the mechanics of the process of respiration, it is necessary first to review very briefly the respiratory organs—the nose, the mouth, the pharynx, the larynx, the trachea, the bronchi, and the lung tissue itself. The accessory organs that make breathing possible are the thorax (or box in which the lungs are held), the ribs, and the diaphragm.

Air enters the body either through the nose or the mouth, passes through the larynx (which is the "voice box"), into the trachea, and then into the lungs. This may

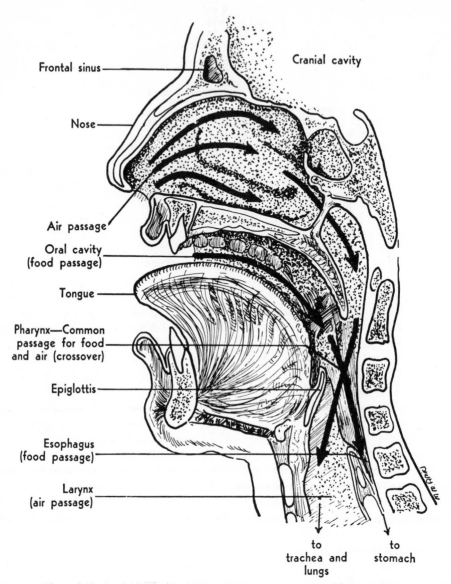

Figure I-12. Structure involved in breathing and swallowing (Shilling 1965).

be graphically traced by looking at Figures I-12 and I-13. As the air passes through the nose it is partially saturated with moisture and some of the particulate matter is caught. As seen in Figure I-12, there is a crossover of the air passage and the food passage. As a rule everything works well, and when food or water is swallowed, the larynx is lifted up (one can notice or feel the "Adam's apple" move up), and the epiglottis, or lid for the air passage, comes down to close off the normally open air passage so that the food or water will flow past without getting into the trachea or windpipe.

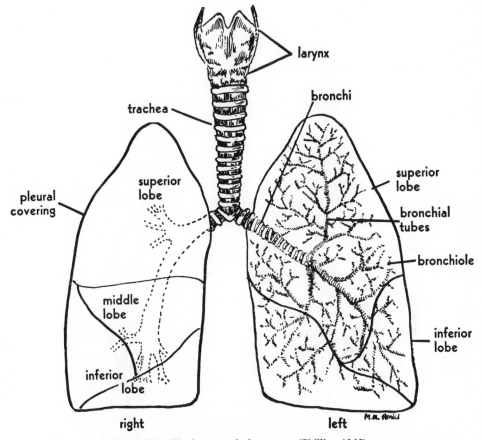

Figure I-13. The lungs and air passage (Shilling 1965).

The trachea leads to the bronchi, which, as can be seen from Figure I-13, lead to the lungs. These latter organs are like two large sacks divided into lobes or compartments, each of which contains innumerable tiny alveoli. Thus, the lungs are divided and redivided by partitions which form a vast multitude of these tiny air pockets. It is estimated there are some 300 million alveoli in each human lung; and, because all of them are in contact with tiny subdivisions of the blood vessels, they have exposed to the air a blood surface of about 750 ft^2! This large blood surface is necessary because the diffusion of a gas depends upon the exposed surface area.

The mechanism of breathing can best be understood by studying the diagrams in Figure I-14. In the action of inspiration the diaphragm contracts and moves downward, the ribs are elevated, and a larger cavity is formed. Thus negative pressure is produced inside the chest cavity and air is drawn into the lungs in order to equalize this pressure and satisfy the vacuum. In exhaling, the diaphragm relaxes and moves up, and the weight and elasticity of the chest walls cause the chest to return to its original size, thus reversing the process and expelling the air from the lungs.

There are a number of terms used in discussing respiratory function or respiratory mechanics, which are illustrated in Figure I-15. The four pulmonary lung "volumes"

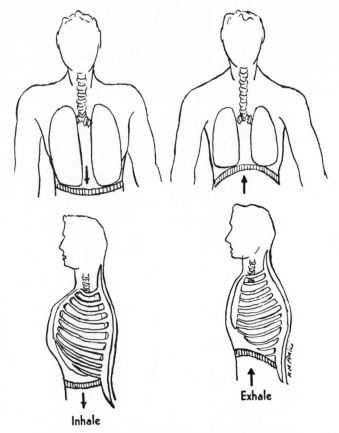

Inhale

Exhale

Figure I-14. The mechanics of breathing (Shilling 1965).

are: inspiratory reserve volume; tidal volume; expiratory reserve volume; and residual volume. Together they represent the maximum volume to which the lungs can be expanded. Whenever it is needful to consider two or more of the pulmonary volumes together, the combination is termed "capacity" (Guyton 1971).

1. *Inspiratory capacity* (inspiratory reserve and tidal volumes). The amount of air that a person can breathe beginning at the normal expiratory level and distending his lungs to the maximum amount.

2. *Expiratory capacity*. The amount of air that a person can exhale at the deepest possible expiration.

3. *Functional residual capacity*. The amount of air remaining in the lungs at the end of normal expiration, sometimes referred to as the "resting expiratory level."

4. *Vital capacity* (inspiratory and expiratory reserve volumes plus tidal volume). The maximum amount of air which can be expelled from the lungs after a maximum inspiration.

5. *Total lung capacity*. The maximum volume to which the lungs can be expanded with the greatest possible inspiratory effort.

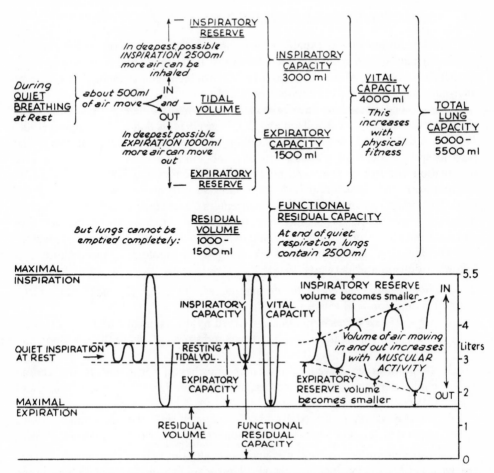

Figure I-15. Capacity of the lungs. [McNaught and Callander (1970); reprinted from *Federation Proceedings* 9:602 (1950).]

The adult male breathes in and out about 16 times per minute and the amount of air breathed at rest is about 500 ml; thus, 500 ml × 16 = 8000 ml or eight liters. This is the *respiratory minute volume*, or *pulmonary ventilation*. During exercise this value may go up to as much as 200 liters. (These values are about 25% lower in women.)

Average normal respiratory rate is 14–18 cycles/min, but varies greatly with exercise and with disease. The rate of respiration is governed by the respiratory center in the brain (see the thermostat, p. 16), which is activated primarily by the accumulation of carbon dioxide and the consequent acidity of the blood. Thus, the more oxygen burned and, consequently, the more carbon dioxide produced, the faster and deeper one breathes. This leads to the paradoxical situation that a person who has stopped breathing, such as an apparently drowned person, should be resuscitated with a gas mixture rich in carbon dioxide as well as in oxygen, even though his breathing normally eliminates carbon dioxide.

G. *The Information Receptors*

The human machine is capable of sensing four forms of energy in the environment: mechanical, thermal, chemical, and photic. It has many highly developed sense organs for obtaining information from the environment and is uniquely capable of analyzing and storing the information obtained.

The sensory subsystems have been classified in many different ways. One of the classifications reported by Adolfson and Berghage (1974) is that of Sherrington (1906, 1948). It is based on the source of the stimulus and the location of the receptor. "The proprioceptors, found in muscles, tendons and joints, and in the labyrinth, give information concerning the movements and position of the body in space. The exteroceptors, the sense organs of the skin, give information on changes in the immediate external environment. The interoceptors transmit impulses from the visceral organs. The teleceptors, or distance receptors, are the sense organs of the eyes, ears and nose and give information concerning changes in the more remote environment."

The clinicians classify the senses according to morphological considerations, as follows (Adolfson and Berghage 1974):

1. *Special senses* served by the cranial nerves: (1) vision, (2) audition, (3) taste, (4) olfaction, and (5) vestibular.
2. *Superficial* or *cutaneous sensations* served by the cutaneous branches of spinal and certain cranial nerves: (1) touch–pressure, (2) warmth, (3) cold, and (4) pain.
3. *Deep sensations* served by muscular branches of spinal nerves and certain cranial nerves: (1) muscle, tendon, and joint sensibility, or position sense, and (2) deep pain and deep pressure.
4. *Visceral sensations* served by fibers conducted with the autonomic nervous system: (1) organic sensation (e.g., hunger and nausea), and (2) visceral pain.

However, according to Aristotle, and classically throughout the centuries, there are five special senses: vision, hearing, smell, taste, and touch. Current popular usage also speaks of the five senses, but of recent years the research psychologists and physiologists have spoken of several additional senses. This has been brought about quite largely by subdividing touch, or *feeling* as it is often called.

Some of the additional senses for which independent status has been claimed are pressure, contact, deep pressure, prick pain, deep pain, warmth, heat, cold, muscular pressure, articular or joint pressure, appetite, hunger, thirst, cardiac sensation, pulmonary sensation, and vibration.

In this discussion we will consider the classical five—vision and the eye; hearing and the ear; taste and the tongue; smell and the nose; touch, or feeling, and the skin—and add two others: kinesthetic and organic sensibilities, and equilibrium and the inner ear.

1. *Vision and the Eye*

Of all the senses by which the human machine receives information concerning the world around it, the visual sense is by far the richest. It has been said that 80% of our knowledge comes to us by way of the eye (Shilling 1965). It becomes even more important when man enters a strange environment, like the underwater world, whether he is wearing a face mask or goggles or whether he is a free swimmer or in a submerged vehicle. Because it is so important that engineers and underwater workers better their understanding of the normal visual process as well as the changes that take place in an underwater environment, normal anatomy and physiology of the eye are discussed here and underwater vision is the subject of a special section in Chapter V.

Figure I-16 represents an anatomical cross section of the human eye. The functioning of the eye, in many ways similar to that of a camera, is diagrammed in Figure I-17. The basic parts of the eye used for vision are the cornea, the lens, and the retina (the light-sensitive layer of the eyeball composed of photoreceptors—the rods and the cones). The fovea, at the center of the macula lutea, is the point of maximum acuity. It is used for fine focus, in reading, etc., and is composed entirely of cones. Rods predominate in the periphery of the retina and are extremely important in night vision.

Three types of cones are responsible for perceiving color—light radiation corresponding to those responsive primarily to red, green, and blue, respectively. They are stimulated only in fairly high illumination (greater than 10^{-3} mL) and are used in detecting fine detail. The rods are able to give vision of only black, white, and shades of gray, are sensitive to lower levels of illumination (minimum 10^{-5} mL), and can only detect gross characteristics of configuration.

Focusing of incoming light is accomplished by the eye due to refraction or bending of light rays. The majority of the refraction occurs at the cornea. The light

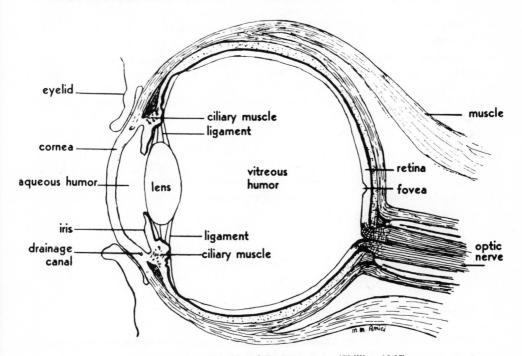

Figure I-16. Cross section of the human eye (Shilling 1965).

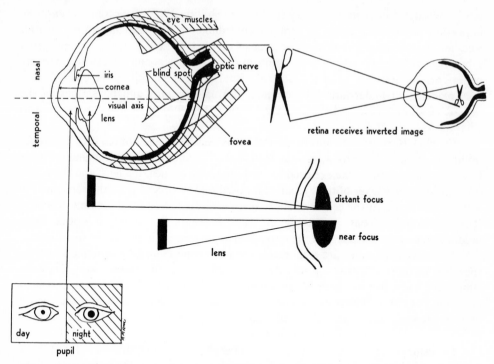

Figure I-17. Diagrammatic representation of left eye (horizontal cross section) (Shilling 1965).

rays then pass to the lens, which is elastic and via specialized muscles and ligaments can be made more convex (to converge light coming from nearby) or less convex (to converge light coming from a distance) for the purpose of accommodating focus for objects at varying distances. By constriction or dilation of the iris (the colored pigmentation encircling the pupil) the amount of light entering the eye can also be varied. Thus, the pupil dilates when one enters a darkened room and constricts on exposure to bright sunlight.

Since the two eyes are separated slightly, they give slightly different views of a scene. These slight differences are used by the brain as a cue for depth perception (stereoscopic vision), at least for distances less than about 20 ft, beyond which the differences in the images furnished by the two eyes become too small to be effective as depth cues. Another binocular depth cue is furnished by the fact that the angle formed by the two eyes and a point on which both are focused depends on the distance of the object.

Monocular depth cues include shadows, overlapping contours, gradations in color and visual texture, linear perspective, and familiarity with sizes.

2. Hearing and the Ear

In terms of information-gathering, the ear is second only to the eye as a sense organ. Hearing depends on: the ear and its associated nerve pathways; the sound

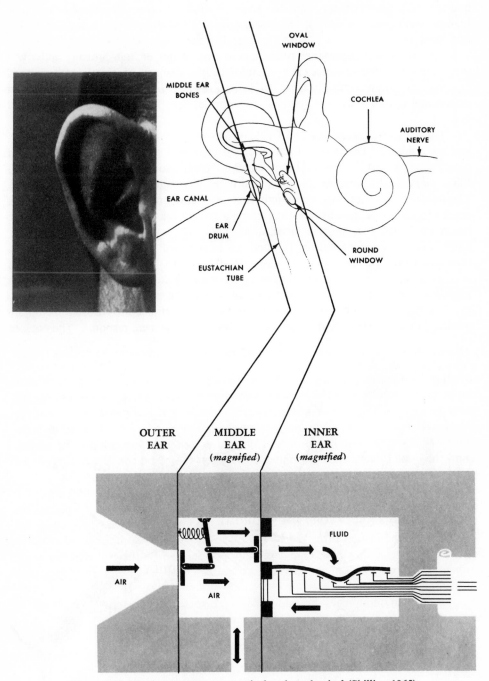

Figure I-18. The human ear—anatomical and mechanical (Shilling 1965).

source (a vibrating body, such as the human larynx, a bongo drum, or a jet airplane); and the medium, which transmits the pressure vibrations from the sound source (air, high-pressure gases, or water). The first is termed the auditory system and the second and third are combined and called the sound system.

Because of the importance of the ear to underwater or high-pressure activity, a section in Chapter V discusses the physiological problems of hearing in a pressurized environment and a section in Chapter X discusses hearing as it relates to speech and underwater communications systems. The normal function of the ear is considered here.

The ear (see Figures I-18 and I-19) consists of three parts: the outer, the middle, and the inner ear. The shape of the outer ear enables it to capture sound waves and direct them toward the canal. The ear canal directs the sound toward the eardrum—a thin, tough, slightly conical membrane situated between the outer and middle ear. The eardrum vibrates with the fluctuation of sound pressure, transmitting it to a chain of small bones (the ossicles) stretching across the cavity of the middle ear. These bones are called the malleus, incus, and stapes, or more commonly, the hammer, anvil, and stirrup. One end of the malleus is attached to the eardrum, the bones themselves are attached, and the stapes is attached to the oval window. The oval window is covered with a membrane and communicates with the inner ear. This anatomical structure causes infinitely small fluctuating pressures at the eardrum to produce pistonlike vibrations at the oval window. The transmitted vibrations initiate motion in the cochlear fluid, which in turn elicits the hearing response. The cochlea

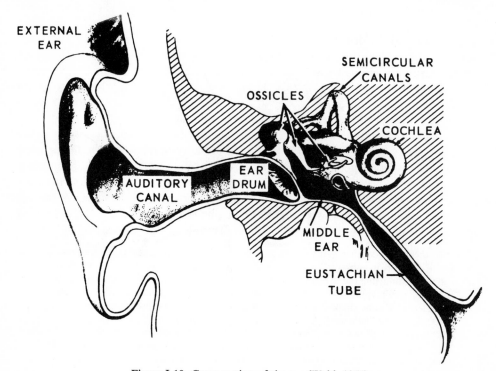

Figure I-19. Cross section of the ear (Webb 1964).

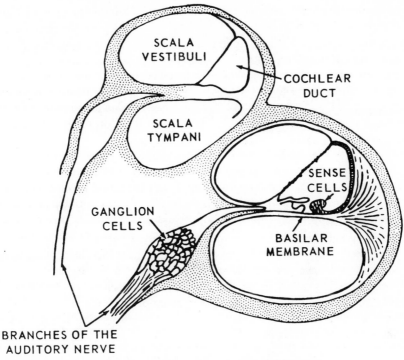

Figure I-20. Cross section of the cochlea (Webb 1964).

is a spiral canal divided into three fluid-filled channels (see Figures I-19 and I-20). Sound waves displace the partition called the basilar membrane so that each frequency creates maximum resonance at its particular place. The primary receptors for hearing are rows of tiny hair cells along the basilar membrane. When these receptors are stimulated, nerve impulses are generated and carried by way of the auditory nerve to the auditory center of the brain.

The eustachian tube, which connects the tympanum (middle-ear cavity) with the nasopharynx (nose and throat) is of especial importance in subaquatic and hyperbaric activity. It has bony portion leading from the tympanum down toward the naso-pharynx, where it becomes cartilaginous, and opens in the throat right behind the nose. In its narrowest portion it is only about 2–3 mm high and 1–1.5 mm wide, but the nasopharyngeal opening is somewhat larger. Figure I-21 is a model of the eustachian tube as diagrammed by Flisberg *et al.* (1963). They described their model as follows:

> ... the model consists of a "middle ear" chamber (ME) whose top opening is closed with a rubber membrane, i.e., the "tympanic membrane" (TM). At ambient pressure inside and outside the ME this occupies a volume of 1 ml. If needed, a "rigid air cell system" (AC), volume 9 ml, can be airtightly connected to the ME-space. The "tube" consists of a cone-shaped semicircular rod, whose narrowest part is fastened to the ME, i.e. at the "isthmus." A thin rubber film is distended over the convex of the rod, while it is airtightly fastened to the back side. The tension of the rubber film over the convexity is assumed equal along the whole of the rod at atmospheric pressure. At the "isthmus" a small portion of the rubber film (ETM) leaves the rod and is air-tightly fixed to that part of the circumference of the ME tube which is not connected

to the rod. An airtight rigid chamber (TCH) is mounted around the "tube" in such a way that "the rhinopharyngeal orifice of the tube" is open to the atmosphere at RH. A main homogeneous layer of mucus is assumed to be applied between the rubber film and the convex surface of the rod. Thus under normal pressure conditions, the lumen of the "tube" is closed. The lumen will be opened by applying a negative pressure inside the TCH, which causes the rubber film to be sucked away from the rod surface. Since mucus has been applied between the rod and the rubber film, surface forces acting at the liquid air interface may contribute to the elastic recoil of the rubber film. Now it is well known that the smaller [the] radius of curvature, the greater the pressure opposing surface tension must be, i.e. the amount of pressure needed for opening must increase in the aural direction of the "tube."

Since the oxygen in the air in the middle ear is always being absorbed by the mucosal lining of the walls, there is usually a slight negative pressure of -20 mm Hg in the middle ear with respect to the throat. There are three small muscles attached to the nasopharyngeal end of the tube to open it for the passage of air. Although it is normally closed, Riu *et al.* (1969) found that in the usual waking state the eustachian tube opens about once per minute, and about once per 5 min in sleep. Although swallowing, yawning, or phonation are usually said to open the tube, they found that phonation did not open the tube, only swallowing or yawning.

For additional information about the action of the eustachian tube under changing barometric pressure see Chapter III.

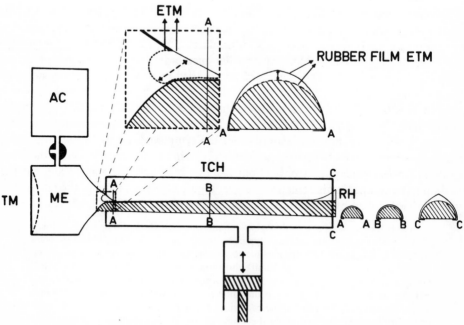

Figure I-21. Ear–eustachian tube model. For details see text. [From Flisberg *et al.* (1963) by permission of *Acta Oto-Laryngologica*, Stockholm.]

3. *Equilibrium and the Inner Ear*

The labyrinthine or vestibular "sense" gives information concerning balance and equilibrium. This proprioceptive function is concerned with the relation of man to his environment, viz. which end is up? As will be seen in the section dealing with vestibular function in Chapter V, the vestibular sense is of major importance in diving, particularly when visibility is poor.

The receptors for the vestibular sense are part of the inner ear and are encased in the dense bone of the skull. These receptors are in the vestibule and in the three semicircular canals which lie approximately at right angles to each other, one for each major plane of the body (see Figures I-22 and I-23). The combination of the three canals is well suited to monitor the movement of the head—and thus the body—at varying degrees of pitch, yaw, and roll. Because of this they have been referred to as the three "spirit levels" of the body.

Within the body labyrinth lies the membranous labyrinth, all parts of which are continuous with each other and with the sacs of the vestibule. A watery fluid, the perilymph, surrounds the entire membranous labyrinth, preventing it from rubbing against its body surroundings. Inside is the endolymph, which plays an important part in stimulation. In response to change in movement it either shifts its position or circulates within the canal.

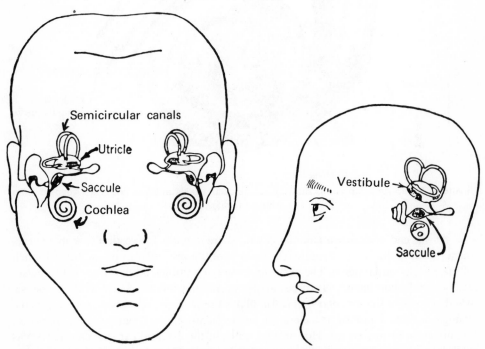

Figure I-22. The primary structures of the inner ear (greatly enlarged in proportion to the size of the head). [Adapted from Camis (1930).]

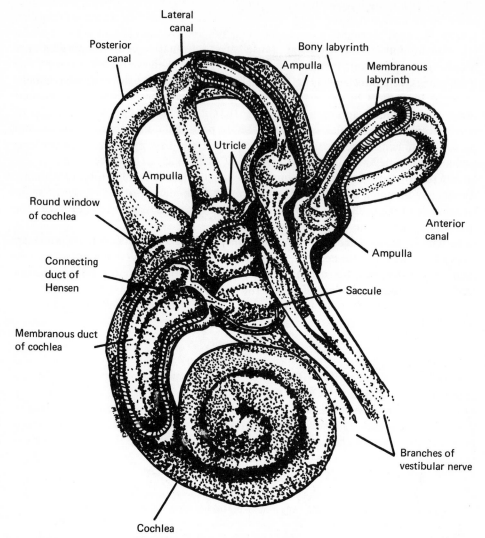

Figure I-23. The right labyrinth, showing semicircular canals and the cochlea. [Adapted from Shilling (1965).]

At one end of each semicircular canal, where it enters the vestibule, is an enlargement (the ampulla) which contains the endings of the vestibular or nonauditory branch of the eighth cranial nerve. The nerve fibers in the vestibule are large so that impulses can be rapidly conducted, as is necessary for emergency reflexes. It is not motion per se, whether straight line or rotary, that stimulates the semicircular canals; rather, it is the change in rate of motion (acceleration or deceleration) and, particularly, change in the direction of movement. Hot or cold water in the external ear (the caloric test) as well as mechanical pressure and electricity can excite them in a similar manner.

The vestibular sense is not necessary for general maintenance of equilibrium and balance under normal conditions when other cues are acting to provide information about one's position in the environment. These include visual, cutaneous, organic, kinesthetic, and, occasionally, auditory cues. When these cues are absent or obscure (as in many underwater activities) or when fast action is necessary, vestibular cues become very important. In any case, to maintain equilibrium and balance on a fine level, particularly during rapid and intricate body movements, the vestibular sense is necessary regardless of the number of other cues available. Here the semicircular canals are of prime importance (Guyton 1971). The utricle can sense changes of position after they have occurred but this is not very useful during walking and particularly during running or in climbing stairs. The canals, however, sense acceleration and therefore small changes of direction (Wendt 1951).

Through early learning the canals take on a "predictive" function. Walking, running, and moving about, the child learns that a signal indicating a change of angular velocity of the head (angular acceleration) must be counteracted or accounted for immediately or else he will fall. For example, to round a corner, he must lean into it or he will not make it around; to stop abruptly, he must lean backward to counteract inertia (although he does not put these things into actual words or physical concepts). Such responses are epitomized in the behavior of cats, which, when dropped, "always land on their feet."

4. Taste and Smell

Foods and beverages, once taken into the mouth, make at least a dual, and usually a triple, appeal. The tissues of the mouth, throat, and nasal cavity are so innervated and so disposed in relation to one another that any or all three sensory systems (pressure, taste, and smell) can go into operation simultaneously in response to the same stimulus. The pressure sensibilities of the mouth region are brought into play and the sense of taste is obviously involved; however, the most elaborate experiences are furnished by olfaction or the sense of smell.

According to Shilling (1965):

> Olfactory receptors, or neurons, are located in a relatively inaccessible crevice high in each of the two nasal cavities. Only a small portion of a gas, vapor, or volatile material inhaled is able to reach and stimulate these receptors. Olfactory receptors are extensions of the brain, as is the retina in the case of vision. [See Figure I-24.]
>
> Gustatory neurons, ending in taste cells embedded in taste buds of the tongue, are the receptors for taste sensation. They respond to various types of chemical stimulation in very low concentrations; however, the solution of the material must actually contact the taste bud in order to excite it. [See Figure I-24.]

Although of general interest to the individual, these special senses are not of paramount importance to underwater activity.

5. Touch

Much of the information about the immediate surroundings of the human machine comes through the medium of the skin. There are many classifications of the

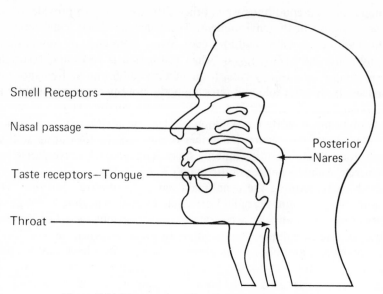

Figure I-24. Taste and smell receptors (Shilling 1965).

various sensations arising from the stimulation of the skin and subcutaneous tissues, but the most satisfactory theory is that the skin has three systems of sensitivity: one for pressure reception, one for pain, and one for response to temperature changes. For the arousal of the sensation of touch, contact, and pressure, any thermally indifferent solid or liquid will suffice. The usual stimulus for arousing pressure sensations is mechanical deformation of the skin and subcutaneous tissue, which produces tension.

On the basis of feel alone, one can make accurate judgments concerning hardness or softness, roughness or smoothness, wetness or dryness, stickiness or oiliness. Movement in some degree is essential, but the amount of relative motion may be very slight and still yield correct impressions. If water temperature is near normal (60–70°F), tactile sensation is not diminished by underwater submersion.

Pain may be elicited by a great range of stimuli—mechanical, thermal, electrical, and chemical. The intensity of the stimulation is the major factor involved; for example, light mechanical stimulation gives a sensation of pressure, while heavy stimulation produces a sensation of pain.

To elicit warm and cold sensations, any device will serve that will conduct heat to or away from the skin. Temperature sensations may also be initiated by various chemicals: by creating rapid interchanges between the skin and surrounding air; by alternating the circulation of blood and thus disturbing the heat equilibrium of the skin tissues; or by direct action on nerve endings.

The sensation of cold is of extreme importance in underwater activity. For example, cold exposure of the hands leads to a distinct drop in tactile sensitivity. (See special section on the effects of cold in Chapter IV.)

6. *Kinesthetic and Organic Sensibilities*

The receptors for the perception of motion and position are located in the muscles, tendons, and joints. They provide impulses which produce coordination of all body parts for complex acts, such as walking. This kinesthetic sense, although concerned primarily with movements of discrete parts of the body, contributes to the proprioceptive function, which is concerned with the relation of man to his environment. For example, when a person lies down, the brain receives information from both the labyrinthine receptors and from the kinesthetic receptors because of movements of the body and sensations of pressure from contact with the environment (bed). As will be seen in the section on vestibular function in Chapter V, the water distorts many of the normally experienced stimuli.

H. *Body Measurement (Anthropometrics)*

For a man–machine complex, the design engineer must have accurate data on the extremes and the mean measurements of the various parts of the body, and of the force different body parts are able to exert in various positions.

To aid the design engineer in development of future diving systems and equipment, a comprehensive anthropometric study was completed by the U. S. Navy Experimental Diving Unit (Beatty and Berghage 1972). Fifty-nine measurements were obtained on 100 divers. The mean for each of the measurements with standard deviation, range skewness, and kurtosis are given in both the metric (a) and the English (b) systems in Table I-4.

I. *Psychological Overlay*

There is an axiom in geometry which states, "The whole is greater than any of its parts and is equal to the sum of all of its parts." But in considering the potential for performance of the human machine, it is obvious that not only is the whole greater than any of its parts, but it is also greater than the sum of all of its parts.

One of the reasons for the increased potential of the whole over the sum of its parts lies in the organization of the parts. As Gerard (1940) said, ". . . the real basis of the difference in behavior of a man and a mouse, of a mussel and a mushroom, is not in the materials (after all only more or less elaborated protoplasm) of which the organisms are built, but rather in the intricacy of organization of these materials."

But organization is not enough, for no matter how it is put together, an elaboration of protoplasm does not describe the total man, the whole person. The human machine, man, is a thinking, emotional personality, and his total potential cannot be realized or measured without taking into consideration many of the nontangible aspects (such as personality, emotional makeup, intelligence, drive, ambition, motivation, social adjustment, leadership ability, education, home life, and religious feelings).

Table I-4
Diver Anthropometrics
(Beatty and Berghage 1972)

		Mean	Standard deviation	Range		Skewness	Kurtosis
1. Weight	(a)	81.52	11.24	60.38–120.09	(kg)	0.66	0.44
	(b)	179.72	24.77	133.11–264.73	(lb)		
2. Height	(a)	176.22	5.98	155.70–188.50	(cm)	-0.13	0.32
	(b)	69.38	2.36	61.30–74.21	(in.)		
3. Suprasternal height	(a)	143.87	5.35	128.00–155.70	(cm)	0.11	-0.24
	(b)	56.64	2.11	50.39–61.30	(in.)		
4. Anterior superior iliac spine height	(a)	107.11	4.90	92.30–119.50	(cm)	0.10	-0.17
	(b)	42.17	1.93	36.34–47.05	(in.)		
5. Tibiale height	(a)	48.82	2.79	41.50–56.50	(cm)	0.21	-0.09
	(b)	19.22	1.10	16.34–33.24	(in.)		
6. Lower leg length	(a)	40.12	2.77	31.90–46.50	(cm)	-0.08	-0.23
	(b)	15.80	1.09	12.56–18.31	(in.)		
7. Biacromial diameter	(a)	42.05	1.93	36.60–46.40	(cm)	-0.16	0.12
	(b)	16.55	0.76	14.41–18.27	(in.)		
8. Biiliocristal diameter	(a)	28.81	1.57	26.00–33.50	(cm)	0.41	-0.33
	(b)	11.34	0.62	10.24–13.19	(in.)		
9. Transverse chest	(a)	30.61	1.81	27.00–36.50	(cm)	0.61	0.57
	(b)	12.05	0.71	10.63–14.37	(in.)		
10. Anteroposterior chest	(a)	21.65	1.62	18.40–25.90	(cm)	0.20	-0.49
	(b)	8.53	0.64	7.24–10.20	(in.)		
11. Chest circumference	(a)	97.96	5.98	87.00–115.00	(cm)	0.55	-0.15
	(b)	38.57	2.36	34.25–45.28	(in.)		
12. Abdominal circumference	(a)	89.81	8.13	73.00–114.00	(cm)	0.65	0.28
	(b)	35.36	3.20	28.74–44.88	(in.)		
13. Thigh circumference	(a)	58.07	4.45	48.60–69.30	(cm)	0.17	-0.40
	(b)	22.86	1.75	10.13–27.28	(in.)		
14. Ankle circumference	(a)	23.27	1.57	19.60–27.70	(cm)	0.50	0.11
	(b)	9.16	0.62	7.72–10.91	(in.)		

No.	Measurement		Mean	S.D.	Range	Unit		
15.	Sitting height	(a)	90.91	3.21	83.60–98.00	(cm)	−0.24	−0.30
		(b)	35.79	1.27	32.91–38.58	(in.)	−4.74	35.56
16.	Bicondylar femur	(a)	9.56	0.68	4.20–10.60	(cm)	−0.07	−0.34
		(b)	3.76	0.27	1.65–5.17	(in.)		
17.	Buttock–knee length	(a)	59.53	2.71	53.00–65.00	(cm)	0.00	−0.33
		(b)	23.44	1.07	20.87–25.63	(in.)		
18.	Total arm length	(a)	75.28	3.44	65.90–83.10	(cm)	0.15	−0.35
		(b)	29.64	1.35	25.94–32.72	(in.)		
19.	Upper arm length	(a)	31.87	1.90	27.40–37.10	(cm)	0.05	−0.33
		(b)	12.55	0.75	10.79–14.61	(in.)		
20.	Forearm length	(a)	26.64	1.55	22.70–30.70	(cm)	−1.24	5.33
		(b)	10.49	0.61	8.94–12.09	(in.)		
21.	Bicondylar humerus	(a)	7.03	0.40	5.00–7.90	(cm)	−0.31	−0.51
		(b)	2.77	0.16	1.97–3.11	(in.)		
22.	Wrist breadth	(a)	5.42	0.34	4.60–6.00	(cm)	−0.43	0.12
		(b)	2.14	0.14	1.81–2.36	(in.)		
23.	Hand breadth	(a)	8.40	0.40	7.20–9.40	(cm)	0.82	1.11
		(b)	3.31	0.16	2.83–3.70	(in.)		
24.	Head length	(a)	19.75	0.73	18.20–22.10	(cm)	0.68	0.85
		(b)	7.78	0.29	7.17–8.70	(in.)		
25.	Head breadth	(a)	15.40	0.62	14.20–17.70	(cm)	−0.17	−0.24
		(b)	6.60	0.25	5.59–6.97	(in.)		
26.	Bizygomatic diameter	(a)	13.36	0.81	11.40–15.30	(cm)	0.06	−0.40
		(b)	5.26	0.32	4.49–6.02	(in.)		
27.	Nasion-gnathion	(a)	12.20	0.55	11.00–13.50	(cm)	0.37	−0.29
		(b)	4.80	0.22	4.33–5.31	(in.)		
28.	Nose height	(a)	5.30	0.36	4.50–6.20	(cm)	0.22	0.43
		(b)	2.09	0.14	1.77–2.44	(in.)		
29.	Nose breadth	(a)	3.38	0.33	2.40–4.30	(cm)	−1.25	5.38
		(b)	1.33	0.13	0.94–1.69	(in.)		
30.	Ear length	(a)	6.39	0.51	3.80–7.50	(cm)		
		(b)	2.52	0.20	1.50–2.95	(in.)		

Table I-4—*Cont.*

		Mean	Standard deviation	Range		Skewness	Kurtosis
31. Ear breadth	(a)	3.83	0.36	3.00–4.80	(cm)	0.26	−0.20
	(b)	1.51	0.14	1.18–1.89	(in.)		0.11
32. Upper face height	(a)	18.31	1.27	15.70–22.40	(cm)	0.34	
	(b)	7.21	0.50	6.18–8.82	(in.)		
33. Minimum frontal diameter	(a)	11.56	0.79	9.70–14.30	(cm)	0.71	0.77
	(b)	4.55	0.31	3.82–5.63	(in.)		
34. Bigonial diameter	(a)	10.76	0.55	9.50–12.60	(cm)	0.20	0.25
	(b)	4.24	0.22	3.74–4.96	(in.)		
35. Mouth width	(a)	5.20	0.40	4.20–6.10	(cm)	−0.06	−0.50
	(b)	2.05	0.16	1.65–2.40	(in.)	−0.06	−0.50
36. Lip thickness	(a)	1.62	0.37	0.50–3.50	(cm)	−0.24	0.16
	(b)	0.64	0.14	0.20–0.98	(in.)		
37. Head height	(a)	14.26	0.75	12.20–16.10	(cm)	−0.08	−0.22
	(b)	5.61	0.30	4.80–6.34	(in.)		
38. Ankle breadth	(a)	7.38	0.39	6.50–8.30	(cm)	0.11	−0.27
	(b)	2.90	0.15	2.56–3.27	(in.)		
39. Foot length	(a)	26.52	1.38	22.80–31.70	(cm)	0.20	1.41
	(b)	10.44	0.54	8.98–12.48	(in.)		
40. Head circumference	(a)	57.02	1.68	54.00–62.00	(cm)	0.43	0.12
	(b)	22.45	0.66	21.26–24.41	(in.)		
41. Neck circumference	(a)	38.80	2.05	34.00–45.00	(cm)	0.24	0.33
	(b)	15.28	0.81	13.39–17.72	(in.)		
42. Upper arm circumference	(a)	30.37	2.49	25.50–37.00	(cm)	0.46	−0.26
	(b)	11.96	0.98	10.04–14.57	(in.)		
43. Upper arm contr. circumference	(a)	33.80	2.67	27.50–40.00	(cm)	0.26	−0.43
	(b)	13.31	1.05	10.83–15.75	(in.)		
44. Forearm circumference	(a)	27.81	1.70	24.30–32.00	(cm)	0.37	−0.34
	(b)	10.95	0.67	9.57–12.60	(in.)		
45. Wrist circumference	(a)	17.32	0.94	15.40–20.50	(cm)	0.65	1.08
	(b)	6.82	0.37	6.06–8.07	(in.)		

#	Measurement		Mean	SD	Range	Units		
46.	Calf circumference	(a)	37.86	2.79	31.00–45.50	(cm)	0.33	−0.16
		(b)	14.91	1.10	12.20–17.91	(in.)		
47.	Triceps skinfold	(a)	10.83	4.51	3.50–24.00	(mm)	0.76	0.21
		(b)	0.43	0.18	0.14–0.94	(in.)		
48.	Subscapular skinfold	(a)	16.58	6.88	6.00–33.00	(mm)	0.52	−0.71
		(b)	0.65	0.27	0.24–1.30	(in.)		
49.	Midaxillary skinfold	(a)	13.76	7.25	2.00–42.00	(mm)	1.29	2.45
		(b)	0.54	0.29	0.08–1.65	(in.)		
50.	Chest skinfold (Juxta-nipple)	(a)	6.42	5.34	2.00–27.00	(mm)	1.99	3.03
		(b)	0.25	0.21	0.08–1.06	(in.)		
51.	Biceps skinfold	(a)	3.44	1.72	2.00–11.00	(mm)	20.6	4.63
		(b)	0.14	0.07	0.08–0.43	(in.)		
52.	Forearm skinfold	(a)	6.08	2.56	3.00–16.00	(mm)	1.39	2.10
		(b)	0.24	0.10	0.12–0.63	(in.)		
53.	Abdomen skinfold	(a)	26.58	11.62	5.00–58.00	(mm)	0.34	−0.32
		(b)	1.05	0.46	0.20–2.28	(in.)		
54.	Suprailiac skinfold	(a)	7.37	5.05	2.00–26.00	(mm)	1.72	2.87
		(b)	0.29	0.20	0.08–1.02	(in.)		
55.	Forced vital capacity		5120.85	663.31	999.00–7120.00	(ml)	0.38	0.45
56.	Forced expiratory volume: 1 sec		4022.68	740.58	999.00–5940.30	(ml)	0.37	−0.60
57.	Specific gravity		1.0652	0.0113	1.0339–1.0886		−0.25	−0.11
58.	Total body fat, %		16.46	5.55	5.20–32.18		0.30	−0.08
59.	Body surface area		1.97	0.14	1.59–2.42	(m^2)	0.39	0.32

This handbook is not the place for a study of psychology, philosophy, or religion. However, a few paragraphs may help in understanding the human machine as a human being.

1. *The Brain and Its Functions*

The human machine is distinguished from all other machines in that it has a brain which with proper education and training allows it to make crucial decisions as to the proper course of action on a continuing basis. The human brain, as shown in Figure I-25, has two main divisions: the cerebrum, which occupies almost all of the skull cavity, and the cerebellum, which is situated beneath the rear portion of the cerebrum.

The cerebrum is composed of the right and left hemispheres, and its outer surface is called the gray matter because the nuclei of the cell bodies make it appear gray. There are at least 10^{10} of these nerve cells, arranged in complex and systematic three-dimensional patterns. Shilling (1965) describes the cerebrum as follows:

> Beneath this outer layer lie the connecting axons (white matter), or the nerve fibers, which form a connecting link between the two hemispheres and continue on down to form the pons, the medulla, and the spinal cord.
>
> As will be noted in Figure [I-25], the activity of the cerebrum is concerned with sensation, thought, memory, judgment, reason, and the initiation or management of any and all activity which we commonly consider to be under the control of the will. This is a large order and is carried out in specialized areas as well as in higher integrating centers. As examples of the specialized areas, the olfactory lobes might be designated the nose-brain; the occipital lobe, the eye-brain; and the hindbrain may be called the ear-brain. As will be seen, there are other areas where the control for special functions is located, such as speech, writing, and general sensory and motor areas. Some of the higher integrating centers have to do with such functions as the control of breathing, heart action, and swallowing. These are the integrating centers for functional coordination. The higher centers dealing with intelligence, memory, judgment, and thought are the ones that differentiate man from all other mammals.
>
> All of this activity is accomplished by interlocking and interrelating of the circuits. That there is ample scope for detail or fantasy in this mechanism is evident from the fact that there are in the order of $10^{10,000}$ possible permutations. The impulse brought in by a single sensory fiber may either directly or indirectly reach many synapses, bringing the fiber into contact with a vast array of other neurons and fibers. Usually an impulse does not activate more than a very small fraction of the fantastic number of possible connections.
>
> *Cerebellum.* The *cerebellum* has as its chief function the maintenance of balance, harmony, and coordination of the motions initiated by the cerebrum. Its principal sensory connections are with the fibers that convey proprioceptive sensitivity (deep muscle and joint sensations), and with the semicircular canals. It plays a major role in making possible the extraordinary precision of muscular movement required for such activity as writing, speaking, piano playing, or even just walking and maintaining posture.
>
> *Pons.* The *pons* acts as a bridge or connecting station between the other three parts of the brain—the cerebrum, the cerebellum, and the medulla.
>
> *Medulla.* The *medulla* is anatomically the lowest part of the brain, situated just above the beginning of the spinal cord. But it is very important, for it has the centers which control heart action, breathing, and other vital processes.

Electroencephalography. The advent of the *electroencephalograph* (E.E.G.) made it possible to accurately study brain action. It became increasingly evident that there was more to the brain than some of the classical philosophers thought when they called it 'the thing in the head.' The E.E.G. records show that there is continuous activity of an extremely complex nature in the brain. By studying carefully the normal electrical discharges, it has been possible to use the E.E.G. tracings of brain activity to identify abnormal and diseased conditions of the brain.

2. *Communication*

Another distinguishing feature of the human machine is that it can communicate through the medium of language, either spoken or written. Voice communication is considered in detail in Chapter X, but some general considerations are presented here. According to Shilling (1965):

> The human machine has an external communication network which may be considered at three distinct levels. For ordinary spoken language, the first level involves the ear and that part of the brain which is connected with the inner ear. This apparatus, when stimulated by sound vibrations in the air or their equivalent in electric circuits, represents the part of the machine concerned with sound, or what is called the phonetic aspect of language.

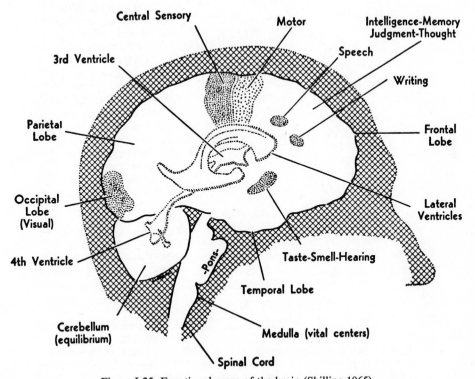

Figure I-25. Functional areas of the brain (Shilling 1965).

The semantic aspect, which is concerned with the meaning of the words, is the second level. Here the receiving apparatus neither receives nor transmits the language word by word, but idea by idea, and this is not by any means a trivial part of the problem of communication.

There is a third level of communication, which consists of actions which may be observed externally. This may be called the behavior level of language. These actions consist of two parts: the direct gross actions of the sort which we also observe in the lower animal, and the coded or symbolic actions which we know as spoken or written language.

3. Man–Machine Relationship

According to Shilling (1965):

Norbert Wiener has contributed greatly to this aspect of the communication problem and his two books, *Cybernetics*, and *The Human Use of Human Beings*, contain a wealth of information on the subject. Cybernetics comes from the Greek word meaning "steersman." The material covers the entire field of control and communication theory, both machine and animal; it is applied to the man–machine, machine–machine, and machine–man communication process. The nervous system of the human machine fits perfectly into this mathematical and electronic setup. For example, the synapse transmission of an impulse from one nerve to another is like a coincidence-recorder and the reflex arc is related to servo-mechanisms.

Feed-back is well illustrated by the simple act of picking up a pencil lying on the table. We decide to pick up the pencil; however, we do not signal consciously to any of the many muscles that are to be used. As we reach toward the pencil we are constantly getting signals back from our eyes and proprioceptive (deep muscle and joint) signals from the arm telling us that our hand is not quite in position yet to pick it up and that we will have to move the hand a little farther. Thus there is a constant feed-back in operation in this simple process, although it is so extremely smooth that we are not aware of the constant interchange of signals concerning the amount of motion which has already been accomplished and how much more is needed.

The first time a man tries to operate the bow-plane of a submerged submarine, the result will be like the first time he tried to ride a bicycle, except instead of wobbling, the submarine will be bobbing like a porpoise. This is a good example of a feedback system and the wide oscillations which occur if it is crudely constructed or ineptly operated. Other mechanical examples of the feed-back are the steering mechanism aboard ship, and the "Christmas Tree" board of signal lights on a submarine that informs one whether or not the hatch or the main induction has been closed in response to an order and the movement of a lever. Another very common example of a purely mechanical feed-back is the thermostat, which receives its signal from the thermometer indicating that the temperature is too low; as a result the thermostat then closes an electric switch which activates the mechanisms of the furnace, thus sending out more heat; this additional heat continues to drive the thermometer up until the desired temperature is reached, whereupon the thermostat cuts the furnace off.

We have noted earlier that the body tends to maintain a constant level of basic chemical operation. This is accomplished by what might be called homeostatic feed-backs, which operate chemically and which tend to be much slower than the nerve feed-backs. However, they do collectively keep the body's operations in balance.

It is obvious from this discussion that there can be both positive and negative feed-back. The positive type provides mechanisms for hunting, scanning, and probing the environment for information; and the negative feed-back maintains stability.

4. *Memory*

The process of memory in the human brain is not yet understood. Computers have been developed which utilize all types of memory devices (including punched cards, punched tape, magnetic tape, phosphorescent substances, telegraphic type repeaters, photography, and vacuum tubes) but none of these devices, no matter how remarkable their refinement, can come anywhere near the memory function of the properly functioning human brain, even though they can accomplish some tasks impossible for man.

5. *Motivation*

Of all the attributes that distinguish man the human being from man the machine, none is more important, and at the same time more difficult to define or measure, than motivation. What drives an individual toward a certain goal? Why does this drive persist in spite of adverse psychological or physical environmental conditions that may even entail marked discomfort? It has been well demonstrated that in time of extreme stress, when the motivation may be a life-saving situation, the human being is able to perform at an almost supernatural level.

A thorough discussion of motivation is not within the province of this handbook. Generally, after divers have been selected and trained, and chosen for a particular task, there are two ways to increase motivation. One is communication. The more the diver or underwater worker knows of the goals of the task, the more likely it is that motivation will be enhanced. Working "in the dark" is not fun. The other approach is to favorably modify the environment. Obviously there are many ways to do this, but one of the most important is to provide the very best life support and operational equipment. In this connection it is worthwhile to look carefully at the cost–reward model (Figure I-26) of the desirability of various environments as depicted by Helmreich (1971).

The effect or consequences of any environment for an individual can be stated in terms of the rewards received and the costs incurred by participation. Acceptability of outcomes can be judged by the comparison level (CL) and the comparison level of alternatives CL(alt). The CL defines whether a person's outcomes are better than, worse than, or equal to his expectations, while the CL(alt) indicates the relative attractiveness of other possible environments. The cost–reward diagram is an attempt to place various situations in a cost–reward matrix. Normally, the farther to the left and above the diagonal, the more desirable are outcomes, and, therefore, the more desirable the situation. The farther to the right and below the diagonal, the less desirable is the situation. The area below the dashed line in the lower right half of the matrix defines situations in which a person would not remain voluntarily. Similarly, outcomes above and to the left of the upper dashed line are those which seldom occur because they normally yield unattainably high outcomes.

This discussion would not be complete without emphasizing two points. The first is that although the human machine has an amazing ability to adapt itself to the requirements of a given environment or situation, it still has certain definite limitations.

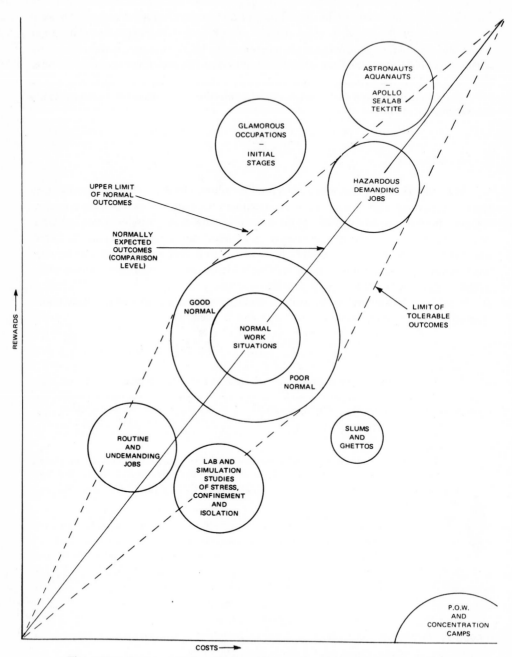

Figure I-26. A cost–reward model for various environments (Helmreich 1971).

Therefore the second point made here should be self-evident. In the complex of man, machine, and environment, it is the machine and the environment which must be modified or changed. Although it seems unnecessary to repeat, it cannot be over-emphasized that the engineer must bear in mind the very firm fact that the human being has only two eyes, and these are located in his head; that he has only two arms, and these are of limited length and power; and that he has only a very definitely limited ability to fit into the rapidly changing environment of modern machines.

If man must be used in the man–machine combination, *the environment must be modified to meet the requirements of man.*

6. Stress

Man experiences stress when he is unable to find an adjustive response to problems confronting him. Stress may occur under conditions variously described as trauma, frustration, or conflict. Thus it follows that man in competition with his environment is inevitably subjected to physiological and psychological stress. Whenever anyone attempts to accomplish the new, the different, or the difficult, he must work against the environment, thereby subjecting himself to stress. The man who does nothing experiences little or no stress, whereas the man who works in a diving rig at 500 ft subjects himself to an enormous amount of strain and the resulting stress. According to Shilling (1965):

> Environmental physical trauma (heat, cold or physical violence) leads to physiological stress upon the human body (burns, freezing, cuts, fractures with their resulting hemorrhage, and shock).
>
> Even though the physiological type of stress is more readily identified, it must never be forgotten that psychological stress is just as real and may be even more devastating. Fear alone without the application of an iota of external physiological stress has been known to kill an individual. This is admittedly a rare occurrence, but far from rare is the development of psychosomatic complaints as a response to an intolerable or overly stressful situation.
>
> It is difficult to evaluate the tolerance of the human organism because of enormous variation between individuals. It is almost equally difficult to evaluate the degree of environmental stress, since 'one man's meat is another man's poison.' Determining the acceptable balance between stress and tolerance in any situation is one of the greatest challenges but at the same time one of the most rewarding.
>
> It is interesting to note that motivation and the resulting effort play a large part both in the breakdown under stress and in the defense against such breakdown. It is axiomatic to say that without some commitment to an objective, some goal toward which effort is expended, there is no such experience as stress. It might be said that without motive, there is no effort; without effort, no strain; without strain, no stress.

References

ADOLFSON, J., and T. BERGHAGE. *Perception and Performance Underwater*. New York, Wiley (1974).
BEATTY, H. T., and T. E. BERGHAGE. Diver anthropometrics. U. S. Navy Exp. Div. Unit, Report NEDU 10-72 (June 1, 1972).

BECKMAN, E. L. A review of current concepts and practices used to control body heat loss during water immersion. U. S. Nav. Med. Res. Inst., MR 005.13-4001.06, Rep. 3 (Sept. 12, 1964).

BLESSER, W. B. *A Systems Approach to Biomedicine*. New York, McGraw-Hill (1969).

BROWN, J. H. U., J. E. JACOBS, and L. STARK. *Biomedical Engineering*. Philadelphia, F. A. Davis (1971).

CAMIS, M. *The Physiology of the Vestibular Apparatus*. Oxford, Clarendon Press (1930). [Translated by R. S. Creed from *Fisiologia dell' apparato vestibolare*. Bologna, Casa Editrice N. Zanichelli (1928).]

FLISBERG, K., S. INGELSTEDT, and U. ORTEGREN. The value and "locking" mechanisms of the eustachian tube. *Acta Oto-laryng. Suppl.* **182**:57–68 (1963).

GERARD, R. *Unresting Cells*. New York, Harper and Brothers (1940).

GUYTON, A. C. *Textbook of Medical Physiology*, fourth edition. Philadelphia, W. B. Saunders (1971).

HELMREICH, R. Evaluation of environments: behavioral observations in an undersea habitat. Austin, Tex., Univ. Tex., Tech. Rep. 16 on Contract N00014-67A-0126-0001 (Aug. 1971).

HEMPLEMAN, H. V. Human problems of open sea diving. In: *Diving Applications in Marine Sciences Research Seminar. 14 Dec. 1971*, pp. 120–126. Godalming, Surrey, U.K., National Institute of Oceanography (1972).

MCNAUGHT, A. B., and R. CALLANDER. *Illustrated Physiology*, second edition. Baltimore, Williams and Wilkins (1970).

RIU, R., L. FLOTTES, R. GUILLERN, R. BADRE, and R. LEDEN. La trompe d'eustache dans la plongee. *Rev. Physiol. Subaquatique Med. Hyperbare* **1**:194–198 (Jan./Mar. 1969).

SHILLING, C. W. *The Human Machine*. Annapolis, Md., U. S. Naval Institute (1965).

WEBB, P. *Bioastronautics Data Book*. Washington, D. C., National Aeronautics and Space Administration (1964) (NASA SP-3006).

WENDT, G. R. Vestibular function. In: Stevens, S. S., ed. *Handbook of Experimental Psychology*, pp. 1191–1224. New York, Wiley (1951).

The Ocean as an Environment

A. *Introduction*

"The ocean environment can be described in terms of the density, temperature, and salinity of water; atmospheric conditions (barometric pressure, air temperature and humidity, winds and contaminants); tides and currents; and bottom topography, water clarity, and marine organisms. All of these factors interact to make an ever-changing and formidable environment for men and material. For the working diver these conditions become more rigorous with each additional fathom" (Battelle 1971).

Approximately 361.1×10^6 km^2 (71%) of the earth's surface of 510.1×10^6 km^2 is covered by water (Riley and Chester 1971). Table II-1 gives the areas, volumes, and depths of the oceans and seas. Both the mean and greatest depth of the oceans make diving to the bottom seem impossible, but there are areas, particularly the extensive continental shelves, where diving to the bottom is feasible.

The ocean environment imposes conditions of high hydrostatic pressure, dynamic forces from waves and currents, limited visibility, and low temperatures, which, as will be seen in other sections of this handbook, severely limit the diver's effectiveness and time on the bottom.

This chapter should not be considered as a text on oceanography. It is designed only to provide information considered to be of interest either from the standpoint of equipment engineering or operational diving.

B. *Physical and Chemical Properties of Sea Water*

A number of properties of water and sea water should be considered by both the diver and the design engineer. Some characteristics which are important to the "fitness" of water for the needs of living organisms are shown in Table II-2. Such properties as pressure and depth, specific gravity, density, salinity, compressibility, electrical conductivity, thermal properties, colligative properties, sound transmission, and light transmission are discussed here, followed by a final section that summarizes

Table II-1

Areas, Volumes, and Depths of Oceans and Seas[a]

Sea or ocean	Area, 10^6 km²	Volume, 10^6 km³	Mean depth, m	Greatest depth, m
Oceans, including adjacent seas				
Atlantic Ocean	106.2	353.5	3,331	8,526[b]
Indian Ocean	74.9	291.9	3,897	7,450[c]
Pacific Ocean	179.7	723.7	4,028	11,034[d]
All Oceans	361.1	1,370.3	3,795	—
Seas				
Mediterranean and Black Sea	2.97	4.32	1,458	4,404
Hudson Bay	1.23	0.16	128	229
Baltic	0.42	0.02	55	463
North Sea	0.58	0.05	94	665
English Channel and Irish Sea	0.18	0.01	58	263
Red Sea	0.44	0.22	491	2,359
Persian Gulf	0.24	0.01	25	84
Japan Sea	1.00	1.36	1,350	3,712

[a] Reprinted from Defant (1961) with permission of Pergamon Press.
[b] Puerto Rico Trough north of Puerto Rico.
[c] Java Trench south of Java.
[d] Mariana Trench (11°N, 143°E) according to sounding by Russian research ship *Vityaz*.

the major properties with brief definitions, symbols, and governing equations in table form (Table II-19).

1. *Pressure and Depth*

There is a direct relationship between pressure under water and the depth in the water at which the measurement is taken. Thus, an object at a depth of 33 ft in sea water or 34 ft in fresh water is at a pressure of 2 atm. This is because, in addition to the 1 atm of pressure from the water overhead, an additional 1 atm of air pressure must be counted. When gauge pressure is used, the gauge is normally calibrated to zero at sea level, so that gauge pressure in 33 ft of sea water would be only 1 atm, or 14.7 psi.

The symbol ATA, commonly used in diving operations, represents 1 atm of pressure at 0°C at sea level, which is the absolute value, or atmosphere absolute (ATA). Table II-3 gives most of the ways of expressing pressure and of converting from one to another. Although there are a glossary of terms and a complete list of conversion units elsewhere in this book, it is well here to note that pressure is at standard when at 0°C the height of a column in a mercury (Hg) barometer is 760 mm (called 1 atm or standard atmospheric pressure), and absolute pressure is the total pressure on a submerged object. Thus a man at 99 ft below the water surface is subject to a pressure of 4 atm absolute (ATA) or 3 atm gauge (atm). For more detailed information on pressure effects on man see Chapter III.

Table II-2

Anomalous Physical Properties of Liquid Water[a]

Property	Comparison with other substances	Importance in physical–biological environment
Heat capacity	Highest of all solids and liquids except liquid NH_3	Prevents extreme ranges in temperature; heat transfer by water movements is very large; tends to maintain uniform body temperatures
Latent heat of fusion	Highest except NH_3	Thermostatic effect at freezing point due to absorption or release of latent heat
Latent heat of evaporation	Highest of all substances	Large latent heat of evaporation extremely important in heat and water transfer of atmosphere
Thermal expansion	Temperature of maximum density decreases with increasing salinity; for pure water it is at 4°C	Fresh water and dilute sea water have their maximum density at temperatures above the freezing point; this property plays an important part in controlling temperature distribution and vertical circulation in lakes
Surface tension	Highest of all liquids	Important in physiology of the cell; controls certain surface phenomena and drop formation and behavior
Dissolving power	In general dissolves more substances and in greater quantities than any other liquid	Obvious implications in both physical and biological phenomena
Dielectric constant	Pure water has the highest of all liquids	Of utmost importance in behavior of inorganic dissolved substances because of resulting high dissociation
Electrolytic dissociation	Very small	A neutral substance, yet contains both H^+ and OH^- ions
Transparency	Relatively great	Absorption of radiant energy is large in infrared and ultraviolet; in visible portion of energy spectrum there is relatively little selective absorption, hence is "colorless"; characteristic absorption important in physical and biological phenomena
Conduction of heat	Highest of all liquids	Although important on small scale, as in living cells, the molecular processes are far outweighed by eddy conduction

[a] Reprinted from Sverdrup et al. (1942) with permission of Prentice-Hall.

Table II-3
Depth–Pressure Conversion Tables
(Berghage and Beatty 1971)

Depth of sea water			Pressure, ATA	Absolute pressure					Equivalent depth of fresh water, ft	Equivalent pressure of fresh water, psi
ft	m	fathoms		psi	mm	in. Hg	bars	kg/cm²		
0.	0.0	0.0	1.000	14.696	760.000	29.921	1.013	1.033	0.0	14.696
10.	3.048	1.667	1.303	19.146	990.303	38.988	1.320	1.346	10.272	19.149
20.	6.096	3.333	1.606	23.596	1220.605	48.055	1.627	1.660	20.545	23.603
30.	9.144	5.000	1.909	28.046	1450.908	57.122	1.934	1.973	30.817	28.056
40.	12.192	6.667	2.212	32.496	1681.211	66.189	2.242	2.286	41.090	32.509
50.	15.240	8.333	2.515	36.946	1911.514	75.256	2.549	2.599	51.362	36.962
100.	30.480	16.667	4.030	59.195	3063.028	120.591	4.084	4.165	102.724	59.229
150.	45.719	24.999	5.545	81.445	4214.539	165.925	5.619	5.730	154.086	81.495
200.	60.959	33.332	7.061	103.694	5366.055	211.260	7.155	7.296	205.448	103.762
250.	76.199	41.665	8.576	125.944	6517.566	256.595	8.690	8.861	256.810	126.028
300.	91.439	49.998	10.091	148.194	7669.082	301.930	10.225	10.427	308.172	148.295
350.	106.678	58.331	11.606	170.443	8820.598	347.264	11.760	11.993	359.534	170.561
400.	121.918	66.664	13.121	192.693	9972.109	392.599	13.296	13.558	410.896	192.828
450.	137.158	74.996	14.636	214.942	11123.625	437.934	14.831	15.124	462.258	215.094
500.	152.397	83.329	16.151	237.192	12275.094	483.267	16.366	16.689	513.618	237.360
600.	182.877	99.995	19.180	281.682	14577.035	573.894	19.435	19.819	616.294	281.871
700.	213.356	116.660	22.209	326.164	16878.977	664.521	22.505	22.949	718.969	326.383
800.	243.836	133.326	25.238	370.646	19180.918	755.147	25.574	26.078	821.645	370.895
900.	274.309	149.992	28.267	415.129	21482.859	845.774	28.643	29.208	924.321	415.407

1000.	304.777	166.657	31.296	459.611	23784.801	936.401	31.712	32.338	1026.996	459.919
1100.	335.246	183.323	34.325	504.094	26086.742	1027.028	34.781	35.468	1129.672	504.431
1200.	365.715	199.989	37.354	548.576	28388.680	1117.655	37.850	38.597	1232.348	548.943
1300.	396.184	216.654	40.382	593.059	30690.621	1208.281	40.919	41.727	1335.023	593.455
1400.	426.652	233.320	43.411	637.541	32992.563	1298.908	43.989	44.857	1437.699	637.967
1500.	457.121	248.986	46.440	682.023	35294.504	1389.535	47.058	47.987	1540.375	682.479
1600.	487.590	266.642	49.469	726.506	37596.445	1480.162	50.127	51.116	1643.050	726.991
1700.	518.059	283.292	52.498	770.988	39898.387	1570.788	53.196	54.246	1745.726	771.503
1800.	548.527	299.942	55.527	815.471	42200.328	1661.415	56.265	57.376	1848.401	816.014
1900.	578.996	316.593	58.556	859.953	44502.270	1752.042	59.334	60.505	1951.077	860.526
2000.	609.465	333.243	61.584	904.436	46804.207	1842.669	62.404	63.635	2053.752	905.038
2100.	639.934	349.894	64.613	948.918	49106.148	1933.295	65.473	66.765	2156.428	949.550
2200.	670.402	366.544	67.642	993.400	51408.090	2023.922	68.542	69.895	2259.104	994.062
2300.	700.871	383.194	70.671	1037.883	53710.031	2114.549	71.611	73.024	2361.779	1038.574
2400.	731.340	399.845	73.700	1082.365	56011.973	2205.176	74.680	76.154	2464.455	1083.086
2500.	761.809	416.495	76.729	1126.848	58313.914	2295.803	77.749	79.284	2567.131	1127.598
2600.	792.277	433.146	79.758	1171.330	60615.855	2386.429	80.818	82.414	2669.806	1172.110
2700.	822.746	449.796	82.787	1215.813	62917.797	2477.056	83.888	85.543	2772.482	1216.622
2800.	853.215	466.446	85.815	1260.295	65219.734	2567.683	86.957	88.673	2875.157	1261.134
2900.	883.684	483.097	88.844	1304.777	67521.625	2658.310	90.026	91.803	2977.833	1305.646
3000.	914.152	499.747	91.873	1349.260	69823.563	2748.937	93.095	94.932	3080.509	1350.158

2. *Specific Gravity*

The specific gravity of a substance is the ratio of its density to that of pure water at 4°C; the specific gravity of pure water at 4°C is thus defined as 1.0 (no units); the density of water, which is the ratio of weight to volume, is 1 g/cm³, or 1000 g/liter. The specific gravity of sea water depends on the salinity and varies to some extent with temperature and depth (see Table II-4). The range for sea water is from 1.020 to 1.030; sea water is high in salinity in such areas as the Red Sea and is relatively low in salinity in the Atlantic Ocean. The importance of specific gravity lies in its direct effect on buoyancy (Miles 1969).

3. *Density*

The density of any material is defined as the mass per unit volume. In the metric system the density is expressed in grams per cubic centimeter (g/cm³). In oceanography the terms density and specific gravity are used interchangeably: The density of sea water is defined as the weight of a given volume of sea water at a specified temperature as compared with the weight of the same volume of fresh water at a temperature of 4°C. As the weight of a given volume of sea water will vary with temperature, it is now the practice of the Coast and Geodetic Survey to adopt 15°C as the standard temperature for sea water for the purpose of comparison and the hydrometers used for the observations are graduated accordingly. The average density of sea water is approximately 1.026 at a temperature of 15°C.

The density of sea water is dependent upon three properties: temperature, salinity, and pressure. The symbol for density in sea water is ρ, but this must be corrected for standard temperature and pressure (STP), and when so corrected the

Table II-4

Specific Gravity Anomaly × 10^5 of Sea Water[a]

S, ‰	Temperature, °C									
	0	2	4	6	8	10	15	20	25	30
0	−13	−3	0	−5	−16	−32	−87	−177	−293	−433
5	397	403	402	394	381	362	301	207	87	−57
10	801	804	799	788	772	750	685	586	462	315
15	1204	1204	1195	1181	1162	1138	1067	964	836	686
20	1607	1603	1589	1573	1551	1532	1450	1342	1210	1057
25	2008	2001	1988	1970	1947	1920	1832	1720	1585	1428
30	2410	2400	2384	2363	2340	2308	2215	2098	1960	1801
32	2571	2560	2543	2521	2494	2464	2364	2250	2110	1950
34	2732	2719	2701	2678	2651	2619	2522	2402	2261	2100
36	2893	2879	2860	2836	2808	2775	2676	2554	2412	2250
38	3055	3040	3019	2994	2965	2931	2830	2707	2563	2400
40	3216	3200	3179	3153	3122	3088	2985	2860	2714	2550
42	3377	3361	3337	3310	3279	3243	3138	3011	2864	2700

[a] Reprinted from Riley and Chester (1971) with permission of Academic Press.

symbol is σ_0. It is defined as $\sigma_0 = (\rho - 1)1000$. Thus if $\rho = 1.02570$, then $\sigma_0 = 25.70$. At atmospheric pressure and a temperature $t°C$, the symbol σ_t is used.

Density is related to chlorinity Cl (see Table II-19) by the equation

$$\sigma_0 = 0.069 + 1.4708Cl - 0.001570Cl^2 + 0.0000398Cl^3.$$

According to Battelle (1971):

> The density of seawater varies with latitude, location, seasons, temperature, dissolved salt content (salinity), and depth (compressibility). Fortunately, the dominating factors, temperature and salinity, tend to counterbalance and, for most practical purposes, a density of 64 lb/ft³ or 0.4444 psi for each foot of depth, can be assumed. The error resulting from such an assumption should be less than 0.5 percent, or 5 feet in 1000 feet of water. The effect of compressibility on density is negligible (0.01 percent at 1500 feet).

4. *Salinity*

Salinity is often used in conjunction with temperature and pressure as a chemical parameter for calculating other properties, such as density. Salinity (S) is defined as the total amount of solid material, in grams, contained in 1 kg of sea water when all the carbonate has been converted to oxide, the bromide and iodine replaced by chlorine, and all organic matter completely oxidized (Horne 1970). The unit of salinity is often given as ‰ (per mille) rather than g/kg. Total salt content of sea water is variable, but major constituents remain *nearly* constant and this is a generally useful assumption for practical and theoretical purposes of the engineer or working diver, if not for the physical or chemical oceanographer.

Determinations of specific gravity have for many years been based on use of such tables as Knudsen's (1901) *Hydrographical Tables*, which give specific gravity as a function of chlorinity, salinity, and temperature. In recent years, however, salinity has been determined in many cases by measuring other physical properties of sea water, such as electrical conductivity. One such study (Cox *et al.* 1970) determined the relationship of specific gravity, salinity, and temperature in natural sea water. The results, shown in Table II-5, agreed well with Knudsen's tables except in the low-salinity region, but indicated that Knudsen's tables are "low by an average of 0.006 in sigma-*t*."

Specific gravities are calculated as follows: According to Knudsen (1901),

$$\sigma_t = (S_t - 1)1000$$

where σ_t is the specific gravity of sea water at $t°C$ referred to pure water at 4°C.

5. *Compressibility*

Table II-6 shows the relationship between specific gravity (density) and compression (percentage reduction in volume relative to the volume at atmospheric pressure) of sea water of salinity 35‰ and temperature 0°C, conditions similar to those of most oceanic deep waters, at pressures of up to 1000 bars. Data on the compression of sea water of various salinities and temperatures under a pressure 1000 dbar are given in Table II-7.

$$\frac{2.54 \text{ cm}}{IN} / 2 \left(30.48 \frac{cm^3}{FT^3}\right)$$

Table II-5

The Specific-Gravity/Salinity/Temperature Relationship in Natural Sea Water[a]

S, ‰	$t,^b$ °C							Source of sample[c]
	0.000	4.998	9.996	14.994	17.494	19.993	24.992	
41.398	—	—	—	—	—	—	28.181	R
41.390	—	—	—	—	30.308	—	—	R
41.379	—	—	—	—	—	29.648	—	R
41.365	—	—	—	—	—	—	28.175	R
40.336	—	—	—	—	29.494	—	—	R
40.315	—	—	—	—	—	28.835	—	R
40.288	32.401	31.906	31.119	30.087	29.510	—	—	R
39.232	31.547	31.056	30.277	29.268	28.672	—	—	M
39.230	—	—	—	—	28.663	—	—	M
39.216	—	—	—	—	—	28.014	—	M
35.027	—	—	—	—	25.433	—	—	A
35.017	—	—	—	—	—	24.799	—	A
35.006	—	—	—	—	—	—	23.376	A
35.004	28.141	27.708	26.992	26.014	25.431	—	—	A
34.919	—	—	—	—	25.352	—	—	B/R
34.905	—	—	—	—	—	24.716	—	B/R
34.890	—	—	—	—	—	—	23.293	B/R
34.885	—	27.624	26.883	25.931	25.334	—	—	B/R
29.723	—	—	—	—	21.376	—	—	B/R
29.721	—	—	—	—	—	20.773	—	P
29.712	—	—	—	—	—	—	19.386	B/R
29.710	—	—	—	—	—	20.774	—	B/R
29.701	—	—	—	—	—	—	19.382	B/P
29.698	23.878	23.514	22.864	21.937	21.372	—	—	B/R
29.729	—	—	—	—	21.384	—	—	B/R
25.445	—	—	—	—	18.137	—	—	B/R
25.440	—	—	—	—	—	17.543	—	B/R
25.439	20.448	20.157	19.551	18.667	18.118	—	—	B/R
25.436	—	—	—	—	—	—	16.189	B/A
20.158	—	—	—	—	—	—	12.227	B/R
20.156	—	—	—	—	—	—	12.226	B/R
20.154	16.194	—	—	—	—	—	—	B/R
20.152	—	—	—	—	14.095	—	—	B/R
20.147	—	—	—	—	—	13.523	—	B/R
20.131	—	—	—	—	—	—	12.227	B/R
20.130	16.194	—	—	—	—	—	—	B/R
20.128	—	15.981	15.445	14.614	14.081	—	—	B/R
15.553	—	—	—	—	10.612	—	—	B/R
15.543	—	—	—	—	—	10.061	—	B/R
15.541	12.492	12.342	11.877	11.098	10.604	—	—	B/R
15.536	—	—	—	—	—	—	8.781	B/R
9.942	—	—	—	—	6.368	—	—	B
9.940	—	—	—	—	—	5.841	—	B
9.896	—	7.944	7.539	—	—	—	—	B
9.629	—	—	—	—	6.123	—	—	B
9.626	—	—	—	—	—	5.598	—	B
9.623	—	—	—	—	—	—	4.361	B
9.621	—	—	—	—	6.124	—	4.363	B
9.618	—	—	—	—	—	5.598	—	B
9.579	7.695	7.694	7.295	6.582	6.118	—	—	B

[a] Reprinted from Cox *et al.* (1970) with permission of Pergamon Press.
[b] IPTS-68.
[c] *Key:* R—Red Sea; B—Baltic Sea; M—Mediterranean Sea; P—Pacific Ocean; A—Atlantic Ocean.

Table II-6

Specific Gravity and Volume Reduction of Sea Water Under Pressure[a]

Pressure, dbar	Specific gravity	Percent decrease in volume
0	1.02812	0.000
100	1.02860	0.046
200	1.02908	0.093
500	1.03050	0.231
1,000	1.03285	0.458
2,000	1.03747	0.902
3,000	1.04198	1.330
4,000	1.04640	1.747
5,000	1.05071	2.150
6,000	1.05494	2.542
7,000	1.05908	2.932
8,000	1.06314	3.294
9,000	1.06713	3.656
10,000	1.07104	4.007

[a] Reprinted from Riley and Chester (1971) with permission of Academic Press. Salinity, 35.0‰; temperature 0°C.

Ekman (1908) has empirically determined the mean compressibility of sea water between varying pressures and from this relationship the true compressibility can be calculated utilizing the equation

$$K = \frac{k + p(dk/dp)}{1 - kp}$$

where p is the pressure in bars and k is the mean compressibility obtained from Ekman's tables.

Table II-7

Percentage Reduction in Volume of Sea Water under a Pressure of 1000 dbar at Various Temperatures and Salinities[a]

S, ‰	Temperature, °C			
	0	10	20	30
0	0.487	0.470	0.451	0.435
10	0.475	0.461	0.443	0.426
20	0.458	0.452	0.435	0.420
30	0.451	0.442	0.425	0.413
35	0.445	0.436	0.420	0.409
40	0.440	0.431	0.418	0.400

[a] Reprinted from Riley and Chester (1971) with permission of Academic Press.

6. *Electrical Conductivity*

It is important to know the exact relationship between the electrical conductivity and the salinity of sea water because the use of salinometers depends on that ratio. Most salinometers are standardized with Standard Sea Water (salinity 35.00 ± 0.01‰). In these instruments the conductivity of the sample is compared with the conductivity of Standard Sea Water. There are comprehensive tables which relate conductivity ratio to salinity at both 15° and 20°; Table II-8 is a sample, covering salinities from 30 to 40‰. There are also tables of corrections available to be used with nonthermostatic salinometers, at other temperatures, to convert their readings to standard (UNESCO 1966).

Correction must be made for the effect of pressure on electrical conductivity of sea water when using salinometers designed to be used under pressure. Table II-9 shows the effect of salinity variations on the pressure coefficient of conductivity is slight, but the effect of temperature is relatively large (Bradshaw and Schleicher 1965).

7. *Thermal Properties of Sea Water*

Water temperatures at depth are important, not only from the standpoint of the survival of man at various ambient temperatures, but also from the standpoint of rates of oxidation (corrosion) of metals and other materials used at depth under the sea.

a. Temperature

The temperature on the moon varies from +134°C at noon of the lunar day to −153°C during the night. In contrast, the temperature of the earth's seas rarely varies by more than 25°C either in a given location or from place to place. Clearly, the earth's seas are a great moderating factor with respect to temperature, making life on this planet possible. Geographically, the surface temperature of the ocean varies roughly

Table II-8

Relative Conductivity of Sea Water
(UNESCO 1966)

S, ‰	15°C	20°C
30	0.87100	0.8713
31	0.89705	0.8973
32	0.92296	0.9232
33	0.94876	0.9489
34	0.97443	0.9745
35	1.00000	1.0000
36	1.02545	1.0254
37	1.05079	1.0506
38	1.07601	1.0758
39	1.10112	1.1008
40	1.12613	1.1257

Table II-9
Effect of Pressure on the Conductivity of Sea Water[a]

Tempera-ture, °C	Pressure dbar	S, ‰			Tempera-ture, °C	S, ‰		
		31	35	39		31	35	39
0	1,000	1.599	1.556	1.512	15	1.032	1.008	0.985
	2,000	3.089	3.006	2.922		1.996	1.951	1.906
	3,000	4.475	4.354	4.233		2.895	2.830	2.764
	4,000	5.759	5.603	5.448		3.731	3.646	3.562
	5,000	6.944	6.757	6.569		4.506	4.403	4.301
	6,000	8.034	7.817	7.599		5.221	5.102	4.984
	7,000	9.031	8.787	8.543		5.879	5.745	5.612
	8,000	9.939	9.670	9.401		6.481	6.334	6.187
	9,000	10.761	10.469	10.178		7.031	6.871	6.711
	10,000	11.499	11.188	10.877		7.529	7.358	7.187
5	1,000	1.368	1.333	1.298	20	0.907	0.888	0.868
	2,000	2.646	2.578	2.510		1.755	1.718	1.680
	3,000	3.835	3.737	3.639		2.546	2.492	2.438
	4,000	4.939	4.813	4.686		3.282	3.212	3.142
	5,000	5.960	5.807	5.655		3.964	3.879	3.795
	6,000	6.901	6.724	6.547		4.594	4.496	4.399
	7,000	7.764	7.565	7.366		5.174	5.064	4.954
	8,000	8.552	8.333	8.114		5.706	5.585	5.464
	9,000	9.269	9.031	8.794		6.192	6.060	5.929
	10,000	9.915	9.661	9.408		6.633	6.492	6.351
10	1,000	1.183	1.154	1.125	25	0.799	0.783	0.767
	2,000	2.287	2.232	2.177		1.547	1.516	1.485
	3,000	3.317	3.237	3.157		2.245	2.200	2.156
	4,000	4.273	4.170	4.067		2.895	2.837	2.780
	5,000	5.159	5.034	4.910		3.498	3.429	3.359
	6,000	5.976	5.832	5.688		4.056	3.976	3.896
	7,000	6.728	6.565	6.402		4.571	4.481	4.390
	8,000	7.415	7.236	7.057		5.045	4.945	4.845
	9,000	8.041	7.847	7.652		5.478	5.369	5.261
	10,000	8.608	8.400	8.192		5.872	5.756	5.640

[a] Reprinted with permission from Bradshaw and Schleicher (1965). Values given are percentage increases compared with the conductivity at 1 atm.

latitudinally; the temperature ranges from 28°C near the equator to −2°C in the polar seas. In shore, if ocean currents run close to the land, the isotherms tend to be parallel to the coast.

Temperature profiles are classified into a dozen or more types which tend to be characteristic of particular ocean regions and seasons. However, the oceans can be divided into three vertical temperature zones as shown in Figure II-1. The uppermost of the vertical temperature zones, with a thickness of 50–200 m, usually consists of well-mixed water having a temperature similar to that of the surface. Beneath this lies a zone in which the temperature decreases rapidly in depth and is known as the permanent thermocline. It occurs at a depth of about 200–300 m in temperate and tropical waters, but is absent from polar waters because there is a net loss of heat

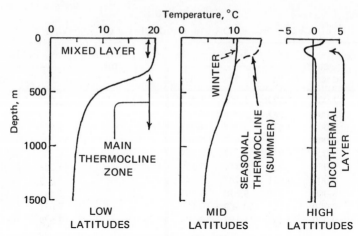

Figure II-1. Typical mean temperature profiles in the open ocean. [Reprinted from Pickard (1964) by permission of Pergamon Press.]

through the surface. Below the thermocline is a deep zone in which temperature falls more gradually.

b. Thermal Expansion

The coefficient of thermal expansion

$$\alpha = (1/V)(\Delta V/\Delta T)$$

increases as pressure, temperature, and salinity increase (Horne 1970). This is shown in Table II-10.

Table II-10

Coefficient of Thermal Expansion of Sea Water at Different Temperatures, Salinities, and Pressures ($e \times 10^6$)[a]

Pressure, dbar	Salinity, ‰	Temperature, °C							
		−2	0	5	10	15	20	25	30
0	0	−105	−67	17	88	151	207	257	303
	10	−65	−30	46	112	170	222	270	315
	20	−27	4	75	135	189	237	282	324
	30	7	36	101	157	200	250	292	332
	35	23	51	114	167	214	257	297	334
2,000	35	80	105	157	202	241	278		
	40	94	118	168	210	248	283		
4,000	35	132	152	196	233	266			
	40	144	162	204	240	272			
6,000	34.85	177	194	230					
8,000	34.85	—	231	260					
10,000	34.85	—	263	287					

[a] Reprinted from Sverdrup et al. (1942) with permission of Prentice-Hall.

c. Thermal Conductivity

According to Sverdrup et al. (1942): "The amount of heat in gram calories per second which is conducted through a surface area 1 cm² is proportional to the change in temperature per centimeter along a line normal to that surface and the coefficient of proportionality γ is called the coefficient of thermal conductivity $(dQ/dt = \gamma \, d\vartheta/dn)$." This coefficient is only valid for water at rest or in laminar motion.

d. Specific Heat

The specific heat of any mass is (in cgs units) the number of calories required to increase the temperature of 1 g of it by 1°C. Specific heat may be expressed with respect to constant pressure or constant volume. The equation developed empirically by Kuwahara (1939) for specific heat at 0°C and atmospheric pressure is

$$C_p = 1.00005 - 0.004136S + 0.0001098S^2 - 0.000001324S^3$$

where S is the salinity. Cox and Smith (1959) have measured the specific heat of sea water at temperatures from 0 to 30°C, showing that specific heat falls as salinity increases. The specific heat of sea water at constant pressure for temperatures from −2° to 30°C and salinities from 0 to 40‰ is given in Table II-11. A probable error of 0.0015 J/g is claimed for these values by Cox and Smith (Cox 1965).

e. Latent Heat of Evaporation

The latent heat of evaporation of sea water is assumed to be very close to that of pure water. The formula for its determination is

$$L = 596 - 0.52(T°C)$$

Table II-11
Specific Heat of Sea Water at Constant Pressure[a]

S, ‰	Temperature, °C						
	0	5	10	15	20	25	30
0	4.217	4.202	4.192	4.186	4.182	4.179	4.178
5	4.179	4.168	4.161	4.157	4.154	4.153	4.152
10	4.142	4.135	4.130	4.128	4.126	4.126	4.126
15	4.107	4.103	4.100	4.099	4.098	4.099	4.100
20	4.074	4.072	4.071	4.071	4.071	4.072	4.074
25	4.043	4.042	4.042	4.043	4.045	4.046	4.048
30	4.013	4.014	4.015	4.016	4.018	4.020	4.023
32	4.002	4.003	4.004	4.006	4.008	4.010	4.013
34	3.990	3.992	3.993	3.995	3.998	4.000	4.003
35	3.985	3.986	3.988	3.990	3.993	3.995	3.999
36	3.979	3.981	3.983	3.985	3.988	3.991	3.994
38	3.968	3.970	3.972	3.975	3.978	3.981	3.985
40	3.957	3.959	3.962	3.965	3.968	3.972	3.976

[a] Reprinted from Cox (1965) with permission of Academic Press. Values given are in absolute J/g.

f. Adiabatic Temperature Changes

When a fluid is compressed, work is performed and a rise in temperature occurs if there is no loss of heat to the surrounding environment. Conversely, a drop in temperature occurs if a fluid is allowed to expand. These are known as adiabatic temperature changes and have little effect on the sea water environment except at extreme depths (greater than 1000 m). Table II-12 illustrates this effect.

8. Colligative Properties of Sea Water

The colligative properties of a substance—e.g., vapor pressure lowering, freezing point depression, boiling point elevation, and osmotic pressure—are properties that depend upon the concentration of particles present and not upon their nature. Although their theoretical values do not pertain directly to sea water because of its complex makeup, departures from these values are directly proportional to one another in all cases. If the magnitude of one colligative property is known, any of the other three may be determined.

a. Vapor Pressure Lowering

The vapor pressure of a solution may be determined from the relationship

$$e/e_0 = 1 - 0.000969Cl$$

where e is the vapor pressure of the sample, e_0 is the vapor pressure of distilled water at the same temperature, and Cl is chlorinity (Figure II-2). Sea water at normal ranges of salt concentration has a vapor pressure about 98% of that of pure water at the same temperature (Sverdrup *et al.* 1942).

Table II-12
Adiabatic Temperature Gradient in the Sea[a]

Depth, m	Temperature, °C								
	−2	0	2	4	6	8	10	15	20
0	0.016	0.035	0.053	0.070	0.087	0.103	0.118	0.155	0.190
1,000	0.036	0.054	0.071	0.087	0.103	0.118	0.132	0.166	0.199
2,000	0.056	0.073	0.089	0.104	0.118	0.132	0.146	0.177	0.207
3,000	0.075	0.091	0.106	0.120	0.133	0.146	0.159	0.188	
4,000	0.093	0.108	0.122	0.135	0.147	0.159	0.170	0.197	
5,000	0.110	0.124	0.137	0.149					
6,000	0.127	0.140	0.152	0.163					
7,000	—	0.155	0.165	0.175					
8,000	—	0.169	0.178	0.187					
9,000	—	0.182	0.191	0.198					
10,000	—	0.194	0.202	0.209					

[a] Reprinted from Sverdrup *et al.* (1942) with permission of Prentice-Hall. Values given are in °C/1000 m at a salinity of 34.85‰.

b. Freezing Point

The freezing point t of sea water was measured by Knudsen (1903), who derived the expression

$$t = -0.0086 - 0.064633\sigma_0 - 0.0001055\sigma_0{}^2$$

which relates the freezing point (t in °C) with specific gravity at 0°C. Table II-13 gives the freezing point at various salinities from 5 to 40‰ computed from this expression.

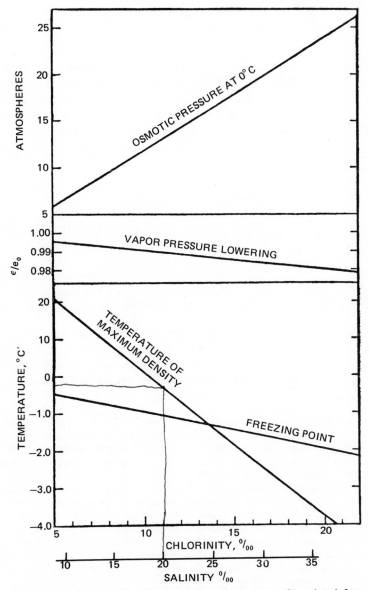

Figure II-2. Some colligative properties of sea water. [Reprinted from Sverdrup *et al.* (1942) by permission of Prentice-Hall.]

Table II-13

The Freezing Point (*FP*) and Osmotic Pressure (*OP*) of Sea Water[a]

Salinity, ‰	FP °C	OP at 0°C, atm.
5	−0.27	3.2
10	−0.53	6.4
15	−0.80	9.7
20	−1.07	13.0
25	−1.35	16.3
30	−1.63	19.7
35	−1.91	23.1
40	−2.20	26.6

[a] Reprinted from Cox (1965) with permission of Academic Press.

c. Osmotic Pressure

Stenius (1904) formulated an expression for computing osmotic pressure at temperature t °C as follows:

$$OP(t°) = OP(0°) \times \frac{273 + t}{273}$$

Values of osmotic pressure at 0°C computed for salinity values from 5 to 40‰ are given in Table II-13.

9. *Transmission of Sound in Sea Water*

Since man is determined to explore, exploit, and fight in the seas it is necessary for him to "see." Since light penetration in sea water (see Section A.10) is feeble and the ocean depths lie in darkness, the visual sense is not adequate (see also the section on vision in Chapter V). But acoustic energy does provide a "window" in the sea. According to Horne (1970), sonar, or the

> . . . technology of underwater sound transmission provides an invaluable research tool for the exploration of the topology and geology of the ocean floor and even sub-floor, while the closely related sister technique of ultrasonics is being increasingly used in fundamental research on the properties and structure of liquids, including water and aqueous electrolytic solutions.
>
> Sonar systems fall conveniently into two types: *Active systems*, in which a sound signal is bounced off an object and the reflected signal detected; and *passive systems*, which simply listen to detect any sound signal originated by the object. In traveling through the seawater the sound signal is subject to degradation and the degradation can also be divided into two types (a) dispersion, scattering, and reflection, which represent deflections or changes in the direction of the sound ray; and (b) absorption which, together with the unavoidable diminution due simply to spherical spreading, represents a reduction of the energy level or intensity of the sound field.
>
> . . . the physical chemistry of sea water and the structural changes they reflect have remarkably little effect on the velocity of sound, but they are directly responsible for

the absorption of sound in sea water. The velocity of sound in a medium depends on the density and the compressibility. Hence those parameters which affect the density and the compressibility, and in the same order of relative importance, namely, temperature, pressure and salinity, will determine the velocity of sound in sea water.

Sound travels at 4700 ft/sec in water, compared to only 1090 ft/sec in air (Bark *et al.* 1964). Generally, the velocity of sound in sea water can be computed on a thermodynamic basis from the expression

$$V_s = \lambda / \rho k$$

where V_s is the sound velocity, λ is the ratio of specific heats C_p/C_v, ρ is the density, and k is the compressibility.

a. Dispersion, Scattering, and Reflection

Even though dissolved gases do not have a significant effect on sound velocity in sea water, large concentrations of bubbles can affect sound velocity by altering density and compressibility (Greenspan and Tschiegg 1956).

As pressure, temperature, and salinity ratios are complex in real situations, so the path of sound rays yields a complex profile. Vertical thermal profiles or deflections from high-pressure and high-temperature regions may provide sound wave guides or channels and affect velocity. Even stronger deflections result from heterogeneities in the sea. Figure II-3 shows examples of sonar paths in the sea. Sound waves vary from place to place and in given locations, chaotically from fluctuations in thermal micro-structure, systematically from periodic processes such as internal waves (Horne 1970).

b. Absorption

Although most acoustic energy loss results from geometric spreading, loss by absorption is also appreciable. Two kinds of absorption processes affect sound velocity in sea water—*normal* absorption, which occurs in pure water as well as sea water, and an additional absorption that occurs in sea water but not in pure water. This additional absorption has been referred to as the *structural* (Horne 1970) or *chemical* (Andersen 1973) *relaxation phenomenon.* According to Andersen (1973):

> In the frequency range of 1 to 1000 kHz, acoustic absorption in sea water exceeds that of fresh water by as much as thirtyfold. This anomaly is principally due to the dissociation of magnesium sulfate ion-pairs, although this compound in undissociated form has a concentration in sea water of only about 5×10^{-3} molar. Magnesium sulfate alone does not completely regulate the excess sound absorption properties of sea water, and it is apparent that other chemical equilibria also contribute. Present research is beginning to indicate that magnesium bicarbonate and hydrated boron also may be significant.

Wilson (1960*a,b*, 1962) has made direct measurements of sound velocity in sea water, which are available for detailed study; tables have been published (U. S. Navy 1961) based on Wilson's measurements, with pressures expressed in decibars and the pressure at the sea surface taken as the arbitrary zero. Brief extracts from this source appear as Tables II-14(a) and II-14(b).

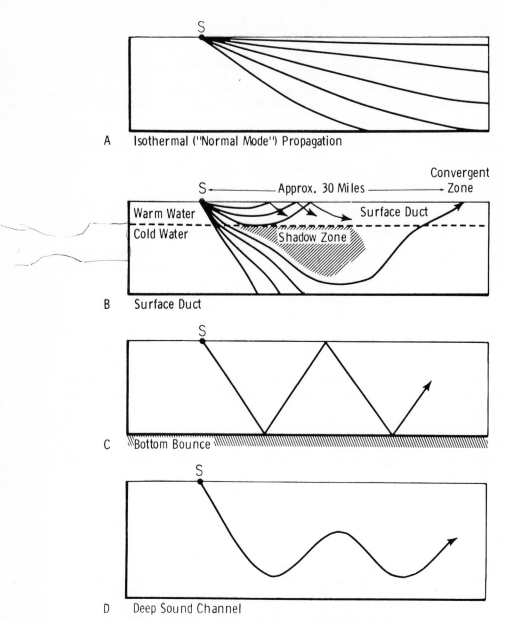

Figure II-3. Some sonar paths in the sea. [Reprinted from Horne (1969) with permission of John Wiley and Sons.]

More recently measurements of sound velocity have been made by del Grosso and Mader (1972) which do not agree with Wilson's data at pressures greater than 1 atm. Unfortunately, it cannot be stated with certainty who is right at the present time.

Table II-14(a)
Velocity of Sound in Sea Water[a]
(U. S. Navy 1961)

Pressure, dbar	Temperature, °C						
	0	5	10	15	20	25	30
0	1449.3	1471.0	1490.4	1507.4	1522.1	1534.8	1545.8
1000	1465.8	1487.4	1506.7	1523.7	1538.5	1551.3	1562.5
2000	1482.4	1504.0	1523.2	1540.2	1555.0	1567.9	1579.2
3000	1499.4	1520.7	1538.6	1555.6			
4000	1516.5	1537.7	1555.2	1572.2			
5000	1533.9	1554.8	1571.9	1588.9			
6000	1551.5	1572.1					
7000	1569.3						
8000	1587.3						
9000	1605.4						
10000	1623.5						

[a] Velocities in m/sec; pressures in decibars above atmosphere. Salinity 35‰. For other salinities see Table II-14(b).

10. *Light Transmission in Sea Water*

The rate at which downward-traveling radiation decreases, an important aspect of light transmission in sea water, is called the extinction coefficient (Petersson and Petersson 1929). The absorption coefficient measures the decrease of intensity of radiation passing through a layer of water because of conversion into another form of energy or lateral scattering. The refractive index is the measure of the bending of light rays occurring when the rays traverse the boundary between two media of different densities, from the atmosphere into sea water in this case.

Table II-14(b)
Effect of Salinity on Sound Velocity[a]
(U. S. Navy 1961)

S, ‰	Temperature, °C						
	0	5	10	15	20	25	30
30	−7.0	−6.7	−6.5	−6.2	−5.9	−5.6	−5.3
32	−4.2	−4.0	−3.9	−3.7	−3.5	−3.4	−3.2
33	−2.8	−2.7	−2.6	−2.5	−2.4	−2.2	−2.1
34	−1.4	−1.3	−1.3	−1.2	−1.2	−1.1	−1.1
35	0	0	0	0	0	0	0
36	1.4	1.3	1.3	1.2	1.2	1.1	1.1
37	2.8	2.7	2.6	2.5	2.4	2.3	2.1
38	4.2	4.1	3.9	3.7	3.6	3.4	3.2
40	7.0	6.8	6.5	6.2	6.0	5.7	5.3

[a] Corrections to be applied to the values in Table II-14(a) for salinities other than 35‰.

a. Extinction Coefficient

When light enters a body of water it experiences in general two types of attenuation. Part of it is subjected to scattering, being reflected in all directions, while another portion is absorbed by being converted into another form of energy. Strictly, therefore, we have to distinguish the true absorption coefficient from the scattering coefficient; but for practical purposes it is sometimes convenient to add them together as the total attenuation or extinction coefficient.

The extinction coefficient is determined from the relationship

$$K_\lambda = 2.30(\log_{10} I_{\lambda, z} - \log_{10} I_{\lambda, (z+1)})$$

where $I_{\lambda, z}$ and $I_{\lambda, (z+1)}$ represent radiation intensities of wavelength λ at depths z and $(z + 1)$ m. Figures II-4 and II-5 show the extinction against wavelength in sea water free of dissolved or suspended organic material. Table II-15 gives the differences between sea water and pure water.

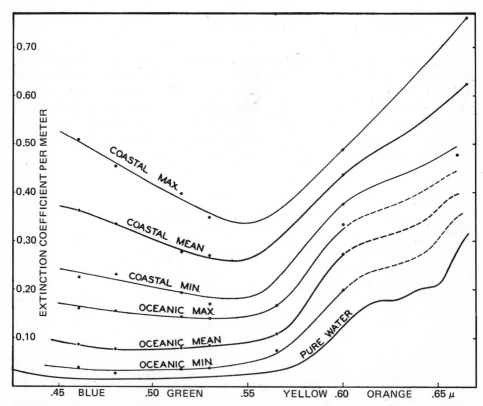

Figure II-4. Extinction coefficients of radiation of different wavelengths in pure water and in different types of sea water. [Reprinted from Sverdrup *et al.* (1970) by permission of Prentice-Hall.]

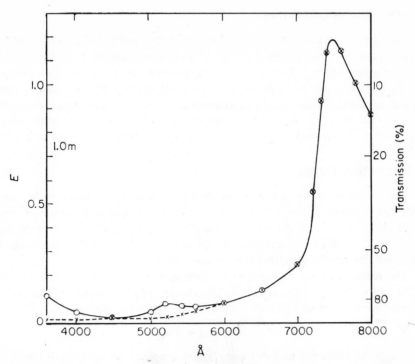

Figure II-5. Extinction (1-m path) against wavelength. Solid line, sea water. Broken line, pure water. [Reprinted from Cox (1965) by permission of Academic Press.]

Table II-15

Differences between the Extinctions of Sea Waters and Pure Water[a]

Sample	Wavelength, Å							
	3600	4000	5000	5200	6000	7000	7500	8000
Artificial sea water	0.010	0.002	0.003	0.005	nil	nil	nil	nil
Ocean water, unfiltered	0.011	0.008	0.005	0.006	0.001	nil	nil	nil
Continental slope water, unfiltered	0.051	0.029	0.009	0.008	0.002	0.010	0.015	0.002
Continental slope water, filtered	0.015	0.009	0.003	0.003	0.002	0.005	nil	nil
Inshore water, unfiltered	0.054	0.041	0.026	0.024	0.025	0.027	0.025	0.015
Inshore water, filtered	0.014	0.009	0.003	0.003	nil	nil	nil	nil

[a] Reprinted from Cox (1965) with permission of Academic Press. Values given are $E_{10\,cm}$(sea water)—$E_{10\,cm}$(pure water) as a function of wavelength.

b. Absorption Coefficient

The absorption coefficient as applied to electromagnetic radiation is a measure of the rate of decrease in intensity of a beam of photons or particles in its passage through a particular substance, in this case water. As pointed out above, we must differentiate between scattering and absorption. It is well to remember that the scattered radiation may still be effective in the same way as the original radiation, but the absorbed portion ceases to exist as radiation or is reemitted as secondary radiation. The absorption for pure water of various wavelengths of electromagnetic radiation is given in Table II-16.

c. Refractive Index

The refractive index is defined as the phase velocity of radiation in free space divided by the phase velocity of the same radiation in a specified medium. It may also be defined as the ratio of the sine of the angle of incidence to the sine of the angle of refraction. Table II-17 shows this latter relationship, and Figure II-6 shows the various factors influencing the penetration of sunlight into the sea.

The refractive index increases with increasing salinity and decreasing temperature; it varies directly with wavelength. Utterback *et al.* (1934) found that the refractive index for sea water could be represented by an expression such as the following:

$$n_t = n_{0,t} + k_t Cl$$

where n_t is the refractive index of a sea water sample at a temperature t in °C, $n_{0,t}$ is that of distilled water at the same temperature, k_t is a constant for that temperature, and Cl is the chlorinity. Miyake (1939) determined the refractive index of sea water from the relationship

$$n = n_0 + \sum (U - n_0)$$

where n_0 is the refractive index of distilled water and U is the refractive index of solutions of single salts having comparable concentrations of these salts in sea water.

Table II-16
Absorption Coefficients per Meter of Pure Water[a]

Wavelength, μm	Absorption coefficient per meter	Wavelength, μm	Absorption coefficient per meter	Wavelength, μm	Absorption coefficient per meter	Wavelength, μm	Absorption coefficient per meter
0.32	0.58	0.52	0.019	0.85	4.12	1.60	800
0.34	0.38	0.54	0.024	0.90	6.55	1.70	730
0.36	0.28	0.56	0.030	0.95	28.8	1.80	1700
0.38	0.148	0.58	0.055	1.00	39.7	1.90	7300
0.40	0.072	0.60	0.125	1.05	17.7	2.00	8500
0.42	0.041	0.62	0.178	1.10	20.3	2.10	3900
0.44	0.023	0.65	0.210	1.20	123.2	2.20	2100
0.46	0.015	0.70	0.84	1.30	150	2.30	2400
0.48	0.015	0.75	2.72	1.40	1600	2.40	4200
0.50	0.016	0.80	2.40	1.50	1940	2.50	8500

[a] Reprinted from Sverdrup (1942) with permission of Prentice-Hall.

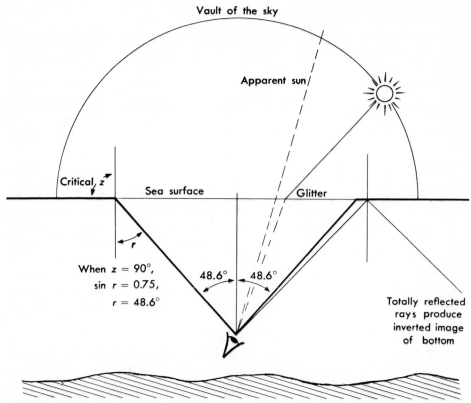

Figure II-6. Factors influencing the radiance of the sky as seen from a point beneath the surface of a glassy sea. [Reprinted from von Arx (1962) by permission of Addison-Wesley.]

Table II-17
Reflectivity of Light Waves as a
Function of Angle of Incidence[a]

Angle of incidence, deg	Percent reflectivity
0	2.0
10	2.0
20	2.1
30	2.1
40	2.5
50	3.4
60	6.0
70	13.4
80	34.8
90	100.0

[a] Reprinted from von Arx (1962) with permission of Addison-Wesley.

Table II-18
Refractive Index of Sea Water[a]

S, ‰	Temperature, °C					
	0	5	10	15	20	25
0	1.3 3395	1.3 3385	1.3 3370	1.3 3340	1.3 3300	1.3 3250
5	3500	3485	3465	3435	3395	3345
10	3600	3585	3565	3530	3485	3435
15	3700	3685	3660	3625	3580	3525
20	3795	3780	3750	3715	3670	3620
25	3895	3875	3845	3805	3760	3710
30	3991	3966	3935	3898	3851	3798
31	4011	3985	3954	3916	3869	3816
32	4030	4004	3973	3934	3886	3834
33	4049	4023	3992	3953	3904	3851
34	4068	4042	4011	3971	3922	3868
35	4088	4061	4030	3990	3940	3886
36	4107	4080	4049	4008	3958	3904
37	4127	4099	4068	4026	3976	3922
38	4146	4118	4086	4044	3994	3940
39	4166	4139	4105	4062	4012	3958
40	(4185)	(4157)	(4124)	(4080)	(4031)	(3976)
41	(4204)	(4176)	(4143)	(4098)	(4049)	(3994)

[a] Reprinted from Riley and Chester (1971) with permission of Academic Press. Values are for sodium D light.

The refractive index of sea water at various salinities between 0 and 41‰ and at various temperatures between 0 and 25°C is given in Table II-18.

11. Summary

Whatever the engineering challenge, understanding the various physical and chemical characteristics of sea water will be a help. For this reason Table II-19 summarizing information concerning basic properties of sea water was designed for ocean engineers by Jolliff (1973) using data of Defant (1961), Groen (1967), Hill (1962–63), Pierson and Neumann (1966), Sverdrup et al. (1942), von Arx (1967), and Williams (1962).

C. Coping with Waves, Tides, and Currents*

1. General Background

Inland divers and particularly scuba divers trained in a swimming pool or a rock quarry must receive additional training before attempting ocean diving. Some of the environmental problems will be presented in the following material, but these written

* All of the material in this section was taken from Somers (1972).

Table II-19

Physical and Chemical Properties of Sea Water[a]

Property	Definition	Symbol	Units	Governing equation
Adiabatic temperature change	Adiabatic temperature changes which accompany pressure changes as a result of the compressibility of sea water	$\dfrac{\delta T}{\delta Z}$	°C/1000 m	$\dfrac{\delta T}{\delta Z} = \beta g \dfrac{T}{JC_p}$
Alkalinity	Quality or state of being alkaline	A	P_H	$A = 0.068\ S\%$ $A = 0.123\ Cl\%$
Chlorinity	The number giving the chlorinity in grams per kilogram of a salt water sample is identical with the number giving the mass in grams of the "atomic weight silver" just necessary to precipitate the halogen in 0.3285233 kg of salt water sample	$Cl\%$	‰	Normal sea water taken as that which has a $Cl\% = 19.4\%$. This technique gives $\sim$0.02% accuracy in salinity calculation; $$Cl\% = \dfrac{S\% - 0.030\%}{1.805}$$
True compressibility coefficient	Sea water's relative change (decrease) of volume $\Delta V/V$ is proportional to the change of pressure ΔP, where K is defined as the proportionality coefficient	K	g/cm²	$\dfrac{\Delta V}{V} = K \dfrac{1}{\Delta P}$ or $K = -\dfrac{1}{\alpha}\dfrac{d\alpha}{dP}$
Mean compressibility of sea water	Since α is a function of S, T, and P, there exists a mean compressibility k of sea water between pressures of 0 and p bars defined as in the governing equation in terms of α	k	cm²/g	$K = \dfrac{k + p(dk/dp)}{1 - kp}$ $\alpha_{STP} = \alpha_{STO}(1 - kp)$ where p = pressure of 0 to p bars
Density	Defined as mass per unit volume	ρ_{STP}	g/cm³	$\rho_{STP} = 35.0 = \underbrace{\epsilon_S + \epsilon_T + \epsilon_{ST} + \epsilon_P + \epsilon_{SP} + \epsilon_{TP} + \epsilon_{STP}}_{C}$ $\rho_{STP} = 1 + 10^{-3}\,\sigma_{T+c}$ at 0° $C\sigma_0 = (\rho_{S00} - 1)10^3$ when $p = 0$ or $\sigma_0 = 0.069 + 1.4708\ Cl - 0.0157\ Cl^2 + 0.0000398\ Cl^3$ $\sigma_0 = -0.093 + 0.8149\ S - 0.000482\ S^2 + 0.0000068\ S^3$

Table II-19—Cont.

Property	Symbol	Units	Governing equation	Definition
Dielectric constant	E	No units (pure scalar)	$E = C_0/C$	Number which expresses how much smaller the electric intensity is in space filled versus in a vacuum if same electric field is present.
Diffusion coefficient (laminar flow)	Δ	$\dfrac{cm^3}{conc\ sec}$	$\dfrac{dM}{dt} = -\Delta\dfrac{dc}{dn}$	In a solution in which the concentration of a dissolved substance varies in space, the amount which diffuses through 1 cm²/sec is proportional to the change in concentration per centimeter along line normal to surface
Dynamic viscosity	μ	$\dfrac{g}{cm\ sec}$	$\tau = \mu\dfrac{du}{dz}$	Proportionality coefficient relating frictional fluid stress to the velocity change $\perp$ to the x direction
Eddy conductivity (turbulent flow)	r	No units (pure scalar)	$\dfrac{dQ}{dt} = -rA\dfrac{dt}{dn}$	Heat flux across unit area and temperature gradient related by coefficient of proportionality r
Electric conductivity coefficient	Σ	1/ohm-cm	$\Sigma = 1/\omega$	The rate of electrical flow in the sea water solution
Elevation of boiling point	Δt_s	°C	$\Delta t_s = 0.0158\ S‰$	Temperature above 100°C at which sea water will boil
Freezing point	T_g	°C	$T_g = -0.003 - 0.0527\ S - 0.00004\ S^2$ $T_g = -0.0966\ Cl - 0.0000052\ Cl^3$ $T_g = 0.0086 - 0.06463\sigma_0 - 0.0001055\sigma_0^2$	Temperature at which sea water will freeze
Kinematic viscosity	v	cm²/sec	$v = \mu/\rho$	Defined as the ratio of μ/ρ
Latent heat of evaporation	L	g/cal	$L = 596 - 0.52\ (T°C)$	Amount of heat needed for producing 1 g of water vapor of the same temperature as the water
Osmotic pressure	OP	dbar or atm	$OP_0 = 12.08\ (T_g°C)$ $OP_t = OP_0(1 + \alpha_{T}) = OP_0\dfrac{273 + T}{237}$	Difference or increase in pressure due to salinity effect on osmosis

Term	Description	Symbol	Units	Formula
Radiation absorption coefficient	Intensity of parallel beams of radiation decreases in direction of beams; decrease in layer proportional to intensity	K_λ	radiation/m	$K_\lambda = \dfrac{2.30}{L}(\log I_{\lambda h} - \log I_{\lambda(h+1)})$
Refractive index	Ratio of the sine of the angle of incidence to the sine of the angle of refraction	η	No units (pure scalar)	$\eta = \eta_0 + \Sigma U - \eta_0$ $\eta = \dfrac{\sin i}{\sin r}$
Salinity	Total amount of solid materials in grams contained in 1 kg of sea water when all carbonate was thought to be oxide, the bromine and iodine replaced by chlorine, and all organic matter completely dissolved	S	‰ (per mille) or g/kg	$S_{\text{total}} = 0.043 + 1.0044\,S$ $S\%_0 = 0.030 + 1.8050\,Cl\%_0$
Sound absorption coefficient	Loss of sound energy in direction of sound propagation	N	cm³/λ²	$N = \dfrac{16\pi^2 \eta}{3\lambda^2 V_s \rho}$
Sound velocity	Speed of sound in sea water	V_s	cm/sec	$V_s = \lambda/\rho k$
Specific heat	Amount of heat required to increase the temperature of 1 g sea water by 1°C	C_p	J/g	$\dfrac{d(C_P)}{d_P} = -\dfrac{T}{J}\left(\dfrac{d\beta}{dt} + \beta^2\right)$ $C_P = C_V + \dfrac{T\beta^2}{\rho KJ}$ $C_P = 1.005 - 0.004136\,S + 0.0001098\,S^2 - 0.000001324\,S^3$ at 0°C and atmospheric pressure
Specific volume	Defined as the reciprocal of density	α	cm³/g	$\alpha_{STP} = 1/\rho_{STP}$ $\alpha_{STO} = 0.97264 + \Delta_{ST}$ where $\Delta_{ST} = 0.02736 - \dfrac{10^{-3}\sigma_T}{1 + 10^{-3}\sigma_T}$
Surface tension	That property due to molecular forces by which the surface film of all liquids tends to bring the contained volume into a form having the least superficial area	T'	dyne/cm²	$T' = 75.64 - 0.144\,T + 0.0399\,Cl\%_0$

Table II-19—*Cont.*

Property	Definition	Symbol	Units	Governing equation
Temperature	Degree of heat or cold measured on a subscribed scale	T	°C or °F	$°F = \frac{9}{5}C° + 32°$
Thermal conductivity coefficient (laminar)	Proportionality coefficient relating heat flux to temperature gradient	γ	$\dfrac{cal}{cm\ sec\ °C}$	$\dfrac{dQ}{dt} = -\gamma g \dfrac{di}{dn}$
Thermal expansion coefficient	Relative change of the specific volume with temperature	β	°C	$\beta = \dfrac{1}{\alpha_{STP}}\dfrac{\delta\sigma_{STP}}{\delta T}$
Turbulent eddy viscosity	Proportionality constant relating shear of observed velocities to the stress	A_t	$\dfrac{W}{cm\ °C}$	$\tau_s = A_t \dfrac{d\bar{N}}{dn}$
Vapor pressure	Pressure of saturated vapor as a function of temperature only	e	atm	$\Delta P/P_0 = 0.538 \times 10^{-3}\ S‰$ $e/e_0 = 1 - 0.000969\ Cl‰$
Pressure	Force per unit area	P	lb/in.²	Pressure in ocean increases by 1 atm for every 33 ft of water depth
Visibility range	Distance from an object at which the contrast is reduced to the threshold value of 0.02	R	m	$R = \dfrac{\ln 50}{a}$
Chlorosity	Property corresponding to chlorinity expressed on a volume basis	Ch	g/20°C/liter	Ch is obtained by multiplying the chlorinity of a water sample by its density at 20°F

[a] Reprinted from Jolliff (1973) with permission of Compass Publications. The symbols used in this table are as follows:

$\delta T/\delta Z$ Ratio of adiabatic temperature change to a vertical distance change of 1000 m
β Coefficient of thermal expansion
g Acceleration of gravity
T Temperature °K, °F, or °C as required
J 4.1862×10^7 ergs/cal (mechanical equivalent of heat)
A Alkalinity
$S‰$ Salinity "per mille"
$Cl‰$ Chlorinity "per mille"
P_H Symbol used to express alkalinity or acidity
V Volume
K Coefficient of compressibility
P Pressure

α Specific volume
k Mean compressibility coefficient
ρ Density
σ_0 Density, corrected for S T P
dM/dt Diffusion of substance per unit time
dc/dn Concentration of substance along normal line to diffusion surface
Δ Diffusion coefficient
τ Shear stress
μ Dynamic viscosity
du/dz Velocity change in vertical direction
E Dielectric constant
C_0 Capacitance in vacuum of a condenser
C Capacitance with dielectric between condenser plates

r	Eddy conductivity coefficient of proportionality
Σ	Electric conductivity coefficient
dQ/dt	Heat flux
A	Surface area, cm²
$d\bar{T}/dn$	Temperature gradient in direction n
Δt_s	Change in elevation of boiling point above 100°C
T_g	Freezing point of sea water
v	Kinematic viscosity
L	Latent heat of vaporization
OP_0	Osmotic pressure at T_g °C
OP_t	Osmotic pressure at any temperature
K_λ	Radiation absorption coefficient
I_λ	Radiation energy
N	Sound absorption coefficient
η	Refractive index
λ	Wavelength
V_s	Propagation of velocity of sound
C_p	Specific heat at constant pressure
C_v	Specific heat at constant volume
Δ_{STP}	Change in specific volume at STP
γ	Thermal conductivity coefficient
τ_s	Shear of vertical velocity
e	Vapor pressure of sea water sample
e_0	Vapor pressure of distilled water
q	Area, cm²
A_t	Turbulent eddy viscosity
T'	Surface tension
ω	Specific resistance of conductor
R	Visibility range
a	Attenuation coefficient per meter
Ch	Chlorinity
V_s	Speed of sound in water

descriptions are not sufficient in themselves to prepare inland divers for an ocean experience.

2. Waves

Waves are a series of undulations generally propagated on the water's surface by the force of the wind. Ocean waves are usually measured in terms of their length, height, and period. *Wave length* is the horizontal distance between successive crests, *height* is the vertical distance between crest and trough, and *period* is the time required for the movement of two successive crests (or troughs) past a given reference point.

Waves are moving forms, a translation of energy from water particle to water particle, with very little mass transport of the water. Waves develop under the influence of winds. The air pressure changes on the surface and the frictional drag of the moving air of these winds develop ripples on the water surface which evolve into waves whose dimensions tend to increase with the wind velocity, duration, and *fetch* (the length of the area over which the wind is blowing). Energy is transferred directly from the atmosphere to the water. These waves, generated locally by a continuing wind, are known as *sea*. This sea persists only as long as the wind velocity continues. When the wind velocity decreases the wave continues as a *swell wave* until it loses its height and steepness and decays. As the wave forms approach shore and move across shallow bottoms, they are reflected, diffracted, and refracted. When a wave encounters a vertical wall, such as a steep, rocky cliff rising from deep water, or a seawall, it is *reflected* back upon itself with little loss of energy (Figure II-7). If the period of the

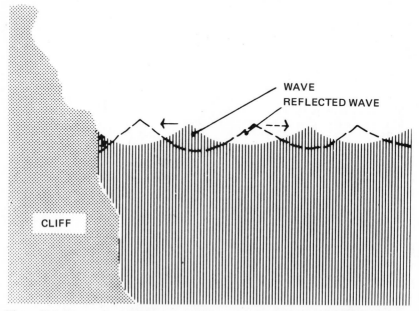

Figure II-7. Wave reflection [Somers (1972); reprinted by permission from the author].

approaching wave train is regular, a pattern of standing waves may be established in which the orbits of the approaching and reflected waves modify each other in such a way that there is only vertical water motion against the cliff and only horizontal motion at a distance out of one-fourth wave length. Submerged barriers, such as a coral reef, will also cause reflections.

When a wave encounters an obstruction, the wave motion is diffracted around it (Figure II-8). As the waves pass the obstruction, some of their energy is propagated sideways due to friction with the obstruction, and the wave crest bends into the apparently sheltered area.

As the wave train moves into shallow water, the friction on the bottom causes it to slow. Since different segments of the wave front are moving in different depths of water, the crest bend and the wave direction constantly change. This is called *refraction* (Figure II-9). Essentially the wave crest or front parallels the contours of the bottom. A simple example of refraction is that of a set of waves approaching a straight shoreline at an angle. The part of each wave nearest shore is moving in shallower water and, consequently, is moving slower than the part in deeper water. Thus the wave fronts tend to become parallel to the shoreline and the observer on the beach will see larger waves coming directly toward him. On an uneven shoreline the effect of refraction is to concentrate the wave energy on points of land and disperse the wave energy in coves or embayments. Submarine depressions, i.e., canyons, also cause the waves to

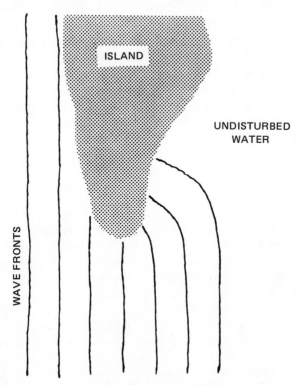

Figure II-8. Wave diffraction [Somers (1972); reprinted by permission from the author].

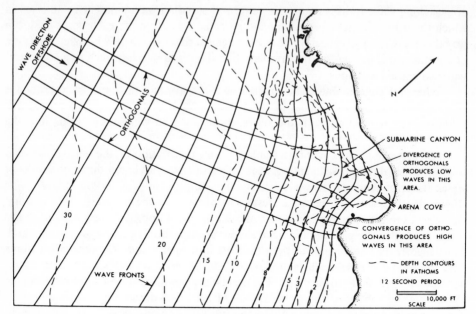

Figure II-9. Wave refraction (Baker *et al.* 1966).

react in a similar fashion. The waves dissipate over the canyon and increase in intensity on the perimeter of the canyon. Any irregularities in bottom topography in shallow waters will cause refraction to some degree.

A knowledge of the behavior of waves as they enter shallow water is of considerable significance to scuba divers when planning entries from shore. By observing wave patterns and by studying the shoreline configuration and bottom topography, the diver can select the locations where wave energy and, consequently, height is least. This will aid entry and nearshore work.

3. *Surf*

As swell, the waves traverse vast expanses of ocean with little modification or loss of energy. However, as the waves enter shallow water, the motion of the water particles beneath the surface is altered. When the wave enters water of depth equal to or less than one-half the wave length, it is said to "feel bottom." As the wave "feels bottom," its wave length decreases and steepness increases. Furthermore, as the wave crest moves into water where the depth is about twice that of the wave height, the crest changes from rounded to a higher, more pointed mass of water. Finally, at a depth of approximately 1.3 times the wave height, when the steepest surface of the wave inclines more than 60° from the horizontal, the wave becomes unstable and the top portion plunges forward. The wave has broken; this is *surf* (Figure II-10). This zone of "white water," where the waves finally give up their energy and where systematic water motion gives way to violent turbulence, is the surf zone. The "white water" is a mass of water with bubbles of entrapped air. Having broken into a mass of turbulent

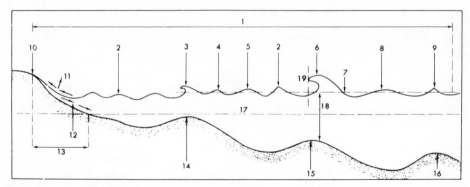

Figure II-10. Schematic diagram of waves in the breaker zone. [Somers (1972); reprinted by permission from the author.]

foam, the wave continues landward under its own momentum. Finally, at the beach face, this momentum carries it into an uprush or swash. At the uppermost limit, the wave's energy has diminished. The water transported landward in the uprush must now return seaward as a backrush or current flowing back to the sea. This seaward movement of water is generally not evident beyond the surface zone or a depth of 2–3 ft. This backrush is *not* to be considered as an undertow. Undertow is one of the most ubiquitous myths of the seashore. These mysterious mythical currents are said to flow seaward from the beach along the bottom and "pull swimmers under." There are currents in the surf zone and other water movements which may cause trouble for swimmers, but not as just described. Once the wave has broken, if the water deepens again, as it does where bars or reefs lie adjacent to shore, it may reorganize into a new wave with systematic orbital motion. The new wave is smaller than the original one and it will proceed into water equal to 1.3 times its height and also break. A diver may use the presence of waves breaking offshore an an indicator for the location of rocks, bars, etc., and plan his entry or approach to shore accordingly (Figure II-11).

4. Currents

Currents are caused primarily by the influence of surface winds, changing tides, and rotation of the earth. They are essentially flowing masses of water within a body of water. Divers must always take currents into account in planning and executing a dive, particularly a scuba dive.

a. Currents in Oceans and Large Lakes

Large ocean currents, such as the Gulf Stream of the Atlantic and Japan current of the Pacific, flow continuously, although there may be local variations in magnitude and location. Local wind-derived currents are common throughout the oceans and on large lakes. The current velocity may exceed 2–3 knots. Attempts to swim against this type of current may result in severe fatigue. Sometimes in the Gulf of Mexico as well as other portions of the ocean, there may be no noticeable current at the surface with

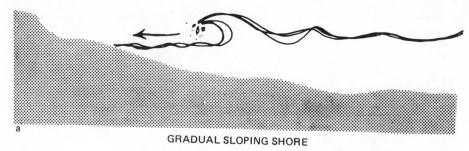

GRADUAL SLOPING SHORE

SHORE BREAKER   REEF BREAKER

orbital
motion

b

REEF

Figure II-11. (a) Gradual sloping shore; (b) reef. [Somers (1972); reprinted by permission from the author.]

a 1–2-knot current at a depth of 10–20 ft, or there may be a current at the surface and no current at 10–20 ft down.

b. Rip Currents

In and adjacent to the surf zone, currents are generated by waves approaching the bottom contours at an angle and by irregularities in the bottom, but these currents seldom exceed a speed of one knot. These currents move seaward at weak points (gaps in a bar or reef, submarine depressions perpendicular to the shore) and form a *rip current* through the surf (Figure II-12). The knowledgeable diver will use modest rip currents to aid seaward movement. An unsuspecting swimmer, when caught in a rip, should ride the current and swim to the side, not against the current. Outside the surf zone the current widens and slackens. He can then enter the beach at another location. The rip current dissipates a short distance from the shore.

c. Tides and Tidal Currents

The tidal phenomenon is the periodic motion of the ocean waters in response to the variations in attractive forces of various celestial bodies, principally the moon and sun, upon different parts of the rotating earth. On the seacoasts this motion is evidenced by a rhythmic, vertical rise and fall of the water surface called the tide, and by horizontal movements of the water called tidal currents. Essentially, tides are long-

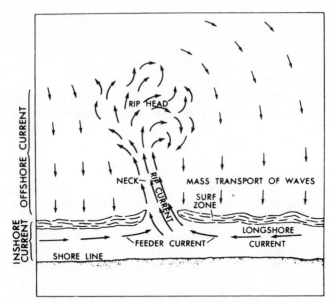

Figure II-12. Nearshore current system (Baker *et al.* 1966).

period waves having a period of 12 hr and 25 min and a wave length equal to one-half the circumference of the earth. The tidal cycle is 24 hr and 50 min.

Tidal current, the periodic horizontal flow of water accompanying the rise and fall of the tide, is of considerable significance to the diver who must work in restricted bay-mouth areas, channels, etc. Offshore, where the direction of flow is not restricted by any barriers, the tidal current flows continuously, with the direction changing through all points of the compass during the tidal period. In rivers or straits, or where the direction of flow is more or less restricted to certain channels, the current reverses with the rise and fall of the tides. In many locations there is a definite relationship between times of current and times of high and low water. However, in some localities it is very difficult to predict this relationship. Along channels or waterways the relationship will change as the water progresses upstream.

Tidal current movement toward shore or upstream is the *flood*; the movement away from shore or downstream is the *ebb*. At each reversal of current, a short period of little or no current exists, called *slack water*. During flow in each direction, the speed will vary from zero at the time of slack to a maximum, called *strength of flood or ebb*, about midway between the slack periods.

d. Diving in Currents

Divers are encouraged to consult local tide tables, confer with local authorities, and make personal evaluations of the water movements in order to determine time of slack water, and consequently, the best time to dive. Tide tables and specific information are contained in various forms in many navigational publications. Tidal current tables, issued annually, list daily predictions of flood and ebb times, and of the times of intervening slacks.

In some channels or straits safe diving time will be limited to 10–20 min of slack water. Specific precautions must be taken when divers work in these areas. Dives must be planned and timed precisely. Scuba diving may be least desirable. Surface-supplied diving equipment, with heavy weighted shoes, may be required for the diver to work in the currents. He should not attempt to swim against the tidal current. If he is caught in a current, he should surface, inflate his lifejacket, and swim perpendicular to the current toward shore or signal for pick up by the safety boat.

When working in large ocean or wind currents the diver should observe the following precautions to minimize the hazards:

1. The diver should always wear a personal flotation device.

2. The diver should be in good physical condition when working in currents.

3. A safety line at least 100 ft with a float should be trailed over the stern of a boat anchored in a current during diving operations. Upon entering the water, a diver who is swept away from the boat by the current can use this line to keep from being carried far down current.

4. Descent should be made down a weighted line placed at the stern or, if unavoidable, down the anchor line. Free-swimming descents in currents should be avoided. If the diver stops to equalize pressure, he may be swept far down current. Furthermore, if a diver has to fight a current all the way to the bottom, he will be fatigued, a hazardous situation under water. Ascent should also be made up a line.

5. When a bottom current is encountered at the start of the dive, the diver should always swim into the current, not with it. This will facilitate easy return to the boat at the end of the dive. He should stay close to the bottom and use rocks if necessary to pull himself along in order to avoid overexertion. If the diver wants to maintain position, he should grasp a rock or stop behind a rock, not attempt to swim. The same technique should be used by a fatigued diver to rest.

6. A qualified assistant should stay on the boat at all times. This will facilitate rescue of a diver swept down current.

e. Effects of Currents on Undersea Operations

The tragedy in the Johnson Sea-Link is a vivid reminder of the importance of evaluating currents before embarking upon any type of underwater operation. The submersible made its dive near Fort Pierce to inspect the marine life in the vicinity of a sunken destroyer, which acts as an artificial reef. In this area currents influenced by both tides and the Gulf Stream combine to produce currents of varying velocities and directions. At that particular time, the current was running at a higher velocity than had been realized, thus preventing the submersible from making headway and forcing it into the hull of the destroyer, where it became entangled in cables. By the same token, the strength of the current greatly hampered and delayed rescue operations, with the result that the two men in the ambient pressure lock-out compartment died from heat loss and carbon dioxide poisoning (Booda 1973).

5. *Shoreline Diving Entries*

Most shorelines are not straight features. Irregularities in the form of coves, bays, points, etc., affect the incoming waves, tidal movements, and the resultant current patterns. When preparing for a dive where beach entries and exits are necessary, the diver must take wave approach, shoreline configuration, and currents into account. Entries and exits should be planned to avoid high waves, as on the windward side of points, and to take maximum advantage of current movements. It is wise to avoid dives that require swimming against the current. A dive should never be undertaken from an ocean beach without considering these factors. Hypothetical beach configuration, wave approach, and current diagrams are included in Figure II-13 to aid the diver in the concepts of planning beach-entry dives.

a. Sand Beach Entry

The width of surf zone and the severity of the breaking waves will be influenced significantly by the slope of the beach. On a gradually sloping beach the surf zone will be wide since the wave will break, reform, and break again (Figure II-14). The diver

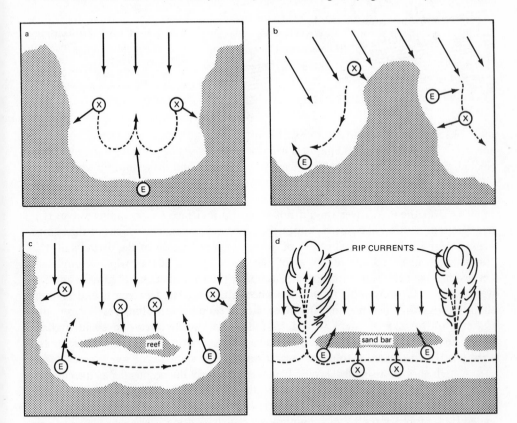

Figure II-13. (a) Small deep coves; (b) points; (c) rocky cove—reefs; (d) sand bar—sandy beach—rip current. [Somers (1972); reprinted by permission from the author.]

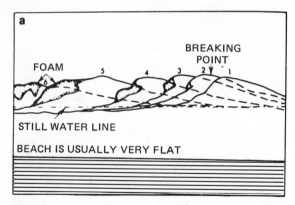

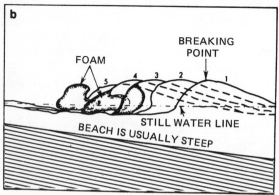

Figure II-14. (a) General character of spilling breakers. (b) General character of plunging breakers. [Somers (1972); reprinted by permission from the author.]

must observe the wave pattern and surf beat in order to time his movement into the surf zone. The best entry technique is usually for the diver to get outfitted completely (including fins), select the best time (least wave height), and move into the zone backward while watching the oncoming waves. As soon as the water is deep enough, the diver should start swimming. He must swim under the oncoming waves, not attempt to swim over them. A diver should not stand up and face an oncoming wave. If a float is used, it should be towed, not pushed into the waves. The weight of the equipment, the shift of the normal center of gravity, the restriction of the diving suit, the cumbersome fins, and the fogginess of the mask are all factors which complicate entries through surf. A diver can compensate for the shift in center of gravity and the weight of the tank by moving with his knees slightly bent, feet apart, and leaning slightly forward. When moving, the diver should slide his feet along the bottom and not attempt to take big steps. If a diver falls or is knocked down, even in shallow water, he should not attempt to stand and regain his footing. He should conserve his energy and swim or crawl to deeper water (or back to shore).

A high surf on a steeply sloping beach is extremely dangerous for a diver in full equipment. The waves will break violently directly on the beach, with a very narrow surf zone (only a few feet wide) [Figure II-14(b)]. A diver wearing fins may be up-ended by the force of the water running down the steep slope after a wave has broken. The diver must evaluate both the shoreline and the surf conditions to determine if safe entry is possible. Under severe conditions, the best judgment may be to abort the dive. To make the entry, the diver should move as close to the water's edge as possible, select the proper time (smallest wave), and move into the water under the oncoming wave as soon as possible.

When exiting through surf, the diver should stop just seaward of the surf zone and evaluate wave conditions. The exit should be timed so that the diver rides the back of the last large wave of a series as far up the beach as possible. At a point where the diver can stand, he should turn his back toward the beach, face the oncoming waves, and move toward the beach with his body positioned to retain his balance. If the oncoming waves are still at chest level or higher, the diver should dive head first into the wave and stand up as soon as possible when the breaking part of the wave has passed. If the wave is below chest level, the diver should simply lie on top of the wave, keep his feet under him, and ride the wave toward shore. A fatigued diver should not attempt to regain his footing, but ride the wave as high up the beach as possible and crawl out on his hands and knees. On exits through the surf, the float should be pushed in front of the diver and released if necessary to avoid injury or entanglement.

b. Rock Shore Entry

When entering surf from a rocky shore, the diver should not attempt to stand or walk. A fall can be extremely hazardous. The diver should evaluate the wave conditions, select the backwash of the last large wave of a series, and crawl into the water. the backwash will generally carry the diver through the rocks. Once the diver is moving, he should not attempt to stop or slow down. If the diver retains a prone swimming position and faces the next oncoming wave, he can grasp a rock or kick to keep from being carried back toward the shore. He can then kick seaward after the wave passes. Floats should be towed behind the diver.

When exiting on a rocky shoreline, the diver must stop outside the surf zone and evaluate the wave conditions. Exit toward the beach is made on the backside of the last large wave of a series. As he loses momentum the diver should grasp a rock or kick in order to avoid being carried seaward by backwash. The diver will finally find it necessary to crawl from the water. When exiting through the surf, the diver should always look back in order to avoid surprise conditions.

References

ANDERSEN, N. R. Personal communication (July 1973).

BAKER, B. B. Jr., W. R. DEEBEL, and R. D. GEISENDERFER, eds. *Glossary of Oceanographic Terms*, SP-35, second edition. Washington, D. C., Oceanographic Analysis Division, Marine Sciences Department, U. S. Naval Oceanographic Office (1966).

BARK, L. S., P. P. GANSON, and N. A. MEISTER. *Tables of the Velocity of Sound in Sea Water*. New York, Macmillan (1964).

BATTELLE COLUMBUS LABORATORIES. *U. S. Navy Diving Gas Manual*, second edition. Washington, D. C., U. S. Navy Supervisor of Diving (1971) (Navships 0994-003-7010).

BERGHAGE, T. E., and H. T. BEATTY. Computer generated depth-pressure conversion tables. U. S. Navy Exp. Diving Unit, Rep. NEDU-RR-3-71 (Apr. 1971).

BOODA, L. L. Tragedy in the Johnson Sea Link. *Sea Technol.* 14:17, 28, 58 (July 1973).

84 Chapter II

BRADSHAW, A., and K. E. SCHLEICHER. The effect of pressure on the electrical conductance of sea water. *Deep Sea Res.* **12**:151 (Apr. 1965).

COX, R. A. The physical properties of sea water. In: Riley, J. P., and G. Skirrow, eds. *Chemical Oceanography*, pp. 73–120. London–New York, Academic Press (1965).

COX, R. A., and N. D. SMITH. The specific heat of sea water. *Proc. Roy. Soc. London A* **252**:51–62 (Aug. 25, 1959).

COX, R. A., M. J. McCARTNEY, and F. CULKIN. The specific gravity/salinity/temperature relationship in natural sea water. *Deep Sea Res.* **17**:679–689 (Aug. 1970).

DEFANT, A. *Physical Oceanography*. Volume I, Oxford, U. K., Pergamon Press (1961).

DEL GROSSO, V. A., and C. W. MADER. Speed of sound in sea water samples. *J. Acoust. Soc. Amer.* **52**:961–974 (Sept. 1972).

EKMAN, V. W. Die Zusammendruckbarkeit des Meerwassers. *Cons. Perm. Int. Explor. Mer, Pub. Circonstance* **43** (1908).

GREENSPAN, M., and C. E. TSCHIEGG. Effect of dissolved air on the speed of sound in water. *J. Acoust. Soc. Amer.* **28**:501 (May 1956).

GROEN, P. *Waters of the Sea*. Princeton, N. J., Van Nostrand (1967).

HILL, M. N. *The Seas*. Vol. I. New York, Wiley–Interscience (1962–63).

HORNE, R. A. *Marine Chemistry*. New York, Wiley (1970).

JOLLIFF, J. V. Physical and chemical properties of sea water. In: *Under Sea Technology Handbook/Directory*, Chapter 2, pp. A7–A13. Arlington, Va., Compass Publications (1973).

KNUDSEN, M. *Hydrographical Tables*. Copenhagen, Tutein og Koch (1901).

KNUDSEN, M. Gefrierfurkttabelle fur Meerwasser, *Cons. Perm. Int. Explor. Mer, Pub. Circonstance* (1903).

KUWAHARA, S. Velocity of sound in sea water and calculation of the velocity for use in sonic sounding. *Hydrogr. Rev.* **16**(2):123–140 (Nov. 1939).

MILES, S. *Underwater Medicine*. Philadelphia, J. B. Lippincott (1969).

MIYAKE, Y. Chemical studies of the western Pacific Ocean, III. Freezing point, osmotic pressure, boiling point and vapour pressure of sea water. *Bull. Chem. Soc. Japan* **14**(3):58–62 (1939).

PETERSSON, H., and O. PETERSSON. Methods for the determination of the density and salinity of sea water. *Svenska Hydrogr.-Biol. Komm. Skr. N.S. Hydrografi* **3**:1–4 (1929).

PICKARD, G. L. *Descriptive Physical Oceanography*. Oxford, U. K., Pergamon Press (1964).

PIERSON, W. J., and G. NEUMANN. *Physical Oceanography*. Englewood Cliffs, N. J., Prentice Hall (1966).

RILEY, J. P., and R. CHESTER. *Introduction to Marine Chemistry*. London and New York, Academic Press (1971).

SOMERS, L. H. *Research Diver's Manual*. Ann Arbor, University of Michigan (1972) (Sea Grant TR-16).

STENIUS, S. Ofrers, *Finska Vetensk Soc., Forh.* **46**:6 (1904).

SVERDRUP, H. U., M. W. JOHNSON, and R. H. FLEMING. *The Oceans, Their Physics, Chemistry and General Biology*. Englewood Cliffs, N. J., Prentice-Hall (1942).

UNESCO. *International Oceanographic Tables*. Wormby, U. K., National Institute of Oceanography (1966).

U. S. NAVY. Tables of the velocity of sound in sea water. Bureau of Ships ref. NObsr 81564 S-7001-0307. Washington, D. C. (1961).

UTTERBACK, C. L., T. G. THOMPSON, and B. D. THOMAS. Refractivity–chlorinity–temperature relationships of ocean waters. *Cons. Perm. Int. Explor. Mer, J. Cons.* **9**:35–38 (1934).

VON ARX, W. S. *An Introduction to Physical Oceanography*. Reading, Mass., Addison-Wesley (1962, 1967).

WILLIAMS, J. *Oceanography, An Introduction to Marine Science*. Boston, Little, Brown (1962).

WILSON, W. D. Equation for the speed of sound in sea water. *J. Acoust. Soc. Amer.* **32**:1357 (Dec. 1960a).

WILSON, W. D. Speed of sound in sea water as a function of temperature, pressure, and salinity. *J. Acoust. Soc. Amer.* **32**:641–644 (June 1960b).

WILSON, W. D. Extrapolation of the equation for the speed of sound in sea water. *J. Acoust. Soc. Amer.* **34**:866 (June 1962).

Man in the Ocean Environment: Physical Factors

A. *Basic Concepts*

Under normal atmospheric conditions man seldom thinks about the air he breathes, its composition, or the fact that it is under pressure. He may note some shortness of breath at high altitudes, but it is not until he goes under increased barometric pressure either in a chamber or as a diver that he is really made aware of pressure changes.

The physical characteristics of the ocean include its pressure, specific gravity, viscosity, chemical composition, temperature, and changes in the conduction of light and sound waves. All of these characteristics have been considered in Chapter II, the Ocean as an Environment. For the underwater worker, plain physical pressure is probably the most important of the ocean environmental factors.

Pressure is the amount of force applied per unit area. In underwater or pressure-chamber activity the pressure units commonly used are atmospheres (atm), pounds per square inch (lb/in.^{2} or psi), and millimeters of mercury (mm Hg). Atmospheric pressure is the amount of pressure or force exerted by the earth's atmosphere and varies with elevation, but for the diver at sea level it is equal to 7×14.7 lb/in.^{2} or 760 mm Hg. Frequently used pressure terms are gauge pressure, absolute pressure, and ambient pressure. Since most gauges are calibrated to read zero at normal atmospheric pressure, gauge pressure is the increase in the pressure being measured above atmospheric pressure. Absolute pressure is gauge pressure plus atmospheric pressure. Ambient pressure is the absolute pressure surrounding an object. Under water the 760 mm Hg may be expressed in several ways:

$$
\begin{aligned}
760 \text{ mm Hg} &= 29.9 \text{ in. Hg} \\
&= 34 \text{ ft fresh water} \\
&= 33 \text{ ft salt water} \\
&= 14.7 \text{ lb/in.}^2 \\
&= 1033.3 \text{ g/cm}^2 \\
&= 1 \text{ atm gauge} \\
&= 1013 \text{ mbar}
\end{aligned}
$$

For each 33 ft depth of salt water there is an increase of 760 mm Hg. Thus, at 33 ft there would be 1 atm gauge, and 2 atm absolute (ATA); and at 99 ft the pressure would be 3 atm gauge and 4 ATA. It is not easy for the uninitiated to understand how a human being in 500 ft of water could withstand 237 lb of pressure per square inch of his body surface, or about 339 tons. But the diver is quite unaware of this and he functions normally. The reason for this is found in the laws which apply to liquids. (Air-filled parts such as the lungs, middle ear, and sinuses, and gas in the gastrointestinal tract are considered later.) The most important law is that water in its liquid form (for purposes of this chapter, its form as steam or ice is not considered) is incompressible; i.e., it is a liquid that does not change significantly in volume or other characteristics due to changes in either temperature or pressure. The further laws which apply to liquids and which can be applied to "fluid man" are (Miles 1969):

1. If pressure is applied to the surface of an enclosed fluid, it is transmitted to all parts of that fluid.
2. The pressure at any point in a fluid is the same in all directions if the fluid is at rest.
3. In a homogeneous liquid the pressure is the same at all points in the same horizontal plane.

B. *Gases*

Air, in comparison to a liquid, has a very low density (it weighs about 0.081 lb/ft^3), is compressible, and its behavior is governed by simpler laws of physics. The temperature, pressure, and volume relationships are usually expressed in terms of an imaginary substance called an "ideal" or "perfect" gas.

In underwater activity it is impossible to divorce air or gas pressure from water pressure, since man must continue to breathe air or some oxygen-containing gas mixture in order to survive. It is equally important to take into account the effect of pressure on the air contained in the lungs, respiratory passages, middle ears, sinuses, and viscera. Underwater the human being must continue to breathe, and in order that his respiratory system may function, the air or gas mixture breathed must be under pressure equal to the depth of the water. Thus under normal diving conditions no pressure difference will be built up between the air in the body and the surrounding tissues, which are, in turn, under the pressure of the water. (For information on breathing mixtures, see Chapter IV.)

Air and other gases are governed by the "gas laws," which are the following:

1. Boyle's—At a constant temperature the volume of a perfect gas varies inversely as the pressure, and the pressure varies inversely as the volume.
2. Charles'—At a constant pressure the volume of a given mass of perfect gas varies directly with the absolute temperature.
3. Dalton's—In a mixture of gases the pressure exerted by one of those gases is the same as it would exert if it alone occupied the same volume.

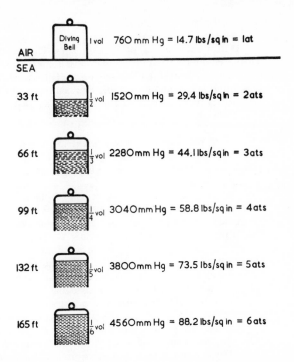

AIR	Diving Bell	1 vol 760 mm Hg = 14.7 lbs/sq in = 1 at
SEA		
33 ft		$\frac{1}{2}$ vol 1520 mm Hg = 29.4 lbs/sq in = 2 ats
66 ft		$\frac{1}{3}$ vol 2280 mm Hg = 44.1 lbs/sq in = 3 ats
99 ft		$\frac{1}{4}$ vol 3040 mm Hg = 58.8 lbs/sq in = 4 ats
132 ft		$\frac{1}{5}$ vol 3800 mm Hg = 73.5 lbs/sq in = 5 ats
165 ft		$\frac{1}{6}$ vol 4560 mm Hg = 88.2 lbs/sq in = 6 ats

Figure III-1. Boyle's law and diving. $PV = K$. [Reprinted from Miles (1969) by permission of the author and Staples Press.]

4. Henry's—At a constant temperature, the amount of a gas which dissolves in a liquid, with which it is in contact, is proportional to the partial pressure* of that gas.

All of these laws are involved as man breathes air or an oxygen-containing gas mixture under increased pressures. Failure to fully understand and heed these laws will lead to trouble. Boyle's law is of fundamental importance in underwater physiology. Miles (1969) has described and illustrated this law so clearly, that we quote:

> The pressure of a given quantity of gas whose temperature remains unchanged varies inversely as its volume.
> $PV = K$, when P = pressure, V = volume and K = constant.
> This simply means that if the pressure of a given amount of gas is doubled it must be compressed to half its volume. An enclosed volume of gas cannot be subjected to increased pressure unless it occupies a smaller space.
> The importance of this under water is well illustrated in Figure [III-1], which represents the changes taking place if an inverted bell, or bucket, containing air is pushed under water.
> The changes illustrated form the basis of underwater medicine and their understanding is of paramount importance. It is worth while therefore at this stage mastering the important implications, especially as the inverted bell can, to all intents and purposes, be taken as representing the air contained in the lungs of a man who goes under water.
> At the surface the bell contains a given mass of air which occupies the volume of the bell and exerts a pressure of 1 atmosphere.

* Partial pressure is the pressure exerted by one component in a system, usually one gas or vapor in a mixture.

If the bell is now lowered 33 ft through the water its lower end, being open, is subjected to the increased water pressure, i.e. a pressure of the atmosphere plus that of 33 ft of water, a total of 2 atmospheres. The pressure on the air in the bell has in fact been doubled and therefore it must contract to half its volume. This is achieved by the water level rising half way up the bell. At 33 ft therefore the volume of air in the bell is halved and its pressure doubled though its mass (the number of molecules present) is unchanged. (It should be pointed out that no account is being taken of slight variations which may occur as a result of changes in the amount of air dissolving in the water.)

If the bell is now lowered a further 33 ft to a depth of 66 ft, a further 33 foot's-worth of water pressure is added giving now a total of 3 atmospheres of pressure on the confined air which must therefore be compressed to $\frac{1}{3}$ of its original volume by water rising $\frac{2}{3}$ the way up.

Similarly if the bell is lowered to 99 ft the air would be compressed to $\frac{1}{4}$ of the original volume, at 132 ft to $\frac{1}{5}$ of the volume and so on until at 627 ft, for example, it would be reduced to $\frac{1}{20}$ of the original volume when the pressure of the water and the air at that depth would be 20 atmospheres. This is of course a pressure of 20 atmospheres "absolute" i.e. it includes the one atmosphere of the air at sea-level. A pressure gauge at this depth would read only 19 atmospheres "gauge."

It may however, in practice, be necessary to maintain air under water at a constant volume, i.e. occupying an unchanging space. How can this be done?

Two methods are available. The first and simplest [Figure III-2A] is to contain the air in a rigid case, i.e. in the present example to seal off the bottom of the bell. In this case both pressure and volume of the air are unchanged but as depth increases the container is subjected to an increasing external pressure. At 627 ft, for example with only one atmosphere of pressure inside, the case would need to be strong enough to withstand a pressure of 19 atmospheres, i.e. about 280 lb per sq. inch. It would be essential, therefore, to construct the chamber of material of sufficient strength to withstand the pressure difference of the proposed operating depth . . .

The second method of maintaining a constant volume of air under the increasing pressure of depth is to keep pumping in more air to maintain the pressure at that of the surrounding water pressure [Figure III-2B]. In the case of the diving bell the bottom may be left open provided a pipe from an air pump is attached to supply air at the required pressure. To keep the bell full of air at 33 ft it would have to be pumped up to a pressure of 2 atmospheres (1 atmosphere on the pump gauge). At 99 feet an air pressure of 4 atmospheres would be needed and at 627 feet 20 atmospheres. (The pump gauge reading in each case is 3 atmospheres and 19 atmospheres, respectively.) In practice a pressure a little above those given would cause bubbles of air to escape from the edges of the bell and give an assurance to those working the pump that an adequate supply was being provided to maintain the volume. This method is indeed that used to supply air to the diver who being in a flexible suit needs air at the pressure of his surroundings. The supply may be obtained from a source of compressed air on the surface or gas cylinders carried by the diver himself.

C. *Solubility and Partial Pressures*

In addition to the physical effects of pressure and volume it is necessary to consider the effect pressure has on the movement in and out of solution of the gases in the breathing mixture. This problem is discussed in depth in Chapter IV in the sections on respiration and breathing mixtures. Here a very general statement is included in order to present a complete picture of the physical factors involving man in the ocean environment.

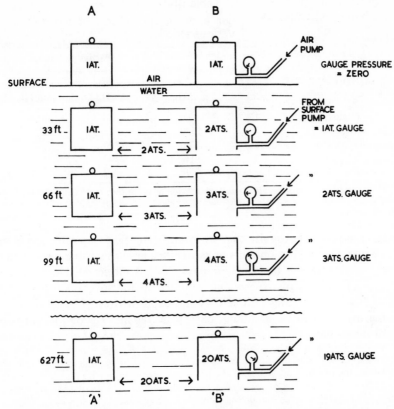

Figure III-2. Maintaining constant gas volume under water. [Reprinted from Miles (1969) by permission of the author and Staples Press.]

In diving one works with a mixture of gases rather than a single pure gas, and thus we must consider partial pressure, which is explained by Dalton's and Henry's laws. Partial pressure computations are necessary for mixed-gas diving and are useful for understanding diving physiology. The partial pressure p_X of a given gas in a mixture may be calculated by the formula

$$p_X = P_t X_\%$$ (1)

where P_t is the total pressure of the gas mixture (absolute) and $X_\%$ is the percent of gas X by volume in the mixture. Disregarding the trace elements and the 0.04% carbon dioxide, air is a mixture of nitrogen (79%) and oxygen (21% approximately). At normal atmospheric pressure the 760 mm Hg pressure is shared by these two gases:

$$p_{nitrogen} = 79\% \text{ of } 760 = 600 \text{ mm Hg}$$

$$p_{oxygen} = 21\% \text{ of } 760 = 160 \text{ mm Hg}$$

These values are partial pressures of the two gases.

The partial pressure of oxygen in the air at sea level may also be calculated by

$$p_{O_2} = 14.7 \times 0.21 \text{ or } 3.1 \text{ lb/in.}^2$$ (2)

Table III-1

Partial Pressures of Oxygen and Nitrogen in an Inverted Bell at Various Depths[a]

Depth, ft	Total pressure		Partial pressure	
	atm	mm Hg	Nitrogen, mm Hg	Oxygen, mm Hg
0 (surface)	1	760	600	160
33	2	1520	1200	320
66	3	2280	1800	480
99	4	3040	2400	640
132	5	3800	3000	800

[a] Reprinted from Miles (1969) with permission of the author and Staples Press.

It is important to recognize that when air or another respirable gas is breathed under pressure, the varying amounts of gases diffusing through and dissolving in body fluids and tissues create far-reaching physiological effects. Table III-1 gives the partial pressure values of oxygen and nitrogen in the inverted bell illustrated in Figure III-1. It is interesting to note that at a depth between 99 and 132 ft (actually 124 ft) the diver would be breathing oxygen at 760 mm Hg or the same as 100% oxygen at the surface.

D. *Buoyancy*

The buoyant effect of liquids is expressed by Archimedes' principle, which may be stated as: Any object wholly or partially immersed in a liquid is buoyed up by a force equal to the weight of the liquid displaced. The laws of flotation based on Archimedes' principle can be summarized as follows (U. S. Navy 1970):

1. A body sinks in a fluid if the weight of fluid it displaces is less than the weight of the body.
2. A submerged body remains in equilibrium, neither rising nor sinking, if the weight of the fluid it displaces is exactly equal to its own weight.
3. If a submerged body weighs less than the volume of liquid it displaces, it will rise and float with part of its volume above the surface.

After a deep breath most individuals are positively buoyant; i.e., they remain on the surface. In fresh water about 10% will sink; i.e., they are negatively buoyant. In sea water the figure is only about 2%.

In a self-contained outfit the diver is buoyant on the surface because of air trapped between his suit and his body, and he may have to use weights to descend. But as he descends the trapped air is compressed and he becomes heavier and heavier.

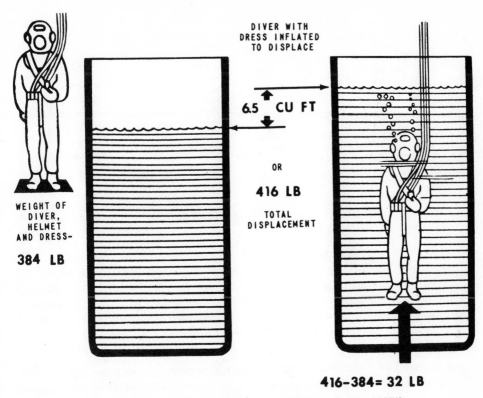

Figure III-3. Buoyancy—Archimedes' principle (U. S. Navy 1970).

In a standard "hard hat" diving outfit buoyancy can be regulated by air flow and inflation of the suit. The Archimedes' principle is illustrated in Figure III-3.

The diver, with his helmet and dress, weighs 384 pounds. If he inflates his dress so that he displaces 6.5 cubic feet of water, he will be buoyed up by a force equal to the weight of 6.5 cubic feet of water. Because sea water weighs 64 pounds per cubic foot, the buoyant force acting on this diver would be 6.5 × 64 = 416 pounds. This force is 32 (416 − 384) pounds more than his total weight. Such an excess of buoyancy force is called positive buoyancy. In this example, the diver would actually float with half a cubic foot of his volume out of water. The volume of water displaced would then be 6 cubic feet, the weight of which (6 × 64 = 384 pounds) would equal his own weight. To give himself neutral buoyancy, with which he would neither rise nor sink in the water, the diver could either exhaust one-half of a cubic foot of air from his dress or wear an additional 32 pounds of weight. If he required negative buoyancy (the state of being heavy in the water), he would have to add still more weight or let out more air (U. S. Navy 1970).

E. *Effects of Increased Pressure*

In the underwater environment there is always increased pressure on the diver. As he moves up and down in the water this pressure changes, and so does any gas

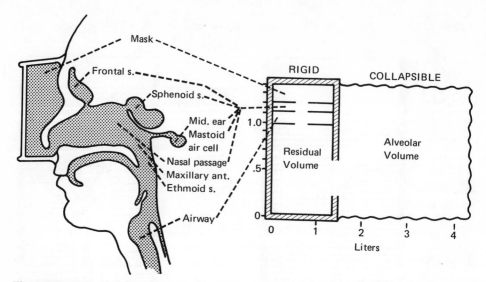

Figure III-4. Schematic presentation of the various rigid chambers in man which must be properly pressure-compensated during a dive. The figures for a typical Ama give relative proportions. [Reprinted from Rahn and Yokoyama (1965), National Academy of Sciences–National Research Council, Washington, D. C.]

volume exposed to the pressure. These changes in pressure and in gas volume as they effect divers may be considered as *barotrauma*; here we include all of the physical effects of pressure and volume change except the physiological reactions such as oxygen poisoning, nitrogen narcosis, and decompression sickness, which are discussed separately.

Gas cavities of the body must adjust to large and rapid changes in pressure during diving. It is worthwhile to think of the gas volume system of the body as divided into the rigid and the collapsible chambers, as illustrated in Figure III-4. Since the rigid chambers cannot collapse to accommodate pressure changes, air must be able to flow into them or there will be congestion and bleeding to accommodate the relative vacuum situation.*

The diver, carefully selected and trained (see Chapter XI, Selection and Training) and carefully checked prior to diving (see Chapter VIII, discussion of predive conditions), can normally accommodate to the pressure changes, but situations may arise where the pressure of air or gas in a body cavity becomes different from the surrounding tissues, and then trouble begins.

The resulting conditions and situations, listed alphabetically below, vary greatly

---

* This problem has been humorously presented by Perrault (1972): "Since man has not yet completed his evolution into a scuba diver, he is troubled by air spaces in his head called sinuses . . . these sinuses are, fortunately connected to the outdoors by passages opening to the diver's general ventilating system . . . [but the designer] goofed up both the ears and the sinuses by making their drainage passages much too narrow and by lining them with some kind of crazy tissue that swells like mad whenever exposed to even the most insignificant irritant. As a result, divers often have problems balancing the pressure inside their heads with the pressure outside."

in severity and frequency of occurrence, but are included because they are due to physical factors related to changes in pressure.

1. *Aural Barotrauma*

The most common complaint of anyone experiencing increased air pressure changes is difficulty in "clearing" or "popping" the ears. As noted in Chapter I, Section G2 (Hearing and the Ear), the middle ear is an air-filled cavity sealed on the outer side by the eardrum and open to the throat through the eustachian tube. Normally equalization of air pressure in the middle ear is accomplished during the act of swallowing, and there is no trauma to the eardrum, or to the middle ear. Chronically in some individuals, however, and in the presence of upper respiratory infection in most individuals, the eustachian tube does not open freely. Difficulty may thus ensue when the ambient pressure changes, which it does rather dramatically in hyperbaric activity. It is important to remember that "it is the difference between the pressure in the middle ear and the ambient pressure, not the absolute value of either pressure that calls forth the series of tissue changes known collectively as aerotitis media" (Shilling *et al.* 1946). The damage or pathological changes vary from slight congestion to rupture of the drum, and have been classified by Teed (1944) as follows:

Grade 0. Normal appearance.
Grade 1. Retraction of eardrum with redness in Shrapnell's membrane and along the manubrium.
Grade 2. Retraction with redness of entire eardrum.
Grade 3. Same as Grade 2 plus evidence of fluid in the middle ear.
Grade 4. Blood in the middle ear or perforation of the eardrum, or both.

The condition called aerotitis media in the quote above is most frequently referred to in the literature as otitis media, but is also known as aero-otitis, aerotitis, aural barotrauma, otitic barotrauma, otic barotrauma, and aerosalpingotympanitis, among divers as "ear squeeze," and among aviators simply as "aviator's ear."

The incidence of otitis media has been studied extensively in men undergoing submarine escape training, where a 50-lb pressure test is required prior to actual training using the submarine escape appliance. Shilling and Everley (1942), Teed (1944), Shilling (1944), Schulte (1957), and Alfandre (1965) reported a 26.9–36.2% incidence in prospective submarine personnel undergoing the pressure test for the first time. Many of these were able to return and complete the test, but some were permanently disqualified.

Regardless of the pathology or the incidence, this aural barotrauma may be classified as due to ascent or descent and according to the part of the ear affected (Edmonds *et al.* 1973).

a. Aural Barotrauma of Ascent

Little difficulty is experienced when the pressure increases in the middle ear because of decreasing ambient pressure either by ascent in air or water. In this case,

Table III-2
Effects of Difference between Middle Ear and Ambient Air Pressure

Effect	Pressure differential	Reference
Tube opens	15 mm Hg	Armstrong and Heim 1937
	20 mm Hg	Flisberg *et al.* 1963
	4–10 mm Hg	Coles 1964
	20–25 bars	Riu *et al.* 1969
Tube locked	−80 to −90 mm Hg	Armstrong and Heim 1937
	−90 mm Hg	Keller 1958
	−30 to −50 mm Hg	Flisberg *et al.* 1963
	−100 mm Hg	Coles 1964
	−60 to −90 bars	Riu *et al.* 1969
Pain	−60 mm Hg	Armstrong and Heim 1937
	−60 mm Hg	Keller 1958
	−10 lb/in.^{2}	Shilling and Everley 1942
Rupture of the drum	−100 to −500 mm Hg	Armstrong and Heim 1937
	−200 to −400 mm Hg	Perlman 1943
	−20 to −30 lb/in.^{2}	Vail 1929
	−25 lb/in.^{2}	Keller 1958

as the ambient pressure decreases, the trapped gas in the middle ear expands according to Boyle's law and is simply forced out through the eustachian tube when the differential pressure reaches about 10–20 mm Hg (see Table III-2). This type of barotrauma does not usually present any severe problem, although there may be a sensation of pressure or pain in the affected ear and dizziness may occur (alternobaric vertigo). It appears that some individuals can sense a pressure change of 0.040 psi, an altitude change of about 75 ft in air (Williams and Cohen 1972).

b. Aural Barotrauma of Descent

Negative pressure in the middle ear created artificially because of increased ambient pressure is an entirely different matter. Because the nasopharyngeal (throat) end of the eustachian tube is normally closed and because of the construction of this end of the tube, the "flutter valve" opening will only seal more firmly under excess pressure. The tube can be opened by various maneuvers (see following material) if the maneuvers are performed before the pressure becomes too great. In various studies locking of the tubal opening has been shown to occur between −30 and −100 mm Hg. In the case of the −100 mm Hg the subjects were experienced divers. (See Table III-2.)

If pressure continues without relief, the relative vacuum in the middle ear increases and the eardrum is pressed inward by the ambient pressure, causing pain, hemorrhage, and ultimately perforation (rupture) of the drum as noted above.

Perforation of the eardrum is a relief from pain for the diver, and actually is not dangerous unless it occurs in cold water. Cold water flowing into the middle ear may cause severe dizziness (sometimes complicated by nausea and vomiting), and the diver may become confused and unable to find his way back to the surface. Dizziness passes when the water in the ear warms to body temperature.

c. External-Ear Barotrauma

This may occur during descent and is known as "external-ear squeeze" or "reversed ear." If the external-ear canal is blocked by an earplug or a snug-fitting diving hood, a negative pressure will develop in the canal and the eardrum will be drawn out as the ears are "cleared" to equalize pressure in the middle ear. A depth of 30–50 ft (120–150 mm Hg) has been shown to cause trauma (Jarrett 1961). Prevention of this condition is easy—do not wear ear plugs underwater, and adjust the hood so that there is an air channel to the outer ear.

d. Prevention of Aural Barotrauma

In general, prevention of aural barotrauma depends on the proper selection of men who can pass the pressure test and in not allowing divers to go underwater with an upper respiratory infection.

Divers should also be aware of the various maneuvers that may help pressure equalization, the simplest being swallowing and yawning. Frequent clearing of the ears during descent is important so that the differential pressure does not become too great. The most commonly used method is called the Valsalva maneuver and consists of a forced expiration with the mouth and nose closed. A force of 130 cm H_2O (Riu *et al.* 1969) is the average pressure required to force gas up the tube by pulmonary and thereby nasopharyngeal pressure. Because of the associated rise in right arterial and pulmonary capillary pressure, and the possibility of obstruction to the return of venous blood from the brain, care must be used not to exert too much pressure during the Valsalva maneuver.

There is a nontraumatic way to ventilate the middle ear by way of positive pressure in the nasopharynx, now called the Frenzel maneuver in honor of its propounder: voluntarily closing the glottis, closing the mouth and nose, while contracting the muscles of the floor of the mouth and the superior pharyngeal constrictors. With the nose, mouth, and glottis closed, the elevated tongue can be used as a piston to compress the air behind it in the nasopharynx, and thus up the eustachian tube.

In a well-executed Frenzel maneuver, the mass of the tongue is very strongly driven backward. Note that no remotely dangerous pulmonary pressures are generated, the intrathoracic pressure being practically unchanged. Probably since the tubal muscles are active, it takes less nasopharyngeal pressure (only 6 mm Hg instead of about 33) to force air up the tube; also the nasopharyngeal pressures are higher than those possible with the Valsalva maneuver; finally, the Frenzel maneuver can be performed at any stage of inspiration, the Valsalva usually only after a sharp inspiration. Although the Frenzel maneuver is superior to the Valsalva on several important counts, it is difficult to learn and harder to teach, though most subjects can finally learn (Chunn 1960).

Another method of equalization is activation of the pharyngeal muscles by raising the soft palate (velum). Experienced divers can use isolated contraction of the velum, without swallowing or movements of the jaws. Riu *et al.* (1969) studied such movements by X-ray movies and found these maneuvers always had the effect of raising the velum. The muscles of the pharynx were thus shown to be active. These active muscular openings of the tube apparently are not effective when the differential pressure in the middle ear exceeds -60 to -80 μbar, and more forceful maneuvers are necessary to force gases into the middle ear.

Riu *et al.* (1969) also found by survey that while beginners nearly always use the Valsalva maneuver, experienced divers rarely do, relying instead upon isolated contraction of the velum together with jaw movements. They suggest that these often unconscious movements point to a conditioned reflex.

The voluntary opening of the tubes, "beance tubaire voluntaire" (BTV), has been well described by Delonca (1970), who found that about half of experienced divers could master the technique. Table III-3 compares the BTV with the Valsalva, Frenzel, and Toynbee maneuvers. The Toynbee maneuver consists in swallowing with the mouth and nose closed; it is the inverse of the Valsalva maneuver, and is of some limited use in relieving hyperpressure, rather than hypopressure in the middle ear.

Prevention also may be aided by use of decongestant nasal sprays prior to entering the water. Feet-first entry into the water is also recommended since head-first entry may bring on vascular congestion in the eustachian tubes.

Table III-3

Comparative Table of Different Techniques of Clearing the Tube
(Translated from Delonca 1970)

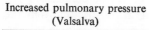

	Increased pulmonary pressure (Valsalva)	Increased rhinopharyngeal pressure (Frenzel)		Voluntary opening of the tubes
	Valsalva	Frenzel	Toynbee[a]	BTV
Nose	Compressed	Compressed	Compressed	Open
Mouth	Closed	Open	Closed	Open
Glottis	Open	Closed	Closed	Open
Action	Exhale through nose	Move the back of the tongue upward and backward	Swallow	Take the open position
Result	Increased pressure of pulmonary origin	Increased pressure of rhinopharyngeal origin	Reduced pressure	Equalization of pressure
Execution	Easy	Fairly difficult	Easy	Difficult
Safety	Fair	Good	Good	Perfect

[a] Maneuver particularly useful in ascending or in leaving a caisson, for clearing the middle ear during pressure.

According to Reuter (1971), 95% of divers' ear problems start in the eustachian tube and he urgently recommends the feet-first entry. Also, he recommends inflating the ears every few feet during descent rather than waiting and then using great force in the Valsalva maneuver, which may cause damage.

It is well to remember that many acute and chronic infectious processes as well as anatomical anomalies may lead to otitis media.

Treatment is generally unnecessary but in case of rupture of the eardrum it is important to put nothing in the ear and to refrain from diving until a doctor says the eardrum is healed. Recommendations for treatment and prevention of aural baro-trauma [described in the *U. S. Navy Diving Manual* (1970) as "middle-ear squeeze" and "external-ear squeeze"] are summarized in Table III-4.

The older literature reported little if any permanent hearing loss associated with aural barotrauma, but the evidence is accumulating that severe and permanent hear-ing loss may occur. (For a detailed review of hearing loss see Chapter V, section on hearing.)

2. Blowup

One of the most serious, but, fortunately, not very common, accidents associated with diving in suit and helmet is blowup. This occurs when the diver becomes positively buoyant. As he moves toward the surface, water pressure decreases and positive buoyancy increases and the diver may blow to the surface. The excess air pressure in the suit causes it to inflate to the maximum and become so rigid that the diver is "spread-eagled" and unable to move. See Table III-4 for a description of causes and methods of prevention.

3. Gastrointestinal Barotrauma

During ascent any gas trapped in the gastrointestinal tract will expand and may cause discomfort or even colicky pains. The condition is rarely severe and may be relieved by belching or by flatus. Carbonated beverages and heavy meals should be avoided before and during exposure to hyperbaric conditions. Treatment involves either slowing the rate of ascent or stopping for a time, and in severe cases even descending again.

4. Pulmonary Barotrauma

Pulmonary barotrauma or tissue damage and its sequelae resulting from an imbalance between pressures in physiological gas spaces and body tissue may occur in the lungs as well as the ears and sinuses.

Pulmonary barotrauma of descent is not very common but may occur in breath-hold diving. As the descent continues the lung volume decreases with the increasing pressure, in accordance with Boyle's law. An average normal full lung contains 6 liters

Table III-4

Physical Factors Relating to Changes in Pressure[a]

Conditions	Causes	Symptoms and signs	Treatment	Prevention
Aural barotrauma *Aural barotrauma of ascent*	Expansion of gas trapped in the middle ear as ambient pressure decreases	Sensations of pressure or pain; dizziness (alternobaric vertigo)	Usually none is needed	Do not dive with a head cold or infection that blocks the eustachian tubes; slow ascent when symptoms occur
Aural barotrauma of descent (middle-ear squeeze)	Diving with blocked eustachian tube; failure to "pop" ears on descent	Symptoms: pain in ear during descent; sudden relief of pain if eardrum ruptures. Signs (dependent on extent of damage): redness and swelling of eardrum; bleeding into eardrum or middle-ear space; rupture of eardrum with bleeding; spitting up of blood; bleeding to outside if eardrum is ruptured. May result in otitis media	Report to medical officer. Mild case without rupture: avoid pressure until damage heals and ears can be readily cleared. Ruptured eardrum: no diving until healed (usually about two weeks); keep water and all objects and materials (including medications) out of ear; keep hands away from ears. Return to medical officer at once if pain increases or if drainage appears (these signs may indicate infection requiring antibiotic treatment)	Do not accept men who cannot pass pressure test. "Pop" ears properly during descent—swallow or yawn; move jaw; blow gently against closed nostrils. Do not dive while having a head cold or infection that blocks eustachian tubes. Use nose drops, spray, or inhaler for mild difficulty
External-ear barotrauma (external-ear squeeze)	Suit squeeze; hood sealing over external ear; use of ear plugs	Symptoms: pain on descent even though able to "pop" ears; feels almost same as middle-ear squeeze; pain stops if eardrum ruptures	Same as in middle-ear squeeze (keep hands away from ears); report any signs of infection	If using closed rubber swimsuit be sure to admit air for equalization during descent (pinch face seal at junction with mask to make a channel)

		Signs: drum may have same appearance as in middle-ear squeeze; often see blood blisters on or around eardrum or in canal; eardrum may be ruptured, but bleeding to outside *does not* necessarily mean rupture	Line hood (ear aura) with flannel or porous rubber to prevent sealing
		May result in otitis media	Never use ear plugs
Blowup	In deep-sea or helium-oxygen rig: any mishap or error that causes overinflation of dress, poor adjustment of air-control and exhaust valves, or plugging of exhaust openings; loss of shoe or weights; allowing legs to be higher than body (if legs are not properly laced or shoes are too light); too strong or rapid a pull by tenders, suddenly breaking free from being stuck in mud; strong tide causing diver to lose hold on bottom or descending line, thus sweeping him to surface	Air embolism: if breath is held during blowup even from extremely shallow depth (i.e., 7 ft above diver's head)	First, guard against the causes with careful safety procedures
		Decompression sickness: if diver required decompression stops or was close to non-decompression limits (the rapid ascent of blowup may cause trouble even in a dive not requiring stops)	Prohibit use of controlled blowup as a means of ascent
			Exhale continuously if blowup occurs
		Mechanical injury resulting from striking bottom of boat or other object at surface	Tenders: take in at once all slack in diver's line when diver reaches surface after blowing up
	In self-contained diving: unintentional dropping of weights; accidental over-inflation of breathing bag; excess air in closed dry suit as result of efforts to equalize suit squeeze; failure of vent valves on suit during ascent; unintentional inflation of life-jackets	Squeeze may result from falling back into deeper water after reaching surface and exhausting air from diving dress	Diver: exhaust only enough air to prevent rupture of suit and retain positive buoyancy until tenders have taken in slack
		Drowning not unlikely to occur if suit ruptures at surface	Diagnosis of all the mishaps that may have occurred during blowup and act accordingly
			If diver is unconscious, recompress immediately (probable air embolism)
			If dive did not require decompression and diver appears all right, watch him closely and keep him near recompression chamber; if symptoms of decompression sickness develop, treat according to treatment tables in Chapter VII
			If dive did require decompression, follow procedure described in Chapter VII
			Apply first aid and other measures as required for injuries if any

Table III-4—*Cont.*

Conditions	Causes	Symptoms and signs	Treatment	Prevention
Gastrointestinal barotrauma	Gas trapped in the gastro-intestinal tract during ascent	Abdominal discomfort; collicky pains; rarely severe	Belching or flatus Slowing the rate of ascent or stopping for a time—in severe cases, descending again	Carbonated beverages and heavy meals should be avoided before diving
Pulmonary barotrauma	Imbalance between pressures in physiological gas spaces and body tissues On descent may occur in breathhold diving by descending too deep On ascent (more common), internal air trapping; inadequate exhalation caused by faulty apparatus; panic, water inhalation In scuba diving, throwing off the mask and coming to the surface holding the breath	Symptoms: pulmonary tissue damage; emphysema; pneumothorax; air embolism Signs: difficult breathing; cough; bloody sputum	Adequate respiration with 100% oxygen; drug support for the cardiovascular system; if severe, immediate recompression For pneumothorax: *at surface*: bed rest and supportive treatment and, when necessary, suction to withdraw air and fluid *during decompression*: it may be necessary to put a trocar through the chest wall to release trapped gas For air embolism: immediate recompression	All care should be taken to condition divers to react rationally and with patience when breathing problems occur Equipment should be kept in perfect condition and should never be used without prior checking Do not descend further than 99 ft during breath-hold diving In scuba diving, divers should be trained never to remove their masks to come to the surface holding the breath; in an emergency it seems to be very hard to remember to exhale during ascent

		Symptoms/Signs	Treatment	Prevention
Sinus barotrauma	Blocking of the passage from the sinus to the nose	Symptoms: congestion of the lining of the sinus; hemorrhage into the sinus cavity Signs: blood and mucous may be expelled		Medical screening for abnormal nasal passage or polyps; do not dive when you have a cold or nasal congestion Medical selection to rule out abnormalities of the ear and vestibular system, and routine predive examination for ear, nose, and throat infections "Buddy system"
Squeeze	Face squeeze: when wearing a face mask, failure to equalize internal and external pressure Body squeeze: falling through the water	Tissues will swell and may even bleed In a hard-hat rig, if a diver falls he may be slammed into his helmet and severely injured or killed	Stop ascent, descent, compression, or decompression—whichever is in effect when dizziness occurs—until dizziness passes When there is topside communication, reassurance can sometimes prevent panic; if not, a diving companion may prevent a serious accident by helping the diver cope with his confusion Blowing air into the diver's mask to equalize pressure Emergency rescue operations to achieve medical aid as soon as possible	
Vertigo[b]	Vestibular problems; middle-ear barotrauma of descent; overforceful Valsalva maneuvers; unilateral caloric stimulation	Dizziness; disorientation; nausea		Same as treatment Use self-contained systems whenever possible; take safety precautions against accidents, especially falling off platform

[a] Data from Riu et al. (1969), U. S. Navy (1970), Reuter (1971), Williams and Cohen (1972), Edmonds et al. (1973).

[b] See also Chapter V, section on vestibular function.

of air at the surface, but this is compressed to 1.5 liters at 99 ft. This is approximately the normal residual volume (see Chapter I, section on gas exchange) and further descent could cause pulmonary congestion, edema, and hemorrhage.

Pulmonary barotrauma associated with ascent is a much more serious and common occurrence. Lanphier (1957) believes it to be second only to drowning as a cause of death among scuba divers. Recalling the various pressure phenomena, it becomes evident that in breath-hold diving the air in the lungs of the diver is compressed by the pressure on the wall of the chest and the upward movement of the diaphragm due to pressure on the abdomen during the descent. Then, upon ascent the pressure of the air in the lung returns to the ambient atmospheric pressure, and the diver is not exposed to any danger of lung rupture.

On the other hand, the scuba diver with his self-contained breathing apparatus breathes in air which is compressed to the ambient pressure of the water and thus, at a depth of 99 ft he will be exposed to a pressure of 60 lb/in.² of his body surface and will also have his lungs filled with air at the same pressure. If he continues to breathe normally as he slowly ascends he will have no problem and the external pressure and the internal lung pressure will both be 15 lb at the surface. However, if through panic or for other causes he clamps his glottis shut (laryngospasm) and holds his breath as he ascends, he will have an external pressure of 15 lb at the surface and an internal lung pressure of 60 lb. Rupture of the very thin-walled lung alveoli (see Chapter I) will ensue, causing air embolism, pneumothorax (air in the chest cavity), or emphysema (air in the lung tissues).

According to Edmonds and Thomas (1972):

> Pressure gradients necessary to cause pulmonary barotrauma are approximately 80 mm Hg near the surface. . . . Hence, barotrauma may supervene when the ambient water pressure falls by 80 mm Hg or more below the intrapulmonary pressure—i.e. with an ascent of four feet to the surface. This is more likely to occur after full inspiration before ascent. A diver whose total lung volume is 6 liters at 33 ft will need to exhale 6 liters of gas during ascent in order to maintain his normal 6-liter lung volume at the surface. Precipitating factors include: internal air trapping and inadequate exhalation caused by faulty apparatus, through panic, or because of water inhalation. This is encountered in free ascents and submarine escape training, or emergency ascents.

Pulmonary barotrauma is all too common in scuba diving and occurs primarily as the result of "ducking" the mask and coming to the surface holding the breath. Here, as pointed out earlier, the water pressure on the body decreases, the air in the lungs expands, the diaphragm and chest walls stretch to accommodate the increased volume of gas, and the lung tissue stretches and tears, allowing air to flow into the pulmonary circulation, or intrapleural space. This, in turn, leads to extensive air embolism and often to death. The simple expedient of exhaling the trapped, expanding air during ascent seems to be most difficult to remember in an emergency.

Four conditions may result: pulmonary tissue damage, emphysema, pneumothorax, and air embolism.

Pulmonary tissue damage always precedes either emphysema, pneumothorax, or embolism; it may consist of the rupture of a few or many alveoli and be accompanied by difficult breathing, cough, and bloody sputum. Treatment involves adequate respiration with 100% oxygen to maintain acceptable arterial gas levels, and drug support for the cardiovascular system may be required. Positive-pressure respiration could increase lung damage and should be avoided.

Emphysema of a surgical nature may result following rupture of the alveoli. Gas may track along the loose tissue planes surrounding major blood vessels, finding its way into the mediastinum (space between the lungs where the heart is located), thus embarrassing the heart, or into the neck, where it may cause a feeling of fullness in the throat. If signs and symptoms are mild, no treatment may be necessary but 100% oxygen respiration would be advisable and if the symptoms are severe, recompression at once is indicated.

Pneumothorax may result if the visceral pleura (outer lining of the lung) ruptures, allowing air into the chest cavity. If the diver has reached the surface and shows the signs and symptoms of air in the chest cavity, he may need only bed rest and general supportive treatment and the lung will heal and reexpand; however, more severe cases may require suction to withdraw air and fluid so the lung may expand more readily. If the condition occurs while the diver is under pressure, it may be necessary to put a trocar through the chest wall to allow the expanding trapped gas to flow out during decompression of the patient. All of these suggested treatments are in the province of the doctor or medically trained assistant if no doctor is available.

Air embolism is a most dangerous condition and is the result of gas passing into the pulmonary veins following rupture of the alveoli and then passing into the circulation where it can cause all of the signs and symptoms of air embolism due to decompression sickness, except that in the case of alveoli rupture, the condition is apparent immediately upon surfacing. In this condition, which is often marked by complete collapse, treatment is urgent and recompression to 165 ft (6 ATA) should be instituted at once. Oxygen should be given as soon as it is safe and circulatory and respiratory support should be instituted as recommended by the doctor.

5. *Sinus Barotrauma*

If this condition occurs during descent, it is often referred to as sinus squeeze and results from blocking of the passage from the sinus to the nose. Congestion of the lining of the sinus and hemorrhage into the cavity compensate for the contraction of the air within the sinus cavity, and during ascent blood and mucous may be expelled.

Sinus barotrauma of ascent may occur as a result of occlusion of the sinus openings by folds of tissue or polyps, preventing escape of expanding gases. Painless hemorrhage may follow.

6. *Squeeze*

Face squeeze is due to wearing a face mask and not equalizing pressure by blowing air into the mask during descent. Facial tissues will swell and even bleed as they are forced into the space, external to, but in contact with, the face. Breath-hold diving wearing goggles also results in a similar problem. The relative vacuum caused by the increased external pressure must be satisfied.

Body squeeze is caused by falling through the water without the air pressure being able to keep up. In a hard-hat rig, unless extra gas is added during descent to compensate for the effects of Boyle's law, the suit and occupant may be forced into the helmet, causing bizarre injury and even death. The self-contained set is dangerous only if it fails and the diver falls through the water. The depth of the diver at the time of the fall is most important. If the diver falls off the platform from 33 ft to 66 ft he would be subjected to an increase of pressure from 2 to 3 atm. This would represent a decrease in air volume in suit and respiratory system to two-thirds the original volume, or a $33\frac{1}{3}\%$ change, which would lead to gross crushing. However, if the diver fell from 132 ft to 165 ft—the same distance—the reduction in volume would be only half as much.

7. Toothache

Although not common, pressure changes can cause pain in a tooth. This is almost surely the result of a cavity, and the diver should be sent to his dentist. The condition is not operationally important.

8. Vertigo

Although usually due to other causes (see Chapter V, section on vestibular function) vertigo may also result from middle-ear barotrauma of descent (middle-ear squeeze), overforceful Valsalva maneuvers, and unilateral caloric stimulation (cold water entering through a perforated ear drum) (Edmonds 1971).

Vertigo is a common disorder among divers, and is directly related to the problem of disorientation, which may have serious or even fatal consequences.

Prevention rests in careful examination to rule out acute infection or abnormalities of the ear and to demonstrate patency of the eustachian tube. Frequent "clearing" of the ears during compression is helpful.

In the presence of vertigo it is wise to "freeze" on the line until the dizziness passes, which usually requires only a few minutes.

The "buddy" system (two divers working together) is particularly important in this situation.

References

ALFANDRE, H. J. Aerotitis media in submarine recruits. U. S. Nav. Submar. Med. Cent., Rep 450 (May 16, 1965).

ARMSTRONG, H. G., and J. W. HEIM. The effect of flight on the middle ear. J. Amer. Med. Assoc. 109:417–421 (Aug. 7, 1937).

CHUNN, S. P. A comparison of the efficiency of the Valsalva maneuver and the pharyngeal pressure test and the feasibility of teaching both methods. U. S. Air Force Sch. Aerosp. Med., ACAM thesis (1960).

COLES, R. R. A. Eustachian tube function. J. Roy. Nav. Med. Serv. 50:23–29 (Spring 1964).

DELONCA, G. Considerations sur les manoeuvres sites d'equilibration de l'oreille chez le plongeur. *Bull. Medsubhyp.* **3**:10–14 (Sept. 1970).

EDMONDS, C. Vertigo in diving. Balmoral, New South Wales, Australia, Roy. Aust. Navy, Sch. Underwater Med., Rep. 1/71 (1971).

EDMONDS, C., and R. L. THOMAS. Medical aspects of diving. Part 3, *Med. J. Aust.* **2**:1300–1304 (Dec. 2, 1972).

EDMONDS, C., P. FREEMAN, R. THOMAS, J. TONKIN, and F. A. BLACKWOOD. *Otological Aspects of Diving.* Glebe, New South Wales, Australia, Australasian Medical Publishing Co. (1973).

FLISBERG, K., S. INDELSTEDT, and D. ORTEGREN. The valve and "locking" mechanisms of the eustachian tube. *Acta Otolaryngol. Suppl.* **182**:57–68 (1963).

JARRETT, A. S. Revised-ear syndrome and the mechanism of barotrauma. *Brit. Med. J.* **2**(5250):483–486 (Aug. 19, 1961).

KELLER, A. P. A study of the relationship of air pressure to myringorupture. *Laryngoscope* **68**:2015–2028 (Dec. 1958).

LANPHIER, E. H. Diving medicine. *New Eng. J. Med.* **256**:120–131 (Jan. 17, 1957).

MILES, S. *Underwater Medicine.* Third edition. Philadelphia, L. B. Lippincott Co. (1969).

PERLMAN, H. B. The effect of explosions on the acoustic apparatus. *Trans. Amer. Acad. Opthalmol. Otolaryngol.* **47**:442–453 (July/Aug. 1943).

PERRAULT, P. B. Basic etiquette for divers. *Skin Diver* **21**:50–51, 77 (July 1972).

RAHN, H., and T. YOKOYAMA. *Physiology of Breath-Hold Diving and the Ama of Japan. A Symposium Held in Tokyo, August 31–September 1, 1965.* Washington, D. C., National Academy of Sciences–National Research Council (1965).

REUTER, S. H. 95% of divers' ear problems start in the eustachian tube. *Clin. Trends* **10**:8 (Oct./Nov. 1971).

RIU, R., L. HOTTES, R. GIULLERM, R. BADRE, and R. LEDEN. La trompe d'eustache dans la plongee. *Rev. Physiol. Subaquatique Med. Hyperbare* **1**:194–198 (Jan.–Mar. 1969).

SCHULTE, J. H. Aerotitis media and aerosinusitis in submarine trainees: A prophylactic study. *U. S. Armed Forces Med. J.* **8**:1571–1576 (Nov. 1957).

SHILLING, C. W. Aerotitis media and auditory acuity loss in submarine escape training. *Trans. Amer. Acad. Ophthalmol. Otolaryngol.* **49**:97–102 (Nov./Dec. 1944).

SHILLING, C. W., and I. A. EVERLEY. Auditory acuity in submarine personnel. Part III. *U. S. Nav. Med. Bull.* **40**:664–686 (July 1942).

SHILLING, C. W., H. L. HAINS, J. D. HARRIS, and W. J. KELLY. The prevention and treatment of aerotitis media. *U. S. Nav. Med. Bull.* **46**:1529–1558 (Oct. 1946).

TAYLOR, G. D. The otolaryngolic aspects of skin and scuba diving. *Laryngoscope* **69**:809–858 (July 1959).

TEED, R. W. Factors producing obstruction of the auditory tube in submarine personnel. *U. S. Nav. Med. Bull.* **42**:293–306 (Feb. 1944).

U. S. NAVY, *U. S. Navy Diving Manual.* Washington, D. C., Department of the Navy (1970) (NAVSHIPS 0994-001-9010).

WILLIAMS, D. H., and E. COHEN. Human thresholds for perceiving sudden changes in atmospheric pressure. *Percept. Motor Skills* **35**:437–438 (Oct. 1972).

VAIL, H. H. Traumatic conditions of the ear in workers in an atmosphere of compressed air. *Arch. Otolaryngol.* **10**:113–126 (Aug. 1929).

IV

Man in the Ocean Environment: Physiological Factors

A. *Respiration in a Hyperbaric Environment*

1. *Introduction*

Even though breathing under increased ambient pressure is a recent development in man's history, there is now ample reason and need to put man at the greatest possible depth consistent with his ability to do useful work. Respiratory variables are of crucial importance in determining both the capacity and the safety of working at depth and under high pressure.

All activity and life itself depend upon energy which is produced by the chemical reaction of oxidative metabolism (see Chapter I, section on production of energy). Although the body can store *fuel* in adequate quantities for the chemical reactions, the "oxygen uptake and elimination of carbon dioxide must be matched to the level of activity on an almost moment-to-moment basis. If the rate of oxygen supply or carbon dioxide elimination is restricted by disease or environmental factors, activity will be restricted correspondingly. Life itself may be threatened if the limitation is severe" (Lanphier 1969).

This section is *not* written for the respiratory physiologist, but for the engineer or diving officer, and from the standpoint of problems which may be encountered by a diver at depth or a worker in a hyperbaric environment.

A number of topics closely related to pulmonary function are beyond the scope of this section, while a number of topics are treated in detail elsewhere in this handbook.

For example, the type of breathing apparatus is crucial but is presented in Chapter IX, in the section on operational equipment, under the topic, personal equipment. Also, oxygen toxicity can have drastic effects upon pulmonary function, but this condition is discussed later in this chapter, in Section C. Again, one of the main environmental factors causing deficiency of pulmonary ventilation is increased gas density, which will be mentioned in this section, but is also handled in more detail in

Section E of this chapter, on the physiology of breathing mixtures. The normal anatomy of the respiratory system is presented in Chapter I.

2. *Physiology of Respiration*

a. General Respiratory Function

The act of breathing, or the alternate inspiration and expiration of gas into and out of the lungs, serves to bring oxygen into the lungs and to eliminate carbon dioxide. True respiration, to the biochemist, refers to the ultimate utilization of oxygen by the body cells with the coincident release of carbon dioxide (metabolism). Some speak of internal respiration as the exchange of gases between blood and the cells via the tissue fluids. External respiration is then thought of as the exchange of gases among the ambient atmosphere, the lung alveoli, and the blood in the lung capillary bed.

The significance of lung volumes and capacities to the engineer and equipment designer is such that detailed consideration is required. Figure IV-1 illustrates the several functional subdivisions. It will be noted that there are four primary lung volumes, which do not overlap: tidal volume (TV), inspiratory reserve volume (IRV), expiratory reserve volume (ERV), and residual volume (RV). These are all defined in the accompanying glossary. Also, as seen in Figure IV-1, there are four capacities, which include two or more of the primary volumes: total lung capacity (TLC), inspiratory capacity (IC), vital capacity (VC), and functional residual capacity (FRC). These are also defined in the glossary that follows.

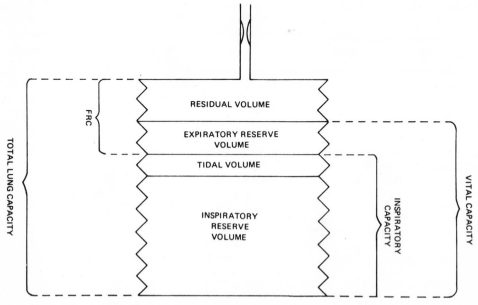

Figure IV-1. Lung volumes and capacities (Bartlett 1973).

Glossary of Terms, Abbreviations, and Symbols Used in Respiration Physiology

Alveolar ventilation   The volume of inspired gas that reaches and ventilates the alveoli, i.e., total ventilation minus dead space ventilation.

Alveolus (pl., alveoli; adj., alveolar)   Small outpocketings of the sacs in the lung. Gas exchange with the blood in pulmonary capillaries takes place through alveolar walls.

Anatomical dead space (ADS)   That area superior to the alveoli comprised of the upper airways where no gas exchange occurs to or from the blood.

Apnea   Cessation of breathing.

ATP   Ambient temperature and pressure.

ATPS   Ambient temperature and pressure and saturated water vapor.

Breathing   The alternate inspiration and expiration of air into and out of the lungs.

BTPS   Body temperature (37°C unless otherwise specified), existing barometric pressure, gas saturated with water vapor at body temperature. BTPS values are about 1.08 times average ambient (ATPS) room-temperature values. STPD values are roughly 0.8 of BTPS values if the barometric pressure is 760 mm Hg.

Conductance   The ratio of flow to driving pressure expressed in liters per second per centimeters of water.

Convective acceleration (Bernoulli effect)   The energy required to accelerate gas particles between two points in a tube with converging boundaries.

Dead space   See anatomical dead space.

Diffusion dead space (DDS)   The fraction of gas in the lung that does not diffuse due to damage to the intrapulmonary diffusion capability (see Albano 1970).

Dyspnea   Difficult or labored breathing, especially if not warranted by physical exertion.

Elastic work   That work performed in the act of breathing to overcome the elastic resistance of the lung and pleural cavity.

ERV (Expiratory reserve volume)   See definition.

Eucapnia   The condition in which the carbon dioxide of the blood is normal.

Expiratory reserve volume (ERV)   The volume of gas that can still be expired by maximum expiration after the end of a normal tidal expiration, normally 1100 ml.

Forced expiratory reserve volume (FEV)   The rate of flow of air per second that can be forcefully expired after the end of the normal tidal expiration.

FRC (Functional residual capacity)   See definition.

Functional residual capacity (FRC)   The volume of gas remaining in the lung after a normal expiration. The sum of ERV and RV (about 2300 ml).

Hypercapnia   Excess of carbon dioxide in the blood.

Hyperoxia   Excess of oxygen in the body.

Hyperventilation   An increase in rate and/or depth of respiration over and above that required to meet the body's respiratory requirement.

Hypocapnia   Deficiency of carbon dioxide in the blood.

Hypoxia   Low oxygen content or tension in the body.

IC (Inspiratory capacity)   See definition.

Inspiratory capacity (IC)   The volume by which the lung can be increased by a maximum inspiratory effort following a normal expiration. It is the sum of TV and IRV (about 3500 ml).

Inspiratory reserve volume (IRV)   The extra volume of gas that can be inspired over and beyond the normal tidal volume (about 3000 ml).

IRV (Inspiratory reserve volume)   See definition.

Laminar flow   Streamlined flow of a viscous fluid or gas where the flow is in layers without large irregular fluctuations. Laminar flow occurs at low Reynolds numbers (see definition).

Maximum breathing capacity (MBC)   The maximum volume of air which can be moved in and out of the lungs.

Maximum expiratory flow rate (MEFR)   Rate of flow of expirate at maximum voluntary effort on the part of the subject.

Maximum inspiratory flow rate (MIFR)   Rate of flow of inspirate at maximum voluntary effort on the part of the subject.

Glossary of Terms—*Cont.*

Maximum voluntary ventilation (MVV)   See maximum breathing capacity.

Minute volume (ventilation)   The total volume of air passing in and out of the lungs in 1 min.

mm Hg   Millimeters of mercury, usually used as measure of pressure of a gas.

Negative-pressure breathing (NPB)   Breathing from a mask, helmet, or the like, where the pressure of the gaseous mixture being breathed is less than the ambient pressure, thus requiring an additional conscious effort to inhale.

Parenchyma   The functional elements of an organ, as distinguished from its framework or stroma.

Perfusion dead space (PDS)   The fraction of gas in the lung that reaches poorly perfused alveoli and where no gas exchange occurs (see Albano 1970).

Pleura   The serous membrane investing the lungs and lining the thoracic cavity, completely enclosing a potential space known as the pleural cavity.

Positive-pressure breathing (PPB)   Breathing from a mask, helmet, or the like, where the pressure of the gaseous mixture being breathed is greater than the ambient pressure external to the airways. This situation eases the work of breathing and can force breathing, as with a respirator.

Pulmonary function   The factors included in the act of breathing, including ventilatory mechanics (mechanics of breathing), alveolar ventilation, and gas exchange between the alveoli and the blood.

Pulmonary ventilation   The total amount of air moved by breathing movements.

Residual volume (RV)   The volume of gas remaining in the lung after a maximum expiratory effort (about 1200 ml).

Respiration   The act or function of breathing.

Respiratory quotient (RQ)   The ratio between the volume of carbon dioxide expired and the volume of oxygen inspired in a given time.

Respiratory rate   The rate at which gases are exchanged between the blood and the tissues; by biochemists, the rate of tissue metabolism; by laymen, the breathing frequency.

Reynolds number   A criterion for a type of fluid motion. It is the ratio between the inertial force $\rho v d$, and the viscous force μ, or $\rho v d / \mu$, where ρ is fluid density, v is the velocity, d is a characteristic length, and μ is fluid viscosity. In a tube, laminar flow normally exists at a Reynolds number less than 2000 (the threshold), and the flow is turbulent when the number exceeds that value.

RQ (Respiratory quotient)   See definition.

RV (Residual volume)   See definition.

STPD   Indicates that a volume has been corrected to standard conditions of temperature (0°C), pressure (760 mm Hg), and dry gas. This correction is universally used for $\dot{V}_{O_2}$ and $\dot{V}_{CO_2}$.

Tidal volume (TV)   The volume of gas inspired or expired during each respiratory cycle.

Timed vital capacity (TVC)   The time required to exhale or inhale completely from a full state of inspiration or expiration, respectively.

Tissue viscous resistance   The frictional resistance to changing deformation of the lung parenchyma.

TLC (Total lung capacity)   See definition.

Total lung capacity (TLC)   The sum of VC plus RV (about 5800 ml); or the sum of all four primary lung volumes: TV, IRV, ERV, and RV.

Transmural pressure   The differential between the pressures acting on the outer and inner surfaces of the tubes in the lungs.

Transpulmonary pressure   The difference between the oral pressure and the pressure exerted on the visceral pleural surface of the lung.

Turbulent flow   The motion of fluids or gases in which local velocities and pressures fluctuate irregularly (such flows occur at high Reynolds numbers).

Turbulent work   That work performed in the act of breathing to overcome the effect of turbulent flow in the lung caused by high gaseous densities and/or rapid breathing.

TV (Tidal volume)   See definition.

V Represents a gas volume.

$\dot{V}$ Represents volume per unit of time, or flow.

VC (vital capacity)   See definition.

Ventilatory capacity (work limits)   A function of maximum breathing capacity (MBC), timed vital capacity (TVC), and maximum expiratory flow rate (MEFR), all of which are maximum-effort dynamic ventilatory measures and all reflect the work limits of the anatomical respiratory apparatus.

Viscous work   That work performed in the act of breathing to overcome the effect of increased viscosity of the inspired gaseous medium.

Vital capacity (VC)   The maximum volume of gas that can be exhaled from the lung following a maximum inspiration. It is the sum of IRV, TV, and ERV (about 4600 ml).

In considering the importance of these lung volumes and capacities for equipment design, the work of Bartlett (1973) is so clear that it is quoted in its entirety:

Functional Residual Capacity. FRC is important in equipment design for two principal reasons. First, when a closed-loop breathing system is used, the total gas volume will be the sum of the gas volumes of the rebreather loop and FRC and the tidal volume. (The tidal volume will be contained in either the lungs or the rebreather bellows or will be distributed between them during inhalatory or exhalatory activity.) This total system gas volume is important in considerations of inert gas dilution, expansions or contractions with pressure change, inert gas washout, etc. Secondly, should the expanding gases exceed tolerable pressures, the designer must assure that a breathing apparatus will not so impede the flow of these rapidly expanding gases as to increase lung pressures above these critical levels.

Residual Volume. Because the residual volume is that portion of lung volume that remains in the lung after a forced exhalation, to reduce lung volume below this level will require compression of the chest/lung complex or an increased trans-lung (intra-pleural pressure to alveolar pressure) pressure gradient as the result of reduced external breathing apparatus acting through the airway. Regardless of the mechanism, lung collapse (atelectasis), lung blood vessel rupture, pulmonary edema, or any combination of these three effects may occur. It is, therefore, important to assure that system operation over every possible range of pressure changes never produces such pressure differentials as to require a lung volume less than residual volume.

Inspiratory Capacity and Inspiratory Reserve Volume. Inspiratory capacity represents the maximum volume of air that can be forced into the lung by changes in system volume, without producing overexpansion of the lungs. The volume of the inspiratory reserve capacity is progressively reduced during an inhalation, and the remaining volume that can be inhaled is referred to as the inspiratory reserve volume. Thus, during a breathing cycle, the permissible increase in lung volume imposed by system pressures is reduced by the tidal volume. When system pressures are such as to require a lung volume increase greater than that represented by the inspiratory capacity (or inspiratory reserve volume, as appropriate), the lungs are overextended, and lung rupture is possible. It is, therefore, important to assure that system pressure changes will never require an increase in lung volume above what could be produced by a maximum inhalation.

Tidal Volume. A closed-loop system must have a rebreather bellows that will accomodate the largest expected tidal volume. Such a tidal volume could, of course, be as large as the vital capacity. But it might not be practical to provide so large a rebreather bellows. In such a case a rebreather bellows of reasonable size should be provided (determined by expected level of exercise, etc.). In this case increased inhalation volumes can be provided from oxygen storage and increased exhalation volumes can be accomodated by venting through a spring loaded exhalation pressure safety valve.

Total Lung Capacity. In an extreme case, the maximum amount of gas the lungs could contain, which must be considered for total system volume calculations and accommodation for gas expansion, is the total lung capacity. Although such a circumstance is unlikely, a conservative design must use this volume as the basis for the design.

Table IV-1

Partial Pressure and Gas Exchange Breathing Air at Sea Level

	Gas partial pressure, mm Hg				
	O_2	CO_2	N_2	H_2O	Total
Inspired air	158	0.3	596	5.7	760
Expired air	116	32	565	47	760
Alveolar air	100	40	573	47	760
Arterial blood	100	40	573	47	760
Venous blood	40	46	573	47	760
Tissues	≤ 30	≥ 50	573	47	760

In order to understand the effect of a hyperbaric environment on respiratory function it is also necessary to look at the normal sea-level partial pressure and gas exchange while breathing air. Transportation is the only active process, i.e., ventilation of the lungs and circulation of the blood. There is no secretory process and the gases move along concentration gradients as shown in Table IV-1.

Respiratory mechanics as related to hyperbaric conditions is the subject of a separate section which follows.

b. Oxygen Requirements

As noted above, the metabolic energy by which man operates depends on oxidation or burning of the *fuel*. The oxidation depends upon a continuous supply of oxygen and the oxygen is supplied through respiration. Also, the body's need for oxygen depends on the level of activity, and oxygen consumption, in turn, provides the most useful index of exertion.

Oxygen is transported under normal atmospheric conditions almost entirely by the hemoglobin contained in the red cells of the blood. Blood leaving the lungs contains approximately 19 ml of O_2 per 100 ml of blood, of which 18.7 ml is in chemical combination with the hemoglobin and only about 0.3 ml is in solution in the blood serum. Under hyperbaric conditions the amount in the dissolved state increases dramatically.

The average oxygen consumption* $\dot{V}_{O_2}$ in liters per minute (STPD†) for the 70-kg (154-lb) "average man" at rest is 0.3 liter/min, while an underwater swimmer doing 40 yards/min with fins, suit, and breathing equipment would have an oxygen consumption of at least 2.5 liters/min. All oxygen consumption values are subject to variation depending on the individual's level of activity, size, and proficiency, and on various environmental conditions. The $\dot{V}_{O_2}$ of any individual at a given moment will obviously be somewhere between his minimum or basal value and his maximum $\dot{V}_{O_2}$ ($\dot{V}_{O_2max}$) for aerobic capacity. For a diver of average size and reasonable fitness a value of $\dot{V}_{O_2\,max}$ above 3 liters/min would be expected and values as high as 6 liters/min have been reported. But such a high level is reached only with vigorous

---

* In standard respiratory symbols V represents a gas volume, while $\dot{V}$ indicates *volume per unit time* or flow. The subscript indicates the volume at issue, here of O_2, taken up by the body.
† See glossary.

activity involving a large proportion of the total skeletal-muscle mass and there is reason to doubt that underwater workers reach that level often since the aqueous medium appears to keep the muscles from reaching their surface-work maximum.

One can go into *debt* for oxygen so far as one's muscles are concerned, but both the central nervous system and the heart will tolerate scarcely any lag in oxygen supply. For example, a man can run 50 yards before he begins to supply the accelerated use of energy by increased respiration, and it takes several minutes after he stops running for breathing to return to its resting rate; but lack of O_2 to the brain can cause unconsciousness in 10–15 sec.

c. Carbon Dioxide Elimination

The by-products of oxidation are water and carbon dioxide. The concentration of water in the body is controlled by osmotic regulation and excess water is eliminated mainly via the kidneys. The elimination of CO_2 is effected by respiration—the greater the ventilation, the greater the amount of CO_2 *washed out* of the body.

Since CO_2 is the end-product of the same metabolic process in which O_2 is consumed, they are closely related variables. The relationship is expressed by the respiratory quotient (RQ):

$$RQ = \frac{CO_2 \text{ production}}{O_2 \text{ consumption}}$$

The value of RQ is determined largely by the diet and can range from 0.7 for a heavy fat diet to 1.0 for a carbohydrate diet. Protein and an average mixed diet yield an RQ of about 0.8.

Alveolar ventilation is treated in more detail later since the exchange is basic to the entire metabolic process. It is certain in this connection, however, that the inspired air should be as free of CO_2 as is possible, because the washout of CO_2 from the body will also depend on the pressure gradient for CO_2 between the alveoli and the ambient atmosphere.

"Fresh air contains about 0.04% carbon dioxide ($F_{ICO_2} = 0.0004$). Even at 10 ATA this would yield a P_{ICO_2} of only about 3 mm Hg. Unfortunately, a diver is seldom certain that he is receiving 'fresh' air, and in many situations he certainly is not. The P_{ICO_2} is almost certain to be above a negligible level in a diving helmet, in much breathing equipment, and in many pressure chambers" (Lanphier 1969). Since the diver is constantly producing CO_2, the problem of adequate ventilation becomes crucial.

d. Control of Respiration

The mechanisms concerned with the regulation of respiration are beyond the scope of this Handbook, but cannot be completely neglected because most of the problems of pulmonary function are more closely related to the regulation of pulmonary ventilation than to local pulmonary effects. "The regulation of respiration at rest does not present many difficulties in diving. However, divers are not often at rest; and when they are, most of their problems of pulmonary function largely disappear.

Most problems are in connection with respiratory control during exertion. This, unfortunately, has been one of the most controversial topics in respiratory physiology" (Lanphier 1969).

There seems now, however, to be general agreement that there is both a *neurogenic* and a *humoral* factor. The neurogenic factor is a fast component with almost instantaneous increase in ventilation at the onset of dynamic exercise caused by nerve impulses from muscles and joints; the humoral factor is a slower change caused by the accumulation of carbon dioxide and change in acidity of the body during exercise.

e. Breath-Hold Diving

Breath-hold diving, sometimes called skin diving, is the oldest form of diving and is still practiced in many parts of the world, e.g., the Korean amas who dive for pearls. The length of time the breath can be held is the time to the breaking point, which may be defined as "the voluntary termination of breath-holding in response to the development of a net ventilatory stimulus too strong to be further resisted by voluntary effort" (Mithoefer 1965).

By far the strongest part of the stimulus to ventilation comes from raised P_{CO_2} and the breaking point is affected by such factors as the initial lung volume, oxygen breathing before breath-hold diving, and hyperventilation prior to diving.

Figure IV-2 illustrates that initial lung volume has a marked effect on breath-holding time. Figure IV-3 shows the effect of different concentrations of oxygen on the breaking point; at vital capacity the breath-holding time increases by 75% using 100% oxygen as compared to air, while at low lung volumes the effect of oxygen is much greater. Hyperventilation is so important that it is treated in more detail next.

f. Hyperventilation

Hyperventilation may be defined as a voluntary or involuntary increase in rate and/or depth of respiration over that necessary to maintain the homeostatic (i.e., normal) condition in the body. If prolonged, it will produce an excessive washout of carbon dioxide, which can result in hypocapnia with the concomitant disruption of

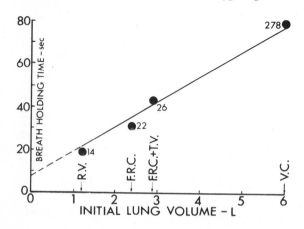

Figure IV-2. Relationship between initial lung volume and breath-holding time when the inspired gas is air. Number of observations indicated at each point. [From Mithoefer (1965) by permission of the American Physiological Society.]

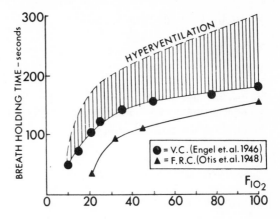

Figure IV-3. Effect of inspired oxygen concentration P_{IO_2} on breath-holding time at two levels of inspired volume, VC and FRC. Broken line indicates predicted elevation of VC curve that would be produced by 2 min of hyperventilation. [From Mithoefer (1965) by permission of the American Physiological Society.]

the acid–base balance, further resulting in finger tingling, tetany, and finally loss of consciousness (Cotes 1968, Miles 1969).

Hyperventilation is frequently caused by anxiety on the part of the diver, but may also be caused by the simple act of donning a mouthpiece and nose clip, as illustrated in Figure IV-4. Notice the drop in minute volume which occurs automatically after removal of the equipment to allow carbon dioxide tension in the blood to rise back to normal.

Hyperventilation is sometimes practiced deliberately in an effort to extend the underwater diving time. Deep breathing can wash out so much CO_2 that the alveolar tension is as low as 15 mm Hg, instead of the normal 40 mm Hg. At the same time the alveolar oxygen tension may be raised from the normal 100 mm Hg to 140 mm Hg. This maneuver does give the diver a longer breath-hold time because of the much reduced CO_2 stimulus to respiration. However, the exertion of swimming uses up the available oxygen supply and the pressure of O_2 may decrease to a dangerous level

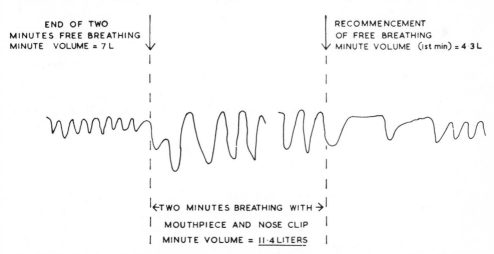

Figure IV-4. Effect of mouthpiece and nose clip on breathing pattern. [From Miles (1969) by permission of Granada Publishing, London.]

before the breaking point is reached. Unconsciousness and drowning may be, and in many cases have been, the result.

It should be remembered that "when trying for records, or to better one's own best efforts in the water, one should never do so alone. It is most important to have a buddy who knows what one is trying to do, and who is watching. This is the rule skin divers use and should apply to all water activities, even in pools" (Dumitru and Hamilton 1963). But of paramount importance is that hyperventilation before a breath-hold underwater swim is extremely dangerous.

3. Alveolar Ventilation

a. Dead Space

Albano (1970) classified the dead space in the human lung into three entities—the anatomical dead space (ADS) made up of the upper airways superior to the alveoli; the perfusion dead space (PDS), which comprises the fraction of gas which reaches poorly perfused alveoli; and the diffusion dead space (DDS), which comprises those areas in the lung where intrapulmonary diffusion is obstructed. The relation between pulmonary ventilation and alveolar ventilation may be expressed by

$$\dot{V}_E = f \times \dot{V}_D + \dot{V}_A \qquad (1)$$

where $\dot{V}_E$ is the expired ventilation in liters/min, f is the frequency of breaths per minute, $V_D = ADS + PDS + DDS =$ functional dead space in liters, and $\dot{V}_A$ is the alveolar ventilation in liters/min.

Elliott (1970) postulates that the airways in the lungs are "incompletely flushed" by the denser gaseous mixtures at depth, which leaves a layer of exhausted gases lining the alveoli. This lining, which obstructs the diffusion of oxygen as well as carbon dioxide, adds to the diffusion dead space (DDS).

b. Ventilation and Carbon Dioxide

The level of alveolar ventilation necessary to maintain the body is dependent upon the energy expenditure with its associated oxygen consumption and carbon dioxide production, and, in practice, the efficiency of alveolar ventilation is evaluated on the basis of the efficiency of exchange of carbon dioxide. Rossier and Mean (1943) derived the formula

$$\dot{V}_A = \frac{863.5\dot{V}_{CO_2}}{P_{A_{CO_2}}} \qquad (2)$$

where $\dot{V}_A$ is the alveolar ventilation in ml/min, $\dot{V}_{CO_2}$ is the CO_2 produced in ml/min, 863.5 is the transformation coefficient in mm Hg, and $P_{A_{CO_2}}$ is the average partial pressure of CO_2 in alveolar gas. This equation is represented graphically in Figure IV-5 with $\dot{V}_A$ on the vertical scale and $\dot{V}_{CO_2}$ on the horizontal scale.

The diagonal lines represent the indicated values of $P_{A_{CO_2}}$. For example, the $P_{A_{CO_2}} = 40$ mm Hg line gives all the possible combinations of $\dot{V}_{CO_2}$ and $\dot{V}_A$ that yield $P_{A_{CO_2}} = 40$. Shaded areas indicate values of $P_{A_{CO_2}}$ so far above or below the normal

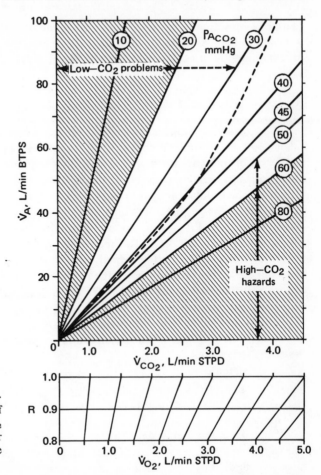

Figure IV-5. Relationship determining $P_{A_{CO_2}}$ as a function of alveolar ventilation and CO_2 production. [From Lanphier (1969) by permission of Bailliere and Tindall, London.]

range as to present distinct problems. The dashed line shows typical relationships for increasing levels of exertion. At lower values of $\dot{V}_{CO_2}$ a certain $P_{A_{CO_2}}$ (here 40 mm Hg) is maintained quite constantly, but at some level (usually corresponding to about 70% of the individual's $\dot{V}_{O_2 max}$) $\dot{V}_A$ begins to increase out of proportion to further increases in $\dot{V}_{CO_2}$. As a result, $P_{A_{CO_2}}$ must fall. The scales below the $\dot{V}_{CO_2}$ scale represent corresponding values of $\dot{V}_{O_2}$ at different values of R, where R is the apparent RQ at any given time and is defined as $\dot{V}_{CO_2}/\dot{V}_{O_2}$. Find the value of $\dot{V}_{O_2}$ on the appropriate R scale, then read the $\dot{V}_{CO_2}$ directly above this point. The graph as a whole can be used in place of calculation for several types of determinations and permits the relationships to be visualized (Lanphier 1969).

c. Ventilation and Oxygen

As with carbon dioxide, there are upper and lower limits for the inspired partial pressure of oxygen $P_{I_{O_2}}$. Partial pressures of inspired oxygen in excess of 0.42 ATA for prolonged periods of time at any density of the mixture result in chronic or acute hyperoxia (see Section C on oxygen toxicity). On the other hand, partial pressures below 0.21 (pressure of the ambient air at sea level) result in hypoxia. To preclude exceeding these limits when diving to great depths, the quantity of oxygen in the mixture must be reduced to maintain the $P_{I_{O_2}}$ within the range 0.21–0.42 (see Section E on breathing mixtures) (Chouteau and Corriol 1971).

d. Exertion at Depth

At sea level the limiting factor on the amount of work a man can perform is considered to be the cardiovascular system, whereas at depth the limiting factor may be the degree of tolerance the diver has to a high level of CO_2 in the alveoli. Because of the lowering of breathing efficiency, already discussed, the amount of air reaching the alveoli may be insufficient to clear out the CO_2 being produced. Figure IV-6 represents breath-by-breath CO_2 measurements made at increasing pressure. There was a slight increase in end-tidal P_{CO_2}, from 36 to 41 mm Hg between 1 and 6 ATA, and a dramatic rise to 50 mm Hg at 7.8 ATA. This increase in CO_2 level indicates that the subject was having difficulty in ventilating, which subsequently resulted in a higher retention of CO_2 and forced discontinuation of the experiment (J. N. Miller *et al.* 1971).

In experiments conducted on four men in a hyperbaric chamber, Jarrett (1966) confirmed the fact that the high alveolar P_{CO_2} effects during exercise under pressure are directly attributable to the reduction of alveolar ventilation (Table IV-2). He concluded that reduced alveolar ventilation is most probably caused by increased breathing resistance.

A simulated dive to 47 bars (1500 ft) was conducted at Alverstoke in 1971 in a helium–oxygen environment. The alveolar P_{CO_2} showed little change at rest, but increased markedly under a mild work load of 300 kg-m/min (Figure IV-7) (Morrison 1971).

Table IV-3 represents data showing the effects of higher levels of exercise to 900 kg-m/min down to a depth of 600 ft. Bradley *et al.* (1971) observed that one of the most striking findings of their tests was the wide variation in the ventilatory response to exercise. One subject required almost twice the volume of air over another when exercising at 900 kg-m/min at all depths.

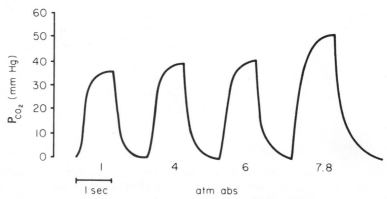

Figure IV-6. End-tidal P_{CO_2} records during 220-W exercise at various depths. [From Miller *et al.* (1971) by permission of Academic Press, New York.]

Table IV-2

CO_2 Retention under Various Ambient Pressures and Work Loads
(Means for Four Subjects) [a]

Work rate, kg-m/min	Pressure, ATA	P_{CO_2}, mm Hg	Peak $\dot{V}_I$, liters/min	$\dot{V}_A$ liters/min	Respiration rate/min	Heart rate/min
300	1	47.4	68.5	19.8	14.8	111
	2	47.0	70.5	19.2	17.3	111
	3	48.8	68.0	20.5	14.0	108
	4	48.6	68.0	21.1	12.8	103
573	1	49.1	102.2	31.1	14.5	132
	2	51.4	89.5	30.0	17.0	127
	3	54.2	89.5	28.3	13.8	118
	4	58.2	67.8	25.7	13.8	118
846	1	48.3	126.5	46.0	17.5	159
	2	54.8	119.3	41.1	17.4	151
	3	59.3	115.0	36.6	19.0	146
	4	62.3	113.0	36.6	15.8	143

[a] Reprinted from Jarrett (1966) with permission of the American Physiological Society.

e. Gas Exchange

Hyperbaric pressures appear to have little direct effect on the amounts or rates of exchange of gases between the alveoli and the arterial blood, except as they relate to the ventilation of the alveoli. Flynn (1971) found that by immersing the body in water a significant impairment was produced in the exchange of oxygen, with a con-

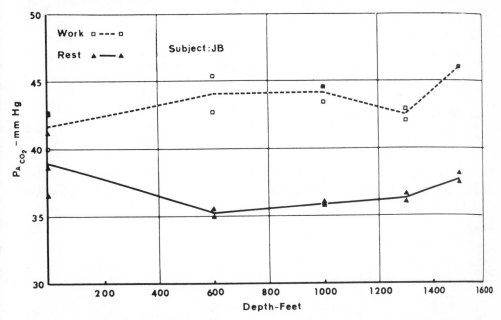

Figure IV-7. Alveolar carbon dioxide partial pressure $P_{A_{CO_2}}$ measured at rest and at a work load of 300 kg-m/min to depths of 1500 ft. [From Morrison (1971) by permission of the Royal Naval Physiological Laboratory, Hants, England.]

Table IV-3

Oxygen Consumption, Carbon Dioxide Consumption, Respiratory Quotient, Alveolar Ventilation, Alveolar Carbon Dioxide Tension, and Respiration Rate at Varying Work Loads down to 600 ft[a]

	Depth, ft	Rest	450 kg-m/min	900 kg-m/min
O_2 consumption,	Sea level	0.296	1.272	2.174
liters/min, STPD	150	0.303	1.159	2.203
	300	0.315	1.334	2.300
	450	0.310	1.230	2.332
	600	0.319	1.324	2.408
CO_2 production,	Sea level	0.244	1.153	2.043
liters/min, STPD	150	0.246	1.105	2.039
	300	0.285	1.235	2.230
	450	0.247	1.210	2.162
	600	0.282	1.216	2.319
Respiratory quotient	Sea level	0.84	0.91	0.94
	150	0.81	0.96	0.93
	300	0.87	0.93	0.97
	450	0.80	0.95	0.92
	600	0.90	0.94	0.95
$\dot{V}_A$, liters/min, BTPS	Sea level	5.19	22.22	41.14
	150	5.29	22.53	44.08
	300	5.89	23.22	45.41
	450	5.13	23.00	42.77
	600	5.82	23.92	44.00
$P_{A_{CO_2}}$, mm Hg	Sea level	38.8	44.1	42.5
	150	38.5	43.0	41.6
	300	39.4	44.7	42.4
	450	39.8	44.3	43.5
	600	40.0	43.5	44.5
f, breaths/min	Sea level	9.9	16.1	22.6
	150	7.0	13.3	20.1
	300	7.7	14.8	18.7
	450	7.7	12.5	20.2
	600	6.6	14.4	21.1

[a] Reprinted from Bradley *et al.* (1971) with permission of Academic Press. Values are the means of four subjects breathing an He–O_2 mixture. The changes in respiratory frequency and alveolar ventilation are statistically significant: p less than 0.05.

comitant increase in the alveolar to arterial difference, $(A-a)D_{O_2}$ (Table IV-4). This effect, however, does not produce a significant decrease in the oxygen content of the blood.

4. Respiratory Mechanics

a. General Considerations

According to Mead and Milic-Emili (1964):

The breathing mechanism would be characterized by an engineer as a reciprocating bellows pump. The walls of the bellows have two concentric parts: the lungs and the chest wall. The lungs are specialized to disperse the gas over the gas exchange surface,

Table IV-4

Summarized Mean Results of Tests on Two Subjects Dry at Sea Level and Simulated
Depth of 5 fsw
(Flynn 1971)

	Sea level			Simulated depth of 5 fsw		
	Dry	Immersed	% Change	Dry	Immersed	% Change
$\dot{V}_E$, m/min, BTPS	7.8	7.2	−7.7	9.5	10.8	+13.8
f, breaths/min	11.2	11.2	0.0	17.0	18.8	+10.6
V_T, ml, BTPS	701	647	−7.7	553	571	+3.3
$\dot{V}_{O_2}$, liters/min, TSPD	0.302	0.290	−3.9	0.351	0.355	+1.1
P_{IO_2}, mm Hg	169.3	168.5	−0.5	170.2	168.4	−1.0
P_{ACO_2}, mm Hg	38.7	40.4	+4.4	37.2	38.5	+3.5
P_{AO_2}, mm Hg	103.5	90.3	−12.7	112.6	99.3	−11.7
$(A\text{-}a)D_{O_2}$, mm Hg	16.3	28.3	+73.7	13.3	28.9	+117.3
V_D, ml, MTPS	237	201	−15.2	153	150	−1.9

and the chest wall includes the muscles that power the pump. The bellows chamber is made up of the respiratory bronchioles, alveolar ducts, alveolar sacs, and the alveoli, which contain some 95 percent of the gas within the pump. The remainder of the gas is in the tracheobronchial tree connecting the chamber to the outside. This system of tubes has an intrathoracic portion extending well inside the bellows—an unusual feature for a pump, with important implications to gas-flow resistance, as well as an extrathoracic portion (the upper trachea, larynx, glottis, nasopharynx, mouth and nose).

From the standpoint of function, most pumps have three parts: (1) an energy source, (2) passive elements that couple the energy source to the fluid to be pumped, and (3) the fluid itself, along with the channels through which it is moved. In many pumps the energy source, coupling elements, and fluid are physically distinct. It is clear that physical separation in the respiratory system is along other lines: the chest wall intermingles motor and coupling functions, whereas the lungs intimately combine coupling and fluid conduction. Still it is possible to accomplish to a limited but useful extent separate measurements of these functions.

The reciprocating bellows pump concept is well illustrated by Figure IV-8, which shows the balance of forces at midposition and the mechanism of passive exhalation. Normal breathing motion is produced by active inhalation (increasing the lung volume above the midposition) and subsequent passive exhalation. In inhalation the expansion of the chest in the lateral diameter is produced by movement of the ribs, which also increases the anterior/posterior diameter, and the chest is increased in the head–foot dimension by downward movement of the diaphragm. Exhalation is passive through relaxation of the inhalation muscles, which permits the forces of gravity and the elastic recoil energy produced during inhalation to return the lung–chest complex to normal midposition.

The normal anatomy of the respiratory system is presented in Chapter I, in the section on gas exchange. The pressure effects on the components and functions of the respiratory system are presented in the section dealing with respiratory mechanics, for they are generally physical in nature.

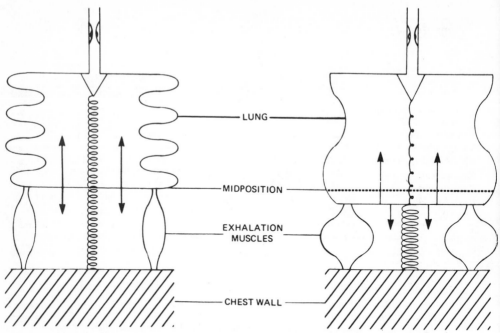

Figure IV-8. Balance of forces at midposition and mechanism of passive exhalation (Bartlett 1973).

b. Ambient Pressure Differences

One of the most dramatic effects of ambient pressure (pressure external to the lungs) is on the volume of gases in the lungs during a breath-hold dive, as shown in Table IV-5.

It should be noted that at a depth of 99 ft the total lung capacity has been reduced to the residual volume. Thus a diver entering the water in full inspiration will be in a state equivalent to full expiration without having lost any air when he has dived to 99 ft. If the diver continued to go deeper, (1) the air in the lungs would remain at constant volume; (2) pressure and the pressure differential in the alveoli would increase, causing a transfer of fluid from the alveolar capillaries into the lungs (pulmonary edema); (3) as the descent was continued, vessels would burst, resulting in pulmonary hemorrhage; and finally, (4) ribs would crack and the chest wall would

Table IV-5

The Effects of Pressure at Various Depths on Total Lung Capacity during a Breath-Hold Dive

		Average man, liters	Small man, liters	Big man, liters
Surface	Vital capacity + residual air	$4\frac{1}{2} + 1\frac{1}{2} = 6$	$3 + 1 = 4$	$6 + 2 = 8$
At 33 ft	Lung contains	3	2	4
At 66 ft	Lung contains	2	$1\frac{1}{3}$	$2\frac{2}{3}$
At 99 ft	Lung contains	$1\frac{1}{2}$	1	2

cave in (Miles 1969). Therefore, a human diving to depths greater than 99 ft must be protected by a positive-pressure breathing apparatus of some kind.

c. Pressure Differences Caused by Immersion

Agostoni *et al.* (1966) reported that immersion of the seated human body in water up to the xiphoid process (top of the abdomen) produces an appreciable reduction in the lung volume due to craniad displacement of the abdomen caused by the difference in ambient pressure on the upper and lower chest, and the lesser effect of gravity on the submersed abdomen. Even further reduction occurs during submersion to the neck—the expiratory reserve volume is reduced from about 36% to about 11% of the vital capacity in air.

Where pressure was measured across the rib cage, the diaphragm, and the abdominal wall of two subjects sitting in air and in water at the first level (the xiphoid process), the pressure on the abdominal side of the diaphragm increased by 6 cm H_2O while the pressure on the lung side of the diaphragm increased only 1.5 cm H_2O. This produced a pressure difference of 4.5 cm H_2O, which caused the diaphragm to move upward, decreasing the volume of the lungs (Figure IV-9). (Measurements were made at the end of spontaneous expiration with respiratory muscles in a relaxed condition.) Further submersion of the thorax to the neck increased the difference in pressure across the diaphragm to 14 cm H_2O.

Negative-pressure breathing experiments conducted on the same subjects in a seated position determined that breathing from a tank at -20.5 cm H_2O produced the same reduction of lung volume as submersion to the neck.

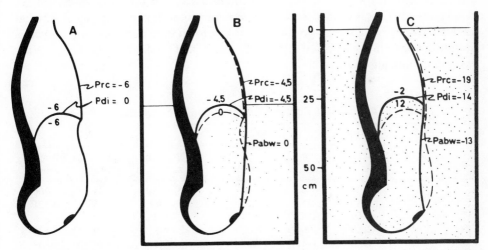

Figure IV-9. Drawings illustrating the thoraco-abdominal pressures at the end of spontaneous expirations of subjects sitting in air (A), in water up to the xiphoid process (B), and in water up to the neck (C). The broken lines in B and C indicate the profile of condition (A) for comparison. The values of pressure in cm H_2O are the mean values recorded in two subjects. Prc, Pdi, and Pabw indicate, respectively, the average pressures across the rib cage, the diaphragm and rib cage, and the diaphragm and the abdominal wall. [From Agostoni *et al.* (1966) by permission of the American Physiological Society.]

Thompson *et al.* (1967) investigated the effect of transpharyngeal pressure gradients caused by the difference in ambient pressure between the lungs and the mouth when a diver is underwater in the upright position. One effect is that it causes discomfort by stretching the pharynx, which is relatively unsupported. In experiments with eight subjects seated immersed to the level of the auditory meatus in water at normal skin temperature (33–34°C), Thompson *et al.* found that subjects breathing through a mouthpiece or facemask subjectively preferred to breathe at a pressure that was negative relative to the sternal notch rather than choosing a pressure that would seem to be physiologically sound, a pressure equal to the mean external pressure on the thorax. Perhaps the choice is a compromise between the overstretching of the upper respiratory tract and underinflation of the lower tract. They concluded that whether mouthpiece, facemask, or helmet is used, "a wide range of transthoracic pressure gradients (−30 to +40 cm H_2O) is subjectively more comfortable than a slight increase in the transpharyngeal gradient (up to 7.5 cm H_2O), suggesting that during immersion intrapulmonic pressures are selected by the subject to minimize the transpharyngeal pressure gradient" (Thompson *et al.* 1967).

d. Lung Volume Alterations Caused by Postural Changes

Postural changes from one position to another at atmospheric pressure have a significant effect on vital capacity (VC) and expiratory reserve volume (ERV), as well as other relative volumes in the lungs (Table IV-6) (Moreno and Lyons 1961).

Investigating the effects of postural changes on tidal volume and vital capacity under several breathing pressures, Thompson *et al.* (1967) found that (1) tidal volume was unaffected; (2) negative-pressure breathing during vertical immersion reduced vital capacity (VC); (3) vertical immersion decreased total lung capacity and functional residual capacity but a supine posture partially restored these decreases; and, (4) positive-pressure breathing increased total lung capacity and residual volume in the sitting position in both air and water.

Table IV-6

Analysis of Postural Changes for 20 Subjects[a]

	Mean absolute values			Mean difference		
	Sitting	Supine	Prone	Sitting to supine	Supine to prone	Sitting to prone
IC	2562	3002	2768	+440	−234	+206
ERV	1345	759	824	−586[b]	+65	−521[b]
VC	3907	3762	3594	−145	−168	−313
FRC	2689	1904	2053	−785[b]	+149	−636[b]
RV	1292	1143	1229	−149	+86	−63
TLC	5193	4900	4799	−293	−101	−394
RV/TLC × 100	25%	23%	25%			
FRC/TLC × 100	52%	39%	43%			

[a] Reprinted from Moreno and Lyons (1961) with permission of the American Physiological Society. IC, inspiratory capacity; ERV, expiratory reserve volume; VC, vital capacity; FRC, functional reserve capacity; RV, residual volume; TLC, total lung capacity.
[b] Statistically significant difference.

5. *The Work of Breathing*

a. General Considerations

Not only the most prominent, but the most important disorder of pulmonary function in diving is inadequate ventilation of the lungs, especially during exertion. The mechanics of breathing have already been covered, but some aspects will be reemphasized here.

The work of breathing can be analyzed in terms of the mechanical work done against forces that resist the movement of air and those forces that oppose changes in lung volume. Thus the most important aspects are those related to resistance to gas flow in the airways and to elasticity of the chest–lung system (Lanphier 1969). In hyperbaric work with the increased gas density and the hydrostatic pressure differences, both these factors are affected.

b. The Breathing Mixture, Gas Density

It was pointed out in the introduction to this section that the density of a gas increases with pressure. Section E on breathing mixtures contains an in-depth treatment of respirable gas mixtures. However, for the purposes of discussing the work of breathing as it relates to the density of the gas breathed, it is important to point out here that the relative density is directly related to the type of gas in the mixture.

It might logically be assumed that values for helium at greater depths could be predicted simply on the basis of relative density between air and helium and known values for air at various depths. But, as shown in Figure IV-10 and discussed by Lanphier (1967), "although pure helium is less than one-seventh as dense as air, the relative density increases markedly with addition of oxygen or nitrogen. For example, an 80% He–20% O_2 mixture is almost exactly one-third as dense as air."

Figure IV-11 shows a comparison of the densities of various helium–oxygen mixtures at equivalent depths to 40 ATA with the density of air at 1 ATA (Workman 1967). Obviously, it is possible to vary the density of the inspired gas by changing either its composition or the ambient pressure. Man's inefficiency in the water environment is compounded by progressive restriction of pulmonary ventilation due to increasing gas density with depth. The use of a lighter gas helps, but, as greater depths are reached, the restrictive properties of man's air pump are again encountered.

Increased air density may disturb intrapulmonary gas exchange because of an alteration in the regional distribution of ventilation as shown in Figure IV-12. Wood and Bryan (1971) found that at low flow rates, the distribution of ventilation was normal for all the gas densities; that is, alveoli in the lower lung were better ventilated than those near the apex. As flow rate increased, this distribution was reversed at higher gas densities. Thus, many different patterns of ventilation–perfusion may exist and it is possible, especially during exercise at depth, that gas exchange may be impaired.

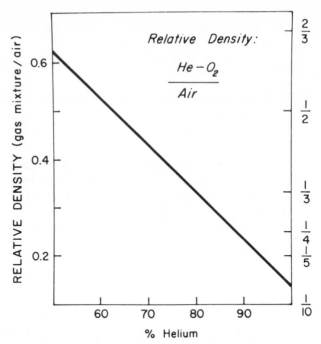

Figure IV-10. Density of various He–O$_2$ mixtures relative to that of air at the same ambient pressure. The density of such mixtures is greatly influenced by the proportion of the heavier gases. [From Lanphier (1967) by permission of Williams & Wilkins.]

c. Flow Rates

According to Murphy *et al.* (1969):

The flow of gas through a tube is laminar below a critical velocity and turbulent above it, the transition occurring when the Reynolds number exceeds 2,000. Since this quantity is defined as (density × velocity × diameter)/viscosity, the critical velocity will be (2,000 × viscosity)/(density × diameter). As the viscosity of 21% oxygen in helium is 1.1 times, and the density .34 times that of air, the critical velocity for He–O$_2$ must be approximately three times that of air. When ventilation is increased from a resting value, the flow of both gases must, therefore, be laminar at first, then turbulent with air but still laminar with He–O$_2$, and finally turbulent with both gases.

During laminar flow the flow-resistive pressure, being proportional to the viscosity and velocity, should be almost the same for the two gases. With turbulent flow, however, the resistive pressure is related to the density and the square of the velocity, so that the value for air must rise rapidly above that for He–O$_2$ as turbulence develops. Finally with turbulent flow in both gases, airway resistance must increase in a parallel manner in both gases, the values differing in proportion to the densities.

It follows that, since the density of the gas flowing in the tracheobronchial tree is one of the determinants of airway resistance, an increase in depth (ambient pressure) or a change in gas composition will affect the rate at which gases can be inhaled and exhaled.

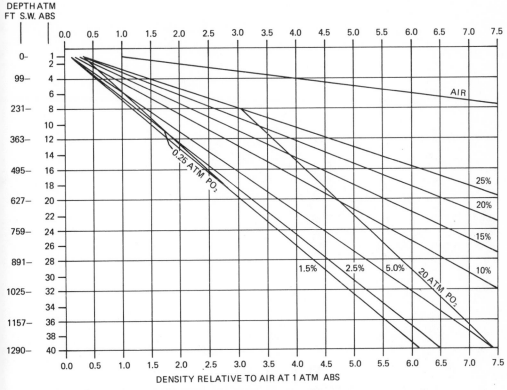

Figure IV-11. Comparison of densities of various mixtures of oxygen in helium with air at depths equivalent to 40 ATA. [From Workman (1967) by permission of Williams & Wilkins.]

d. Maximum Breathing Capacity and Flow Rates

The maximum breathing capacity (MBC) is the maximum volume of air which can be voluntarily moved in and out of the lungs in a certain amount of time. By measuring MBC and two other factors, the timed vital capacity (TVC) and the maximum expiratory flow rate (MEFR), the ventilatory (work) limits of the respiratory system can be determined. Miles (1969) observed that the MBC varies inversely with the square root of the mass, which is directly proportional to the density of air. This observation was made assuming that the work of breathing is constant at every depth. Thus, MBC and MEFR vary inversely and exponentially with the density of air.

The relationship of the average forced expiratory volume at 1 sec (FEV_1) to relative gas density is shown in Figure IV-13, where it can be seen that FEV_1 falls markedly as the relative gas density (RGD) increases.

The decrease in the average maximum expiratory flow rates in relation to relative gas density is presented in Figure IV-14.

Vorosmarti et al. (1971) found that the average work done against airway resistance at various minute ventilations and relative gas densities was described by Figure IV-15: "At resting ventilations gas density has little effect on the work required

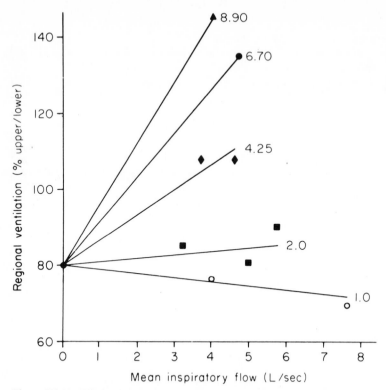

Figure IV-12. Effect of flow rate of upper and lower lung regions in one subject breathing gases of different densities. The relative gas densities are indicated adjacent to each line. [From Wood and Bryan (1971) by permission of Academic Press.]

to overcome airway resistance. At higher ventilations the general effect of increasing gas density was to increase the work done against airway resistance. Although these subjects were not attempting to perform a maximum breathing capacity maneuver during hyperventilation and no definite end point was established at each relative gas density (RGD), the theoretical minute ventilation decreased with increasing gas density to 50 liters per minute at RGD 15."

The maximum inspiratory flow also varies inversely as the square root of air density. Figure IV-16 shows that, even at moderate flow rates under pressure, very large pleural pressure increments are required to produce moderate increases in flow. This means that inspiratory work increases significantly at depth (Wood et al. 1971).

e. Resistance to Breathing in Self-Contained Diving

Considerable depth can be reached without density being a serious problem if the individual is in a free-breathing situation (such as a chamber, habitat, or rigid helmet diving dress) where only the diver's own respiratory resistance need be con-

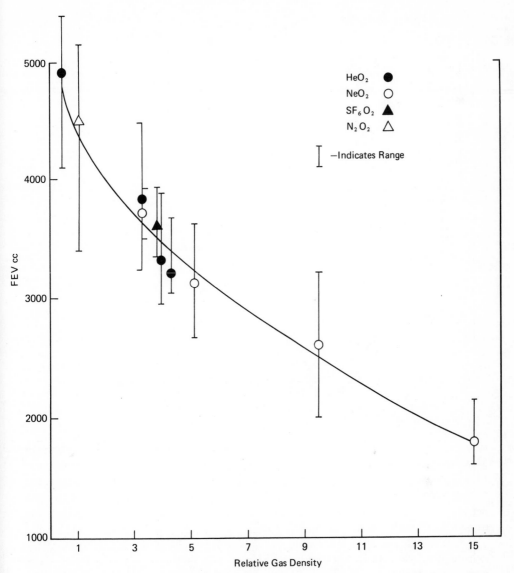

Figure IV-13. FEV$_1$ at various relative gas densities (70% N$_2$–30% O$_2$ = 1). [From Vorosmarti *et al.* (1971) by permission of the authors.]

sidered. However, with self-contained diving equipment there may be added the additional resistance of mouthpiece, tubes, valves, or even a carbon dioxide-absorbing canister.

Just as depth causes increase in gas density and adds to personal respiratory resistance, so depth will increase any equipment-breathing resistance.

Personal resistance at atmospheric pressure varies with individuals, but may be

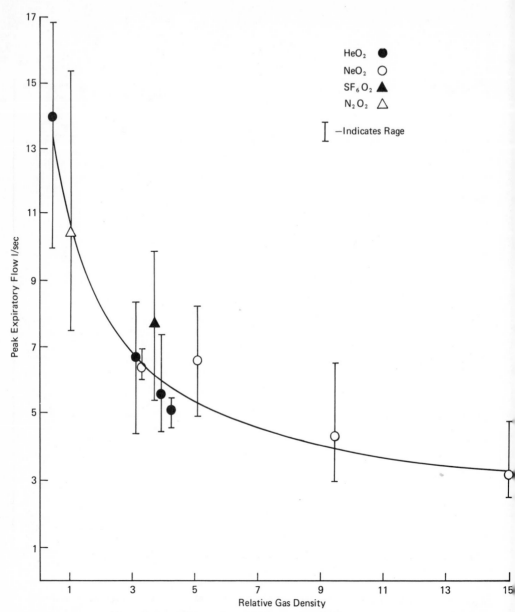

Figure IV-14. Peak expiratory flow at various relative gas densities (70% N_2–30% O_2 = 1). [From Vorosmarti *et al.* (1971) by permission of the authors.]

taken as 4 cm of water per liter per second (4 cm H_2O liter^{-1} sec^{-1}). Measurement of resistance in a breathing set was also found to be 4 cm H_2O liter^{-1} sec^{-1}, so the overall resistance when used at atmospheric pressure is 8 cm H_2O liter^{-1} sec^{-1}. While the volume of air moved is proportional to the inverse of the square root of the density, other factors being equal, the resistance may be taken to be proportional to the square

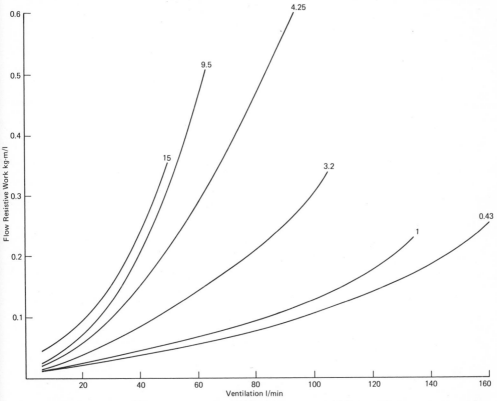

Figure IV-15. Work done against flow resistance at increasing ventilatory rates at various relative gas densities (70% N_2–30% O_2 = 1). [From Vorosmarti *et al.* (1971) by permission of the authors.]

root of the density, $R \propto \sqrt{D}$. This is shown graphically in Figure IV-17. From results of experimental work a desirable limit is 12 cm H_2O liter^{-1} sec^{-1}. This does not change with depth and is thus shown as a vertical line.

In addition to the effect of density, the postural position of the diver has a substantial effect on the work of breathing with an underwater breathing apparatus. It has been shown that divers in the horizontal swimming position with certain scuba diving rigs (Mark II Diving System) were positive-pressure breathing at +15 cm H_2O, while in the vertical position they were negative-pressure breathing (NPB) at −10 cm H_2O. Both of these factors make the work of breathing more difficult for the diver (Bradley 1970, Agostoni *et al.* 1966).

The several scientists whose work has been presented all agree that when man breathes a gas denser than air at sea level, his pulmonary flow resistance and work of breathing are increased and, as he goes under pressure, the situation becomes aggravated. It has been suggested that some sort of respiratory assistance is desirable but what form the assistance should take is not decided. The least that can be done is not to add additional resistance to breathing in the development of underwater life support equipment, if it can be avoided.

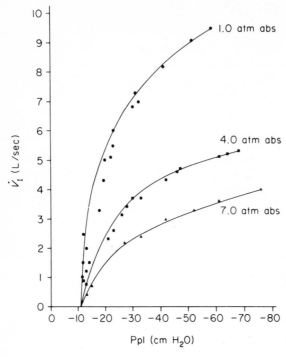

Figure IV-16. Inspiratory volume-pressure curves at 50% VC. Note the large increase in Ppl required to produce moderate flow increases at 2–3 liters/sec at 4.0 and 7.0 ATA. [From Wood *et al.* (1971) by permission of Academic Press.]

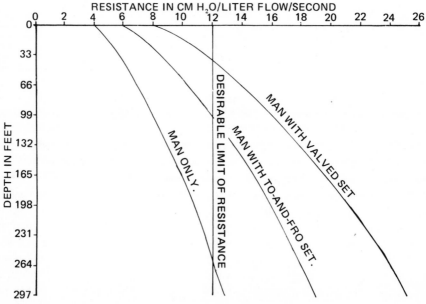

Figure IV-17. Resistance to breathing at depth. Surface density is represented as unity, 2 at 33 ft, 2 at 66 ft, etc. [From Miles (1969) by permission of Granada Publishing, Ltd., London.]

f. Respiratory Rate and Hyperbaric Environments

As shown in Table IV-7, there is an almost linear relationship between the average decrease in respiration rate and increase in ambient pressure down to 4.0 ATA with the subject at rest breathing air. Concomitantly, a correlation between the increasing tidal volume and the increasing ambient pressure is evident. These phenomena are believed to be brought about by the increase in air density and thus breathing resistance, and to a lesser degree by the increase in inspired P_{O_2} (Hesser and Holmgren 1959).

Although ventilatory responses to exercise vary widely from subject to subject, significant trends can be seen at various work loads at various depths. Table IV-8 shows that at any given work load, as depth increases, the respiration rate is decreased and the tidal volume is increased relative to the values determined at sea level. Table IV-9 extends these data to a depth of 1000 ft, showing a clear decreasing trend in respiration rate with increasing pressure.

These phenomena are similarly attributed to the increase in density of the gaseous environment as well as the increase in turbulence brought about by the more labored breathing when exercising (Bradley *et al.* 1971, Salzano *et al.* 1970).

6. *Conclusion*

As density of the gas mixture increases with depth, so efficiency decreases and, since this is unavoidable, the work rate must be reduced. "At atmospheric pressure in

Table IV-7

Effects of Breathing in an Air, 100% Oxygen, and 5% Oxygen–Nitrogen Environment at Various Hyperbaric Pressures on Eight Subjects[a]

Pressure	Air					Oxygen		5% O_2 in N_2	
	1.0 atm	1.3 atm	2.0 atm	3.0 atm	4.0 atm	1.0 atm	1.6 atm	3.0 atm	4.0 atm
Respiration rate, Breaths/min	9.60	9.21	8.65	7.73	6.89	8.65	8.81	8.57	7.73
Standard deviation	3.71	3.73	3.93	3.43	3.48	3.71	3.75	3.71	3.97
Percent of control[b]	100	95.7	88.4	80.5	72.8	90.0	91.1	85.9	79.9
Standard deviation	0.0	11.6	13.9	16.4	22.9	13.1	12.6	21.9	23.4
Tidal volume Milliliters, BTPS	808	801	868	895	1019	893	885	869	964
Standard deviation	411	384	486	396	417	419	439	533	505
Percent of control[b]	100	99.9	105.4	113.1	133.1	111.7	110.0	109.9	121.1
Standard deviation		9.6	10.3	44.7	44.7	17.7	15.1	21.6	44.5

[a] Reprinted from Hesser and Holmgren (1959) with permission of *Acta Physiologica Scandinavica*.
[b] Control: air-breathing at sea-level pressure.

Table IV-8

Mean Values (Statistical Significance : $p < 0.05$) of Respiratory Frequency and Tidal Volume for Four Subjects during Rest and Exercise at Work Loads of 450 and 900 kg-m/min from Sea Level to 600 fsw[a]

Work, kg-m/min	f, breaths/min					V_T, liters, BTPS				
	Sea level	150	300	450	600	Sea level	150	300	450	600
Rest	9.9	7.0	7.7	7.7	6.6	1.176	1.805	1.694	1.294	1.699
450	16.1	13.3	14.8	12.5	14.4	2.327	3.196	2.525	2.735	2.504
900	22.6	20.1	18.7	20.2	21.1	2.656	2.877	3.266	2.834	2.785

[a] Reprinted from Bradley *et al.* (1971) with permission of Academic Press. All depths are in fsw.

a healthy individual, muscular effort is limited by cardiac output. No physical activity can be great enough to need a respiratory minute volume anything like the maximum breathing capacity" (Miles 1969). Work rates described by Silverman *et al.* (1951) of 1660 kg-m/min on a bicycle ergometer require 3500 ml of O_2/min and would presumably require a minute volume of at least 70 liters. It would produce great stress to attempt such work under pressure:

> At an equivalent depth of 100 feet this ventilatory requirement would be almost equivalent to the maximum breathing capacity. A situation would, therefore, arise where a task which was within the capacity of the individual's muscles and cardiac output could not be performed because of an inability to achieve adequate ventilation. Breathing capacity now becomes the limiting factor to muscular effort, a situation foreign to the healthy individual. Under such circumstances an accustomed task may be accompanied by unaccustomed respiratory embarrassment. The inexperienced diver may find this respiratory distress alarming and his anxiety may lead to hyperventilation when he has given up his work with its attendant disturbances and sometimes dangerous results (Miles 1969).

Table IV-9

Pulmonary Ventilation and Respiratory Rate during Exercise at Normal and Elevated Ambient Pressure[a]

Work load, kpm/min	Atm	Number of experiments	Pulmonary ventilation $\dot{V}_E$, liters/min, BTPS	Respiration rate, breaths/min
275	1.0	18	23.8 (22.1–26.9)	18.9 (16.0–21.0)
	10.7	6	Not recorded	17.5 (13.0–21.3)
	31.3	6	25.1 (22.0–28.6)	15.3 (13.5–17.0)
582	1.0	18	41.7 (34.7–47.6)	24.7 (15.5–30.0)
	10.7	6	Not recorded	20.8 (15.0–26.5)
	31.3	6	40.6 (38.5–43.6)[b]	17.8 (15.0–20.0)
735	1.0	18	52.6 (47.1–62.1)	26.2 (22.0–30.0)
	10.7	6	Not recorded	20.9 (15.0–25.0)
	31.3	6	51.1 (47.0–58.7)	23.0 (17.0–30.0)

[a] Reprinted from Salzano *et al.* (1970) with permission of the American Physiological Society. Values are means of three subjects. Values in parentheses are the extreme values.
[b] Based on four experiments by two subjects.

B. *Cardiovascular Factors*

Cardiovascular functions of man in the ocean environment vary with the type of activity, such as simple immersion to the neck, breath-hold diving, scuba diving, or saturation diving. Basically, any change in cardiovascular functions may be associated with one or more of the following three factors: mechanical, neural, and humoral (also hormonal or chemical). In the ocean environment, man is always under the influence of hydrostatic pressure, which is directly proportional to the depth. Exposure to a pressure higher than 1 ATA (atmosphere absolute) results in certain hazards, particularly if the changes in pressure are rapid. Although many of these hazards involve respiratory functions (see the discussion of the physiology of respiration), there are certain attendant changes in cardiovascular functions. In addition to this pressure (mechanical) factor, the ocean environment presents a considerable thermal stress to man. Because of the higher thermal conductivity of water, which is about 25 times greater than that of air, there is a net loss of body heat when the water temperature is below 35°C or so. Cardiovascular functions are closely coupled to thermoregulatory functions and thus the continuous body heat loss is associated with certain cardiovascular disturbances through either neural or humoral mechanisms. In this section, changes in cardiovascular functions in the ocean environment will be described, with special reference to the role of environmental pressure and temperature. For convenience, the overall cardiovascular changes with the type of activity in the ocean will be described.

1. *Immersion to the Neck*

The first step of any activity in the ocean necessarily involves immersion in the water surface up to the neck. In this case, many subtle changes in pressure distribution are brought about which affect the pulmonary, cardiovascular, and renal functions.

a. Negative-Pressure Breathing

When one breathes in an ordinary air environment, the pressure within the lungs is equal to that surrounding the body. This is no longer true when the body is immersed up to the neck. Since the subject still keeps his head above the water and breathes the outside air, the pressure of the gas phase within the lungs (the intrapulmonary pressure) must be equal to 1 ATA. On the other hand, the body below the neck is under the influence of the hydrostatic pressure, which is greater than 1 ATA. In other words, the intrapulmonary pressure is lower than the surrounding pressure, and hence the subject performs so-called "negative-pressure breathing." (In physiology, a sub-atmospheric pressure is called "negative pressure.") As a result, the functional residual capacity (FRC) is greatly reduced due to the compression of the chest (or thoracic) cage. This reduction in the FRC is largely due to a reduction in the expiratory reserve volume (Hong *et al.* 1969*b*), but there is evidence that the residual volume also decreases during immersion to the neck (Agostoni *et al.* 1966, Craig and Ware 1967).

The latter finding has been attributed to an increase in the intrathoracic blood volume due to the increase in the absolute pressure of the extrathoracic vascular system in the face of unchanged intrathoracic pressure. Such a pressure gradient would drive more blood from the periphery into the thorax, resulting in an increase in venous return. Because the lungs cannot be compressed to below the residual volume during a breath-hold dive, the reduction in the residual volume during immersion would extend the maximal depth to which a diver goes. It should be mentioned, however, that the intrathoracic vessels are largely protected from the changes in the absolute pressures to which all the extrathoracic vessels are exposed. The rib cage and the highly collapsible veins provide an efficient barrier to blood and prevent excessive accumulation within the thorax (see below for a more detailed discussion).

b. Density

The next important mechanical factor to be considered is the density of the medium in which the extrathoracic vessels are supported. In an air environment the pressure at the bottom of a 100-cm column of blood is 103,370 dyn/cm² (or 103 cm H_2O), as compared to 2800 dyn/cm² (or 2.8 cm H_2O) in sea water (see Table IV-10). Such a huge pressure difference between a column of blood and air is due to the extremely low density of air, which is about 0.0012 g/cm³ at 15°C as compared to 1.056 g/cm³ for blood. Since the density of water is only slightly lower than that of blood, the pressure difference between a column of blood and supporting media virtually disappears during immersion in water. This condition is analogous to a gravity-free state where blood pooling in the peripheral veins, usual in an air environment, is essentially eliminated. In other words, a redistribution of the peripheral blood volume, particularly the venous blood, contributes to the increase in the intrathoracic blood volume.

c. Water Temperature

In addition to mechanical factors, there is a neural factor which plays a very important role in water colder than thermoneutral. As stated earlier, whenever the water temperature is below 35°C, there is a net loss of body heat, resulting in lowering of deep body temperature. As a defense against this body heat loss, the blood flow through the periphery, particularly the skin, decreases greatly by inducing a constriction of blood vessels in this region. This is a reflex phenomenon initiated by the

Table IV-10

Pressure Differences ΔP between a Column of Blood (100 cm) and Supporting Media[a]

Medium	Height (cm)	Acceleration $(cm \cdot sec^{-2})$	$(\rho_{blood} - \rho_{media})$ $(g \cdot cm^{-3})$	$= \Delta P$ $(dyn \cdot cm^{-2})$
Air	100 ×	980	× (1.056 − 0.0012) =	103,370
Fresh water	100 ×	980	× (1.056 − 0.999) =	5,586
Sea water	100 ×	980	× (1.056 − 1.027) =	2,842

[a] Modification of an original table by Rahn (1965). ρ is density.

stimulation of cold receptors distributed all over the body surface. The information received by these receptors is relayed to the central nervous system via sensory nerves, where the vasoconstriction center is located. According to the data reported by Hong *et al.* (1969a), the blood flow through the lower arm and finger decreases to virtually zero within 1 hr of immersion even in water at 30°C. Naturally, such a reduction in the peripheral blood flow is accompanied by a corresponding reduction in the heat loss from the surface of these regions to the surrounding water. At any rate, it is important to note that the intense peripheral vasoconstriction would also contribute to the central pooling of blood at the expense of the peripheral blood volume.

d. Intrathoracic Blood Volume

Accurate measurement of the magnitude of the increase in intrathoracic blood volume during immersion is rather difficult, and thus the values reported in the literature are highly variable. However, it is conservatively estimated to be about 300–500 ml (Hong *et al.* 1969b, Rahn 1965).

Cardiovascular consequences of the increase in the intrathoracic blood volume during immersion in 35°C water to the neck have been studied only recently by Arborelius *et al.* (1972). Their data are summarized in Table IV-11. In the absence of any significant changes in the heart rate (see the following sections for more comprehensive discussions on the heart rate), the cardiac output increases 30% during immersion, indicating that the stroke volume (cardiac output/heart rate) must be increased similarly. The calculated stroke volume increases from 84 ml in air to 108 ml during immersion. The right atrial pressure is usually slightly negative (i.e., slightly below the atmospheric pressure) in air but increases to +16 mm Hg during immersion. These findings can be explained by the redistribution of blood, in which the

Table IV-11

Cardiovascular Functions of Human Subjects Immersed in 35°C (95°F) Water up to the Neck[a]

	Control (in air)	Immersion
Heart rate, beats/min	73	71
Cardiac output, liters/min	6.0	7.7
Stroke volume, ml	84	108
Right atrial pressure, mm Hg	−2	16
Pulmonary arterial pressure, mm Hg		
Systolic	13	28
Diastolic	3	20
Mean	5	22
Brachial arterial pressure, mm Hg		
Systolic	114	128
Diastolic	71	79
Mean	86	98
Vascular resistance of systemic circulation, mm Hg/liter·min^{-1}	15.2	10

[a] Data of Arborelius *et al.* (1972).

intrathoracic blood volume is increased in the immersed subject. Evidently, the heart (and the intrathoracic blood vessels) is distended as a result of the pronounced blood shift and may be under considerable strain. In this connection, it is important to note again that the large veins collapse as they enter the chest in immersed subjects, thus preventing an excessive congestion of intrathoracic vessels and of the heart (Rahn 1965).

The increase in cardiac blood volume implicated by increases in the right atrial pressure and in the cardiac output seems to be causally related to the so-called "immersion diuresis" (a temporary increase in urine flow) experienced by divers (McCally 1965). This diuresis is known to be due to suppression of antidiuretic hormone (ADH), which is responsible for conservation of water. The left atrial wall contains a special structure, called "volume receptors," that responds to an increase in the volume of the left atrium by suppressing the ADH system (Gauer *et al.* 1954). However, the mechanism of "immersion diuresis" seems to be more complicated than that described above, and there is at least one additional contributing factor. Recently Epstein and Saruta (1971) reported significant decreases in plasma renin activity and in urinary aldosterone excretion during immersion in 34°C water for 2 hr. The renin–aldosterone system is very important in conserving salt and water, and, hence, a suppression of this system would increase the excretion of salt and water. At any rate, there is a net loss of water and salt during prolonged immersion and this could lead to a disturbance of cardiovascular functions.

Table IV-11 also indicates that both the pulmonary and brachial arterial pressures are elevated during immersion to the neck. Interestingly enough, the magnitude of increase in the mean brachial arterial pressure is somewhat lower than the hydrostatic pressure increase induced by the external water column. This may be somehow related to a slight reduction in the vascular resistance of systemic circulation, also shown in Table IV-11. However, it is not understood how these changes are brought about during immersion.

A summary of events taking place during immersion up to the neck is shown in Figure IV-18. One is impressed by the complexity of these cardiovascular changes induced by a simple act of immersion. Many of these changes are still not well understood but have a great impact on human performance in the ocean.

2. Breath-Hold Diving

More than a century ago Bert in France observed a pronounced bradycardia (slowing of the heart rate) in the diving duck. Since then, this has been confirmed in every diving animal. Largely due to the excellent pioneering work of Irving and Scholander, the physiological implication of this fascinating bradycardial response to breath-hold diving has been beautifully documented. In essence, Irving and Scholander view this diving bradycardia as a reflex phenomenon which is accompanied by an intense peripheral vasoconstriction and by a drastic reduction in the cardiac output [for review, see Scholander (1961–62)]. (Only the blood circulation to the brain and the cardiac tissue is maintained.) The arterial blood pressure P is determined by the

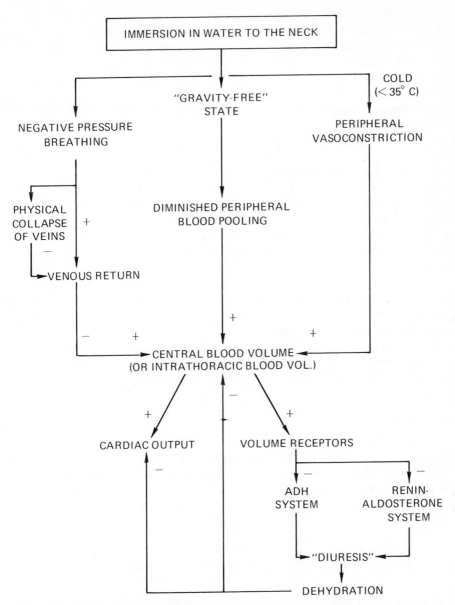

Figure IV-18. Cardiovascular changes during immersion up to the neck (+ indicates stimulation or increase; − indicates inhibition or decrease).

product of the cardiac output $\dot{Q}$ and the peripheral resistance R, in analogy to Ohm's
law:

$$P = \dot{Q}R$$

The resistance term R is mainly determined by the geometry of blood vessels, especially
the radius r: $R \propto 1/r^4$. In other words, R is expected to increase greatly whenever a

marked vasoconstriction develops as in the case of diving. However, the arterial blood pressure during diving does not increase in diving animals. This indicates that the reduction in the cardiac output offsets the pressure-increasing effect of intense vasoconstriction during diving. Such a reflex plays a vital role in conserving oxygen during the dive. Because of the marked vasoconstriction in all organs other than the brain and the heart, the oxygen consumption decreases drastically during the dive, which helps prolong the duration of dive in these animals. Since this remarkable cardiovascular readjustment develops rather instantaneously with the onset of breath-hold diving, Irving and Scholander maintain that a neural reflex mechanism is involved.

a. Diving Bradycardia in Man

Whether or not such a cardiovascular reflex mechanism exists in man has been the topic of exhaustive investigations during the last decade. Conveniently enough, one can elicit a diving bradycardia in man by a breath-hold face immersion in water and hence most of the current knowledge on this subject is derived from such face immersion studies.

There are major differences in diving bradycardia in man and in diving animals (Table IV-12). The onset of bradycardia is usually slow in man, reaching maximum in 20–30 sec, in contrast to 5–10 sec in many diving animals. Moreover, the degree of bradycardia in man is not as large as in diving animals. In some diving animals—such as beavers, seals, and whales—it is not unusual to observe a 90% reduction in the heart rate during diving (Strauss 1970). In man, the degree of maximal bradycardia is less than 50% of the predive control heart rate (Song *et al.* 1969).

b. Water Temperature

The most interesting aspect of diving bradycardia in man is its dependence on the water temperature. Regardless of whether one is engaged in actual breath-hold diving or breath-hold face immersion, the degree of bradycardia increases as the water temperature decreases (Kawakami *et al.* 1967, Corriol and Rohner 1968, Hong *et al.* 1967, Song *et al.* 1969). For instance, the magnitude of bradycardia is equivalent to

Table IV-12

Comparison of Cardiovascular Responses to Breath-Hold Diving between Diving Animals and Man

Response	Diving animals	Man
Bradycardia		
Onset	Fast	Slow
Maximal reduction, % of control heart rate	50–90%	10–30%
Temperature dependence	Absent	Present
Cardiac arrhythmia	Absent	Present
Arterial blood pressure	No change	Increase
Cardiac output	Large decrease	Slight decrease
O_2 conservation	Present	Questionable

only about 10% of the normal heart rate during face immersion in water of 30°C but increases to about 35% of the normal heart rate during face immersion in water of 0°C.

However, the relationship between the two variables is not linear (see Figure IV-19). It is only when the water temperature is below 15°C that the degree of brady-cardia increases linearly with temperature reduction. It should also be pointed out that the heart rate decreases even when the subject breathes through a snorkel during face immersion in cold water (Brick 1966, Paulev 1969) or when the subject is immersed in cold water up to the neck without breath-holding (Craig and Dvorak 1969, Paulev 1969). However, the degree of bradycardia in the latter case is less than 50% of that observed during immersion plus breath-holding. It is also of interest to note that the degree of bradycardia observed during breath-hold face immersion is not different from that during whole-body immersion in water of comparable temperature (Paulev 1969, Moore *et al. in press*). This means that the cold receptors (nerve cells sensing the temperature lower than the prevailing skin temperature) on the skin surface of the face play a very important role in the development of diving bradycardia.

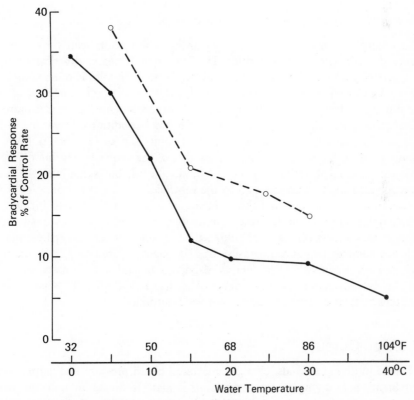

Figure IV-19. Bradycardial response to breath-hold face immersion in water of various temperatures.

c. The Role of Cold Receptors on the Face

It is still not resolved whether the face should be wet during breath-holding to elicit the maximal response. It has been shown that a dry, breath-hold face immersion (with the face covered with polyvinyl sheet to prevent wetting) elicits the same degree of bradycardia as a wet breath-hold face immersion when the water temperature is as low as 5°C (Moore *et al. in press*). However, wetting seems to be necessary in water of more moderate temperature (Stromme *et al.* 1970).

Whether a particular region of the face is directly involved in triggering this response has been questioned by several investigators. Although some have claimed that the nasal region is involved, others have failed to confirm such a finding. Central to this argument is the degree of bradycardia during breath-hold face immersion with and without a facemask which covers the eyes and the nose. Although there are some disagreements, the difference in the degree of diving bradycardia with and without a facemask seems to be negligible (Sasamoto 1965, Moore *et al. in press*). In contrast to these equivocal findings in man, the nasal region seems to play a key role in triggering the bradycardial response in the duck (Andersen 1966). Curiously enough, however, this response, initiated by wetting the nasal region of the duck, is not dependent on the temperature of the water (Andersen 1966).

d. Cardiac Arrhythmia

In diving animals, the diving bradycardia is not complicated by any other electrocardiogram (ECG) irregularities. The heart rate decreases drastically but each beat is perfectly normal. In contrast, diving bradycardia in man is often complicated by many ECG irregularities (called "cardiac arrhythmia") (Scholander *et al.* 1962, Sasamoto 1965, Hong *et al.* 1967), indicating that the effect on the heart is more than just decreasing the frequency of beats. Under normal conditions, frequency of heart beat is paced by the rhythmic excitation of the sino-atrial node (S-A node). Moreover, the contraction of the atrium is followed by that of the ventricle. During diving, not only is the pacemaker activity of the S-A node altered, but sometimes the cardiac contraction also seems to be initiated by the ventricle. From a practical point of view, it is important to point out that the incidence of cardiac arrhythmias during diving increases significantly when the water temperature is lower. In a study conducted on Korean women divers (Hong *et al.* 1967), the incidence of cardiac arrhythmias was 43% in the summer (water temperature of 27°C) as compared to 72% in the winter (water temperature of 10°C). In other words, diving in cold water is associated with a greater bradycardia and a higher incidence of cardiac arrhythmias. To what extent this contributes to fatal diving accidents is not yet determined.

e. Cardiac Output

Unlike in diving animals, changes in arterial blood pressure and cardiac output during breath-hold diving are not remarkable in man. In diving animals, the increase in peripheral vascular resistance due to the intense vasoconstriction during diving is counterbalanced by the reduction in the cardiac output, and hence, the arterial blood pressure remains unchanged. Similarly, an intense vasoconstriction of peripheral vessels develops in man during diving (Brick 1966, Elsner and Scholander 1965, Song *et al.* 1969), but this is accompanied by an increase in arterial blood pressure (Kawakami *et al.* 1967, Hong *et al.* 1970). This suggests that the cardiac output may be decreased only slightly, if at all, during diving. Because of technical difficulties involved in the measurement of cardiac output during diving, no data are available in the literature to evaluate the above notion. However, the data obtained during simple breath-holding or breath-hold face immersion in water indeed indicate that the reduction in the cardiac output during these maneuvers is negligible or rather slight (about 20% reduction) (Kawakami *et al.* 1967, Paulev 1969, Hong *et al.* 1971). As expected, the degree of peripheral vasoconstriction and the magnitude of increase in blood pressure increase as the water temperature decreases (Kawakami *et al.* 1967, Song *et al.* 1969, Hong *et al.* 1970). It is also possible that cardiac output may decrease more significantly during cold water diving, but no data are available.

All of these cardiovascular adjustments during diving are geared to conserve oxygen, which serves to prolong the duration of diving. There is no doubt this is the case in diving animals. However, this matter has not been fully resolved for man. There is no unequivocal experimental evidence to indicate that oxygen consumption is reduced in man during either actual diving or breath-hold face immersion (Raper *et al.* 1967, Heistad and Wheeler 1970). If there is any conservation of oxygen in man during diving, it is perhaps too small to detect.

f. Venous-Return Mechanism

There are quantitative as well as qualitative differences in the changes in cardiovascular functions during diving in man and in diving animals. The adjustment mechanism in animals is most likely a neural reflex triggered by a breath-hold immersion of the nasal regions. A similar mechanism seems to play a definite role in initiating diving bradycardia in man; however, additional mechanisms are involved. One of these is the role of cold receptors on the face, as discussed earlier. The other is a mechanical factor related to the venous return. It has been shown that the degree of breath-hold face immersion bradycardia can be manipulated by changing the level of intrathoracic pressure (the pressure in the potential space between the lung surface and the chest wall). For a given breath-hold face immersion, the degree of bradycardia decreases with an increase in the intrathoracic pressure, which, in turn, increases the heart rate through a reflex mechanism mediated by the pressure-sensitive receptors (baroreceptors) located in the carotid and aortic sinus. In the case of immersion in water to the neck, the venous return increases, as discussed in a previous section of this chapter, and this could potentiate the diving bradycardia.

Another factor is concerned with the lung volume at the start of diving. Somehow, the degree of diving bradycardia is proportional to the lung volume at the start of diving, and some investigators postulate the existence of certain thoracic receptors which mediate this lung-volume-dependent response (Song *et al.* 1969).

g. Effect of Changes in Ambient Pressure

So far, the possible effect of changes in ambient pressure on cardiovascular functions has not been discussed, mainly because of the scarcity of information. There are two main factors to be considered when the pressure undergoes a cyclic change during a breath-hold dive: One is the effect of hydrostatic pressure itself and the other the effect of attendant changes in blood gas pressures (Lanphier and Rahn 1963, Hong *et al.* 1963). The only cardiovascular function studied systematically along these lines is the degree of diving bradycardia. It has been shown that the time course of change in heart rate during actual breath-hold diving is not different from that during either breath-hold surface swimming or breath-hold, whole-body immersion at the surface (Hong *et al.* 1967). This indicates that the cyclic changes in pressure during breath-hold diving do not seem to have any effect on bradycardial response.

In order to ascertain this particular point, Moore *et al.* (*in press*) carried out breath-holding experiments on scuba divers at depths to 18.3 m (60 ft). At various depths, the degree of bradycardia was significantly less than at the surface (see Figure IV-20). Attenuation of bradycardial response to breath-hold face immersion in dry hyperbaric air environments was even more remarkable, as also shown in Figure IV-20. Since the oxygen pressure is elevated in hyperbaric air environments, the observed attenuation of bradycardial response is most likely due to the increase in the oxygen pressure of blood perfusing the chemoreceptors located in the carotid and aortic body. In fact, in subsequent studies, these authors proved the above hypothesis (Moore *et al.* 1973). It is of interest to note that these authors failed to observe any effect of higher CO_2 pressure on the diving bradycardia. At any rate, if the higher oxygen pressure is responsible for attenuation of bradycardial response, one would expect the same degree of attenuation in both wet- and dry-dive experiments. Since the attenuation at a given depth is greater in dry dives as compared to wet dives, one may speculate that the water immersion itself overcomes the effect of high oxygen pressure to some extent. How this is done is entirely unknown at present.

In summary, the most consistent change in cardiovascular function during breath-hold diving in man appears to be bradycardia. The current state of knowledge concerning the mechanism of this response is schematically illustrated in Figure IV-21. The bradycardia is triggered by the act of breath-holding, which through some unknown mechanisms activates the cardioinhibitory center in the central nervous system and then the vagus nerve. This basic response to breath-holding is subject to either potentiation or attenuation by neural, mechanical, and chemical factors as discussed so far. Unlike in diving animals, this bradycardia is not associated with the corresponding reduction in oxygen consumption and this is one of the reasons why man is a poor breath-hold diver.

3. *Air Scuba Diving*

Unlike breath-hold divers, scuba divers are engaged in respiratory movements and consequently can dive longer and deeper. Theoretically, the imbalance of pressures between the lungs and the surrounding water (as noted in the discussion of immersion

RELATIVE HR, % OF CONTROL

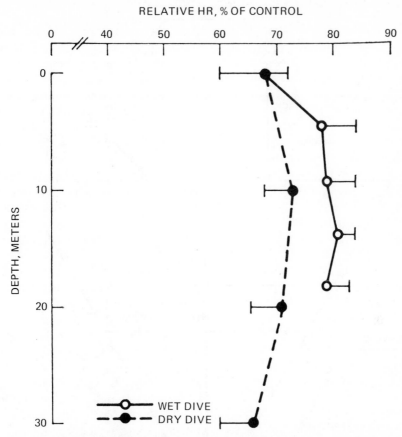

Figure IV-20. Relative heart rate responses to 60-sec breath-holding as a function of depth in ocean (wet dive) and with face immersion in 25°C (77°F) water in a recompression chamber (dry dive). [From Moore *et al.* (1972) by permission of the American Physiological Society.]

to the neck) can be eliminated in scuba diving if a demand valve is positioned properly. The classical work in this area has been carried out by Paton and Sand (1947). They report that underwater divers felt most comfortable when a demand valve is placed at the level of the suprasternal notch in all positions. However, it is still not clear whether the above balance point represents the mean (integrated) pressure to which the thorax is subjected. If it is, one would expect an imbalance of pressures whenever a demand valve is positioned either high or lower than the level of the suprasternal notch. When it is higher, the intrapulmonary pressure would be lower than the pressure of the surrounding water (i.e., negative pressure breathing), a situation analogous to immersion in water up to the neck. Physiological implications of negative intrapulmonary pressure have been discussed with regard to immersion to the neck. The opposite situation develops when the position of a demand regulator is lower than the level of the suprasternal notch. In this case, the intrapulmonary pressure will

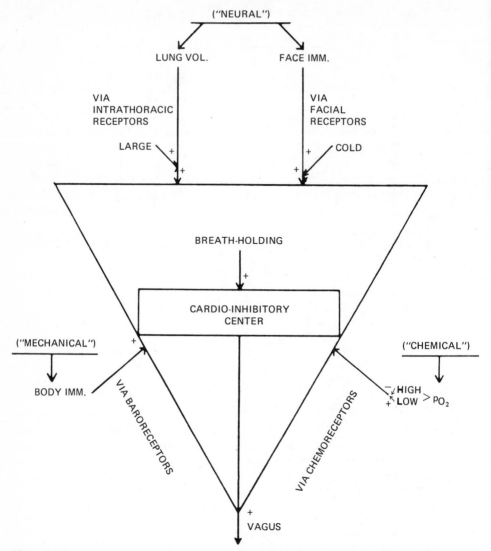

Figure IV-21. Schematic model of factors modifying diving bradycardia (+ indicates potentiation; − indicates attenuation of bradycardial response). [From Moore *et al.* (1973) by permission of the American Physiological Society.]

be higher than the surrounding pressure and hence the diver would engage in "positive-pressure breathing."

No studies dealing with cardiovascular functions during underwater positive pressure breathing are available in the literature. However, studies on positive pressure breathing in air indicate profound cardiovascular changes. In general, the return of venous blood from the periphery to the heart is inhibited and consequently the cardiac output decreases significantly (Lenfant and Howell 1960). Moreover, the urine flow decreases while the plasma ADH level increases, which has been attributed in part to a possible reduction in the intrathoracic blood volume (Khambatta and Baratz 1972). As previously discussed, in an ocean environment the high density of the medium and cold water stress lead to the central pooling of blood, which should partly offset the effect of positive pressure breathing on the venous return.

a. Effects of Oxygen at High Pressure

In scuba diving on air, there is an increase in oxygen partial pressure of the breathing gas in proportion to depth. Although less well defined and less dramatic than the effects on the lungs or the central nervous system, there are certain cardio-vascular changes induced by oxygen at high pressure (OHP). For instance, when the oxygen pressure is as high as 3.5 ATA the cerebral blood flow decreases by 25%, which is accompanied by a 55% increase in cerebral vascular resistance (Lambertsen *et al.* 1953). Saltzman *et al.* (1964) also reported a 46% decrease in the size of retinal blood vessels in man within 5 min of exposure to 3 ATA oxygen. In contrast, the changes in pulse rate and blood pressure induced by OHP are equivocal. Some investigators report a slight decrease, while others report an increase. Recently Young (1971) measured the heart rate of divers working in dry hyperbaric environments breathing a gas mixture containing about 85–90% O_2. He found that heart rate at a given level of exercise is lower by 10–20 beats/min at 35 and 47 fsw than at 26 and 29 fsw.

b. Effects of Depth

In one of the rare studies conducted in open waters (Clear Lake, Oregon), the heart rate of scuba divers (on air) either resting or working at depths to 60 ft in 5.5°C water was reported to increase slightly as a function of depth (Russell *et al.* 1972) (Table IV-13). However, the heart rate increase at a given depth is considerably less than the oxygen consumption increase; hence, oxygen pulse (milliliters of oxygen consumption per heart beat) increases with depth. A similar phenomenon was found in scuba divers in a laboratory water tank (water temperature of 15–30°C) at 1 and 2 ATA (Moore *et al.* 1970, Lally *et al.* 1971). This means that the increased demand

Table IV-13

Heart Rate and Oxygen Consumption of Scuba Divers [a]

	Depth, ATA	Rest	Work
Heart rate, beats/min	1	87.7 ± 2.1 [b]	135.1 ± 4.1
	2	90.7 ± 2.2	139.7 ± 3.9
	3	94.3 ± 2.4	142.4 ± 4.2
Oxygen consumption, ml/min	1	512 ± 25	1,657 ± 52
	2	643 ± 23	1,809 ± 82
	3	735 ± 25	1,935 ± 104
Oxygen pulse, ml O_2/heart beat	1	5.8 ± 0.4	12.4 ± 0.6
	2	7.1 ± 0.4	13.0 ± 0.7
	3	8.1 ± 0.4	13.4 ± 0.8

[a] Data of Russell *et al.* (1972).
[b] Mean ± SE of data from ten divers. Water temperature 5.5°C (43°F).

for oxygen during scuba diving in cold water is met most probably by a larger stroke volume. As discussed earlier, this may be attributed to an increased return of blood from the periphery due to intense vasoconstriction in cold water.

4. Mixed-Gas Saturation Diving

In order to circumvent oxygen poisoning, nitrogen narcosis, and other hazardous problems associated with hyperbaric air, most chamber dives of the past decade have used a helium–oxygen breathing mixture. Usually, oxygen concentration is such that the oxygen pressure is maintained at 0.3–0.4 ATA as compared to 0.2 ATA in a 1-ATA air environment. The balance between the total ambient and oxygen pressure is made up mostly by helium. For some dives, neon has also been used successfully (Lambertsen *in press*). The simulated depth of these dives has been recently extended to 2001 ft in a French dive "Physalie VI" (Fructus 1972).

a. Changes in Heart Rate

Because of technical difficulties involved in determining certain cardiovascular functions during these dives, the only cardiovascular data available in the literature are on heart rate and arterial blood pressure. The most impressive changes are found in the heart rate. In the majority of dives, a definite bradycardia is found in both resting and exercising divers (Hamilton *et al.* 1966, Raymond *et al.* 1968, Schaeffer *et al.* 1970, Salzano *et al.* 1970, Bühlmann *et al.* 1970, Moore *et al.* *in press*, Bennett 1972*b*). According to Hamilton *et al.* (1966) and Moore *et al.* (*in press*), this bradycardia is most conspicuous when the divers are on the bottom and seems to disappear during the decompression phase. However, in studies reported by Bühlmann *et al.* (1970) and Schaeffer *et al.* (1970) the degree of bradycardia tended to increase during decompression. As summarized in Table IV-14, average heart rate reduction during saturation diving amounts to about 15% of the predive heart rate. It is also important to note in this table that the magnitude of bradycardia is not correlated with the depth over a wide range of 140–1150 ft.

In contrast to such a clear-cut bradycardia observed in many dives, changes in the heart rate are equivocal in other studies (Bradley *et al.* 1971, Strauss *et al.* *in press*) in which bradycardia was observed in some divers but not in others. Bradycardia was more apparent in divers whose predive resting heart rate was above 60 per min. The heart rate did not decrease in divers whose predive heart rate was close to 50 per min. This may explain why bradycardia was consistently observed in dives shown in Table IV-14, in which the average predive heart rate was above 60 per min in all cases. However, it does not explain the fact that in the Sealab II dive the average heart rate *increased* from a predive level of 71 to 85 per min during the saturation dive to 200 ft (Hock *et al.* 1966).

b. Effects of Gas Density

Albano (1970) describes a series of hyperbaric experiments in which changes in heart rate were carefully studied. He found a usual bradycardia in 10-ATA oxygen–nitrogen environment but found a slight tachycardia (increase in the heart rate) in

Table IV-14

Resting Heart Rate of Divers during Saturation Diving[a]

Number of subjects	Depth, ft	Heart rate per min at		% Reduction at depth	Reference
		Surface	Depth		
2	650	70	59	16	Hamilton *et al.* 1966
5	480	79	65	18	Raymond *et al.* 1968
	390	79	62	22	
	300	79	68	14	
	240	79	64	19	
	190	79	72	9	
	140	79	71	10	
4	800	80	65	19	Schaeffer *et al.* 1970
3	1000	90	79	12	Salzano *et al.* 1970
	320	90	81	10	
3	1150	73	62	15	Bühlmann *et al.* 1970
	1000	73	59	19	
6	500	64	57	11	Moore *et al.* 1972
	250	64	54	16	
2	1000	?	51	?	Bennett 1972

[a] Data from Salzano *et al.* (1970).

10-ATA oxygen–helium environment. Since the density of an oxygen–nitrogen mixture is considerably greater than that of an oxygen–helium mixture, Albano speculates that the heart rate decreases with an increase in gas density. However, according to the data of Strauss *et al.* (*in press*), there is no correlation between resting heart rate and gas mixture density over a range of 5–25 g/liter. Data summarized in Table IV-14 also indicate that the degree of bradycardia is more or less independent of depth. The latter findings are at variance with the view of Albano and suggest that the gas density does not play a major role in inducing hyperbaric bradycardia.

c. Temperature

A factor which does seem to play a major role is temperature of the environment. Although chamber temperature and changes in body temperature were not recorded in many studies, available data indicate that hyperbaric bradycardia may be, in part, related to cold stress. For instance, in the Sealab II dive (Hock *et al.* 1966) the chamber temperature was kept at 30–31°C and the oral temperature of divers was elevated by 0.7°C; as stated earlier, these divers did not show bradycardia. In the studies reported by Strauss *et al.* (*in press*) in which one of two divers did not show any bradycardia, the early morning rectal temperature was also elevated by about 1°C (Webb, *in press*), In contrast, the oral and/or skin temperatures were lowered in a hyperbaric environment in studies reported by Moore *et al.* (*in press*) and Raymond *et al.* (1968), who observed a significant bradycardia. The latter studies indicate that the rate of body heat loss increases significantly in hyperbaric heliox environments even when the ambient temperature is relatively warm. In other words, a certain degree of cold stress seems to exist in a hyperbaric heliox environment of moderate temperature which would be perfectly comfortable to man in the ordinary air environment (Webb, *in press*).

A further suggestion that the cold stress associated with hyperbaric heliox environment may be causally related to hyperbaric bradycardia is supported by the work of Moore *et al.* (1972), who observed an attenuation of bradycardia upon raising the ambient temperature from 27.8 to 29°C at 500 ft depth. It is tempting to speculate that the cold stress provided by helium at high pressure induces both a peripheral vasoconstriction and an increase in the blood pressure (particularly the diastolic pressure), resulting in bradycardia through a reflex mechanism mediated by the pressure-sensitive receptors located in the carotid and aortic sinus. However, the changes in the blood pressure reported in saturation diving studies are not consistent enough to warrant the above hypothesis.

d. Effects of Helium

Recently, a very important effect of helium on the sympathetic nervous system has been reported. Using anesthetized dogs, Raymond *et al.* (1972) found that breathing 75% helium–25% oxygen instead of 75% nitrogen–25% oxygen at 1 ATA reduced the occurrence of cardiac arrhythmias induced by ligation of the coronary artery (which perfuses the cardiac tissue). Moreover, helium reduces the baseline heart rate and the level of plasma catecholamines, which mediate the heart rate-increasing action of the sympathetic nervous system. More recently, Hong *et al.* (1973) found that the magnitude of bradycardia induced by either simple breath-holding or face immersion becomes significantly greater in hyperbaric heliox environments as compared to that in 1 ATA air. The latter finding could be explained on the basis of suppression of the effect on the sympathetic nervous system in the heliox environment, as suggested by the finding of Raymond *et al.* (1972). These findings strongly suggest that the presence of helium at high pressure may play an important role in inducing hyperbaric bradycardia.

As stated earlier, heart rate during exercise is maintained lower in hyperbaric heliox environments than in 1 ATA air, a situation analogous to that found in scuba diving on air (see the section on air scuba diving, this chapter). However, this difference in the heart rate between hyperbaric and 1-ATA air environments tends to disappear during heavy exercise (Salzano *et al.* 1970, Strauss *et al. in press*). This means that, in the presence of a high demand for oxygen, all the mechanisms underlying hyperbaric bradycardia seem to disappear.

It is thus evident that the mechanism of hyperbaric bradycardia is yet to be solved. It is almost impossible to single out a factor responsible for the bradycardia. It is most likely that the combined effects of high gas density, cold stress, the level of the vagal tone (which determines the resting heart rate) of the divers, and the presence of helium at high pressure may be responsible.

C. Oxygen Toxicity

To understand the physiological problems caused by breathing various gases and combinations of gases, it is necessary to understand the principle of partial pressure.

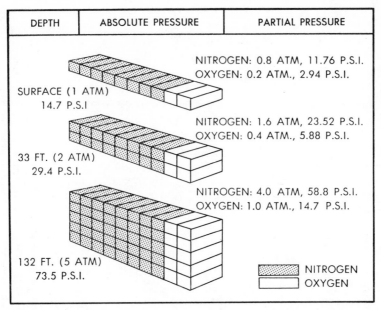

DEPTH	ABSOLUTE PRESSURE	PARTIAL PRESSURE

SURFACE (1 ATM)
14.7 P.S.I

NITROGEN: 0.8 ATM, 11.76 P.S.I.
OXYGEN: 0.2 ATM., 2.94 P.S.I.

33 FT. (2 ATM)
29.4 P.S.I.

NITROGEN: 1.6 ATM, 23.52 P.S.I.
OXYGEN: 0.4 ATM., 5.88 P.S.I.

132 FT. (5 ATM)
73.5 P.S.I.

NITROGEN: 4.0 ATM, 58.8 P.S.I.
OXYGEN: 1.0 ATM., 14.7 P.S.I.

NITROGEN
OXYGEN

Figure IV-22. Partial pressures of gases in air at various depths. [From Lee (1967) (illustrated by Elaine Grant) by permission of Doubleday and Co.]

In any breathing mixture under pressure, each component exerts its share of the total pressure, i.e., its partial pressure. Figure IV-22 illustrates how partial pressures increase while the percentage volume remains the same. It can be seen that the partial pressure of oxygen in air at 132 ft is about the same as that of pure oxygen at the surface.

Prolonged exposure to increased partial pressure of oxygen can result in toxic effects which become progressively more severe as the inspired partial pressure and/or duration of exposure is increased. The most dramatic of these are toxic effects on the respiratory system (the Lorraine–Smith effect) and on the central nervous system (the Paul Bert effect). Effects include destruction of red blood cells and the neurosensory tissues of the eye (Lambertsen 1965).

In continuous exposure of many hours at oxygen partial pressures not exceeding 2.0 atm, the respiratory system is the first to be affected. Oxygen at partial pressure less than about 0.5 atm is unlikely to have a toxic effect regardless of the percentage composition in inhaled mixtures, but above 0.5 atm, depending upon degree of partial pressure and duration of exposure, irreversible pulmonary damage can result (Clark and Lambertsen 1971a).

Exposure to oxygen partial pressures of several atmospheres and higher affects brain and spinal cord, i.e., the central nervous system (CNS) first, and can culminate in convulsions, paralysis, and possible death. Although the CNS is the first to be noticeably affected during shorter exposures to several atmospheres and higher, pulmonary effects occur concurrently with the development of CNS effects (Clark and Lambertsen 1971a).

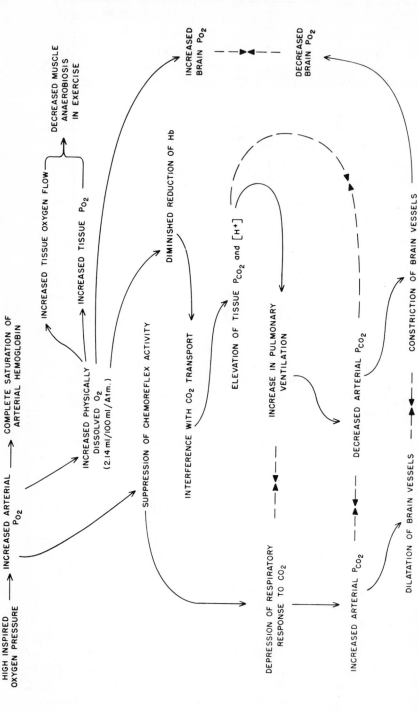

Figure IV-23. Sequence of acute physiological effects of oxygen in normal men. As oxygen at ≥ 1 atm pressure is administered, an interrelated sequence of changes occurs. Each of the effects shown has been demonstrated at 1 atm and, with the exception of chemo-reflex suppression, at 3.0–3.5 atm. Each should be considered a physiological (or pharmacological) rather than a toxic effect and to be spontaneously reversible on returning to normal levels of inspired oxygen. In a stable state of oxygen breathing by a normal individual at rest, the magnitude of the changes from normal are proportional to the dose of oxygen. As shown by the five sets of opposing arrows, the conflicting physiological effects lead to a new state of dynamic balance. [From Lambertsen (1965) by permission of the American Physiological Society.]

There are also nontoxic effects which occur during exposure to increased oxygen partial pressure. These are distinguishable from toxic effects in that they do not endanger life nor impose serious limitations, and are characterized by a rapid onset as well as a prompt and complete reversal. These effects, illustrated in Figure IV-23, have been pointed out because they are in operation at increased partial pressures of inspired oxygen, but they will not be discussed in depth as they are not generally of pathogenic significance.

1. *Pulmonary Oxygen Toxicity*

a. Toxic Effects upon Lung Pathology

The harmful effects of oxygen on the lung membranes and function are due both to chemical actions related to the increase in inspired partial pressure of oxygen and to the physical consequences of excluding the inert carrier gas from the pulmonary passage (Lambertsen 1966b).

(i) Chemical Effects of Oxygen on the Lungs. The rate of development and degree of chemical damage to the respiratory passages are proportional to both the amount of oxygen and the duration of exposure (Lambertsen 1966b). Effects of a threshold dose on the respiratory tract appear only after a certain latent period; damage significantly increases in rapidity as the partial pressure is increased above 1 atm (Figure IV-24). Although the advantage of intermittent exposures in delaying convulsions and death of animals from oxygen poisoning has been demonstrated (Lambertsen 1955), it is possible that pulmonary effects of very frequent exposure to increased partial pressures of oxygen may be cumulative (Lambertsen 1966b).

(ii) Physical Effects of Oxygen on the Lungs. Although breathing 100% oxygen produces no biochemical toxicity when the total pressure is maintained low enough so that the alveolar oxygen level is not greater than normal, there is a physical effect

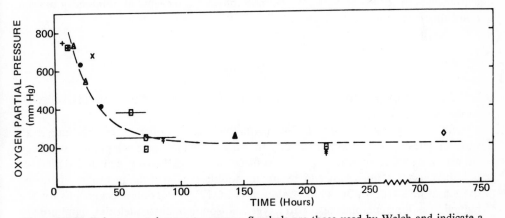

Figure IV-24. Pulmonary tolerance to oxygen. Symbols are those used by Welch and indicate a number of different studies described in a review (Welch *et al.* 1963). Some represent the average time of onset of symptoms and some the earliest indication of pulmonary irritation. [From Lambertsen (1965) by permission of the American Physiological Society.]

due to the rapid and complete absorption of the gas phase (Clark and Lambertsen 1971a). This rapid absorption effect of pure oxygen in combination with random blockage of bronchioles, even by normal secretion, can lead to diffuse, progressive, and eventually severe pulmonary atelectasis (collapse of alveoli).

(iii) Interaction of Physical Factors. Pulmonary atelectasis remains as a possible physical complication of oxygen breathing when the inspired oxygen partial pressure is high enough to produce a chemical pulmonary irritation or when another factor, such as pulmonary infection, obstructs the terminal pulmonary passages. The occurrence of atelectasis in oxygen-poisoned lungs may be a primary effect related to the chemical action of oxygen upon alveolar surfactant, a secondary effect of airway obstruction caused by pulmonary edema (slowed diffusion of gases between alveoli and blood) and subsequent gas absorption, or a combination of both factors (Clark and Lambertsen 1971a). Pulmonary edema is one of the most common pathologic manifestations of pulmonary oxygen toxicity; it first appears as a widening of interstitial spaces and progresses, in most severe cases, to form massive pleural effusions.

Chemical effects on the pulmonary membranes, leading to congestion, edema, and loss of fluid into air passages, also predispose the individual to pulmonary atelectasis by contributing to obstruction of the alveoli.

(iv) End Results of Pulmonary Oxygen Toxicity. While short periods of breathing oxygen at elevated partial pressures produce no demonstrable harm, continuous exposure for long periods leads to severe damage. Limitation of alveolar ventilation (froth in airways), atelectasis, passage of blood through nonventilated alveoli, and pulmonary edema all interfere with pulmonary gas exchange. This will also be true when oxygen at high partial pressures is breathed. The pulmonary damage leads to a lowering of arterial oxygen, elevation of arterial and tissue carbon dioxide, and death from anoxia and acidosis (reduced alkali reserve in the blood and other body fluids, caused by increased carbon dioxide). Because the inspired and alveolar oxygen are initially high, the pulmonary damage may lead to carbon dioxide retention and acidosis, even while arterial oxygenation is sustained above the normal for sea level existence. A proposed sequence of the pathophysiologic events in pulmonary oxygen poisoning is shown in Figure IV-25.

b. Effects upon Pulmonary Function

Effects of oxygen toxicity upon pulmonary function become more marked with its increasing severity.

(i) Pulmonary Mechanics. Change in the elastic properties of the lung is an early manifestation of oxygen toxicity in man. Dynamic lung compliance has been shown to decrease by about 15% in subjects who breathed oxygen at 2.0 atm for 6–11 hr (Fisher *et al.* 1968) and by about double that amount in others breathing oxygen at 0.98 atm for 30–48 hr (Caldwell *et al.* 1966). Possible mechanisms for the reduction in lung compliance due to oxygen poisoning include atelectasis, pulmonary edema and congestion, asymmetric narrowing of the airways, decrease in alveolar surfactant, and change in retractile properties of pulmonary tissue elastic elements (Clark and Lambertsen 1971a).

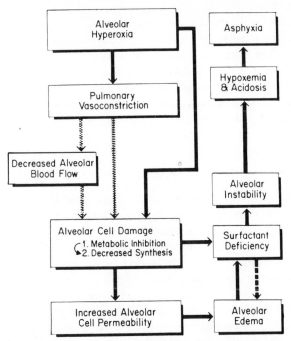

Figure IV-25. Sequence of interrelated pathophysiological events leading to pulmonary oxygen toxicity. [From Wittner and Rosenbaum (1966) by permission of the American Physiological Society.]

(ii) Lung Volume. Decrease in vital capacity (see the section on respiration, this chapter) has been produced by exposure of normal men to oxygen partial pressures ranging from 0.75 to 2.0 atm (Clark and Lambertsen 1967). Reduction of vital capacity is usually progressive throughout the oxygen exposure and can be described accurately by the early part of a dose–response curve (Clark and Lambertsen 1971*b*). After prolonged exposure to 2.0 atm of oxygen, vital capacity decreased for the first few hours after exposure was discontinued (Clark and Lambertsen 1967). This indicates that the consequences of oxygen poisoning can progress in severity beyond the time of the exposure to the direct toxic process. Recovery of vital capacity is not immediate. It usually occurs within two to three days after the oxygen exposure, but occasionally requires several weeks. Vital capacity has been observed to return to normal even after reductions as great as 40% of the control volume. Possible mechanisms for reduction of vital capacity in pulmonary oxygen toxicity include chest pain, atelectasis, pulmonary edema, decreased lung compliance, and decreased force of inspiration (Clark and Lambertsen 1971*a*).

(iii) Pulmonary Gas Exchange. Pulmonary diffusing capacity is adversely affected by high inspired partial pressures of oxygen, but a pathological response requires at least several hours to manifest itself. The reduction of diffusing capacity in pulmonary oxygen poisoning could be caused by atelectasis, lengthening of the diffusion path across the alveolar–capillary tissue barrier, uneven ventilation–diffusion

relationships, decrease in pulmonary capillary blood flow, and alterations of the pulmonary vasculature (Clark and Lambertsen 1971a).

It is possible that the nature of the initial pathological changes responsible for reduction of diffusing capacity in early pulmonary oxygen poisoning may vary at different levels of inspired partial pressures of oxygen, with interstitial edema and increase in thickness of the alveolar membrane appearing initially during prolonged exposures at relatively low partial pressures, and severe vasoconstriction and capillary destruction occurring first during exposure to very high, more acutely toxic levels of oxygen.

c. Signs and Symptoms of Pulmonary Oxygen Poisoning

Symptoms appear to be those of a tracheobronchitis. Beginning as a mild throat irritation and occasional coughing, the tracheal symptoms become progressively more intense and continuous until each inspiration is painful and coughing is uncontrollable (Clark and Lambertsen 1971a). Pulmonary symptoms have appeared after 6, 4, and 3 hr of oxygen breathing at partial pressures of 0.83, 1.0, and 2.0 atm, respectively. Severity of symptoms in most cases rapidly diminishes within the first few hours of the postexposure period and the sensations of pulmonary irritation completely disappear over the following one to three days. Dyspnea (subjective distress in breathing) at rest is produced by severe exposures and dyspnea on exertion continues for the first few days of the postexposure period. Other reported symptoms include nasal congestion, earache, headache, joint and muscle pains, paresthesias ("pins and needles"), and giddiness.

d. Tolerance

A knowledge of the tolerance of the human lung to oxygen toxicity is essential to safe diving and decompression procedures. To maintain an objective basis for a definition of pulmonary tolerance in man, change in the vital capacity has been used as a quantitative index (Clark and Lambertsen 1967). A series of curves has been derived, using such an index and based upon the theoretical assumptions that pulmonary oxygen tolerance in man can be described by families of rectangular hyperbolas with asymptotes at zero time and at an inspired oxygen partial pressure of 0.5 atm. The curves in Figure IV-26 define the development rate of pulmonary oxygen poisoning in 50% of normal individuals exposed to increased partial pressures of oxygen. Figure IV-27 shows the variation in susceptibility to a uniform degree of pulmonary oxygen poisoning represented by a 4% decrease in vital capacity. The assumption that an inspired partial pressure of oxygen of 0.5 atm is a practical horizontal asymptote for pulmonary oxygen tolerance curves implies that normal men can breathe oxygen at this partial pressure for an indefinite period of time, with the occurrence of only a minor degree of pulmonary oxygen poisoning.

The degree of pulmonary toxicity equivalent to a 2% decrease in vital capacity is completely reversible and asymptomatic and is considered as a reasonable maximum limit of oxygen exposure for treatment of uncomplicated decompression sickness (Wright 1972).

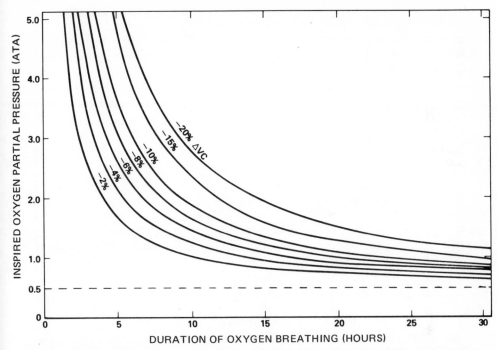

Figure IV-26. Pulmonary oxygen tolerance curves for normal men based on vital capacity changes in 50% of the individuals exposed to increased partial pressures of oxygen. The curves describe rate of development of pulmonary oxygen poisoning in an individual with "average" susceptibility (Wright 1972).

When elevated pressures of oxygen are used in the treatment of more severe decompression sickness, it may be reasonable to accept a greater degree of pulmonary toxicity in order to facilitate better treatment. The degree of pulmonary oxygen toxicity that produces a 10% decrease in vital capacity is associated with moderate symptoms of coughing and pain in the chest on deep inspiration. This degree of impairment is reversible within a few days following cessation of exposure. Greater oxygen exposures may not be reversible. Therefore, it is suggested that an oxygen exposure that produces a 10% decrement in vital capacity (Figure IV-26) be chosen as the extreme limit for hyperbaric exposure.

e. Mechanisms

Although the specific biochemical target sites in the lung have not yet been satisfactorily identified, the direct toxic effects of oxygen in the lung tissue must originate from an inactivation of essential enzymes and the resulting disruption of cellular metabolism. These initial biochemical effects start a complex series of events that terminate in the overall pathological changes of pulmonary oxygen poisoning. Pathological results show that the entire lung can be damaged by prolonged exposure to oxygen at increased partial pressures. Other studies show that the mechanisms responsible for the total syndrome of pulmonary oxygen poisoning include multiple

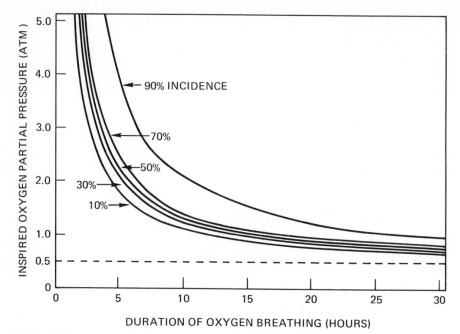

Figure IV-27. Pulmonary oxygen tolerance curves for normal men based on a 4% decrease in vital capacity. The curves describe varying susceptibility to a uniform degree of pulmonary oxygen poisoning. [From Clark and Lambertsen (1971a) by permission of Williams & Wilkins.]

interacting factors, some of which are related to toxic effects of oxygen at extrapulmonary sites (Clark and Lambertsen 1971a).

f. Treatment

Because severe damage and death will result from failure of pulmonary gas exchange, preventive measures such as artificial ventilation and prophylactic use of antibiotics should aid victims of mild oxygen toxicity (Lambertsen 1966b). In choosing the gas to ventilate the lungs, a concentration of oxygen not exceeding 60% at sea level should be used, as higher tensions of oxygen would presumably aggravate the existing pulmonary damage.

If toxicity proceeds to the point that hypoxemia develops, a dilemma exists—removal to an air-breathing situation will result in further pulmonary damage. In this case, gradual reduction of oxygen partial pressure in conjunction with drug therapy (i.e., control of acid–base balance) will be touch-and-go, and if the condition is severe enough, the use of extrapulmonary oxygenation may have to be employed.

g. Prevention

(i) Calculation of Cumulative Oxygen Dose. Most diving and oxygen therapy procedures include exposures to oxygen at several different partial pressures. Although one can estimate the degree of lung toxicity that may be expected from a given partial

pressure of oxygen for a given time period from Figure IV-26, use of this graph to estimate the cumulative damage from an interrupted exposure to oxygen at several different partial pressures can become somewhat complex. Wright (1972) describes a method for calculating the cumulative pulmonary toxicity that may be expected following exposure to oxygen at a variety of pressures and times.

The procedure introduces the concept of the "unit pulmonary toxic dose" or UPTD. One UPTD is defined as the degree of pulmonary toxicity incurred by breathing 100% oxygen at a pressure of 1 atm for 1 min. It is assumed that this same toxicity is achieved by exposures to various other combinations of partial pressures and durations of exposure and that the pulmonary effects are additive. Wright (1972) recommends that oxygen exposures should be planned so as not to exceed the following limits:

> During decompression and for treatment of mild decompression sickness, the total oxygen exposure should be limited to that which yields a UPTD of 615 or less.
> In the use of oxygen for medical therapy or treatment of serious decompression sickness, which is responding poorly, an extreme limit of oxygen exposure which yields a UPTD of 1425 or less should be planned.

Having derived a formula for calculating UPTD, Wright developed a simplified arithmetic method and constructed a table which may be used to calculate oxygen exposure from 0.6 to 5.0 atm.

(a) *Arithmetic method.* At any constant oxygen partial pressure, UPTD is a linear function of the time of exposure to that P_{O_2}, as can be seen by Wright's formula

$$\text{UPTD} = t^{-1.2} \sqrt{\frac{0.5}{P - 0.5}} \tag{3}$$

which reduces to

$$\text{UPTD} = k_p t \tag{4}$$

when P is held constant.

At any P_{O_2}, then, there exists a factor k_p, which, when multiplied by the time of exposure to that P_{O_2}, yields the UPTD for that exposure:

$$k_p = t^{-2.2} \sqrt{\frac{0.5}{P - 0.5}} \tag{5}$$

A list of these k_p factors is given in Table IV-15. To calculate the UPTD for a given exposure:

1. Convert the partial pressure of oxygen breathed at each depth to P_{O_2} in atmospheres.
2. Select the corresponding k_p from Table IV-15.
3. Multiply the time of exposure at that P_{O_2} by the corresponding k_p to get the UPTD for that depth.
4. Add the UPTD's for each P_{O_2} in the complete exposure together to get the total UPTD for the exposure.

(b) *Tabular Method.* The UPTD's resulting from various exposures to different partial pressures are recorded in Table IV-16. Oxygen exposures are given from 0.6 to

Table IV-15

k_p Factors for Calculating UPTD

(Wright 1972)

P_{O_2}	k_p	P_{O_2}	k_p	P_{O_2}	k_p	P_{O_2}	k_p
0.50	0.00	1.70	2.07	2.90	3.70	4.10	5.18
0.60	0.26	1.80	2.22	3.00	3.82	4.20	5.30
0.70	0.47	1.90	2.36	3.10	3.95	4.30	5.42
0.80	0.65	2.00	2.50	3.20	4.08	4.40	5.54
0.90	0.83	2.10	2.64	3.30	4.20	4.50	5.66
1.00	1.00	2.20	2.77	3.40	4.33	4.60	5.77
1.10	1.16	2.30	2.91	3.50	4.45	4.70	5.89
1.20	1.32	2.40	3.04	3.60	4.57	4.80	6.01
1.30	1.48	2.50	3.17	3.70	4.70	4.90	6.12
1.40	1.63	2.60	3.31	3.80	4.82	5.00	6.24
1.50	1.78	2.70	3.44	3.90	4.94		
1.60	1.93	2.80	3.57	4.00	5.06		

5.0 ATA in increments of 0.1 ATA. One ATA is not given because 1 min equals 1 UPTD at that pressure. Time and UPTD are given in the same unit—if time is measured in seconds, UPTD is in seconds; if time is in minutes or hours, UPTD is in minutes or hours. To use the table:

1. Convert the partial pressure of oxygen breathed at each depth in P_{O_2} in atmospheres.
2. Select the appropriate P_{O_2} table.
3. Enter the table in the "time" column at the time corresponding to the duration stop.
4. Read the corresponding UPTD.
5. Add the UPTD's for each depth together to get the total UPTD for the exposure.

According to Wright, one may convert a time longer than 60 min into hours and minutes, enter the table twice, and add the results, being careful to keep minutes and hours separate. "Due to rounding errors in the program which generated the table, conversions from hours to minutes and vice versa may be inaccurate by small fractions of a minute. It is also possible to break a single exposure to the same P_{O_2} down into partial times small enough to be found in the table and then add each of the corresponding UPTD's to find the total UPTD of the stop" (Wright 1972).

(ii) Fluctuating Oxygen Exposure. One of the most practical measures available for extending the use of high oxygen partial pressures is carefully scheduled alternations of exposure to high and normal levels of oxygen (Clark and Lambertsen 1971a). Such interruptions of exposure markedly extend oxygen tolerance in animals and indicates that the recovery rate from direct toxic effects of oxygen is considerably greater than the development rate of overt oxygen toxicity (Lambertsen 1966b). This is consistent with the concept (Haugaard 1955) that oxygen poisoning results from oxidation of critical enzymes and cofactors, which can be rapidly resynthesized when a nontoxic oxygen level is restored.

The increase in tolerance achieved by alternating exposures appears to depend, at least partially, on how much oxygen intoxication develops before the inspired pressure is reduced (Clark and Lambertsen 1971a). Although this method offers a practical approach to increasing the duration of the total exposure to a particular high oxygen pressure within a given time period, the optimal relationships between the duration of exposure and the length of the interruption of exposure have not yet been determined in man.

(iii) Pharmacologic. Development rate of pulmonary oxygen poisoning in intact animals is influenced by a wide variety of conditions, procedures, and drugs. The factors summarized in Table IV-17 have been shown to have a significant effect upon the degree of pulmonary pathology, or the survival time, or both, in animals exposed to toxic levels of oxygen. Little information is available regarding the quantitative effects of such agents upon oxygen poisoning in the human lung. It is reasonable to assume, however, that agents that clearly modify pulmonary oxygen poisoning in animals should have a similar effect in man. It is very likely that persons with elevated levels of catecholamines, adrenocortical or thyroid hormones, or who are febrile will have an increased susceptibility to pulmonary oxygen toxicity. Hypercapnia (increased level of carbon dioxide) is another factor which definitely enhances pulmonary oxygen poisoning when it occurs concurrently with convulsive levels of hyperoxia.

Observations that pulmonary oxygen tolerance is decreased in vitamin E-deficient rats (Kann *et al.* 1964) and increased by administration of antioxidant drugs (Jamieson and van den Brenk 1964) suggest that such drugs may be beneficial and the hypothermia may serve as a protective measure where brief exposures to hyperbaric oxygen are required. Succinate (Sanders *et al.* 1965), disulfiram, and trisaminomethane (THAM) (McSherry and Veith 1968) may also prove to be useful in this respect.

2. CNS Oxygen Toxicity

Breathing oxygen at raised partial pressures can also lead to toxic effects on the central nervous system (CNS). Figure IV-28 illustrates borderline limitations for oxygen use. It can be seen that the CNS is the first to be noticeably affected at oxygen partial pressures of several atmospheres and above. CNS tolerance to oxygen toxicity is assumed to be borderline at 2.0 atm of inspired oxygen since subjects exposed for 10–11 hr produced no detectable CNS effects (Clark and Lambertsen 1967).

a. Convulsions

The overt expression of CNS oxygen toxicity in vertebrates is a generalized convulsion. In man the convulsive seizure resembles that of grand mal epilepsy and is usually, but not always, preceded by localized muscular twitching, especially about the eyes, mouth, and forehead. The convulsive episode has been well described by Lambertsen (1965). Small muscles of the hands may be involved and incoordination of diaphragm activity in respiration may occur. These phenomena increase in severity over a period which may vary from a few minutes to nearly an hour with essentially clear consciousness being retained. Eventually, an abrupt spread of excitation occurs

Table IV-16

UPTD Resulting from Various Exposures to Different Partial Pressures
(Wright 1972)

Time	UPTD for given P_{O_2} exposure										
	0.60	0.70	0.80	0.90	1.10	1.20	1.30	1.40	1.50	1.60	1.70
1.00	0.26	0.47	0.65	0.83	1.16	1.32	1.48	1.63	1.78	1.93	2.07
2.00	0.52	0.93	1.31	1.66	2.33	2.65	2.96	3.26	3.56	3.86	4.15
3.00	0.78	1.40	1.96	2.49	3.49	3.97	4.44	4.90	5.35	5.79	6.22
4.00	1.05	1.86	2.61	3.32	4.66	5.29	5.92	6.53	7.13	7.72	8.30
5.00	1.31	2.33	3.27	4.15	5.82	6.62	7.40	8.16	8.91	9.65	10.37
6.00	1.57	2.80	3.92	4.98	6.98	7.94	8.88	9.79	10.69	11.57	12.44
7.00	1.83	3.26	4.57	5.81	8.15	9.27	10.36	11.42	12.47	13.50	14.52
8.00	2.09	3.73	5.23	6.64	9.31	10.59	11.84	13.06	14.25	15.43	16.59
9.00	2.35	4.19	5.88	7.47	10.48	11.91	13.32	14.69	16.04	17.36	18.67
10.00	2.62	4.66	6.53	8.30	11.64	13.24	14.79	16.32	17.82	19.29	20.74
11.00	2.88	5.13	7.19	9.13	12.80	14.56	16.27	17.95	19.60	21.22	22.82
12.00	3.14	5.59	7.84	9.96	13.97	15.88	17.75	19.58	21.38	23.15	24.89
13.00	3.40	6.06	8.49	10.79	15.13	17.21	19.23	21.22	23.16	25.08	26.96
14.00	3.66	6.52	9.15	11.62	16.30	18.53	20.71	22.85	24.95	27.01	29.04
15.00	3.92	6.99	9.80	12.45	17.46	19.85	22.19	24.48	26.73	28.94	31.11
16.00	4.18	7.46	10.45	13.29	18.63	21.18	23.67	26.11	28.51	30.87	33.19
17.00	4.45	7.92	11.11	14.12	19.79	22.50	25.15	27.74	30.29	32.79	35.26
18.00	4.71	8.39	11.76	14.95	20.95	23.83	26.63	29.38	32.07	34.72	37.33
19.00	4.97	8.85	12.41	15.78	22.12	25.15	28.11	31.01	33.85	36.65	39.41
20.00	5.23	9.32	13.07	16.61	23.28	26.47	29.59	32.64	35.64	38.58	41.48
21.00	5.49	9.79	13.72	17.44	24.45	27.80	31.07	34.27	37.42	40.51	43.56
22.00	5.75	10.25	14.37	18.27	25.61	29.12	32.55	35.90	39.20	42.44	45.63
23.00	6.02	10.72	15.03	19.10	26.77	30.44	34.03	37.54	40.98	44.37	47.71
24.00	6.28	11.18	15.68	19.93	27.94	31.77	35.51	39.17	42.76	46.30	49.78
25.00	6.54	11.65	16.33	20.76	29.10	33.09	36.99	40.80	44.54	48.23	51.85
26.00	6.80	12.12	16.99	21.59	30.27	34.41	38.47	42.43	46.33	50.16	53.93
27.00	7.06	12.58	17.64	22.42	31.43	35.74	39.95	44.06	48.11	52.09	56.00
28.00	7.32	13.05	18.29	23.25	32.59	37.06	41.42	45.70	49.89	54.01	58.08
29.00	7.58	13.51	18.95	24.08	33.76	38.39	42.90	47.33	51.67	55.94	60.15
30.00	7.85	13.98	19.60	24.91	34.92	39.71	44.38	48.96	53.45	57.87	62.22
31.00	8.11	14.45	20.25	25.74	36.09	41.03	45.86	50.59	55.24	59.80	64.30
32.00	8.37	14.91	20.91	26.57	37.25	42.36	47.34	52.22	57.02	61.73	66.37
33.00	8.63	15.38	21.56	27.40	38.41	43.68	48.82	53.86	58.80	63.66	68.45
34.00	8.89	15.84	22.21	28.23	39.58	45.00	50.30	55.49	60.58	65.59	70.25
35.00	9.15	16.31	22.87	29.06	40.74	46.33	51.78	57.12	62.36	67.52	72.60
36.00	9.42	16.78	23.52	29.89	41.91	47.65	53.26	58.75	64.14	69.45	74.67
37.00	9.68	17.24	24.17	30.72	43.07	48.98	54.74	60.38	65.93	71.38	76.74
38.00	9.94	17.71	24.83	31.55	44.24	50.30	56.22	62.02	67.71	73.31	78.82
39.00	10.20	18.17	25.48	32.38	45.40	51.62	57.70	63.65	69.49	75.23	80.89
40.00	10.46	18.64	26.13	33.21	46.56	52.95	59.18	65.28	71.27	77.16	82.97
41.00	10.72	19.11	26.79	34.04	47.73	54.27	60.66	66.91	73.05	79.09	85.04
42.00	10.98	19.57	27.44	34.87	48.89	55.59	62.14	68.55	74.84	81.02	87.11
43.00	11.25	20.04	28.09	35.70	50.06	56.92	63.62	70.18	76.62	82.95	89.19
44.00	11.51	20.50	28.75	36.53	51.22	58.24	65.10	71.81	78.40	84.88	91.26
45.00	11.77	20.97	29.40	37.36	52.38	59.56	66.58	73.44	80.18	86.81	93.34
46.00	12.03	21.44	30.05	38.19	53.55	60.89	68.05	75.07	81.96	88.74	95.41
47.00	12.29	21.90	30.71	39.02	54.71	62.21	69.53	76.71	83.74	90.67	97.49
48.00	12.55	22.37	31.36	39.86	55.88	63.54	71.01	78.34	85.53	92.60	99.56
49.00	12.82	22.83	32.01	40.69	57.04	64.86	72.49	79.97	87.31	94.53	101.63
50.00	13.08	23.30	32.67	41.52	58.20	66.18	73.97	81.60	89.09	96.45	103.71
51.00	13.34	23.77	33.32	42.35	59.37	67.51	75.45	83.23	90.87	98.38	105.78
52.00	13.60	24.23	33.97	43.18	60.53	68.83	76.93	84.87	92.65	100.31	107.86
53.00	13.86	24.70	34.63	44.01	61.70	70.15	78.41	86.50	94.44	102.24	109.93
54.00	14.12	25.16	35.28	44.84	62.86	71.48	79.89	88.13	96.22	104.17	112.00
55.00	14.38	25.63	35.93	45.67	64.02	72.80	81.37	89.76	98.00	106.10	114.08
56.00	14.65	26.10	36.59	46.50	65.19	74.12	82.85	91.39	99.78	108.03	116.15
57.00	14.91	26.56	37.24	47.33	66.35	75.45	84.33	93.03	101.56	109.96	118.23
58.00	15.17	27.03	37.89	48.16	67.52	76.77	85.81	94.66	103.34	111.89	120.30
59.00	15.43	27.49	38.55	48.99	68.68	78.10	87.29	96.29	105.13	113.82	122.38
60.00	15.69	27.96	39.20	49.82	69.85	79.42	88.77	97.92	106.91	115.75	124.45

Table IV-16—*Cont.*

Time	UPTD for given P_{O_2} exposure										

Time	1.80	1.90	2.00	2.10	2.20	2.30	2.40	2.50	2.60	2.70	2.80
1.00	2.22	2.36	2.50	2.64	2.77	2.91	3.04	3.17	3.31	3.44	3.57
2.00	4.43	4.72	5.00	5.27	5.55	5.82	6.08	6.35	6.61	6.87	7.13
3.00	6.65	7.08	7.49	7.91	8.32	8.72	9.13	9.52	9.92	10.31	10.70
4.00	8.87	9.43	9.99	10.54	11.09	11.63	12.17	12.70	13.23	13.75	14.27
5.00	11.09	11.79	12.49	13.18	13.86	14.54	15.21	15.87	16.53	17.19	17.83
6.00	13.30	14.15	14.99	15.82	16.64	17.45	18.25	19.05	19.84	20.62	21.40
7.00	15.52	16.51	17.49	18.45	19.41	20.36	21.29	22.22	23.15	24.06	24.97
8.00	17.74	18.87	19.98	21.09	22.18	23.26	24.34	25.40	26.45	27.50	28.54
9.00	19.96	.00	22.48	23.72	24.95	26.17	27.38	28.57	29.76	30.94	32.10
10.00	22.17	23.58	24.98	26.36	27.73	29.08	30.42	31.75	33.07	34.37	35.67
11.00	24.39	25.94	27.48	29.00	30.50	31.99	33.46	34.92	36.37	37.81	39.24
12.00	26.61	28.30	29.98	31.63	33.27	34.90	36.50	38.10	39.68	41.25	42.80
13.00	28.82	30.66	32.47	34.27	36.04	37.80	39.55	41.27	42.99	44.68	46.37
14.00	31.04	33.02	34.97	36.91	38.82	40.71	42.59	44.45	46.29	48.12	49.94
15.00	33.26	35.38	37.47	39.54	41.59	43.62	45.63	47.62	49.60	51.56	53.50
16.00	35.48	37.74	39.97	42.18	44.36	46.53	48.67	50.80	52.90	55.00	57.07
17.00	37.69	40.09	42.47	44.81	47.14	49.43	51.71	53.97	56.21	58.43	60.64
18.00	39.91	42.45	44.96	47.45	49.91	52.34	54.76	57.15	59.52	61.87	64.21
19.00	42.13	44.81	47.46	50.09	52.68	55.25	57.80	60.32	62.82	65.31	67.77
20.00	44.34	47.17	49.96	52.72	55.45	58.16	60.84	63.50	66.13	68.74	71.34
21.00	46.56	49.53	52.46	55.36	58.23	61.07	63.88	66.67	69.44	72.18	74.91
22.00	48.78	51.89	54.96	57.99	61.00	63.97	66.92	69.85	72.74	75.62	78.47
23.00	51.00	54.24	57.46	60.63	63.77	66.88	69.96	73.02	76.05	79.06	82.04
24.00	53.21	56.60	59.95	63.27	66.54	69.79	73.01	76.20	79.36	82.49	85.61
25.00	55.43	58.96	62.45	65.90	69.32	72.70	76.05	79.37	82.66	85.93	89.17
26.00	57.65	61.32	64.95	68.54	72.09	75.61	79.09	82.54	85.97	89.37	92.74
27.00	59.87	63.68	67.45	71.17	74.86	78.51	82.13	85.72	89.28	92.81	96.31
28.00	62.08	66.04	69.95	73.81	77.63	81.42	85.17	88.89	92.58	96.24	99.87
29.00	64.30	68.40	72.44	76.45	80.41	84.33	88.22	92.07	95.89	99.68	103.44
30.00	66.52	70.75	74.94	79.08	83.18	87.24	91.26	95.24	99.20	103.12	107.01
31.00	68.73	73.11	77.44	81.72	85.95	90.15	94.30	98.42	102.50	106.55	110.58
32.00	70.95	75.47	79.94	84.35	88.73	93.05	97.34	101.59	105.81	109.99	114.14
33.00	73.17	77.83	82.44	86.99	91.50	95.96	100.38	104.77	109.12	113.43	117.71
34.00	75.39	80.19	84.93	89.63	94.27	98.87	103.43	107.94	112.42	116.87	121.28
35.00	77.60	82.55	87.43	92.26	97.04	101.78	106.47	111.12	115.73	120.30	124.84
36.00	79.82	84.91	89.93	94.90	99.82	104.69	109.51	114.29	119.04	123.74	128.41
37.00	82.04	87.26	92.43	97.53	102.59	107.59	112.55	117.47	122.34	127.18	131.98
38.00	84.25	89.62	94.93	100.17	105.36	110.50	115.59	120.64	125.65	130.62	135.54
39.00	86.47	91.98	97.42	102.81	108.13	113.41	118.64	123.82	128.96	134.05	139.11
40.00	88.69	94.34	99.92	105.44	110.91	116.32	121.68	126.99	132.26	137.49	142.68
41.00	90.91	96.70	102.42	108.08	113.68	119.23	124.72	130.17	135.57	140.93	146.25
42.00	93.12	99.06	104.92	110.72	116.45	122.13	127.76	133.34	138.87	144.36	149.81
43.00	95.34	101.41	107.42	113.35	119.23	125.04	130.80	136.52	142.18	147.80	153.38
44.00	97.56	103.77	109.91	115.99	122.00	127.95	133.85	139.69	145.49	151.24	156.95
45.00	99.78	106.13	112.41	118.62	124.77	130.86	136.89	142.87	148.79	154.68	160.51
46.00	101.99	108.49	114.91	121.26	127.54	133.77	139.93	146.04	152.10	158.11	164.08
47.00	104.21	110.85	117.41	123.90	130.32	136.67	142.97	149.22	155.41	161.55	167.65
48.00	106.43	113.21	119.91	126.53	133.09	139.58	146.01	152.39	158.71	164.99	171.21
49.00	108.64	115.57	122.40	129.17	135.86	142.49	149.06	155.57	162.02	168.43	174.78
50.00	110.86	117.92	124.90	131.80	138.63	145.40	152.10	158.74	165.33	171.86	178.35
51.00	113.08	120.28	127.40	134.44	141.41	148.30	155.14	161.91	168.63	175.30	181.91
52.00	115.30	122.64	129.90	137.08	144.18	151.21	158.18	165.09	171.94	178.74	185.48
53.00	117.51	125.00	132.40	139.71	146.95	154.12	161.22	168.26	175.25	182.17	189.05
54.00	119.73	127.36	134.89	142.35	149.72	157.03	164.27	171.44	178.55	185.61	192.62
55.00	121.95	129.72	137.39	144.98	152.50	159.94	167.31	174.61	181.86	189.05	196.18
56.00	124.16	132.07	139.89	147.62	155.27	162.84	170.35	177.79	185.17	192.49	199.75
57.00	126.38	134.43	142.39	150.26	158.04	165.75	173.39	180.96	188.47	195.92	203.32
58.00	128.60	136.79	144.89	152.89	160.82	168.66	176.43	184.14	191.78	199.36	206.88
59.00	130.82	139.15	147.38	155.53	163.59	171.57	179.48	187.31	195.09	202.80	210.45
60.00	133.03	141.51	149.88	158.16	166.36	174.48	182.52	190.49	198.39	206.23	214.02

Table IV-16—*Cont.*

Time	\multicolumn{11}{c}{UPTD for given P_{O_2} exposure}
	2.90	3.00	3.10	3.20	3.30	3.40	3.50	3.60	3.70	3.80	3.90
1.00	3.70	3.82	3.95	4.08	4.20	4.33	4.45	4.57	4.70	4.82	4.94
2.00	7.39	7.65	7.90	8.15	8.40	8.65	8.90	9.15	9.39	9.64	9.88
3.00	11.09	11.47	11.85	12.23	12.61	12.98	13.35	13.72	14.09	14.46	14.82
4.00	14.78	15.29	15.80	16.31	16.81	17.31	17.80	18.30	18.79	19.28	19.76
5.00	18.48	19.12	19.75	20.38	21.01	21.64	22.26	22.87	23.48	24.09	24.70
6.00	22.17	22.94	23.70	24.46	25.21	25.96	26.71	27.45	28.18	28.91	29.64
7.00	25.87	26.77	27.65	28.54	29.42	30.29	31.16	32.02	32.88	33.73	34.58
8.00	29.57	30.59	31.61	32.62	33.62	34.62	35.61	36.59	37.58	38.55	39.52
9.00	33.26	34.41	35.56	36.69	37.82	38.94	40.06	41.17	42.27	43.37	44.46
10.00	36.96	38.24	39.51	40.77	42.02	43.27	44.51	45.74	46.97	48.19	49.40
11.00	40.65	42.06	43.46	44.85	46.23	47.60	48.96	50.32	51.67	53.01	54.34
12.00	44.35	45.88	47.41	48.92	50.43	51.92	53.41	54.89	56.36	57.83	59.28
13.00	48.04	49.71	51.36	53.00	54.63	56.25	57.86	59.47	61.06	62.65	64.22
14.00	51.74	53.53	55.31	57.08	58.83	60.58	62.31	64.04	65.76	67.47	69.16
15.00	55.44	57.35	59.26	61.15	63.03	64.91	66.77	68.61	70.45	72.28	74.11
16.00	59.13	61.18	63.21	65.23	67.24	69.23	71.22	73.19	75.15	77.10	79.05
17.00	62.83	65.00	67.16	69.31	71.44	73.56	75.67	77.76	79.85	81.92	83.99
18.00	66.52	68.83	71.11	73.38	75.64	77.89	80.12	82.34	83.70	86.74	88.93
19.00	70.22	72.65	75.06	77.46	79.84	82.21	84.57	86.91	89.24	91.56	93.87
20.00	73.91	76.47	79.01	81.54	84.05	86.54	89.02	91.49	93.94	96.38	98.81
21.00	77.61	80.30	82.96	85.61	88.25	90.87	93.47	96.06	98.64	101.20	103.75
22.00	81.31	84.12	86.91	89.69	92.45	95.19	97.92	100.64	103.33	106.02	108.69
23.00	85.00	87.94	90.87	93.77	96.65	99.52	102.37	105.21	108.03	110.84	113.63
24.00	88.70	91.77	94.82	97.85	100.86	103.85	106.82	109.78	112.73	115.66	118.57
25.00	92.39	95.59	98.77	101.92	105.05	108.18	111.28	114.36	117.42	120.47	123.51
26.00	96.09	99.41	102.72	106.00	109.36	112.50	115.73	118.93	122.12	125.29	128.45
27.00	99.78	103.24	106.67	110.08	113.46	116.83	120.18	123.51	126.82	130.11	133.39
28.00	103.48	107.06	110.62	114.15	117.67	121.16	124.63	128.08	131.51	134.93	138.33
29.00	107.18	110.89	114.57	118.23	121.87	125.48	129.08	132.66	136.21	139.75	143.27
30.00	110.87	114.71	118.52	122.31	126.07	129.81	133.53	137.23	140.91	144.57	148.21
31.00	114.57	118.53	122.47	126.38	130.27	134.14	137.98	141.80	145.61	149.39	153.15
32.00	118.26	122.36	126.42	130.46	134.47	138.46	142.43	146.38	150.30	154.21	158.09
33.00	121.96	126.18	130.37	134.54	138.68	142.79	146.88	150.95	155.00	159.03	163.03
34.00	125.66	130.00	134.32	138.61	142.88	147.12	151.33	155.53	159.70	163.84	167.97
35.00	129.35	133.83	138.27	142.69	147.08	151.45	155.79	160.10	164.39	168.66	172.91
36.00	133.05	137.65	142.22	146.77	151.28	155.77	160.24	164.68	169.09	173.48	177.85
37.00	136.74	141.47	146.17	150.84	155.49	160.10	164.69	169.25	173.79	178.30	182.79
38.00	140.44	145.30	150.13	154.92	159.69	164.43	169.14	173.82	178.48	183.12	187.73
39.00	144.13	149.12	154.08	159.00	163.89	168.75	173.59	178.40	183.18	187.94	192.67
40.00	147.83	152.94	158.03	163.08	168.09	173.08	178.04	182.97	187.88	192.76	197.61
41.00	151.53	156.77	161.98	167.15	172.30	177.41	182.49	187.55	192.58	197.58	202.55
42.00	155.22	160.59	165.93	171.23	176.50	181.74	186.94	192.12	197.27	202.40	207.49
43.00	158.92	164.42	169.88	175.31	180.70	186.06	191.39	196.70	201.97	207.22	212.43
44.00	162.61	168.24	173.83	179.38	184.90	190.39	195.84	201.27	206.67	212.03	217.38
45.00	166.31	172.06	177.78	183.46	189.10	194.72	200.30	205.84	211.36	216.85	222.32
46.00	170.00	175.89	181.73	187.54	193.31	199.04	204.75	210.42	216.06	221.67	227.26
47.00	173.70	179.71	185.68	191.61	197.51	203.37	209.20	214.99	220.76	226.49	232.20
48.00	177.40	183.53	189.63	195.69	201.71	207.70	213.65	219.57	225.45	231.31	237.14
49.00	181.09	187.36	193.58	199.77	205.91	212.02	218.10	224.14	230.15	236.13	242.08
50.00	184.79	191.18	197.53	203.84	210.12	216.35	222.55	228.72	234.85	240.95	247.02
51.00	188.48	195.00	201.48	207.92	214.32	220.68	227.00	233.29	239.54	245.77	251.96
52.00	192.18	198.83	205.43	212.00	218.52	225.01	231.45	237.86	244.24	250.59	256.90
53.00	195.87	202.65	209.38	216.07	222.72	229.33	235.90	242.44	248.94	255.40	261.84
54.00	199.57	206.48	213.34	220.15	226.93	233.66	240.35	247.01	253.64	260.22	266.78
55.00	203.27	210.30	217.29	224.23	231.13	237.99	244.81	251.59	258.33	265.04	271.72
56.00	206.96	214.12	221.24	228.31	235.33	242.31	249.26	256.16	263.03	269.86	276.66
57.00	210.66	217.95	225.19	232.38	239.53	246.64	253.71	260.74	267.73	274.68	281.60
58.00	214.35	221.77	229.14	236.46	243.74	250.97	258.16	265.31	272.42	279.50	286.54
59.00	218.05	225.59	233.09	240.54	247.94	255.29	262.61	269.88	277.12	284.32	291.48
60.00	221.74	229.42	237.04	244.61	252.14	259.62	267.06	274.46	281.82	289.14	296.42

Table IV-16—*Cont.*

	UPTD for given P_{O_2} exposure									
Time	4.00	4.10	4.20	4.30	4.40	4.50	4.60	4.70	4.80	4.90	5.00
1.00	5.06	5.18	5.30	5.42	5.54	5.66	5.77	5.89	6.01	6.12	6.24
2.00	10.12	10.36	10.60	10.84	11.08	11.31	11.55	11.78	12.02	12.25	12.48
3.00	15.18	15.54	15.90	16.26	16.62	16.97	17.32	17.67	18.02	18.37	18.72
4.00	20.24	20.73	21.20	21.68	22.16	22.63	23.10	23.57	24.03	24.50	24.96
5.00	25.31	25.91	26.51	27.10	27.69	28.28	28.87	29.46	30.04	30.62	31.20
6.00	30.37	31.09	31.81	32.52	33.23	33.94	34.65	35.35	36.05	36.75	37.44
7.00	35.43	36.27	37.11	37.94	38.77	39.60	40.42	41.24	42.06	42.87	43.68
8.00	40.49	41.45	42.41	43.36	44.31	45.25	46.20	47.13	48.07	49.00	49.92
9.00	45.55	46.63	47.71	48.78	49.85	50.91	51.97	53.02	54.07	55.12	56.16
10.00	50.61	51.81	53.01	54.20	55.39	56.57	57.74	58.92	60.08	61.24	62.40
11.00	55.67	56.99	58.31	59.62	60.93	62.23	63.52	64.81	66.09	67.37	68.64
12.00	60.73	62.18	63.61	65.04	66.47	67.88	69.29	70.70	72.10	73.49	74.88
13.00	65.79	67.36	68.91	70.46	72.00	73.54	75.07	76.59	78.11	79.62	81.12
14.00	70.86	72.54	74.21	75.88	77.54	79.20	80.84	82.48	84.12	85.74	87.36
15.00	75.92	77.72	79.52	81.30	83.08	84.85	86.62	88.37	90.12	91.87	93.60
16.00	80.98	82.90	84.82	86.72	88.62	90.51	92.39	94.27	96.13	97.99	99.84
17.00	86.04	88.08	90.12	92.14	94.16	96.17	98.17	100.16	102.14	104.12	106.08
18.00	91.10	93.26	95.42	97.56	99.70	101.82	103.94	106.05	108.15	110.24	112.32
19.00	96.16	98.45	100.72	102.98	105.24	107.48	109.71	111.94	114.16	116.37	118.56
20.00	101.22	103.63	106.02	108.40	110.78	113.14	115.49	117.83	120.17	122.49	124.81
21.00	106.28	108.81	111.32	113.82	116.31	118.79	121.26	123.72	126.17	128.61	131.05
22.00	111.35	113.99	116.62	119.24	121.85	124.45	127.04	129.62	132.18	134.74	137.29
23.00	116.41	119.17	121.92	124.66	127.39	130.11	132.81	135.51	138.19	140.86	143.53
24.00	121.47	124.35	127.22	130.08	132.93	135.76	138.59	141.40	144.20	146.99	149.77
25.00	126.53	129.53	132.53	135.50	138.47	141.42	144.36	147.29	150.21	153.11	156.01
26.00	131.59	134.72	137.83	140.92	144.01	147.08	150.14	153.18	156.21	159.24	162.25
27.00	136.65	139.90	143.13	146.34	149.55	152.74	155.91	159.07	162.22	165.36	168.49
28.00	141.71	145.08	148.43	151.76	155.09	158.39	161.68	164.96	168.23	171.49	174.73
29.00	146.77	150.26	153.73	157.18	160.62	164.05	167.46	170.86	174.24	177.61	180.97
30.00	151.83	155.44	159.03	162.60	166.16	169.71	173.23	176.75	180.25	183.73	187.21
31.00	156.90	160.62	164.33	168.02	171.70	175.36	179.01	182.64	186.26	189.86	193.45
32.00	161.96	165.80	169.63	173.44	177.24	181.02	184.78	188.53	192.26	195.98	199.69
33.00	167.02	170.98	174.93	178.86	182.78	186.68	190.56	194.42	198.27	202.11	205.93
34.00	172.08	176.17	180.23	184.29	188.32	192.33	196.33	200.31	204.28	208.23	212.17
35.00	177.14	181.35	185.54	189.71	193.86	197.99	202.11	206.21	210.29	214.36	218.41
36.00	182.20	186.53	190.84	195.13	199.40	203.65	207.88	212.10	216.30	220.48	224.65
37.00	187.26	191.71	196.14	200.55	204.93	209.30	213.66	217.99	222.31	226.61	230.89
38.00	192.32	196.89	201.44	205.97	210.47	214.96	219.43	223.88	228.31	232.73	237.13
39.00	197.38	202.07	206.74	211.39	216.01	220.62	225.20	229.77	234.32	238.85	243.37
40.00	202.45	207.25	212.04	216.81	221.55	226.27	230.98	235.66	240.33	244.98	249.61
41.00	207.51	212.44	217.34	222.23	227.09	231.93	236.75	241.56	246.34	251.10	255.85
42.00	212.57	217.62	222.64	227.65	232.63	237.59	242.53	247.45	252.35	257.23	262.09
43.00	217.63	222.80	227.94	233.07	238.17	243.24	248.30	253.34	258.36	263.35	268.33
44.00	222.69	227.98	233.25	238.49	243.71	248.90	254.08	259.23	264.36	269.48	274.57
45.00	227.75	233.16	238.55	243.91	249.24	254.56	259.85	265.12	270.37	275.60	280.81
46.00	232.81	238.34	243.85	249.33	254.78	260.22	265.63	271.01	276.38	281.73	287.05
47.00	237.87	243.52	249.15	254.75	260.32	265.87	271.40	276.90	282.39	287.85	293.29
48.00	242.93	248.71	254.45	260.17	265.86	271.53	277.17	282.80	288.40	293.97	299.53
49.00	248.00	253.89	259.75	265.59	271.40	277.19	282.95	288.69	294.40	300.10	305.77
50.00	253.06	259.07	265.05	271.01	276.94	282.84	288.72	294.58	300.41	306.22	312.01
51.00	258.12	264.25	270.35	276.43	282.48	288.50	294.50	300.47	306.42	312.35	318.25
52.00	263.18	269.43	275.65	281.85	288.02	294.16	300.27	306.36	312.43	318.47	324.49
53.00	268.24	274.61	280.95	287.27	293.55	299.81	306.05	312.25	318.44	324.60	330.73
54.00	273.30	279.79	286.26	292.69	299.09	305.47	311.82	318.15	324.45	330.72	336.97
55.00	278.36	284.97	291.56	298.11	304.63	311.13	317.60	324.04	330.45	336.85	343.21
56.00	283.42	290.16	296.86	303.53	310.17	316.78	323.37	329.93	336.46	342.97	349.45
57.00	288.48	295.34	302.16	308.95	315.71	322.44	329.14	335.82	342.47	349.10	355.69
58.00	293.55	300.52	307.46	314.37	321.25	328.10	334.92	341.71	348.48	355.22	361.93
59.00	298.61	305.70	312.76	319.79	326.79	333.75	340.69	347.60	354.49	361.34	368.17
60.00	303.67	310.88	318.06	325.21	332.33	339.41	346.47	353.50	360.50	367.47	374.42

Table IV-17

Factors Influencing Rate of Development of Pulmonary Oxygen Poisoning[a]

Decrease survival time or increase severity of pulmonary pathology	Increase survival time or decrease severity of pulmonary pathology	
Adrenocortical hormones	Adrenal medullectomy	Gluthathione
Adrenocorticotrophic hormone	Adrenalectomy	Glycine
Aspartic acid	Adrenergic blocking drugs (dibenamine, SKF 501, dibenzyline, dehydrobenzoperidol)	Glycylglycine
Atropine	Alanine	Hibernation
Carbon dioxide	Altitude acclimatization	Histamine
Chorionic gonadotropin with progesterone	Anesthesia (barbiturates, alpha-chlorolose, propylene glycol, urethane)	Hyopohysectomy
Epinephrine	Antihistamines (promethazine, metyramine, thiazinanium)	Hypothermia
Estrogen	Antioxidants (alpha-tocopherol polyethylene glycol 1000 succinate, ascorbic acid, N,N'-diphenyl-p-phenylenediamine, hydroquinone, methylene blue, porpyl gallate, trihydroxyphenone antioxidants, beta-aminoethylisothiuronium, nordihydroquiaretic acid)	Hypothyroidism (propylthiouracil)
Glutamic acid		Immaturity
Dextroamphetamine		Intermittent exposure
Dihydroxyphenylalanine		Methacholine
Histamine	Arginine	Oxytyramine
Hyperthermia	Carbachol	Reserpine
Inert gas (inspired at high ambient pressure)	Chlorpromazine	Serotonin
Insulin	Cobalt (II)	Sodium bicarbonate
Norepinephrine	Coenzyme A	Sodium diethyldithiocarbamate
Thyroid hormones	Convulsions	Sodium lactate
Vitamin E deficiency	Cystamine	Starvation
X-irradiation	Cysteamine	Saccinate
	Cysteine	Thionrea
	Dimercaprol	Thyroidectomy
	Ethanol	Tris aminomethane (THAM)
	gamma-Aminobutyric acid	Tryptamine
	Ganglionic blocking drugs (hexamethonium, tetraethylammonium)	Vitamin E

[a] Reprinted from Clark and Lambertsen (1971a) with permission of Williams and Wilkins.

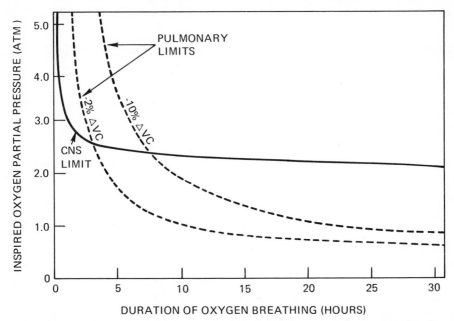

Figure IV-28. Toxicity limits of oxygen use: inspired partial pressure of oxygen vs. duration of exposure. From Wright (1972) and Institute for Environmental Medicine (1970).

and the rigid tonic (continuous muscular contraction) phase of the convulsion begins. Respiration ceases at this point and does not begin again until muscular coordination returns. The tonic phase usually lasts for about 30 sec and is accompanied by an abrupt loss of consciousness. Vigorous clonic (rapid succession of contraction and relaxation of muscle) contractions of the muscle groups of the head, neck, and limbs then occur, becoming progressively less violent over about 1 min. As the uncoordinated motor activity stops, respiration can again proceed. Following the convulsion, hyperpnea (a deeper and more rapid than normal breathing pattern) is marked due to accumulation of metabolic products during breath-holding. Respiration is complicated by soft tissue obstruction and by the extensive secretions which result from what is probably an autonomic component of the CNS convulsive activity. Since a high alveolar oxygen partial pressure persists during convulsive breath-holding, the individual remains well oxygenated throughout the convulsion itself. Actually, due to the high arterial carbon dioxide partial pressure, brain oxygenation should increase during the breath-holding period. This is in sharp contrast to the epileptic patient, who convulses while breathing air at ambient pressure. Electroencephalographically, the seizures caused by high partial pressures of oxygen are indistinguishable from grand mal epilepsy; the EEG shows high-amplitude, fast spikes which break off rather sharply to low-amplitude, fast activity or to a relatively isoelectric state (Stein 1955).

(i) Early Warning Signs. The onset of convulsions is usually preceded by other symptoms of CNS oxygen poisoning, both objective and subjective in nature. Some of

them include the following (Behnke 1955, Lambertsen 1965, Gillen 1966, Young 1971):

Localized muscular twitching, especially about the face
Incoordination of diaphragm activity in respiration
Nausea
Paresthesia ("pins and needles")
Dizziness
Incoordination
Light-headedness
Euphoria

Dyspnea (subjective distress or difficulty in breathing)
Confusion
Unusual fatigue
Visual symptoms
Hiccups
Dilated pupils
Bradycardia (slowing of the heart rate)

However, convulsions are not always preceded by early warning signs; early warning signs are sometimes rapidly—and in some cases immediately—followed by the onset of convulsions, and convulsions can still occur after oxygen breathing has been discontinued (Institute for Environmental Medicine 1970).

(ii) *Latency or Tolerance.* Convulsions due to oxygen toxicity occur only after a "safe latent period," the length of which is inversely proportional to the inspired oxygen partial pressure. This is not a linear change with pressure, but decreases sharply with increasing partial pressure. Figure IV-29 shows the relationship between the pressure (oxygen) in atmospheres and the time required to develop symptoms of CNS oxygen toxicity in a dry pressure chamber. It can be seen that at 60 fsw pressure in a dry chamber none of the 20 subjects breathing pure oxygen developed symptoms in the 2-hr exposure. At 80 fsw, 50% of the subjects developed symptoms in about 60 min, and at 100 fsw, 50% of the subjects developed symptoms in 25 min.

The CNS tolerance, or latency, differs markedly between individuals. The oxygen tolerance test (Gillen 1966), which requires breathing pure oxygen for 30 min at 60 fsw pressure in a dry compression chamber, is designed to detect unusually susceptible individuals. Definite preconvulsive signs such as twitching of the lips or limbs constitute "failure" on the test. In addition to the latent period variation among individuals, there is also a considerable variance from day to day within a single individual. This is illustrated in Figure IV-30.

At a particular oxygen partial pressure, the latent period for development of convulsions in the intact animal is less than the time required for enzymic inhibition to alter metabolism *in vitro* (Davies and Davies 1965). This may indicate that extremely subtle chemical changes can affect the electrical activity of the highly organized central nervous system and that the effects of oxygen on membranes, not directly paralleled by changes in the oxidative metabolism of cell systems, may possibly be responsible for the convulsions (Lambertsen 1965).

(a) *Effect of carbon dioxide.* The latent period of oxygen convulsions is prolonged by hyperventilation and shortened by administration of low concentrations of carbon dioxide. These effects appear in part to be indirect results of the influences of hypocapnia and hypercapnia upon brain circulation and hence upon the dose of oxygen delivered to the brain cells (Lambertsen 1963). The level of arterial carbon dioxide accompanying high arterial oxygen is of greater practical importance in that changes

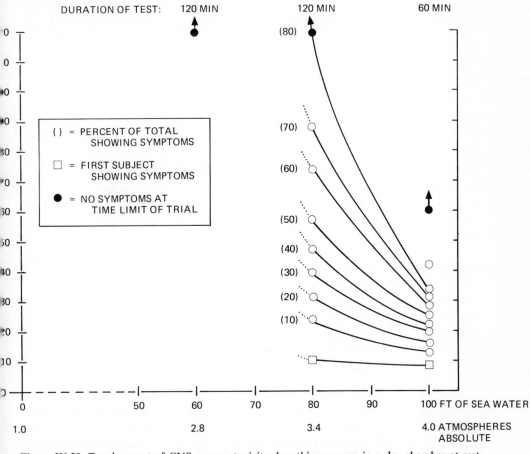

Figure IV-29. Development of CNS oxygen toxicity, breathing oxygen in a dry chamber at rest. [From Lambertsen (1955) by permission of the National Academy of Sciences–National Research Council.]

in brain flow are accompanied by large alterations of oxygen in the capillaries of the CNS. Addition of carbon dioxide to the inspired oxygen overcomes cerebral vaso-constriction, speeds the flow of blood through the brain capillaries, and the consequent extreme rise in capillary and venous oxygen can be presumed to increase the dose of oxygen to which the brain cells are exposed (Lambertsen 1965). In contrast to the situation while breathing air at 1 atm pressure, it would be expected that breath-holding when the respiratory gas is oxygen at high pressure would increase the dose of oxygen delivered to the brain.

(b) *Effect of exercise*. The latent period of CNS oxygen toxicity is decreased by exercise. This effect is illustrated in Figure IV-31 for men breathing 100% oxygen at various pressures. Although symptoms of CNS toxicity have been reported in men breathing 100% oxygen as shallow as 20 fsw when working to exhaustion (Young 1971), the work load of swimming underwater does not appear to cause problems

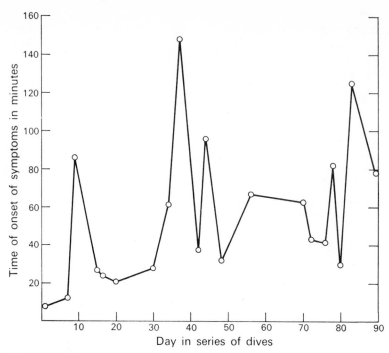

Figure IV-30. Variation in time of onset of symptoms of oxygen toxicity in the same diver exposed repeatedly to 3.12 ATA (70 fsw) over a period of 90 days. The symptoms in the vast majority of cases was lip twitching. [From Donald (1947) with permission from Bailliere and Tindall, London.]

until about 30 fsw. The tolerance to oxygen during underwater swimming becomes sharply reduced at depths greater than 30 fsw (Lambertsen 1955).

It is difficult to determine the exact nature of the influence of exercise on oxygen tolerance. The increased electrical activity of cerebral, cortical, and other central neurons during exercise and the bombardment of the reticular activating system by efferent impulses are factors of possible but uncertain importance (Lambertsen 1965). It is possible that a superimposed carbon dioxide retention results from such factors as interference with alveolar ventilation due to increased gas flow resistance in respiratory passages, or in the external breathing apparatus, or (when closed-circuit breathing systems are used) depression of alveolar ventilation produced by physiological effects of inadequate methods of carbon dioxide absorption. It is not yet certain whether the neurophysiological effects of muscular exercise or some consequent interference with carbon dioxide elimination is responsible for the shortened latent period.

(c) *Other factors affecting latency or tolerance.* Still other factors are known to modify the duration of the safe latent period before the development of oxygen convulsions.

1. Immersion. Immersion (in the vertical position) appears to shorten the latent period but whether this is an experimental artifact related to a circulatory adjustment

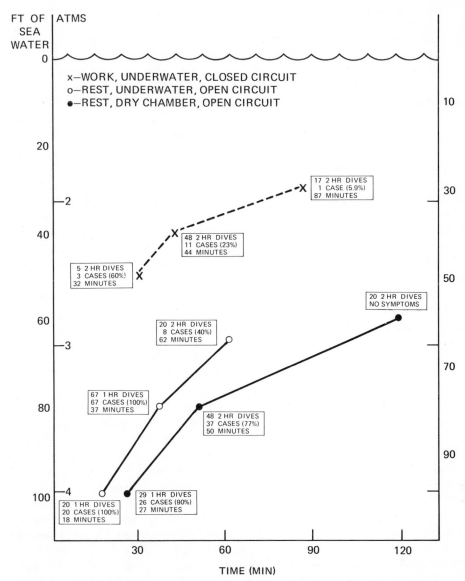

Figure IV-31. Effect of exercise underwater upon CNS oxygen tolerance. [From Lambertsen (1955) by permission of the National Academy of Sciences–National Research Council.]

due to the external hydrostatic forces or is secondary to ventilatory changes is not known (Lambertsen 1965).

2. Hypothermia. Cold-blooded animals have a long latent period at low body temperatures, but become more susceptible to oxygen toxicity when their body temperature is raised. This could have important implications for man in that hypothermia should delay and hyperthermia should accelerate the onset of convulsions (Lambertsen 1965).

b. Mechanisms

The effects of high-pressure oxygen on metabolite and electrolyte transport at neuronal membranes may be the key to understanding the events that culminate in oxygen convulsions (Wood 1969). Generally, high-pressure oxygen tensions interfere with the oxidation of glucose and with the use of oxygen itself. The effect may well be, at least in part, the result of an excessive rate of oxidation of essential cellular constituents. High-pressure oxygen-induced lipid peroxidation (Jerett *et al.* 1973) has been suggested as disrupting the integrity of cellular membranes. Loss of potassium from cerebral cortex slices under high-pressure oxygen conditions has been reported (Kaplan and Stein 1957), and studies (Watkins 1965) have suggested the possible involvement of gamma-aminobutyric acid (GABA) in the etiology of oxygen convulsions. CNS epinephrine has been proposed as a possible chemical which determines a receptor tone in the CNS, and thus the threshold of susceptibility to seizure (Hof *et al.* 1972). It has also been proposed that X-irradiation and oxygen poisoning produce some of their effects through a common mechanism, the formation of oxidizing free radicals (Gerschman *et al.* 1954*a*). Further study of the biochemistry of high-pressure oxygen is obviously needed.

It should be stressed that the varied nature of the physiological and biochemical events which are related to high-pressure oxygen indicates that oxygen toxicity involves various mechanisms, many of them interconnected.

c. Treatment

Due to rapid oxygen metabolism in the cell, the oxygen level at a particular intracellular site should fall to a normal, nontoxic level within a few minutes of a pulmonary "oxygen washout time" (Lambertsen 1966*b*). When any of the early warning signs of oxygen toxicity is apparent, the return to normal partial pressure of inspired oxygen should begin as soon as possible. However, due to the frequently rapid onset of convulsions, this return cannot always be accomplished quickly enough.

If a convulsion occurs, the first step is to remove the mask or oxygen source. It is necessary to prevent the subject from injuring himself. Excessive restraint should be avoided. At the onset of the convulsion, the head is hyperextended and the lower jaw is strongly depressed so that the jaws are separated. During this period of about 10 sec, a soft but firm object such as a padded tongue depressor can be easily inserted in the mouth to prevent chewing of the tongue during the subsequent clonic phase of the convulsion. Fingers should not be used for this purpose. Due to severe interference with pulmonary ventilation, it is extremely important to avoid decompression during any part of the convulsion. Expanding pulmonary gas could rupture the lung and produce a possibly fatal pulmonary embolism. Rhythmic breathing returns as the clonic convulsion ceases. It is at this stage that attention to the airway is important.

Consciousness usually returns gradually over a period of 5–10 min. The subject may be irrational and will require reassurance and gentle restraint. In some cases consciousness returns abruptly and the patient shows surprising mental clarity. Headaches or nausea may occur and muscular fatigue is to be expected.

d. Residual Effects

If the conditions of pressure and duration of an oxygen exposure do not cause acute manifestations of CNS oxygen toxicity (i.e., convulsions), it is unlikely that important residual harmful effects will result. Even convulsions do not necessarily produce CNS damage under laboratory or treatment conditions. No physical, physiological, or psychometric evidence of harm, either acute or residual, has been detected in normal subjects (Lambertsen 1965) or in several schizophrenic patients who were subjected to repeated, biweekly oxygen convulsions at an inspired oxygen pressure of 4 atm. The convulsion is apparently a sensitive and early manifestation of central nervous system toxicity. Most likely, the development of toxic effects within central neurons has not reached an irreversible stage by the time the diffuse electrical discharge of the convulsion occurs.

Even if it does not produce neuronal damage, a convulsion can conceivably result in the same forms of physical trauma associated with epileptic seizures or convulsions produced by electroshock. Although recovery of consciousness is normally prompt, an instance of persistent convulsions has been reported (Workman 1963). Residual neurological damage and paralysis (Kydd 1964) have been shown to occur in rats following severe oxygen poisoning, but have not been described in larger animals or man.

e. Prevention

(i) Avoidance. Effective diving procedures take CNS oxygen toxicity into account and essentially eliminate it as a problem by maintaining oxygen partial pressures below the convulsive threshold. A maximum of 2.2 ATA will avoid virtually all risk, while exposures of up to 3.0 ATA are used by some groups under the proper conditions, such as low carbon dioxide, etc.

(ii) Fluctuating Oxygen Exposure. It has been observed that the rate of recovery from the effects of oxygen toxicity is faster than the rate of development. In World War II when twitching developed during oxygen-diving, the diver would surface and symptoms would dissipate even without discontinuing oxygen breathing. When the diver went down again, a long latent period once more existed before symptoms occurred again (Lambertsen 1966*a*). This finding led to the principle of alternating high and low inspired oxygen pressures to permit greater exposure to increased pressures of oxygen within a given time period without producing oxygen toxicity. This principle has been quantitatively demonstrated in animals and, although the optimal durations of oxygen exposure and interruption have not yet been developed for man (Lambertsen 1965), a practical procedure for oxygen exposure in the treatment of decompression sickness is well established (Goodman 1967).

(iii) Hyperventilation. Hyperventilation has been shown to extend CNS oxygen tolerance in animals. This effect is most likely due to reduced cellular oxygen tensions in the brain, since hyperventilation leads to arterial hypocapnia and cerebral vasoconstriction, which acts to lower mean brain–oxygen levels and decrease the mass of brain tissue exposed to the toxic oxygen tensions. This should not be interpreted as an increase in tolerance, but merely indicates that the development of convulsions can be delayed (Lambertsen 1966*b*).

(iv) Pharmacologic. A wide variety of substances have been investigated for their prophylactic properties against oxygen poisoning in animals and, although a number of compounds have afforded some protection, a compound giving significant protection that is safe for human use without undesirable side effects has still not been found. Bennett (1972*a*) recently reviewed pharmacologic agents with possible use in diving. A selection of substances affording partial protection to animals against CNS oxygen toxicity is listed in Table IV-18.

Although the development of convulsions can be prevented or delayed by the use of barbiturates and other central depressants and neuromuscular blocking agents (Davies and Davies 1965), the metabolic disturbance and cellular damage will continue to develop and may result in permanent damage to the central nervous system.

3. *Other Toxic Effects*

While the toxic effects of increased oxygen on pulmonary function and the central nervous system are the most obvious, they are not restricted to these areas but can manifest themselves in most of the body cells that contain susceptible enzymes (Lambertsen 1966*a*).

a. The Eye

Oxygen at high pressure can cause constriction of the retinal arterioles and veins and of the visual peripheral field. Despite this, hyperbaric oxygen appears to increase

Table IV-18

Protective Agents against CNS Oxygen Toxicity in Animals[a]

Substance	Protection against	Reference
Tris buffer (THAM)	Convulsions, lung damage	Bean 1961
Vitamin E	Convulsions, RBC destruction	Taylor 1956, Fischer and Kimzey 1971
GABA	Convulsions, lung damage	Wood *et al.* 1963, 1965
Arginine	Convulsions	Gershenovich and Krichevskaya 1960
Antioxidants	Convulsions, lung damage, paralysis	Jamieson and van den Brenk 1964
Anesthetic	Convulsions, lung damage	Bean and Zee 1965
2-Mercaptoethylamine	Mortality	Gerschman *et al.* 1954
Dimercaprol	Convulsions, mortality	Van Tassel 1965
Vitamin K	Mortality	Horne 1966
Glutathione	Mortality	Gerschman *et al.* 1958, Sanders *et al.* 1972
2:4 Dinitrophenol	Spasticity and paralysis	van den Brenk and Jamieson 1964
5-Hydroxytryptamine	Spasticity and paralysis	van den Brenk and Jamieson 1964
Succinate	Convulsions, lung damage, paralysis, and mortality	Sanders *et al.* 1965, Sanders and Curie 1971, Sanders *et al.* 1972

[a] Reprinted from Wood (1969) with permission of Bailliere and Tindall.

the reservoir of oxygen accessible to the retina. This assumption is based on the demonstration that vision persists for a longer interval in hyperoxygenated man than is normal after inducing retinal ischemia (Carlisle *et al.* 1964). Below an alveolar oxygen partial pressure of 2 atm this increase in persistence time is relatively small, but above that level the time increases in direct proportion to the increase in oxygen and may exceed 50 sec at 4 atm (at ambient pressures of air, normal vision usually persists for only about 4 sec when the intraocular tension is raised). Although these visual effects of oxygen are slow in onset and rapidly reversible in normal men, the effects may be quite different in individuals with ocular manifestations of disease processes (Nichols and Lambertsen 1971). A more detailed discussion of the physiological and pathological effects of high-pressure oxygen on the visual system appears in Chapter V.

b. The Blood

Acceleration of hemoglobin formation at lower than normal oxygen levels raises the possibility that high tensions of inspired oxygen lead to depression of red blood cell formation and hemoglobin synthesis (Lambertsen 1966b). However, a recent review by Fischer and Kimzey (1971) indicates not that hemoglobin synthesis is depressed, but that hemolysis (alteration, dissolution, or destruction of red blood cells) may be accelerated. They also concluded that:

(1) The increase in physically dissolved O_2 contributes more to the chemical toxicity of hyperoxia than that which is chemically bound. Therefore, any elevation in O_2 tension above 160 mm Hg may precipitate the death of susceptible cells.

(2) The deleterious effect of O_2 on RBC [red blood cells] leads to a decrease in the circulating RBC mass. This reduction results either directly from the inactivation of essential glycolytic enzymes [promoting the breakdown of sugars into simpler compounds] by oxidation of SH [sulfhydryl] groups, or indirectly by the formation of lipid peroxides from the lipoproteins of the RBC membrane and the subsequent inactivation of SH-bearing enzymes.

D. *Physiological Aspects of Nitrogen Narcosis*

The physiological and psychological changes in man due to breathing air under hyperbaric conditions comprise a syndrome of neurological and physiological dysfunctions manifested primarily as decreased cognitive and psychomotor ability and behavioral and neurological disturbances. It is the nitrogen in the air under increased pressure that acts as a narcotic agent displaying a general depressant effect upon the body. Behnke *et al.* (1935) were the first to attribute the detrimental effect of hyperbaric air upon cognitive and psychomotor performance to the raised partial pressure of nitrogen. The signs and symptoms have often been compared to those of alcohol intoxication, and their severity depends primarily on the pressure, or depth, at which the air is breathed. Intoxication begins at approximately 100 fsw.

At moderate depths (100–200 fsw) a person breathing compressed air exhibits a delayed response to auditory and visual stimuli and concentration is difficult (Behnke *et al.* 1935). There is a tendency toward idea fixation and a loss of clear thinking accompanied by impaired neuromuscular coordination (Case and Haldane 1941). Subjective effects seem to appear before objective changes in performance. Few divers can work very effectively beyond 200 fsw, and only very exceptional or well-adapted individuals can accomplish useful work at 300 fsw. At 400 fsw symptoms include euphoria, manic or depressive states, a sense of levitation, disorganization of the time sense, other psychosensory phenomena, and in some cases psychotic behavior (Adolfson 1967; Hamilton 1973). The effects appear to reach a maximum within 2–3 min (Case and Haldane 1941; Barnard *et al.* 1962; Bennett and Glass 1961), do not appear to increase with time at constant pressure, and dissipate almost immediately upon return to ambient, or near ambient, pressure (Kiessling and Maag 1962). Individuals exhibit considerable difference in susceptibility (Cousteau 1953) and frequent exposure affords some adaptation (Shilling and Willgrube 1937).

Saturation diving procedures and the use of undersea habitats have resulted in the prolonged exposure of numerous divers to air or nitrogen–oxygen environments at saturation pressures as great as 120 fsw. Adaptation of divers to these environments is of interest in two aspects—the effect of the nitrogen environment on day-to-day performance, and the possible protection afforded against exposure to increased nitrogen pressures on excursions. A recent experiment approached both these questions (Schmidt *et al. in press*).

1. Normoxic nitrogen–oxygen habitats over prolonged periods do not appear to result in any degradation of cognitive and psychomotor function. Rather, adaptation appears to be almost total, at least at pressures up to 100 fsw. [This is in agreement with an earlier laboratory study conducted at the University of Pennsylvania by Elcombe and Teeter (1973).]

2. Performance on compressed air excursions from nitrogen habitats is significantly improved over performance on *bounce* dives made from the surface to the same depth. This effect is especially pronounced at pressures equivalent to 200 and 250 fsw, from saturation in the range of 60–90 fsw.

The following discussion relates primarily to the physiological aspects of inert gas narcosis. Those aspects related to performance decrement will be discussed in Chapter VI.

1. *Narcotic Potency and Possible Mechanisms*

Many gases otherwise chemically *inert* can produce narcosis or anesthesia. They have been the subject of many studies and theories of this effect. Attempts have been made to correlate the narcotic potency of gaseous and liquid anesthetics with various physical properties. Among them are lipid solubility (Meyer 1899, Overton 1901), partition coefficients and molecular weight (Behnke and Yarbrough 1939), adsorption coefficients (Case and Haldane 1941), thermodynamic activity (Ferguson 1939, Brink and Posternak 1948), and the formation of clathrates (Miller 1961, Pauling 1961). Featherstone and Muelbaecher (1973) have stressed that the majority of the above physical properties are simply reflections of van der Waals forces—that is, the relatively weak attractive forces operative between neutral atoms and molecules arising because of the electric polarization induced in each of the particles by the presence of other particles.

By far the most satisfactory correlation is with lipid solubility, which has come to be known as the Meyer–Overton hypothesis. In effect, it states that the more soluble an agent is in lipid, the more potent it will be as an anesthetic. Excellent correlation between the anesthetic potency of gaseous or volatile anesthetics and their lipid solubility have been obtained, both with the inert gases (Bennett 1969) shown in Table IV-19, and with a variety of other gases and vapors, including clinical anesthetics (Miller *et al.* 1967). Since nerve cells are richer in lipid, the narcotic may gain access to nerve tissue by virtue of its lipid solubility (Bennett 1966). Wulf and Featherstone (1957) and Sears (1962) have suggested that the inert gases produce narcosis by an effect on the lateral spacing of the lipid molecules, causing an interference with the permeability of ions across the cell membranes.

Clements and Wilson (1962) studied the affinity of six narcotic gases for lipoprotein and fatty acid monolayers and concluded that there was a correlation between a standard effect of anesthesia and a decrease in the interfacial tension in biological systems. Bangham *et al.* (1965) investigated the penetration of phospholipid monolayers by *n*-alkyl alcohols, ether, and chloroform from an aqueous substrate and related the degree of such penetration to an increase in the permeability of phospholipid bilayers to cations. Bennett *et al.* (1967) then established that both nitrogen and argon penetrate lipid membranes at high pressures and that the extent of the absorption, related to narcosis, was in agreement with the findings of Clements and Wilson (1962). These findings indicate the possibility that raised partial pressures of inert gases may penetrate membranes in lipid areas of the body (particularly nerve cells), causing the membranes to swell.

Such findings are compatible with the recently observed phenomenon of the "pressure reversal of anesthesia" (Johnson and Miller 1970; Lever *et al.* 1971), in which animals in a narcotic or anesthetized state return to a spontaneous, non-anesthetized condition when subjected to sufficient increases in pressure without a change in the dose of anesthesia or narcotic gas. A hypothetical etiology for such a

Table IV-19

Correlation of Narcotic Potency of the Inert Gases with Lipid Solubility and Other Physical Characteristics[a]

Gas	Molecular weight	Solubility in lipid	Temp., °C	Oil–water solubility ratio	Relative narcotic potency
He	4	0:015	37	1:7	4:26[b]
Ne	20	0:019	37.6	2:07	3:58
H₂	2	0:036	37	2:1	1:83
N₂	28	0:067	37	5:2	1
Ar	40	0:14	37	5:3	0:43
Kr	83.7	0:43	37	9:6	0:14
Xe	131.3	1:7	37	20:0	0:039[c]

[a] Reprinted from Bennett (1969) with permission of Bailliere and Tindall.
[b] Least narcotic.
[c] Most narcotic.

phenomenon is that in the anesthetized state, certain membranes in the central nervous system are in an expanded state (caused by the absorption of molecules of an inert substance) and that the application of pressure compresses the membrane back toward the unanesthetized, unexpanded state. The degree of expansion necessary to produce anesthesia and the pressure necessary to reverse the effects have been quantitatively studied by Miller *et al.* (1973) and their conclusion has been termed the "critical volume" hypothesis. Miller *et al.* show that observed data fit their model better than they do the simpler Meyer–Overton model.

Another recent theory or model, the lipid "free volume" hypothesis, has been proposed by Stern and Frisch (1973). According to this hypothesis, narcosis occurs when the inert gas dissolved in the lipid phase causes the free volume to exceed a specific threshold value. The anesthetic potency of a gas depends not only on its lipid solubility, but also on the thermal expansivity and compressibility of the lipid phase as well as on the environmental temperature and hydrostatic pressure. This model agrees with the pressure reversal of anesthesia as well as the Meyer–Overton hypothesis.

2. Interactions

Although it has been demonstrated that the increased partial pressure of nitrogen is a principal cause of compressed air effects, other factors will potentiate the effect of the nitrogen upon human physiology and performance. Few, if any, of these factors operate alone. Most of them are influenced unfavorably when others are operative and a vicious cycle of interactions may possibly ensue.

a. Carbon Dioxide

When a diver uses breathing equipment or occupies a confined space, carbon dioxide may become toxic and cause serious problems if allowed to accumulate. An excess of carbon dioxide can result in (among other things) confusion, an inability to think clearly, drowsiness, and loss of consciousness.

That the combined effects of compressed air and carbon dioxide are much more severe than those of either alone was demonstrated by Case and Haldane (1941). They found that 3–4% carbon dioxide caused no deterioration in manual or arithmetic skill at atmospheric pressure. When air containing about 0.4% carbon dioxide was breathed at 10 ATA (and therefore with a carbon dioxide partial pressure equivalent of about 4%), there was a marked deterioration in manual dexterity and a good deal of confusion among the subjects. When breathing carbon dioxide at partial pressures equivalent to 6.6–9.7% in air at 10 ATA, eight subjects lost consciousness in 1–3 min. The authors concluded that carbon dioxide in air at 10 ATA should be kept below 0.3%. It was evident that carbon dioxide had a marked synergistic effect upon the symptoms of compressed air, but there was no conclusive evidence that this synergism was with the nitrogen.

(i) Rate of Compression. In general, the effects of compressed air will appear sooner and be of greater severity if compression rate is rapid. Albano (1962) described

12 cases where the diver had to stop his descent and return to the surface due to intoxication at depths which had formerly been attained without any difficulty. The one common factor in all these cases was the especially rapid rate of descent. This may have resulted in an unusually large increase in alveolar and cerebral carbon dioxide, and the resultant synergistic effect with the nitrogen. Further support for this contention was obtained by Cabarrou (1964), who measured alveolar carbon dioxide during rapid compression to 5.9 atm. An increase of alveolar carbon dioxide to 7.5–8% was found, which returned to 5.3–5.4% within 4–8 min at pressure. If compression was slow (approximately 30 min), no such alveolar carbon dioxide was found.

(ii) Exercise. Since the high density of gases under hyperbaric conditions increases the work of breathing and thus limits the ventilatory response for exercise, it follows that exercise would increase compressed air dysfunction. Adolfson (1967) has demonstrated that this is indeed the case. Figures IV-32 and IV-33 show the effect of exercise on manual dexterity and arithmetic calculating ability in a dry compression chamber as a function of compressed air pressure.

b. Hypoxia and Hyperbaric Oxygen

Although many of the symptoms of compressed air dysfunction are similar to those of hypoxia (idea fixation, etc.), hypoxia is not likely to be the cause of compressed air effects since air at 10 ATA has an oxygen partial pressure of 2 ATA. Oxygen at

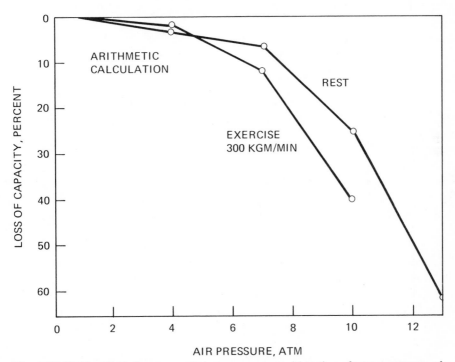

Figure IV-32. The effect of compressed air pressure upon arithmetic performance at rest and during exercise. [From Adolfson (1967) by permission of Almqvist and Wiksell, Stockholm.]

raised partial pressures has a toxic effect upon the body, including chronic changes in lung pathology, and increased excitability of the nervous system, which can culminate in convulsions. The extra oxygen in 10 ATA of air is equivalent to 100% oxygen at 2 ATA, but 2 ATA of oxygen produces none of the signs and symptoms of breathing air at 10 ATA.

The body effect of nitrogen and oxygen was evaluated by Frankenhaeuser *et al.* (1963). They varied the partial pressures of the two gases independently and determined the resultant effect upon performance. An increase in air pressure from 1 to 5 ATA caused only a slight tendency toward impaired performance. However, when a similar rise in inspired partial pressure of nitrogen was combined with a greater rise in inspired partial pressure of oxygen, a more pronounced impairment was induced in two of the performance variables (simple reaction time and mirror drawing error score). At a constant level of nitrogen pressure of 3.9 ATA the simple and choice reaction times tended to show slower responses with increasing oxygen pressure.

In performance tests the observed synergistic relationship between hyperbaric oxygen and nitrogen is probably not due to any direct synergism between the two, but rather to the oxygen excess enhancing carbon dioxide retention. This is thought to

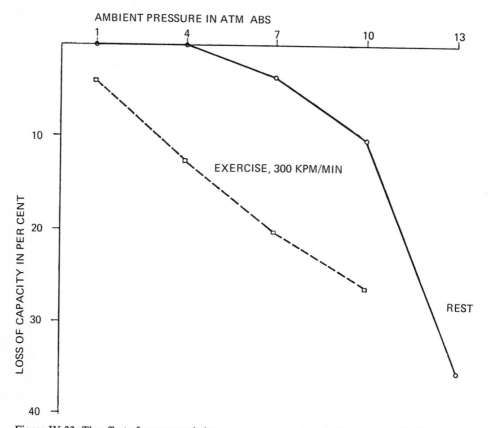

Figure IV-33. The effect of compressed air pressure on manual dexterity at rest and during exercise. [From Adolfson (1967) by permission of Almqvist and Wiksell, Stockholm.]

occur possibly at the ventilatory level, or more likely, by interference with carbon dioxide elimination from tissues due to hemoglobin saturation. This resultant increase in retained carbon dioxide would, in turn, potentiate the effects of nitrogen. If so, the effects of increased oxygen would tend to be more severe at higher partial pressures of nitrogen. Such an effect was observed by Frankenhaeuser *et al.* (1963).

3. *Neurological Measurements*

a. EEG Activity

The electroencephalograph (EEG) is a measurement of frequency displays of cerebral activity recorded through the intact skull. Frequency is designated in cycles per second (Hz) or by Greek letters indicating broad wave bands—the delta band (0.5–3 Hz), the theta band (4–8 Hz), the alpha band (8–13 Hz), and the beta band (14 Hz and higher). A normal, alert subject exhibits continuous spectra of frequencies starting below 1 Hz and becoming gradually imperceptible beyond 40–50 Hz. Relative energy diminishes as frequency increases. Tracings obtained while a subject's eyes are closed ordinarily show a major peak in the alpha band (the principal resting rhythm of the brain) and a minor peak in the beta band. This is shown in Figure IV-34. The alpha rhythm is readily disrupted by visual attentiveness and may be blocked continuously while the eyes are open. Theta and delta waves are also generally inconspicuous in tracings of normal, alert subjects (Kooi 1971).

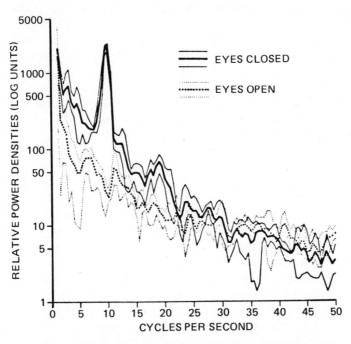

Figure IV-34. EEG frequency spectrum of normal adult. The alpha and beta peaks are shown clearly in the eyes-closed condition. [From Kooi (1971) by permission of Harper and Row.]

Roger *et al.* (1955) reported EEG changes in men exposed to 10 ATA of air. An increase in frequency and decrease in amplitude were noted. Bennett and Glass (1961) reported similar observations in men breathing air at 10 ATA. Although changes vary in details with different agents and different subjects, the most consistent changes in EEG potentials resulting from the administration of anesthetic agents are progressively decreasing frequency and increasing amplitude (Kooi 1971). Indeed, breathing of normoxic nitrogen at 10 ATA produces a decrease in frequency and an increase in voltage with the appearance of theta (slow) wave activity (Albano 1962, Albano and Criscuoli 1962). Thus, the observed EEG changes due to air at 10 ATA appear to be opposite to those produced by nitrogen and anesthetic agents in general. Albano and Criscuoli (1971) expressed the opinion that the neurophysiological effects of exposure to compressed air appear to be like a polymorphous syndrome, with the symptoms of nitrogen narcosis appearing at lower pressures and the neurotoxic effects of oxygen predominating at greater pressures.

The effect of excess oxygen, and of convulsants in general, on the EEG is a desynchronization of alpha activity with appearance of rhythmic theta activity of increasing intensity with a preepileptic phase characterized by the emergence of bursts of slow-wave, sharp discharges, or spike-wave complexes (Kooi 1971, Albano and Criscuoli 1971). Excess carbon dioxide likewise produces a slowing of EEG frequency. Thus EEG changes due to breathing of air at 10 ATA, increased frequency and decreased amplitude, are different from the decrease in frequency due to increased partial pressures of nitrogen or oxygen (or excess carbon dioxide).

That nitrogen plays a primary role in the psychophysiological effects of hyperbaric air is undisputed. However, as the body undergoes simultaneous increases in nitrogen, oxygen, carbon dioxide, and hydrostatic pressure, the net effect of the simultaneous presence of all these environmental factors on the EEG is different from that produced by any of them alone.

Schmidt (1973) noted that certain psychotropic drugs produce an effect upon the EEG similar to that due to hyperbaric air at 10 ATA—attenuation of alpha activity and acceleration of alpha frequency—and Adolfson (1967) compared the effects of air beyond 300 fsw to effects of these same drugs.

b. Alpha Blocking and Critical Fusion Frequency (CFF)

A normal, alert subject usually exhibits a major component of alpha activity in his EEG when he is resting with his eyes closed. This alpha activity is readily disrupted by visual attentiveness and may be blocked continuously while the eyes are open.

The effect of hyperbaric air on EEG and alpha blocking was studied by Bennett and Glass (1961). Subjects who exhibited a clear alpha blocking response during the performance of mental calculations at ambient pressure were compressed to pressures ranging from 2.5 to 7 ATA. At 7 ATA loss of the alpha blocking response occurred immediately and continued while the subjects were at pressure. When subjects were returned to atmospheric pressure alpha blocking was not as intense as before compression, but eventually returned to normal. At pressure, arithmetic problem-solving ability showed a significant decrease. At pressures less than 7 ATA, time from start of compression until abolition of the alpha blocking response varied inversely with the pressure.

The effect of compressed air on brain function may be detected by other means than the EEG. Bennett (1958) demonstrated that simultaneously with abolition of alpha blocking there is a change in the frequency at which a subject perceives a flickering light to be steady. This frequency, called the critical fusion frequency (CFF), was found to increase with increasing air pressure.

c. Evoked Brain Responses

The evoked brain response (EBR) is obtained, like the EEG, by attaching recording electrodes to the scalp; it measures cerebral reaction to sensory stimulation. Recently attempts have been made to utilize potentials evoked in the brain by stimuli such as a sharp sound (the auditory evoked response or AER) and a flash of light (the visual evoked response or VER) in the investigation of compressed air dysfunction.

A typical AER recording, shown in Figure IV-35, consists of four major components. An initial deflection P_1, with a latency of about 50 msec, is followed by a negative deflection N_1 and a second positive deflection P_2. A second negative deflection N_2 and a further positive deflection are sometimes found. For a given individual, the form of these potentials is usually consistent and reproducible.

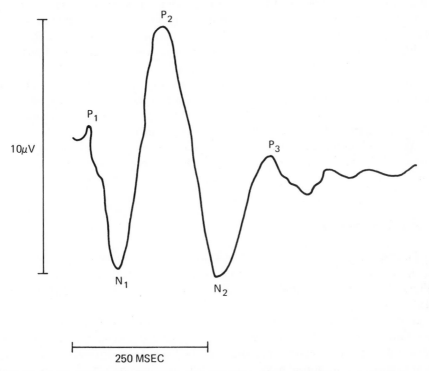

Figure IV-35. A typical auditory evoked response. [From Bennett *et al.* (1969) by permission of Bailliere and Tindall, London.]

Bennett *et al.* (1969) reported on the exposure of divers to compressed air in a hyperbaric chamber. The number of arithmetic problems attempted and the number correct, together with the spike height of the N_1P_2 complex of the AER, were obtained at 50-fsw increments from 100 to 300 fsw. Since the mean percentage decrement of all of these parameters increased with increasing pressure, Bennett concluded that the magnitude of the decrement of the N_1P_2 complex of the AER was a valid measurement of inert gas narcosis.

Using visual responses evoked 12–16 times per second, Kinney and colleagues (Kinney and McKay 1971, Kinney *et al.* 1971) found a consistent reduction in amplitude and regularity of waveform when subjects were exposed to compressed air corresponding to increasing depths. Little or no decrease was seen in equivalent exposure to helium–oxygen.

Ackles and Fowler (1971) used both the AER and the VER to examine further the validity of the EBR as a measure of inert gas narcosis. They included the VER because the decrement noted by Bennett in the AER was possibly due to peripheral attenuation of the auditory stimulus in the hyperbaric conditions. Tests were performed at 4 and 7 ATA, using compressed air and a corresponding oxyargon mixture (argon having about twice the narcotic potency of nitrogen). They found that the number of arithmetic errors increased with increasing pressure for both breathing mixtures, the greatest decrement occurring with the argon mixture. On the other hand, although both the AER and VER depressions increased with increasing pressure, no differentiation between the two breathing mixtures was obtained. They concluded that there was no correlation (calculated using individual values) between the arithmetic test and the EBR depression for either gas mixture. Kinney *et al.* (1972) suggest that EBR responses may approach an asymptote and that Ackles and Fowler were beyond the sensitivity of the method.

Langley (*in press*) commented that stimulation of the auditory and visual systems under hyperbaric conditions is not free of peripheral, non-CNS factors, which may change the effective stimulus intensity perceived by the subject, and proposed the elicitation of EBR's through stimulation of the somatic system.

We are left with the conclusion that evoked brain responses, alpha blocking, and perhaps critical fusion frequency are all potentially useful tools for objective assessment of narcosis, but that they must be used carefully and interpreted critically.

4. Predisposition

There exists considerable individual susceptibility to hyperbaric air dysfunction. One diver may not be affected at 6 ATA while another may be severely affected at 4 ATA. There is evidence that this predisposition may be due to both physiological and psychological factors.

In 1933 the British Admiralty conducted a number of physiological tests, including a breath-holding test (Hill *et al.* 1933). They concluded that divers who exhibited a predisposition to compressed air dysfunction had some physical imperfection, and it was suggested that a thorough physical examination be required of all diving candidates. However, there was also evidence that physical health was not the sole factor. A breath-holding test showed some correlation with the susceptibility of an individual. Besides being a physiological test, this is also a test of will power and may relate to psychological factors. Cousteau (1953) observed that low mental ability and emotional instability enhance an early and extreme reaction to hyperbaric air.

5. *Adaptation*

There is evidence that frequent exposure to hyperbaric air affords considerable adaptation to its effects. An experienced diver described the following (Zinkowski 1971):

> A commonly observed phenomenon relating to nitrogen narcosis is that you apparently build up a resistance to its effects during consecutive exposures. Many divers have commented that after not having dived for a week or so, they are definitely groggy during their first dive to 200 feet. On the second dive the effects are noticeably less and after the third or fourth dive they are unaware of any deleterious effects whatsoever.

The effect of acclimatization on mental performance under hyperbaric air was first quantitatively studied by Shilling and Willgrube (1937). Their results are shown in Figure IV-36. Although there is no quantitative data on the rate of adaptation (or just as important, rate of loss of adaptation), Miles (1965) reports that weekly dives are necessary to maintain adaptation, and the USSR *Manual of Scuba Diving* (Bulenkov *et al.* 1968) recommends conditioning in a chamber at 7 ATA once or twice a month.

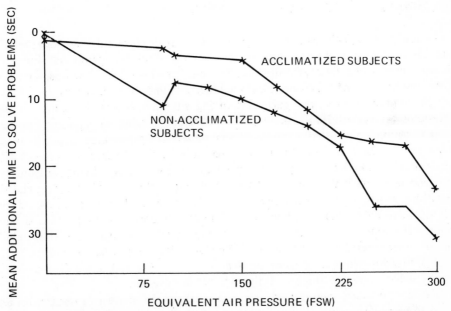

Figure IV-36. Mean additional time required to solve arithmetic problems as a function of air pressure for both acclimatized and nonacclimatized subjects. Data from Shilling and Willgrube (1937).

E. *The Physiology of Breathing Mixtures*

This section is concerned with the elements involved in the choice of breathing mixtures for diving. Two aspects are obligatory from the beginning: A diver's breathing gas must be supplied under pressure approximately equal to the pressure in the diver's upper respiratory tract; and all breathing mixtures must contain a high enough partial pressure of oxygen to maintain effective life. Beyond this the tradeoffs begin, and the ideal choice of gas is a compromise of several factors. These factors—*oxygen toxicity, metabolism, inert gas narcosis, HPNS, density, voice distortion, thermal properties, decompression, fire safety, cost,* and *logistics*—are covered by topic in this section.

It should be noted that this section relates to the physiological factors and does not cover the quantitative guidelines needed in order to plan a dive, design equipment, or operate a dive. There is no presentation of type of equipment, or of the precautions or procedures for carrying out gas mixing or handling. All of this material is presented in detail in two U. S. Navy publications: The *U. S. Navy Diving-Gas Manual,* Second edition (1971), prepared by Battelle Columbus Laboratories for the U. S. Navy Supervisor of Diving (NAVSHIPS 0994-003-7010), and the *U. S. Navy Diving Manual,* 1970 edition (NAVSHIPS 0994-001-9010). They are available from the Superintendent of Documents, U. S. Government Printing Office, Washington, D. C. 20402.

1. *General Planning Considerations*

Careful planning is necessary for each mission since current diving modes require the selection of optimal breathing mixtures for each of several operational situations. These include scuba diving, the use of closed and semiclosed breathing rigs, the use of hose-supplied gas, and the use of both submersible and deck chambers. In addition to the equipment used, the particular diving situation affects the choice of gas. Physiological factors also are involved; these include duration of exposure, temperature, work load, and whether the diver is immersed in a gas or breathing it by mask. Figure IV-37 shows a sequential block diagram indicating the information and decisions required to adequately plan the gas requirements for any diving mission. The first consideration is the working depth, which affects the type of diving rig and the breathing gas to be used. Compressed air is used in shallow water missions, and $He-O_2$ mixtures at depths below 190 ft. The next important consideration is manhours, which must be estimated as accurately as possible, especially if bottled helium or any other expensive gas is to be used. The breathing gas selected depends upon the type of breathing equipment to be used and this in turn depends upon the location of the hyperbaric activity: shipboard divers to habitat dwellers.

It is particularly important to decide in advance the number of inspections and trials to be made, for these use gas. Careful calculations should be made so as to have sufficient on hand. Lastly, it is necessary to consider all of the items in the *operations plan* block.

A helpful aid in the planning of any diving mission is a checkoff list to ensure that all areas of significant gas consumption are adequately planned for. Such a list is shown in Figure IV-38. It is obvious that not all listed factors will apply to each mission, and some factors may have been omitted in the case of special missions. The point is that planning and checking are most important.

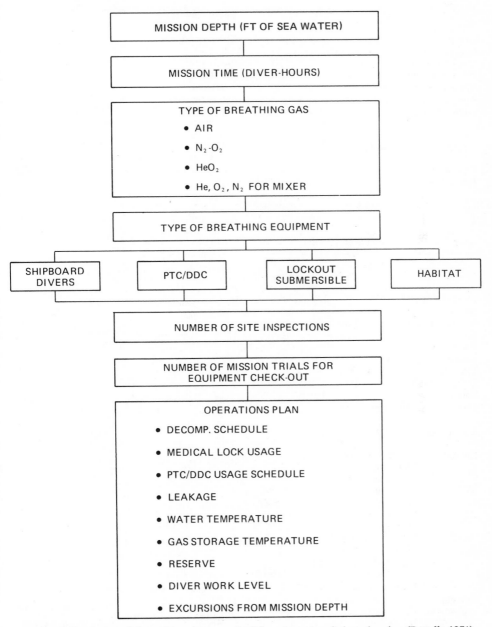

Figure IV-37. Sequential considerations required for adequate mission planning (Battelle 1971).

OPERATIONAL PARAMETERS: FOR EXAMPLE PROBLEM

Operational depth, ft	450
Excursion depth (max), ft	450
Type of breathing equipment	
Open circuit-free flow	
Venturi recirculating	
Open circuit demand	
Semiclosed circuit	
Closed circuit	X
Gas storage pressure, psia	3000
Gas supply pressure (min), psia	700
Gas storage temperature, °F	70
Gas usage temperature, °F	70
Allowable gas storage leak rate	
Allowable environment leak rate (He)	1 fsw per hr

PERSONNEL PARAMETERS

Number of divers	2
Hours to complete mission	144 + 94 Decomp.
Metabolic oxygen provisions	3 SCF hr/div.

DIVER GAS REQUIREMENTS, lb

Mission tasks	Usage rate	Time	O_2 He N_2 Air
Site inspections			
Sea trials			
Mission diving	6 days	72	
Support diving			
Safety reserve			
Cleanup & recovery			
Decompression	94 hr	47	
Treatment reserve	94 hr	70.5	
Totals		189.5	

EQUIPMENT VOLUMES, ft³

Habitat	Living chamber	
	Wet room	
	Adsorption scrubber	
PTC	Chamber	143.8
	Trunk	8.2
DDC	Living chamber	2 at 321
	Medical lock	2 at 1.6
	Adsorption scrubber	2 at 2.9
EL	Trunk to No. 1 chamber	3
	Trunk to No. 2 chamber	3
	Lock chamber	103.8
	Trunk to atmosphere	9.3
Recomp chamber	Inner lock	
	Outer lock	
	Medical lock	

EQUIPMENT GAS REQUIREMENTS, lb

Mission tasks	Number used	O_2	He	N_2	Heliox mixture	Air
Test breathing equipment	—	—	—	—	—	—
Charge and test PTC	1		129.5			
Charge and test DDC	1					24
Charge and test habitat						
Precharge mission equipment						
Safety reserve gas supply	1		108.2			
Recompression chamber						
DDC medical lock use	16/day		35.8			
Habitat leakage						
DDC leakage			71.3			
Supply gas leakage						
Decompression reserve gas supply	1		34.9			
PTC leakage						
Absorbent changing						
PTC mating	6		14.7			
Sampling and bleed gas			8.4			
Totals			402.8			24

Figure IV-38. Example program—
mission planning (Battelle 1971).

2. Properties of Gas

The physical properties of the various gases are intimately involved in the choice of the best ones to use for diving. Gas properties are summarized in Tables IV-20 and IV-21. Detailed listings of properties of the gases and gas mixtures traditionally used in diving, as well as details of gas mixing and the use of breathing apparatus, are contained in the *U. S. Navy Diving Gas Manual* (Battelle 1971). Conversion factors for the units of pressure and volume used in gas computation are given at the end of the book.

3. Factors Relating to Choice of Gas

a. Oxygen: Metabolic Needs and Toxicity

The fundamental requirement in a breathing gas is for oxygen, to meet the metabolic needs of the body. These needs are met by supplying a breathing mixture containing an adequate partial pressure of oxygen.

The partial pressure of oxygen P_{O_2}, considered "normoxic" and to which man is adapted, is 0.21 atm inspired. A healthy person can maintain blood oxygenation at an inspired oxygen partial pressure of about 0.16 atm; below this point a relative hypoxia (called by some "anoxia") will prevail—this is discussed further below. There was concern for a time that dense mixtures (e.g., 80 atm He–O_2, or 12 times the density of air at sea level) could cause a diffusion limitation in the lung ("the Chouteau effect," Chouteau *et al.* 1967) and that this could render a P_{O_2} of 0.21 atm inadequate. Subsequent experiments, however, involving densities twice this value showed no evidence of this limitation (Strauss *et al. in press*).

The upper toxicity limit for oxygen breathing is discussed in detail later in this section. In practical terms oxygen limits can be set for in-water work, for chambers, and for habitats. Where decompression is to follow, it appears that an advantage lies in maintaining the highest possible P_{O_2} during the pressure exposure. For a lone diver in deep water and working hard, a limit of 1.2 atm P_{O_2} is a reasonable compromise between toxicity and optimal decompression. Under less strenuous conditions up to 2.0 atm may be tolerable. These are practical limits which consider such factors as mixing variations and analytical errors. In-water work at higher P_{O_2} than this may be risky. Some experienced diving operators avoid the use of pure oxygen in the water at all.

For a habitat the same lower limits (0.16 atm) should prevail; pressures between 0.25 and 0.4 atm seem to be an optimal operational level for indefinite periods, while pressures up to 1.0 atm may be maintained safely for a few hours. During treatment for decompression sickness using the U. S. Navy Oxygen Treatment Schedules (see Chapter VII) a diver is exposed to 2.8 atm for several 20-min periods. This has been found to be tolerable to most divers and hence operationally suitable, but a certain risk of convulsion is nevertheless present. Figure IV-39 gives in graphic form the ranges of oxygen that are acceptable for diving.

Table IV-20
Critical Properties of Gases

	Hydrogen H₂	Helium He	Neon Ne	Nitrogen N₂	Oxygen O₂	Argon Ar	CF₄	SF₆
Molecular weight	2.016	4.003	20.183	28.016	32.000	39.944	88.01	146.07
Density at 0°C, 1 atm, g/liter	0.0056 (lb/ft³)	0.1784	0.9004	1.251	1.429	1.784	—	6.139 (70°F)
Viscosity at 0°C, 1 atm, micropoise	89.2 (28.1°C)	194.1	311.1	175.0	201.8 (19.1°C)	221.7	—	145 (2.11°C)
Thermal conductivity at 0°C, 1 atm, cal/°C-cm²-sec	39.7×10^{-5}	34.0×10^{-5}	11.04×10^{-5}	5.66×10^{-5}	5.83×10^{-5}	3.92×10^{-5}	—	—
Specific volume, 70°F, 1 atm, ft³/lb	192	96.7	19.2	13.8	12.08	9.67	4.4	2.5
Specific heat C_p, cal/mole deg	3.39 BTU/lb °F (60°F)	4.968	4.968	6.95	6.97	4.968	0.132 cal/g °C (−80°C)	22.93
Solubility in water at 38°C, cm³/1000 g	16	8.6	9.7	13	28.9 (25°C)	26	4.3	5 (25°C)
Solubility in oil at 38°C, cm³/1000 g	50 (40°C)	15	19	61	120 (40°C)	140	72	250

Table IV-21

Comparative Properties of Suggested Gases for Use in Diving
(Billings 1973)

Properties	Argon	Nitrogen	Neon	Helium	Hydrogen
Decompression	Not good—can only be used under special circumstances	Good in short dive; slow return from long dives	Fairly easy to eliminate in long dive; builds up slowly in short dive	Easily eliminated from body in long dives; builds up fast in short dive	About the same as He
Narcosis	Very narcotic in nominal diving range	Narcotic beyond about 200 fsw	No narcosis	No narcosis	Slight narcosis at great depths
Voice distortion	Makes voice deep	Normal	Nearly normal	Large distortion	Large distortion
Thermal conductivity	Good insulation	Fair to good insulation	Fair insulation	Poor insulation	Very poor insulation
Breathability	Hard to breathe	Hard to breathe at depth	Relatively easy to breathe	Can be breathed to 5000 fsw	Same as He
Cost and worldwide availability	Low cost, readily available	Lowest cost, available anywhere	High cost, depending on purity; available many places	Moderate cost, available only in certain places	Low cost, readily available

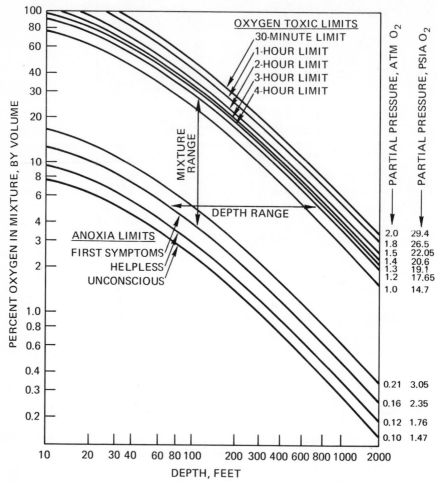

Figure IV-39. Percentage of oxygen in breathing mixture as a function of depth and oxygen partial pressure. Percent O_2 = partial pressure O_2 (atm) $\times$ 100/sea water pressure (atm).

Another aspect of the oxygen requirement of a diver, in addition to a suitable partial pressure range, is the amount of oxygen needed to supply metabolic needs. In the diving situation where oxygen is usually supplied at tensions near the upper limits of tolerance and where CO_2 elimination is usually the predominant factor controlling breathing, metabolic needs are easily met. Semiclosed breathing rigs and conventional helmets, as well as all closed chambers, must be supplied with oxygen in accordance with the physiological demands of the diver. The oxygen actually consumed by the body varies from about $\frac{1}{2}$ liter/min at rest, to about 1 liter/min for moderate work, to 3 or more liters/min for very heavy work. The requirements for supplying the various types of gas with the appropriate oxygen flow are given in detail in the *U. S. Navy Diving Gas Manual* (Battelle 1971).

Fire risks of high-oxygen environments must be considered and are discussed briefly later in this section, and in more detail in Chapter VIII, under fire safety.

b. Inert Gas Narcosis

As with oxygen, the matter of inert gas narcosis is of sufficient importance and complexity to merit a separate section (see Section D, this Chapter). The gas of primary concern here is nitrogen, or in particular, air. There is no established upper limit for the safe breathing of nitrogen; a pressure of 13 ATA has been shown to cause severe narcosis but is not incompatible with survival (Adolfson 1967).

It is generally considered safe to dive to about 200 fsw breathing air. This produces a noticeable narcosis on the first exposure, but a great deal of accommodation is seen in the next few dives to the same depth if they follow no later than a day or two. A diver living in an undersea habitat acquires a resistance to narcosis related to the partial pressure of nitrogen in the habitat.

Nitrogen is believed to *interact* with high levels of both oxygen and carbon dioxide; these gases apparently increase the susceptibility of the diver to the narcotic effects of nitrogen.

Helium and neon are nonnarcotic. Hydrogen apparently causes a narcosis at very high pressures that is balanced by the hydrostatic pressure effect. All other gases are more narcotic than nitrogen; this category includes oxygen, but the narcotic properties of oxygen are not manifest until well beyond the usual CNS toxicity limits. Argon has been used in diving; it is about four times as narcotic as nitrogen.

c. HPNS and Hyperbaric Arthralgia

The high pressure nervous syndrome is an increasingly important consideration as depths exceed about 600 fsw. It is invoked by gases with low lipid solubility, hence its early designation, "helium tremors." Helium and, to a slightly lesser extent, neon are associated with HPNS because they allow high hydrostatic pressures to be applied without a compensating narcotic effect. In situations where HPNS is a problem, its effects may be counteracted by including in the breathing mixture a gas with sufficient lipid solubility to counterbalance the compression effect (presumably in the lipid component of cell membranes.) The optimal doses are not yet established but it has been shown that rapid compressions to 1000 fsw cause far less tremor and other manifestations of the syndrome when 12–25% nitrogen is included in the inert component of the breathing mixture. Lesser percentages of gases that have higher anesthetic potency might also be used, with the effect on decompression the deciding factor.

Another direct pressure effect is hyperbaric arthralgia. The mechanism of this phenomenon is likely to be different from that of HPNS—perhaps due to osmosis—and the mitigating effect of narcotic gases has not been demonstrated. Slow rates of compression seem to be the best procedure for managing the problem or, for short jobs, the opposite approach of a rapid compression and faster decompression. In any case, until hyperbaric arthralgia is better understood the choice of gas will most generally be determined by other operational considerations.

d. Density and Viscosity

As a diver's depth increases, the density of his breathing gas increases as an almost linear function of his absolute pressure. High gas densities act to limit the diver's ability to ventilate his lungs and to increase the work required; this effect acts on both the diver and his equipment. The viscosity of a gas, however, does not change appreciably with depth, and the viscosities of most respiratory gases are similar (Lanphier 1969). Breathing a mixture of dense gases at sea level has been shown experimentally to be essentially identical to breathing a lighter gas at increased pressure, provided that the densities are equivalent (Maio and Farhi 1967). Respiratory mechanics are covered in Chapter I. The salient facts are that during laminar flow the pressure required to cause gas to flow is a linear function of viscosity, but that during turbulent flow pressure must increase approximately with the square of the flow. Normal breathing at rest involves some turbulent flow and consequently the work of breathing is increased when denser gases are breathed. This effect is exaggerated during exercise, when respiratory flow rates are increased.

In diving the physiological effect of a limitation in breathing capability is slightly different from that at sea level in that oxygen pressures are usually more than adequate to meet the metabolic needs. Restrictions on lung ventilation, therefore, cause only a buildup in CO_2. The rate of removal of CO_2 is entirely a function of ventilation volume flow $\dot{V}_E$, whereas oxygen can be delivered at very low rates of ventilation.

Another important factor in respiratory mechanics is the concept of airway collapse. In a forced expiration there is a maximum possible rate of gas flow beyond which extra effort will not provide extra flow, but instead will force temporary constriction of the airways. This situation is reached more easily with denser gases and, although a person does not require these high flow rates under normal circumstances, at high densities the maximum flow capacity and minimum ventilatory requirement tend to converge. The result is that at increasing pressures a limit will be encountered of the amount of work which a diver can do. Artificial respiratory assistance is not likely to be very helpful other than in doing the extra work of overcoming the resistance of the breathing apparatus.

Gas densities as great as 25 times that of sea-level air have been breathed experimentally (Lambertsen *in press*); it was found that work was restricted to less than three-fourths the sea-level maximum for the individual, but that this amount of work could be accomplished at densities up to about 18 times normal. These experiments were conducted with equipment having low resistance and low dead space; with standard diving gear, problems will be encountered at much lower densities, especially if gas-delivery lines are long.

The same gas properties prevail in breathing equipment as in man and both density and viscosity of the gases to be used must be considered in engineering design. It is worth pointing out that equipment must be designed to provide the maximum desired *peak flow* during a breathing cycle, rather than just the average minute volume.

Ordinarily, breathing-gas density is not a limitation in the use of air or nitrogen–oxygen mixtures (or argon) since narcosis becomes a definite problem before density becomes limiting. This may not be the case where deeper depths can be tolerated as a result of adaptation to nitrogen in a hyperbaric habitat. Also, the density of air at

200–300 fsw is enough to be difficult to breathe through inadequate equipment or at inadequate supply pressures. One other difficulty with air is that any CO_2 accumulation which might occur can interact unfavorably with the effects of both the nitrogen and the oxygen of air.

One of the most desirable characteristics of helium is its low density and the consequent ease with which it is breathed. The same applies to hydrogen. Experiments at the University of Pennsylvania have shown that density equivalent to that of a helium–oxygen mixture at 5000 fsw can be tolerated at light work levels (Lambertsen *in press*). Because narcosis does not limit the depth at which it can be used, neon mixtures can pose density limitations to both equipment and man. A good practical depth limit for Neon 75 (75% neon, 25% helium) using current equipment is 650 fsw.

e. Voice Distortion

Voice distortion is encountered when light gases (e.g., helium and hydrogen) are breathed, and the distortion increases in severity with increasing depth. (See Chapter X, discussion of communication.) This is due to an increase in the speed of sound in the gas, and the effect of this increase on the phonetic properties of the resonant cavities used in speech (Sergeant 1969). (Some distortion occurs when *air* is breathed at depth, but this is a minor problem.) Helium speech involves a linear increase in the frequency of important formants (the frequency bands which make up the phonetic quality of vowel sounds) by a factor of about 2.25 (Gertsman *et al.* 1966). Electronic devices (helium speech unscramblers) which restore these formants to their original frequency will restore the intelligibility of helium speech, although quality must necessarily be sacrificed if this is to be done in *real time*. A practical aspect of the frequency shift is that many formants are shifted out of the frequency range of ordinary communications gear, so that a really useful system requires suitable microphones and preamplifiers as well as an unscrambler.

Hydrogen causes slightly greater distortion than helium (Sergeant 1972) but is reasonably well translated by a good helium-speech unscrambler. Neon causes much less distortion than helium, preserving a high level of intelligibility throughout the depth range to 600 fsw.

Many diving operators add 10 or 15% nitrogen to a He–O_2 breathing mixture in order to improve voice intelligibility (as well as to help keep the diver warm). The addition of small amounts of very dense gases (such as SF_6 or CF_4) has also been suggested.

f. Thermal Properties

Another property which has a prominent effect on the success of a gas used for deep diving is the way it influences the loss of heat from a diver's body. Divers lose heat by two routes, the skin and the respiratory tract. An unprotected diver in shallow water has most of his heat loss to the water through his skin, but even here the tremendous heat exchange capabilities of the lungs are significant. At greater pressures, especially where helium is used, the proportion of heat lost through the breathing system increases with the increase in density of the breathing gas, until at 20 atm this may equal the metabolic heat production of the body (Webb and Annis 1966). Exercise to produce more heat is no help because respiratory gas exchange increases almost linearly with increased metabolism.

Most deep-diving experience is based on helium and it has been found that respiratory heat loss, where helium-based mixtures are used as breathing gas, makes it practically impossible to carry out prolonged deep dives (beyond 1 hr) in cold water without heating the breathing gas (Webb 1973a). This problem is complicated by the pressure effect on insulation. Virtually all thermal insulation relies on vacuum or *dead* gas space; at pressure, conventional insulations are compressed to a fraction of their original dimensions and, to make matters still worse, the little remaining space may be filled with a thermally conductive gas. Insulation containing tiny, noncompressible spheres is effective, or a new gas space can be created by addition of pressurized gas at depth (but helium is not much help there; air, nitrogen, Freon, and carbon dioxide are more effective insulators).

As in the case of voice improvement, small proportions of a denser (hence less thermally conductive) gas added to a helium–oxygen breathing mixture may reduce heat loss enough to be worth the added density and decompression complications. Neon has been used in larger proportions (up to 75%) and, subjectively, appears to help.

Another situation where heat loss to a gaseous medium may occur is when a transfer bell or SDC (submersible decompression chamber) is used. This is a particularly difficult problem since a diver may be required to decompress in the bell for several hours after leaving the water, already in a chilled condition and without a chance to rewarm.

Heat loss is an important aspect of undersea habitats and saturation chambers. Divers in helium atmospheres are comfortable at about 30°C. Though they feel comfortable, there is an extra energy burden which may approach 2000 kcal/day in the case of helium-habitat dwellers (Webb 1973a).

g. Decompression Considerations and Counterdiffusion

A discourse on the relative merits of the different inert gases with regard to their effect on decompression will, if it sticks to experience with human divers, be almost entirely concerned with the relative merits of helium versus nitrogen. (For a detailed discussion of decompression, see Chapter VII). The properties that presumably govern the decompression characteristics of a gas are solubility in water, solubility in fat, diffusivity, thermodynamic properties, and derived factors such as oil–water solubility ratio.

Helium was first suggested as a diving gas on the basis of its low solubility, with regard also to its low density and high diffusivity. In small animals the advantages of helium are substantial, resulting in a reduction of safe decompression time by several times as compared with nitrogen. These advantages prevail in humans also, but to a lesser extent. Because helium is taken up much faster than nitrogen, there are situations where this advantage is lost.

A convenient way of comparing the decompression properties of these two gases is to look at the limiting half-times for nitrogen, which are 2–3 times as long as those of helium.

It has been suggested, on the basis of the independent partial pressures of gases in a mixture, that a *porridge* of gases could be prepared which would permit immediate decompression. If seven different inert gases were used, each having a partial pressure of 1 atm, then a dive of 200 fsw would involve no more than 1 atm of each gas, hence no supersaturation. In practice this does not work; partial pressures are additive with respect to tendency to form a gas phase.

Whether or not all gases in solution act independently, there is evidence that changing the gases being breathed during decompression can substantially influence the decompression (Workman 1969, Bühlmann 1969). Standard practice in deep commercial helium–oxygen (mixed-gas) diving involves a shift of air as the breathing gas during the later stages of decompression.

Other gases considered here from a decompression point of view are oxygen, neon, and hydrogen. Argon is appreciably more soluble than even nitrogen, and thus must be used sparingly. Neon has a low solubility, of the same order as that of helium, both in fat and water, but resembles nitrogen more in diffusion. Hydrogen is slightly more soluble than helium but also diffuses more readily. How each of these factors affects whether one gas is *better* than another depends on how each gas is used and in what manner a comparison is made. For example, though helium is *unloaded* about three times as fast as nitrogen, certain deep, short-dive profiles result in a shorter decompression time with nitrogen than with helium (Workman 1969).

The metabolic gases CO_2 and O_2 also play a role in decompression. Oxygen displaces inert gas and, if time is allowed for it to be consumed in the tissue, it is not likely to be involved in bubble formation; however, oxygen acts as a vasoconstrictor at high pressures and as such no doubt influences gas transport. On the other hand, carbon dioxide acts as a vasodilator, and likewise will affect gas transport. In tunnel work a high level of CO_2 has been observed to be associated with an increased bends incidence (End 1938, Kindwall *in press*).

Different theorists in decompression computation treat the matter of different gases in different ways. The basic Haldane approach was used for development of the U. S. Navy Helium Tables. Here a 2.15:1 ratio was used, with the longest half-time 75 min; by contrast, the exceptional-exposure air tables used a 240-min longest half-time and a 2:1 ratio (Workman 1969).

The same approach, essentially, is used by Bühlmann in a slightly different way; he sums the nitrogen and helium partial pressures in equivalent compartments ($t_{1/2} = 180$ min for helium and 480 min for nitrogen) and uses a Haldane ratio which is reduced as a function of total pressure (Bühlmann 1969). Schreiner and Kelley (1971) use a more pragmatic yet physiological modification of this method in which the half-times are determined for any gas on the basis of the oil–water solubility ratio and blood-perfusion assumptions. All gases are summed in each compartment and ascent is controlled by a matrix of M values (Workman 1965) which reflects total depth. Hempleman's (1952) single-tissue concept considers helium's properties, since it is based on the rate of linear diffusion into a slab of tissue. Hills' (1969) thermo-dynamic model regards gas transport as diffusion-limited and, consequently, considers the diffusion properties of the gases in question.

Under certain conditions it is possible to evoke many aspects of the syndrome associated with decompression sickness without a change in pressure. This phenomenon, known as counterdiffusion, results when a diver is immersed in one gas (e.g., helium) while breathing a denser, more slowly diffusing gas (e.g., nitrogen). The most apparent effect is intense itching of the exposed skin (skin covered with foil or blanketed in the gas being breathed does not itch) (Blenkarn *et al.* 1971), but symptoms similar to vestibular decompression sickness have been seen (Idicula *et al. in press*). For the helium–nitrogen system a *soak* of at least 2 hr in helium is needed at a pressure of at least several atmospheres before breathing the nitrogen; symptoms then appear in about 20 min. Lesions seen in animals are clearly bubbles of gas. An *in vitro* model system has been devised which demonstrates that supersaturation can occur when two gases of different diffusivities (and perhaps solubilities) are diffusing in opposite directions through a lipid layer (Graves *et al.* 1973).

The most obvious relevance of this phenomenon to diving is in the laboratory, where such exposures have taken place and where a new tool is now available for the study of gas transport. The severity of the symptoms seen makes intentional human exposures not advisable. Whether the effects will appear when a diver immersed in nitrogen breathes helium is not apparent; even if skin lesions are not seen, disorders in the ear, for example, are possible. A diver saturated in a nitrogen habitat who wishes to make a deep excursion breathing helium–oxygen may be more susceptible to decompression sickness than he otherwise would be.

h. Fire Safety

The choice of breathing gas has a significant effect on fire safety if it results in a chamber having an oxygen level which will support combustion readily (see Chapter VIII). It is the percentage of oxygen more than the partial pressure which determines its flame-propagating character; mixtures with less than 6% oxygen are safe. It is more difficult to heat materials to the ignition temperature in a helium–oxygen atmosphere than in an equivalent one with nitrogen, but once ignited, the burning rate is higher.

Effects of other possible diving gases on flame propagation have not been determined, with the exception of hydrogen. Here, of course, the fire safety situation is of a higher order of magnitude. Mixtures containing less than 5 or 6% hydrogen will neither burn nor support combustion, but making these mixtures safely and coping with other general handling problems with hydrogen are formidable tasks (Edel 1972).

i. Cost and Logistics

This section is concerned primarily with the economics of obtaining, storing, and delivering the various gases to the dive site. For hardware and equipment considerations, the *U. S. Navy Diving-Gas Manual* (Battelle 1971) and the *U. S. Navy Diving Manual* (U. S. Navy 1970) are available.

Air is, of course, the least expensive gas available, requiring only clean compression—perhaps with proper filtering—provided a smog-free atmosphere is accessible. The components of air—nitrogen and oxygen—are also inexpensive, and handling, mixing, analyzing, storing, etc., make up the bulk of the cost of using nitrogen–oxygen mixtures. These costs apply to some extent to any gas used except air.

Nitrogen–oxygen mixtures may have a role in diving which is yet unexploited: their use in habitats from which excursions are made to deeper work sites (Hamilton *et al.* 1973). The narcosis that limits safe use of air or N_2–O_2 mixtures can be tolerated at greater depths on excursions from saturation than in bounce dives from the surface (Schmidt *et al. in press*). By using the normal oxygen consumption of the divers to *breathe down* the oxygen in a habitat to a tolerable range, such operations can be conducted entirely by means of compressors and gas available at the site. [Consider the situation of a habitat having 125 ft³ volume per man (a small one to be sure) at a depth of 100 fsw: If each man consumes 25 ft³ of oxygen per day the P_{O_2} could be brought down from its initial 0.8 atm to 0.6 atm in one day and 0.4 atm in two. This regime would be well below the toxicity limits for oxygen.]

The term *mixed gas* in its broadest interpretation as applied to diving refers to any gas being used for diving except air (and perhaps the rarely used pure oxygen). In its general usage, however, the term refers specifically to helium–oxygen mixtures, which may or may not contain a little nitrogen.

Helium, although abundant in the universe, has been found in commercial quantities in only certain places on earth. It is usually found with natural gas in practical percentages of from 1 to as much as 8%. Until recent years the only helium production was in the U. S. but sources have now been found in Canada, Poland, and the Soviet Union. The U. S. has had laws controlling export and requiring conservation of helium. These were relaxed in 1969; helium stockpiling by the Bureau of Mines was also discontinued and the Helium Research Unit at Amarillo, Texas was closed.

In the U. S., helium is not particularly expensive; a typical bulk price (in 1972 dollars) is about 6¢/ft³.

This can appreciate to a cost of 15–25¢/ft³ for gas delivered to a North Sea Diver on the sea floor. The expense of using helium in diving operations in remote locations is dictated by the cost of handling, which includes mixing, shipping, and storing. The basic container is the conventional gas cylinder, varieties of which hold 200–300 standard ft³ of gas at pressures of 2000–3000 psi. The empty weight may be 150–200 lb. For easier shipboard handling these are often manifolded in *quads* of four or six cylinders. Larger quantities of gas are handled in trailers holding 10,000–150,000 ft³.

Much mixed-gas diving is done using gas mixtures prepared on shore and shipped to the dive site as mixtures. In addition to the bulk shipping problem, this practice involves other inefficiencies. Depth of use (hence optimum oxygen percentage in the mixture) must be anticipated well in advance of need, and not only must the *steel* be shipped back, but in deep-diving operations a considerable amount of gas remains unusable in the quads when its pressure falls below the appropriate delivery pressure. However, this system has certain advantages in that large initial capital investments can be replaced by distributed demurrage costs, and there is less dependence on highly trained technicians at the dive site.

With the addition of compression, storage, and analytical equipment, gases can be mixed to order, invoking the obvious advantages. Still another method of preparing diving-gas mixtures is by means of on-line mixing devices. Using a system of appropriate regulators and flowmeters, mixtures can be *dialed* as needed from bulk storage of the pure components (Gilardi 1972).

As the final step in mixed-gas logistics, it is relevant to consider the methods of gas conservation, recovery, and reuse. The basic breathing apparatus is an open-circuit demand system, whereby each inspiration is taken from a compressed source and each expiration is lost *overboard*. Some breathing systems use partial or total recirculation of the breathing gas, removing the accumulated CO_2 with a chemical absorber. A *push–pull* hookah system returns expired gas from the diver to the bell atmosphere, where it is scrubbed and recirculated to the diver. On decompression the gas from the bell can be used to pressurize the desk chamber, with the divers transferring after equalization has taken place. Simple buffering, scrubbing, and recompressing systems can be assembled to permit reuse of a large fraction of the mixture used in a deep dive (Schmidt *et al.* 1973). More sophisticated systems have been developed which actually re-refine helium, separating pure helium from all other components and contaminants in a mixture. These systems are complex, expensive, and require a supply of liquid nitrogen, but where usage is great and supply lines long they can be cost-effective (Slack 1973).

Along with the development of devices and procedures for conserving helium there has been a continual push to develop alternatives to the use of helium for mixed-gas diving. This effort has been to circumvent the physiological, economical, and operational disadvantages of helium as well as to avoid dependence on a politically sensitive commodity. The two gases considered as serious alternatives to helium are neon and hydrogen.

Neon is a product of the distillation of atmospheric air, being present at about 18 ppm (air also contains 5 ppm helium). Purified neon is too expensive to consider for ordinary diving operations, but mixtures of neon and helium produced as by-products of the air distillation process are usable (Schreiner *et al.* 1972). From the uncondensed fraction in an air-distillation column (normally discarded) a mixture of neon and helium can be obtained in the proportion of 72–78% neon, 22–28% helium. It is called "first-run" neon or Neon 75. This mixture is relatively inexpensive; it could be available at 10–20¢/ft^3 if bought in large quantities. The significant fact about neon is that, as a product of the atmosphere, it is available anywhere there is sufficient industrialization to require an air-reduction plant.

Neon has another logistic advantage: Because of its thermodynamic properties, it is the diving gas most easily transported and stored as a liquid.

Hydrogen is available by electrolysis of water and, as such, rivals air in its worldwide availability and low cost. Hydrogen's high flammability and diffusivity add to its cost—perhaps substantially—by increasing the difficulty of handling and mixing it safely (Edel 1972). Increasing interest in the use of hydrogen as a low-pollution fuel should result in future improvements in handling methods and equivalents.

4. Conventional Breathing Mixtures

This section considers the various breathing mixtures in general use, their history, uses, and advantages and disadvantages.

a. Air and Nitrogen–Oxygen Mixtures

Until the introduction of helium to diving in the later 1930's the history of diving was the history of air diving. Although most early diving exploits were breath-holding or snorkel dives, Alexander the Great is reported to have descended in an air-filled bell, and a compressed-air bell was designed in 1690 by Edmond Halley (of comet fame). Fréminet first used a bellows to pump air down to a diver in the 1770's, but diving as we know it began in 1819 with the invention of a diving suit by Auguste Siebe (Larson 1959). This suit, supplied with compressed air, had all the essential features of the classical diving gear still in use today (Goodman 1962).

Air, because of its availability and suitability for shallow diving, is the gas of choice for nearly all diving to about 150 fsw. Beyond this depth narcosis becomes a factor, which may limit safe and effective diving to divers adapted to nitrogen and well trained for the job at hand. If suitable support (e.g., a diving partner, life lines, air supply, communications, etc.) is available, the safe depth can be extended to 250 or perhaps 300 fsw for short times (less than $\frac{1}{2}$ hr). For short dives in the deeper air range decompression may be easier with air than with helium–oxygen mixtures.

Air is the gas of choice for undersea habitats to a depth of at least 70 fsw; at some depth beyond this, oxygen toxicity may cause lung discomfort after a few days; an appropriate depth–duration tradeoff for this limit has not yet been established.

Nitrogen–oxygen mixtures having a greater oxygen fraction than air are sometimes used, the added oxygen for the purpose of displacing inert gas and hence reducing decompression.

b. Helium (Mixed-Gas Diving)

The record shows that C. J. Cooke applied in 1919 for a patent on the use of helium–oxygen mixtures to be supplied to men under pressure, but he apparently never tried it (End 1939). Credit for development of helium should go to the eminent chemist, Dr. Joel H. Hildebrand, who, with R. R. Sayers and W. P. Yant of the U. S. Bureau of Mines, tried helium decompressions with small animals (Sayers *et al.* 1925). It was on the basis of its low solubility that these investigators proposed helium, and they reasoned correctly that this property would improve decompression time in comparison with nitrogen. It was not until End tried breathing helium, and, with Nohl, actually tried helium on a 420-fsw dive (End 1938, 1939) that the real advantage of helium was noted—its lack of narcotic properties. Shortly thereafter Momsen, with Behnke and Yarbrough, at the U. S. Navy Experimental Diving Unit, explored the use of helium to the pressure equivalent of 500 fsw (Ellsberg 1939) and in 1941 it was successfully used in dives to 440 fsw (Behnke 1942). The salvage of the submarine SQUALUS could not have been accomplished safely without helium.

The use of helium in diving has exceeded the expectation of its early proponents, with the 1000 fsw mark surpassed at sea and depths twice that reached in the laboratory. These dives have not been without problems, however. Serious HPNS symptoms may result if compression is rapid, and long exposure to hyperbaric arthralgia and extended decompression are consequences of slow compression rates.

Helium is presently the diluent gas of choice for virtually all diving beyond 150 fsw. Where immediate decompression is to follow, oxygen is kept as high as possible in the mixture to produce a bottom-gas oxygen partial pressure of 1.2–2.0 atm; the higher the oxygen in the breathing mixture, the lower the exposure to inert gas. For saturation, oxygen is kept between 0.2 and 0.6 atm, but it should be raised during decompression. Nitrogen is sometimes added to helium–oxygen mixtures for the purpose of improving voice communication and mitigating slightly the chilling effect. The range used for this purpose is 5–15%; higher doses might be used, but even 15% is enough to have a detrimental effect on decompression. Another purpose for adding nitrogen is to counteract the effects of HPNS. Here 12–25% might be used; this technique is still experimental. Use of larger percentages of nitrogen for diving in the 200–300 fsw range has been suggested, where conservation of helium is more important than hasty decompression.

To summarize the properties of helium as they affect its use as a diving gas: Helium is light and very easy to breathe; it has low solubility and high diffusivity, which enable it to be easily eliminated during most decompressions. The same low solubility gives it no narcotic properties, but allows HPNS to become manifest in rapid compressions. Although not expensive, helium can only be obtained in certain locations. When kept under pressure helium has a high leak rate, and it can leak into devices stored in it, such as a TV tube. One of the most disturbing factors about helium is its sonic velocity; when a diver breathes helium his speech is difficult to understand.

c. Oxygen

With the advent of fully closed, mixed-gas rebreathing systems that produce no bubbles and allow a long bottom time on a single charge of gas, the incentive to use pure oxygen underwater has disappeared. The main use of pure-oxygen rebreathing systems has been for clandestine operations where the presence of bubbles was unacceptable. The Italian Navy used closed-circuit oxygen rebreathing early in World War II, and the LARU (Lambertsen amphibious respiration unit) was introduced in the U. S. Navy shortly thereafter (Larson 1959).

Because of the possibility of CNS oxygen toxicity (convulsions), the use of such units is restricted to a maximum depth of about 25 fsw.

5. Experimental Breathing Mixtures

This section deals with the more unusual diving gas mixtures—those not in routine use but on which some experience has been accumulated. A qualitative approach is used in this discussion, with attention directed primarily at the components of a mixture, not the proportions.

a. Hydrogen

Although highly explosive when mixed with air in certain proportions, hydrogen can be used in diving because, at increased pressures, enough oxygen to meet physiological needs can be added and the mixture can still remain below flammability limits (Dorr and Schreiner 1969). Physiologically it appears to be inert.

Hydrogen was used in experimental dives with animals as far back as 1914, before helium had been discovered in enough abundance to merit consideration of its use in diving. More recently Zetterstrom made successful dives with hydrogen–oxygen mixtures, but unfortunately died in a diving accident (Zetterstrom 1948, Bjürstedt and Severin 1948). His death, although unrelated to the use of hydrogen, had the effect of setting back work with this gas.

In addition to its low cost and ready availability—advantages partially offset by its explosive properties—hydrogen has a physiological advantage which might make it the gas of choice for deep diving: The narcotic properties of hydrogen apparently counteract the neurological disturbances due to high hydrostatic pressures (Brauer and Way 1970). Hydrogen is about one-fourth as potent as nitrogen in causing narcosis. Further, its low density makes hydrogen mixtures the easiest ones to breathe at great depths.

The thermal properties of hydrogen seem comparable to those of helium, and it causes a voice distortion of equal proportions. Helium unscramblers, however, seem to work with hydrogen, too.

Short diving exposures to hydrogen conducted by Edel have revealed no contraindications against the eventual use of hydrogen in diving (Edel *et al.* 1972). Decompression times appear to be generally comparable to helium, perhaps slower. Hydrogen offers an intriguing possibility for exploring the question of whether perfusion or diffusion plays the limiting role in decompression; it has a diffusion comparable to helium and an oil–water solubility ratio like nitrogen. In practice hydrogen appears to fall between helium and nitrogen in its decompression efficiency.

Not all reports on the use of hydrogen are favorable. Experiments conducted with rabbits at pressures of 280 m (917 ft) showed reduction in respiratory and motor activity; this, however, may be due to the peculiar thermal susceptibility of rabbits (LeBoucher 1970).

b. Neon

The high cost of neon has limited the study of this gas for use in diving. Numerous experiments with animals and other biological models have shown no detrimental effects (Schreiner *et al.* 1962; K. W. Miller *et al.* 1967*a*); human exposures have verified this and also have shown that neon has virtually no tendency to produce narcosis to at least 1200 fsw (Hamilton *in press*).

Because of the expense of pure neon, most experiments have concentrated on the use of a mixture of neon and helium, Neon 75, which is obtained by distillation of air. Neon, therefore, has worldwide availability in this form.

Experiments designed to evaluate neon as a diving gas in side-by-side comparison with nitrogen and helium (Schreiner *et al.* 1972) have shown it to be equivalent to helium in its effect on mental and psychomotor processes, and to present no particular problems in decompression. Voice is less distorted by neon than by helium or hydrogen, and the lower thermal conductivity of neon might prove to be advantageous when diving in cold water. Divers who have used neon at sea report some subjective reduction in heat loss but definite measurements of this factor remain to be made.

It appears that the optimal use of neon will be in the range 150–600 fsw. Deeper depths result in enough of an increase in gas density that Neon 75 mixtures may be hard to breathe by a diver doing heavy work and limited by breathing equipment designed for helium.

c. Argon

The properties of argon are not beneficial with respect to its use as a diving gas. It has a high lipid solubility and is consequently narcotic and a problem in decompression. Further, its density makes it difficult to breathe at high pressures, but it does allow effective voice communication and also acts as a fair insulator against heat loss.

Argon has been used as a decompression gas, with the purpose of reducing the inspired partial pressures of both helium and nitrogen (Keller 1967, Keller and Bühlmann 1965). The real benefits of argon in this situation have not been systematically explored, but its use seems to be effective (Schreiner 1969).

Argon is about twice as narcotic as nitrogen, causing an equivalent decrement in performance tests at about half as much pressure (Ackles and Fowler 1971).

d. Other Gases

Few other biologically inert gases exist which seem to be practical as the major component of a diving-gas mixture, and scant data exist on the ones that do exist. Heavier gases exhibit the same deficiencies as argon, having high lipid solubilities, which result in both narcosis and difficult decompression, and high densities, which cause breathing resistance.

Of the light gases, deuterium is not available in sufficient quantities to be considered; it should have properties similar to hydrogen and helium, and probably offers no particular advantage. Methane is sufficiently inert and lighter than most other gases, but its lipid solubility should make it reasonably narcotic. Like hydrogen, it is not flammable if the oxygen is kept low (below about 4%).

Certain gases have some value when added as only a small fraction of a breathing mixture. Gases known to be narcotic may reduce symptoms of HPNS. A heavy gas having a relatively low solubility may improve voice communication; such a gas is CF_4, tetrafluoromethane (Airco patent). Other fluorinated hydrocarbons (Freons) are used as fire extinguishing agents. These Freons are apparently not completely *inert*, in that they exhibit certain toxic characteristics (Call 1973; D. G. Smith and Harris 1973).

One additional use of gases in diving is to simulate other conditions in the laboratory. Examples are the use of nitrous oxide to cause narcosis similar to that of hyperbaric nitrogen (Brauer and Way 1970, Hamilton 1973). Nitrous oxide is 30–40 times more narcotic than nitrogen. Sulfur hexafluoride has a molecular weight of 146, making it about five times as dense as air. This property has been used to study breathing resistance as a function of density (Uhl *et al.* 1972; Anthonisen *et al.* 1971).

e. Liquid Breathing

As diving depths increase, the available choices diminish but the factors relating to gas characteristics become progressively more important. One method of eliminating breathing gas problems is to eliminate the breathing gas altogether—by breathing a liquid.

In the early 1960's, the Dutch physician Johannes A. Kylstra was looking for a better dialysis technique for patients with kidney failure. He reasoned correctly that the lung might be an effective exchange medium for removing toxic products and maintaining electrolyte balance. Kylstra was fortunate enough to have been associated with one of the first hyperbaric medicine installations, and this inspired him to experiment with supplying oxygen dissolved under pressure while flushing the lungs with fluid. He found that mammalian lungs could indeed obtain sufficient oxygen from physiological saline solutions if they were first equilibrated with oxygen under a pressure of at least 6 atm (Kylstra 1962, 1967). In subsequent experiments he found that CO_2 removal was a limiting factor, since hyperbaric techniques could do nothing to increase the rate of CO_2 elimination. Another problem was the washing out of the lung surfactant by the lavage; this was found to be avoidable by eliminating bubbles in the lavage fluid.

Other researchers (L. C. Clark and Golden 1966) found that fluorocarbon fluids would dissolve enough oxygen at sea-level pressures to meet oxygen-delivery requirements in the lungs.

Supplying oxygen and ventilating of the lungs with a liquid eliminates two significant problems of deep diving. With no need for an inert gas diluent, inert gas narcosis and decompression are no longer problems. Speech communication is not possible, and carbon dioxide elimination assumes a distinctly restrictive role.

As with the very dense gas mixtures, there are definite limits as to how fast a fluid will move through the airways. For a given level of CO_2 production, maintaining a proper CO_2 level is essentially a function of lung ventilation rate. Delivery of more CO_2 to the ventilating fluid can be accomplished by increasing the level in the blood with which this fluid (or gas) is equilibrated, and a certain increase can be tolerated. (Many experienced divers allow their CO_2 to increase from a normal 40 mm Hg to as much as 55 mm Hg when conserving air). Recent experiments in Kylstra's laboratory have shown that a light work rate should be tolerated by a liquid-breathing diver if he is breathing an emulsion of fluorocarbon fluid and a buffer solution (THAM) (Kylstra *et al.* 1973; Schoenfisch and Kylstra 1973).

6. *Breathing-Gas Purity Standards*

The composition of breathing gas mixtures and the quantity required are determined by the physiological characteristics of the human body both at atmospheric pressure and at elevated pressures. The breathing mechanisms that serve to regulate partial pressures of oxygen and carbon dioxide in the blood at sea level serve equally well under hyperbaric conditions, providing the partial pressures of oxygen, nitrogen, and carbon dioxide, as well as the respiratory volume, are kept near those at sea level. *It is the partial pressure of the gases breathed that is the controlling factor.*

a. Purity Standards

The U. S. Navy Bureau of Medicine and Surgery has established purity standards for diver's breathing air which should be met by all apparatus used to supply breathing air (Battelle 1971). The requirements are as follows:

Oxygen	20–22% by volume
Carbon dioxide	300–500 ppm (0.03–0.05%) by volume
Carbon monoxide	20 ppm maximum
Oil, mist, and vapor	5 mg/m³ maximum
Solid and liquid particles	Not detectable except as noted above under oil, mist, and vapor
Odor	Not objectionable

Oxygen purity is covered by Federal Specification BB-0-925, and there are three grades: (A) aviator's breathing; (B) industrial and medical; and (C) technical. Grades A and B differ only in moisture content and are tested as specified by the U. S. Pharmacopeia for acidity or alkalinity, carbon dioxide, other oxidizing substances, halogens, and carbon monoxide; they both must contain not less than 99.5% O_2.

Helium is produced by the Federal Government and four grades are listed, but only A and D are currently being produced. Grade A helium is approximately 99.999% pure and is free of oil and moisture and thus suitable for diving.

b. Contaminants

Diving gases supplied by cylinder from a reputable manufacturer are most unlikely to contain deleterious contaminants, since the manufacturing processes used have very few steps in which contamination can occur. Mixing errors are possible but introduction of alien gas contaminants is improbable (Hamilton and Erb 1970; Gilardi 1972). Contaminants can, however, be introduced in handling. These might be solvents used in cleaning, such as Freons or other halogenated hydrocarbons, or petroleum distillates.

By far the main contaminants of diving gases are those found in compressed air. These are primarily CO, CO_2, oxides of nitrogen, oil mist, and gaseous hydrocarbons. Their presence would be an indication of a gas from an improper source, an overheating compressor, or improper filtering.

In a habitat or breathing rig the major sources of contamination are the divers themselves, and the major contaminant is CO_2. For long-duration missions with semiclosed life-support systems divers will also generate CO, methane, H_2S, and hydrogen, and their activities may produce acroleins, other hydrocarbons, ammonia, SO_2, and chlorine. Freons passing through catalytic burners or a fire can break down into toxic fluorine compounds. Solvents used in glue, paint, etc., may be given off for a long time.

The importance of these contaminants is not completely clear. The effect of CO is definite and clearly detrimental, as is CO_2 present in physiologically active amounts.

Table IV-22
Typical Contaminant Exposure Limits
(Bishop 1973)

Substance	8-hr weighted average limit	Ceiling concentration	Comments
Ammonia	50 ppm	—	—
Carbon dioxide	5000 ppm	—	—
Carbon monoxide	50 ppm	—	—
Freon-12	1000 ppm	—	—
Hydrogen chloride	—	5 ppm	—
Hydrogen fluoride	3 ppm	5 ppm	10 ppm for max 30 min
Mercury	—	$0.1 \ mg/m^3$	—
Nitric oxide	25 ppm	—	—
Nitrogen dioxide	5 ppm	—	—
Oil mist	$5 \ mg/m^3$	—	—
Ozone	0.1 ppm	—	—
Phosgene	0.1 ppm	—	Freon decomposition
Stibene	0.1 ppm	—	Lead-acid battery
Sulfur dioxide	5 ppm	—	—

Oil mist can cause lung damage, probably by inactivating the surfactants. Freon-breakdown products, oxides of nitrogen, SO_2, and other irritant gases can cause serious responses if present in large enough amounts. What constitutes too much of some of these products has been established by the Occupational Safety and Health Administration and may serve as a guide here (Table IV-22). These exposure limits are for an 8-hr work day and 40-hr work week at sea level, but they are also computed to allow a 40-yr career of exposure without damage. Appropriate limits for the occasional concentrated exposure of a diving chamber have not been agreed upon.

c. Testing

There are two basic methods of preparing compressed-gas mixtures. One involves mixing the component gases by volume in a large gas holder at normal barometric pressure and then compressing the mixture into cylinders. Considerable accuracy is possible with this method because analysis and adjustment of percentage can be accomplished before compression.

The only practical methods for field use are those in which the mixing is done with the gases under pressure in cylinders. Here pressure provides the usual basis for proportioning. In both types of mixing, samples must be carefully analyzed and percentage adjustment of the component gases is often necessary. For the pressure-transfer method an oxygen-transfer pump is sometimes needed.

When gas mixtures are used for diving, an accurate determination of the oxygen concentration is essential, for any significant deviation from the planned oxygen content may lead to serious difficulties. Both chemical and physical methods of analysis are employed. In the U. S. Navy both the Model C and the Model D Beckman oxygen analyzers are used, the Model D being smaller and more suited to field activity. Of paramount importance in all gas analysis is to obtain a sample that is free of any contamination.

In diving, analysis of gas for its carbon dioxide content may be done to check for CO_2 as a contaminant in a breathing mixture, or to determine the CO_2 level in a habitat or closed or semiclosed scuba or other equipment. It is well to remember that low concentrations could cause trouble in deep diving. For example, 1% CO_2 breathed at 132 ft would have the same physiological effect as 5% CO_2 breathed at the surface. Reasonably accurate analysis can be accomplished with several types of analyzers. The one carried on every submarine is the Dwyer CO_2 analyzer and is reasonably accurate when the content is over 1 or 2%.

A small portable carbon monoxide detector is readily available that is simple to operate, sufficiently sensitive, and accurate. Since the maximum allowable concentration of CO is only 10 ppm it is important to analyze the breathing gas, especially compressed air from a compressor at the site of diving.

In general other gases must be analyzed by special laboratory techniques not suitable for field diving operations.

7. Physiology of the Wrong Gas

When a person is exposed to an atmosphere that differs very much from the one in which he evolved—ample air at pressure of 1 atm—he is generally first warned by certain physiological signs and symptoms and may, if the alien atmosphere is improper, suffer a decrement in performance, or injury, loss of consciousness, and perhaps even death. Certain other stresses (e.g., cold, decompression sickness, etc.) may cause symptoms similar to those caused by an improper breathing mixture. This section is an attempt to provide a set of guidelines on what to expect from various situations.

a. Low Oxygen (Hypoxia)

(i) Situation. Brain or tissue hypoxia (sometimes incorrectly called anoxia) may result from any disturbance in the oxygen-transport mechanism; the concern here is with an inadequate oxygen partial pressure P_{O_2} in the lungs. (But see also *carbon monoxide* below.) This can be caused by lowered total pressure or reduced fraction of oxygen in the ambient gas. Commonly encountered in high-altitude flying, hypoxia is not common in diving because of the prevalence of high oxygen partial pressures. It can result from consumption of the oxygen in a confined space (e.g., a bell, submarine, or rebreathing system), improper computation or analysis of a mixture, using it at the wrong depth, failure of a control system, incomplete mixing (e.g., of oxygen with helium), or ascending to lower pressure while breathing a mixture that is correct for a greater depth.

(ii) Physiology. Hypoxia can result from several other problems: A constricted airway or laryngospasm can reduce lung ventilation; or pneumonia, water in the lung, or pneumothorax can reduce lung function. Restrictions in delivery of oxygen can be due to shortage of functioning hemoglobin—as might result from anemia, hemorrhage, or carbon monoxide poisoning—or inadequate general or tissue blood flow. Tissue utilization of oxygen can be poisoned, as by cyanide. Like many other physiological functions, rate of onset has a bearing on the effects, as does degree of physical exertion; rapid onset and exercise tend to exaggerate symptoms.

(iii) Symptoms. Mild hypoxia causes headache, fatigue, listlessness, and a measurable deterioration of night vision. There may be varying degrees of tingling, paresthesia, and formication (*crawling ants*). As the condition progresses, judgment is impaired, accompanied by euphoria which may result in high self-confidence and a disregard for the sensory indications of hypoxia. Coordination becomes poor (handwriting shows obvious deterioration) and vision is impaired; a distinct characteristic is *tunnel* or *gun-barrel* vision, where peripheral vision grays out. As the condition worsens, these symptoms intensify, leading to eventual unconsciousness. Rapid onset of hypoxia may result in loss of consciousness with little or no warning; warning signs may be disregarded. Not all individuals show the same responses. Cyanosis—blue lips and fingernail beds—may be noticeable.

(iv) What to Do. Correct the oxygen deficiency. Administer higher oxygen in the breathing gas—1–2.5 atm P_{O_2}.

b. High Oxygen (Hyperoxia)*

(i) Situation. Chronic oxygen toxicity can be invoked after hours or days of exposure to as little as 0.5 atm P_{O_2}; acute oxygen toxicity requires more than 1 atm P_{O_2} and hence cannot be induced at normal barometric pressure. Most people can tolerate oxygen at rest at 60 fsw (2.8 atm) for many minutes. However, in exercise these symptoms may be encountered at less than 2 atm P_{O_2}. Acute oxygen toxicity may be encountered when pure oxygen is breathed at increased pressure for decompression, treatment of decompression sickness, or hyperbaric therapy. It may also be encountered as a result of an improper breathing mixture. Habitation of an environment with elevated oxygen may lead to lung symptoms after several hours or days.

(ii) Physiology. It is possible to detect high oxygen by a physiological mechanism. Finger pressure applied to the corner of the eye will result in a gray spot in the visual field on the opposite side of that eye. This takes about 10 sec to develop under normal conditions. This time is increased when breathing hyperbaric oxygen, to as long as 50 sec at 3 atm P_{O_2} (Carlisle *et al.* 1964, Anderson 1968).

(iii) Symptoms. Chronic or lung effects, gradual in onset, may result in chest pain (substernal); coughing, especially on taking a deep breath; dry or sore throat; and general breathing discomfort. These may be confused with or complicated by the effects of breathing dry gas by mask. Vital capacity is reduced (Clark and Lambertsen 1971*b*), even before other symptoms are noted.

Acute or CNS symptoms culminate in convulsions or seizures. These may be accompanied by dizziness, nausea, lightheadedness or confusion, euphoria, and pupil dilation. Twitching of lips and facial muscles—possibly not noticeable to the affected person—are common (Young 1971).

(iv) What to Do. Remove source of high oxygen. This is helpful even if only intermittent reduction of P_{O_2} is possible. In case of convulsion, spontaneous and complete recovery will occur if P_{O_2} is reduced; patient should be protected from mechanical injury.

---

* See also the discussion of oxygen toxicity in this chapter.

c. Lack of Inert Gas

(i) Situation. This occurs when pure oxygen is breathed for an extended period. This is possible only at normal or reduced total pressure.

(ii) Physiology. Pure oxygen is absorbed more quickly and completely from body spaces than a mixture containing inert gas. This may result in a relative *squeeze* in the middle ear or sinuses, particularly if the individual is asleep and therefore not ventilating the middle ear. If absorption is from the lung, alveoli are allowed to collapse (atelectasis).

(iii) Symptoms. The patient may awake with ear squeeze. Lung atelectasis may cause coughing, substernal pain, and difficulty in taking a deep breath.

(iv) What to Do. For the ears, ventilate through eustachian tubes several times. For the lungs, take a few deep breaths, preferably with slight positive pressure.

d. High Carbon Dioxide (Hypercapnia)

(i) Situation. Elevated carbon dioxide levels encountered in diving may be generated by the divers themselves or by the technicians managing the dive. This may occur in a habitat or closed space, in a diving suit or breathing rig, or in the lungs or cardiopulmonary system. Levels increase any time rate of removal falls behind rate of generation.

(ii) Physiology. The only way CO_2 can be removed from the body is by ventilation of the lungs. To maintain the normal 40 mm Hg P_{CO_2} in the arterial blood, CO_2 must be removed from the body at the same rate as it is being produced. Expired air is essentially equilibrated with the blood, which means it contains a partial pressure of 40 mm Hg. At sea level this is about 5%, but at depth the partial pressure remains the same and the percentage consequently decreases. Each breath of expired gas contains a given *amount* of CO_2 based on the fraction of CO_2 and the volume. Unloading CO_2 is then a simple matter of material transport. The significant factor is that the same ambient volume of gas must be moved regardless of the depth and, hence, regardless of the density or total number of molecules of gas. During exercise, as production in the body goes up—with a constant P_{CO_2} in the blood—the ventilation must increase by almost exactly the same degree as the increase in production. This relation holds even in liquid breathing. It is modified if CO_2 is present in the inspired air, the effective (alveolar) ventilation being then based on the difference between inspired and expired gas.

Flushing a chamber which contains no chemical scrubber is the same process; to maintain a given level at a constant rate of production, an equal amount must be removed per minute, and that is equal to the fraction in the vented gas multiplied by the volume.

Carbon dioxide is a normal constituent of the body; not only is it continually produced by all oxidative metabolism, it plays a major role in many body functions, especially as a chemical messenger and regulator of acid–base balance. There is no particular reason why heroic and expensive efforts must be mounted to prevent the breathing of this product.

(iii) Symptoms. The effects of abrupt, short-term exp
range from no detectable effect to unconsciousness and conv
the concentration inspired. Presence of 1, 2, or, in most norm;
air at sea level produces no recognizable respiratory or c
(Lambertsen 1960). Presence of 4, 5, or 6% inspired CO_2 pro
tory stimulation of increasing degree, with an accompanyi
stimulation to the degree of true dyspnea occurs with 7 ai
deterioration of mental competence becomes evident (Dripps and Comroe 1947,
Lambertsen 1971).

At still higher inspired carbon dioxide levels, as with 15–20%, respiratory distress
is abrupt and violent in onset and is accompanied by rapid loss of consciousness and by
spasmodic neuromuscular twitching. With 20–30% inspired CO_2, convulsions occur
within 1–3 min after the beginning of the exposure. Inspired CO_2 can reduce tolerance
to heavy work, but high levels are required—as much as 40 mm Hg (Clark 1973). A
summary of acute exposure symptoms is given in Table IV-23.

A number of biochemical changes have been detected in chronic exposures to low
levels of CO_2 (up to 30 mm Hg) (Clark 1973, Schaefer 1964, Clark *et al.* 1971) but
these are all adaptive responses, and no evidence has been found that exposures of this
sort are detrimental. (The changes are no greater than those resulting from exposures
to high altitude, a stress considered socially and medically acceptable.) Respiration
increases, more at first, then settles down to a constant level (Clark *et al.* 1971).

(iv) What to Do. Removing the diver from the CO_2-rich environment should be
as quick as is conveniently possible, without involving the rescuer as part of the
problem. Headache, if not already present, will often result on withdrawal and can be
treated with aspirin. Sodium bicarbonate will hasten return to normal acid–base
balance. If an adequately oxygenated individual is known to be exposed to CO_2
levels high enough to cause unconsciousness, a gradual (over 5–10 min) reduction of

Table IV-23

Acute Effects of Exposure to Increased P_{CO_2}
(Clark 1973)

Inspired P_{CO_2}, mm Hg	Acute effects
0–15	Barely detectable; ventilation increases slightly; little or no effect on work capacity
15–21	Aware of increased ventilation; severe exercise subjectively more difficult; headache in sensitive individuals
21–30	Easily detectable; ventilation approximately doubled; moderate exercise subjectively more difficult; increased incidence of headaches
>50	More rapidly developing and vigorous respiratory stimulation, dizziness, dyspnea, restlessness, faintness, severe headache, and dulling of consciousness
>100	Unconsciousness, muscular rigidity, and tremors
150–200	Generalized convulsions, death

is advisable; abrupt reduction from high levels may cause heart fibrillation due potassium release (Brown and Miller 1952).

Many present standards for CO_2 exposure call for reducing the inspired level to no more than 0.5% at all times. This choice is obviously not based on physiological principles; a more realistic value would be 2.0% (Clark 1973).

e. Low Carbon Dioxide (Hypocapnia)

(i) Situation. Hypocapnia results from hyperventilation performed either voluntarily prior to breath-holding or involuntarily due to other causes.

(ii) Physiology. Rapid ventilation *washes* CO_2 out of the lungs. The primary urge to breathe is due to CO_2. By reducing the CO_2 level in lungs and blood, a skin diver can hold his breath longer. (This is further increased by breathing oxygen.) The diver hyperventilating on air does not increase his oxygen load significantly. If he then exercises vigorously, it is possible for him to pass out as a result of brain hypoxia occurring before he reaches the *breaking point* of breath-holding. This "shallow water blackout" (Dejours 1965) is dangerous and is responsible for many drowning deaths each year.

Involuntary hyperventilation can result from the stimulus of breathing equipment, or claustrophobia and other apprehensions. The reduction of blood CO_2 decreases brain circulation and may result in cerebral hypoxia.

(iii) Symptoms. In the breath-hold situation the diver has a much reduced urge to breathe and may experience no warning before passing out abruptly. In continued hyperventilation the symptoms are similar to those of hypoxia—dizziness, poor coordination, tingling, paresthesia, headache, and nausea. In extreme cases muscle spasms occur in jaws and wrists.

(iv) What to Do. In the breath-hold case, limit hyperventilation to 2 min, or to a third of what it takes to feel symptoms. When the diver is apprehensive, he must force himself deliberately to slow his breathing.

f. Inert Gas Narcosis*

(i) Situation. Exposure to air in the depth range beyond 100 fsw, or to other narcotic gases at appropriate depths, may cause symptoms.

(ii) Symptoms. Narcosis resembles hypoxia in loss of coordination, dizziness, and euphoria. Narcosis also causes numbness of the lips, which is less common in hypoxia. Hypoxia, however, causes more visual disturbances, especially gun-barrel vision. Tingling is more common in hypoxia or hyperventilation than in nitrogen narcosis.

(iii) What to Do. Reduce exposure to hyperbaric narcotic gas. Tolerance can be improved by daily diving or by habituation (Hamilton *et al.* 1973).

g. Carbon Monoxide

(i) Situation. Carbon monoxide may be present in compressed air from compressors whose intake air contains it or, in some cases, from compressors allowed to

* See also the discussion of physiological aspects of nitrogen narcosis in this chapter.

overheat (as might result from improper lubrication). It results from incomplete combustion of carbonaceous fuels, such as charcoal, from engine exhausts, from fires, and from smoking. It is also produced by the body; a man generates 0.3–1.0 ml/hr, a consideration in closed atmospheres. Effects of CO are according to its partial pressure.

(ii) Physiology. Carbon monoxide reacts with hemoglobin in the blood, displacing oxygen; hemoglobin has an affinity for this gas some 200 times greater than for oxygen. The carboxyhemoglobin (COHb) which is formed shifts the oxygen dissociation curve (to the left) in such a way as to reduce drastically the blood's ability to transport oxygen, resulting in tissue hypoxia. Effects of CO resemble those of other forms of hypoxia. Inspiring 35 ppm (0.0035%) produces a COHb of 5%, while 400 ppm converts over 40%. A one-pack-per-day smoker will have 5–6% COHb (Lassiter 1972).

(iii) Symptoms. Acute exposures resulting in COHb levels as low as 5% have been shown to affect performance in a detrimental way. Like hypoxia, higher exposures cause headache, nausea, vomiting, dizziness, drowsiness, and collapse. Since COHb is a brighter red than Hb or oxygenated Hb, the nail beds, cheeks, and lips may look rosy or cherry red (in contrast with the blue or purple color of reduced hemoglobin, which results from general hypoxia).

(iv) What to Do. Provide uncontaminated air or, preferably, oxygen or hyperbaric oxygen (2–2.5 atm P_{O_2} for up to 2 hr). Use supportive means, if needed, to maintain breathing and heart rate.

F. *The High Pressure Nervous Syndrome and Other High-Pressure Effects*

Many of the problems of exposing man to a high-pressure environment depend overtly on the properties of gases in the gaseous phase—such as the various squeezes, voice distortion, thermal effects of helium, density effects, and decompression sickness. Other problems relate directly to the properties of a specific gas—such as oxygen or carbon dioxide toxicity and nitrogen narcosis. Still another set of phenomena appear to be related to pressure per se, or to changes in pressure. These are the subjects of this section.

The distinction is an arbitrary one, since both hyperbaric arthralgia and the high pressure nervous syndrome may, in fact, be gas-related problems and there is no doubt that the nature of the gas present plays a secondary role; however, for convenience of presentation, we will treat them as independent of the nature of the gas.

1. *History*

In 1964 and 1965, a series of simulated dives were conducted at the Royal Naval Physiological Laboratory (RNPL) in which men were exposed to helium–oxygen at pressures of 600 and 800 fsw for 1–4 hr (Bennett 1965, 1967). Compressions were rapid, from 20 to 100 ft/min. The men experienced dizziness, nausea, vomiting, and a marked tremor of the arms, hands, and occasionally the whole body. Performance decrements of 25% on a test of fine manual dexterity at 600 fsw were twice as severe at 800 fsw. Within 90 min subjects had returned to normal. It was considered on the basis of these results that man might be severely incapacitated at 1000 fsw. Bennett (1965) called this phenomenon "helium tremors." At about the same time a series of dives was undertaken by Ocean Systems, only one of which has been reported in the literature (Bennett 1965, Hamilton and MacInnis 1965, Hamilton et al. 1966). Dives were to sea-water equivalents of 500–650 ft and, except for the saturation dives, involved compression rates of 100 ft/min. It required 1 hr to reach 650 fsw. The rapidly compressed subjects reported hand tremor, dizziness or disorientation, difficulty in focusing and tracking with the eyes, and joint stiffness. No nausea sufficient to cause vomiting occurred. Divers reported that their eyes felt as if they lacked lubricant. This was attributed to the overwhelming presence of hot, dry helium (Hamilton and MacInnis 1965).

Rapid-compression dives conducted by Bühlmann et al (1970) invoked the same set of responses, and a new one called "microsleep" by Brauer (1968). Subjects seemed constantly on the verge of sleep unless stimulated or busy on a task.

The first saturation dive to 1000 fsw was conducted at Duke University in 1968 with a compression time of 24 hr, a rate appreciably slower than in any previous dive. Only mild symptoms were seen—transient joint stiffness and mild tremor (Summitt et al. 1971). In the same year, COMEX conducted a simulated dive to 1189 fsw (Physalie III) but with a total compression time of 123 min (Brauer 1968). Marked tremors accompanied by dizziness and nausea occurred during compression. In addition, significant EEG changes (theta waves) and periods of somnolence ("microsleep") during the 4 min at 1189 fsw were sufficiently severe to warrant abandonment of the dive. Brauer postulated that there might be a physiological barrier to deep diving at about 1200 fsw; he referred to the combination of signs and symptoms as the "high pressure nervous syndrome."

The 1200-fsw barrier was surpassed in 1970 during a simulated dive to 1500 fsw at the RNPL (Bennett 1970, 1971). The dive's success was attributed to a significantly slow compression rate and interim stops at 600, 1000, 1100, 1200, 1300, and 1400 fsw. Symptoms of HPNS were present upon arrival at each of the interim depths, but they dissipated or were on the decline during periods spent at constant pressure.

Finally, a series of dives by the Compagnie Maritime d'Expertise (COMEX), Marseilles, employing a compression rate governed by an equation which takes into account the necessity of maintaining a certain gradient between the dissolved inert gas tension in the "fastest" and "slowest" body tissues (Fructus et al. 1972), attained a depth of 1700 fsw in 1970 (COMEX 1970) and 2001 fsw in 1972 (COMEX 1972). Symptoms of HPNS were present at depths beyond 1000 fsw and were observed to be more pronounced after sleeping at constant depth than during compression phases.

2. High Pressure Nervous Syndrome

The "high pressure nervous syndrome" or HPNS (Brauer et al. 1969, Fructus and Agarate 1971) is a syndrome of neurological and physiological dysfunctions which appear in men compressed to high hydrostatic pressures. Onset and severity seem to be a function of both compression rate and absolute pressure. The condition has been characterized by motor disturbances (such as tremor and uncontrolled muscle jerks); somnolence; significant EEG changes (which may signal imminent convulsions); and visual disturbances, dizziness, and nausea.

The HPNS should not be confused with inert gas narcosis. In fact, the two phenomena seem to act in opposition to one another, with HPNS causing a general excitation in the central nervous system (CNS), and narcosis acting as a manifestation of a less active CNS (Lever *et al.* 1971).

In animals, at pressures greater than those yet attained by man, the syndrome develops into motor seizures and convulsions accompanied by EEG changes characterized by paroxysmal, or spiking, discharges which increase in frequency as the syndrome builds up toward generalized epileptiform seizures (Brauer 1972a).

Compression rate has an influence; the onset and severity of the syndrome are enhanced by fast compression but can be minimized by slow compression with stops at interim depths. Absolute pressure is also important in that symptoms become more intense as absolute pressure is increased. Also, as depth increases, the sensitivity to rate also increases, such that even small changes in pressure must be made quite slowly. Symptoms appear progressively and some divers are more rapidly and/or severely affected than others. Symptoms are not usually accompanied by decreases in mental performance. Hand tremors may hamper the performance of certain tasks, but they can be overcome by intentional, concentrated effort (Zaltsman 1968). Generally, divers accommodate in time and recover from the symptoms after stays at constant pressure or during recompression. At depths beyond 2000 fsw, it appears that there may be a point beyond which increased time at depth results in a deterioration of condition rather than recovery.

a. Tremor

By far the most common effect of rapid compression is tremor. This section discusses the physiology of tremor, both normal and as a result of HPNS.

(i) Normal Tremor. Tremors of various sorts are present in the body at all times, and are considered to be a part of normal physiology. One classification of tremor is given in Table IV-24.

1. *Rest tremor.* Normal rest tremor or microtremor seen in relaxed muscles is of very low amplitude, 10–80 μm displacement, and usually has its greatest amplitude at a frequency of 6–13 Hz. Normal rest tremor is probably a cardiorespiratory phenomenon but it may result from continual contraction of individual muscle fibers (Brumlik and Yap 1970).

2. *Postural tremor.* When a hand or finger is held in a fixed position, it is the result of isometric contraction of antagonistic muscles. The tremor associated with this condition, in contrast to that of relaxed muscles, is known as postural tremor. Normal postural tremor has a predominant frequency about the same as rest tremor, 7–12 Hz, but involves excursions of up to fractions of a millimeter, and can be seen visually. The physiological origin of this type of tremor is believed to be neural, resulting from

Table IV-24

Characteristics of Normal and Abnormal Tremor[a]

Clinical examples	Peripheral characteristics		
	Frequency, cycles/sec	Amplitude[b]	Waveform
I. Normal tremor			
A. Normal rest tremor (NRT) — BCG	8–12	80 μm	BCG
B. Normal postural tremor (NPT) — Postural effort (isometric contraction)	8–12	1–5 mm	Irregular, continuous
C. Normal intentional tremor (NIT) — Intentional effort (isotonic contraction)	8–12	1–5 mm	Irregular, continuous
II. Abnormal tremor			
A. Abnormal rest tremor (ART) — Extrapyramidal disease (e.g., the tremor of parkinsonism)	3–8	Variable, mm–cm	Regular, rhythmic
B. Abnormal postural tremor (APT)			
1. Continuous — a. Alcoholism, anxiety, thyrotoxicosis	8–12	mm	Variable, regular to irregular
b. Extrapyramidal diseases (parkinsonism) and cerebellar diseases	3–8	Variable, mm–cm	Regular, rhythmic
2. Intermittent (shuddering) — Shivering, tremorine	8–12	mm	"Bursts" of regular tremor
C. Abnormal intentional tremor (AIT) — Cerebellar diseases (e.g., the tremor in multiple sclerosis)	3–8	Variable, mm–cm	Regular, rhythmic

[a] Reprinted from Brumlik and Yap (1970) with permission of Charles C Thomas.
[b] Approximate amplitude, dependent on, among other variables, the transducers employed. NRT is not visible to the unaided eye; NPT may or may not be visible, whereas the other tremors described are clinically apparent.

alternating impulses from either the spinal cord or the brain directed to opposing muscles, or from servoloop mechanisms between the muscles and the spinal cord. These tremors are affected by changes in mass and hence the resonant frequency of the extremity (Stiles and Randall 1967).

3. *Intention tremor.* Another type of tremor, intention tremor, is manifest when the individual is making a definite muscular movement. In contrast to postural tremor, contraction is isotonic rather than isometric, and there is a continual change of tension between antagonistic muscles (Brumlik and Yap 1970).

(ii) Abnormal Tremor. All normal tremors (rest, postural, and intention) tend to show a large frequency component between 8 and 12 Hz, while abnormal tremors (such as those found in parkinsonism and cerebellar disease) are usually between 3 and 8 Hz.

(iii) Tremor Measurement. Measurements of tremor usually are directed toward both amplitude and frequency, and may be made with force, displacement, or acceleration transducers, using various electrical, mechanical, optical, or magnetic means (Bennett 1971, Stiles and Randall 1967, MacInnis 1970, Bachrach *et al.* 1971, Langley 1970).

Performance-oriented measurement of tremor—manipulating small ball-bearings with a pair of tweezers (Bennett 1965)—reflects the subject's ability to perform tasks requiring fine neuromuscular coordination, whereas quantitative analysis of tremor by the use of power spectral density analysis (Randall 1967) indicates that each individual has his own characteristic rhythm or tremor "signature" (Bachrach *et al.* 1971). This has possible applications for monitoring slight changes in neuromuscular performance. Typical tremor "signatures" are shown in Figure IV-40.

(iv) HPNS Tremor. A number of investigators have reported the occurrence of marked tremors of the hands and body during actual and simulated dives to depths of 500 fsw and greater (Bennett 1965, 1970, 1971; Brauer 1968; Bühlmann *et al.* 1970; Hamilton *et al.* 1966). They may affect one diver but not another under identical conditions. Mild tremors may be intentionally overcome, especially during the performance of coordinated movements. They may dissipate spontaneously or precede other HPNS symptoms. These tremors not only interfere with diver coordination and performance, but may also serve as an early warning sign of more serious neuromuscular disorders.

The HPNS condition can be compared with the akinetic-rigid syndrome arising in parkinsonism as a result of impaired functioning of certain parts of the brain. In both cases the tremor is stereotyped; has a frequency of 5–8 Hz; is rhythmic; is propagated to the extremities, torso, and head; is most clearly expressed in the hands; and decreases during spontaneous movements (Zaltsman 1970). The effects observed grossly resemble the type of tremor brought about by lesions in the cerebellum. Rhythmic tremors with a frequency of 5–8 Hz afflicting the upper extremities, especially the hands, torso, and sometimes the lower jaw, were observed in subjects breathing a 10–14% oxygen-in-helium mixture at pressures of 14 and 16 atm in a compression chamber (Zaltsman 1970). At a constant pressure the tremor disappeared after 3–5 min. Acts of will suppressed the tremor to a considerable degree, but the tremor hindered work performance requiring fine coordination. Bachrach and Bennett have reviewed the role of tremor in diving (1973).

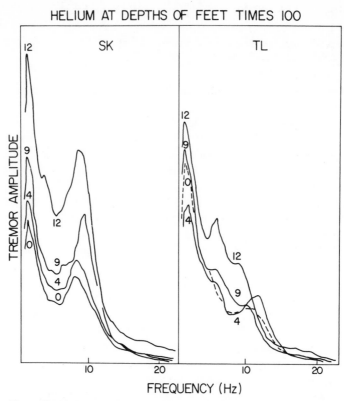

Figure IV-40. Power spectra of microtremor recorded from two different subjects under the same conditions. Numbers show equivalent depth in hundreds of feet during a chamber experiment. Curves are composite averages of nine samples each. [From Thorne *et al.* (*in press*) by permission of the authors.]

b. Convulsions

Brauer (1972*a*) reported that experiments using rhesus monkeys to study the relative narcotic potencies of helium and hydrogen had found, contrary to expectations, that the monkeys developed convulsions and paroxysmal (sudden) EEG discharges similar to grand mal epileptic seizures instead of narcosis. Later studies determined that this phenomenon was not a result of temperature, oxygen and carbon dioxide concentrations, or noise. The conclusion reached was that the hyperexcitability of the central nervous system had to be attributed either to the high pressure as such, or to a change in the pressure over many atmospheres.

The convulsion syndrome as seen on the EEG consists of generalized paroxysmal (sudden) discharges of the cerebral cortex, frequently preceded by episodes of focal spiking, simulating epileptiform seizures. The motor symptoms show two characteristic stages. An early stage is unaccompanied by gross brain wave abnormalities and is characterized by tremors increasing in severity with increasing pressure. The second, or motor-seizure, phase ranges in severity from isolated involuntary muscle jerks to maximal tonic or clonic seizures, and is correlated with the generalized paroxysmal activity of the EEG. The early tremor stage is not associated with any recognizable changes in the EEG. Early myoclonic (muscle tensing) episodes tend to be associated with focal spiking, and these spikes tend to increase in frequency as the syndrome builds up toward generalized seizures.

Studies so far seem to indicate that the epileptiform seizures are not of the autonomic type, since dilantin, which is commonly used to treat epilepsy, has no effect in raising the convulsive threshold in animals (Brauer 1972b).

In studies using animals, tremors have always preceded convulsions. The convulsion threshold appears to be a highly stable characteristic of individual species (Brauer 1972a). Under the conditions of a constant compression rate of 24 atm/hr (with an oxygen partial pressure of 0.5 atm, and ambient temperature of 30–33°C), the mean convulsive threshold (P_c in atm) in all species was closely correlated with the mean tremor threshold (P_t in atm) (Brauer 1971). The data are described by the regression equation

$$P_c = 0.943P_t + 41.5 \qquad\qquad (6)$$

with a correlation coefficient of 0.82. These findings indicated that the determination of the tremor threshold in other species would provide a rational approach to the prediction of the convulsive threshold.

In a series of ten human dives performed as part of the series Physalie (Brauer et al. 1969) (compression rate 24 atm/hr; oxygen partial pressure 0.5 atm; and ambient temperature 30–33°C) the tremor thresholds varied from 22 to 27 atm, with a mean value of 26.4 atm. Based upon the previous equation, Brauer (1971) postulated that the convulsion threshold for human subjects under the same conditions would be 66.3 ± 7.8 atm, or about 2300 fsw.

Studies on monkeys and mice indicate that the mean convulsion thresholds are substantially independent of oxygen partial pressure over a range of values from 0.2 to 1.0 atm, but that the thresholds are significantly correlated with compression rate. Keeping all other conditions constant, the relation between the convulsion threshold P_c and compression rate for these two species is given by a common regression equation

$$\bar{P}_c/\bar{P}_c{}^{24} = 1.2 - 0.17 \log \dot{P} \qquad\qquad (7)$$

where $\bar{P}_c{}^{24}$ is the mean convulsion threshold at a compression rate of 24 atm/hr, and $\dot{P}$ is the compression rate in atm/hr (Brauer 1971).

If this same relation is applicable to man, it would predict that the mean convulsion threshold at a compression rate of 24 atm/hr can be increased measurably by slowing the compression rate. At 1 atm/hr the expected mean convulsion threshold in man would be approximately 81 atm, while at the fast rates of compression characteristic of some forms of submarine escape, the same threshold might fall as low as 46 atm (Brauer 1972a). Figure IV-41 is a graph of predicted convulsion thresholds in man as a function of compression rate as predicted by the previous equations.

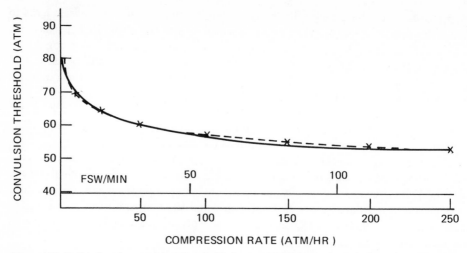

Figure IV-41. Predicted convulsion thresholds in man (solid line) as a function of compression rate as predicted by Brauer (1971). The curve is calculated on the basis of the relationship of tremor and convulsive thresholds in animals, and the observed tremor threshold in man.

c. Somnolence and EEG Changes

In 1968 a series of simulated dives (Physalie) were performed by the COMEX organization (Brauer 1968, Fructus *et al.* 1971), the objective of which was to bring man as close as possible to the depth at which it was believed convulsions might commence. The three deepest dives of the series ranged from 1100 to 1189 fsw, and the compression rates were such that maximum depths were attained within a little over 2 hr. Tremors were noticed in both subjects on each of the dives, the onset pressure ranging from 22 to 28 atm. The effect increased in severity as the pressure was increased even though compression rates were progressively decreased. The tremors became rather severe beyond 25 atm, and included gross motor difficulties, suggesting an increasing prominence of spastic movements underlying generalized fasciculation. EEG changes began to appear at pressures close to 30 atm, the threshold being consistently lower for one subject (RV) than for the second (RB) in the three exposures to maximum pressure attained by each. The EEG changes were characterized by the appearance of slow waves in the theta range, with a frequency of approximately 5–6 Hz. These increasingly displaced the alpha waves in the EEG pattern of the resting subject. Such symptoms increased in severity as the pressure was increased, and above 33 atm, were associated with behavioral effects, including intermittent bouts of somnolence ("microsleep") occupying an increasing portion of the time, associated with intermittent stage 1, later stage 2, sleep patterns in the EEG. Once again, these changes were noted sooner and more severely in RV than in RB. As pressure exceeded 35 atm, theta wave activity persisted even while the divers were actively performing tasks. As 36 atm, RV appeared to be asleep 50% of the time. Transition from sleep to wakefulness seemed to be almost imperceptible: A word or a light touch sufficed to initiate the transition, and withdrawal of stimulation sufficed to permit return to a somnolent state. Due to the severity of the behavioral and EEG changes, the dive was terminated after 4 min at a maximum pressure of 36.5 atm (1189 fsw). EEG changes persisted during decompression for approximately 10 hr in one subject monitored during Physalie II, and for more than 12 hr in both subjects during Physalie III. These early phases of decompression were associated subjectively with fatigue and a degree of depression. A strikingly similar description of the "micro-sleep" phenomenon came from a dive to 1000 fsw in Bühlmann's laboratory (Schreiner 1969).

A more recent simulated dive to 1500 fsw at RNPL made stops at interim depths, including 24-hr periods at constant pressures of 600, 1000, and 1300 fsw (Bennett 1971, Bennett and Towse 1971). On reaching stable pressure, the EEG showed a continuation of the changes initiated by compression. With eyes either open or shut on arrival at 600 fsw in subject PS, theta activity and, to a lesser extent, delta activity continued to increase for 6 hr. These changes were seen most readily when the subject's eyes were open. A 25% peak increase in theta was reached, which gradually returned to normal after an additional 12 hr at pressure. Compression to 1000 and 1300 fsw initiated similar cycles, with the theta activity increasing to 75% above control level during the first 6 hr at 1000 fsw, and 90% at 1300 fsw, tending to return to lower values during the subsequent period at constant pressure. On compression to 1500 fsw there was another increase in theta, but not as severe as occurred at 1000 and 1300 fsw. These changes are illustrated in Figure IV-42. In the other subject, changes in slow activity were not so clear, probably due to large amounts of delta and theta activity in the controls. On return to the surface, the delta activity had recovered to normal and there was a significant recovery of all activities other than theta, which remained high when the eyes were shut. These EEG changes were preceded by remission of the signs and symptoms well before reaching the surface.

The HPNS somnolence appears outwardly very much like narcolepsy; both are characterized by the subject falling asleep if unstimulated. However, there are significant differences between the two. Normal sleep consists of two different states. One is the phase of sleep associated with rapid eye movements (REM); the other, characterized by a typical EEG of slow waves, is called nonrapid eye movement (NREM) sleep. In normal individuals, NREM sleep always precedes REM sleep. HPNS somnolence is accompanied by NREM (slow wave) EEG patterns, but they persist even while the subject is apparently awake and spontaneous. In patients affected by narcolepsy, however, REM sleep is almost instantaneous, with no preceding phase of NREM sleep (Merck Manual 1972).

d. Dizziness and Nausea

Symptoms of dizziness, vertigo, nausea, and vomiting have been reported during compression to, and during constant pressure at, depths greater than 500 fsw (Bennett 1965, Bennett 1971, Brauer et al. 1969, Bühlmann et al. 1970, Chouteau et al. 1971). These symptoms, which tend to dissipate with time at constant pressure, are usually seen secondary to tremor and the other aspects of HPNS.

There are other causes of dizziness and nausea in diving and in the absence of other HPNS symptoms it is likely that some other cause may be involved. Other possible factors might be alternobaric vertigo or other vestibular effects of changing

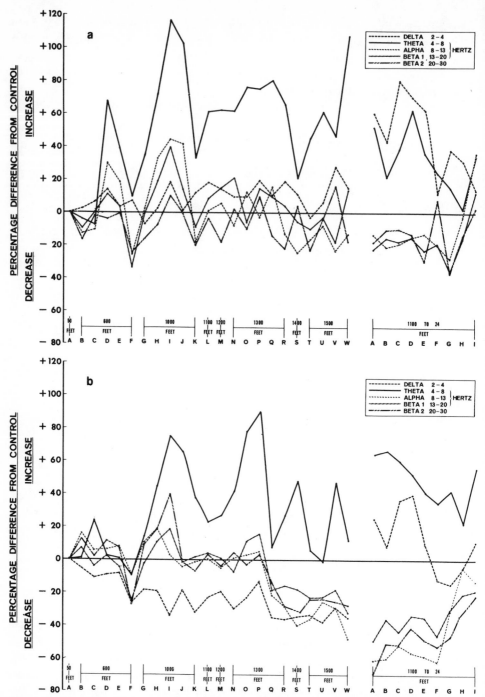

Figure IV-42. Subject PS. Analysis of percentage change in the delta, theta, alpha, beta 1, and beta 2 EEG activity at various stable stages of a dive to 1500 fsw, as compared to control measurement at 50 fsw. (a) EEG eyes open; (b) EEG eyes closed. [From Bennett (1971) by permission of the Marine Technology Society and the author.]

pressure, buildup of CO_2 due to inadequate ventilation during rapid compression, cold water in the external ear canal, and motion sickness (see Chapter V, discussion of vestibular function).

3. Hyperbaric Arthralgia

The same dramatic increases in diving depth that led to the discovery of the high pressure nervous syndrome disclosed another new problem of deep diving, hyperbaric arthralgia. This is a term used to describe an ill-defined set of symptoms ranging from vague discomfort to outright pain in various joints, following and as a direct result of being compressed to high pressures. Some authors (Fructus *et al. in press*, Chouteau *et al.* 1971) classify hyperbaric arthralgia as a component of HPNS, and rightly point out that slow compression works to alleviate both conditions, but since HPNS appears primarily to affect the nervous system and the pain of hyperbaric arthralgia seems to have a mechanical origin, they are considered separately here.

During the Ocean Systems 650-fsw chamber dive in 1965, both divers experienced joint pains. Compression took 1 hr to reach 650 fsw. At 300 fsw one diver became aware of stiffness and mild pain in both wrists, and shortly thereafter the other diver began to complain of pain in his lower back. The wrist pain gradually subsided after maximum pressure was reached, while the back pain virtually disabled the diver for one day and did not go away for still another (Hamilton *et al.* 1966). In the same laboratory, during a series of short (up to 40-min) compressions followed by direct decompression without saturation, these and other divers had experienced the wrist stiffness, as well as stiffness in shoulders, elbows, and knees. They called it "no joint juice," a term with which other descriptions agree, and one which may in fact be related to the cause. Interestingly, in more than 40 man-dives to greater than 500 fsw, all at a compression rate approaching 100 fsw/min, that laboratory experienced only one case of debilitating pain, and that was in the dive that took 1 hr to compress (Hamilton and MacInnis 1965).

In both SEALAB experiments similar symptoms were observed, and all the men in the Conshelf III dwelling reported pain in the joints. Similar symptoms have been reported during the Centre d'Etudes Marines Avancees (CEMA) Saturation I and II dives, with arthralgias and muscular pains becoming more severe the morning after arriving at 250 m (825 fsw). During a simulated dive by AIRCO, International Underwater Contractors, and the U. S. Naval Submarine Medical Research Laboratory, one of the divers experienced pain in the hip region shortly before arriving at 180 m (594 fsw) and both divers complained of lower back pains during compression and shortly after arrival at 240 m (800 fsw). This pain disappeared the next morning, about 11 hr after arrival. During the 1500-fsw simulated dive at RNPL, divers reported creaking in the joints precisely on arrival at the stops at 183 m (604 fsw) and 305 m (1007 fsw), and the symptoms became less severe during following stops. These and other relevant experiences have been reviewed by Chouteau *et al.* (1971).

In a comprehensive review of over 200 U. S. Navy helium–oxygen dives covering the depth range between 100 and 1000 fsw, Bradley and Vorosmarti (1972) reported that divers described hyperbaric arthralgia as a sensation of the joints being "dry and grainy," "a slight sprained feeling," and "a deep, sharp pain with sudden movement." Intensity of symptoms increased with increasing depth, as did involvement and persistence of the pain. About 25% of the divers had symptoms by 250 fsw, 50% by 450 fsw, 75% by 725 fsw, and 95% by 1000 fsw. Joint "creaking" as a result of strain or rapid movement was also seen to increase with depth.

The shoulder was most frequently affected and, in decreasing order, knees, wrists, hips, and back. Involvement was apparently unrelated to age of the diver. Symptoms correlated significantly with compression rate, with more involvement and more pain on dives having faster compressions. This analysis, however, considered all compressions faster than 2.5 fsw/min as "fast."

One aspect not discussed in Bradley and Vorosmarti's report is the time required for the syndrome to develop—it may not be fully developed for many minutes following the start of compression. The type of commercial diving conducted in support of offshore oil production often involves short tasks at depths in excess of 600 fsw. These are routinely accomplished by rapid compression (100 fsw/min) dives with short (up to 1 hr) bottom time. Hyperbaric arthralgia is not unheard of in this diving mode, but it does not constitute an operational limitation.

In saturation diving, however, hyperbaric arthralgia is a serious operational problem. Current procedures for saturation dives at depths beyond about 400 fsw use slow compression rates, generally in combination with "hold" periods during which compression is halted for perhaps several hours.

4. Hyperbaric Bradycardia

The early literature on diving physiology contains the almost unanimous observation that pulse rate is slowed under hyperbaric conditions (Hoff 1948). This effect can be partially explained by the attendant increases in oxygen partial pressure. Recently, however, exposures to much greater pressures under more or less normoxic conditions have shown that there seems to be a pressure-related component. This appears to be a real effect, but its mechanism is still undetermined. More important, perhaps, is the fact that this phenomenon seems to have little operational importance.

A 12–18% decrease in heart rate was reported during a 650-fsw simulated dive at Ocean Systems (Hamilton *et al.* 1966). Heart rates of the Sealab II subjects remained reduced as compared to baseline levels, but with a rather marked increase seen on the third day (Hock *et al.* 1966). During a 23-atm dive in Zurich (Waldvogel and Bühlmann 1968) heart rates were in the lower range of the norm while at pressure (accredited possibly to lack of physical activity) but were markedly low at night. A 1500-fsw simulated dive at RNPL (Bennett 1970, 1971) reported a bradycardia in both divers at 600 fsw, which became more pronounced at 1000 fsw. Thereafter the bradycardia was a little less pronounced and rates were no slower at 1500 fsw. Bradycardia was also seen during a rapid-compression simulated dive at 1000 fsw at Ocean Systems (Langley 1970).

That this effect on cardiac activity is a result of pressure, rather than any effect of helium on the body, is supported by the findings that mice breathing oxygenated fluorocarbon liquid may exhibit a decrease in heart rate that varies almost linearly with increasing pressure (Ornhagen and Lundgren *in press*).

Hyperbaric bradycardia is not seen in all cases of exposure to pressure, but since heart rate is subject to many additional factors, reports of the occurrence of brady-cardia may depend on how carefully and consistently measurements are made (Bradley *et al.* 1971). It is important to distinguish between hyperbaric bradycardia and the diving reflex, a distinct bradycardia seen primarily in diving animals, but also in humans and related to immersion of the face in water.

5. *Etiology and Mechanisms*

a. HPNS

Originally associated with deep helium–oxygen dives and hence referred to as "helium" tremor, HPNS is clearly related to both rate of compression and the total pressure attained. Though the association with helium remains, the specific effect of helium as an irritant [as suggested by Zaltsman (1968)] is probably not an essential element, except perhaps for what helium does not do, rather than for what it does. More about this below.

One specific mechanism advocated by numerous investigators as a possible cause of HPNS and hyperbaric arthralgia is that of gas osmosis (Kylstra *et al.* 1968, Chouteau *et al.* 1971, Bradley and Vorosmarti 1972). Clearly, gas molecules in solution exhibit colligative properties. The compression-rate and total-pressure sensitivity and the tendency for symptoms to dissipate with time at constant pressure all support the idea that gas osmotic differences in tissues or cells could cause fluid shifts and changes in membrane permeability and hence the observed symptoms. Difficulties with this mechanism are that rather small osmotic effects of the relatively insoluble gases (e.g. helium) in comparison with the rather high osmotic effects of the usual solutes found in body fluids, the lack of knowledge that a suitable membrane exists to allow diffusion of water and not gas, and the fact that HPNS is seen in fluid-breathing animals not subjected to increased gas pressure (Kylstra 1967).

Another theory is that HPNS symptoms are in some way due to direct mechanical effects of pressure, pressure per se. Pressure has been shown to have detrimental effects on biological functions in general, including muscle and nerve function, and it is reasonable that some of these effects may be involved. The possibility that HPNS tremor may be due to shivering is unlikely. HPNS tremor frequency is predominantly 5–8 Hz, while that of shivering is 8–12 Hz. Further, compression produces heat, not chilling.

An elegant theory which draws on both the presence and the nature of the inert gas as well as effects of hydrostatic pressure is currently the best candidate to explain the observed phenomena. This theory deals with a problem broader than HPNS, that of general anesthesia. It is related to earlier ideas, such as the classic Meyer–Overton theory and Sears'observations on lipid solubility and resulting volume changes (Sears and Fuller 1968). Called the "critical volume hypothesis" by Miller *et al.* (1961) and "lipid free volume" by Stern and Frisch (1973), it holds that the narcotic potency of "inert" and anesthetic gases is related to the lipid volume increase (presumably cell membranes) which the gases cause. When dissolved in the lipids this volume increase can be counteracted by hydrostatic pressure, and pressures great enough to counteract the volume change of an anesthetic gas are found also to counteract the anesthesia

(Lever *et al.* 1971, Brauer *et al. in press*). An extension of this idea is that compression of the appropriate lipid in the absence of enough of a suitably soluble gas (or anesthetic) should cause a hyperexcitable membrane, and hence HPNS. The interaction of narcotic gases and HPNS is discussed below under alleviation.

b. Hyperbaric Arthralgia

Direct pressure effects may also be involved in the mechanism of hyperbaric arthralgia. A joint involves multiple layers of tissue of different properties. If the different layers have different compressibility properties, then hydrostatic pressure acting on the entire joint might well distort the tissue, causing shearing forces and pain (Workman 1969). It is not known whether a man subjected to high pressures in the absence of high partial pressures of gas would be spared this arthralgia, nor has the mitigating effect of a mixture of a narcotic gas been quantified.

The load-bearing surface of a joint is a "spongy" matrix whose spaces are filled with a gel. A very thin layer (0.01 μm) of the gel is squeezed into the joint space to "lubricate" the joint (Faber *et al.* 1967). The formation of this thin layer could be disturbed by high hydrostatic pressures, which have been shown to affect sol–gel relationships (Fenn 1969). Cavitation of joints has been verified at normal pressure (Unsworth *et al.* 1971); the cracking of joints observed at depth may cause pain as a result of cavitation (Bradley and Vorosmarti 1972).

Osmosis has also been proposed as an explanation for the effects of hyperbaric arthralgia, since osmotic gradients could result in dehydration of articular surfaces (Kylstra *et al.* 1968, Bradley and Vorosmarti 1972).

6. *Alleviation*

Regardless of whether HPNS and hyperbaric arthralgia are caused by the same or different mechanisms, similar procedures seem to be useful in reducing the presence and intensity of the symptoms.

a. Compression Schedule

For short working dives of less than 1 hr, at depths less than 700 or 800 fsw, compression rates of up to 100 ft/min may be used. For compression to depths beyond 700 fsw, slower compression rates (less than 10 ft/min) and compression stops (of one to many hours) should be used for postponement of HPNS symptoms. In general, for deeper dives, increasingly slower compression rates and increasingly longer stops at intermittent depths are required. For example, during the RNPL dive to 1500 fsw, 24-hr stops at 600, 1000, and 1300 fsw were used (Bennett 1970, 1971).

Two mathematical models of compression have been developed which take into account the necessity of maintaining a specific gradient between the dissolved inert gas tension in the "fastest" and "slowest" body compartments; this gradient is a function of depth. In general, both these models must be accomplished by a continual compression at a progressively decreasing rate, or compression at a uniform rate interrupted by progressively longer stops at interim depths.

(i) COMEX Model. If P is the depth, $t_{1/2}$ is the half-time of the tissue, and $G(P)$ is the proposed gradient function, the compression speed is given by

$$\frac{dP}{dt} = \frac{(\log 2)G}{t_{1/2}(1 - G')} \tag{8}$$

where G' is equal to dG/dP (Fructus *et al. in press*). Using this relationship, simulated dives to pressures equivalent to 1700 fsw (COMEX 1970) and 2001 fsw (COMEX 1972) were attained. Although symptoms of HPNS were present, they appeared to be within a tolerable level.

(ii) CNEXO–CEMA Model. Inert gas dynamics in the body can be described by the following gas transport equation (Chouteau *et al.* 1971):

$$d\pi/dt = k(P - \pi) \tag{9}$$

which has a special solution where the rate of change of pressure is constant (including zero), or $dP/dt = C$:

$$\pi = P_{I_0} + C\left(t - \frac{1}{k}\right) - \left(P_{I_0} - \pi_0 - \frac{C}{k}\right)e^{-kt} \tag{10}$$

where π is the partial pressure of dissolved inert gas; t is time; k is the specific time constant of inert gas exchange, $k = (\ln 2)/t_{1/2}$; P_I is the partial pressure of inspired inert gas; C is the rate of change of P as a function of time; P_{I_0} is the partial pressure of inspired inert gas at $t = 0$; and π_0 is the partial pressure of dissolved inert gas at $t = 0$.

These will be recognized as the equations used for computation of decompression schedules (Schreiner and Kelley 1967, 1971).

The tension π of an inert gas dissolved in a tissue or compartment of half-time $t_{1/2}$ is never in equilibrium, during compression, with the partial pressure of the inert gas in the mixture inhaled. Thus there remains a pressure gradient ΔP which is equal to the difference between the partial pressure of the inert gas in the inhaled mixture P_{I_0} and the tension π of the inert gas dissolved in the compartment of half-time $t_{1/2}$. The equation for the pressure gradient is

$$P_I - \pi = \Delta P = \frac{C}{k}(1 - e^{-kt}) + (P_{I_0} - \pi_0)e^{-kt} \tag{11}$$

Chouteau *et al.* (1971) have made comparisons of the results of various animal and human experiments with the calculated values of the gradients of the inert gas tension (particularly helium) in tissues, and have found a constant and significant correlation between symptoms of HPNS and a maximum value of the given gradient ΔP. They determined that if compression with helium was made slowly enough so that a ΔP of 8–10 atm could be maintained in the 120-min compartment ($t_{1/2} = 120$), then serious HPNS manifestations could be avoided.

It is significant that this method is not consistent with the most promising model describing the mechanism of HPNS (Lever *et al.* 1971, Miller *et al.* 1973, Stern and Frisch 1973), but, like the decompression equations on which it is based, it works.

b. Pharmacological

As a converse of the pressure reversal of anesthesia, the lipid free volume theory (Johnson and Miller 1970, Lever *et al.* 1971) indicates that HPNS may be at least partially reversed by adding narcotic gases to the breathing mixtures or by administering appropriate drugs.

(i) Narcotic Gases. Brauer (1971, 1972*a*) and his colleagues have performed quantitative determinations on the use of narcotic gases to combat HPNS, using animals in compressed-gas mixtures containing various proportions of hydrogen, nitrogen, and nitrous oxide in the basic helium–oxygen gas mixture. The tremor as well as the convulsion thresholds were found to vary linearly with the narcotic potency of the gas mixture and mixtures of equivalent narcotic potency showed equal mean convulsion and tremor thresholds regardless of the chemical composition or density of the gas mixtures. This is illustrated in Figure IV-43. It was found that adding as little as 0.3% nitrous oxide to helium protected mice and monkeys to the extent that apparently conscious and functionally intact animals could be subjected to pressures as least 30% higher than those attainable in pure helium–oxygen.

It may prove to be particularly advantageous that hydrogen seems to exert enough narcotic effect to permit deeper diving free of HPNS since its low density would permit hydrogen to be breathed at depths greater than would be possible with helium.

(ii) Drugs. Findings have led to the exploration of a variety of barbiturates and of certain muscle relaxants (of the oxazolamine type) with predominantly central nervous system action sites. These compounds have been used to raise convulsion thresholds in monkeys by about 50% or more above values in untreated animals (Brauer 1972*a*, 1972*b*).

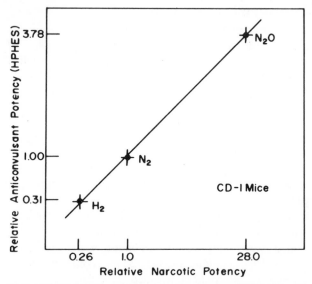

Figure IV-43. Relative anticonvulsant potencies of N_2O, N_2, and H_2 in He–O_2 against HPNS convulsions as a function of narcotic potency in CD1 mice. [From Brauer *et al.* (1972) by permission of Doin, Paris.]

In general, inert gas anesthetics are relatively more effective against tremor than against the convulsion phase of the syndrome, whereas barbiturates affect convulsion thresholds rather than tremor (Brauer *et al. in press*).

Although oxygen convulsions can be suppressed by barbiturates and other CNS depressants (Davies and Davies 1965), cellular and enzymatic oxygen poisoning probably continues to develop (Lambertsen 1966). It is possible, but less likely, that permanent neurological damage may result from exposure to pressure in the ranges that would normally cause convulsions but which are suppressed by artificial means.

Finally Brauer (1971) has reported that sodium pentobarbital, in doses as small as 20% of those producing surgical anesthesia, can provide a great deal of protection against high-pressure tremor and convulsions. However, animals compressed under these conditions ultimately develop motor seizures which differ markedly in important respects from the epileptiform convulsions seen in untreated animals at much lower pressures. He suggests that these changes may represent a new phase of HPNS.

7. Areas in Need of Further Study

The principal explanations of the cause of HPNS are pressure per se and rate of change in pressure. Both are indicated in that the compression rate becomes more critical with increasing compression rate. Whether this is due to two separate effects superimposed upon one another, or whether they are directly interrelated, is not known.

Although symptoms can be delayed by slower compression rates, there must be a physiological limit yet unknown to the depth attainable by man. This limit may be pragmatically unattainable due to the time required to reach and return from such great depths.

Although the etiological explanations implying hypoxic and circulatory mechanisms do not appear to be the sole primary causes of HPNS symptoms, their exact role should be determined. The effect of such drugs as Levodopa (L-dopa, used in the treatment of parkinsonism) on HPNS tremor should be explored.

Although high-pressure convulsions have not occurred in man, and other symptoms (such as major EEG changes and somnolence) appear to be limiting factors encountered before convulsions, the possibility of convulsions must not be disregarded, especially in view of the fact that in animals the distance between the onset of tremor and that of convulsions decreases with decreasing rates of compression.

It is important to investigate whether dizziness and nausea are due to direct pressure effects on middle-ear function, or to effects on the central or autonomic nervous systems. Also, specific values of rapid compression and physiologically safe bottom times should be determined. Evaluation should be made of the tradeoff between rapid compression followed by immediate decompression and slow compression and decompression. Adding narcotic gases to the breathing mixture may help to alleviate the cause of HPNS symptoms; their use should be further explored.

G. Cold

Of all man's natural enemies, none is more awesome than the cold. And nowhere is cold a more relentless foe than in the oceans, whose temperature is too low in most places for man to survive unprotected for more than a little while. The need for life-support systems that can reliably provide proper skin and breathing-gas temperatures is still a stumbling block to man's optimal performance in the oceans. It is vital that underwater bioengineers understand this problem.

1. Thermoregulation in Man*

For the purposes at hand, it is necessary to condense the extensive literature on thermoregulation. Detailed discussions are presented in the standard textbooks of physiology. A recent symposium (Senay 1973) dealt with some of the newer concepts, and Bligh (1973) has summarized the current information and its interpretation in an excellent book.

a. The Role of Hypothalamus

The neurons of the primate hypothalamus have the ability both to sense and to respond to deviations in temperature from the usual value (37°C for healthy men). By so sensing and responding, the hypothalamus acts to maintain its own temperature at a stable value and to provide fairly constant temperatures in many other vital tissues. It is able to accomplish this regulatory job over a wide range of environmental conditions of temperature, wind speed, and water vapor pressure (humidity).

The hypothalamus regulates temperature in two distinct but related ways. Most important, it causes thermal perturbations to be sensed as discomfort, leading to behavioral changes (clothing, position, room thermostat setting). Second, it sets off a series of functional adjustments on the part of the body to offset the impending thermal imbalance.

(i) Behavioral Response. Although Bligh (1973) properly calls attention to the importance of behavioral thermoregulation, it is clear that behavioral adjustments are much less available to the cold-water diver than to his landward counterpart. To a great degree, the diver waives many of his *physiological rights* when he dives in cold water (or cold hyperbaric helium–oxygen). If such exposures are prolonged, his behavioral options may be limited to the use or rejection of whatever protective devices are available and functioning, and to possible changes in positioning and area-to-mass relationships (huddling) as may be required by emergency circumstances.

(ii) Autonomic Processes. The effector pathways responsible for the non-behavioral mechanisms of thermal homeostasis are not completely understood. However, they involve changes in sweat-gland function, circulation to the body surface, and, in the case of cold, shivering and nonshivering forms of increased heat generation.

---

* The standard terms and units published by Gagge *et al.* (1969) have been used. Conversion tables in this book will allow engineers to translate these terms to more familiar ones when necessary.

Shivering will occur in most normal persons when the skin is cooled to an average value of 30°C (86°F), but thin males often shiver at a mean skin temperature $\bar{T}_s$ 1–2°C warmer (Craig and Dvorak 1966, Hong 1963). The $\bar{T}_s$ threshold for augmenting thermogenesis is clearly affected by deep body temperature (Benzinger *et al.* 1961), as well as by the insulation of subcutaneous fat, and by other individual characteristics.

The nonshivering form of thermogenesis is a complex metabolic response to cold, thought to be mediated by the neurotransmitters epinephrine and norepinephrine. The other central mechanisms of thermoregulation also involve the release of neurotransmitters. An elegant demonstration of the hormonal nature of central thermoregulation has been presented by Myers and Sharpe (1968). Although the relevance of these findings to the problems of body temperature regulation during immersion may not be apparent, this topic needs further study to elucidate how the body responds to this form of environmental stress, as well as to develop better protective equipment.

b. Maintaining Body Temperature

As in other regulated systems, the body maintains its temperature by striking a balance between heat generation and dissipation, and by using its limited capacity for heat storage. The following equation expresses this balance:

$$M \pm S = C_s + C_r + E + R \tag{12}$$

where M is the thermal equivalent of metabolic activity, S is heat storage, and the four modes of heat dissipation are: C_s, skin convection; C_r, respiratory-tract convection; E, evaporation; and R, radiative heat transfer from the skin to the surrounding surfaces. For a resting subject not undergoing thermal stress, M is about 60 W/m² of body surface. During cold stress, resting values of M may rise to 3–6 times normal (Behnke and Yaglou 1951, Reeves *et al.* 1965). When M is balanced by the sum of these processes, a thermal steady state exists and storage is zero. When this steady state does not exist, S has a positive or negative value which will be reflected in a rise or fall in mean body temperature. When the discrepancy between heat generation and dissipation is so great as to cause deep body temperature in normal humans to exceed 39°C or to fall below 35°C, thermal stress is present and corrective steps should be taken.

c. Effects of Thermal Stress

Even within the 35–39°C limits, thermal stress may exist on a local level in regions of the body exposed to high thermal drain or influx—such as the face or extremities—when they are inadequately protected from temperature extremes. When surface temperatures of the fingers or toes remain below 10°C (50°F) longer than 1 or 2 hr, numbness and tingling sensations may arise that last days or weeks afterwards. Prolonged exposures at colder temperatures are likely to lead to pathological changes in the exposed parts. Even more serious to the diver, however, is the fact that acute effects such as loss of motor and sensory function may quickly render a diver completely helpless and thus cause death by drowning or asphyxia, well before deep body temperature has fallen to dangerous levels.

When the decrease in deep body temperature proceeds unchecked, deterioration in judgment, thinking, and motor performance will occur in many subjects when the rectal temperature falls below about 35°C. Unconsciousness will follow if the exposure continues and, at deep body temperatures below 30°C, death from abnormal cardiac rhythm or circulatory failure will probably occur.

2. Problems of Cold Exposure in Underwater Work

The nature, intensity, and duration of cold exposure in diving depend on details of operational and local conditions, such as temperature, humidity, and wind or water velocity. Response of the exposed personnel will be greatly affected by their individual characteristics and conditioning, as well as by whatever protective equipment they may employ. For one example, the diving women of Korea (Hong 1963) are able to alter their bodies in ways which favor maintenance of central body temperature in the face of seasonal swings in water temperature. For another, such seasonal swings in surface water temperature are important in many latitudes, but not in polar regions. At the poles, of course, the water temperature is often higher than that of the ambient air since sea water freezes at about −2°C (29°F). In such polar diving, a constant-volume dry suit has been reported to afford very good protection (MacInnis 1972) in comparison to other types of protection available (Covey 1972).

The main categories of operations where cold is a potential hazard are immersion and surface swimming; shallow and deep air or He–O₂ diving; saturation diving with dry chambers; and saturation diving with open sea or "wet-pot" chambers. From this list, it is apparent that each category of diving has its own set of problems related to cold. These problems will be discussed next.

a. Immersion and Surface Swimming

This category encompasses the overwhelming majority of exposures to cold, since it includes recreational and accidental exposures, as well as many which occur in commercial diving and in military maneuvers. It is the simplest to analyze and, therefore, has received the most study. Recent reviews (Beckman 1967, Keatinge 1969, Bullard and Rapp 1970) have emphasized man's vulnerability to thermal stress, even in waters of mild temperature. Because their emphasis differs, these reviews are all recommended for bioengineers concerned with man's function in the water. For example, Beckman (1967) analyzes the effects of water temperature, insulation, and the metabolic rate needed to maintain thermal homeostasis during immersion. In this analysis, he has combined the nomographic estimates of Smith and Hames (1962) with his own studies (Beckman et al. 1966, Goldman et al. 1966) on the performance of thermal protection in air and water (Table IV-25). He also considers the theoretical and practical aspects of providing supplementary heating for dives in very cold water (0–10°C).

Mainly because water conducts heat much more readily than does air, the temperature of unprotected skin rapidly approaches the temperature of the water. Thus, even water as warm as 25°C (77°F) can cause thermal stress in many subjects when exposure time is prolonged much beyond 1 hr and thermal protection is not

Table IV-25

Comparative Insulation Value of Some Protective Suits[a]

Suit	Insulation value in air, clo[b]	Insulation value in water, clo[b]
Manikin, nude	0.8	0.1
Dry suit (USN Mk 5a)	2.0	0.6
¼-in. Wet suit, 1 ATA	1.5	0.8
¼-in. Wet suit, 3 ATA	1.0	0.4
¼-in. Wet suit, 6 ATA	—	0.2

[a] Reprinted from Goldman *et al.* (1966) with permission of Aerospace Medical Association. The values for hyperbaric conditions are based on the unpublished findings of Dr. John Betts of the British Sub-Aqua Club's Scientific and Technical Group.

[b] 1 clo = $0.18°C \cdot kcal^{-1} \cdot m^{-2} \cdot hr^{-1}$ = that amount of insulation which will transfer $5.56 \ kcal \cdot m^{-2} \cdot hr^{-1} \cdot °C^{-1}$.

worn. This is true despite the fact that the body reduces blood flow to the peripheral tissues as soon as skin temperature drops below 33°C (Behnke and Willmon 1941). Like most compensatory mechanisms, vasoconstriction in response to cooling is an imperfect adjustment.

In waters colder than 25°C, thermal protection must usually be provided. For example, immersion in still water at 20°C (68°F) causes rectal temperature T_r in resting men clad in swim trunks to fall at a rate of 1.2, 0.7, or 0.3°C/hr, for thin, average, and obese subjects, respectively (Beckman *et al.* 1966). From Keatinge's work (1969) it is evident that exercising subjects lose heat even faster in 20°C water unless they are quite obese. More recently, Sloan and Keatinge (1973) studied deep body cooling of young swimmers in 20°C water. The rate of fall in T_r was closely correlated with overall surface fat thickness, and ranged from 0.02 to 0.05°C/min. Thus, in the absence of obesity or thermal protection, such exposures will rapidly chill the skin, produce shivering, and lower T_r. These effects will cause changes in many vital functions (see discussion below). If unchecked, these changes will hinder performance, produce severe discomfort, and ultimately threaten the health and safety of the divers.

b. Shallow and Deep Dives Using Air or Helium–Oxygen

(i) Shallow Dives. Breath-hold dives and dives that use air or helium–oxygen but do not require decompression are referred to here as shallow dives. They range from dives of unlimited duration at depths less than 10 m (31 ft) and those of up to 5 hr at 11 m (35 ft), to dives with bottom times of only 5 min at depths exceeding 43 m (140 ft). For dives lasting 30 min or more, it is likely that thermal stress will become a problem when the water temperature is less than 25°C (77°F).

As discussed above, muscular activity will hasten the fall in body temperature in nonobese subjects, unless protective suits are worn. This is true for two reasons. First, skin convection C_s will be increased by forms of activity that disturb the layer of still water in contact with the skin. Second, respiratory-tract convection C_r will tend to rise due to (1) the hyperventilation which accompanies exercise, (2) an increase in the density of the breathing mixture with depth (Webb and Annis 1966), and (3) the use of helium, because of the high specific heat of this gas. Figure IV-44 shows the rectal

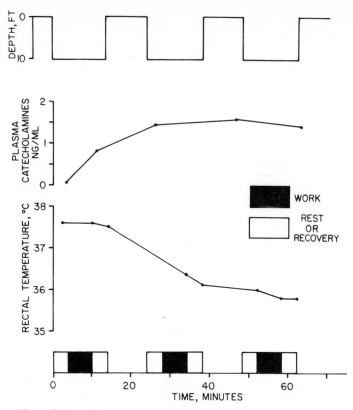

Figure IV-44. Effect of work–rest cycles (in a single subject who exercised in 25°C water at depth of 10 ft) upon deep body temperature and blood levels of catecholamines. When the same work was done in air at 25°C, there was no change in body temperature and the rise in blood catecholamines was much less. These results are typical of lean, athletic subjects, e.g., 1.8 m height, 70 kg weight, 1.9 m² body surface area.

temperature of one subject who breathed air during three successive, brief rest–work–recovery cycles at a depth of 3 m (10 ft) in unstirred 25°C water. The work, repetitive leg extension against calibrated springs, was of moderate intensity [oxygen consumption 1.7 liters/min, STPD (standard temperature and pressure, dry)]. The subject wore only swim trunks and, as shown, he surfaced after each cycle in order to have blood samples taken from a heparinized venous cannula. Rectal temperature fell almost 2°C over the 1-hr period of observation. The prompt and sustained rise in the plasma concentration of total catecholamines (epinephrine plus norepinephrine) provides an index of the stress imposed by the combination of immersion, cold, and moderate work (Raymond, Langworthy *et al. in preparation*). Responses such as these are moderated by conditioning, sex, body composition, and protective equipment, as stated earlier.

   (ii) Deep Dives. Dives that require decompression stops yet are shallow enough to be done without recourse to saturation diving techniques have generally been done

with *hard-hat* diving dress. Such equipment, worn with several sets of long underwear, provides marginal protection in cold water. (Equipment and diving dress for thermal protection underwater is discussed in detail in Chapter IX.) At present, new developments in equipment suggest that hard-hat dress will be replaced by the constant-volume dry suit for routine as well as experimental diving operations. Physiological evidence in support of the advantages afforded by the new dry suit has not yet been published, however.

When helium is present in the breathing mixture, thermal stress becomes more severe (Behnke and Yarbrough 1938). (See also following discussion of deep saturation diving, dry chambers.) This is due to an increase in both C_s and C_r when a helium-rich gas mixture is in contact with the skin and the surface of the respiratory tract. Helium's great ability to transfer heat is magnified by increasing pressure, an effect of density which is true to a lesser degree of any pressurized atmosphere (Raymond 1966). However, no systematic studies of this problem are available except for helium (Hoke *et al. in press*, Goodman *et al.* 1971).

c. Deep Saturation Diving, Dry Chambers

In order to avoid nitrogen narcosis, helium is the most generally used breathing gas diluent at depths beyond 90 m (298 ft). Because of the high thermal conductivity and specific heat of helium–oxygen atmospheres, warmer ambient temperatures (than in air) are required to maintain thermal homeostasis even at normal pressure (Cook *et al.* 1951, Leon and Cook 1960). At increased pressure, helium's heat transfer properties are accentuated, so that ambient temperature T_a must be further increased to provide a desirable skin temperature T_s and ensure comfort (Raymond *et al.* 1968, Varene *et al. in press*). Webb (1970, 1973a) has emphasized the impact of the high convective character of helium at great pressures, requiring divers to have a large caloric intake and a T_a of 36°C at 50 ATA to maintain thermal balance in prolonged exposures. A later study (Raymond, Thalmann *et al. in preparation*) indicated that a T_a of 32°C was high enough for thermal comfort at this pressure. A mean caloric intake of 2400 kcal/day afforded adequate metabolic intake under these conditions. In two subjects studied in detail, resting oxygen consumption was 311–378 ml/min, STPD, which was within the normal range for their surface area. The difference between these findings and those of Webb suggests a need for further study, with special attention to overall metabolic requirements and cutaneous convection.

The relationship between the rate of skin heat transfer C_s and the difference between skin and ambient temperatures can be stated by the equation

$$C_s = h_c(\bar{T}_s - T_a)A \tag{13}$$

in which h_c is the convective conductance (W/m² °C) and A is the surface area. The relationship between h_c and the ambient pressure P of the helium-rich atmosphere was described by Raymond, Thalmann *et al.* (1973) using the equation

$$h_c = 1.08P^{0.78} \tag{14}$$

in which P is in atmospheres absolute and h_c is in W/m² °C. Of two subjects, the man with greater subcutaneous fat had a consistently lower value of both h_c and of conductance between body core and skin, confirming the importance of individual differences in body composition mentioned earlier. The form of Eq. (14) is similar to that given by Varene *et al.* (*in press*), who studied French males over a similar range of pressures.

Since it is not always possible to provide a high T_a during month-long hyperbaric exposures, it may be useful to estimate the changes in body temperature that might be expected in the event of power failure or other accident. Equation (12) can again be used to express the balance between body heat production and loss. Approximate values of evaporative heat transfer E and radiant heat transfer R can be used since they do not change greatly with pressure in hyperbaric helium [6 and 8 W/m², respectively; Raymond *et al.* (1968)]. With special reference to evaporation heat loss, this assumes that relative humidity is about 70%, and that atmospheric motion is minimal. C_s can be calculated from Eqs. (12) and (13), assuming the difference between $\bar{T}_s$ and T_a is 2°C. To compute the convective heat transfer from the respiratory tract C_r, the following equation can be employed (Webb 1970, Hoke *et al. in press*, Goodman *et al.* 1971):

$$C_r = \rho c_p (T_E - T_a)\dot{V}_E \qquad (15)$$

where ρ is the density of the breathing mixture (a function of gas composition and ambient pressure) and c_p is the specific heat (1.24 cal/g °C can be assumed for a mixture with a helium concentration above 95%). T_E is the expired gas temperature, for which a value of 35°C is a satisfactory estimate for resting subjects. $\dot{V}_E$ is the volume of gas exhaled per unit time, at the ambient temperature and pressure, and saturated with water vapor (ATPS). A $\dot{V}_E$ of 10 liters/min may be representative of resting normals undergoing thermal stress in hyperbaric helium, but no direct measurements have been published.

If one applies the above equations and assumptions to the case of unprotected subjects at any desired pressure, it is possible to calculate values for S, the stored heat, for any ambient temperature of interest. For $T_a < 30$°C, values of S would generally be negative in helium–oxygen atmospheres, since the skin would be likely to cool to within 2°C of T_a soon after T_a began to fall from its initial "comfort level." With continuing exposure, $-S$ would increase in magnitude in accordance with Eq. (12). From $-S$, a rate of fall in deep body temperature could then be calculated.

Example. *Given:* Hyperbaric helium chamber, $P = 49.6$ ATA (488 m = 1600 ft). Power failure causes T_a to fall to 20°C (68°F). No thermal protection; diver's height 1.8 m, weight 70 kg, surface area 1.9 m².

Question: At what rate will diver's T_r decrease?

Solution: Using the above assumptions, Eq. (12) is written as follows:

$$60 \pm S = C_s + C_r + 6 + 8$$

With Eq. (13), C_s is calculated as 45 W/m², using the values $h_c = 22.5$ W/m² °C [Eq. (14)] and $\bar{T}_s - T_a = 22 - 20 = 2$°C. Equation (15) provides a value of $C_r = 1488$ cal/min based upon $\rho = 8.0$ g/liter when $P = 49.6$ ATA and $c_p = 1.24$ cal/g °C, and $T_E - T_a = 35 - 20 = 15$°C. Converting C_r to W/m², $C_r = 54$ W/m², and solving

Eq. (12), we find that $S = -53$ W/m². This "negative storage" will result in cooling of the body core, the more peripheral tissues being cooled directly and rapidly on contact with the cool helium atmosphere to about 22°C. It is probably correct to assume that the entire magnitude of S will be effective in reducing deep body temperature. Since these deeper tissues comprise about two-thirds of the body mass, and have a specific heat of 0.83, it is possible to equate a value of $S = -53$ W/m² to a rate of change of $T_r = 1.5$°C/hr for the subject described at the start of this example. As the example implies, a subject so exposed might well be rendered helpless by central hypothermia within about 2 hr, since any increases in M which might tend to offset hypothermia would also be likely to increase heat dissipation—for example, by increasing $\dot{V}_E$ and thus increasing C_r. Whether shivering would change h_c or $\overline{T}_s$ has not been studied, but such changes might also increase C_s.

Using the above equations, and considering the effect of atmospheric velocity upon h_c by analogy to the "wind-chill" expression of Siple (1949), it is possible to show what combinations of temperature and velocity in hyperbaric helium may cause severe thermal stress in unprotected subjects (Raymond 1971). A qualitative estimate of such thermal stress for 305 m (1000 fsw) is shown in Figure IV-45. In the event that conditions become severe during a dive the obvious step—removing personnel to a warmer environment—may be impossible. The peril of such a situation is clear from the fact that decompression from this pressure (31.3 ATA) takes several weeks. While awaiting emergency measures to restore T_a to a level above 30°C, steps to protect the divers might include additional hot drinks, food, and protective suits and extra blankets. Other protective equipment, such as the foamed-neoprene wet suit, is limited in value when compressed by increased ambient pressure (Table IV-25). This loss in insulation value is more marked at higher pressures, such as those of the above example, especially in helium-rich atmospheres, which transfer heat almost as well as water does (Webb 1970). Nevertheless, any available form of protection would be worth trying, in the hope that it might provide a boundary layer able to retard the heat flux, which would otherwise go unimpeded. Localized heating of chamber surfaces by infrared sources or hot water would be very helpful, if logistics permitted. Failing these measures, the risk of death from hypothermia would mount with each hour of exposure.

d. Saturation Diving, Open-Sea or "Wet-Pot" Chambers

Although still somewhat experimental, open-sea or "wet pot" saturation attracts great interest because it enables operations to be conducted at great depths while keeping costs within the realm of economic feasibility. Among the many difficulties encountered by saturation-diving operators and their divers, cold can be a consistently troublesome one (Cousteau 1966, Rawlins 1972).

As Othmer and Reels (1973) point out, "Sea water is always cold in the deeps, and often it approaches the temperature of its maximum density, near the freezing point." Although such extreme water temperatures can be endured for short periods in a laboratory setting at 1 atm (Figure IV-46), they would be rapidly fatal in deep-sea operations. Calculations similar to those in the previous example show that the large

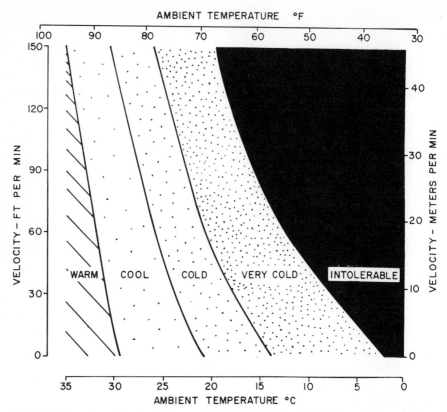

Figure IV-45. Schematic "comfort chart" for unprotected humans in helium–oxygen atmospheres at 1000 fsw (31.3 ATA). The temperature zones which represent thermal comfort, discomfort, and hazard are much closer to one another than in air at normal pressure, especially if there is substantial atmospheric motion. The theoretical basis of the figure is given in the text. No systematic studies of its accuracy are available.

C_r, coupled with severe surface chilling, would rapidly immobilize the average unprotected diver. At a depth of 305 m (1000 fsw) and a temperature of 0°C, it is likely that survival time would be measured in minutes rather than hours, even if the diver were equipped with a standard neoprene wet-suit. [This form of protection appears to have been quite satisfactory in a shallow, nitrogen–oxygen saturation dive at water temperatures of 21–24°C, according to Beckman and Smith (1972).]

Considerable progress appears to have been made in the quest for more effective thermal protective devices (Beckman 1967). For example, the constant-volume dry suit mentioned above has been used in combination with electrically heated underwear and a heater for the breathing gas (Jegou 1972, Jenkins 1973). Each of the heating devices has a capacity of 1000 W, and is designed to keep a diver in thermal balance at an ocean depth of 305 m. Studies of performance and physiological function in divers with such protection will be of great interest. Given the existing equipment, and the cold temperatures found at this depth in many parts of the world, 305 m (1000 fsw) may be the maximum safe working depth (Harvey 1966). Several successful hyperbaric

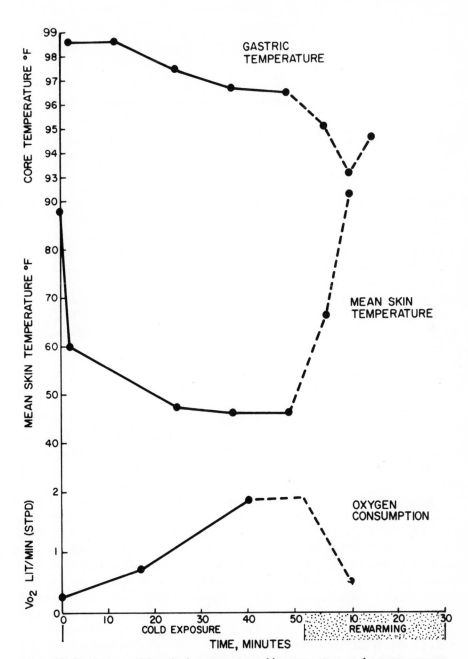

Figure IV-46. Changes of deep body temperature, skin temperature, and oxygen consumption in a resting subject immersed in still water at 43°F. Rewarming took place in 102° water. The subject of these observations had considerable subcutaneous fat, unlike the subject in Figure IV-44. [Modified from Behnke and Yaglou (1951). Used by permission from the American Physiological Society.]

Table IV-26

Minimum Safe Temperature of Inspired Gas as a
Function of Depth[a]

Depth		Gas temperature	
fsw	msw	°C	°F
600	182	−1.0	30.2
700	212	4.0	39.2
800	242	7.8	46.1
900	273	10.8	51.5
1000	305	13.3	55.9

[a] Adapted from Braithwaite (1972).

chamber dives have been described to depths of 364 m (1200 fsw) (Lambertsen *in press*) to 515 m (1700 fsw) (Naquet and Rostain *in press*), and a more recent dive to 485 m (1600 fsw) showed that men can do a moderate amount of work at these depths (Spaur 1973). The degree to which these studies can be applied to future open-sea operations will depend to a large degree on how well thermal balance can be maintained. Initially, cold-water "wet-pot" chamber studies are necessary to investigate heat loss, especially respiratory heat loss (Hoke *et al. in press*), as well as thermal clothing and equipment. Because of the magnitude of respiratory heat loss, the U. S. Navy has published minimum acceptable temperatures of inspired gas for helium–oxygen dives of various depths (Table IV-26).

3. Methods of Measurement

It is beyond the scope of this work to present a critique of the methodology for thermal physiology in the various environments encompassed in underwater bio-engineering. On the other hand, some of the imperfections in the current methods are more likely to be overcome if they are pointed out to those whose expertise may help make the improvements. The measurements are elementary ones: temperatures of the body and the environment, humidity and atmospheric motion, volume of pulmonary ventilation, and the exchange of oxygen and carbon dioxide. The problems arise from the variety of circumstances in which the measurements must be made and the differing purposes for which the information is to be used. Skin temperature is a good example (Teichner 1958). This fundamental quantity reflects to a great degree the state of comfort, shivering, and performance of the diver. If one is studying thermal balance in a dry hyperbaric chamber, it would be desirable to know T_s and T_a to an accuracy of $\pm 0.2°C$. To achieve this accuracy for T_s, it would be best to use some method which would not require skin contact, since the temperature probe and its means of contact would provide some insulation at the site of measurement, altering the value. A non-contact method would also overcome the use of connecting wires on the subject, which interfere greatly with activity when multiple temperatures are required. Such devices as infrared bolometers are available, but none has been adapted for use at the pressures currently under study. The same statement applies to most of the other instruments used in thermal physiology. Some off-the-shelf instruments, such as thermistors and telethermometers, can be used with safety and accuracy at pressures up to 50 ATA, while others are either unsafe or inoperative under even a few atmospheres. Further work is needed to develop more effective means of measuring body temperatures in diving and other hyperbaric work, and the same is true of environmental humidity and velocity, and of the ventilation and metabolism of the diver. In addition, as Webb (1973b, *in press*) has pointed out, further research is needed to define better thermal criteria for the various levels of stress and the adequacy of its reversal.

4. *Physiological Effects of Cold in Underwater Operations*

Up to this point, the discussion has emphasized *prevention* of cold stress in diving, because unchecked cold exposure can threaten the diver in a number of ways. Despite the most careful planning, however, diving operations can go awry, the best protective equipment can fail, and the most skillful team can be incapacitated. It is therefore necessary to consider the ways in which accidental hypothermia can impair the diver's functions and well-being. An overview of the problem is presented in Table IV-27, adapted from Webb (*in press*), who intended it as a general guide rather than an exact predictor. This table summarizes the deteriorations in function which should be anticipated when cold exposure lowers deep body temperature to the levels shown.

This problem will be considered in additional detail, by analyzing the effects of cold on the vital functional systems of the body. The systems which have been studied in some detail are the nervous system, including diver performance and musculo-skeletal function; the cardiovascular–renal system, including effects on body salt and water balance; the respiratory system; and the endocrine system.

In addition, cold can alter body function in ways not related to any single organ system, and such effects will be considered in a miscellaneous category after the above systems.

Table IV-27

Effects of Reduced Body Temperature on Body Function[a]

T_r, °C	Impairment of function
36–37	Cold sensations, cutaneous vasoconstriction; increase in oxygen consumption and in muscle tension by electromyogram
35–36	Sporadic shivering, suppressed by voluntary movements; bouts of shivering give way to uncontrollable shivering; oxygen consumption rises to 200–500% of resting value; *decreasing will to struggle increases risk of drowning*
34–35	Amnesia and poor articulation; sensory and motor dysfunction
33–34	Clouding of consciousness, hallucinations, and delusions
32–33	Cardiac abnormalities
30–32	Motor performance grossly impaired; no response to pain; familiar persons not recognized

[a] Adapted from Webb (*in press*).

a. Nervous System

Exposure to cold in the undersea situations described in the first part of this section can change neurological function in ways that range from pain and shivering to confusion and amnesia, culminating in loss of consciousness (Table IV-27). A diver so affected would soon find his performance degraded to a negative quantity as the operation turned into a rescue mission. When viewed with academic detachment, changes in neurological and musculoskeletal function are perhaps the most challenging effects of cold for the physiologist. Because they occur in a very complex multi-component system, their precise localization is often difficult. The study of man's performance in cold water is a good illustration (see Chapter VI).

(i) Performance. To analyze the effects of cold on performance one must deal with the perception and integration of specific information in the midst of numerous other stimuli. As pointed out by Vaughan and Andersen (1973), cold can interfere with performance in several ways—distraction, discomfort, or dysfunction—depending on the intensity of the thermal stress. One might add *distortion* to that list, since impaired perception is one of the best documented effects of cold. [For example, Mackworth (1953) found a progressive loss in the tactile discrimination of the finger tip as it was cooled below 25°C, thus giving a quantitative meaning to the phrase "numb with the cold."] Cold also impairs mental function (Weltman *et al.* 1971). (See Chapter VI, Man in the Ocean Environment: Performance.)

Many studies have documented the fact that tasks requiring gross or fine motor skills are performed less well in cold water than on dry land or in warm water. In discussing this problem, Poulton (1970) drew attention to the work of Kay (1949), who related the loss in handgrip strength and motor skills directly to the skin temperature of the hand exposed to cold air (Figure IV-47). Function was markedly impaired at temperatures below 50°C. Similar results have been reported by Bowen (1968) and by Stang and Weiner (1970). As Kay pointed out, a part of such decrements is obviously due to impaired perception (numbness), but some of the loss in function may also be due to shivering, stiffness, and, when central hypothermia is allowed to occur, to the diver's altered mentation. When protective gloves or mittens are worn for severe exposures, these will add some functional impairment of their own. These factors obviously need attention in the difficult tasks of designing tools for underwater use and planning underwater operations (see Chapter IX, Operational Equipment).

(ii) Hypothermia. All who have studied man in the cold are familiar with the euphoria, amnesia, and other changes in higher intellectual function which hypothermia can produce. While such changes are often humorous (Beckman 1963), they may be the signs of impending collapse. Terminating the exposure, while mandatory in an experiment or an operation, does *not* terminate the risk. It is in the *first few minutes of rewarming that the deep body temperatures often reach the lowest values* (Figure IV-46), and this is the time the subject needs the closest observation. This phenomenon, which Behnke and Yaglou (1951) termed the "afterdrop," is probably due to the return of still cold blood from the peripheral tissues, which remain at a low temperature for a long time during rewarming. Incomplete rewarming can produce another neuromuscular effect of cold: delayed shivering. This can recur many hours after the exposure, interfering with sleep. Webb (1973*b*) pointed out the lack of any

Figure IV-47. The average deterioration in various manual tasks produced by a fall in the temperature of skin of the hand. Efficiency is expressed as a percentage of efficiency in the warm. The top line shows the reliable diminished strength of handgrip measured with a hand ergometer. The middle line shows the reliable reduced number of 0.6-in. hexagonal nuts threaded onto screws in a fixed time. The bottom line shows the reliable reduced number of 0.2-in. screws put into a metal plate. The results are from six people. [From Kay (1949) by permission from the Royal Navy Personnel Research Committee.]

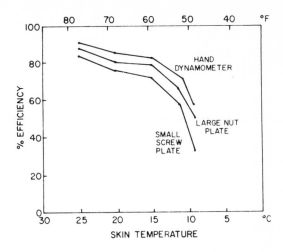

objective (or subjective) criterion to indicate when rewarming is achieved, falling back on the empirical practice of using sweating as an indication that the process has been overdone.

(iii) Shivering. Shivering is so common among divers that it may be an accepted part of the dive, even when protective suits are worn. When such shivering is violent or prolonged, it is often followed by muscle pain and tenderness lasting about two days. Muscle structure or function after prolonged shivering in humans has not been studied systematically but abnormalities would probably be found. Rats exercised in cold air (1.7°C) showed evidence of muscle damage, and two-thirds died during a 9-hr exercise period that was nonfatal to rats that had been acclimated to cold (Dieter *et al.* 1970).

b. Cardiovascular-Renal System

Cold has a number of important effects on cardiovascular and renal function. Many facets germane to underwater bioengineering are discussed in this chapter under cardiovascular factors. In addition brief discussions of the effects of cold on "immersion diuresis," diving bradycardia, and heart rate are presented here. Except for work on temperature effects on immersion diuresis (Reeves *et al.* 1965) and studies discussed by Keatinge (1969) which involved rather profound hypothermia, there does not seem to be any relevant work on the effects of cold-water diving on renal function.

(i) Immersion Diuresis. A temporary increase in urine flow is potentiated by low water temperatures, which also increase the loss of sodium, potassium, calcium, and other physiologically important ions (Reeves *et al.* 1965). If not replaced during prolonged exposures, such losses could lead to cardiovascular collapse. Immersion for 12 hr at 23.9°C is also accompanied by a striking rise in the excretion of creatinine and urea, raising the question of whether the prolonged violent shivering is associated with cellular damage.

(ii) Diving Bradycardia. Sudden death sometimes occurs in apparently normal persons following abrupt immersion in cold water. The suddenness suggests that it is due to an abnormality of cardiac rhythm. An illustration of the kind of abnormality that might lead to sudden death is shown by Keatinge (1969), whose article contains a brief but excellent discussion of the problem. Since apneic bradycardia (the "diving reflex") has been shown by Hong *et al.* (1969*a*) and others to be potentiated by cold, it seems reasonable to expect that "escape rhythms" of the heart might supervene as the normal heart beats become infrequent. It would seem, then, that serious cardiac irregularities would actually be more common. Whatever the explanation, persons with heart disease should avoid cold water. Cold increases the resistance against which the heart must pump and thus increases its need for oxygen and metabolic fuels (Epstein *et al.* 1969). Since cold can also cause spasm of the coronary arteries, the heart may be placed at a double disadvantage, which is especially risky for persons who already have diseased heart muscle.

(iii) Heart Rate. Analysis of the effect of prolonged cold-water diving on heart rate is complicated by the presence of multiple independent variables having opposite effects. Facial immersion slows down heart rate, especially when the water is cold. On the other hand, many workers have found that immersing the whole body in cold water increases heart rate. At greater depths, however, it is generally believed that the increased hydrostatic pressure will decrease heart rate (Flynn *et al.* 1972). Thus, the prediction of whether heart rate will increase, fall, or stay unchanged in a given dive would depend on a knowledge of several variables, and of the individual's response to them. From the data of Russell *et al.* (1972), it appears that most recreational diving done in cold (5.5°C) water involves a considerable increase in heart rate, oxygen consumption, and minute ventilation—even at rest—when conventional wet suits are worn.

c. Respiratory System

The respiratory system, which comprises the upper and lower airways, the alveoli and pulmonary capillaries, and the musculoskeletal and neural structures that drive gas flow, is quite vulnerable to the effects of cold in underwater environments. (For detailed discussion of the respiratory system in the underwater environment see this chapter, on the physiology of respiration.) One immediate effect is on the rate and volume of respiratory exchange, discussed below.

The respiratory vulnerability stems from two main factors. In most shallow diving, gas comes to the respiratory system directly from a mouthpiece and, in very deep diving, the rather high resistance of the nose makes mouth breathing almost obligatory, especially during exercise. Thus the normal warming, humidifying, and filtering role of the nose is eliminated. The second aspect is the great capacity for heat transfer which the breathing mixture possesses in diving situations.

(i) Respiratory Exchange. The most dramatic effect of cold on respiration is the intense hyperventilation which immediately attends the exposure of an unprotected man to an ice-cold shower (Keatinge *et al.* 1964). The response may seem appropriate to the drama of the experiment but, indeed, some degree of hyperventilation is usually present when cold stress causes a less acute hypothermia. For example, the work of

Russell *et al.* (1972) showed that minute ventilation $\dot{V}_E$ was greater than normal at 1, 2, and 3 ATA in wet-suited divers exposed to 5.5°C water. However, the increase in $\dot{V}_E$ was probably due to the increased metabolic rate of the chilled divers, whereas in the Keatinge *et al.* (1964) experiment, hyperventilation was disproportionate to metabolic needs in that it lowered alveolar carbon dioxide levels.

No other data on the acid–base status of thermally stressed divers at any pressure are available at present. Keatinge (1969) has summarized the information from animal studies and from therapeutic hypothermia in patients, but most of these data consider central body temperatures below 30°C an extreme degree of hypothermia in the present context. Arterial blood gas and *p*H values would be quite valuable, since they give a nearly complete view of the adequacy of the gas exchange function of the respiratory system.

(ii) Heat Transfer. The capacity of respiratory heat transfer is proportional to the product of gas density and specific heat, for a given difference between inspired and expired gas temperatures [see Eq. (15), $C_r = \rho c_p (T_E - T_a)\dot{V}_E$]. Since density increases linearly with pressure, the heat capacity increases directly with depth. At pressures much greater than 6 ATA (50 m or 165 fsw), helium must be substituted for nitrogen in most diving and its great specific heat offsets its lesser density, leading to still greater heat transfer ability. Furthermore, as depth increases, water temperature generally becomes lower. The net effect of these changes is that the upper and probably the lower airways are exposed to a gas stream which is considerably colder than what their surfaces are accustomed to. The gas stream is relatively unhumidified, in contrast to the normal condition, wherein air is both warmed and humidified well before it reaches the lower airways (Christie and Loomis 1933). Whereas Mather *et al.* (1953) found no evidence that blood in the pulmonary capillaries was losing heat to the alveolar gas—despite an inspired gas temperature of −18°C—the work of Hoke *et al.* (*in press*) makes it seem most unlikely that such homeostasis is maintained during deep diving, especially during helium–oxygen breathing and exercise.

(iii) Effects of Respiratory Heat Loss. What is exposure to cold, unhumidified dense gas of high heat capacity likely to do to the respiratory system? One adverse effect seen by Hoke *et al.* (*in press*) was marked respiratory distress, due to profuse secretions. Even in the absence of such secretions, cold facial stimulation (Josenhans *et al.* 1969) and cold air breathing (Guleria *et al.* 1969) can increase breathing resistance in normal men at sea level. Since respiratory resistance is already high in diving (see discussion of physiology of respiration, this chapter), factors that further increase it are clearly undesirable. If the above facts are true, one may wonder that so much cold-water diving is accomplished without greater evidence of respiratory failure occurring. The absence of respiratory failure speaks more for the great reserve capacity of the respiratory system in healthy, screened subjects, than it does to minimize the nature of the stress.

The great magnitude of respiratory heat loss has already been mentioned (Rawlins 1972), as well as the fact that it has led to the setting of limits for how cold a helium–oxygen mixture can safely be breathed at various depths (Braithwaite 1972). However, Cooper and Scarratt (1973) recently reported that extreme cold hindered operations at 46–76 m (150–250 ft) in 3.5°C water, raising the possibility that heating of such gas mixtures may be advisable in dives considerably shallower than currently suggested.

(iv) Other Respiratory Functions Affected by Cold. In addition to the function of its airways in conducting gas in and out of alveoli, the respiratory system has other known and unknown functions which might be affected by exposure to cold. These include the following:

1. Mucociliary clearance—the removal of foreign particles, including infectious agents.
2. Killing of infectious agents, once they have been removed from the gas stream.
3. Removal (or addition) of some blood constituents by the pulmonary capillaries.
4. Maintenance of structural and functional integrity of alveolar–capillary units, to promote gas exchange.

Some of these functions have only recently come under study in the normal environment, so data relevant to the present discussion cannot be expected in the immediate future. Some clues are already available, however. Several studies, such as that of Won and Ross (1971), have demonstrated an increased susceptibility of mice to bacterial infections of the lung when the animals are exposed to cold. Boyd (1972) has shown that a cold climate causes tissues of the respiratory system to become dehydrated; this may adversely affect mucociliary function and other lung defenses. With regard to other changes in respiration due to cold, Hong *et al.* (1969*b*) found that immersion changes lung mechanics in man at normal pressure even when no thermal stress is present. No complementary studies are yet available for cold water, but it is likely that hypothermia would increase the shift in fluid volume from the periphery into the thorax. How this might affect lung function is unknown.

d. Endocrine System

(i) Adrenal Responses. When compared to other forms of environmental stress, diving has attracted surprisingly little attention from endocrine physiologists. Cold stress in diving has recently been studied by several groups, however. Davis *et al.* (1972) studied eight men diving in British coastal waters to depths of 3 and 30 m. The water temperature at these depths was 12.5 and 11.5°C, respectively, and despite the use of a dry suit worn over a neoprene wet suit, the divers lost about 1.5°C in rectal temperature over a 1-hr period (about half of which was actually spent in the water) on each dive. The divers' adrenal responses showed clear increases in cortisol and epinephrine, and the increases were more marked on the day of the deeper dive. However, the predive values were also higher than "control" levels, again more so on the day of the deeper dive. The authors concluded that the adrenal responses were due more to anxiety than to thermal stress. Reeves *et al.* (1965) also found increased adrenal steroid excretion in a laboratory exposure to cold water, where thermal stress presumably overshadowed the anxiety often present in open-water dives. Increases in plasma corticosterone, epinephrine, and norepinephrine also occur in rats exposed to -20°C air (Chin *et al.* 1973). Exercise-conditioning at room temperature was able to prevent most of the hypothermia found in nonconditioned rats, whose rectal temperature fell to 31.9°C by the end of a 3-hr cold exposure. Exercise conditioning also diminished the magnitude of the hormonal responses to cold air, but the mean values were all

still significantly higher than those of rats kept at room temperature. The authors concluded that physical training provided an increased metabolic capacity which enabled the trained rats to forestall the development of hypothermia in these acute exposures.

Other studies in normal humans provide support for the view that adrenal cortical and medullary function is stimulated by exposure to cold (Arnett and Watts 1960, Suzuki *et al.* 1967, Budd and Warhaft 1970, Wilson *et al.* 1970), although negative results have been reported for steroid hormones (Thorn *et al.* 1953, Golstein-Golaire *et al.* 1970) and for catecholamines (Keatinge *et al.* 1964).

(ii) Thyroid Function. Function of the thyroid gland has not been directly studied in cold-water diving. The increased metabolic rates of Korean women divers during colder months (Hong 1963, 1973) could be due to increased thyroid activity, but could also be explained by other effects of cold, such as increased catecholamine secretion or increased sensitivity to catecholamines. Studies of acute exposure of humans to cold air have given contradictory results. Fisher and Odell (1971) have summarized the negative results found in adults. In pointing out that adolescents and young adults exposed to 2–4°C air show an increased blood level of the thyroid-stimulating hormone (TSH), they reason that the adult hypothalamus has become refractory to acute cold as a stimulus, so that it no longer directs the pituitary to release TSH to drive the thyroid. However, Fisher and Odell also gave results on more chronic exposure of adult men to Arctic cold, which clearly resulted in higher serum TSH concentrations. The latter work may be more applicable to diving exposures to cold, which are often repeated or chronic, especially in saturation diving (Webb 1973*a*). Other workers (Suzuki *et al.* 1967, Golstein-Golaire *et al.* 1970) have found some evidence of increased thyroid activity in humans exposed to cold air, a finding generally true in numerous studies with small mammals.

(iii) Glucose Homeostasis. The relative lack of endocrine studies in diving hypothermia extends even to the question of glucose homeostasis. While one of three subjects studied by Reeves *et al.* (1965) exhibited progressively lower blood glucose concentrations during immersion in 29.4°C (85°F) and 23.9°C (75°F) water, he showed a similar lowering during a 35°C (95°F) study. Thus the trend may reflect factors other than thermal stress in that subject. However, the glucose concentrations at 8 and 12 hr in the coldest water had reached levels of only 48 and 53 mg/100 ml. This degree of hypoglycemia suggests the need for additional study of this issue, even though the other two subjects' glucose levels remained well within the normal range in the above study. Keatinge (1969) cites some earlier, unpublished data of Alexander (1946) that blood glucose *rises* during hypothermia and falls during rewarming, making it advisable to administer glucose during the latter process. In 1-hr exposures of unprotected men to 25°C water, Raymond, Langworthy *et al.* (*in preparation*) found no changes in the serum concentrations of glucose, insulin, and growth hormone and thus concluded that glucose homeostasis was unimpaired by the acute stress. With the increasing availability of radioimmunoassay techniques for the determination of steroid, peptide, glycoprotein, and other hormones, it is likely that a more complete understanding of the endocrine responses to cold stress in diving will be available in the next few years.

e. Miscellaneous Effects of Cold

In addition to effects upon specific functional systems mentioned above, cold can affect the diver's well-being in ways not attributable to a single system, because it is acting on several systems at once. Two examples of this are: possible effects of cold on safe decompression schedules, and damage to inadequately protected extremities.

The first example concerns a matter of speculation. One can find unsupported statements to the effect that cold divers are more likely to develop decompression sickness than they would be without the added thermal stress, but we know of no published evidence that this is true. Its instinctive appeal is reinforced by the knowledge (Behnke and Willmon 1941) that the body surface becomes very poorly perfused, even to the point that cutaneous gas transfer drops off, when the surface is cooled much below normal. Thus, any inert gas deposited, for example, in subcutaneous adipose tissue with normal temperature and perfusion would probably be released more slowly if tissue perfusion were impaired by hypothermia. Such an idea finds support from a recent abstract (Balldin and Lundgren 1971), where it is stated that "temperature may influence nitrogen elimination by inducing circulatory changes."

With regard to the second example, it is paradoxical that cold water can kill, often in the absence of frostbite or other forms of damage to the extremities that are associated with severe exposures in cold air. However, cold water can produce long-lasting functional changes even when the exposure lasts less than 1 hr and the temperature is 8°C above the freezing point of sea water, −1.9°C (Behnke and Yaglou 1951). Cold-water immersion is not free from the risk of pathological changes, including freezing, at this temperature. The damage involves nerves, muscles, blood vessels, and skin, and has been discussed by Keatinge (1969). His discussion of the cold-injured extremity suggests that treatment should be limited to rewarming in a 40°C (104°F) bath, maintenance of the patient's fluid volume, treatment of pain, and the protection of the injured site from further trauma, infection, and excessive moisture.

A third miscellaneous effect of cold worth mentioning is one directly related to equipment, rather than to any direct effect on the body. In the more specialized breathing devices used in some types of diving, breathing gas is sometimes recirculated. This requires some means for carbon dioxide removal, and metallic hydroxides are commonly used in the form of a granular absorption bed. At lower temperatures, these materials become less effective, with the possible consequence of carbon dioxide accumulation in the body, especially during exercise. This problem has recently been discussed by J. G. Smith (1973).

H. *Diet and Metabolism*

The diver, carefully selected to be a normal, healthy human being, should eat a basic diet that contains proteins, carbohydrates, fats, minerals, and vitamins (see Chapter I). If he eats such a balanced diet there is no reason to worry about deficiencies in relation to any underwater activity. However, one problem in the past is that dietary regulation has rarely been enforced, either in simulated or actual deep dives. The diver in most cases has been allowed to request whatever he wishes to eat, and no record has been kept of intake. In many cases the diver eats only part of what is furnished, since the food does not taste or smell right to him. In fact reduction or severe impairment of the senses of taste and smell has been reported under long-term deep-dive situations. It is in this situation that the cook becomes the most important person on the team.

For man in an underwater habitat there are a host of problems involving food processing, packaging, storage, final preparation, and waste disposal. Cans with any appreciable air space distort to an extent that makes them difficult to open. Rigid foam containers and packaging material crush. The excellent thermal conductivity of pressurized helium makes cooking difficult. Fumes and toxic gases in a galley tax the life-support system. Environmental commitment and the necessity to protect swimmers from scavenger fish dictate hygienic containment and disposal of food and excretory waste. Strict control and monitoring is required during the shipment, storage, and transfer of frozen food. Food that has been thawed and refrozen presents a potential health hazard.

From existing data in the literature it is possible to estimate the minimal nutritional requirements for man at rest or working at or near sea level and breathing air and, thus, to determine how much water and macro-micronutrients are required daily per man. But the literature is indeed scarce on the nutritional requirements of man breathing a helium–oxygen gas mixture in a hyperbaric, cold, dark, underwater environment. Some work has been done over the years in nutritional research involving submarine crews, and extensive research has dealt with dietary needs of the astronauts (K. J. Smith *et al.* 1966). Although most of the astronaut and submarine personnel work is not transferable to the situation of the aquanauts, a few general conclusions may apply.

At least 2 hr should elapse between the last meal and a diving operation. Certainly a heavy ingestion of fat should be avoided since evidence from animal experiments indicates that lipemia (fat or lipids in the blood) may lead to thrombi and blockage of the circulation (Philp *et al.* 1967). During a prolonged, heavy exercise effort it is wise to provide a supplementary diet heavy in carbohydrates (fuel). This was successfully tried in a long, underwater scuba swim where the supplementary, or in this case, replacement diet consisted of 18% protein, 72% carbohydrates, and 10% fat (Hunt *et al.* 1964).

Effects of severe cold should be taken into account when planning the diet since evidence shows that a marked increase in the caloric intake is necessary to counteract the energy expended to maintain body temperature. In one experiment (Burton 1971), eight aquanauts spent up to 10 days in the Helgoland Underwater Laboratory at a depth of only 65 ft but the water temperature ranged from 54 to 57°F and they worked outside for up to 3 hr at a time. Even though they consumed some 6000 cal/day, they lost up to 11 lb per man in the 10-day period.

Interestingly, the lack of food has been incriminated in oxygen syncope and collapse (Miles 1969). Some people faint easily, but the situation is much worse among those who "take a last meal between 6 and 7 P.M. in the evening and are content with a cup of tea and a cigarette for breakfast. Forenoon diving with such a routine is not to be recommended." The underwater worker definitely should not be overweight (see Table IV-28). Also, animal work indicates that unregulated dietary regimens

Table IV-28

Weight Standards for Male Officers—Navy and Marine Corps [a]

(Shilling 1955)

Height, in.	17–20			21–25			26–30			31–35		36–40		41–45		46–50		51–64	
	Min.	Std.	Max.	Min.	Std.	Max.	Min.	Std.	Max.	Std.	Max.	Std.	Max.	Std.	Max.	Std.	Max.	Std.	Max.
60	105	117	146	108	120	150	110	122	153	125	157	128	160	131	164	133	166	135	169
61	107	119	149	110	122	153	112	124	155	127	159	130	163	133	166	135	169	137	171
62	109	121	151	112	124	155	113	126	158	129	161	132	165	153	169	137	171	139	174
63	111	124	155	113	126	158	115	128	160	131	164	134	168	137	171	139	174	141	176
64	113	127	159	115	128	160	118	131	164	134	168	137	171	140	175	142	178	144	180
65	115	130	163	119	132	165	121	135	169	138	173	141	176	144	180	146	183	148	185
66	117	133	166	122	136	170	125	139	174	142	178	145	181	148	185	150	188	152	190
67	121	137	171	126	140	175	129	143	179	146	183	149	186	152	190	154	193	156	195
68	125	141	176	130	144	180	132	147	184	150	188	153	191	156	195	158	198	160	200
69	129	145	181	133	148	185	136	151	189	154	193	157	196	160	200	162	203	164	205
70	133	149	186	137	152	190	139	155	194	158	198	161	201	164	205	166	208	168	210
71	137	153	191	140	156	195	143	159	199	162	203	165	206	168	210	170	213	172	215
72	141	157	196	145	161	201	148	164	205	167	209	170	213	173	216	175	219	177	221
73	145	161	201	149	166	208	152	169	211	172	215	175	219	178	223	180	225	182	228
74	149	165	206	154	171	214	157	174	218	177	221	180	225	183	229	185	231	187	234
75	153	169	211	158	176	220	161	179	224	182	228	185	231	188	235	190	238	192	240
76	157	173	216	163	181	226	166	184	230	187	234	190	238	193	241	195	244	197	246
77	161	177	221	167	186	232	170	189	236	192	240	195	244	198	248	200	250	202	253
78	165	181	226	172	191	239	175	194	242	197	246	200	250	203	254	205	256	207	259

Weight according to age and height

[a] The standard weight for each height for the age group 26–30 is the ideal one to maintain thereafter. For age after this group, the minimal allowance will be that for the age group 26–30.

conducive to alimentary hyperlipidemia (fat in the blood) are known to be hazardous for animals exposed to a compression–decompression cycle (Behnke 1971).

If the diver actually eats a well-balanced diet, there is no reason to worry about vitamin deficiency; however, there is evidence that in mice vitamin E (or tocopherol) deficiency is associated with hemolysis when the mice are exposed to hyperoxia (Goldstein and Menzel 1969). A direct relationship between vitamin E deficiency and the toxicity of hyperbaric oxygen has been proven in animals, premature infants, and astronauts but, according to Money and Strong (1972), there has been no reference in the literature to the importance of vitamin E intake to the underwater diver and they feel that this oversight should be remedied.

Although an underwater habitat has specific food-management problems in common with the pressurized space capsule, life in an underwater habitat differs in that contact with the outside world is relatively easy and, thus, it is even possible to send down hot meals. For this reason some underwater studies have investigated dietary management using foods that have not been specially processed or prepared. Investigating food preferences during a saturation dive at 650 ft (Hamilton *et al.* 1966), researchers sent food and drink "to the divers through the small medical lock. Each man was given considerable freedom of choice in the selection of his meals which, for the most part, were provided by a local restaurant. Both men exhibited normal appetites except that, as the dive progressed, they preferred a lighter diet (e.g., fruits, salads) to heavy, rich foods. Fluid intake for the period of study remained within normal limits. Both divers showed preference for coffee and fruit juice."

During the conduct of TEKTITE II, data were collected and evaluated on food preferences and on food management (Miller *et al.* 1971). Food preference data obtained during TEKTITE II provided a basis for designing food systems for SkyLab and subsequent manned spaceflights. Data revealed that food monotony could be delayed by providing a variety of food types (frozen, dehydrated, and thermostabilized) in a food system. And a negative psychological effect associated with the use of disposable plastic eating utensils led to the conclusion that stainless steel utensils would be preferable for the SkyLab program, even though additional work is required for cleansing and sanitizing.

During the Janus II experiment, a study was made of the alimentary behavior of three divers and, as precisely as possible, of calcium intake. The normal 3500-cal diet of the athlete was increased to 4000 cal for divers, because of the heat-loss factor. During decompression slightly less was consumed (3000 cal).

The diet was composed of 120 g of protein, 160 g of lipids, and 500 g of glucides. Green vegetables were limited to 200 g and fruits to 400 g because of the variability of their calcium content; more cereals and potatoes were included than would normally be the case, in the interest of stabilizing the calcium intake. Four meals were given— breakfast, lunch, a snack at 6 P.M., and dinner at 10 P.M. Breakfast and the snack, which were given 1 hr before a working dive, contained a high proportion of glucides. The diet was simple, containing no sauces or fried foods, except for an occasional tomato sauce, fried potatoes, or sauteed fish. Alcoholic beverages were eliminated, since they contribute neither energy nor thermal balance, and might aggravate narcotic phenomena or HPNS. Carbonated beverages were also eliminated because of the gas. A minimum of two liters of liquid was recommended because of increased diuresis.

Table IV-29

Reference Values for Blood and Urine Data[a]

Blood component	Units	Deficient	Low	Acceptable	High	Normal range
Vitamin C, plasma	mg/100 ml	<0.10	0.10–0.19	0.20–0.39	>0.40	—
Vitamin C, plasma	mg/100 ml	—	—	—	—	0.4–2.0
Vitamin A, plasma	μg/100 ml	<10	10–19	20–49	>50	—
Carotene, plasma	μg/100 ml	—	20–39	40–99	>100	—
Total protein, serum	g/100 ml	<6.0	6.0–6.4	6.5–6.9	>7.0	—
Albumin	g/100 ml	<3.0	3.0–3.9	4.0–4.9	>5.0	—
Globulin	g/100 ml	<1.4	1.4–1.9	2.0–2.9	>3.0	—
A/G ratio	unitless	—	—	—	—	1.4–3.0
Hemoglobin, blood	g/100 ml	<12.0	12.0–13.9	14.0–14.9	15.0	—
Erythrocyte transketolase						
Brin method	Brin units[b]	<800	—	—	—	850–1000
Warnock method	IU[c]	5.0–7.5[d]	—	—	—	8.5–13.0
Na$^+$, plasma	meq/liter	—	—	—	—	135–155
K$^+$, plasma	meq/liter	—	—	—	—	3.6–5.5
Ca^{++}, plasma	mg/100 ml	—	—	—	—	9.0–11.0
Mg^{++}, plasma	mg/100 ml	—	—	—	—	1.8–3.1
Zn^{++}, plasma	μg/100 ml	—	—	—	—	50–120
Cu^{++}, plasma	μg/100 ml	—	—	—	—	70–140
Thiamin	μg/g creatinine	<27	27–65	66–129	>130	—
Riboflavin	μg/g creatinine	<27	27–79	80–269	>270	—
Vitamin B$_6$	μg/g creatinine	—	—	—	—	3.5–120
Vitamin B$_6$	μg/g creatinine	—	<19	>19	—	—
Niacin	mg/g creatinine	—	—	—	—	0.087–0.87
N^1-methylnicotinamide	mg/g creatinine	<0.5	0.5–1.59	1.6–4.29	>4.3	—
Folacin	μg/g creatinine	—	—	—	—	1.3–13

[a] Reprinted from Frattali and Robertson (1973) with permission of the Aerospace Medical Association.
[b] Brin units = μg hexose produced/ml hemolyzed red blood cells/hr at 38°.
[c] International units = μmole ribose-5-phosphate utilized/ml blood/hr at 37°.
[d] Marginal thiamin deficiency for adult humans.

It was found, contrary to expectation, that the food lost little of its characteristic flavor and the sense of taste of the divers remained acute enough so that they refused soup which had accidentally been oversalted. However, textures of most foods changed noticeably: bread became elastic, thick soup coagulated, fried foods diminished by one-third in volume, rice became lumpy, apples and pears bruised, and bananas spoiled at once. But oranges underwent little change. During decompression, changes were less severe. It was found that circumstances permitted a calcium study as accurate as had been intended, but studies of blood and urine indicated that phosphocalcic metabolism was not disturbed during the experiment. Appetites remained good, and weight was about the same at the beginning and at the end of the experiment (Segui and Conti 1972).

A biochemical evaluation of the nutritional status of two divers was performed by the U. S. Navy (Frattali and Robertson 1973) during a saturation helium–oxygen dive to a simulated depth of 850 ft with two excursions to 1000 ft. Dietary records of food consumption permitted estimation of caloric and nutrient intake, and blood and urine samples provided information for evaluation of vitamin status and other indicators of their nutritional state. Most of the analyses gave normal results, but there was a marked decrease in thiamin excretion and also a marked decrease in erythrocyte transketolase activity. Since thiamin requirement is directly related to energy expenditure, it is possible that hyperbaric stress, exercise, and cold stress created a greater thiamin demand. In the same article a complete reference guide is given for interpretation of blood and urine data, which should be useful for future biochemical nutritional studies (Table IV-29).

References

ACKLES, K. N., and B. FOWLER. Cortical evoked response and inert gas narcosis in man. *Aerosp. Med.* **43**:1181–1184 (Nov. 1971).

ADOLFSON, J. Human performance and behavior in hyperbaric environments. *Acta Psychol. Gothoburgensia* **6**:1–74 (1967).

AGOSTONI, E., G. GURTNER, G. TORRI, and H. RAHN. Respiratory mechanics during submersion and negative pressure breathing. *J. Appl. Physiol.* **21**:251–258 (Jan. 1966).

ALBANO, G. Influenza della velocita di discesa sulla latenza dei disturbi neuropsichici da compressa nel lavoro subacqueo. Presented at the 25th National Congress of Medicine, Taormina, Italy (October 1962).

ALBANO, G. Principles and observations on the physiology of the scuba diver. (Translation). Off. Nav. Res. Rep. ONR-DR-150 (1970).

ALBANO, G., and P. M. CRISCUOLI. La sindrome neuropsichia di profondita. Note 4. *Boll. Soc. Ital. Biol. Sper.* **38**:754 (Aug. 15, 1962).

ALBANO, G., and P. M. CRISCUOLI. Neuropsychological effects of exposure to compressed air. In: Lambertsen, C. J., ed. *Underwater Physiology: Proceedings of the Fourth Symposium on Underwater Physiology*, pp. 193–204. New York, Academic Press (1971).

ALEXANDER, L. Treatment of shock from prolonged exposure to cold especially in water. Combined Intelligence Objectives Sub-committee, Item No. 24, File No. 26-37 (1946).

ANDERSEN, H. T. Physiological adaptation in diving vertebrates. *Physiol. Rev.* **46**:212–243 (Apr. 1966).

ANDERSON, B. Ocular effects of changes in oxygen and carbon dioxide tension. *Trans. Amer. Ophthalmol. Soc.* **66**:423–474 (1968).

ANTHONISEN, N. R., M. E. BRADLEY, J. VOROSMARTI, and P. G. LINAWEAVER. Mechanics of breathing with helium–oxygen and neon–oxygen mixtures in deep saturation diving. In: Lambertsen, C. J., ed. *Underwater Physiology. Proceedings of the Fourth Symposium on Underwater Physiology*, pp. 339–345. New York, Academic Press (1971).

ARBORELIUS, M., JR., U. I. BALLDIN, B. LILJA, and C. E. G. LUNDGREN. Hemodynamic changes in man during immersion with the head above water. *Aerosp. Med.* **43**:592–598 (June 1972).

ARNETT, E. L., and D. T. WATTS. Catecholamine excretion in man exposed to cold. *J. Appl. Physiol.* **15**:499–500 (May 1960).

BACHRACH, A. J., and P. B. BENNETT. Tremor in diving. *Aerosp. Med.* **44**:613–623 (June 1973).

BACHRACH, A. J., D. R. THORNE, and K. J. CONDA. Measurements of tremor in the Makai Range 520 foot saturation dive. *Aerosp. Med.* **42**:856–860 (Aug. 1971).

BALLDIN, U., and C. LUNDGREN. Nitrogen elimination in man during immersion, shifts in temperature and body position. In: *Abstracts of the Twenty-Fifth Congress of Physiological Sciences Satellite Symposium: Recent Progress in Fundamental Physiology of Diving*. Marseille, France (July 1971), pp. 59–60 (unpublished).

BANGHAM, A. D., M. M. STANDISH, and K. MILLER. Cation permeability of phospho-lipid model membranes: effect of narcosis. *Nature* **208**:1295 (Dec. 25, 1965).

BARNARD, E. E. P., H. V. HEMPLEMAN, and C. TROTTER. Mixture breathing and nitrogen narcosis. Alverstoke, U. K., Roy. Nav. Personnel Res. Comm., Med. Counc. Rep. (1962).

BARTLETT, R. G., JR. Respiratory system. In: Parker, J. F., and V. R. West, eds. *Bioastronautics Data Book*, pp. 489–531. Washington, D. C., National Aeronautics and Space Administration (1973) (NASA SP-3006).

BATTELLE COLUMBUS LABORATORIES, *U. S. Navy Diving-Gas Manual*. Second edition. Washington, D. C., U. S. Navy Supervisor of Diving (1971) (NAVSHIPS 0994-003-7010).

BEAN, J. W. Tris buffer, carbon dioxide and sympatho-adrenal system in reactions to oxygen at high pressure. *Amer. J. Physiol.* **201**:737–739 (Oct. 1961).

BEAN, J. W., and D. ZEE. Metabolism and the protection by anesthesia against toxicity of oxygen at high pressure. *J. Appl. Physiol.* **20**:525–530 (1965).

BECKMAN, E. L. Thermal protection during immersion in cold water. In: *Proceedings of the Second Symposium on Underwater Physiology*, pp. 246–266. Washington, D. C., National Academy of Sciences–National Research Council (1963).

BECKMAN, E. L. Thermal protective suits for underwater swimmers. *Milit. Med.* **132**:195–209 (Mar. 1967).

BECKMAN, E. L., and E. M. SMITH. Tektite II: Medical supervision of scientists in the sea. *Texas Rep. Biol. Med.* **30**:1–204 (Fall 1972).

BECKMAN, E. L., E. REEVES, and L. W. RAYMOND. Unpublished observations (1966).

BEHNKE, A. R. Employment of helium in diving to new depths of 440 feet. *U. S. Nav. Med. Bull.* **40**:65–67 (Jan. 1942).

BEHNKE, A. R. Oxygen decompression. In: Goff, L. G., ed. *Proceedings of the Underwater Physiology Symposium*, pp. 61–73. Washington, D. C., National Academy of Science, National Research Council (1955), Publ. 377.

BEHNKE, A. R. The Harry G. Armstrong lecture. Decompression sickness: advances and interpretations. *Aerosp. Med.* **42**:255–267 (Mar. 1971).

BEHNKE, A. R., and T. L. WILLMON. Cutaneous diffusion of helium in relation to peripheral blood flow and the absorption of atmospheric nitrogen through the skin. *Amer. J. Physiol.* **131**:627–632 (Jan. 1941).

BEHNKE, A. R., and C. P. YAGLOU. Physiological responses of men to chilling in ice water and to slow and fast rewarming. *J. Appl. Physiol.* **3**:591–602 (Apr. 1951).

BEHNKE, A. R., and O. D. YARBROUGH. Physiologic studies of helium. *U. S. Nav. Med. Bull.* **36**:542–558 (Oct. 1938).

BEHNKE, A. R., and O. D. YARBROUGH. Respiratory resistance, oil water solubility and mental effects of argon compared with helium and nitrogen. *Amer. J. Physiol.* **126**:409–415 (June 1939).

BEHNKE, A. R., R. M. THOMPSON, and E. P. MOTLEY. The psychologic effects from breathing air at 4 atmospheres pressure. *Amer. J. Physiol.* **112**:554–558 (July 1935).

BENNETT, P. B. Flicker fusion frequency and nitrogen narcosis. A comparison with EEG changes and the narcotic effect of argon mixtures. Alverstoke, U. K., Roy. Nav. Personnel Res. Comm., Med. Res. Counc., Underwater Physiol. Sub-comm. Rep. 176 (1958).

BENNETT, P. B. Psychometric impairment in men breathing oxygen–helium at increased pressures. Alverstoke, U. K., Roy. Nav. Personnel Res. Comm., Rep. 251 (1965).

BENNETT, P. B. *The Aetiology of Compressed Air Intoxication and Inert Gas Narcosis.* New York, Pergamon Press (1966).

BENNETT, P. B. Performance impairment in deep diving due to nitrogen, helium, neon and oxygen. In: Lambertsen, C. J., ed. *Underwater Physiology, Proceedings of the Third Symposium on Underwater Physiology,* pp. 327–340. Baltimore, Williams and Wilkins (1967).

BENNETT, P. B. Inert gas narcosis. In: Bennett, P. B., and D. H. Elliott, eds. *The Physiology and Medicine of Diving and Compressed Air Work,* pp. 155–182. Baltimore, Williams and Wilkins (1969).

BENNETT, P. B. Pressure physiology. Interim report on some physiological studies during 1500-foot simulated dive. Alverstoke, U. K., Roy. Nav. Physiol. Lab., Rep. 1/70 (1970).

BENNETT, P. B. Simulated oxygen–helium saturation diving to 1500 ft and the helium barrier. In: *1971 Offshore Technology Conference, April 19–21, Houston, Texas. Preprints,* vol. II, pp. 195–210. Published by the Conference (1971).

BENNETT, P. B. Review of protective pharmacological agents in diving. *Aerosp. Med.* **43**:184–192 (Feb. 1972*a*).

BENNETT, P. B. Some physiological measurements during human saturation diving to 1,500 ft. In: Fructus, X., ed. *Third International Conference on Hyperbaric and Underwater Physiology,* pp. 35–43. Paris, Doin (1972*b*).

BENNETT, P. B., and A. GLASS. Electroencephalographic and other changes induced by high partial pressures of nitrogen. *Electroencephalogr. Clin. Neurophysiol.* **13**:91–98 (1961).

BENNETT, P. B., and E. J. TOWSE. The high pressure nervous syndrome during a simulated oxygen–helium dive to 1500 ft. *Electroenceph. Clin. Neurophysiol.* **31**:383–393 (1971).

BENNETT, P. B., D. PAPAHADJOPOULOS, and A. D. BANGHAM. The effect of raised pressure of inert gas on phospholipid membranes. *Life Sci.* **6**:2527–2533 (Dec. 1, 1967).

BENNETT, P. B., K. N. ACKLES, and V. J. CRIPPS. Effects of hyperbaric nitrogen and oxygen on auditory evoked responses in man. *Aerosp. Med.* **40**:521–525 (May 1969).

BENZINGER, T. H., A. W. PRATT, and C. KITZINGER. The thermostatic control of human metabolic heat production. *Proc. Nat. Acad. Sci. USA* **47**:730–739 (May 1961).

BILLINGS, C. E. Barometric pressure. In: Parker, J. F., Jr., and V. R. West, eds. *Bioastronautics Data Book,* pp. 1–34. Washington, D. C., National Aeronautics and Space Administration (1973).

BISHOP, R. P. OSHA breathing gas purity standards. In: Battelle Columbus Laboratories. *Proceedings, 1973 Divers' Gas Purity Symposium, Nov. 27–28, 1973,* pp. V-1–V-10. Washington, D. C., Navy Supervisor of Diving (1973).

BJÜRSTEDT, H., and G. SEVERIN. The prevention of decompression sickness and nitrogen narcosis by the use of hydrogen as a substitute for nitrogen. (The Arne Zetterstrom method of deep sea diving.) *Milit. Surg.* **103**:107–116 (Aug. 1948).

BLENKARN, G. D., C. AQUADRO, B. A. HILLS, and H. A. SALTZMAN. Urticaria following the sequential breathing of various inert gases at a constant pressure of 7 ATA. A possible manifestation of gas-induced osmosis. *Aerosp. Med.* **42**:141–146 (Feb. 1971).

BLIGH, J. *Temperature Regulation in Mammals and Other Vertebrates.* New York, American Elsevier Publishing Co. (1973).

BOWEN, H. M. Diver performance and the effects of cold. *Hum. Factors* **10**:445–463 (Oct. 1968).

BOYD, E. M. *Respiratory Tract Fluid.* Springfield, Ill., Charles C Thomas (1972).

BRADLEY, M. E. The interaction of stresses in diving and adaptation to these stresses. In: Scripps Institute of Oceanography. *Human Performance and Scuba Diving. Proceedings of the Symposium on Underwater Physiology, La Jolla, Calif., April 10–11, 1970,* pp. 63–69. Chicago, Ill., The Athletic Institute (1970).

BRADLEY, M. E., and J. VOROSMARTI. The nature of hyperbaric arthralgia during dives from 100 to 1000 feet. Presented at the 1972 annual meeting of the Aerospace Medical Association, Bal Harbor, Fla., 8–11 May 1972.

BRADLEY, M. E., N. R. ANTHONISEN, J. VOROSMARTI, and P. G. LINAWEAVER. Respiratory and cardiac responses to exercise in subjects breathing helium–oxygen mixtures at pressures from sea level to 19.2 atmospheres. In: Lambertsen, C. J., ed. *Underwater Physiology. Proceedings of the Fourth Symposium on Underwater Physiology*, pp. 325–345. New York, Academic Press (1971).

BRAITHWAITE, W. R. The calculation of minimum safe inspired gas temperature limits for deep diving. U. S. Navy Exp. Diving Unit, Rep. NEDU 12-72 (July 1972).

BRAUER, R. W. Narrative account of a series of pressure chamber dives reaching to 1190 ft. in search of special physiological effects attributable to depths in excess of 1000 feet, Series Physalie, Wrightsville, Marine Biomedical Laboratory and Compagnie Maritime d'Expertises, Marseilles, France, May/June 1968. Wilmington, North Carolina, Wrightsville Marine Biomedical Laboratory (24 July 1968).

BRAUER, R. W. Current studies on physiology of extremely deep diving. *Mar. Technol. Soc. J.* **5**:31–32 (Nov./Dec. 1971).

BRAUER, R. W. Studies concerning the high pressure hyperexcitability in the squirrel monkey. Wilmington, N. C., Wrightsville Marine Biomedical Lab., Final report on ONR contract N00014-69-C-0341 (July 13, 1972a).

BRAUER, R. W. Probing deep for keys to convulsive disorders. *Biomed. News* (July 1972b).

BRAUER, R. W., and R. O. WAY. Relative narcotic potencies of hydrogen, helium, nitrogen, and their mixtures. *J. Appl. Physiol.* 23–31 (July 1970).

BRAUER, R. W., S. DIMOV, X. FRUCTUS, P. FRUCTUS, A. GOSSET, and R. NAQUET. Syndrome neurologique et electrographique des hautes pressions. *Rev. Neurol.* **121**:264–265 (Sept. 1969).

BRAUER, R. W., M. R. JORDAN, and R. O. WAY. The high pressure neurological syndrome in the squirrel monkey, Saimari sciureus. In: *Third International Conference on Hyperbaric and Underwater Physiology*, pp. 23–30. Paris, Doin (1972).

BRAUER, R. W., M. R. JORDAN, R. W. BEAVER, and S. M. GOLDMAN. Interaction of the high pressure neurological syndrome with various pharmacologic agents. In: *Proceedings of the Fifth Symposium on Underwater Physiology, Freeport, Bahamas, 21–25 Aug. 1972. In press.*

BRICK, I. Circulatory responses to immersing the face in water. *J. Appl. Physiol.* **21**:33–36 (Jan. 1966).

BRINK, F., and J. M. POSTERNAK. Thermodynamic analysis of the relative effectiveness of narcosis. *J. Cell. Comp. Physiol.* **32**:211 (Oct. 1948).

BROWN, E. B., JR., and F. MILLER. Ventricular fibrillation following a rapid fall in alveolar carbon dioxide concentration. *Amer. J. Physiol.* **169**:56–60 (Apr. 1952).

BRUMLIK, J., and C. B. YAP. *Normal Tremor.* Springfield, Ill., Charles C Thomas (1970).

BUDD, G. M., and N. WARHAFT. Urinary excretion of adrenal steroids, catecholamines and electrolytes in man, before and after acclimatization to cold in Antarctica. *J. Physiol. (London)* **210**:799–806 (Nov. 1970).

BÜHLMANN, A. A. The use of multiple inert gases in decompression. In: Bennett, P. B., and D. H. Elliott, eds. *The Physiology and Medicine of Diving and Compressed Air Work*, pp. 357–385. Baltimore, Williams and Wilkins (1969).

BÜHLMANN, A. A., H. MATTHYS, G. OVERRATH, P. B. BENNETT, D. H. ELLIOTT, and S. P. GRAY. Saturation exposures at 31 ATA in an oxygen–helium atmosphere with excursions to 36 ATA. *Aerosp. Med.* **41**:394–402 (Apr. 1970).

BULENKOV, S. YE., et al. Illnesses peculiar to underwater swimming and adverse effects due to them. Nitrogen narcosis. In: *Manual of Scuba Diving*, p. 185. Moscow, Publishing House of the Ministry of Defense (1968).

BULLARD, R. W., and G. M. RAPP (1970). Problems of body heat loss in water immersion. *Aerosp. Med.* **41**:1269–1277 (Nov. 1970).

BURTON, R. Helgoland underwater laboratory. *Sea Frontiers* **17**:335–341 (Nov./Dec. 1971).

CABARROU, P. L'ivresse des grandes profondeurs. *Presse Med.* **72**:793–797 (Mar. 14, 1964).

CALDWELL, P. R. B., W. L. LEE, JR., H. S. SCHILDKRAUT, and E. R. ARCHIBALD. Changes in lung volume, diffusing capacity, and blood gases in men breathing oxygen. *J. Appl. Physiol.* **21**:1477–1483 (Sept. 1966).

CALL, D. W. A study of Halon 1301 (CBrF₃) toxicity under simulated flight conditions. *Aerosp. Med.* **44**:202–204 (Feb. 1973).

CARLISLE, R., E. H. LANPHIER, and H. RAHN. Hyperbaric oxygen and persistence of vision in retinal ischemia. *J. Appl. Physiol.* **19**:914–918 (Sept. 1964).

CASE, E. M., and J. B. S. HALDANE. Human physiology under high pressure. *J. Hyg.* **41**:225–249 (Nov. 1941).

CHIN, A. K., R. SEAMAN, and M. KAPILESHWARKER. Plasma catecholamine response to exercise and cold adaptation. *J. Appl. Physiol.* **34**:409–412 (Apr. 1973).

CHOUTEAU, J. Respiratory gas exchange in animals during exposure to extreme ambient pressures. In: Lambertsen, C. J., ed. *Underwater Physiology. Proceedings of the Fourth Symposium on Underwater Physiology*, pp. 385–397. New York, Academic Press (1971).

CHOUTEAU, J., and J. H. CORRIOL. Physiological aspects of deep sea diving. *Endeavour* **30**:70–76 (May 1971).

CHOUTEAU, J., J. Y. COUSTEAU, J. ALINAT, and C. F. AQUADRO. Sur les limites physiologiques d'utilisation de melange oxygen–helium pour la plongee profonde et les sejours prolonges sous pression. *C. R. Acad. Sci. (Paris) (D)* **264**:1731–1734 (Mar. 29, 1967).

CHOUTEAU, J., J.-M. OCANA DE SENTUARY, and L. PIRONTI. Theoretical, experimental, and comparative study of compression as applied to intervention dives and saturation dives at great depths. Marseilles, Centre d'Etudes Marines Avancees, Rep. CEMA1-71 (Mar. 25, 1971). (Translated by M. E. Hashmall, Biological Sciences Communication Project, The George Washington University Medical Center, Washington, D. C.)

CHRISTIE, R. V., and A. L. LOOMIS. The pressure of aqueous vapor in the alveolar air. *J. Physiol. (London)* **77**:35–48 (1933).

CLARK, J. M. Tolerance and adaptation to acute and chronic hypercapnia in man. In: Battelle Columbus Laboratories. *Proceedings, 1973 Divers' Gas Purity Symposium, Nov. 27–28, 1973*, pp. I-1–I-20. Washington, D. C., U. S. Navy Supervisor of Diving (1973).

CLARK, J. M., and C. J. LAMBERTSEN. Pulmonary oxygen tolerance and the rate of development of pulmonary oxygen toxicity in man at 2 atm inspired oxygen tension. In: Lambertsen, C. J., ed. *Underwater Physiology. Proceedings of the Third Symposium on Underwater Physiology*, pp. 439–451. Baltimore, Williams and Wilkins (1967).

CLARK, J. M., and C. J. LAMBERTSEN. Pulmonary oxygen toxicity: a review. *Pharmacol. Rev.* **23**:38–133 (June 1971*a*).

CLARK, J. M., and C. J. LAMBERTSEN. Rate of development of oxygen toxicity in man during oxygen breathing at 2 atm. *J. Appl. Physiol.* **30**:739–752 (Nov. 1971*b*).

CLARK, J. M., R. D. SINCLAIR, and B. E. WELCH. Rate of acclimitization to chronic hypercapnia in man. In: Lambertsen, C. J., ed. *Underwater Physiology. Proceedings of the Fourth Symposium on Underwater Physiology*, pp. 399–408. New York, Academic Press (1971).

CLARK, L. C., and L. GOLDEN. Survival of mammals breathing organic liquids equilibrated with oxygen at atmospheric pressure. *Science* **153**:1755–1756 (June 24, 1966).

CLEMENTS, J. A., and K. M. WILSON. The affinity of narcotic agents for interfacial films. *Proc. Nat. Acad. Sci. USA.* **48**:1008–1014 (June 1962).

COMEX: New world record simulated deep dive (Physalie V experiment). Marseilles, France, Hyperbaric Research Center, COMEX (1970).

COMEX: "Physalie VI" 2001 feet. Marseilles, France, Hyperbaric Research Center, COMEX (1972).

COOK, S. F., F. E. SOUTH, JR., and D. R. YOUNG. Effect of helium on gas exchange of mice. *Amer. J. Physiol.* **164**:248–250 (Jan. 1951).

COOPER, R. A., and SCARRATT. Evaluation of lock-out submarine Deep Diver for in situ biological work in boreal waters. *Helgoland Wiss. Meeresuntersuch* **24**:82–90 (Mar. 1973).

CORRIOL, J., and J. J. ROHNER. Role de la temperature de l'eau dans la bradycardie d'immersion de la face. *Arch. Sci. Physiol.* **22**(2):265–274 (1968).

COTES, J. E. *Lung Function: Assessment and Application in Medicine*. Second edition. Philadelphia, F. A. Davis Co. (1968).

COUSTEAU, J. Y. *The Silent World*. London, Reprint Society (1953).

COUSTEAU, J.-Y. Working for weeks on the sea floor. *Nat. Geogr.* **129**:498–537 (Apr. 1966).

COVEY, C. W. Unisuit takes the chill out of diving. *Undersea Technol.* **13**:39–42 (Sept. 1972).

CRAIG, A. B., JR. Effects of submersion and pulmonary mechanics on cardiovascular function in man. In: Rahn, H., and T. Yokoyama, eds. *Physiology of Breath-Hold Diving and the Ama of Japan*, pp. 295–302. Washington, D. C., National Academy of Sciences–National Research Council (1965), Publ. 1341.

CRAIG, A. B., JR., and M. DVORAK. Thermal regulation during water immersion. *J. Appl. Physiol.* **21**:1577–1585 (Sept. 1966).

CRAIG, A. B., JR., and M. DVORAK. Comparison of exercise in air and in water of different temperatures. *Med. Sci. Sports* **1**:124–130 (Sept. 1969).

CRAIG, A. B., JR., and D. E. WARE. Effect of immersion in water on vital capacity and residual volume of the lungs. *J. Appl. Physiol.* **23**:423–425 (Oct. 1967).

DAVIES, H. C., and R. E. DAVIES. Biochemical aspects of oxygen poisoning. In: Fenn, W. O., and H. Rahn, eds. *Handbook of Physiology: Section 3, Respiration*, Vol. II, pp. 1047–1058. Washington, D. C., American Physiological Society (1965).

DAVIS, F., M. CHARLIER, R. SAUMAREZ, and V. MULLER. Some physiological responses to the stress of aqualung diving. *Aerosp. Med.* **43**:1083–1088 (Oct. 1972).

DEJOURS, P. Hazards of hypoxia diving. In: Rahn, H., and T. Yokoyama, eds. *Physiology of Breath-Hold Diving and the Ama of Japan*, pp. 183–193. Washington, D. C., National Academy of Sciences–National Research Council (1965).

DIETER, M. P., P. D. ALTLAND, and B. HIGHMAN. Tolerance of unacclimated and cold-acclimated rats to exercise in the cold: serum, red and white muscle enzymes, and histological changes. *Can. J. Physiol. Pharmacol.* **48**:723–731 (Oct. 1970).

DONALD, K. W. Oxygen poisoning in man. *Brit. Med. J.* **1**:667 (May 17, 1947).

DORR, V. A., and H. R. SCHREINER. Region of non-combustion, flammability limits of hydrogen–oxygen mixtures, full scale combustion and extinguishing tests and screening of flame-resistant materials. Tonawanda, N. Y., Ocean Systems, Inc., (May 1, 1969) (AD 689,545).

DRIPPS, R. D., and J. H. COMROE, JR. The respiratory and circulatory response of normal man to inhalation of 7.6 and 10.4 percent CO_2 with a comparison of the maximal ventilation produced by severe muscular exercise, inhalation of CO_2 and maximal voluntary hyperventilation. *Amer. J. Physiol.* **149**:43–51 (Apr. 1947).

DUMITRU, A. P., and F. G. HAMILTON. A mechanism of drowning. *Anesth. Analg. (Cleveland)* **42**:170–176 (Mar./Apr. 1963).

EDEL, P. O. Mixing hydrox safely. *Oceanol. Int.* **7**:31–33 (Jan. 1972).

EDEL, P. O., J. M. HOLLAND, C. L. FISHER, and W. B. FIFE. Preliminary studies of hydrogen–oxygen breathing mixtures for deep sea diving. In: *The Working Diver 1972. Symposium Proceedings, February 1972, Columbus, Ohio*, pp. 257–270. Washington, D. C., Marine Technology Society (1972).

ELCOMBE, D. D., and J. H. TEETER. Nitrogen narcosis during a 14-day continuous exposure to 5.2% O_2 in N_2 at pressure equivalent to 100 fsw (4 ata). *Aerosp. Med.* **44**(7, Sec. II):864–869 (July 1973).

ELLIOTT, D. H. Man underwater. IV: His limitations as a submersible. *Underwater Sci. Technol. J.* **2**:69–73 (June 1970).

ELLSBERG, E. Diving gas. *Collier's* **103**:22,26,28 (Apr. 15, 1939).

ELSNER, R., and P. F. SCHOLANDER. Circulatory adaptations to diving in animals and man. In: Rahn, H., and T. Yokoyama, eds. *Physiology of Breath-Hold Diving and the Ama of Japan*, pp. 281–294. Washington, D. C., National Academy of Sciences–National Research Council (1965), Publ. 1341.

END, E. The use of new equipment and helium gas in a world record dive. *J. Ind. Hyg. Toxicol.* **20**:511–520 (Oct. 1938).

END, E. The physiological effects of increased pressure. In: *Proceedings of the Sixth Pacific Science Congress of the Pacific Science Association*, pp. 91–97. Berkeley, California, University of California Press (1939).

EPSTEIN, M., and T. SARUTA. Effect of water immersion in reninaldosterone and renal sodium handling in normal man. *J. Appl. Physiol.* **31**:368–374 (Sept. 1971).

EPSTEIN, S. E., M. STAMPFER, G. D. BEISER, R. E. GOLDSTEIN, and E. BRAUNWALD. Effects of a reduction in environmental temperature on the circulatory response to exercise in man. *New Engl. J. Med.* **280**:7–11 (Jan. 2, 1969).

FABER, J. J., G. R. WILLIAMSON, and N. T. FELDMAN. Lubrication of joints. *J. Appl. Physiol.* **22**:793–799 (1967).

FEATHERSTONE, R. M., and C. A. MUEHLBAECHER. The current role of inert gases in the search for anesthesia mechanisms. *Pharmacol. Rev.* **15**:97–121 (Mar. 1973).

FENN, W. O. The physiological effects of hydrostatic pressures. In: Bennett, P. B., and D. H. Elliott, eds. *The Physiology and Medicine of Diving and Compressed Air Work*, pp. 36–57. Baltimore, Williams and Wilkins (1969).

FERGUSON, J. The use of chemical potentials as indices of toxicity. *Proc. Roy. Soc. London (B)* **127**:387 (July 4, 1939).

FISCHER, C. L., and S. L. KIMZEY. Effects of oxygen on blood formation and destruction. In: Lambertsen, C. J., ed. *Underwater Physiology—Proceedings of the Fourth Symposium on Underwater Physiology*, pp. 41–47. Academic Press (1971).

FISHER, A. B., R. W. HYDE, R. J. M. PUY, J. M. CLARK, and C. J. LAMBERTSEN. Effect of oxygen at 2 atm on the pulmonary mechanics of normal man. *J. Appl. Physiol.* **24**:529–536 (Apr. 1968).

FISHER, D. A., and W. D. ODELL. Effect of cold on TSH secretion in man. *J. Clin. Endocr. Metab.* **33**:859–862 (Nov. 1971).

FLYNN, E. T. Effect of immersion on the exchange of oxygen in the lung. U. S. Navy Exp. Diving Unit, Rep. NEDU 1-71 (Jan. 31, 1971).

FLYNN, E. T., T. E. BERGHAGE, and E. F. COIL. Influence of increased ambient pressure and gas density on cardiac rate in man. U. S. Navy Exp. Diving Unit, Rep. NEDU 4-72 (Aug. 1972).

FRANKENHAEUSER, M. V., V. GRAFF-LONNEVIG, and C. M. HESSER. Effects on phychomotor functions of different nitrogen–oxygen mixtures at increased ambient pressures. *Acta Physiol. Scand.* **59**:400–409 (Dec. 1963).

FRATTALI, V., and R. ROBERTSON. Nutritional evaluation of humans during an oxygen–helium dive to a simulated depth of 1000 feet. *Aerosp. Med.* **44**:14–21 (Jan. 1973).

FRUCTUS, X. Physalie VI: 610 metres. Nouvelle performance mondiale de plongee profonde en caisson realisee a Marseille, au Centre Experimental Hyperbare de la COMEX, du 16-5 au 2-6 1972. *Med. Sport* **46**(3):180–182 (1972).

FRUCTUS, X., and C. AGARATE. The high pressure nervous syndrome. *Med. Sport.* **24**:272–278 (Nov. 1971).

FRUCTUS, X. R., R. W. BRAUER, and R. NAQUET. Physiological effects observed in the course of simulated deep chamber dives to a maximum of 36.5 atm in helium–oxygen atmospheres. In: Lambertsen, C. J., ed. *Underwater Physiology. Proceedings of the Fourth Symposium on Underwater Physiology*, pp. 545–550. New York, Academic Press (1971).

FRUCTUS, X., C. AGARATE, and F. SICARDI. Postponing the "high pressure nervous syndrome" (HPNS) down to 500 meters and deeper. In: *Proceedings of the Fifth Symposium on Underwater Physiology, Freeport, Bahamas, 21–25 August 1972. In press.*

GAGGE, A. P., J. D. HARDY, and G. M. RAPP. Proposed standard system of symbols for thermal physiology. *J. Appl. Physiol.* **27**:439–446 (Sept. 1969).

GAUER, O H , J. P. HENRY, H. O. SIEKER, and W. E. WENDT. The effect of negative pressure breathing on urine flow. *J. Clin. Invest.* **33**:287–296 (Feb. 1954).

GERSCHMAN, R., D. L. GILBERT, S. W. NYE, P. DWYER, and W. O. FENN. Oxygen poisoning and irradiation: a mechanism in common. *Science* **119**:623–626 (1954*a*).

GERSCHMAN, R., S. W. NYE, D. L. GILBERT, P. DWYER, and W. O. FENN. Oxygen poisoning: Protective effect of beta-mercaptoethylamine. *Proc. Soc. Exp. Biol. Med.* **85**:75–77 (1954*b*).

GERSCHMAN, R., D. L. GILBERT, and D. CACCAMISE. Effect of various substances on survival time of mice exposed to different high oxygen tensions. *Am. J. Physiol.* **192**:563–571 (1958).

GERSHENOVICH, Z. S., and A. A. KRICHEVSKAYA. The protective role of arginine in oxygen poisoning. *Biokhimiya* **25**:790–795 (1960).

GERTSMAN, J. L., G. R. GAMERTSFELDER, and A. GOLDBERGER. Breathing mixture and depth as separate effects on helium speech. *J. Acoust. Soc. Amer.* **40**:1283A (Nov. 1966).

GILARDI, R. C. Saturation diving gas logistics. In: *The Working Diver 1972. Symposium Proceedings, Feb. 1972, Columbus, Ohio*, pp. 9–22. Washington, D. C., Marine Technology Society (1972).

GILLEN, H. W. Oxygen convulsions in man. In: Brown, I. W., Jr. and B. G. Fox, eds. *Proceedings of the Third International Conference on Hyperbaric Medicine*, pp. 217–223; Washington, D. C., National Academy of Sciences–National Research Council (1966).

GOLDMAN, R. F., J. R. BRECKENRIDGE, E. REEVES, and E. L. BECKMAN. "Wet" versus "dry" suit approaches to water immersion protective clothing. *Aerosp. Med.* **37**:485–487 (May 1966).

GOLDSTEIN, J. R., and C. E. MENZEL. Hemolysis in mice exposed to varying levels of hyperoxia. *Aerosp. Med.* **40**:12–81 (Jan. 1969).

GOLSTEIN-GOLAIRE, J., L. VANHAELST, O. D. BRUNO, R. LECLERCQ, and G. COPINSCHI. Acute effects of cold on blood levels of growth hormone, cortisol, and thyrotropin in man. *J. Appl. Physiol.* **29**:622–626 (Nov. 1970).

GOODMAN, M. W. The syndrome of decompression sickness in historical perspective. U. S. Nav. Med. Res. Lab., Rep. NMRL368 (June 1962).

GOODMAN, M. W. Minimal-recompression, oxygen-breathing method for the therapy of decompression sickness. In: Lambertsen, C. J., ed. *Underwater Physiology, Proceedings of the Third Symposium on Underwater Physiology*, pp. 165–182. Baltimore, Williams and Wilkins (1967).

GOODMAN, M. W., N. E. SMITH, J. W. COLSTON, and E. L. RICH, III. Hyperbaric respiratory heat loss study. Annapolis, Md., Westinghouse Electric Corp., Ocean Res. Eng. Cent., Final Rep. on contract N000-4-71-0099 (Oct. 31, 1971).

GRAVES, D. J., J. IDICULA, C. J. LAMBERTSEN, and J. A. QUINN. Bubble formation in physical and biological systems. A manifestation of counterdiffusion in composite media. *Science* **179**:582–584 (Feb. 9, 1973).

GULERIA, J., J. TALWAR, O. MALHOTRA, and J. PANDE. Effect of breathing cold air on pulmonary mechanics in normal man. *J. Appl. Physiol.* **27**:320–322 (Sept. 1969).

HAMILTON, R. W., JR. Comparative narcotic effects in performance tests of nitrous oxide and hyperbaric nitrogen. *Fed. Proc.* **32**(3, PP.2):682 (Mar. 1973).

HAMILTON, R. W., JR. Psychomotor performance in normoxic neon and helium at 37 atmospheres. In: *Proceedings of the Fifth Symposium on Underwater Physiology, August 1972, Freeport, Bahamas. In press.*

HAMILTON, R. W., JR. and E. ERB. Are special purity standards for divers' breathing gas really needed? In: Battelle Memorial Institute. *Purity Standards for Divers' Breathing Gas. Proceedings of a Symposium, Columbus, Ohio, July 1970*, pp. I-1–I-10. Columbus, Ohio, Battelle Memorial Institute, Rep. 6-70 (July 1970).

HAMILTON, R. W., JR. and J. B. MACINNIS. Unpublished observations (1965).

HAMILTON, R. W., JR., J. B. MACINNIS, A. D. NOBLE, and H. R. SCHREINER. Saturation diving to 650 feet. Tonawanda, N. Y., Ocean Systems, Inc., Tech. Mem. B-411 (1966).

HAMILTON, R. W., JR., D. J. KENYON, M. FREITAG, and H. R. SCHREINER. NOAA OPS I and II: Formulation of excursion procedures for shallow undersea habitats. Tarrytown, N.Y., Union Carbide Corp., Rep. UCRI-731 (1973).

HARVEY, H. W. *The Chemistry and Fertility of Sea Waters*, pp. 14–19. London, Cambridge University Press (1966).

HAUGAARD, N. Effect of high oxygen tensions on enzymes. In: Goff, L. G., ed. *Proceedings of the Underwater Physiology Symposium*, pp. 8–12; Washington, D. C., National Academy of Science–National Research Council (1955), Publ. 377.

HAUGAARD, N. The scope of chemical oxygen poisoning. In: Lambertsen, C. J., ed. *Underwater Physiology, Proceedings of the Fourth Symposium on Underwater Physiology*, pp. 1–7. New York, Academic Press (1971).

HEISTAD, D. D., and R. C. WHEELER. Simulated diving during hypoxia in man. *J. Appl. Physiol.* **28**:652–656 (May 1970).

HEMPLEMAN, H. V. Investigation into the diving tables. Report III. Alverstoke, U.K., Roy. Nav. Personnel Res. Comm., Rep. 131 (1952).

HESSER, C. M., and B. HOLMGREN. Effects of raised barometric pressures on respiration in man. *Acta Physiol. Scand.* **47**:28–43 (1959).

HILL, L., R. H. DAVIS, R. P. SELBY, A. PRIDHAM, and A. E. MALONE. Deep diving and ordinary diving. U. K., Report of Committee appointed by the British Admiralty (1933).

HILLS, B. A. Thermodynamic decompression: an approach based upon the concept of phase equilibrium in tissue. In: Bennett, P. B., and D. H. Elliott, eds. *The Physiology and Medicine of Diving and Compressed Air Work*, pp. 319–356. Baltimore, Williams and Wilkins (1969).

HOCK, R. J., G. F. BOND, and W. F. MAZZONE. Physiological evaluation of Sealab II: Effects of two weeks exposure to an undersea 7-atmosphere helium–oxygen environment. Anaheim, Calif., Nortronic (Dec. 1966).

HOF, D. G., W. H. CLINE, JR., J. D. DEXTER, and C. E. MENGEL. CNS epinephrine tone, a possible etiology for the threshold in susceptibility to oxygen toxicity seizures. *Aerosp. Med.* **43**:1194–1199 (Nov. 1972).

HOFF, E. C. A bibliographic sourcebook of compressed air, diving and submarine medicine. Washington, D. C., Department of the Navy, Bureau of Medicine and Surgery (February 1948) (NAVMED 1191).

HOKE, B., D. L. JACKSON, J. M. ALEXANDER, and E. T. FLYNN. Respiratory heat loss and pulmonary function during cold gas breathing at high pressure. In: Lambertsen, C. J., ed. *Proceedings of the Fifth Symposium on Underwater Physiology, Freeport, Bahamas, August 1972. In press.*

HONG, S. K. Comparison of the diving and nondiving women of Korea. *Fed. Proc.* **22**:831–833 (May/June 1963).

HONG, S. K. Patterns of adaptation in women divers of Korea (ama). *Fed. Proc.* **32**:1614–1622 (May 1973).

HONG, S. K., H. RAHN, D. H. KANG, S. H. SONG, and B. S. KANG. Diving pattern, lung volumes, and alveolar gas of the Korean diving women (Ama). *J. Appl. Physiol.* **18**:457–465 (May 1963).

HONG, S. K., S. H. SONG, P. K. KIM, and C. S. SUH. Seasonal observations on the cardiac rhythm during diving in the Korean Ama. *J. Appl. Physiol.* **23**:18–22 (July 1967).

HONG, S. K., C. K. LEE, J. K. KIM, S. H. SONG, and D. W. RENNIE. Peripheral blood flow and heat flux of Korean women divers. *Fed. Proc.* **28**:1143–1148 (May/June 1969a).

HONG, S. K., P. CERRETELLI, J. C. CRUZ, and H. RAHN. Mechanics of respiration during submersion in water. *J. Appl. Physiol.* **27**:535–538 (Oct. 1969b).

HONG, S. K., T. O. MOORE, G. SETO, H. K. PARK, W. R. HIATT, and E. M. BEQNAUER. Lung volumes and apneic bradycardia in divers. *J. Appl. Physiol.* **29**:172–176 (Aug. 1970).

HONG, S. K., Y. C. LIN, D. A. LALLY, B. J. B. YIM, N. KONIRNAMI, P. W. HONG, and T. O. MOORE. Alveolar gas exchanges and cardiovascular functions during breath-holding with air. *J. Appl. Physiol.* **30**:540–547 (Apr. 1971).

HONG, S. K., T. O. MOORE, D. A. LALLY, and J. F. MORLOCK. Heart rate response to apneic face immersion in hyperbaric heliox environment. *J. Appl. Physiol.* **34**:770–774 (June 1973).

HORNE, T. Protective action of some vitamin K analogues against the toxic action of hyperbaric oxygen. *Biochem. J.* **100**:11p (July 1966).

HUNT, H., E. REEVES, and E. L. BECKMAN. An experiment in maintaining homeostasis in a long distance underwater swimmer. U. S. Nav. Med. Res. Inst., Rep. 2 on MR005.13-4001.06 (July 23, 1964).

IDICULA, J., D. J. GRAVES, J. A. QUINN, and C. J. LAMBERTSEN. Bubble formation resulting from steady counterdiffusion of two inert gases. In: *Proceedings of the Fifth Symposium on Underwater Physiology, August 1972, Freeport, Bahamas. In press.*

INSTITUTE FOR ENVIRONMENTAL MEDICINE. *Practical Aspects of Oxygen Tolerance and Oxygen Toxicity.* Philadelphia, University of Pennsylvania (April 1970).

JAMIESON, D., and H. A. S. VAN DEN BRENK. The effects of antioxidants on high pressure oxygen toxicity. *Biochem. Pharmacol.* **13**:159–164 (Feb. 1964).

JARRETT, A. S. Alveolar carbon dioxide at increased ambient pressures. *J. Appl. Physiol.* **21**:158–162 (Jan. 1966).

JEGOU, A. Deep diving and cold water, some practical results. In: *The Working Diver, 1972. Symposium Proceedings, February 1972, Columbus, Ohio*, pp. 127–143; Washington, D. C., Marine Technology Society (1972).

JENKINS, W. T. Personal communication (1973).

JERETT, S. A., D. JEFFERSON, and C. E. MENGEL. Seizures, hydrogen peroxide formation and lipid peroxides in brain during exposure to oxygen under high pressure. *Aerosp. Med.* **44**:40–44 (Jan. 1973).

JOHNSON, S. M., and K. W. MILLER. Antagonism of pressure and anesthesia. *Nature* **228**:75–76 (Oct. 3, 1970).

JOSENHANS, W. T., G. N. MELVILLE, and W. T. ULMER. The effect of facial cold stimulation on airway conductance in healthy man. *Can. J. Physiol. Pharmacol.* **47**:453–457 (May 1969).

KANN, H. E., JR., C. E. MENGEL, W. SMITH, and B. HORTON. Oxygen toxicity and vitamin E. *Aerosp. Med.* **35**:840–844 (Sept. 1964).

KAPLAN, S. A., and S. N. STEIN. Effects of oxygen at high pressure on the transport of potassium, sodium, and glutamate in guinea pig brain cortex. *Amer. J. Physiol.* **190**:157–162 (July 1957)

KAWAKAMI, Y., B. H. NATELSON, and A. B. DuBOIS. Cardiovascular effects of face immersion and factors affecting diving reflex in man. *J. Appl. Physiol.* **23**:964–970 (Dec. 1967).

KAY, H. Report on arctic trials on board HMS Vengeance February–March 1949. London, England, Med. Res. Counc., Roy. Nav. Pers. Res. Comm., Rep. 534 (1949).

KEATINGE, W. R. *Survival in Cold Water.* Oxford and Edinburgh, Blackwell Scientific Publications (1969).

KEATINGE, W. R., M. B. McILROY, and A. GOLDFIEN. Cardiovascular responses to ice-cold showers. *J. Appl. Physiol.* **19**:1145–1150 (Nov. 1964).

KELLER, H. Use of multiple inert gas mixtures in deep diving. In: Lambertsen, C. J., ed. *Underwater Physiology. Proceedings of the Third Symposium on Underwater Physiology,* pp. 267–274. Baltimore, Williams and Wilkins (1967).

KELLER, H., and A. A. BÜHLMANN. Deep diving and short decompression by breathing mixed gases. *J. Appl. Physiol.* 1267–1270 (Nov. 1965).

KHAMBATTA, H. J., and R. A. BARATZ. IPPB, plasma ADH, and urine flow in conscious man. *J. Appl. Physiol.* **33**:362–364 (Sept. 1972).

KIESSLING, R. J., and C. H. MAAG. Performance impairment as a function of nitrogen narcosis. *J. Appl. Psychol.* **46**:91–95 (Apr. 1962).

KINDWALL, E. P. Medical aspects of commercial diving and compressed air work. In: Zenz, C., ed. *Occupational Medicine: Principles and Practical Applications.* Chicago, Medical Yearbook Publications. *In press.*

KINNEY, J. A. S., and C. L. McKAY. The visual-evoked response as a measure of nitrogen narcosis in Navy divers. *U. S. Nav. Submar. Med. Cent.,* Rep. SMRL 664 (Apr. 21, 1971).

KINNEY, J. A. S., C. L. McKAY, A. MENSCH, and S. M. LURIA. The visual-evoked response as a measure of stress in naval environments: methodology and analysis. *U. S. Nav. Submar. Med. Cent.,* Rep. SMRL 669 (June 25, 1971).

KINNEY, J. A. S., C. L. McKAY, and S. M. LURIA. Visual evoked responses for divers breathing various gases at depths to 1200 feet. *U. S. Nav. Submar. Med. Cent.,* Rep. SMRL 705 (Mar. 23, 1972).

KOOI, K. *Fundamentals of Electroencephalography.* New York, Harper and Row (1971).

KYDD, G. H. Observations on acute and chronic oxygen poisoning. *Aerosp. Med.* **35**:1176–1179 (Dec. 1964).

KYLSTRA, J. A. Breathing fluid. *Experientia* **18**(2):68 (1962).

KYLSTRA, J. A. Hydraulic compression of mice to 166 atm. *Science* **158**:793–794 (1967).

KYLSTRA, J. A., I. S. LONGMUIR, and M. GRACE. Dysbarism: osmosis caused by dissolved gas. *Science* **161**:289 (1968).

KYLSTRA, J. A., W. H. SCHOENFISCH, J. M. HERRON, and G. D. BLENKARN. Gas exchange in saline-filled lungs of man. *J. Appl. Physiol.* **35**:136–142 (July 1973).

LALLY, D. A., T. O. MOORE, and S. K. HONG. Cardiorespiratory responses to exercise in air and water at 1 and 2 ATA. Honolulu, Univ. Hawaii, Sch. Med., Dept. Physiol., Rep. UNIHI-SEAGRANT-TR-71-04 (Dec. 1971).

LAMBERTSEN, C. J. Respiratory and circulatory actions of high oxygen pressure. In: Goff, L. G., ed. *Proceedings of the Underwater Physiology Symposium,* pp. 25–38, Washington, D. C., National Academy of Science–National Research Council (1955), Publ. 377.

LAMBERTSEN, C. J. Carbon dioxide and respiration in acid–base homeostasis. *Anesthesiology* **21**:642–651 (Nov./Dec. 1960).

LAMBERTSEN, C. J. Physiological effects of oxygen. In: Lambertsen, C. J., and L. J. Greenbaum, Jr. eds. *Proceedings of the Second Symposium on Underwater Physiology*, pp. 171–187. Washington, D. C., National Academy of Sciences–National Research Council (1963).

LAMBERTSEN, C. J. Effects of oxygen at high partial pressure. In: Fenn, W. O., and H. Rahn, eds. *Handbook of Physiology*, Section 3: *Respiration*, Vol. II, pp. 1027–1046. Washington, D. C., American Physiological Society (1965).

LAMBERTSEN, C. J. Discussion. In: Brown, I. W., and B. G. Cox, eds. *Proceedings of the Third International Conference on Hyperbaric Medicine*, p. 207. Washington, D. C., National Academy of Sciences–National Research Council (1966a).

LAMBERTSEN, C. J. Oxygen toxicity. In: Committee on Hyperbaric Oxygenation. *Fundamentals of Hyperbaric Medicine*, pp. 21–40; Washington, D. C., National Academy of Sciences–National Research Council (1966b).

LAMBERTSEN, C. J. Chemical control of respiration at rest. In: Mountcastle, V. B., ed. *Medical Physiology*, pp. 713–763. St. Louis, Mosby (1968).

LAMBERTSEN, C. J. Therapeutic gases: Oxygen, carbon dioxide, and helium. In: DiPalma, J. R., ed. *Drill's Pharmacology in Medicine*, fourth edition, pp. 1145–1179. New York, McGraw-Hill (1971).

LAMBERTSEN, C. Collaborative investigation of limits of human tolerance to pressurization with helium, neon and nitrogen. Simulation of density equivalent to helium–oxygen respiration at depths to 2,000, 3,000, 4,000, and 5,000 feet of sea water. In: *Proceedings of the Fifth Symposium on Underwater Physiology, Freeport, Bahamas, August 1972. In press.*

LAMBERTSEN, C. J., R. H. KOUGH, D. Y. COOPER, G. L. EMMEL, H. H. LOESCHCKE, and C. F. SCHMIDT. Oxygen toxicity. Effects in man of oxygen inhalation at 1 and 3.5 atmospheres upon blood gas transport, cerebral circulation and cerebral metabolism. *J. Appl. Physiol.* **5**:471–486 (Mar. 1953).

LANGLEY, T. D. Neurophysiological investigation of inert gas depression of the central nervous system. Tarrytown, N. Y., Ocean Systems, Inc., Ann. Prog. Rep. on ONR contract N00014-69-C-0405 (April 30, 1970).

LANGLEY, T. D. Somatic and auditory evoked brain responses in many breathing mixtures of normoxic helium, nitrogen and neon at pressures to 37 atmospheres. In: *Proceedings of the Fifth Symposium on Underwater Physiology, Freeport, Bahamas, August 1972. In press.*

LANPHIER, E. H. Interactions of factors limiting performance at high pressures. In: Lambertsen, C. J. *Underwater Physiology. Proceedings of the Third Symposium on Underwater Physiology*, pp. 375–385. Baltimore, Williams and Wilkins (1967).

LANPHIER, E. H. Pulmonary function. In: Bennett, P. B., and D. H. Elliott, eds. *The Physiology and Medicine of Diving and Compressed Air Work*, pp. 58–112. Baltimore, Williams and Wilkins (1969).

LANPHIER, E. H., and H. RAHN. Alveolar gas exchange during breath-hold diving. *J. Appl. Physiol.* **18**:471–477 (May 1963).

LARSON, H. E. A history of self-contained diving and underwater swimming. Publication 469. Washington, D. C., National Academy of Sciences–National Research Council (1959).

LASSITER, D. V., ed. Occupational exposure to carbon monoxide. HSM 73-11000. Washington, D. C., Department of Health, Education, and Welfare (1972).

LEBOUCHER, F. Status report on equipment developments in the French Navy. In: *Equipment for the Working Diver. Symposium Proceedings, February 24–25, 1970, Columbus, Ohio*, pp. 345–351. Washington, D. C., Marine Technology Society (1970).

LEE, O. *The Complete Illustrated Guide to Snorkel and Deep Diving.* New York, Doubleday (1967).

LENFANT, C., and B. HOWELL. Cardiovascular adjustments in dogs during continuous pressure breathing. In: Rahn, H., ed. *Studies in Pulmonary Physiology.* WADD Tech. Rep. 60-1 (1960).

LEON, H. A., and S. F. COOK. A mechanism by which helium increases metabolism in small animals. *Amer. J. Physiol.* **199**:243–245 (Aug. 1960).

LEVER, M. J., K. W. MILLER, W. D. PATON, and E. B. SMITH. Pressure reversal of anesthesia. *Nature* **231**:368–371 (June 11, 1971).

MacINNIS, J. B. Some comparisons of U. S. Navy and civilian diving programs. In: *Progress into the Sea. Transactions of the Symposium, October 1969*, pp. 129–134. Washington, D. C., Marine Technology Society (1970).

MacINNIS, J. B. Arctic diving and problems of performance. In: *The Working Diver, 1972. Symposium Proceedings, February 1972, Columbus, Ohio*, pp. 159–174. Washington, D. C., Marine Technology Society (1972).

MACKWORTH, N. H. Finger numbness in very cold winds. *J. Appl. Physiol.* **5**:533–543 (Mar. 1953).

MAIO, D. A., and L. E. FARHI. Effect of gas density on mechanics of breathing. *J. Appl. Physiol.* **23**:687–693 (Nov. 1967).

MATHER, G. W., G. G. NAHAS, and A. HEMINGWAY. Temperature changes of pulmonary blood during exposure to cold. *Amer. J. Physiol.* **173**:390–392 (June 1953).

McCALLY, M. Body fluid volumes and the renal response to immersion. In: *Physiology of Breath-Hold Diving and the Ama of Japan*, pp. 253–269. Washington, D. C., National Academy of Sciences–National Research Council (1965), Publ. 1341.

McSHERRY, C. K., and F. J. VEITH. The relation between the central nervous system and pulmonary forms of oxygen toxicity: Effect of THAM administration, *Surg. Forum* **19**:33–35 (1968).

MEAD, J., and J. MILIC-EMILI. Theory and methodology in respiratory mechanics with glossary of symbols. In: Fenn, W. O., and H. Rahn, eds. *Handbook of Physiology*, Section 3, Vol. I, pp. 363–376. Washington, D. C., American Physiological Society (1964).

Merck Manual of Diagnosis and Therapy. Twelfth edition. Rahway, N. J., Merck Sharp and Dohme Research Laboratories (1972).

MEYER, H. H. Theorie der Alkoholnarkose. I. Mittwelche Eigenschaft der Anasthetika bedingt ihre narkotische Wirking. *Arch. Exper. Path. Pharmakol.* **42**:109 (May 1899).

MILES, S. *Underwater Medicine.* London, Staple Press (1965).

MILES, S. *Underwater Medicine.* Third edition. Philadelphia, J. B. Lippincott (1969).

MILLER, J. N., O. D. WANGANSTEEN, and E. H. LANPHIER. Ventilatory limitations on exertion at depth. In: Lambertsen, C. J., ed. *Underwater Physiology. Proceedings of the Fourth Symposium on Underwater Physiology*, pp. 317–323. New York, Academic Press (1971).

MILLER, J. W., J. G. VAN DERWALKER, and R. A. WALLER, eds. *TEKTITE 2. Scientists-in-the Sea.* Washington, D. C., U. S. Department of the Interior (August 1971).

MILLER, K. W., W. D. M. PATON, W. B. STREET, and E. B. SMITH. Animals at very high pressures of helium and neon. *Science* **157**:97–98 (July 7, 1967a).

MILLER, K. W., W. D. M. PATON, and E. B. SMITH. The anesthetic pressures of certain fluorine-containing gases. *Brit. J. Anaesth.* **39**:910–917 (Dec. 1967b).

MILLER, K. W., W. D. M. PATON, R. A. SMITH, and E. B. SMITH. The pressure reversal of anesthesia and the critical volume hypothesis. *Mol. Pharmacol.* **9**(2):131–143 (1973).

MILLER, S. L. A Theory of gaseous anesthetics. *Proc. Nat. Acad. Sci. U. S.* **47**:1515 (Sept. 15, 1961).

MITHOEFER, J. C. Breath-holding. In: Fenn, W. O., and H. Rahn, eds. *Handbook of Physiology*, Section 3, Vol. II, pp. 1011–1025. Washington, D. C., American Physiological Society (1965).

MONEY, D. F. L., and P. J. STRONG. Underwater diving, oxygen poisoning and vitamin E. *N. Z. Med. J.* **75**:34–35 (Jan. 1972).

MOORE, T. O., E. M. BERNAUER, G. SETO, Y. S. PARK, S. K. HONG, and E. M. HAYASHI. Effect of immersion at different water temperatures on graded exercise performance in man. *Aerosp. Med.* **41**:1404–1408 (Dec. 1970).

MOORE, T. O., Y. C. LIN, D. A. LALLY, and S. K. HONG. Effects of temperature, immersion, and ambient pressure on human apneic bradycardia. *J. Appl. Physiol.* **33**:36–41 (July 1972).

MOORE, T. O., R. ELSNER, Y. C. LIN, D. A. LALLY, and S. K. HONG. Effects of alveolar PO_2 and PCO_2 on apneic bradycardia in man. *J. Appl. Physiol.* **34**:795–798 (June 1973).

MOORE, T. O., J. F. MORLOCK, D. A. LALLY, and S. K. HONG. Thermal cost of saturation diving: Respiratory and whole body heat loss at 16.1 ATA. In: *Proceedings of the Fifth Symposium on Underwater Physiology, Freeport, Bahamas, 1972. In press.*

MORENO, F., and H. A. LYONS. Effect of body posture on lung volumes. *J. Appl. Physiol.* **16**:27–29 (Jan. 1961).

MORRISON, J. B. Respiratory function. In: Hempleman, H. V., ed. Experimental observations on men at pressures between 4 bars (100 ft) and 47 bars (1500 ft), pp. 34–59. Alverstoke, U. K., Roy. Navy Physiol. Lab., Rep. 1-71 (1971).

MURPHY, T. M., W. H. CLARK, I. P. B. BUCKINGHAM, and W. A. YOUNG. Respiratory gas exchange in exercise during helium–oxygen breathing. *J. Appl. Physiol.* **26**:303–307 (Mar. 1969).

MYERS, R. D., and L. G. SHARPE. Temperature in the monkey: transmitter factors released from the brain during thermoregulation. *Science* **161**:572–573 (Aug. 9, 1968).

NAQUET, R., and J. C. ROSTAIN. Postponing the "High pressure nervous syndrome" (H.P.N.S.) down to 500 meters and deeper. The evolution of H.P.N.S. with depth, compression rate and bottom time. In: Lambertsen, C. J., ed. *Proceedings of the Fifth Symposium on Underwater Physiology, Freeport, Bahamas, 1972. In press.*

NICHOLS, C. W., and C. J. LAMBERTSEN. Effects of oxygen upon ophthalmic structures. In: Lambertsen, C. J., ed. *Underwater Physiology, Proceedings of the Fourth Symposium on Underwater Physiology*, pp. 57–66. New York, Academic Press (1971).

ORNHAGEN, H. C., and C. E. G. LUNDGREN. Hydrostatic pressure tolerance in liquid breathing mice. In: *Proceedings of the Fifth Symposium on Underwater Physiology, Freeport, Bahamas, August 1972. In press.*

OTHMER, D. F., and O. A. REELS. Power, fresh water, and food from cold, deep sea water. *Science* **182**:121–125 (Oct. 12, 1973).

OVERTON, E. *Studies uberdie Narkose.* Jena, East Germany, Fischer (1901).

PATON, W. D. M., and A. SAND. The optimum intrapulmonary pressure in underwater respiration. *J. Physiol. (London)* **106**:119–138 (June 1947).

PAULEV, P. E. Respiratory and cardiovascular effect of breath-holding. *Acta Physiol. Scand. Suppl.* 324 (1969).

PAULING, L. A molecular theory of general anesthesia. *Science* **134**:15–21 (July 7, 1961).

PHILP, R. B., C. W. GOWDEY, and M. PRASAD. Changes in blood lipid concentration and cell counts following decompression sickness in rats and the influence of dietary lipid. *Can. J. Physiol. Pharmacol.* **45**:1047–1059 (Nov. 1967).

POULTON, E. C. *Environment and Human Efficiency.* Springfield, Ill., Charles C Thomas (1970).

RAHN, H. The physiological stresses of the Ama. In: Rahn, H., and T. Yokoyama, eds. *Physiology of Breath-Hold Diving and the Ama of Japan*, pp. 295–302. Washington, D. C., National Academy of Sciences–National Research Council (1965), Publ. 1341.

RANDALL, J. E. Analog and digital computers in the study of physiologic tremor. *Arch. Phys. Med. Rehab.* **48**:463–466 (1967).

RAPER, A. J., D. W. RICHARDSON, H. A. KONTOS, and J. L. PATTERSON, JR. Circulatory responses to breath holding in man. *J. Appl. Physiol.* **22**:201–206 (Feb. 1967).

RAWLINS, J. S. P. Thermal balance in divers. *J. Roy. Nav. Med. Serv.* **58**:182–188 (Winter 1972).

RAYMOND, L. W. Temperature problems in multiday exposures to high pressures in the sea. Thermal balance in hyperbaric atmospheres. In: Lambertsen, C. J., ed. *Underwater Physiology. Proceedings of the Third Symposium on Underwater Physiology. 23–26 March, 1966, Washington, D. C.*, pp. 138–147. Baltimore, Williams and Wilkins (1967).

RAYMOND, L. W. The thermal environment for undersea habitats. In: *Human Factors 1970, Symposium of the American Society of Heating, Refrigerating and Air Conditioning Engineers, San Francisco, January 1970*, pp. 22–24 (1971).

RAYMOND, L., W. H. BELL II, K. R. BONDI, and C. R. LINDBERG. Body temperature and metabolism in hyperbaric helium atmospheres. *J. Appl. Physiol.* **24**:678–684 (May 1968).

RAYMOND, L., R. B. WEISKOPF, M. J. HALSEY, A. GOLDFIEN, E. I. EGER III, and J. W. SEVERINGHAUS. Possible mechanisms for the antiarrhythmic effect of helium in anesthetized dogs. *Science* **176**:1250–1252 (June 16, 1972).

RAYMOND, L. W., H. C. LANGWORTHY, J. SODE, and J. BLOSSER. Metabolic responses to work in 25°C water at 3 meters. *In preparation.*

RAYMOND, L. W., E. D. THALMANN, W. H. SPAUR, W. R. BRAITHWAITE, J. H. CROTHERS, and H. C. LANGWORTHY. Thermal homeostasis of man in helium–oxygen at 1–50 atmospheres absolute. *In preparation.*

REEVES, E., J. W. WEAVER, J. J. BENJAMIN, and C. H. MANN. Comparison of physiological changes during long term immersion to neck levels in water at 95°, 85° and 75°F. U. S. Nav. Med. Res. Inst., Rep. 9 on MF011.99-1001 (1965).

ROGER, A., P. CABARROU, and H. H. GASTAUT. EEG changes in humans due to changes in surrounding atmospheric pressure. *Electroencephalogr. Clin. Neurophysiol.* **7**:152 (Feb. 1955).

ROSSIER, P. H., and H. MEAN. L'insuffisance pulmonaire. *Schweiz. Med. Wochenschr.* **73**(11):327–332 (March 13, 1943).

RUSSELL, C. J., A. McNEIL, and E. EVONUK. Some cardiorespiratory and metabolic responses of scuba divers to increased pressure and cold. *Aerosp. Med.* **43**:998–1001 (Sept. 1972).

SALTZMAN, H. A., L. HART, B. ANDERSON, E. DUFFY, and H. O. SIEKER. The response of the retinal circulation to hyperbaric oxygenation. *J. Clin. Invest.* **43**:1283 (June 1964).

SALZANO, J., D. C. RAUSCH, and H. A. SALTZMAN. Cardiorespiratory responses to exercise at a simulated seawater depth of 1,000 feet. *J. Appl. Physiol.* **28**:34041 (Jan. 1970).

SANDERS, A. P., and W. D. CURRIE. Chemical protection against oxygen toxicity. In: Lambertsen, C. J., ed. *Underwater Physiology. Proceedings of the Fourth Symposium on Underwater Physiology*, pp. 35–40. New York, Academic Press (1971).

SANDERS, A. P., I. H. HALL, and B. WOODHALL. Succinate: Protective agent against hyperbaric oxygen toxicity. *Nature* **150**(3075):1830–1831 (1965).

SANDERS, A. P., R. M. GELIEN, JR., R. S. CRAMER, and W. D. CURRIE. Protection against the chronic effects of hyperbaric oxygen toxicity by succinate and reduced glutathione. *Aerosp. Med.* **43**:533–536 (May 1972).

SASAMOTO, H. The electrocardiogram pattern of the diving Ama. In: Rahn, H., and T. Yokoyama, eds. *Physiology of Breath-Hold Diving and the Ama of Japan*, pp. 271–280. Washington, D. C., National Academy of Sciences–National Research Council (1965), Publ. 1341.

SAYERS, R. R., W. P. YANT, and J. H. HILDEBRAND. Possibilities in the use of helium–oxygen mixtures as a mitigation of caisson disease. Serial No. 2670. Washington, D. C., U. S. Department of the Interior, Bureau of Mines (Feb. 1925).

SCHAEFER, K. E. Environmental physiology of submarines and aircraft (atmospheric requirements of confined spaces). *Arch. Environ. Health* **9**:320–331 (Sept. 1964).

SCHAEFFER, K. E., C. R. CAREY, and J. DOUGHERTY, JR. Pulmonary gas exchange and urinary electrolyte excretion during saturation–excursion diving to pressures equivalent to 800 and 1,000 feet of seawater. *Aerosp. Med.* **41**:856–864 (Aug. 1970).

SCHMIDT, T. C., D. J. KENYON, M. FREITAG, and R. W. HAMILTON, JR. Recovery and reuse of diving gas. In: Battelle Columbus Laboratories. *Proceedings, 1973 Divers' Gas Purity Symposium, November 27–28, 1973*, pp. XV-1–XV-12. Washington, D. C., U. S. Navy Supervisor of Diving (1973).

SCHMIDT, T. C., R. W. HAMILTON, JR., G. MOELLER, and C. P. CHATTIN. Diver performance during NOAA OPS I and II: Cognitive and psychomotor performance during nitrogen saturation exposures of 7-day duration at 2, 3, 4, 5, atm and air excursions to pressures up to 10 atm. U. S. Nav. Submar. Med. Cent., Rep. *In press.*

SCHOENFISCH, W. H., and J. A. KYLSTRA. Maximum expiratory flow and estimated CO_2 elimination in liquid ventilated dogs' lungs. *J. Appl. Physiol.* **35**:117–121 (July 1973).

SCHOLANDER, P. F. Physiological adaptation to diving in animals and man. *Harvey Lecture Ser.* **57**:93–110 (1961/1962).

SCHOLANDER, P. F. The master switch of life. *Sci. Amer.* **209**:92–106 (Dec. 1963).

SCHOLANDER, P. F., H. T. HAMMEL, H. LeMESSURIER, E. HEMMINGSEN, and W. GAREY. Circulatory adjustment in pearl divers. *J. Appl. Physiol.* **17**:184–190 (Mar. 1962).

SCHREINER, H. R. Advances in decompression research. *J. Occup. Med.* **11**:229–237 (May 1969).

SCHREINER, H. R., and P. L. KELLEY. Computation methods for decompression from deep dives. In: Lambertsen, C. J. ed. *Underwater Physiology. Proceedings of the Third Symposium on Underwater Physiology*, pp. 275–299. Baltimore, Williams and Wilkins (1967).

SCHREINER, H. R., and P. L. KELLEY. A pragmatic view of decompression. In: Lambertsen, C. J., ed. *Underwater Physiology. Proceedings of the Fourth Symposium on Underwater Physiology*, pp. 205–219. New York, Academic Press (1971).

SCHREINER, H. R., J. A. LAURIE, and R. C. GREGOVIE. The effect of helium and the rare gases on cellular growth. *Physiologist* **5**:210 (Aug. 1962).

SCHREINER, H. R., R. W. HAMILTON, JR., and T. D. LANGLEY. Neon: An attractive new commercial diving gas. In: *1972 Offshore Technology Conference, May 1–3, Houston, Texas. Preprints*, Vol. I, pp. 501–516. Published by the Conference (1972).

SEARS, D. F. Mechanisms of anesthesia. III Role of lipid molecules in anesthesia and narcosis. In: *Proceedings of the Twenty-Second International Congress of Physiological Sciences, Leiden, Holland* London, Excerpta Medica Foundation (1962).

SEARS, D. F., and E. L. FULLER. Volume changes of polar and non-polar liquid hydrocarbons exposed to pressures of gases. *Resp. Physiol.* **5**:175–186 (1968).

SEGUI, G., and V. CONTI. Comportement alimentaire de tiors oceanauties au cours d'une experience de vie a saturation. *Bull. Medsubhyp* **7**:15–18 (Oct. 1972).

SENAY, L. C., JR. Body temperature regulation: a critical commentary on selected topics. In: Iberall, A. S., and A. C. Guyton, eds. *Regulation and Control in Physiological Systems. Conference Proceedings, Rochester, N. Y., August 1973*, pp. 207–211. Pittsburgh, Pa., Instrument Society of America (1973).

SERGEANT, R. L. Distortion of speech. In: Bennett, P. B., and D. H. Elliott, eds. *The Physiology and Medicine of Diving and Compressed Air Work*, pp. 211–225. Baltimore, Williams and Wilkins (1969).

SERGEANT, R. L. The intelligibility of hydrogen-speech at 200 fsw equivalent. U. S. Naval Submarine Medical Research Lab., Rep. NSMRL 701 (Mar. 1, 1972).

SHILLING, C. W. *The Human Machine.* Annapolis, Md., U. S. Nav. Inst. (1955).

SHILLING, C. W., and W. W. WILLGRUBE. Quantitative study of mental and neuromuscular reactions as influenced by increased air pressure. *U. S. Nav. Med. Bull.* **35**:373–380 (Oct. 1937).

SILVERMAN, L., G. LEE, T. PLOTHIN, L. H. SAWYER, and A. R. UANGY. Air flow measurements on human subjects with and without respiratory resistance at several work rates. *Arch. Ind. Hyg. Occup. Med.* **3**:461 (1951).

SIPLE, P. A. In: Newburgh, L. H., ed. *Physiology of Heat Regulation and the Science of Clothing*, pp. 422–424. New York, Hafner (1949).

SLACK, D. S. Recycling systems for helium. In: Professional diving symposium, New Orleans, November 1972. *Mar. Technol. Soc. J.* **7**:13–16 (Mar./Apr. 1973).

SLOAN, R. E. G., and W. R. KEATINGE. Cooling rates of young people swimming in cold water. *J. Appl. Physiol.* **35**:371–375 (Sept. 1973).

SMITH, D. G., and D. J. HARRIS. Human exposure to Halon 1301 ($CBrF_3$) during simulated aircraft cabin fires. *Aerosp. Med.* **44**:198–201 (Feb. 1973).

SMITH, G. B., JR. and E. G. HAMES. Estimation of tolerance times for cold water immersion. *Aerosp. Med.* **33**:834–840 (July 1962).

SMITH, J. G. Low temperature performance of CO_2 scrubber systems. In: Battelle Columbus Laboratories. *Proceedings, 1973 Divers' Gas Purity Symposium, November 27–28, 1973*, pp. II-1–II-34. U. S. Navy Supervisor Diving, Nav. Ships Syst. Comm., Rep. 2-73 (1973).

SMITH, K. J., E. W. SPECKMAN, and R. L. HEIN. Selected bibliography on the sustenance of man in aerospace systems. Wrightsville Air Force Base, Aerosp. Med. Res. Lab., Rep. (May 1966).

SONG, S. H., W. K. LEE, Y. A. CHUNG, and S. K. HONG. Mechanism of apneic bradycardia in man. *J. Appl. Physiol.* **27**:323–327 (Sept. 1969).

SPAUR, W. H. 1600 ft dive, 20 April–20 May, 1973. U. S. Navy Exp. Diving Unit, U. S. Nav. Med. Res. Inst., Bur. Med. Surg., Rough draft (May 20, 1973).

STANG, P. R., and E. L. WEINER. Diver performance in cold water. *Hum. Factors* **12**:391–399 (Aug. 1970).

STEIN, S. N. Neurophysiological effects of oxygen at high partial pressure. In: Goff, L. G. ed. *Proceedings of the Underwater Physiology Symposium*, pp. 20–24. Washington, D. C., National Academy of Sciences–National Research Council (1955), Publ. 377.

STERN, S. A., and H. L. FRISCH. Dependence of inert gas narcosis on lipid "free volume." *J. Appl. Physiol.* **34**:366–373 (Mar. 1973).

STILES, R. N., and J. E. RANDALL. Mechanical factors in human tremor. *J. Appl. Physiol.* **23**:324–330 (1967).

STRAUSS, M. B. Physiological aspects of mammalian breath-hold diving: a review. *Aerosp. Med.* **41**:1362–1381 (Dec. 1970).

STRAUSS, R. H., W. B. WRIGHT, R. E. PETERSON, M. J. LEVER, and C. J. LAMBERTSEN. Respiratory function in exercising subjects breathing nitrogen, helium or neon mixtures at pressures from 1 to 37 atmospheres absolute. In: *Proceedings of the Fifth Symposium on Underwater Physiology, Freeport, Bahamas, August 1972. In press.*

STROMME, S. B., D. KEREM, and R. ELSNER. Diving bradycardia during rest and exercise and its relation to physical fitness. *J. Appl. Physiol.* **28**:614–621 (May 1970).

SUMMITT, J. K., J. S. KELLY, J. M. HERRON, and H. A. SALTZMAN. 1000-ft helium saturation exposure. In: Lambertsen, C. J. ed. *Underwater Physiology. Proceedings of the Fourth Symposium on Underwater Physiology*, pp. 519–527. New York, Academic Press (1971).

SUZUKI, M., T. TONOUE, S. MATSUZAKI, and K. YAMAMOTO. Initial response of human thyroid, adrenal cortex, and adrenal medulla to acute cold exposure. *Can. J. Physiol. Pharmacol.* **45**:423–432 (May 1967).

TAYLOR, D. W. The effects of vitamin E and methylene blue on the manifestations of oxygen poisoning in the rat. *J. Physiol.* **131**:200–206 (Jan. 1956).

TEICHNER, W. H. Assessment of mean body surface temperature. *J. Appl. Physiol.* **12**:169–176 (Mar. 1958).

THOMPSON, L. J., M. MCCALLY, and A. S. HYDE. The effects of posture, breathing pressure and immersion in water on lung volumes and intrapulmonary pressures. U. S. Air Force, Aerosp. Med. Res. Lab., Rep. AMRL-TR-66-201 (May 1967).

THORN, G. W., D. JENKINS, and J. C. LAIDLAW. The adrenal response to stress in man. *Rec. Prog. Horm. Res.* **8**:171–215 (1953).

THORNE, D. R., A. J. BACHRACH, and A. W. FINDLING. Muscle tremors under helium, neon, nitrogen, and nitrous oxide at 1 to 37 atm. *J. Appl. Physiol. In press.*

UHL, R. R., C. VAN DYKE, R. B. COOK, R. A. HORST, and J. M. MERZ. Effects of externally imposed mechanical resistance on breathing dense gas at exercise: mechanics of breathing. *Aerosp. Med.* **43**:836–841 (Aug. 1972).

UNSWORTH, A., D. DOWSON, and V. WRIGHT. Cracking joints. A bioengineering study of cavitation in the metacarpophalangeal joint. *Ann. Rheum. Dis.* **30**:348 (1971).

U. S. NAVY. *U. S. Navy Diving Manual.* Washington, D. C., Department of the Navy (1970) (NAVSHIPS 0994-001-9010).

VAN DEN BRENK, H. A. S., and D. JAMIESON. Brain damage and paralysis in animals exposed to high pressure oxygen; pharmacological and biochemical observations. *Biochem. Pharmacol.* **13**:165–182 (Feb. 1964).

VAN TASSEL, P. V. Effect of dimercaprol on oxygen toxicity in rats. *J. Appl. Physiol.* **20**:531–533 (May 1965).

VARENE, P., J. TIMBAL, H. VIELLEFOND, H. GUENARD, and L. HUILLIER. Energetic balance of man in simulated dive from 1.5 to 31 ATA. In: Lambertsen, C. J., ed. *Proceedings of the Fifth Symposium on Underwater Physiology, Freeport, Bahamas, August 1972. In press.*

VAUGHAN, W. S., JR. and B. G. ANDERSEN. Effects of long-duration cold exposure on performance of tasks in naval inshore warfare operations. Landover, Md., Oceanautics, Inc., Tech. Rep. on Contract N00014-72-C-0309 (1973).

VOROSMARTI, J. JR., M. E. BRADLEY, and N. R. ANTHONISEN. The effects of increased gas density on pulmonary mechanics. Unpublished paper presented at the twenty-fifth congress of physiological sciences satellite symposium: Recent progress in fundamental physiology of diving. Marseille, France (23–24 July, 1971).

WALDVOGEL, W., and A. A. BÜHLMANN. Man's reaction to long-lasting overpressure exposure. Examination of the saturated organism at a helium pressure of 21–22 ATA. *Helv. Med. Acta* **34**:130–150 (Mar. 1968).

WATKINS, J. C. Pharmacological receptors and general permeability phenomena of cell membranes. *J. Theor. Biol.* **9**:37–50 (July 1965).

WEBB, P. Body heat loss in undersea gaseous environments. *Aerosp. Med.* **41**:1282–1288 (Nov. 1970).

WEBB, P. The thermal drain of comfortable hyperbaric environments. *Nav. Res. Rev.* **26**:1–7 (Mar. 1973*a*).

WEBB, P. Rewarming after diving in cold water. *Aerosp. Med.* **44**:1152–1157 (Oct. 1973*b*).

WEBB, P. Thermal stress in undersea activity. In: *Proceedings of the Fifth Symposium on Underwater Physiology, Freeport, Bahamas, August 1972. In press.*

WEBB, P., and J. F. ANNIS. Respiratory heat loss with high density gas mixtures. Final report on contact Nonr-4965(00). Yellow Springs, Ohio, Webb Associates (May 31, 1966).

WELCH, B. E., T. E. MORGAN, and H. G. CLAMANN. Time–concentration effects in relation to oxygen toxicity in man. Symposium on respiratory physiology in manned spacecraft. *Fed. Proc.* **22**:1053 (1963).

WELTMAN, G., G. H. EGSTROM, M. A. WILLIS, and W. CUCCARO. Underwater work measurement techniques. Los Angeles, Univ. Cal., Sch. Eng. Appl. Sci., Rep. UCLA-ENG-7140 (July 1971).

WILSON, O., P. HEDNER, S. LAURELL, B. NOSSLIN, C. RERUP, and E. TOSENGREN. Thyroid and adrenal response to acute cold exposure in man. *J. Appl. Physiol.* **28**:543–548 (May 1970).

WITTNER, M., and R. M. ROSENBAUM. The physiological pathology of pulmonary oxygen toxicity. In: Brown, I. W., Jr. and B. G. Cox, eds. *Proceedings of the Third International Conference on Hyperbaric Medicine, Duke University, November 1965*, pp. 319–321. Washington, D. C., National Academy of Sciences–National Research Council (1966).

WON, W. D., and H. ROSS. Relationship of low environmental temperature to mouse resistance to infection with Klebsiella pneumoniae. *Aerosp. Med.* **42**:642–645 (June 1971).

WOOD, J. D. Oxygen toxicity. In: Bennett, P. B., and D. H. Elliott, eds. *The Physiology and Medicine of Diving and Compressed Air Work*, pp. 113–114. Baltimore, Williams and Wilkins (1969).

WOOD, J. D., N. E. STACEY, and W. J. WATSON. Pulmonary and central nervous system damage in rats exposed to hyperbaric oxygen and protection therefrom by gamma-aminobutyric acid. *Can. J. Physiol. Pharmacol.* **43**:405–410 (May 1965).

WOOD, J. D., W. J. WATSON, and F. M. CLYDESDALE. Gamma-aminobutyric acid and oxygen poisoning. *J. Neurochem.* **10**:625–633 (Sept. 1963).

WOOD, L. D. H., and A. C. BRYAN. Mechanical limitations of exercise ventilation at increased ambient pressure. In: Lambertsen, C. J., ed. *Underwater Physiology. Proceedings of the Fourth Symposium on Underwater Physiology.* New York, Academic Press (1971).

WORKMAN, R. D. Discussion. In: Lambertsen, C. J., and L. J. Greenbaum, Jr., eds. *Proceedings of the Second Symposium on Underwater Physiology.* Washington, D. C., National Academy of Sciences–National Research Council (1963), Publ. 1181.

WORKMAN, R. D. Calculation of decompression schedules for nitrogen–oxygen and helium–oxygen dives. U. S. Navy Exp. Diving Unit, Res. Rep. 6-65 (May 26, 1965).

WORKMAN, R. D. Underwater research interests of the U. S. Navy. In: Lambertsen, C. J., ed. *Underwater Physiology. Proceedings of the Third Symposium on Underwater Physiology*, pp. 4–15. Baltimore, Williams and Wilkins (1967).

WORKMAN, R. D. American decompression theory and practice. In: Bennett, P. B., and D. H. Elliott, eds. *The Physiology and Medicine of Diving and Compressed Air Work*, pp. 252–290. Baltimore, Williams and Wilkins (1969).

WRIGHT, W. B. Use of the University of Pennsylvania, Institute for Environmental Medicine procedure for calculation of cumulative pulmonary oxygen toxicity. U. S. Navy Exp. Diving Unit, Rep. NEDU 2-72 (1972).

WULF, R. J., and R. M. FEATHERSTONE. A correlation of Van der Waals constants with anesthetic potency. *Anesthesiology* **18**:97–105 (Jan./Feb. 1957).

YOUNG, J. M. Acute oxygen toxicity in working man. In: Lambertsen, C. J., ed. *Underwater Physiology. Proceedings of the Fourth Symposium on Underwater Physiology*, pp. 67–76. New York, Academic Press (1971).

ZALTSMAN, G. L., ed. *Hyperbaric Epilepsy and Narcosis.* Leningrad, Sechenov Institute of Evolutionary physiology and biochemistry, USSR Academy of Sciences (1968). (Translated as JPRS 51714. Washington, D. C. Joint Publications Research Service Nov. 4, 1970.)

ZETTERSTROM, A. Deep-sea diving with synthetic gas mixtures. *Milit. Surg.* **103**:104–106 (1948).

ZINKOWSKI, N. B. *Commercial Oilfield Diving.* Cambridge, Md., Cornell Maritime Press (1971).

V

Man in the Ocean Environment: Psychophysiological Factors

A. *Vision*

1. *Basic Processes of the Eye*

a. Introduction

Besides providing man a wide range of information about his environment, the visual sense provides confirmatory data for information received through the other senses. In an underwater environment certain changes take place in the visual process due to the change of the visual medium from air to water. Understanding these changes is relevant to the effective engineering of underwater tools and equipment and the orientation of the underwater worker who must perform in a strange environment.

The basic structure and anatomy of the eye has been covered in Chapter I. It may be helpful, however, to briefly examine the way in which the eye functions.

b. Functioning of the Eye

The human eye contains two fairly separate sensing systems; one of these is active at ordinary daylight levels of illumination and is called the photopic system, while the other, sensitive at low or night-time levels of illumination, is called the scotopic system. Photopic vision is mediated by the cones, the only receptors found in the fovea or central part of the retina. Rods, the scotopic receptors, are located in the periphery of the eye, along with the cones.

Each of these two systems has its own inherent capabilities and limitations. The daylight or cone system, for example, is capable of appreciating colors and of resolving or seeing extremely fine detail, but it becomes blind when the illumination drops below a certain level. On the other hand, the night-time or rod system is very sensitive to minute amounts of visible radiation but requires very large features for discrimination and is completely color-blind.

271

These differences can be described in terms of the sensitivity of the two systems to different ranges of wavelengths of electromagnetic energy. The retina of the eye as a whole is sensitive to only a narrow band of wavelengths; the visible spectrum consists of wavelengths lying between the long infrared waves and the short ultraviolet waves. Within this range, not all wavelengths are equally effective in stimulating the retina. At daytime levels of illumination, the cones are most sensitive to electromagnetic radiation of wavelength 555 nanometers (nm), a wavelength that corresponds to a visual sensation of yellow-green. Sensitivity to wavelengths on either side of this peak is less. At very low light levels, sensitivity is shifted toward the shorter wavelengths and the peak is at 510 nm rather than 555 nm. The result is that at scotopic light levels the eye is relatively more sensitive to wavelengths in the blue region and less sensitive to wavelengths in the red region than it is at photopic light levels. This is illustrated in Figure V-1, which shows the relative spectral sensitivity of both systems.

Human beings rely heavily on the photopic system of vision and most of our common visual abilities (judging colors, sensing objects at a distance, reading, scanning, coordinating hand movements with visual stimuli, etc.) are the result of activity of

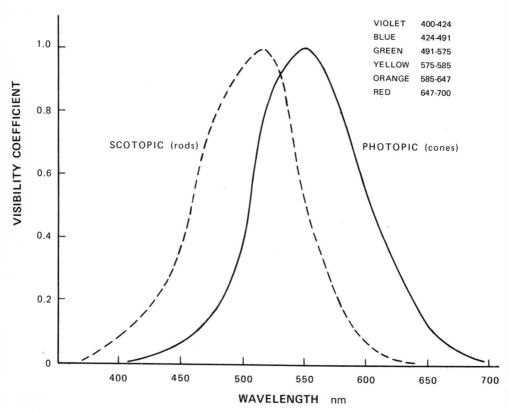

Figure V-1. The relative sensitivity of the eye to different wavelengths of energy at scotopic and photopic light levels. The curves are adjusted to a common scale by making the maximum of each curve equal to one. On an absolute scale of intensities, the rods are much more sensitive than the cones (Woodson 1951).

this system. In fact civilized man is so used to artificial light that he commonly requires special experience or instruction to perform with maximum efficiency at night. This is, of course, not true of animals, some of whom rely exclusively on rod or night vision.

Under water, under good conditions, adequate illumination is present for activity of the cone system, but with increasing depth or turbidity, the illumination levels fall below this point and the diver must rely on rod vision. Therefore, in the sections below, the attributes of both systems are discussed, both as they perform normally, in air, and as they are specifically influenced by the water environment.

2. *Photopic or Daylight Vision*

a. Visual Acuity in Air

Good visual acuity, or the ability to see fine details, is a function of the photopic system. This capability is measured by determining the size of the detail that can be discriminated. Since the smaller the detail, the better the acuity, the formal definition of acuity is the reciprocal of size, in minutes of arc, that can be seen:

$$\text{visual acuity} = 1/\text{visual angle in minutes}$$

A number of factors affect the size of this detail; of primary importance are the position of the target in the visual field, the amount of illumination falling on the target, and the contrast of the target and its background.

Best visual acuity is found with central fixation, that is, by looking directly at the object to be seen. This allows the image of the object to fall on the fovea, the region of the retina that contains closely packed cones and no rods. If the image of the object falls in the periphery of the retina, away from the fovea by only a few degrees, acuity drops dramatically. Thus, in order to see clearly, one must direct the fovea to the object to be seen; this action is so automatic in human beings that it is almost reflexive.

The degree of acuity is also directly affected by the amount of illumination falling on the target; the function, illustrated in Figure V-2, is a linear increase in acuity with a log unit increase in illumination; thus the more light available, the smaller the detail that can be seen.

A third major variable is the amount of contrast: the greater the contrast between an object and its background, the finer the detail that can be discriminated. Typical data (Figure V-3) show that higher acuities are found only with high-contrast targets (black on white or the reverse); gray targets on a gray background can become very difficult to see. In fact there is a rough rule of thumb with regard to human vision that the contrast between an object and its background must be at least 2% or the object will be invisible, no matter how large it is. Such low contrasts are rare in air; thus size rather than contrast usually determines the limit of vision; in water, as is discussed below, this is not the case.

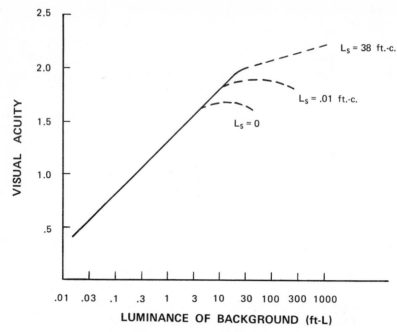

Figure V-2. Acuity as a function of the amount of light falling on the target. L_s refers to the background level of illumination in the booth (Lythgoe 1932).

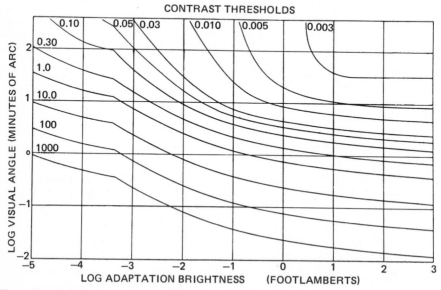

Figure V-3. Relation between stimulus area and luminance for stimuli of varying contrasts. [From Blackwell (1946) by permission.]

b. Visual Acuity under Water

(i) Light Transmission and Its Effect on Acuity. The greatest problem in seeing under water is insufficient light. Only about 5% of the light traveling from air to water is reflected off the surface; the rest enters the water. As one travels deeper, however, the amount of light diminishes very quickly because more light is lost traveling a comparable distance in water than in air. Although propagation of light energy through a medium follows the same principle in water as in air, and the same physical equations are employed, the constants differ. The following equation describes the transmission of energy through a medium:

$$p = p_o e^{-\alpha d} \tag{1}$$

where p is the radiant power reaching a given distance; p_o is the radiant power at the original point; e is the base of the natural logarithmic system; α is the extinction or attenuation coefficient; and d is the distance involved (Duntley 1963). The difference between the situations in air and water is in the size of α. In air it is extremely small, while in water its value is larger by a factor of 1000 times or more. Alternatively, this difference can be described in terms of the units by which distance is measured. Figure V-4 illustrates the amount of light transmitted plotted as a function of distance for an α of 0.05. For air, an α of 0.05 (close to that of pure air) corresponds to distance being measured in *miles*; for water, an α of 0.05 (that of distilled water) corresponds to distance being measured in *meters*: For example, 61% of the original light would remain after traveling through 10 miles of clear air, and less than 1% would still be left after 100 miles. For clear water, about 61% and 1% of the light is left after 10 and 100 m, respectively (Kinney 1970).

This dramatic reduction in available light has severe consequences for visual acuity since, as was noted for air, visual acuity depends directly on the amount of light available. Table V-1 gives the average target size that would be required to be seen at each light level. Comparison with Figure V-4 shows that a target just visible with bright light of 100 mL, would have to be four times as large to be seen after the light had traveled through 100 m of the clearest water.

A further complication occurs since the loss of light transmitted through water is the result of two factors, the absorption and the scattering of light energy by the water. In fact, α is the sum of two coefficients: a, the volume absorption coefficient, and s, the volume scattering coefficient (Duntley 1963). Since light is transformed primarily into thermal energy by absorption, the results are an overall loss of light and a change in the wavelength characteristics. The latter is discussed under color vision.

Scattering refers to the change of direction of light energy or photons as they hit small particles suspended in the water. The particles are generally transparent biological organisms, but in more turbid water may include silt, mineral, and many kinds of suspended particulate matter. Normally in air (excepting such conditions as smoke and fog) light rays travel directly from the object to the eye of the observer. In water, light rays scattered by particles in the water are deflected out of the line of sight and as a result light is diffused, outlines are blurred, and the natural contrast between the object and its background is reduced—the effect is somewhat similar to looking through frosted glass. The more turbid the water, the more particles suspended in it, the

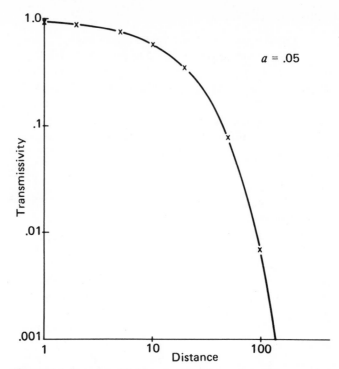

Figure V-4. Amount of light transmitted by various distances of air and water. For clean air, an α of 0.05 would involve distances measured in miles; for clean water, an α of 0.05 would require distances to be measured in meters. [From Kinney (1970) by permission.]

greater the loss of contrast. In extreme conditions, so much scatter is produced that the object and its background do not differ by the required 2% and the object becomes invisible. This, in fact, is the major obstacle to vision under water; as objects recede from the observer in the water they generally become invisible due to the lack of contrast long before they are reduced in size sufficient to make them invisible.

Backscatter, a condition that occurs when a diver attempts to hold his own light

Table V-1

Relative Size of Targets That Would Be Required
to Be Seen at Various Light Levels
(Kinney *et al.* 1968*b*)

	Light level, mL	Target size min of arc
Photopic	100	0.5
	10	1.0
	1	2.0
	0.01	6.0
Scotopic	0.001	17.0
	0.0001	45.0

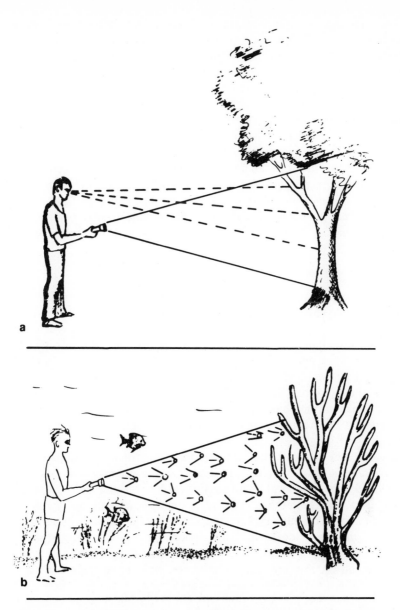

Figure V-5. Normal reflectance from a surface in air compared with backscatter from particles in water. [From Kinney (1970).]

source, can seriously hamper underwater vision (Figure V-5). In air, ordinarily part of the energy of traveling from the source to the reflecting surface is reflected to the observer's eye without undue interference. In water, however, not only does scattering deflect light energy before it reaches the reflecting object, but part of this scattered light returns to the observer's eye, making the foreground brighter than the object and perhaps obliterating the object completely.

The relationship between contrast and particulate matter or amount of turbidity has been extensively investigated (Duntley 1952, 1960, 1963). Formulas and nomographs are available for predicting visibility ranges for various lines of sight, angles of incident light, and kind and size of particles in the water. Although continuous attempts are being made to reduce scattering from artificial light sources, scattering remains an increasingly severe problem as our waters—even the oceans—contain more and more suspended particles.

(ii) Effect of Refraction. Man can see very little under water if his eyes are immersed in it; the reasons are optical and are discussed in the section on facemasks. To circumvent this problem, an air space must be provided between the eye and the water, which, in turn, produces the complication of refraction at the air/water interface.

Refraction or bending of light rays occurs when they pass from one medium to another in which their speed is different. The amount of bending depends upon the angle of incidence of the light rays and can be calculated from the general relation

$$\frac{\sin \angle_1}{\sin \angle_2} = \text{constant (the index of refraction)} \qquad (2)$$

where $\angle_1$ is the angle of incidence and $\angle_2$ is the angle of refraction. This is a common physical principle of great importance in the manufacture of lenses.

When light passes from water into air (for example, the air in a diver's facemask) the index of refraction is 0.75. Illustrations of the effect are shown in Figure V-6. The consequences of refraction for vision are many. For a diver looking from air into water through a facemask, a virtual image is formed at 3/4 of the distance between the object and the interface; thus the light rays appear to emanate from a point at 3/4 of the real distance to the object. The image on the retina corresponding to this virtual image is larger, by about 4/3, than it would have been if produced by the same object in air.

Acuity under water should be better than it is for the same sized target in air, due to the magnified retinal image. Indeed, this has been shown to be true (Kent and Weissman 1966); under ideal conditions of clear water and short viewing distances, acuity under water is better than in air by the amount predicted by the refractive index.

Unfortunately, the enhanced acuity has few practical applications since conditions under water are rarely ideal. Normally, absorption and scattering, discussed above, dominate and produce acuities much poorer than they would be for comparable distances in air. A few individuals, however, will derive some benefit: Since the refraction acts as a negative lens, some myopes (near-sighted individuals) who require a simple negative lens for correction will see better under water than they do, uncorrected, in air.

c. Depth Perception

Even though the eye forms only a two-dimensional pattern of stimulation on the retina, we perceive the world in three dimensions, or in depth. Objects look solid rather than flat, and they appear to be located at various distances which usually corresponds to their physical locations. Many types of stimulus information enable us

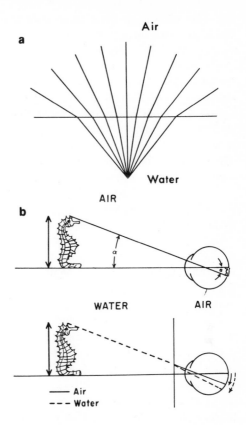

Figure V-6. (a) Classic example of light rays bending as they pass from water to air (Kinney *et al.* 1970a). (b) The effect of water-to-air refraction on the size of the retinal image from an underwater object. (By permission J. A. S. Kinney.)

to perceive depth. Stimuli such as derived from linear perspective and texture gradients, patterns of light and shadow, interposition, relative size, and relative clarity or contrast are quite well known. Artists learned long ago how to simulate these stimuli in order to create the impression of depth in paintings. Another factor for the perception of depth is motion parallax. When an observer moves his head, objects at different distances appear to move relative to one another, thus providing information as to their actual locations. Also important are accommodation and convergence, the eye–muscle actions that enable our eyes to focus and fixate properly. These oculomotor cues are effective only at relatively near distances. Finally, binocular or stereoscopic vision, requiring the use of both eyes, provides an impression of depth even in the absence of all other information. It is this mechanism which enables man to detect very small depth differences.

There are actually two aspects to depth perception, *relative* depth, the separation in depth between two or more objects; and *absolute* depth, the distance between an object and the observer. The discussion below will be limited to absolute depth. A discussion of stereoacuity (detection of relative depth differences) will follow in a later section.

(i) Distance Perception in Air. The nature of absolute distance perception in air is fairly well known. When an observer estimates the distance of a series of targets,

the perceived distance is mathematically related to the physical distance by a power function of the form

$$y = ax^b \tag{3}$$

where y is the perceived distance, x is the physical distance, a is a constant, and b is an exponent which determines the curvature of the relationship: An exponent of 1.0 indicates a linear relation and exponents greater or less than 1.0 indicate upward or downward curvature, respectively. The curve for judging distance in air (outdoors) is nearly linear, with an exponent of about 0.99 (Teghtsoonian and Teghtsoonian 1970). The constant a also tends to be less than 1.0, so that there is a general tendency to underestimate the physical distance (Ferris 1972).

(ii) Distance Perception under Water. Since distance information is distorted under water, it is not surprising that distance perception is also altered. As previously discussed, optical distortion due to wearing a facemask produces a virtual image of an object at a distance which is about three-fourths of the physical distance. Thus the eyes will converge and accommodate for this nearer distance, and one might predict that the object's distance will be underestimated. Underestimation does in fact occur if the water is extremely clear (Ross 1967), but observers tend to overestimate distance if the water is turbid (Luria *et al.* 1967, Kinney *et al.* 1969a, Ferris 1972). Water turbidity increases the magnitude of the estimates, an effect which increases with increased distance. It produces a loss of object brightness and contrast, a distance cue which in air is indicative of relatively large distances. Furthermore, turbidity tends to obscure the entire visual field, thereby limiting the effectiveness of the other stimuli for distance.

The combined effects of optical distortion and varying amount of turbidity produce the distance curves illustrated in Figure V-7. In extremely clear water, optical distortion produces considerable underestimation. Judgments are even less than the theoretical optical distance, due to the added effect of the underestimation tendency which occurs in air. Even in the clearest water, however, the line has an upward curvature, indicative of an exponent slightly greater than 1.0. In moderately turbid water, the exponent increases to about 1.2, and estimates eventually exceed the optical, air, and physical distance values. In very turbid water the exponent may exceed 1.3. The points at which estimates exceed the optical, air, and physical values are quite close to the observer, and underestimation now occurs at very near distances. From Figure V-7 it is apparent that divers will rarely estimate distance correctly. Furthermore, the kinds of errors which will be made are highly dependent on water conditions.

(iii) Stereoacuity in Air. Stereopsis refers to the perceptual experience of three dimensions which results from the fact that our eyes do not have exactly the same view of the scene before us, particularly of those objects that are close by. The two slightly different two-dimensional images are then somehow fused into a single three-dimensional percept. Stereopsis thus allows us to perceive directly that an object is closer or farther than another one without other information, such as relative size, interposition, perspective, gradients, shadows, etc. The measure of how great this difference between the two distances must be before an individual can perceive a difference is called his stereoacuity.

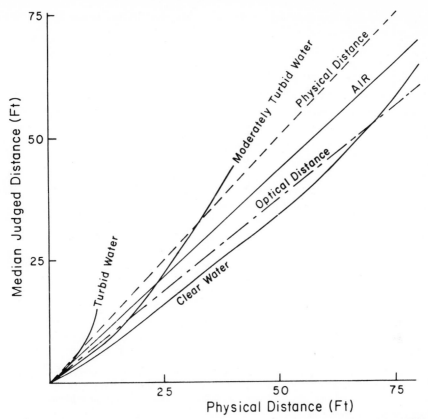

Figure V-7. Summary of data on perception of absolute distance in air and in turbid, moderately turbid, and clear water. [Reprinted from Ferris (1972) with permission of author and publisher.]

The geometry of binocular perception is diagrammed in Figure V-8. When an observer views two objects that are equidistant from him, the angles formed by the two sets of lines of sight are equal. When the two objects are not the same distance away, these angles are unequal. Stereoacuity is the smallest difference in these angles (generally stated in seconds of arc) that results in a judgment of a difference in distance.

Although there are some people who are grossly deficient in this ability, stereo-acuity in most people is remarkable. It is quite common for a disparity of a ¼ in. between two objects 15 ft away to be perceived, although this corresponds to a disparity of the retinal image far smaller than the diameter of a single neural receptor (Kling and Riggs 1972).

(iv) Stereoacuity under Water. In the water, stereoacuity is much worse than it is in air (Ross 1967, Luria and Kinney 1968). Typically, this is the result simply of the turbidity of the water. Since stereoacuity, like resolution acuity, is related to target contrast, any reduction in the visibility of the target, either through loss of contrast or dimming of the level of illumination, will diminish stereoacuity. Surprisingly, however,

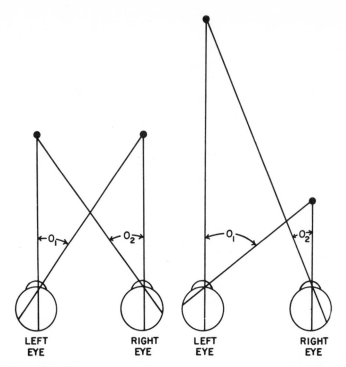

Figure V-8. The geometry of stereoacuity, comparing the equal angles produced by objects at the same distance from the two eyes with the unequal angles from objects at different distances. (By permission J. A. S. Kinney.)

stereoacuity is typically two or three times worse even in the clearest water when the targets are as clearly visible to the diver as in air. There are three major causes of this decline of stereoacuity in clear water.

1. *Lack of visual stimuli.* Stereoacuity declines in the typical underwater environment due to the relative lack of visual stimulation, particularly in the periphery of the visual field. In the air the field of view is usually filled with visual stimuli; there is something to see everywhere, although one is usually not conscious of all the stimulation. In the water, there is relatively little to look at. The underwater scene is generally hazy and undefined. There are few clearly visible objects. Those that appear are attended to, leaving the periphery of the visual field relatively blank. Recent experiments in the air have shown that as the peripheral field is progressively screened from view, stereoacuity becomes increasingly worse (Luria 1969). Introducing a few objects into an otherwise empty visual field improves stereoacuity (Luria 1971). However, where these objects are placed turns out to be important because of the next factor involved.

2. *Accommodation of the lens.* The second cause of the decline of stereoacuity in clear water is the increased state of accommodation of the eye in water. Accommodation refers to the state of the lens in the eye: When it is focused for a near object, the

lens is said to be accommodated; when it is focused for distant objects, it is said to be relaxed.

For reasons which are not understood, stereoacuity declines as the state of accommodation increases. In the water, accommodation for a target at a given distance is greater than would be expected and often greater than is actually necessary. There are two reasons for this. The first is the refraction of the light rays, which was already discussed. In addition, if there are two objects in the field of view at different distances, the eye seems to attempt to focus on both at once. Since this is impossible, the level of accommodation becomes a compromise between what would be appropriate for each object in turn. Thus, if one attempts to focus on a distant object when there is a nearby object also clearly in view, the level of accommodation is for an intermediate distance (Hennessy and Leibowitz 1971). This problem is greatly exacerbated in the water by the fact that nearby objects are always more highly visible than more distant objects. Thus they exert a disproportionate effect on the compromise level of accommodation. When attempting to focus on the more distant object, the diver's level of accommodation is unduly influenced by the more visible nearby object, and stereoacuity suffers (Luria and Kinney 1972). When attempts are made to improve stereoacuity in the water by introducing objects in the peripheral field of view, it is necessary that these objects be about the same distance from the observer as the target. If they are too close, they will do more harm than good (Luria and Kinney 1973).

3. *Reduction of the field of view.* The third cause of the decline in stereoacuity is the reduction in the field of view produced by the typical facemask. This is, of course, related to the first factor, the lack of peripheral stimulation, but it may also be related to the second factor, since the rim of the mask may act as a nearby stimulus on which the eye tries to focus. There is evidence of such a phenomenon in the discovery of "optical instrument myopia," the tendency of the eye to be greatly overaccommodated when looking through microscopes and telescopes, etc. (Schober *et al.* 1970). It follows from these findings that stereoacuity in the water will be improved to some extent by facemasks that provide a wide field of view and by situations in which there is an appreciable amount of peripheral stimulation at the same distance from the diver as the target he is interested in. Often little can be done to provide better viewing conditions, and the diver must simply be aware that his judgments of relative distance will be much worse that in the air.

d. Perception of Size, Shape, Position, and Color

(i) Size Perception in Air. While it is obvious that the subjective perception of the size of an object must be somehow related to its image size on the retina, there is a vast body of evidence which shows that retinal image size, by itself, does not determine subjective or perceived size. The phenomenon of size constancy is a good example; this refers to the fact that objects remain the same apparent size at different distances from the viewer (with the subsequent different retinal image sizes). Thus a person does not appear to shrink in size as he walks away from the viewer even though his image on the retina of the viewer may decrease by a factor of 100 or 1000. Size constancy

depends upon the adequate perception of depth; it has been shown that the size of objects will be perceived correctly as long as the viewer can adequately assess how far away the object is. If the stimuli for depth perception are artificially removed, size constancy breaks down and perceived size tends more toward retinal image size (Epstein 1967).

(ii) Size Perception under Water. The effect of refraction in the underwater viewing situation, as noted above, is to increase the size of the retinal image by a factor of about 1.3 over what it would be for the same object in air. From this it might be assumed, correctly, that the perceived size of objects under water would be magnified. Magnification can easily be experienced by viewing an underwater object through a facemask half in and half out of the water. The portion of the object seen through the underwater half appears much larger and closer than the portion seen through the half that is in air.

Normally, however, a diver does not have simultaneous air and water images to compare. Furthermore, in air, retinal size is not the determining factor but rather retinal size in relation to distance. Thus it becomes an empirical question as to how divers will judge the sizes of objects under water. Several experiments have shown that objects under water are indeed judged to be larger than they are physically (Kinney and Luria 1970, Kinney *et al.* 1970*a*, Ross 1965, Ross *et al.* 1969). For naive subjects judging sizes at close viewing distances, the amount of magnification experienced is almost exactly what would be predicted from refraction. Divers with considerable underwater experience do somewhat better but they, too, see objects as magnified under water. In fact, this distortion plagues divers, no matter how much their diving experience; the fish speared under water is always a disappointment when brought out into the air.

(iii) Perception of Position and Shape under Water. Since the amount of bending or refraction of light rays at an interface is dependent upon the angle of incidence of the rays (the greater the angle, the greater the refraction), the shapes and positions of optical images on the retina will be transformed. Examples of shape distortion are shown in Figure V-9; straight lines may appear curved and squares misshapen into pincushions. Furthermore, since objects are not located physically where they appear to be optically, the hand–eye coordinations and visual motor skills of a diver may be disrupted.

Several studies have shown that predicted distortions do take place under water. Measures of the apparent curvature of straight lines under water show the perceived distortion to be almost as large as predicted from the optical image (Ross 1970). Other tests of hand–eye coordination show that novice divers misreach by an amount predicted by the refracted image (Kinney *et al.* 1968*a*). Such distortion of shape and position is often a problem to the diver, particularly the inexperienced one. The bottom may appear to be sloping up, thus getting shallower, when it is in fact level or even sloping slightly down. Divers frequently reach for objects, only to miss, since the objects are actually farther away than they appear to be.

(iv) Color Vision in Air. The only severe limitation to man's ability to see colors in air is the amount of illumination. If it falls below the level required for photopic or cone vision, man becomes completely color-blind, and the entire scene appears in shades of black, white, and gray. This occurs at approximately 0.01 ft-L

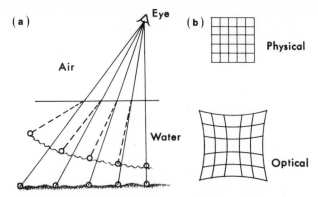

Figure V-9. Examples of distorted appearances of objects under water. Because of refraction (a) a flat bottom may appear to slope upward, (b) a square object appears curved (Kinney *et al.* 1970*a*).

(0.03 cd/m²), about the amount of light provided by a full moon. Normally, levels of illumination outside during daylight hours, no matter how cloudy a day, are well above this level and provide sufficient light for full color vision.

(v) Color Vision under Water. In water, the amount of illumination commonly falls below the levels required for photopic vision, due to extensive absorption by the water. For example, on an overcast day, 200 ft-L might fall on the ocean surface; this would be reduced to 0.01 ft-L (and loss of color vision) by only 200 m of the clearest ocean water. Turbid water would cause even more extreme reduction, of course. Thus divers in clear, deep water or shallow, turbid water frequently do not have enough light for color vision, unless they carry their own light sources.

In addition to overall loss of light, transmission of light energy by the water changes underwater color vision. Water does not absorb the different wavelengths of light equally, transmitting some wavelengths rather well and others very poorly.

The transmission curve for distilled water has a peak at 480 nm, absorbing more strongly all wavelengths on either side of this peak. The same curve applied to extremely clear ocean water is illustrated in Figure V-10. Thus the only color remaining in deep, clear ocean water is blue-green. The loss of the long wavelengths, which correspond to red, in relatively shallow water is pronounced. An interesting consequence is the loss of flesh color; divers in only 20–30 ft of clear water have the chalky-white appearance of a dead man.

The absorption curves for other bodies of water differ greatly from the curve for distilled water, depending on amount of silt, decomposition of plant and animal material, pollution, and living organisms. A common source of change is plankton in the water; plankton absorb short wavelengths of energy much more than long wavelengths, with the result that the peak of the transmittance curve is shifted to the yellow-green portion of the spectrum. The amount of shift depends on the density of the plankton (Kinney 1970).

The relative visibility of different colors will vary greatly in different bodies of water, depending upon which colors are transmitted by the water and which are

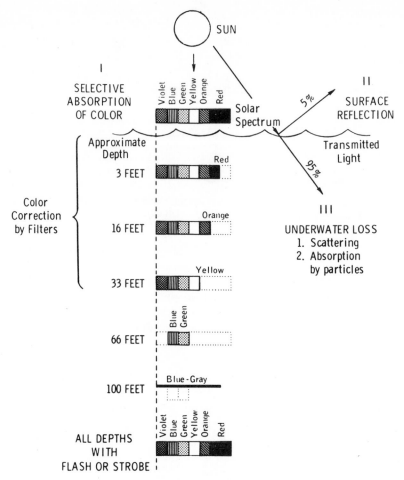

Figure V-10. Selective absorption of colors by clear ocean water at different depths. Reds, oranges, and yellows are filtered out at relatively shallow depths, leaving only blue-green at deeper levels. [From Reuter (1971) by permission.]

absorbed. In general, blues and greens are most visible in clear ocean water, greens and yellows in more turbid coastal water, and yellows, oranges, and reds in turbid river and harbor waters (Kinney *et al.* 1967).

Fluorescent paints are often an effective way of improving visibility under water since these paints convert short-wavelength energy (violets and blues), to which the eye is relatively insensitive, into longer wavelength energy (greens, yellows, oranges), to which the eye is more sensitive. The converted energy is added to the reflected energy, thus increasing the brightness and contrast of the painted object.

As noted above, the total illumination under water is often insufficient for color vision and the diver must provide his own light source. None of the common underwater light sources have the same spectral energy distribution as daylight and this, too, will affect the visibility of colors under water. Tungsten or incandescent sources emit larger amounts of energy in the long wavelengths and have little output at the short end of the visible spectrum. Mercury, on the other hand, has little or no long-wavelength energy, even with thallium additives; its dominant output is in the yellow and green portions of the spectrum (Kinney *et al.* 1969*b*).

The most visible colors for daylight and for tungsten and mercury sources under water in various types of water are summarized in Table V-2. In turbid water, in daylight, oranges and reds are most visible because they are transmitted best by the water; with a tungsten source there is the additional advantage of a high-energy output for these wavelengths. With a mercury light source, in the same water, yellow and yellow-greens are superior due to the emission of the lamp in this region. One caution should be noted. It is sometimes reasoned that specific colored filters should be used under water to take advantage of these results; for example, if the given body of water transmits green best, that green filters should be placed over the light source for maximum visibility. This of course is not true, since the filter merely reduces the amount of energy available, in the same way that the water does. The water itself acts as a colored filter for a white light source.

For color-coding under water, additional factors must be considered. White, while it is always highly visible, tends to take on the color of the water; for this reason it is the easiest to confuse and should not be used for color coding. In general, colors close to each other in the spectrum are the hardest to discriminate. It is best to use only two colors, one from each end of the spectrum, with black as a possible third choice. The choice of colors depends on the body of water and the type of illumination, but

Table V-2

The Most Visible Colors in Various Kinds of Water with Different Light Sources[a]

| | Natural illumination | | Artificial illumination | | | |
| | | | Incandescent light | | Mercury light | |
Type of water	Fluorescent paint	Regular paint	Fluorescent paint	Regular paint	Fluorescent paint	Regular paint
Rivers and harbors; turbid water: visibility 2–6 ft	Yellow Orange Red	Yellow Orange White Red	Yellow Orange Red	Yellow Orange Red	Yellow-green Yellow	Yellow White
Coastal; Moderate turbidity: visibility 10–20 ft	Green Yellow Yellow-green Orange	White Yellow Green	Yellow Orange Red	Yellow Orange Red	Green Yellow-green Yellow Orange	Yellow White
Clear spring or ocean water: visibility 50 ft or more	Green Blue-green Blue	White Yellow Blue Green	Green Yellow Orange	Yellow Orange	Green Yellow-green Yellow	Yellow White

[a] Fluorescent paint superior to regular paint except in turbid water under incandescent light source. White is very visible, but is easily seen as another color; thus it should not be employed in color coding. Black and gray are rarely visible on a horizontal viewing path against a water background. (Data from Kinney *et al.* 1967, 1968*a*, 1968*b*).

the use of green, orange, and black is suggested, under most conditions. If four colors are needed, red and yellow may be substituted for orange in murky water in daylight, or in any water with a tungsten (incandescent) light; blue may be used in clear water in daylight. However, with a mercury light, a fourth color that was not confused with one of the other three could not be found (Kinney *et al.* 1969*b*).

3. *Scotopic Vision*

a. Scotopic Vision under Normal Conditions

When the light level drops below that required for the cones to operate, the rods take over, with consequent dramatic changes in vision. The most obvious losses are color vision and acuity; in the latter case, objects have to be 10–100 times larger to be seen by the rod or scotopic system than by the cones.

Less obvious but equally important is the loss of foveal or central vision. Since this area of the retina contains only cones, which are inoperative in dim light, the normal response of looking directly at objects to see them clearly is inappropriate. Indeed, this natural habit leads to the object's being centered on a blind area in which it cannot be seen at all. The appropriate response, at night, is to purposely look to one side of the object to be seen. During wartime, training was provided for military men to help them use their night vision effectively; divers undoubtedly need the same kind of assistance.

On the plus side, vision is possible with the rod system to light levels four log units below the amount required by the cones; that is, it is possible to see, using the scotopic system, with 0.0001 the amount of light required for the photopic system. This tremendous increase in sensitivity does not occur immediately but requires some time in the dark, a process called dark adaptation (Graham 1965). The amount of time required for dark adaptation depends on many factors, the most important being intensity of illumination prior to darkness. If preexposure light is very bright (i.e., sunlight) the rods will not respond at all for at least 10 min after darkness is imposed and complete adaptation can take up to 45 min. If preexposure light is dim, adaptation will begin immediately and can take a much shorter time. This is illustrated in Figure V-11A.

Color of the illumination prior to darkness also affects time required to adapt. This is illustrated in Figure V-11B, which shows that red preillumination requires less recovery time than does blue, green, or white. This fact has been practically applied in lighting cockpits and ships' control rooms; these areas are often red-lighted to allow quicker dark adaptation plus sufficient light for good task performance before adaptation (Kinney *et al.* 1968*b*).

Since the rod system is not as sensitive to long wavelengths as is the cone system, the use of red lighting thereby decreases the preexposure brightness to the rods. It does, however, allow visual acuity equally as well as does white light of the same intensity, since acuity depends on the cone system. Red light or goggles enable someone to start to adjust his eyes for night vision and at the same time to carry on activity in a well-lighted room. Adaptation to red light is not equivalent to dark adaptation, but produces only an intermediate level of adaptation.

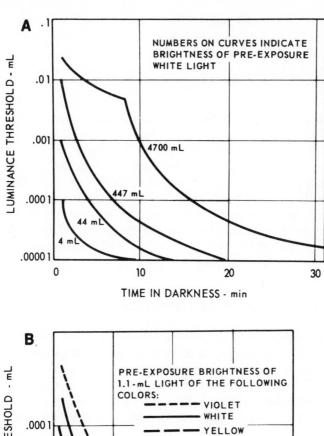

Figure V-11. The process of dark adaptation as indicated by the smaller quantities of light that can be seen after given periods of time in the dark. (a) The effect of different quantities of pre-exposure to white light. [Reprinted from Haig (1941) by permission from publisher and author.] (b) The effect of different colors of preexposure (Peskin and Bjornstad 1948).

b. Scotopic Vision under Water

A diver will often find himself forced to use scotopic vision, even in the daytime, because of insufficient light under water due to turbidity or depth. This is a problem which has not received as much attention as it deserves and divers undoubtedly should be instructed in the use of off-center vision and effective means of dark adaptation.

The use of red lighting or red goggles prior to diving can provide some degree of dark adaptation while allowing the diver light by which to get ready for the dive. An experimental test of this principle showed clearly that the prior wearing of red goggles was effective, allowing divers much more vision when reaching a 170-ft depth than was available without them (Everley and Kennett 1946). One problem arises under water that does not occur in air, however; that is the absorption of long wavelengths or red light by the water. If the diver continues to wear his red goggles while descending to his work site, he will become effectively blind at some depths on the way down (he is wearing goggles which transmit only red light when there is none). The alternative, removing the goggles at the surface, allows short-wavelength light to reach his eyes, destroying to some extent the benefits of prior red adaptation.

4. Effect of Underwater Experience

Many problems discussed in previous sections are worse for the novice diver and can be alleviated, to a greater or lesser extent, by underwater experience. For example, illusions of shape and position, errors in judgment of size and distance, and awkwardness resulting from poor hand–eye coordination can be decreased by diving experience. Many studies have investigated the effect of underwater experience on various visual abilities (Kinney *et al.* 1970b, Ross 1970, Ross *et al.* 1969, Ono and O'Reilly 1971).

There is always some degree of improvement in visual performance with time spent in the water. This may reflect a simple process of adjustment to what might be a frightening experience—with time, the diver learns to pay attention to the visual world about him rather than to the process of breathing. On the other hand, the improvement may reflect knowledge of underwater effects and a conscious attempt to correct for them. Thus the diver may be taught that objects appear too large under water and may consciously adjust his estimates for this knowledge. Finally, the improvement may result from "adaptation," or a real change within the individual in how he responds to the outside world. Such adaptation to transformed stimulation has been extensively studied in air, by having subjects wear prisms or lenses which optically distort the image (Epstein 1967, Held 1965). The results show that adaptation does occur and that individuals can adjust to even the most drastic distortion so that they can move around skillfully without falling or bumping into objects, etc. Unfortunately, the adaptation process may be quite time-consuming, taking weeks or months of continuous wearing of the lenses.

Comparison of different underwater visual tests of men with varying amounts of diving experience reveals that some types of visual performance are much more subject to improvement than others. Specifically, those tasks involving hand–eye coordination or visual–motor skills show real adaptive improvements. The visual–motor performance of novice subjects is completely disrupted; that is, they rely on the distorted optical image. The responses of men with extensive time under water are quite adequate for the physical conditions; they have learned to adjust or have adapted to the transformed underwater stimulation. In between these two extremes there is a regular progression of improvement in visual–motor skills with increasing familiarity with the water. Yet, even within the select group of qualified Navy divers, there are sizable individual differences—the more competence under water, the better the scores on tests of hand–eye coordination (Kinney *et al.* 1970*a*).

This result suggests that training procedures might be devised to assist student divers in making a rapid, adaptive adjustment to the underwater environment. Training procedures have indeed been evolved which improve a novice's performance on tasks of hand–eye coordination with only 15–30 min of practice on visual–motor skills under water (Kinney *et al.* 1970*b*). Even so, this improvement is not complete, and attempts to increase adjustment have failed. At the present time, it appears that prolonged exposure over the course of many years may be necessary to achieve full adaptation in most subjects (McKay *et al.* 1971).

This adaptive improvement which is possible in visual–motor performance is, unfortunately, not evident in tasks that test purely visual phenomena. Distance judgments of experienced divers are better than those of novices, but the evidence suggests that this may result from a conscious correction rather than a change in how distances are perceived. On tests of size perception, experienced divers do very little better than novices; all individuals rely heavily on the distorted optical image and overestimation occurs. This suggests that divers must be taught to consciously correct their size judgments.

Despite these advantages of underwater experience, there are some problems that cannot be changed. For example, underwater experience cannot assist the diver to see when there is too little light, too much scattering, or a change in the color of the light energy. Extensive underwater experience does, however, teach a diver to rely less on vision. Primarily a visual creature, man relies almost exclusively and reflexively on visual information about the outside world. If this is denied him, as is so often the case under water, panic is the frequent result, and the experienced guard against it.

5. Underwater Optical Equipment

a. Masks

For the eye to see clearly, light rays must be in focus on the retina, which requires some 60–65 diopters of focusing power. At least two-thirds of this refraction or bending of light occurs naturally at the corneal surface of the eye in a normal atmosphere, due to the different refractive indices of air outside the eye and fluids inside the eye. When the eye is immersed in water, however, the water and these fluids have about the same refractive index and no bending or focusing takes place. The result is a

Table V-3

Basic Methods for Providing Divers with an Air Space
between the Eye and the Water

Method	Description	Advantages or disadvantages
Hard hat	Made of spun copper with brass fitting and has several round or ellipsoidal port holes (average $4\frac{1}{2}$ in. diameter) covered by plano glass	Severely restricts visual field
Goggles	Generally made of plastic lenses, approximately 44 mm in diameter, with rubber side walls	Presure cannot be equalized; visual fields are contracted to 60° (Figure V–12)
Face masks	Covers eyes and nose only	Air pressure can be equalized by exhaling air through the nose; lateral vision and lower visual fields are greatly restricted
Standard		
Oval	A plain faceplate with a finger recess below the nose to aid in pressure equalization	Offers the widest peripheral field around the brow (Figure V–13); almost whole field is binocular
Kidney	A kidney-shaped faceplate containing a central purge valve for water expulsion	Offers the widest peripheral field almost at eye level; the purge valve assembly cuts into its monocular fields, paring the lower binocular field and producing two "wings" which represent the remaining view over the bridge of the nose (Figure V–13)
Recess kidney	A mask with a deep receptacle for the nose to permit a minimum distance between eye and glass	Appears to combine the individual advantages of the other two above the lateral axis, and increases the peripheral field downward and to the side; the nose receptacle allows a narrow binocular field in a large total field (Figure V–13)
Wraparound	Covers eyes and nose only, but has a faceplate of bent glass extending around to the sides of the head	Distortion from the curved surfaces leaves only 50% of the lateral field undisturbed; double images in the near periphery may also appear (Figure V–14)
Full-face	Covers eyes, nose, and mouth, to facilitate underwater communication	There is a danger of breaking the air supply if the mask comes loose or breaks; it does provide a larger lower field of vision than other facemasks and is sometimes preferred for close work or visual search tasks (Figure V–14)
Rebikoff Visiorama	Covers eyes and nose only, but has two lenses, thereby eliminating the $\frac{1}{4}$ displacement for distance and 33% magnification underwater	The binocular lateral field is contracted to 80°; there is a marked distortion in the periphery of the lens system, which causes vertigo, and a minimum interpupillary distance of 63 mm, so that decentering is not applicable to the lens system

Table V-3—*Cont.*

Underwater contact lenses	Contact lenses that place a flat surface and an air space in front of the eye (Figure V–15)	Supply great field of vision, but some individuals find them irritating underwater and some feel unprotected; the upper face is unprotected (which may be uncomfortable in cold water); individual fitting is necessary, resulting in high cost; most importantly, the contact lens cannot be dislodged, even as the result of a strong blow

45-diopter hyperope (or an individual who is far-sighted by this amount), a disability much more severe than is ever seen in human vision. To restore the focusing power of the cornea it is thus essential that an air space be placed between the eye and the water. Basic methods of providing this air space are shown in Table V-3. Peripheral limits of vision using various methods are given in Table V-4.

b. Underwater Contact Lenses

Skin Diver Contact Air Lenses (SCAL) are an alternative to goggles and masks. They, too, require the placing of a flat surface and an air space in front of the cornea. Such a lens is shown in Figure V-15. It is molded to the eye, and in front of it is an air space with a plane surface. It has been suggested that the fenestration through which contact lenses normally allow the cornea to "breathe" be opaque to eliminate the interference of light rays that would otherwise enter through that point.

Vision with the SCAL is excellent in air as well as under water, except for poor visibility in turbid water. The lenses have been worn to depths of 200 ft with no particular effects of pressure. Tolerance is good, although mild irritation of the eyelids develops with some individuals. This is due more to contact with water than to the lens itself. In fresh water there are few complaints but many individuals experience

Table V-4
Peripheral Limits of Vision[a]

	Upper	Lower	Lateral
Normal	60–70°	80°	200° (binocular)
Standard mask	55°	8°	97°
Full-face mask	Greater than standard		97°
Wraparound mask	Same as standard		97°
Goggles	Same as standard		180° (but only 90° undistorted)
Skin diver contact lenses	Only slightly restricted		120°–157

[a] Data from Weltman *et al.* (1965).

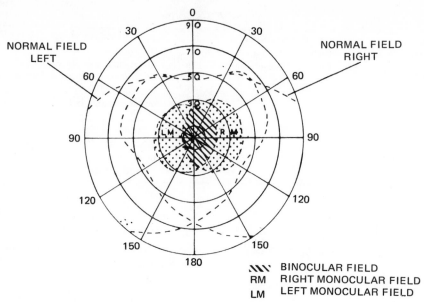

Figure V-12. Visual fields in water with goggles. [From Miles (1969) by permission.]

irritation in chlorinated water and most individuals cannot tolerate sea water very long because their eyes become very irritated and vision is blurred. Other drawbacks include loss of protection of the upper part of the face, which in cold water may become extremely uncomfortable. There may also be a psychological effect of insecurity from this exposure, but this should disappear with training and repeated exposure. There is also the necessity of individual fitting, with the resulting high cost of about $300 (Mosse 1964, Miles 1969, Williamson 1970).

The contact lens does have several advantages over masks or goggles. The field of vision is considerably enlarged, although it is slightly less than the normal field, because of the plane surface. Correction of ametropia is possible by grinding the wearer's prescription into the scleral contact lens before the air space is attached to its anterior surface. The upper part of the face, and especially the nose, is liberated, which may be desirable to some divers. The most important advantage over masks or goggles is that contact lenses cannot be dislodged, even as a result of a strong blow (Williamson 1970, Mosse 1964, Miles 1969).

Soft contact lenses are very comfortable, both in air and water, and no adaptation time is necessary. Visual acuity remains the same following submersion in sea water or chlorinated water, but the lens cannot prevent a burning sensation if the eye comes into direct contact with these types of water. A mask must be worn over the lenses for normal diving maneuvers but the lenses will not dislodge, even if the eye is held open under water; this is applicable when the mask becomes dislodged. The lenses do not cause further restriction of peripheral vision besides that of the mask, and acuity and comfort are unaffected when a facemask is worn to a depth of 200 ft. Several methods are available to allow the use of prescription lenses under water. The reader is referred to Williamson (1970) or to a qualified optometrist.

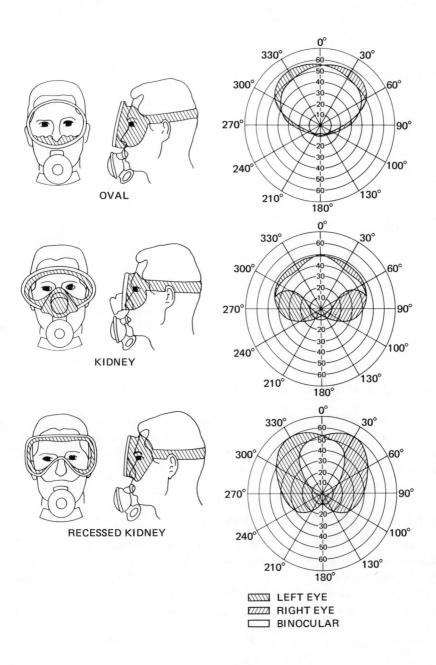

Figure V-13. Median visual fields for the standard masks. [From Weltman *et al.* (1965) by permission.]

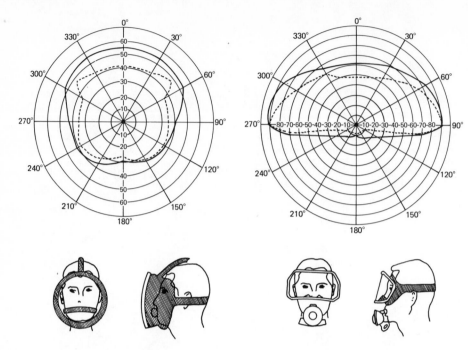

Figure V-14. Total visual fields of the full-face mask and the wraparound mask (two subjects). [From Weltman *et al.* (1965) by permission.]

6. *Other Factors Affecting Vision under Water*

The items discussed below describe some potentially hazardous conditions which would not be encountered by divers in routine diving. However, they are included and should be considered by those extending the frontiers of diving or planning long-term or unusual diving operations.

a. Effect of Confining Quarters on Acuity

Men confined for long periods of time to small quarters—submarine crews, for example—have shown greater loss of visual acuity with age, both far and near, than their counterparts in the normal population, although their vision was better than average at the start of their careers. There is also an indication of increased esophoria (a tendency of the eyes to assume some degree of convergence in the resting position). It is suggested that visual changes found with extended confinement result from lack of opportunity to relax accommodation and convergence (Weitzman *et al.* 1966).

Changes in acuity, stereoacuity, and refractive power occurring during a single submarine patrol are negligible. There is, however, a significant increase in near esophoria. There may be reversals of this shift in phoria when the men return to shore,

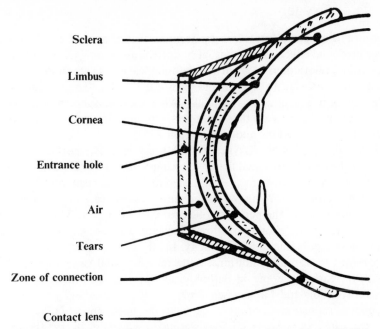

Sclera

Limbus

Cornea

Entrance hole

Air

Tears

Zone of connection

Contact lens

Figure V-15. Principle of the underwater contact lens. [From Mosse (1964) by permission.]

since long-term effects are not as great as would be predicted from short-term findings. Permanent changes in phoria may occur under several conditions: (1) some cumulative change with additional time in submarine; (2) in an older crew member, increasing magnitude of shifts during a patrol; (3) a decreasing ability of an aging visual system to effect the reversal after returning to shore (Luria *et al.* 1970).

b. Effects of Pressure

Since man normally must breathe various artificial mixtures of oxygen, nitrogen, and helium when diving to great depths, it is difficult to separate the effects of pressure and breathing mixture. Nonetheless, two conclusions can be drawn. First is the fact that no adverse effects of pressure on vision have been found as long as partial pressure of oxygen and nitrogen are kept at or near the levels encountered at sea level. For example, a variety of visual functions were tested in a series of dives at the U. S. Naval Experimental Diving Unit. Depths from 15 to 25 atm were simulated with divers breathing a mixture of 0.3 atm oxygen, 1.5 atm nitrogen, and the rest helium. No adverse effects were found on 15 different visual tests (Kelley *et al.* 1968).

Second is the fact that if either oxygen or the nitrogen partial pressure is allowed to rise beyond certain limits, severe and possibly incapacitating effects may be found. When breathing air at depth, both partial pressures are of course increased, but their effects can be studied separately by mixing oxygen, nitrogen, and helium in various proportions. These are discussed separately below.

c. Effects of Oxygen

Exposure to oxygen at pressures greater than that of normal air has been shown to produce profound effects; in fact visual symptoms are among the early indicators of oxygen poisoning. Both physiological and pathological effects are found. Physiological effects are those not due to irreversible chemical toxicity. They occur as soon as oxygen breathing begins and reverse quickly when there is a return to the normal O_2 tension. Pathological effects are those that persist or appear even after terminating the exposure to high P_{O_2} (Nichols and Lambertsen 1969).

(i) Physiological Effects. Oxygen at high pressure causes constriction of retinal vessels and of the peripheral visual field. No changes in visual acuity, visual field, ERG (electroretinographic) measurements, and stereopsis have been found with O_2 exposures of 1 ATA for up to 24 hr. When the pressure is increased to 3 ATA, a bilateral symmetric contraction of the peripheral fields occurs after 3 hr. The visual fields will contract to as little as 10° around fixation, but vision does not entirely disappear until unconsciousness occurs due to CNS toxicity. In all cases, recovery occurs in less than 1 hr after return to normal conditions. A decreased retinal vascular response to O_2 occurs with increasing age and in pathological states such as diabetes mellitus and hypertension (Nichols and Lambertsen 1969).

A symmetric constriction of the visual fields has been reported at lower pressures (Nichols *et al.* 1969). In one case, etiology was unknown and recovery was normal, but, in the other, impairment evidently was aggravated by a previously existing pathological condition of retrobulbar neuritis. Complete recovery from blurring and formation of scotoma took several weeks. Apparently these acute visual changes were due to an altered sensitivity to the metabolic effects of hyperoxia on the part of neuronal enzymes and/or vascular systems. The retrobulbar neuritis seemed to recur, although whether it was induced by oxygen administration or was independent and coincidental was not resolved. This finding does show the dangers involved in exposing patients with a previous history of ocular disease to increased oxygen pressure.

Restriction of visual fields with prolonged exposure to increased pressure is not likely to be due to a paradoxical decrease in the oxygen supply to the tissues. However, since the areas with the lowest blood supply are the first to be affected, it is possible that vasoconstriction may decrease the blood supply to such a degree that essential nutrients (other than oxygen) are not supplied rapidly enough to maintain the high metabolic rate of the retina (Nichols and Lambertsen 1969).

(ii) Pathological Effects. Irreversible damage may occur if sufficiently high partial pressures of oxygen are breathed for sufficient time. A variety of studies performed on animals have been summarized by Nichols and Lambertsen; the effects include visual cell death, retinal detachment, and lesions. The partial pressures employed to produce these effects are generally high—for example, 100% O_2 at 3 ATA for many hours. They are thus beyond the range of values encountered by ordinary divers. However, in deep diving these levels could be obtained if air were employed. Similarly in long-term saturation diving, possibly dangerous amounts could accumulate since the adverse effects build up over time. For these reasons the percentage of oxygen breathed at deeper depths or under saturated conditions is routinely reduced from that of air to prevent oxygen poisoning.

The limits, in time and depth, for saturated air diving are at present unknown. This is the reason for a series of air dives being conducted at NSMRL. In the SHAD series (shallow habitat air diving) a battery of visual and many other tests were administered to men living for 30 days at 50 and 60 ft. No adverse effects on vision have been found under conditions to date.

d. Effects of Nitrogen

The effects of nitrogen under pressure on human beings have been known for years and are commonly referred to as *rapture of the deep, nitrogen narcosis,* or *inert gas narcosis.* The major symptoms are a loss of motor coordination, reduction of mental ability, and changes in mood, often in the direction of euphoria (Jennings 1968). Since any one of these may incapacitate a diver, air is generally not used as a breathing mixture for dives beyond 200–300 ft.

While not primarily a visual phenomenon, nitrogen narcosis can obviously impair visual performance in any test that involves the diver's reasoning, his motivation, or his motor adjustments. Visual hallucinations also are sometimes reported (Adolfson 1967).

Because of the difficulty in separating purely visual changes from the more general symptoms of narcosis, several investigators have turned to the visual evoked response (VER) as an indication of narcosis. The VER, an electroencephalographic response of the human visual system to a visual stimulus, requires no effort from the subject other than that he look at a target. Decrements in evoked responses for divers breathing air at depths of 200–300 ft have been reported by many investigators (Bennett *et al.* 1969, Bennett and Dossett 1970, Bevan 1971, Hamilton and Langley 1971, Kinney and McKay 1971, Larson *et al.* 1971, Ackles and Fowler 1971, Kinney *et al.* 1972, Bartus 1973, Kinney *et al.* 1973). Thus the narcotic symptoms of nitrogen can definitely be expanded to include deficits in the processing of visual information in man.

e. Effects of Drugs

Results from animal experiments performed in connection with the Navy SEALAB project indicated there is no change in the potentiation of a series of commonly used drugs under 20 atm pressure. Further experiments indicated that there was no difference between the action of a variety of drugs used during a 31-atm exposure and the action of the same drugs when administered on the surface. Drugs tested included aspirin, heparin, xylocaine, and neosynephrine as well as a topical anesthetic (0.5% Proparacaine) and a mydriatic drug (17% Tropicamide) (Kelley and Anderson 1969).

B. *Hearing*

In the underwater and high-pressure environment it cannot be assumed that man's sense organs will function with the same efficiency as in normal atmospheric conditions.

In fact, we know that in several ways man's sense organs are less efficient when under water. This section will consider the effects of high-pressure gases and the underwater environment on auditory acuity: What reliance can the underwater worker put on his hearing ability? Such aspects of the problem as absolute thresholds of underwater hearing, maximum intensity for nontraumatic underwater hearing, underwater auditory discriminations, localization of underwater sounds, and cases of cochlear/-vestibular lesions associated with underwater activity showing hearing loss will be presented in this section.

The anatomy and general plan of operation of the outer, middle, and inner ear are covered in Chapter I, under information receptors. Pressure effects on the ear (aural barotrauma) are covered in Chapter III. Speech distortion and the general problem of underwater speech communication are treated in Chapter X. Vestibular dysfunction associated with underwater activity is covered in detail in this section under vestibular function—equilibrium and balance.

For those interested in more detail we recommend Adolfson and Berhage (1974; Chapter III, Audition), Edmonds *et al.* (1973), and Taylor (1959).

1. Underwater Hearing and Air- and Bone-Conducted Hearing

As pointed out in Chapter I, hearing depends on (a) the ear and its associated nerve pathways, (b) the sound source (a vibrating body), and (c) the medium that transmits the pressure vibrations from the sound source.

In air, sound is normally transmitted to the cochlea by way of air conduction (AC). However, one can make a reasonable assumption that underwater hearing is a special case of bone-conducted (BC) hearing. Of course, water-borne acoustic energy would impinge on the whole submerged head at all points rather than at a single point on the skull, as in the case of the bone-conduction vibrator. Thus sound would be transmitted by way of compressional waves through the bones of the skull and the contents of the endocranium to the cochlear capsule.

Acoustic energy in water finds no special barrier at a water/head interface. Since the acoustic impedances of water and of the human body are similar, most of the acoustic energy is admitted and transmitted through the head as waves of condensations–rarefactions. It has been demonstrated (Bekesy 1960) that bone-conducted and air-conducted stimuli are analyzed in exactly the same way by the auditory system once the energy enters the cochlea by any route or routes.

If a diver had a bubble of air in his ear canal, acoustic energy would ensonify that bubble, as it would the air in all air-filled cavities in his head, and that airborne energy could enter the cochlea through the usual route—from the eardrum to the chain of small bones (the ossicles) to the round window and in to the endolymph.

The best estimates available to date are that AC hearing exceeds underwater hearing by 30–60 dB, somewhat greater at the higher frequencies. This appears to indicate that underwater hearing is less efficient since the threshold sound pressure level in air is much less than in water. However, when one reflects that AC and underwater hearing are about equally sensitive to the same vibratory amplitudes—the one of the eardrum and ossicles, the other of the skull—underwater hearing is seen to be just as efficient. Zwislocki (1957) used special ear plugs to insulate human ears from AC sound and took audiograms in air by free-field loudspeaker. The difference with and without the plugs was taken to be BC threshold, 45–68 dB below AC threshold. This difference is similar to the difference in AC and underwater hearing and indicates further that underwater hearing is bone-conducted.

2. *Absolute Thresholds of Underwater Hearing*

a. General

A number of investigators have immersed divers in water and taken some sort of behavioral audiograms. In several of these studies the AC audiogram was not reported, so that one cannot judge how nearly the experimental group's hearing approximates that of an average of young, healthy ears. More to the point, in only two studies were the BC audiograms reported, although the normality of the BC reading should properly determine a subject's suitability for inclusion in a study of underwater hearing.

The unit of sound pressure level is the decibel (dB). Since the dB is a relative unit, a standard reference of 0.0002 μbar (2×10^{-4} dyn/cm^2) has been adopted and it is understood, ordinarily (and in this section), when a sound pressure level is stated. It may be helpful to know, however, that *underwater* sound pressure is often referenced to an underwater hearing threshold of 1 μbar, or 1 dyn/cm^2, or 74 dB re 2×10^{-4}.

b. Measurement of Threshold Values

A number of investigators (Sivian 1943a, 1943b, Reysenbach de Haan 1956, Hamilton 1957, Wainwright 1958, Brandt and Hollien 1967, and Smith 1969) submerged subjects to relatively shallow depths, ranging from 1 to 25 ft of water, and measured the resulting elevated hearing thresholds of these subjects at various frequencies. Unfortunately, not all of the subjects had normal hearing at all frequencies tested in the air; Table V-5 is a composite of these threshold values. Since the conditions for each set of experiments were not the same, no attempt is made to determine mean values for each frequency.

Montague and Strickland (1961) found minor differences in the underwater hearing thresholds in young men vs. older men over the frequency range of 250–6000 Hz. Two of the older men showed a moderate hearing loss in air at 3000–4000 Hz. This evidently had little effect on the underwater thresholds, which are primarily dependent on bone conduction.

This is borne out by the work of Smith (1965), who studied underwater thresholds in eight subjects reclining in a swimming pool with their external ear canals (meatus) filled with water. In addition, both AC and BC hearing thresholds in air were recorded for each subject. The study demonstrated a significant correlation between individual AC audiometric losses vs. air conduction minus underwater (AC − U/W) differences; the more the AC loss, the less the AC − U/W differences. (This is explainable only if

Table V-5
The Effect of Submergence on Hearing Thresholds

Frequency, Hz	Sound pressure level, dB	Investigator[a]
125	67–88	4–6
250	60–75	4–6
500	52–73	3–6
1000	30–68	1–6
2000	53–81	2–6
3000	49	1
4000	54–85	2–6
6000	73	5
8000	72–73	2, 6
15000	104	2

[a] 1, Sivian 1943b; 2, Reysenbach de Haan 1956; 3, Hamilton 1957; 4, Wainwright 1958; 5, Brandt and Hollien 1967; 6, Smith 1969.

underwater hearing is largely BC hearing.) This study is sufficiently important to quote it directly: "It may be concluded on the basis of these results that hearing in water is primarily mediated by bone conduction. The threshold of hearing underwater is approximately 74 dB re 0.0002 dyn/cm² at 500 cps [Hz]. Threshold sensitivity decreases at a rate of 3 dB/octave up to 8000 cps [Hz]."

At 1000 Hz, Smith (1969) found that divers with normal hearing can detect pure tones having a sound pressure level of about 61–64 dB above 0.0002 dyn/cm². Brandt

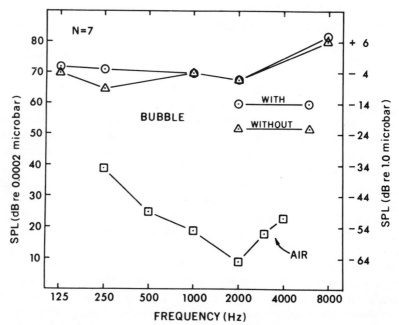

Figure V-16. Mean threshold sound pressure level (dB re 0.0002 μbar) as a function of test frequency in air and water with and without air bubbles (Hollien *et al.* 1969a).

and Hollien (1967) had earlier made the important observation that the peak sensitivity was at a lower frequency underwater (500–1000 Hz) than it is in air (1000–4000 Hz).

c. Effect of Air in External Auditory Meatus

Using three divers, Reysenbach de Haan (1956) found that a bubble of air in the meatus degraded threshold at 1 kHz by about 21 dB, but improved threshold at all higher frequencies used, up to 15 kHz. Hollien *et al.* (1969a) immersed four male and three female experienced divers to 12 ft and collected underwater thresholds with care to retain or exclude a bubble of air in the canals. Figure V-16 shows the mean AC and underwater data. The effect of the bubble was negligible except at 250 Hz, where threshold was better by 6 dB *without* the bubble. This may eventually tell us something about the compliance of the eardrum and middle ear at 250 kHz or it may have been a chance reading. Hollien and Brandt suggest the data mean that underwater hearing is exclusively by (compressional) BC and that the middle ear does not contribute.

d. Effect of Depth

Brandt and Hollien (1969) pursued the matter of underwater hearing in four male and two female experienced divers, adding the information that water depth from 35 to 105 ft was not significant. Their results (Figure V-17) corroborate earlier data as to absolute underwater thresholds and AC − U/W differences. They explain the somewhat elevated underwater data at 105 ft as first-session practice effects.

For effects of depth on hearing in chamber diving see the discussion of human auditory response to air pressure changes.

e. Effect of Wearing a Hood

As was noted by Smith (1969), "Wet suit hoods reduce underwater sensitivity by 25 to 33 decibels over the frequency range 1 to 8 kilohertz." Also, when Hollien and Feinstein (1972) tested a full 3/16 in. wet suit with a 3/16-in. hood (Figure V-18), "the thresholds were significantly lower in the middle and high frequencies for the no-hood condition." A shift in threshold of 20 dB or more occurs at frequencies above 1000 Hz when the diver wears the arctic hood (Montague and Strickland 1961).

3. Underwater Auditory Discriminations

Differential thresholds for frequency and intensity have not been collected as such, but intelligibility for waterborne speech has been assessed. It is common knowledge that near-perfect underwater speech communication is easily achieved with high-quality equipment and good environments in relatively shallow water on pure oxygen or compressed air.

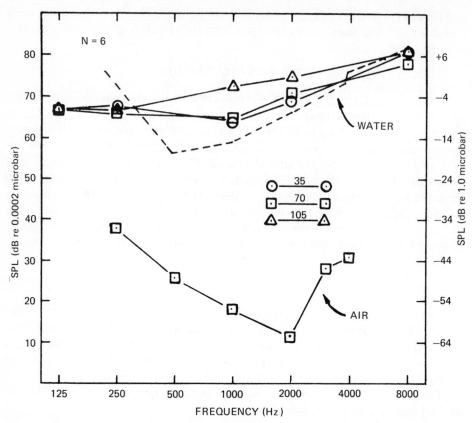

Figure V-17. Air conduction and underwater audiograms, the latter at different depths (Hollien *et al.* 1969*b*). Similar data on seven divers at 30 ft depth, from Hollien and Feinstein (1972), are represented by the dashed line.

Using a Navy J9 underwater loudspeaker at a depth of 35 ft Brandt and Hollien (1968) found 95% correct responses for monosyllable intelligibility at 30 dB above speech reception threshold (SRT). The SRT was determined individually by adjusting the speech signal until the diver achieved a 50%-correct response. The mean SRT of the 16 divers was 80 dB re 2 × 10^{-4} μbar, about 13–15 dB higher than the underwater thresholds for pure tones at the same speech frequencies. (This underwater pure tone vs. speech difference is about the same as the difference in AC hearing and underwater hearing.)

Such clear intelligibility would not be obtained if underwater hearing suffered much greater distortion than AC hearing. But the safety factor in voice communications is such that at least usable communications can be achieved with quite "noisy" systems. Unprocessed voice communications by underwater telephone (AN/UQC-1) through the water between two submerged submarines is usable over a range of several miles (Murry and Strand 1971). For further details on the problem of communication in an underwater or hyperbaric environment see Chapter X.

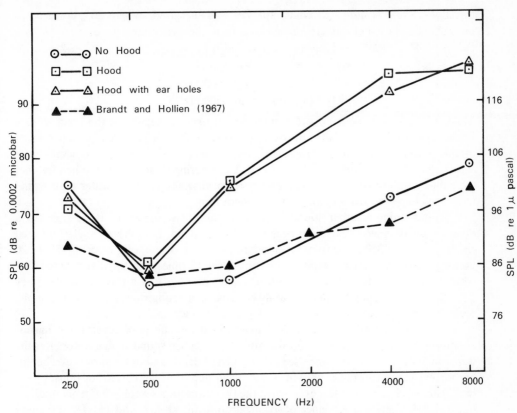

Figure V-18. Threshold in dB re 0.0002 μbar or 1 μ pascal at an ear depth of 30 ft as a function of head covering for seven listeners (Hollien and Feinstein 1972).

4. Localization of Underwater Sounds

The consensus among swimmers and divers is that underwater sound localization is poor. Several authors have even concluded that it is impossible because binaural cues, which allow us to localize sound in air, are removed underwater. However, there are several good reasons to doubt these conclusions. It has been known since 1900 that monaural localization in air can be very accurate indeed. It has also been known that sea-going mammals have excellent directionality, so that it would seem odd if the human head were altogether incapable of sound localization. The most conclusive evidence is a series of recent experiments which indicate that divers do have at least some ability to locate a sound source.

Hollien and his group conducted a series of experiments on underwater sound localization (Hollien 1969, Hollien *et al.* 1970), using a total of 23 subjects immersed in 50 ft of water with bodies held fixed but with heads free to move. Underwater loudspeakers were positioned at 0°, ±45°, and ±90° from the midline and were energized with trains of pulses at different frequencies at intensities of 40 dB over threshold. Subjects indicated which speaker was energized. Where 20% correct was pure chance, the ranges of the mean scores from the two studies were:

53 –57.3% correct with broadband noise
50 –52.7% correct at 250 Hz
37.3–39.0% correct at 1000 Hz
24.0–34.0% correct at 6000 Hz

Most of the errors were no more than 45° in extent. Repeating the trials with a knowledge of previous results improved the score by 30% on the average. When seven of the subjects repeated the trials with their heads fixed, most of the scores were similar except for a drop of 37% at 250 Hz. Considering their results, Hollien felt that the subjects exhibited a far greater ability to localize sound under water than would have been expected from theoretical considerations.

To determine whether sound localization ability would make it possible for a diver to use an emergency sound beacon to summon aid, Leggiere *et al.* (1970) tested six divers and in 350 trials found a standard deviation of pointing error of 58°. In 20 homing trials, the subject swam to the transducer 12 times, but the divers noted that when they were oriented to the bottom plane, homing was always possible. They "concluded that a weak binaural ability exists with scuba divers and that a sonic, scuba emergency beacon is possible and probably practical."

Anderson and Christensen (1969) determined the effects of reverberant harbor conditions versus the open sea on the ability to localize sound under water. Results were much the same in the open sea at a 6-m head depth and in the more reverberant harbor enclosure (porpoise pen) at a 3-m head depth. Mean correct responses were hardly above chance at 2kHz, even when the two transducers were ±90° from midline, but at 1 kHz there was evidence of localization, while at 4, 8, and 16 kHz there was ability slowly increasing with frequency. The best diver achieved 70% correct at 1, 8, and 16 kHz even when the transducers were only ±10° off midline (20° minimum audible angle).

Norman *et al.* (1971) arranged underwater projectors in 30° steps around persons submerged in a reverberant pool. Where localization of clicks at chance was 14%, two persons yielded 42% correct with bare heads, and 40% with hoods with ear-holes, but only 27% with small, Neoprene ear-patches fitted over the external ear. One subject with only one ear-patch fell to chance performance.

5. Maximum Intensity for Safe Underwater Hearing

It is within the capability of present amplifier and transducer technology to create pure tones under water that are loud enough to damage hearing permanently. This sonic hazard is in addition to the effect of underwater explosions, which create steep-front pressure waves and can easily damage body tissues. In one study Montague and Strickland (1961) found that, when divers were facing the transducer, about 50% of them signaled that a tone of 1500 Hz at about 172 dB re. 0.0002 μbar was so annoying that they did not want it increased further. Above 165 dB all subjects reported some distortion of the visual field. A diver's hood created about a 10-dB attenuation, but a hole in the hood as small as 2 in. at the forehead destroyed this protection.

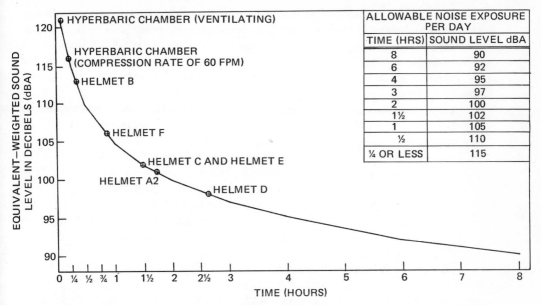

Figure V-19. Relationship between the currently accepted noise exposure limits and the highest levels recorded in individual helmets and in the hyperbaric chamber under the conditions described (Summitt and Reimers 1971).

Noise as a hazard to divers and hyperbaric chamber personnel was studied by Summitt and Reimer (1971) and they concluded that, "Obviously, diving and chamber personnel are frequently exposed to noise which can injure the unprotected ear in a relatively short period of time. Indeed, the recompression chamber during compression or ventilation can have noise intensities greater than the maximum acceptable level even for short exposures." Figure V-19 illustrates the allowable noise exposure limits currently accepted for hearing conservation programs in the United States Navy. Superimposed on the limit curve are the highest noise levels recorded by these experimenters in various diving helmets available for industrial or military use and in a recompression chamber. The result for diving helmet B, for example, is only slightly lower, allowing an exposure time of approximately 20 min.

6. Human Auditory Responses to Air-Pressure Changes

In Chapter I the pressure-equalizing mechanism for the ear is presented and in Chapter III the physical effects of a failure to properly equalize pressure are described. Here we look only at the hyperbaric effects on auditory acuity.

a. Pressure Equalized across the Eardrum

Even when the eustachian tube is patent and adequately ventilates the middle ear, changes occur in the conductive aspects of hearing when the atmospheric pressure is altered. If it is decreased as in an airplane ascent, or the atmosphere is rendered less dense by substituting helium for air, some slight changes in hearing may occur. If it is rendered much denser, as in a hyperbaric chamber or by water immersion at depth, more drastic changes may occur. In this regard as in others, the problems are common to subaquatic and to hyperbaric chamber medicine. Obviously the density of the bubble of air against the eardrum will partially govern the eardrum's response and thus the hearing of the individual studied.

Audiograms were taken on 26 experienced men exposed to 11 ATA in a hyperbaric chamber (Adolfson and Fluur 1965). BC audiometry with a vibrator, affected in no way by pressure, revealed no sensorineural shifts, but AC audiometry (the earphone response carefully corrected for the pressures used, with a Bruel and Kjaer microphone itself calibrated for pressures) showed conductive losses increasing rather regularly with pressure and amounting to 30–40 dB in the speech frequencies.

Farmer et al. (1971) studied six divers down to 19.2 ATA breathing a He–O_2 mixture. AC and BC hearing levels were measured at 0.25–4 kHz at five different depths during the 12-hr descent, twice during the six-day stay at 600 ft, and at six depths during the seven-day decompression. Pressures were always equalized across the eardrum. No BC changes were seen, and no AC changes at less than 100 ft, but at 600 ft, conductive losses averaged 4 $\pm$ 11 bB for 0.25, 0.5, and 1 kHz, and 25 $\pm$ 13 dB for 2 and 4 kHz. Data were less variable after six days on the bottom, but the mean losses at 0.25, 0.5, and 1 kHz increased to 26 $\pm$ 8 dB. The authors explain such changes as due to an upward shift of the ear resonance frequency and greater impedance mismatch with helium.

Using 33 divers, Thomas et al. (1973) took AC and BC hearing levels on one saturation chamber dive to 300 ft, four dives to 600 ft, two dives to 850 ft, and one dive to 1000 ft, all on helium–air mixtures. The audiometer was permanently wired into the hyperbaric chamber and the earphone had been calibrated* down to 1000 ft (31 ATA) in helium–air. Sensorineural loss and also temporary threshold shifts (TTS) can be ruled out by noting that BC never changed under any hyperbaric condition. Since the shallowest depth at which audiograms were collected was 100 ft, by which depth significant conductive losses (15 dB) appeared at 3 and 4 kHz, the authors conclude that a conductive loss may develop at even shallower depths, perhaps at only 1 ATA. Figure V-20 shows the development of mean losses taken at various depths from the surface to 1000 ft and back again. A trend existed at 0.5, 1, 3, and 4 kHz for mean hearing level to decline, and then return, depending on depth. No trend existed for hearing to improve or to worsen during 200 hr at 600 ft.

The mean losses at 1000 ft are somewhat less than the 30–40 dB losses reported by Adolfson and Fluur (1965). The improvement in hearing levels at 6 kHz and the lack of change at 2 kHz mean that the losses probably have rather complicated interactions of more than one effect, involving impedance changes plus shifts in resonance frequencies. See the discussion of speech discrimination in Chapter X.

b. Mild Differential Pressures across the Eardrum

Experimenting with cats, Thompson et al. (1934) inserted a needle through the eustachian tube into the middle ear and recorded cochlear potentials to show there

---

* The ear phone was calibrated by Thomas according to the method described in Thomas et al. (1972).

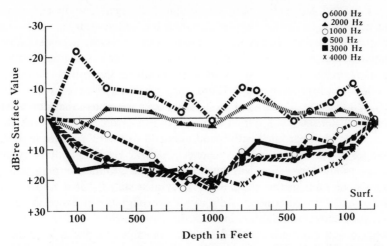

Figure V-20. Mean thresholds for six different frequencies as a function of depth. All thresholds are measured relative to surface threshold (Thomas *et al.* 1973).

was no hearing loss until differential pressure across the eardrum reached ± 5 mm Hg. Then conductive losses increased progressively to the pressure limit of ± 30 mm Hg.

When differential pressures of ± 50 mm Hg were introduced across the eardrum in the cat (Wever *et al.* 1948) a loss (presumed conductive) of 20–40 dB was found, especially in the middle frequencies. At lesser pressures, the data for 2 kHz were exceptional, in that a gain was seen, not a loss. This finding foreshadowed the audiograms under pressure of Thomas *et al.* (1973) in which 2 kHz was also anomalous in not showing changes with pressure.

c. Occluded Eustachian Tube

In certain individuals with chronic occluded eustachian tube (a condition frequently met with in diving) high-tone hearing loss, presumably sensorineural, has been noted. Mygind (1931) suggested the organ of Corti might become edematous, and Galloway (1940) that toxic substances collected in the undrained middle ear might enter the inner ear by way of the round window. However, Loch (1942) closed off his own tubes with an inflatable balloon inserted through the nose. After a 75-min block a hearing loss at 2–10 kHz had progressed irregularly to 15–20 dB, all returning to normal with deflation of the balloon. Evidently tubal obstruction first affects high-tone conduction. Since it is most unlikely that the inner ear is changed in any way, a mild high-tone loss cannot always be presumed to be nonconductive in nature.

d. Middle-Ear Involvement

In the course of submarine escape training a group of 6149 prospective submariners were subjected to 50 lb positive pressure in a dry recompression chamber (Shilling *et al.* 1946). They were carefully evaluated from the standpoint of ear, nose, and throat preconditions and any subsequent pathological changes. Middle-ear damage was classified according to Teed (1944) (see Chapter III) and auditory acuity was matched with ear damage. Ear-damage grades 1, 2, and 3 were so nearly normal that no significant comparison could be made. For grade 4 ears the average loss was from 5 to 15 dB but was statistically significant. It was also significant for the subjects in that they were aware of lowered hearing acuity. "There were, however, a few ears with severe and relatively long-lasting impairment of hearing and we noted that those ears were typically characterized by a dark purple or bluish discoloration of the ear drum, indicating the middle ear filled with blood unmixed with air." In these individuals the loss averaged 20–30 dB through frequencies 256–8192 Hz. The authors hypothesized that the loss of auditory acuity was due, in part at least, to the damping action of free serosanguineous fluid in the middle ear. They also stated, "It is apparent that the otopathology of aerotitis [otitis] media as seen through the otoscope is not a perfect or even a good index to the functional efficiency of the ear."

Freeman and Edmonds (1972) described five cases of reduced sensorineural loss generally not recovered—one bilateral, three of high tones only, all occurring during immersion to depth and in such circumstances that a diagnosis of decompression sickness was stated not to have been possible. Two of the five had symptoms of vertigo. In each there was impaired ability to "clear" the ear on the affected side. They concluded that the pressure differential between the middle and the inner ear was considerable, and that stress on the stapes was relatively violent while the divers, all in some pain, were striving to "clear" their ears. Postulated intracochlear hemorrhage or vascular spasm, etc., was ruled out. They recommended that if a diver should experience aural pain during descent, he should rise to a depth at which he could easily equilibrate his tympanum or, at least, pause before continuing descent.

e. Inner-Ear Involvement

A study (Coles and Knight 1960) of 62 divers indicated most hearing loss is reversible. There were many cases of sensorineural hearing loss, but in every such history the loss could have been the result of previous acoustic trauma. Furthermore, a literature search at that time uncovered no case where pre- and postexposure audiograms clearly indicate sensorineural loss due to barotrauma. In spite of this, the distinct possibility of inner-ear damage exists. More recently, Coles (1973) agreed that his original conclusion must now be supplemented by cases of sensorineural loss attributed to barotrauma.

(i) Intracochlear Vascular Accidents (microhemorrhage, vascular spasm, emboli, bubbles). During decompression, bubbles of gas may be released as emboli to occlude the terminal twigs of the internal auditory artery. Vail (1929) was one of the first to report deafness resulting from nitrogen bubbles in the internal ear. Much later Nourrit (1961) distinguished a unilateral or bilateral perceptive deafness. This deafness may arise upon the patient's emergence from a compression–decompression cycle, or it may appear some hours or even days later; it may or may not be accompanied by vestibular symptoms. According to Plante-Longchamps *et al.* (1970), they have never seen an isolated cochlear barotrauma unaccompanied by vestibular symptoms

and they feel that a compressional lesion of the vestibule would simultaneously lead to a cochlear loss because of the continuity of the labyrinthine fluids. They also report cases of post-dive cochlear/vestibular lesion with delayed treatment that led to permanent deafness.

Stucker and Echols (1971) and Gehring and Buhlmann (*in press*) recommend rapid recompression to avoid permanent damage from progressive necrosis of cochlear structures. In 9 out of 12 cases of sensorineural loss during decompression, Farmer (1973) reported that there were no classical symptoms of decompression sickness, yet recompression was an efficacious treatment. It must be noted, however, that sudden deafness may also be due to acute neuritis of the eighth nerve, virus infection, vascular accident, vasomotor neurosis, and acoustic trauma as well as decompression sickness (Harris 1971).

(ii) Loss of Perilymph (round window rupture). Pullen (1972) repaired a ruptured round window in a man who noticed loss of hearing three days following a dive. Seven weeks later hearing was normal, presumably when the normal volume of perilymph was regained.

Audiograms and nystagmograms showed that the effects of middle-ear barotrauma of descent may temporarily or permanently damage the cochlea and/or the vestibular apparatus (Edmonds 1973*a*). Three out of five cases were shown to have surgically correctable lesions by repairing round-window perforations. The histories of the dives were consistent: There was initial difficulty in equalizing pressures within the middle-ear cleft during the ascent, and usually there was an excessive attempt at forceful Valsalva maneuvers. Following the dive, there was usually a history of tinnitus, with or without vestibular and cochlear damage. In the cases of round-window fistula, the cochlear and vestibular damage progressed following the dive and was corrected only after surgery.

(iii) Gas Saturation of Cochlear Fluids. There is still another possibility for inner-ear damage. In a helium dive, the helium in the middle-ear cleft can readily diffuse across the round window, saturating the inner-ear fluids before other tissues receive gas normally through the circulation. (In the cat, the cochlear microphonic voltage remains for as long as 2 hr, fed by oxygen through the round window, but this "death potential" disappears immediately if the middle ear is flooded with helium.) Thus the usual gas dynamics need not apply to the labyrinth.

f. Central Nervous System Involvement

Some sudden losses of auditory acuity may be due to CNS damage in auditory tracts above the level of the cochlea nuclei, which may mimic peripheral (conductive) loss (Coles 1973).

A too-forceful Valsalva maneuver (Fructus and Ricci 1969) can lead to a severe variation of the alveoli capillary gradient with stress at the level of the smaller vessels leading to the formation of pathogenic bubbles at the level of the alveoli, these bubbles passing in the left ventricle by the pulmonary veins, possibly leading to cerebral embolism of very wide distribution.

Another probable cause of loss of auditory discrimination, particularly at greater depths, may be hyperbaric air intoxication or "nitrogen narcosis," which causes a prolonged associative reaction. (See also Chapter IV). After studying 23 divers by means of speech audiometry in a pressure chamber at 4, 7, and 11 ATA, Adolfson and Fluur (1967) showed that even if the sound intensity was raised far above the hearing threshold, the prolonged associative reactions caused by hyperbaric air intoxication led to difficulties too severe for a diver to comprehend simple common words.

7. Incidence of Auditory Involvement

Although not life-threatening, the most common illnesses affecting divers and pressure workers in general are listed by Wilke and Steffl (1965) as barotrauma and otitis media. They might have added otitis externa or fungus infection of the external ear canal. The incidence of such involvement is given for various groups in the following material.

a. Among Swimmers Generally

In a survey of 900 college graduates Hitschler (1949) found no greater incidence of reported hearing loss among the half of his sample who were trained swimmers than among his controls. There seems nothing about fairly frequent exposure to water as such (especially, clean sea water) to lead to auditory problems in those who are trained in the necessary hygiene.

b. Among Amateur Skin and Scuba Divers

"Of all the problems faced by the skin diver, the commonest are those falling within the otolaryngological field." Fields (1958) then goes on to list all of the problems associated with the ear, from simple otitis externa through otitis media, rupture of the drum, otitis interna, vertigo, and loss of hearing. Wilke and Steffl (1965) describe in detail the conditions under which damage to hearing is incurred as a result of diving accidents and find, as do others, that hearing is not permanently impaired in most cases. However, they find that in some cases there is aero-autosclerosis, with an irreversible progressive course, affecting both middle and inner ear.

c. Among Caisson Workers

It has been long known that caisson disease often has an inner-ear/labyrinthine component, and many articles have been written describing this situation. Poli (1909) found that increased work experience in caissons led to an increased incidence of hearing loss (14.6% for the more experienced vs. 7.5% for the less experienced workers). Lang *et al.* (1971), using more sophisticated otoaudiological techniques on 432 workers under 40 yr of age, found that 60% had hearing defects. In fact, two-thirds of the young men between the ages of 20 and 29 yr were showing some sort of hearing defect within the first six months on the job. Pagano (1959) reported that the types of hearing loss in 40 affected caisson workers were conductive (35%), perceptive (50%), and mixed (15%). A large number of these cases had AC losses exceeding 50 dB and BC losses exceeding 30 dB at all frequencies.

d. Among Inexperienced Hyberbaric Chamber Subjects and Diving Trainees

A high incidence of otitis media has been reported among inexperienced subjects. (See middle-ear involvement above.) If pressurization is too rapid, all ears will undergo the symptoms of middle-ear barotrauma. Teed (1944), Shilling *et al.* (1946), and Alfandre (1965) studied altogether several thousand men subjected to a 50-lb pressure test (simulated 100-ft depth) for 3–10 min. Between 25 and 36% showed some degree of otitis media in this first exposure to pressure. Teed felt that this incidence could have been reduced to 2–3% if all of the men could have performed the Valsalva maneuver. Despite this, Shilling *et al.* found that 6–10% could not complete the test because of ear pain.

Bayliss (1968) reported the incidence of aural barotrauma in 530 diving trainees, all medically fit; 20.4% of these men reported "ear trouble" either in the recompression chamber (58.5%) or in the water (41.5%) and were examined by the author himself. On the 1–5 MacFie scale, only one ear ruptured (grade 5), none showed grade 4 (free blood in middle ear), two ears had grade 3 (gross hemorrhage confined to the eardrum but no blood in middle ear). The remainder who complained of ear trouble showed grades 0–2; about half of them recovered completely in three days or less with oronasal decongestants and antibiotics. The relatively low incidence of serious barotrauma was attributed to rigorous physical selection and appropriate management during the early stages of diving training.

e. Among Professional Divers

In those who dive repeatedly, a high incidence of auditory complications can be expected. But symptoms of barotrauma are either often subclinical or else working divers are reluctant to report aural problems, Prazic and Salaj (1971) carefully administered auditory and equilibrium tests to experienced divers who reported no hearing disorders or vertigo, yet 38% of them had hearing disorders or loss of equilibrium sensitivity in various degrees.

Zannini *et al.* (*in press*) studied 160 professional divers for some time and found that there is cumulative insult to the middle-ear apparatus and, further, that much conductive loss presages sensorineural loss, mainly expressed as high-tone loss from lesions in the basal turn of the cochlea. Perhaps the most common condition among all divers is otitis externa or infection of the external auditory or ear canal. This is discussed in detail in Chapter VIII.

f. Among Breath-Hold Divers

Breath-hold diving, the most primitive form of diving, has been the basic diving technique utilized by various groups in the world. Those that have been most carefully studied are the diving women, the Ama, of Korea and Japan. During a single working day, they might dive 60–90 times down to depths of 100 ft after inspiring as much as 85% of the inhalation capacity. As a result, they experience otorhinolaryngological problems of all kinds (chronic otitis media, opaque eardrums, nonpatent eustachian tubes, sinusitis, rhinitis, and hearing loss) as their chief occupational disorders (Hong and Rahn 1967). Kataoka (1954) found that 36 of 74 divers (49%) had severe hearing losses, as against 28% of nondivers from the same villages. Anzai (1960) compared 88 Ama vs. 86 controls and found that only auditory problems distinguished the groups medically, though only 20% of these particular Ama reported loss of hearing (2% in the controls).

Harashima and Iwasaki (1965) performed a complete medical study of 22 Ama aged 29–59, with 12–43 yr of deep diving experience. Of these, 14 complained of tinnitus, 12 of unilateral eardrum rupture, and 6 of spontaneous otalgia. About half reported otitis media, and about half said they did not dive if they had a cold. The authors feel the chief problem of these divers is chronic otitis media leading to deformities and, often, rupture of the eardrum.

Yamamoto (1936) found that permanent hearing losses at the speech frequencies 250–2000 Hz occurred in about 30% of cases in the youngest divers (16–20 yr), and in about 50% in the older divers (51–60 yr). During heavy diving months, greater losses (usually recoverable) occurred.

C. *Vestibular Function*

1. *Introduction*

Vestibular function is of major importance in diving. Dysfunction in this system, indicated primarily by vertigo, nausea, and vomiting, is at least incapacitating and can be life-threatening, particularly to a diver in the water. Vertigo may be a prime cause of accidents with shallow-water scuba diving. Vestibular dysfunction is increasing as the number of deep dives for research and operational purposes increases and as greater depths are reached. Remoteness and/or unfamiliarity with diagnosis and treatment have complicated many cases. Damage to the vestibular system is a serious hindrance to a diver's occupation. Partial and complete recoveries from vestibular problems have occurred, permitting divers to resume normal activities, but permanent damage to the vestibular system can and does occur, the exact frequency of which is unknown. The usual clinical course of acute vestibular injury is one of gradual reduction of symptoms as repair and/or compensation occur. If repair is incomplete, then full preinjury function is not reached, and the compensatory mechanism will allow the individual to be largely asymptomatic except for varying degrees of motion intolerance which, in some cases, might not become evident except during specific vestibular testing or during certain diving situations. Thus, all divers suffering apparent vestibular damage during or related to diving should be evaluated after they become asymptomatic so that more rational estimates of residual function and further suitability for diving can be made. There is a need for a world-wide review of vestibular dysfunction in naval, commercial, and institutional deep diving so that relevant elements of the diagnosis, treatment, and prevention can be clarified. (See Chapter I for a discussion of normal vestibular function.)

Balance and equilibrium are dependent upon the reception of sensory cues in addition to those from the vestibular system; these include visual, organic, cutaneous, kinesthetic, and occasional auditory inputs, which are integrated in the central nervous system to provide the individual with information about his position and motion changes in the environment. When some of these cues are absent or diminished as in many underwater activities, the undisturbed cues become very important. To maintain equilibrium and balance on a fine level, particularly during rapid and intricate body movements, the vestibular sense is thought to be necessary regardless of the number of other cues available.

Thus, in diving, the vestibular sense takes on increasing importance, particularly for the untethered, free-swimming diver. Under conditions of neutral buoyancy, cutaneous and kinesthetic cues of position are reduced or absent. In fact, the diver in a full wet suit with a weight belt around his waist and neutral buoyancy may have a slight tendency to turn upside down. The diver cannot rely on auditory cues since the water medium distorts phase, intensity, and time-of-arrival differences between the two ears (which are the auditory cues to direction in air). Organic cues are still effective since gravity continues to exert its attraction on internal organs, such as the stomach and intestines, but these cues may be minimized by wearing a snugly fitting wet suit, scuba tanks, and weight belt.

Visual cues may assume great importance for a diver's orientation. When there is visibility he may observe aspects of his environment such as the surface of the water, the bottom, and objects on the bottom. Or he may observe phenomena such as the direction of his exhaled bubbles if he is using an open-circuit breathing system. If he suspends a tool or his light from its lanyard, he may observe, or even feel, the direction in which gravity pulls it (if it is negatively buoyant), or the direction in which it floats (if it has positive buoyancy at his particular depth). To determine if he is ascending or descending in the water column, he may rely on his depth gauge or on the sensations of increasing or decreasing pressure in his ears. But without visual cues, as in darkness or very turbid water, he must rely heavily on his vestibular sense.

There are occasions when vestibular cues may be misinterpreted, leading to spatial disorientation and dizziness. When disorientation is mild or transient the diver may be able to disregard his senses long enough to recover, particularly if he can rest on the bottom or grasp a stationary object. Complete disorientation can result in dizziness (vertigo), nausea, and panic, a life-threatening emergency for the diver.

2. Symptoms of Vestibular Problems

a. Motion Sickness

In diving, motion sickness is generally considered to be a potential problem only while aboard ship; however, there are three other instances where it may occur. First, when decompressing in the water, such as "hanging on at the 10-ft stop," vertical wave action may result in sickness. Second, while a diver is under water, lateral surging due to ground swells may cause problems. Third, if the diver surfaces far from his boat, he may be tossed about quite a bit while swimming or waiting to be picked up; in this case, cold and fatigue may act synergistically with motion to produce sickness. In addition, whether in the water or aboard ship, other factors such as anxiety or uncomfortable warmth may predispose a diver to motion sickness.

The most common cause of motion sickness (Wendt 1951) is repetition of recti-linear accelerations, acting along the vertical axis of the head. This has implicated the utricle (and possibly the saccule) rather than the semicircular canals (which respond primarily to angular acceleration). This view is supported, according to Spector (1967a), by the observation that nystagmus is generally not found with motion sickness.

Treatment of motion sickness is based on: (1) stabilization of the environment, (2) preventing passive movement of the head (reclining or resting the head on a solid object) (Spector 1967a); (3) preventing a conflict between ocular and labyrinthine data (seeing a stable environment but feeling motion, as in the cabin of a ship) (Belmont 1967), and (4) the use of Dramamine and other drugs. The use of anti-motion-sickness drugs is often a helpful preventative, but the true effect of drugs for prevention and treatment of motion sickness is debated.

b. Nystagmus

Nystagmus is a repeated involuntary rapid movement of the eyes, usually in the horizontal plane, but may also be vertical or rotary, or a combination of these. It may be spontaneous or induced by optokinetic testing, changes in position, galvanic testing, rotational tests, or caloric testing.

There are two basic types of spontaneous nystagmus. The first type is called pen-dular nystagmus and is of ocular origin. It is characterized by repeated oscillations of the eyes with the speed of movement in each direction being equal. Pendular nystag-mus is usually congenital, but some cases of multiple sclerosis and lesions in the brain stem can demonstrate this type of eye motion.

The second type of spontaneous nystagmus is called vestibular nystagmus and is characterized by repeated slow movements of the eyes in one direction, followed by a rapid return. It results from either dysfunction of the vestibular end organs or dys-function in the brain itself, most commonly, dysfunction of the vestibular–cerebellar complex in the posterior fossa.

Thus it is important to realize that nystagmus is useful as a diagnostic sign of vestibular dysfunction resulting from injury to either the end organs or the central nervous system. It will always be present shortly after vestibular injury. For further discussion of the clinical diagnosis of end organ versus central vestibular disease the reader is referred to the text of Paparella and Schumrick (1973).

c. Vertigo

The term vertigo is generally used synonymously with the term dizziness as describing an unpleasant sensation of a disturbed relationship to the environment, with the essential feature of movement. (Dizziness may not involve the feeling of movement.) This interchangeability of terms has come about because divers often do not distinguish between the two conditions. Vertigo is used to describe sensations of whirling, including objective vertigo (the sensation that the external world is whirling about the person) and subjective vertigo (the sensation that he is whirling, while the environment is stationary).

Vertigo is usually caused by a lesion of the vestibular apparatus. This may be a peripheral lesion, which involves either the semicircular canals or vestibular nerve, or a central lesion, which involves either the vestibular nuclei or tracks in the brain stem, cerebellum, or cerebrum. A peripheral vertigo usually comes on quickly, is episodic, is made worse by changes of posture, lasts only a few minutes to hours, is occasionally accompanied by tinnitus and deafness, and is always accompanied by horizontal or rotary nystagmus. Central vertigo is usually slower in onset and more continuous or prolonged. The associated nystagmus is often vertical as well as horizontal. Frequently there is evidence of other central nervous system involvement (Reisch and Baker 1973).

There is a correlation between the direction of vertigo and nystagmus—the direction of subjective vertigo is in the direction of the quick component of nystagmus, whereas the direction of objective vertigo is in the direction of the slow component of nystagmus. Also, there is a correlation between the severity of vertigo and nystagmus, since the vertigo resulting from a peripheral disturbance is roughly proportional to the intensity of nystagmus.

The severity of vertigo ranges from slight giddiness or lightheadedness to total incapacitation with inability to function at all, particularly when accompanied by repeated attacks of nausea and vomiting. The latter digestive disturbances are often found with vertigo and seem to be due to reflex actions via the vagus nerves with involvement of the emetic center (vomiting center) of the brain stem, resulting from vestibular stimulation. Stimulation of the vestibular system is frequently accompanied by other vegetative disturbances as well, such as facial pallor, perspiration (cold sweating), and decreased blood pressure. Varying amounts of relief in virtually all cases of vertigo of vestibular origin can be found by restricting movements of the head and/or by lying down.

As noted above, vertigo can be caused by either primary disorders of the vestibular system, which may be end organ or central in location, or by systemic disturbances with secondary effects on the vestibular system, such as hypotension, systemic infections, diabetes, and/or hypoglycemia, arterial insufficiency, etc.

3. *Vestibular Dysfunction in Diving*

There are three general classifications of vestibular problems associated with diving: vertigo associated with inadequate middle-ear pressure equilibration on ascent and descent, vestibular decompression sickness, and isobaric vertigo. The first has been described by Lundgren (1965) and others and is believed to result from difficulties with ear clearing during barometric pressure changes giving rise to a relative negative or positive pressure in the middle ear. The second is a type of decompression sickness resulting from gases evolved during or following decompression. The third classification is a relatively new one which resembles decompression sickness in that it is possibly due to evolved gases but occurs without a change in pressure, under circumstances which allow for the counterdiffusion of gases of different properties in different directions. All are discussed below.

a. Vertigo Associated with Inadequate Middle-Ear Clearing

The term alternobaric vertigo was first suggested by Lundgren (1965) in a report of a survey of several hundred sport divers. Of those who had experienced vertigo while either scuba diving or during breath-hold diving, most incidences had occurred when using scuba and three times as many occurred during ascent as during descent. Others, notably, Terry and Dennison (1966) and Freeman and Edmonds (1972), have described similar problems during descent. The major factor seems to be pressure changes between the middle ear and the environment and the inner ear. The exact magnitude of such pressure differences required to produce vertigo is unknown, but it is generally felt that the pressure differences produced by forceful Valsalva maneuver (clearing the ears by holding the nose and blowing forcefully) or during diving with nasal congestion and subsequent eustachian tube obstruction are great enough to cause vertigo. Indeed, in some cases, permanent inner ear damage, both vestibular and cochlear, can occur under these circumstances (Freeman and Edmonds 1972; Edmonds 1973b). In many cases, vertigo during ascent can be relieved by returning to a deeper depth and then ascending more slowly. Some divers who have experienced vertigo during descent or ascent can induce vertigo by a vigorous Valsalva maneuver on dry land. However, occasionally such vertigo occurs during a descent that is rapid and combined with overexertion. The frequent occurrence is with a novice who is in a great hurry to get to the bottom and therefore rushes down the line hand-over-hand as fast as he can. To his surprise, he suddenly experiences sensations of the environment spinning around him. If he does not panic and risk air embolism by a rapid ascent, he will grasp the line, trying to reestablish his equilibrium. Occasionally, this constitutes treatment for his condition and he soon regains spatial orientation if middle-ear pressure equalization has occurred. Some hapless divers, however, panic or immediately blame the problem on regulator failure or "bad air" and consider the surface to be their only recourse. Such vertigo may be a prominent element in many shallow-water scuba accidents and deaths when vertigo results in panic, which, in turn, leads to drowning or rapid ascent with breath-holding and then air embolism.

Treatment and prevention. The treatment of vertigo associated with inadequate ear-clearing *during descent* is to stop the descent, and if symptoms persist, to begin ascending to a shallower depth, or even the surface, where the middle-ear pressure is equal to the water pressure. However, care must be exercised to prevent inducing the converse condition—that is, vertigo from too rapid release of pressure. The primary treatment of such vertigo *during ascent* is the reverse of the above, increasing pressure for relief of symptoms. This may include going back down to the bottom in order to equalize middle-ear pressure.

Prevention of this ailment is the best course of action. It depends on maintaining pressure equalization in the middle ear and therefore may usually be accomplished by adherence to a few principles. The first is the often-quoted saying: "Don't dive when you have a cold." This may also be extended to the period immediately after a cold when it is frequently more difficult than usual to clear the ears. Extra care should be taken during repetitive diving for the same reason. Use of nasal decongestants is often beneficial, but since there is usually a rebound effect after a few hours and long-term continued use may have adverse effects, they should be used only with a physician's supervision. A diver should keep pressure equilibration in step with external pressure changes by frequent gentle Valsalva maneuvers during descent and by swallowing during ascent. It is also important to avoid rapid ascents and descents (some divers limit themselves to about 20–40 ft/min).

In some cases, caloric stimulation of cold water entering one external ear canal before the other may be the cause of vertigo during descent. To prevent this, use of a well-fitting hood is recommended when diving in cold water; this includes those situations where the surface water may not warrant the use of a hood but the dive involves crossing one or more thermoclines.

b. Vestibular Decompression Sickness

By far the majority of cases of vestibular dysfunction during decompression have occurred from relatively deep dives (300 fsw and greater), contrasting with the occurrences of vertigo from inadequate ear clearing during relatively shallow dives. Vestibular decompression sickness (v.d.s.) is occasionally referred to as "vestibular bends" or as a "vestibular hit" in the literature or in discussion. Vestibular problems included in this category are the result of decompression—specifically, inadequate decompression. Frequently, no other signs of classical decompression sickness are present. When symptoms occur during or shortly after a change of breathing gas at a decompression stop, it may be difficult to determine if it is a case of v.d.s. or due to the change of gas (isobaric vertigo). However, all cases of vertigo occurring during decompression from deep dives should be treated as vestibular decompression sickness as promptly as possible. It should be pointed out that of all central nervous system structures the inner ear seems to be the most vulnerable to damage during decompression from deep dives.

Treatment and prevention. Treatment of v.d.s. can be divided into two general categories. The first of these is operational and consists essentially of recompression plus changing gas mixtures. The second category is pharmacological, although there is no specific agent known at this time which will cure this problem. Drugs are effective only in providing symptomatic relief. The key drugs which have been noted to be effective in this regard are Valium and occasionally Compazine.

(i) Operation Procedures. Recompression is definitely the method of choice. Once the symptoms have been confirmed, it is advisable to recompress the injured diver slowly to at least one additional atmosphere of pressure. Some authorities have advocated recompressing such divers back to the depth of the dive instead of just one additional atmosphere from the onset of symptoms. Indeed, if inner-ear bubble formation is the exact cause of this problem, it is likely that on most deep dives, the bubble begins to form shortly after decompression is commenced and at a depth much greater than 1 atm deeper than the onset of symptoms.

Prompt recompression is important. A minimum of 40–60 min may be required for resolution of the problem and improvement of the patient's condition. During this period, quiet explanation of the problem and confirmation of the likelihood of improvement shortly are essential. The injured diver is very anxious to hear that he should recover soon from his disorientation. This confidence is critical.

Some cases of vestibular damage during decompression have occurred soon after the shift from a helium inert gas to a nitrogen inert gas. Shifting back to the original breathing mixture has been beneficial in some cases. Breathing oxygen is another recommended gas change and treatment success is greatly improved by increasing the partial pressure of oxygen to the maximum tolerable for the symptom depth. The best course of action would be to avoid such shifts altogether, particularly relatively sudden shifts at great depths. If the operational needs dictate switching inert gases in spite of knowledge that possible vestibular damage can occur under these conditions, it is wise to attempt approximation of the partial pressures found in the short oxygen decompression schedules of the *U. S. Navy Diving Manual* (U. S. Navy 1970, 1.6.2, p. 173). Increasing pressure by 75–100 fsw while breathing high-oxygen mixtures (2.0–2.5 atm) has been successful.

Probably a slow shift of inert gas during routine decompressions would do much to minimize the incidence of vestibular dysfunctioning. This could be accomplished by slowly filling the chamber with the new inert gas rather than having divers breathe it directly by mask.

(ii) Pharmacological Procedures. For the same reason that it is unwise to use analgesics for the treatment of "pain only" bends, it is also unwise to use drugs for the treatment of vertigo—by eliminating the symptoms one may be eliminating the index of treatment success. However, vertigo and disorientation are far more debilitating to the diver than a moderate pain in the knee. Thus, it is essential to consider the restrained use of pharmacological agents in this type of decompression accident.

Mild tranquilization with such drugs as Valium can be very useful in providing some relief to divers with vertigo. Short of inducing general anesthesia, it is almost impossible to completely eliminate the symptoms in the acute stages. Most academic otolaryngologists agree that no drugs have been demonstrated to have any beneficial effect on vestibular end organ disease except drugs that are given for their tranquilizing or sedation effects. Use of low-molecular-weight dextran and heparin in treating vestibular decompression sickness is experimental at this time. Other useful adjuncts to treatment are:

1. Having the patient lie down with feet slightly elevated.
2. Keeping the patient's head still.
3. Having the patient orient the eyes on a familiar and stationary object.
4. Ensuring that adequate fluid intake is maintained (preferably by solution with appropriate salt and sugar concentrations).
5. Minimizing positional changes due to the movement of the diver himself or the diving bell in which he is contained.

The prevention of v.d.s. lies in adequate decompressions under the various conditions of diving and divers; this is also the case for "bringing back" a diver who has v.d.s. and is being treated for it while still at pressure. As such, the topic of prevention of v.d.s. belongs under the heading of decompression sickness and the reader is directed to Chapter VII for detailed information on both prevention and mechanisms.

c. Isobaric Vertigo

Isobaric vertigo is the name given to a special phenomenon characterized by the onset of symptoms with vestibular injury (nausea, vomiting, vertigo) during diving situations not closely associated with a change in total atmospheric pressure. This has been described in laboratory experiments involving subjects breathing nitrogen while immersed in a helium–oxygen hyperbaric gaseous atmosphere (Sundmaker 1972). Also, skin itching and other skin effects resembling those of decompression sickness have been noted under such conditions (Blenkarn *et al.* 1971). Animal experiments have shown that skin lesions can be caused under these circumstances; for example, the subject breathes a slowly diffusing gas (e.g., nitrogen or argon) while exposed to an external environment containing a more rapidly diffusing gas (e.g., helium) at a pressure of several atmospheres. Model experiments have shown that bubbles can be made to form at an interface between oil and water while different gases are diffusing through the interface in different directions. The phenomenon is postulated to be related to a local zone of supersaturation, which results in the formation of bubbles (Idicula *et al. in press*).

The exact etiology of isobaric vertigo is unknown but it is being studied in a few laboratories. A significant factor is the diffusion that occurs when one gas is breathed while another surrounds the diver's head. It seems to be related to rapid changes in either chamber or breathing-mixture inert gas at relatively deep depths. Therefore, changing the inert gas in the diving chamber slowly and not gas-switching by mask are precautions recommended for necessary gas switching at a stable depth, during decompression from working dives, and for decompression research.

Treatment. When isobaric vertigo does occur, the immediate treatment is to switch back to the original gas. If administration is by mask, as it seems to have been in most cases of isobaric vertigo, this is quick and simple. If returning to the original gas does not relieve symptoms, other measures—increasing chamber pressure, increasing oxygen partial pressure—should be considered. Absolutely *no* drugs are indicated until exact pathophysiology is understood and drugs are shown to be efficacious in changing such pathophysiology.

For additional information on all aspects of vestibular function and dysfunction at sea level, we refer the reader to the book by Spector (1967*a*). It is clinically oriented and therefore contains much information on diagnosis and treatment. Wendt (1951) presents much experimental data, primarily from animals, dealing with the vestibular system. Kennedy (1974) has recently reviewed vestibular function and dysfunction in diving situations, with information concerning experiences of U. S. Navy personnel. The same author has also written an annotated bibliography on the same topic which contains almost 1000 references (Kennedy 1972). The report by Rubenstein and Summitt (1971) is also recommended to those who want more information on vestibular problems in diving, as are a report by Gehring and Bühlmann (*in press*), which deals with experiences of Swiss divers both on working dives and in very deep laboratory exposures, and a report by Vorosmarti and Bradley (1970).

The auditory function is closely related to the vestibular function and since it is also crucial to the diver it is covered both in Chapter I under the information receptors and in this chapter under hearing.

D. *Other Sensory Functions*

1. *Taste and Smell*

The senses of taste and smell (see Chapter I) are of very little importance to the diver but do have significance for both the submariner and the aquanaut, for the evidence indicates a slight decrease in the threshold for both taste and smell when submerged for extended periods in either submarine or habitats.

2. *Kinesthetic, Proprioceptive, and Organic Effects*

As discussed in Chapter I, the kinesthetic, proprioceptive, and organic senses are major components of the system that makes the diver aware of his body and its position and movements in the environment. These senses are also involved in equilibrium and balance. However, in the water, and particularly under conditions of neutral buoyancy, cues of position received from these senses are markedly reduced or absent. Organic cues are still effective since gravity continues to exert its attraction on internal organs, such as the stomach and intestines, but these cues may be minimized by the wearing of a snugly fitting wet suit, scuba tanks, and weight belt.

3. *Touch*

Manual performance is a complex function of which the tactile or cutaneous sense is but one element. Experience and research have revealed that divers may experience two kinds of impairment to their tactile function, one due to being in the water and in diving dress, the other due to the cold. The water effect seems to be a hindrance, mainly, to motor activity due to some combination of instability, incumbrance of equipment, neutral (approximately) buoyancy, viscous resistance, resistance to limb flexion, and reduced sensory input and, thus, impaired sensory functions. The cold effect has both local (peripheral) and general effects. It causes more or less direct losses in sensory and motor function and it distracts from and disrupts cognitive activity, especially when extended in time so that sustained attention and memory are involved (Bowen and Pepler 1967, Bowen 1967).

The work of McKee (1972) and of Bowen (1967) involved many diversified tests under several different conditions. As a result of these Bowen recommends that:

> . . . underwater equipment should not depend critically on the application of considerable force, on tactile sensation, on speed of operation, or on considerable displacement motions. Tasks requiring mental processes should not be made dependent on prolonged undiverted attention or on the accuracy of short term memory. Alternatively the individual must be more protected and insulated from the impingement of the environment so that his normal, dry land, levels of function are preserved . . . To make man really effective under the sea, we need the thematic development of man/suit, man/tool, and man/machine combinations which will allow the full range of human capability to be deployed in this challenging, though often discomforting, environment.

For additional information see Chapter I, discussion of information receptors, and Chapter VI, discussion of tactile sensitivity.

Terminology, Physical Constants, and Definition of Terms

Terminology and Physical Constants

B	luminance: luminous intensity of an object per unit area of surface
B_t	target luminance
B_b	background luminance
C_B	contrast $= (B_t - B_b)/B_b$
nm	nanometer: unit of length, 10^{-9}m
I	luminous intensity measured in candles (cd)
cd	candela or candle: 1 cd = 1 lm per unit solid angle (steradian)
cd/m²	the internationally accepted unit of measurement: 1 cd/m² = 0.314 mL or 0.292 ft-L
lm	lumen: the unit of luminous flux; one standard candle emits 4π lm; also 0.00146 W at 550 nm (1 nm = 10^{-9}m)
E	illumination: density of the luminous flux deposited on a surface
L	lambert: basic unit of luminance: 1 L = 1 lm/cm² or $1/(\pi\ cd \cdot cm^2)$
mL	millilambert, 10^{-3}L
μL	microlambert, 10^{-6}L
ft-L	foot-lambert: 1 ft-L = 1.076 mL
ft-cd	foot-candle = lm/ft²
f	focal length: distance at which focus forms when incident light is parallel
D	diopters: 1/focal length (meters)
$1/f = 1/a + 1/b$	where a is the distance of a point source of light from a lens (object distance) and b is the distance of the focus from the lens (image distance); lens formula
c	speed of light: c = 300,000 km/sec = 186,000 miles/sec in a vacuum

Definitions

Accommodation. Focus of the lens for near objects.

Ametropia. Imperfection in the refractive powers of the eye, so that images are not brought to a proper focus on the retina, producing hypermetropia, myopia, or astigmatism.

Astigmatism. Improper focus of light on retina due to different requirements for correction in different meridians of the eye; one cause is abnormal curvature of cornea; corrected by cylindrical lens.

Binocular vision. Vision using both eyes.

Color vision. The ability to see different hues as red, green, blue, etc.

Concave lens. A lens that diverges light; it corrects myopia.

Convergence. Fixation of both eyes on an object such that the centers of both visual fields coincide, resulting in binocular vision.

Convex lens. A lens that converges light.

Convex cylindrical lens. A lens for which all rays converge to a focal line.

Convex spherical lens. A lens of which all rays converge to a focal point.

Dark adaptation. Process by which visual receptors gradually become more sensitive to light; rate depends on previous degree of exposure to light; cones adapt first; at very low light intensities only the rods function (only black and white vision, with no detailed configurations).

Emmetropia. Normal vision.

Esophoria. Deviation of the eyes toward each other.

Esotropia. Inward deviation of an eye when both eyes are open and uncovered; convergent strabismus.

Exophoria. Deviation of the eyes away from each other.

Hyperopia. Far-sightedness (hypermetropia); light rays focus in back of the retina; corrected by convex lens.

Monocular vision. Vision using one eye.

Myopia. Near-sightedness; light rays focus in front of the retina; corrected by concave lens.

Parallax. The apparent displacement of an object as seen from two different points not on a straight line with the object.

Phoria. Any tendency of deviation of the eyes from the normal.

Photopic vision. Visual function performed by the cones; color vision; fine detail (high acuity); does not function below 10^{-2}mL.

Presbyopia. Inability to focus light from near objects due to loss of elasticity of lens with advancing age.

Refraction. Bending of light rays as they pass from one medium to another in which they have a different speed.

Refractive index. Ratio between the sine of the angle of incidence and the sine of the angle of refraction; can be represented as the ratio of the velocity of light in air to the velocity of light in the medium (refractive index in a vacuum = 1.00).

Resolving power. The ability of an optical system to form distinguishable images of objects separated by small angular distances.

Scotopic vision. Visual functioning at low or night-time levels of illumination; performed by the rods; characterized by poor acuity; no color vision but extreme sensitivity to light.

Strabismus. Deviation of the eye which the patient cannot overcome; the visual axes assume a position relative to each other different from that required by the physiological conditions.

Threshold. The minimum amount of background luminance, size of an object, contrast, or duration of the visual stimulus required to effect a visual reponse.

Visible light. Wavelengths between 400 nm (violet) and 700 nm (red).

Visual acuity. The ability of the eye to see or resolve fine details; expressed in distance for normal eyesight as 20/20; defined as the reciprocal of the visual angle (in minutes) subtended by the detail; minimal visual angle for normal eyesight is approximately 1 min.

Visual angle. A measure of the size of a visual object at the eye; it is represented by the tangent of the size of the object divided by the viewing distance.

Visual field. That portion of the external environment, measured in degrees, which is represented on the retina; a blind spot in the field appears where the optic nerve joins the retina, but is unnoticeable in normal vision.

References

ACKLES, K. N., and B. FOWLER. Cortical evoked response and inert gas narcosis in man. *Aerosp. Med.* **42**:1181–1185 (Nov. 1971).

ADOLFSON, J. Human performance and behavior in hyperbaric environments. In: *Acta Psychologica Gothoburgensia VI.* Stockholm, Almqvist and Wiksell (1967).

ADOLFSON, J., and T. BERGHAGE. *Perception and Performance Underwater.* New York, Wiley, (1974)·

ADOLFSON, J., and E. FLUUR. Horselforandringar i hyperbar miljo. *Forsvarsmedicin* **1**:167–171 (Oct. 1965).

ADOLFSON, J., and E. FLUUR. Hearing discrimination in hyperbaric air. *Aerosp. Med.* **38**:174–175 (Feb. 1967).

ALFANDRE, H. J. Aerotitis media in submarine recruits. U. S. Nav. Submar. Med. Cent., Rep. SMRL 450 (1965).

ANDERSON, S., and H. T. CHRISTENSEN. Underwater sound localization in man. *J. Aud. Res.* **9**:358–364 (Oct. 1969).

ANZAI, S. Ecological study on women divers. *Mie Med. J.* **4**:1181–1191 (1960). (In Japanese; see Harashima and Iwasaki 1965.)

BARTUS, R. T. Nitrogen narcosis and visual evoked responses in the unanesthetized cat. U. S. Nav. Submar. Med. Cent., Rep. NSMRL 757 (Sept. 1973).

BAYLISS, G. J. A. Aural barotrauma in naval divers. *Arch. Otolaryngol.* **88**:141–147 (Aug. 1968).

BEKESY, G. v. Bone conduction. In: Bekesy, G. v. *Experiments in Hearing*, pp. 127–203. E. G. Wever, transl. and ed. New York, McGraw-Hill (1960).

BELMONT, O. Ocular causes of vertigo. In: Spector, M. ed. *Dizziness and Vertigo: Diagnosis and Treatment.* New York, Grune and Stratton (1967).

BENNETT, P. B., and A. N. DOSSETT. Mechanism and prevention of inert gas narcosis and anaesthesia. *Nature* **228**:1317–1318 (Dec. 26, 1970).

BENNETT, P. B., K. N. ACKLES, and V. J. CRIPPS. Effects of hyperbaric nitrogen and oxygen on auditory evoked responses in man. *Aerosp. Med.* **40**:521–525 (May 1969).

BEVAN, J. The human auditory evoked response and contingent negative variation in hyperbaric air. *Electroencephalogr. Clin. Neurophysiol.* **30**:198–204 (Mar. 1971).

BLACKWELL, H. R. Contrast thresholds of the human eye. *J. Opt. Soc. Amer.* **36**:624–643 (Nov. 1946).

BLENKARN, G. D., C. AQUADRO, B. A. HILLS, and H. A. SALTZMAN. Urticaria following the sequential breathing of various inert gases at a constant pressure of 7 ata: A possible manifestation of gas-induced osmosis. *Aerosp. Med.* **42**:141–146 (Feb. 1971).

BOWEN, H. M. Diver performance and the effects of cold. Darien, Ct. Dunlap and Associates, Inc. Report BSD #67–441 (1967).

BOWEN, H. M., and R. D. PEPLER. Studies of the performance capabilities of divers; the effects of cold. Darien, Ct. Dunlap and Associates, Inc. Report SSD #67–399 (1967).

BRANDT, J. F., and H. HOLLIEN. Underwater hearing thresholds in man. *J. Acoust. Soc. Amer.* **42**:966–971 (Nov. 1967).

BRANDT, J. F., and H. HOLLIEN. Underwater speech reception thresholds and discrimination. *J. Aud. Res.* **8**:71–80 (Jan. 1968).

BRANDT, J. F., and H. HOLLIEN. Underwater hearing thresholds in man as a function of water depth. *J. Acoust. Soc. Amer.* **46**(4, Pt. 2):893–894 (Oct. 1969).

COLES, R. R. A. Labyrinthine disorders in divers: some experiences and activities with the Royal Navy (see McCormick 1973).

COLES, R. R. A., and J. J. KNIGHT. Report on an aural and audiometric survey of qualified divers and submarine escape training tank instructors. Med. Res. Counc., Roy. Nav. Pers. Res. Comm., Rep. He.S. 29 (1960).

DUNTLEY, S. Q. The visibility of submerged objects. Cambridge, Mass., Mass. Inst. Technol., Visibility Lab. (Aug. 31, 1952).

DUNTLEY, S. Q. Improved nomographs for calculating visibility by swimmers (natural light). U. S. Navy BuShips Contract NObs–72–39, Task 5, Rep. No. 5-3 (Feb. 1960).

DUNTLEY, S. Q. Light in the sea. J. Opt. Soc. Amer. 53:214–233 (Feb. 1963).

EDMONDS, C. Personal communication (1973a).

EDMONDS, C. Vestibular and auditory problems during compression (1973b) (see McCormick 1973).

EDMONDS, C., and F. A. BLACKWOOD. Investigation of otological disorders in diving. Balmoral, Aust., Roy. Aust. Navy., Sch. Underwater Med., Rep. 2/71 (1971).

EDMONDS, C., P. FREEMAN, J. TONKIN, R. THOMAS, and F. A. BLACKWOOD. Otological Aspects of Diving. Glebe, Australia, New South Wales, Australasian Medical Publishing Co. (1973).

EPSTEIN, W. Varieties of Perceptual Learning. New York, McGraw-Hill (1967).

EVERLEY, I. A., and W. KENNETT. Field test of dark adaptation of divers. U. S. Nav. Submar. Med. Cent., Rep. NSMRL 106 (July 1946).

FARMER, J. C. Vestibular and auditory problems during decompression (see McCormick 1973).

FARMER, J. C., W. G. THOMAS, and M. PRESLAR. Human auditory responses during hyperbaric helium–oxygen exposures. Surg. Forum 22:456–458 (1971).

FERRIS, S. H. Magnitude estimation of absolute distance underwater. Percept. Mot. Skills 35:963–971 (Dec. 1972).

FIELDS, J. A. Skin diving, the physiology and otolaryngological aspect. Arch. Otolaryngol. 68:531–541 (1958).

FREEMAN, P., and C. EDMONDS. Inner ear barotrauma. Arch. Otolaryngol. 95:556–563 (June 1972).

FRUCTUS, X., and G. C. RICCI. Reflexions sur deux cas de maladie de la decompression. Bull. Medsubhyp 1:10–12 (Dec. 1969).

GALLOWAY, T. C. Discussion. Trans. Amer. Otol. Soc. 30:75 (1940).

GEHRING, H., and A. A. BÜHLMANN. So-called vertigo bends after oxy-helium dives. In: Proceedings of the Fifth Symposium on Underwater Physiology, Freeport, Bahamas, August, 1972. In press.

GRAHAM, C. H. Vision and Visual Perception. New York, Wiley (1965).

GUYTON, A. C. Textbook of Medical Physiology, Third edition, pp. 785–801. Philadelphia, W. B. Saunders Company (1966).

HAIG, C. The course of rod dark adaptation as influenced by intensity and duration of preadapting to light. J. Gen. Physiol. 24:735–751 (July 20, 1941).

HAMILTON, P. M. Underwater hearing thresholds. J. Acoust. Soc. Amer. 29:792–794 (July 1957).

HAMILTON, R. W., Jr. and T. D. LANGLEY. Comparative physiological properties of nitrogen, helium and neon: A preliminary report. Presented at the annual symposium of the Undersea Medical Society, Houston, Texas (29 April, 1971).

HARASHIMA, S., and S. IWASAKI. Occupational disease of the Ama. In: Rahn, H., and T. Yokoyama, eds. Physiology of Breath-Hold Diving and the Ama of Japan, pp. 85–98. Washington, D. C., National Academy of Sciences–National Research Council (1965).

HARRIS, J. D. Hearing loss in decompression. In: Lambertsen, C. J., ed. Underwater Physiology. Proceedings of the Fourth Symposium on Underwater Physiology, pp. 271–286. New York, Academic Press (1971).

HELD, R. Plasticity in sensory-motor systems. Sci. Amer. 213:84–94 (Nov. 1965).

HENNESSY, R. T., and H. W. LEIBOWITZ. The effect of a peripheral stimulus on accommodation. Percept. Psychophys. 10:129–132 (Sept. 1971).

HITSCHLER, W. J. The relationship of swimming and diving to sinusitis and hearing loss. Laryngoscope 59:799–819 (July 1949).

HOLLIEN, H. Underwater sound localization: preliminary information (Abstract). J. Acoust. Soc. Amer. 46(1, Pt. 1):124–125 (Jan. 1969).

HOLLIEN, H., and S. FEINSTEIN. The contribution of the external auditory meatus to human under-water auditory sensitivity. Gainesville, Fla., Univ. Fla., Commun. Sci. Lab., Rep. CSL/ONR 40 (July 1972).

HOLLIEN, H., J. BRANDT, and E. DOHERTY. The effects of air bubbles in the external auditory meatus on underwater hearing thresholds. Gainesville, Fla., Univ. Fla., Commun. Sci. Lab., Rep. CSL/ONR 15 (March 1969a); also, *J. Acoust. Soc. Amer.* **46**:384–387 (Aug. 1969a).

HOLLIEN, H., J. BRANDT, and E. DOHERTY. Underwater hearing thresholds in man as a function of water depth. Gainesville, Fla., Univ. Fla., Commun. Sci. Lab., Rep. CSL/ONR 16 (August 1969b).

HOLLIEN, H., J. L. LAUER, and P. PAUL. Additional data on underwater sound localization (Abstract). *J. Acoust. Soc. Amer.* **47** (1, Pt. 1):127–128 (Jan. 1970).

HONG, S. K., and H. RAHN. The diving women of Korea and Japan. *Sci. Amer.* **216**:34–43 (May 1967).

IDICULA, J., D. J. GRAVES, J. A. QUINN, and C. J. LAMBERTSEN. Bubble formation resulting from steady counterdiffusion of two inert gases. In: *Proceedings of the Fifth Symposium on Under-water Physiology, Freeport, Bahamas, August 1972. In press.*

JENNINGS, R. D. A behavioral approach to nitrogen narcosis. *Psychol. Bull.* **69**:216–224 (Mar. 1968).

KATAOKA, Y. A report of the oto-rhino-laryngological examination of Ama, Japanese diving fishermen and women. *Shikoku J. Med.* **5**:9–13 (1954). (In Japanese; see Harashima and Iwasaki 1965).

KELLEY, J. S., and G. ANDERSON. Visual function and ocular drug effects during a saturation exposure to 31 atmospheres pressure. *Aerosp. Med.* **40**:1296–1299 (Dec. 1969).

KELLEY, J. S., P. G. BURCH, M. E. BRADLEY, and D. E. CAMPBELL. Visual function in divers at 15 to 26 atmospheres pressure. *Milit. Med.* **133**:827–829 (Oct. 1968).

KENNEDY, R. S. A bibliography of the role of the vestibular apparatus under water and pressure: content-oriented and annotated. Bethesda, Md., U. S. Nav. Med. Res. Inst. (1972).

KENNEDY, R. S. General history of vestibular disorders in diving. *Undersea Biomed. Res.* **1**:73–81 (Mar. 1974).

KENT, P. R. and S. WEISSMAN. Visual resolution underwater. U. S. Nav. Submar. Med. Cent., Rep. NSMRL 476 (May 1966).

KINNEY, J. A. S. Visibility of colors underwater. In: *Marine Technology 1970. Preprints*, Vol. 1, pp. 627–636. Washington, D. C., Marine Technology Society (1970).

KINNEY, J. A. S., and S. M. LURIA. Conflicting visual and tactual-kinesthetic stimulation. *Percept. Psychophys.* **8**:189–192 (1970).

KINNEY, J. A. S., and C. L. McKAY. The visual evoked response as a measure of nitrogen narcosis in Navy divers. U. S. Nav. Submar. Med. Cent., Rep NSMRL 664 (April 1971).

KINNEY, J. A. S., S. M. LURIA, and D. O. WEITZMAN. Visibility of colors underwater. *J. Opt. Soc. Amer.* **57**:802–809 (June 1967).

KINNEY, J. A. S., S. M. LURIA, and D. O. WEITZMAN. Responses to the underwater distortions of visual stimuli. U. S. Nav. Submar. Med. Cent., Rep. NSMRL 541 (July 1968a).

KINNEY, J. A. S., S. M. LURIA, and D. O. WEITZMAN. Analyses of a variety of visual problems encountered during naval operations at night. U. S. Nav. Submar. Med. Cent., Rep. NSMRL 545 (Aug. 1968b).

KINNEY, J. A. S., S. M. LURIA, and D. O. WEITZMAN. Effect of turbidity on judgments of distance underwater. *Percept. & Mot. Skills* **28**:331–333 (1969a).

KINNEY, J. A. S., S. M. LURIA, and D. O. WEITZMAN. Visibility of colors underwater using artificial illumination. *J. Opt. Soc. Amer.* **59**:624–628 (1969b).

KINNEY, J. A. S., S. M. LURIA, D. O. WEITZMAN, and H. MARKOWITZ. Effects of diving experience on perception under water. U. S. Nav. Submar. Med. Cent., Rep. NSMRL 612 (Feb. 1970a).

KINNEY, J. A. S., C. L. McKAY, S. M. LURIA, and C. L. GRATTO. The improvement of divers' compensation for underwater distortion. U. S. Nav. Submar. Med. Cent. Rep NSMRL 633 (June 1970b).

KINNEY, J. A. S., C. L. McKAY, and S. M. LURIA. Visual evoked responses for divers breathing various gases at depths to 1200 ft. U. S. Nav. Submar. Med. Cent., Rep. NSMRL 705 (March 1972).

KINNEY, J. A. S., S. M. LURIA, and M. S. STRAUSS. Measures of evoked responses and EEGs during shallow saturation diving. U. S. Nav. Submar. Med. Cent., Rep. NSMRL 761 (Sept. 1973).

KLING, J. W., and L. A. RIGGS. *Experimental Psychology*, Vol. 1, *Sensation and Perception*, third edition, pp. 482–494. New York, Holt, Rinehart and Winston (1972).

LANG, J., I. ROZSAHEGYI, and T. TARNOCZY. Kochleovestibulare Befunde bei Caissonbeitern. *Monatsschr. Ohrenheilkd. Laryngorhinol.* 105(1):9–12 (1971).

LARSON, C. R., D. SUTTON, E. M. TAYLOR, and J. D. BURNS. Visual evoked potential changes in hyperbaric atmospheres. Arizona State University, Tempe, Arizona. ONR Contract No. N00014–68–A–150 Work unit No. NR196–077, Rep. 71–02 (Oct. 1971).

LEGGIERE, T., J. MCANIFF, H. SCHENCK, and J. VAN RYZIN. Sound localization and homing of scuba divers. *Mar. Technol. Soc. J.* 4:27–34 (Mar./Apr. 1970).

LOCH, W. E. The effect on hearing of experimental occlusion of the eustachian tube in man. *Ann. Otol. Rhinol. Laryngol.* 51:396–405 (June 1942).

LUNDGREN, C. E. G. Alternobaric vertigo—a diving hazard. *Brit. Med. J.* 2:511–513 (Aug. 28, 1965).

LURIA, S. M. Stereoscopic and resolution acuity with various fields of view. *Science* 164:452–453 (Apr. 25, 1969).

LURIA, S. M. Effect of limited peripheral cues on stereoacuity. *Psychon. Sci.* 24:195–196 (Aug. 25, 1971).

LURIA, S. M., and J. A. S. KINNEY. Stereoscopic acuity underwater. *Amer. J. Psychol.* 81:359–366 (Sept. 1968).

LURIA, S. M., and J. A. S. KINNEY. Peripheral stimuli and stereoacuity under water. *Percept. Psychophys.* 11:437–440 (June 1972).

LURIA, S. M., and J. A. S. KINNEY. Accommodation and stereoacuity. *Percept. Psychophys.* 13:76–80 (1973).

LURIA, S. M., J. A. S. KINNEY, and S. WEISSMAN. Estimates of size and distance underwater. *Amer. J. Psychol.* 80:282–286 (June 1967).

LURIA, S. M., H. NEWMARK, and H. BEATTY. Effect of a submarine patrol on visual processes. U. S. Nav. Submar. Med. Cent., Rep. NSMRL 641 (1970).

LYTHGOE, R. J. The measurement of visual acuity. Med. Res. Coun. Spec. Rep. Ser. 173. His Majesty's Stationery Office, London (1932).

MAYO CLINIC. *Clinical Examinations in Neurology*, third edition. Philadelphia, W. B. Saunders (1971).

MCCORMICK, J. G. Investigations of vestibular interactions in the high pressure neurological syndrome. Presented at the Undersea Medical Society workshop on labyrinthine dysfunction during diving, Durham, N. C. (1–2 Feb. 1973).

MCKAY, C. L., J. A. S. KINNEY, and S. M. LURIA. Further tests of training techniques to improve visual-motor coordination of Navy divers under water. U. S. Nav. Submar. Med. Cent., Rep. NSMRL 684 (Oct. 1971).

MCKEE, D. L. A study of underwater diver tactile sensitivity. U. S. Nav. Postgrad. Sch., Thesis (Mar. 1972).

MILES, S. *Underwater Medicine*, third edition. Philadelphia, J. B. Lippincott (1969).

MONTAGUE, W. E., and J. F. STRICKLAND. Sensitivity of the water-immersed ear to high- and low-level tones. *J. Acoust. Soc. Amer.* 33:1376–1381 (Oct. 1961).

MOSSE, P. Underwater contact lenses. *Brit. J. Physiol. Opt.* 21:250–255 (Oct./Dec. 1964).

MURRY, T., and C. STRAND. Intersubmarine speech intelligibility levels. U. S. Nav. Submar. Med. Cent., Rep. SMRL 658 (Mar. 1971).

MYGIND, S. H. Die nichtsuppurativen Schalleitungsleiden. *Acta Oto- Laryngol.* 16:333–361 (1931).

NICHOLS, C. W., and C. J. LAMBERTSEN. Effects of high oxygen pressures on the eye. *New Eng. J. Med.* 281(1):25–30 (July 3, 1969).

NICHOLS, C. W., C. J. LAMBERTSEN, and J. M. CLARK. Transient unilateral loss of vision associated with oxygen at high pressure. *Arch. Ophthal.* 81:548–552 (Apr. 1969).

NORMAN, D. A., R. PHELPS, and F. WRIGHTMAN. Some observations on underwater hearing. *J. Acoust. Soc. Amer.* **50**(2, Pt. 2):544–548 (Aug. 1971).

NOURRIT, P. Etude clinique du barotraumatisme de l'oreille interne. *Marseille Med.* **98**(8):823–828 (1961).

ONO, H., and J. P. O'REILLY. Adaptation to underwater distance distortion as a function of different sensory-motor tasks. *Hum. Factors* **12**:133–139 (Apr. 1971).

PAGANO, A. *Otopatie e sinusopatie da barotrauma nei lavorale dei cassoni.* Naples, Casa Editrice V. Idelson di e Gnocchi (1959).

PAPARELLA, M., and D. A. SCHUMRICK, eds. *Otolaryngology*, Vol. I. Philadelphia, W. B. Saunders (1973).

PESKIN, J. C., and J. M. BJORNSTAD. The effect of different wavelengths of light on visual sensitivity. Wright-Patterson, AFB, USAF Air Mater. Command, Rep. MCREXD 694–93A (1948).

PLANTE–LONGCHAMP, G., and BLANCHI. Deux nouveaux cas de lesion labyrinthique apres plongees et leur traitement. *Bull. Medsubhyp* **7**:20–21 (Oct. 1972).

PLANTE–LONGCHAMPS, G., P. MAESTRACCI, and H. NICOLAI–HARTER. Deux cas de destruction labyrinthique apres plongees. *Bull. Medsubhyp* **4**:12–15 (Dec. 1970).

POLI, C. Comment in: XII Kongress der italienischen otorhinolaryngologischen Gesellschaft [brief digest]. *Monatschr. Ohrenheilkd. Laryngorhinol.* **43**:312–314 (1909).

PRAZIC, M., and B. SALAJ. Audiologiczny aspekt dyscypliny sportowej nurkowania. *Autolaryngol. Pol.* **25**:605–607 (1971).

PULLEN, F. W. Round window membrane rupture. A cause of sudden deafness. *Trans. Amer. Acad. Ophthal. Otolaryngol.* **76**:1444–1450 (Dec. 1972).

REISCH, J. A., and A. D. BAKER. Basic sciences and related disciplines. In: Paparella, M., and D. A. Schumrick, eds. *Otolaryngology*, Vol. 1, p. 809. Philadelphia, W. B Saunders (1973)

REUTER, S. H. The problems and techniques of underwater photography. *J. Biol. Photogr. Assoc.* **39**(3):145–156 (1971).

REYSENBACH DE HAAN, F. W. Hearing in whales. *Acta Oto-Laryngol. Suppl.* **134**:114 (1956).

ROSS, H. E. The size-constancy of underwater swimmers. *Quart. J. Exp. Psychol.* **17**:329–337 (Nov. 1965).

ROSS, H. E. Stereoscopic acuity underwater. In: Lythgoe, J. N., and J. D. Woods, eds. *Underwater Association Reports*, pp. 61–64; 1966–67. London, T. G. Williams Industrial and Research Promotions (1967).

ROSS, H. E. Adaptation of divers to curvature distortion under water. *Ergonomics* **13**:489–499 (July 1970).

ROSS, H. E., S. S. FRANKLIN, and G. WELTMAN. Adaptation of divers to distortion of size and distance underwater. Los Angeles, Univ. Cal. Los Angeles, Dep. Eng., Biotechnol. Lab. Rep. 45 (Jan. 1969).

RUBENSTEIN, C. J., and J. K. SUMMITT. Vestibular derangement in decompression. In: Lambertsen, C. J. ed. *Underwater Physiology. Proceedings of the Fourth Symposium on Underwater Physiology* pp. 287–292. New York, Academic Press (1971).

SCHOBER, H. A. W., H. DEHLER, and R. KASSEL. Accommodation during observations with optical instruments. *J. Opt. Soc. Amer.* **60**:103–107 (Jan. 1970).

SHILLING, C. W., H. L. HAINES, J. D. HARRIS, and W. J. KELLY. The prevention and treatment of aerotitis media. *U. S. Nav. Med. Bull.* **45**:1529–1558 (Oct. 1946).

SIVIAN, L. J. Exchange of acoustic pressures and intensities in the air–water system. National Defense Research Council memorandum (January 1943*a*). Available as PB–31033, Office of Technical Services, U. S. Department of Commerce, Washington, D. C. [Summarized in: Sivian, L. J. On hearing in water vs. hearing in air, *J. Acoust. Soc. Am.* **19**:461–463 (1947).]

SIVIAN, L. J. On hearing in water vs. hearing in air with some experimental evidence. National Defense Research Council memorandum (March 1943*b*). Available as PB–31034, Office of Technical Services, U. S. Department of Commerce, Washington, D. C. [Summarized in: Sivian, L. J. On hearing in water vs. hearing in air, *J. Acoust. Soc. Am.* **19**:461–463 (1947).]

SMITH, P. F. Bone conduction, air conduction and underwater hearing. U. S. Nav. Submar. Med. Cent., Mem. Rep. 65–12 (1965).

SMITH, P. F. Underwater hearing in man. I. Sensitivity. U. S. Nav. Submar. Med. Cent., Rep. NSMRL 569 (1969).

SPECTOR, M. *Dizziness and Vertigo: Diagnosis and Treatment.* New York, Grune and Stratton (1967a).

SPECTOR, M. History. In: Spector, M., ed. *Dizziness and Vertigo: Diagnosis and Treatment,* Chapter 3. New York, Grune and Stratton (1967b).

STUCKER, F. J., and W. B. ECHOLS. Otolaryngic problems of underwater exploration. *Milit. Med.* 136:896–899 (Dec. 1971).

SUMMITT, J. K., and S. D. REIMERS. Noise. A hazard to divers and hyperbaric chamber personnel. U. S. Navy Exp. Diving Unit, Washington, D. C., Res. Rep. 5–71 (1971); also *Aerosp. Med.* 42:1173–1177 (Nov. 1971).

SUNDMAKER, W. K. H. Vestibular function. Presented at the Special Summary Program—Predictive Studies III, University of Pennsylvania, State College, Pa. (April 5–7, 1972).

TAYLOR, G. D. The otolaryngologic aspects of skin and scuba diving. *Laryngoscope* 69:809–859 (July 1959).

TEED, R. W. Factors producing obstruction of the auditory tube in submarine personnel. U. S. Nav. Med. Bull. 44:293–306 (Feb. 1944).

TEGHTSOONIAN, R., and M. TEGHTSOONIAN. Scaling apparent distance in a natural outdoor setting. *Psychon. Sci.* 21:215–216 (Nov. 25, 1970).

TERRY, L., and W. L. DENNISON. Vertigo among divers. U. S. Nav. Submar. Med. Cent. Spec. Rep. 66–2 (Apr. 8, 1966).

THOMAS, W. G., M. J. PRESLAR, and J. C. FARMER. Calibration of condenser microphones under increased atmospheric pressures. *J. Acoust. Soc. Amer.* 51(1, Pt. 1):6–14 (Jan. 1972).

THOMAS, W. G., J. SUMMITT, and J. C. FARMER. Human auditory thresholds during deep saturation helium–oxygen dives. In: Thomas, W. G., and J. C. Farmer. *Psychoacoustic and Electrophysiologic Studies of Hearing under Hyperbaric Pressure,* pp. 46–55. Chapel Hill, N. C., Univ. N. C. Med. Sch., Auditory Res. Lab. (June 1973).

THOMPSON, E., H. A. HOWE, and W. HUGHSON. Middle ear pressure and auditory acuity. *Amer. J. Physiol.* 110:312–319 (Dec. 1, 1934).

U. S. NAVY. *U. S. Navy Diving Manual.* Washington, D. C., Department of the Navy (1970) (NAVSHIPS 0994–001–9010).

VAIL, H. H. Traumatic conditions of the ear in workers in an atmosphere of compressed air. *Arch. Otolaryngol.* 10:113–126 (1929).

VOROSMARTI, J. and M. E. BRADLEY. Alternobaric vertigo in military divers. *Milit. Med.* 135:182–185 (Mar. 1970).

WAINWRIGHT, W. N. Comparison of hearing thresholds in air and in water. *J. Acoust. Soc. Amer.* 30:1025–1039 (Nov. 1958).

WEITZMAN, D. O., J. A. S. KINNEY, and A. P. RYAN. A longitudinal study of acuity and phoria among submariners. U. S. Navy Submar. Med. Cent., Rep. NSMRL 481 (Sept. 1966).

WEITZMAN, D. O., J. A. S. KINNEY, and S. M. LURIA. Effect on vision of repeated exposure to carbon dioxide. U. S. Nav. Submar. Med. Cent., Rep. NSMRL 566 (Feb. 1969).

WELTMAN, G., R. A. CHRISTIANSEN, and G. H. EGSTROM. Visual fields of the scuba diver. *Hum. Factors* 7:423–430 (Oct. 1965).

WENDT, G. R. Vestibular function. In: Stevens, S. S., ed. *Handbook of Experimental Psychology,* Chapter 31, pp. 1191–1224. New York, Wiley (1951).

WEVER, E. G., M. LAWRENCE, and K. R. SMITH. Effects of negative air pressure in the middle ear. *Ann. Otol. Rhinol. Laryngol.* 57:418–428 (June 1948).

WILKE, V. M. and M. STEFFL. Horstorung bei touchsportlern. *Dtsch. Gesundeitsw.* 20:1149–1156 (June 24, 1965).

WILLIAMSON, D. E. Correction of ametropia in skin and scuba divers. *Eye Ear Nose Throat Mon.* 49:165–171 (Apr. 1970).

WOODSON, W. E. Human engineering guide for equipment designers. U. S. Navy Electron. Lab., Hum. Factors Division (1951).

YAMAMOTO, K. A report of physical examination of Ama. *Nagoya Med. Soc. J.* 43:169–177 (1936). (In Japanese; see Harashima and Iwasaki 1965).

ZANNINI, D., G. ODALGLIA, and G. SPERATI. Audiographic changes in professional divers. In: *Proceedings of the Fifth Symposium on Underwater Physiology, Freeport, Bahamas, August 1972. In press.*

ZWISLOCKI, J. Ear protectors. In: Harris, C., ed. *Handbook of Noise Control*, pp. 8-1-8-27. New York, McGraw-Hill (1957).

VI

Man in the Ocean Environment: Performance

A. *Introduction*

The purpose of this chapter is to present information about man's performance in an ocean environment, particularly as it contrasts with performance in a normal, dry-land, 1-atm, air environment. The kind of information selected for presentation is that based on experimentation as opposed to that from less formal sources. The strength of this choice is that information obtained by the scientific method can be assessed for confidence, reliability, and generality; the weakness is the piecemeal character of the information. Research results typically are not about holistic job performance by operational divers working in the ocean, but about a dimension of performance as affected by a characteristic of the ocean environment. Most often, several performance dimensions are measured as effects of variations in a single environmental characteristic. Researchers tend to work by varying an environmental characteristic, such as cold or atmospheric pressure, and recording a range of performance dimensions, such as manual dexterity, reaction time, and memory. The principal challenge of preparing this chapter, therefore, was to develop a framework for organizing the fragments of performance-oriented research information relevant to man in the sea.

The organizing framework is presented as Table VI-1. The rows of the table contain 14 performance dimensions ordered more or less hierarchically according to increasing levels of complexity. Selection of these categories for describing performance and the designation of specific tests as representative performance in a given category are based on factor analytic studies by Fleishman and by Guilford and others of psychomotor, perceptual, and cognitive abilities. Selection of the six categories used to define the main characteristics of the ocean diving environment is based on the stress conditions most frequently studied by performance-oriented researchers in the field. The classifications of both the performance dimensions and the characteristics of the diving environment evolved over several iterations; and, while the current sets are not incontestable, they are not unreasonable and they have been useful.

Table VI-1

Organization of Research Based on Diver Performance Dimensions and Environmental Stress Characteristics

Performance dimensions and tests	Stress characteristics of the diving environment					
	Hyperbaric air	Hyperbaric heliox	Tractionless/ viscous	Cold	Isolation/ confinement	Generalized stress
Tactile sensitivity Vibration Two-edge Pressure Size and texture	—	Bennett and Towse 1971	—	Weitz 1941 Mackworth 1953 Mills 1956, 1957 Morton and Provins 1960 Bowen 1968 Stang and Weiner 1970	—	—
Force production Grip strength Pushing and pulling Lifting Torqueing Swimming Load carrying	—	—	Christianson et al. 1965 Andersen 1968, 1969 Barrett and Quirk 1969 Norman 1969 Streimer et al. 1969a, 1969b, 1971, 1972 U. S. Navy 1970 Morrison 1973	Horvath and Freedman 1947 Clarke et al. 1958 Bowen 1968 McGinnis et al. 1972 Egstrom et al. 1973 Vaughan and Andersen 1973	—	—
Steadiness and tremor Tremor transducers Marked points Maze tracking Ball bearing Balance rail Statometer	Adolfson et al. 1972	Bennett, P. B., 1965, 1966, 1967 Bühlmann et al. 1970 Bachrach et al. 1971 Bennett and Towse 1971 Bachrach and Bennett 1973	—	Peacock 1956 Lockhart 1968		

Test / Category					
Dexterity and assembly	—	Tresansky 1973; Braithwaite et al. in press; Berghage et al. in preparation	Hunter et al. 1952; LeBlanc 1955; Dusek 1957; Teichner 1957; Gaydos 1958; Gaydos and Dusek 1958; Clark 1961; Lockhart 1966, 1968; Bowen 1968; Stang and Weiner 1970; Weltman, Egstrom et al. 1971a; McGinnis et al. 1972	—	Bowen et al. 1966; Baddeley 1966, 1967; Baddeley and Flemming 1967; Weltman et al. 1968, 1970
Hunter finger flexion	Kiessling and Maag 1962	Baddeley and Fleming 1967			
O'Connor finger dexterity	Adolfson 1965, 1967	Bennett and Towse 1971			
Purdue Pegboard	Adolfson and Muren 1965	O'Reilly 1973			
Minnesota rate of manipulation	Baddeley 1966a, 1966b				
Screwplate	Baddeley et al. 1968				
Bennett hand tool					
Knot tieing					
Block stringing and packing					
Triangle Test					
UCLA Pipe Puzzle					
Tracking and coordination	—	O'Reilly 1973	Kleitman and Jackson 1950; Teichner and Wehrkamp 1954; Teichner and Kobrick 1955; Payne 1955; Bowen 1968; Vaughan and Mavor 1972; Vaughan and Swider 1972	—	Bowen et al. 1966
Rotary pursuit					
Two-hand coordination					
Multidimensional pursuit					
Reaction time and vigilance	Shilling and Willgrube 1937	Parker 1969; Summitt et al. 1969	Horvath and Freedman 1947	—	Weltman and Egstrom 1966

Table VI-1—Cont.

Performance dimensions and tests	Stress characteristics of the diving environment					
	Hyperbaric air	Hyperbaric heliox	Tractionless/ viscous	Cold	Isolation/ confinement	Generalized stress
Simple visual reaction time Choice reaction time Peripheral signal monitoring	Kiessling and Maag 1962	O'Reilly 1973	—	Kleitman and Jackson 1950 Teichner 1958 Weltman and Egstrom 1966 Stang and Weiner 1970 Weltman et al. 1971a Egstrom et al. 1972 Vaughan and Andersen 1973	—	Weltman et al. 1971a
Perceptual processes Orientation Perceptual speed Time estimation Stroop word–color test Hidden patterns test Digit–symbol test	Shilling and Willgrube 1937 Bennett et al. 1967	Bennett et al. 1967 Parker 1969 Biersner and Cameron 1970b Bühlmann et al. 1970 Bennett and Towse 1971 O'Reilly 1973	Mann and Passey 1951 Margaria 1958 Brown 1961 Diefenbach 1961 Nelson 1967 Ross et al. 1969 Ross et al. 1970	Kleitman and Jackson 1950	—	Deppe 1969 Biersner and Cameron 1970b
Cognitive processes Arithmetic computation Memory Free association	Shilling and Willgrube 1937 Kiessling and Maag 1962 Adolfson 1965	Bennett, P. B., 1965, 1966, 1967 Bowen et al. 1966 Baddeley and Flemming 1967	—	Bowen 1968 Egstrom et al. 1972, 1973 Vaughan and Andersen 1973	—	Weltman et al. 1970

Sentence comprehension	Adolfson and Muren 1965	Parker 1969	
Conceptual reasoning	Baddeley and Flemming 1967	Biersner and Cameron 1970a, 1970b	
Problem solving	Baddeley et al. 1969	Bühlmann et al. 1970	
	Bennett et al. 1969	Bennett and Towse 1971	
	Fowler and Ackles 1972	O'Reilly 1973	
Mood and adjustment	—	Summitt et al. 1969	—
Work productivity			Miller et al. 1967
Emotional stability			Radloff and Helmreich 1968
Gregariousness			Radloff et al. 1970
Habitability			Helmreich 1971a, 1971b
			Nowlis et al. 1971
			Weltman et al. 1971a

Entries in the cells of the matrix in Table VI-1 are abbreviated bibliographic references to performance-oriented research that varied the environmental characteristics identified by the column label and measured the performance dimension indicated by the row label. It is one kind of overview of the field. One can note, by following a row, which environmental characteristics have been studied as a determinate of a performance dimension of interest. Examination of the columns shows the concentration of research on the effects of cold and hyperbaric air. Finally, Table VI-1 serves as a guide to the references given at the end of the chapter. The reader interested in a specific topic, e.g., the effects of cold on cognitive processes, can efficiently identify the relevant source materials.

The remainder of this chapter is organized into two main sections. Section B presents a catalog of tests and measures that have been used to assess each performance dimension. Descriptions include the test materials or apparatus, the technical characteristics of the tests (such as factor purity, reliability, and sensitivity to relevant environmental conditions), and the applications made of the test in diver-performance-oriented research. Section C presents summaries of research findings according to the main categories of environmental stress characteristics that define the diving experience.

In some cases, the catalog of tests and measures in this chapter will include examples of performance research findings. These are intended to give the reader a sense of the range over which the tests have been found to be sensitive, rather than as an exhaustive presentation of performance data. The more complete presentation and discussion of these data are found later in this chapter under summaries of performance research in the diving environment. Inevitably, this results in a certain amount of redundancy, considered justified in these cases.

B. *Tests and Measures of Diver Performance*

1. *Tactile Sensitivity*

Touch sensitivity is reduced as the skin temperature is lowered by exposure to cold water. This aspect of performance is of interest generally as a limitation to the transfer of information underwater via the sense of touch and, specifically, as a factor affecting the ability of a diver to discriminate small objects such as dials, knobs, and switches. Skin cooling has been studied as a modifier of several aspects of sensitivity: vibration, two-point or two-edge discrimination, pressure, and a combination of size and texture discrimination.

a. Vibratory Sensitivity

Apparatus for testing vibratory sensitivity is described by Weitz (1941). It consists of a small needle 0.28 mm in diameter, which can be vibrated against the skin at a specified frequency. Vibration amplitude is gradually increased from a subthreshold level until the subject reports the sensation of touch. This test measures the sensitivity of the skin to stimulation by mechanical vibration.

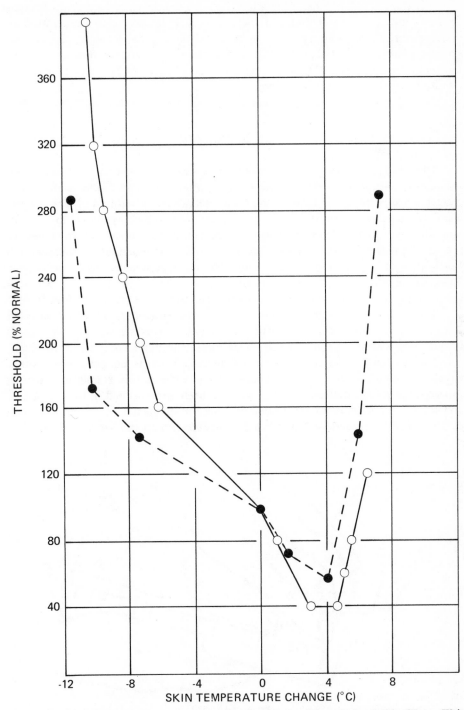

Figure VI-1. Effect of cooling and subsequent warming on vibratory thresholds. [From Weitz (1941) by permission of the American Psychological Association.]

Vibration sensitivity has been studied at finger-skin temperatures in the range 21–41°C, and the general shape of the relationship between skin temperature and vibratory threshold has been mapped by Weitz (1941). Figure VI-1 shows the threshold sensitivity function for 100 and 900 Hz (cycle per sec) vibration frequencies. Vibratory threshold is increased by 400% at a skin temprature of 21°C. At 37°C skin temperature, vibratory sensitivity is at a maximum.

b. Two-Point and Two-Edge Sensitivity

These tests measure the sensitivity of the skin in detecting two-point or two-edge stimulation. Apparatus for two-point discrimination threshold testing is an aesthesiometer whose points are separated by known intervals from trial to trial. The two-edge test device consists of two standard straight-edge rules bolted together at one end and separated by a wedge at the other so that a 13-mm gap separates the two edges, and the gap decreases linearly toward the closed end. This latter test was designed by Mackworth (1953). Test procedure is to have an experimenter press the subject's finger to the edges in an ascending or descending progression and noting where the subject reports the experience of two edges rather than one edge. (See Figure VI-2).

The two indices of sensitivity are closely correlated (Figure VI-3) and sensitivity is a function of finger-skin temperature. Sensitivity decreases gradually as the skin is cooled from a normal value, then degrades sharply at about 8°C.

The Mackworth V-test has been successfully used as an index of sensitivity changes in underwater test environments. The apparatus is simple and the test procedure is easily administered. The test has been administered in the ocean at 10 and 100 ft (Baddeley 1966) and to demonstrate the significant effects of exposure to cold water (Morton and Provins 1960, Bowen 1968, Stang and Weiner 1970).

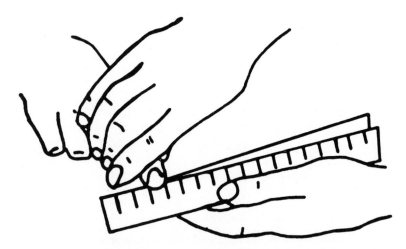

Figure VI-2. The "V" test in use. [From Mills (1957) by permission of the National Academy of Sciences–National Research Council.]

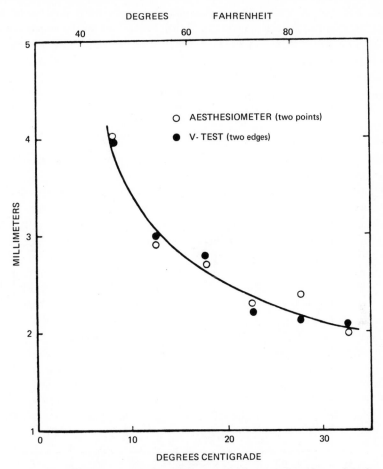

Figure VI-3. Comparison of the two-edge and two-point thresholds as a function of the skin temperature. [From Mills (1957) by permission of the National Academy of Sciences–National Research Council.]

c. Point-Pressure Sensitivity

One test to define the sensitivity of the skin to point pressure uses a small-diameter rod driven against the skin at constant rate (0.25 mm/sec) but at gradually increasing pressure until the subject reports the experience of pressure. Point-pressure sensitivity has been studied under a range of skin temperatures between −5 and 35°C (Mills 1957). Sensitivity is a linear function of skin temperature as shown in Figure VI-4.

d. Pressure Reproduction

Another test defines the ability to reproduce a standard pressure. Apparatus consists of a small brass disk (1.25 cm diameter) connected to a spring steel bar with strain gauges on two sides of the bar (Morton and Provins 1960). The strain gauges

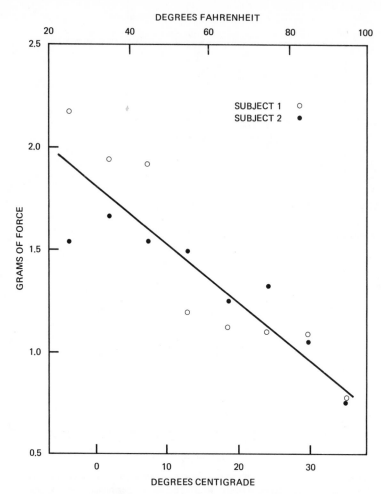

Figure IV-4. Threshold force of a 2-mm stimulus rod on the mid-volar surface of the distal phalanx as a function of the skin temperature at the extreme finger tip. [From Mills (1957) by permission of the National Academy of Sciences–National Research Council.]

are coupled to a pen recorder whose deflection in millimeters is proportional to grams of force applied to the brass disk. At a control temperature of 19°C, mean error in pressure reproduction was 5.9 mm; at −5°C finger-skin temperature, mean error was 12.8 mm. Most of the effects of cold occurred at temperatures below 8°C.

e. Towse Touch Test

A combination of sensitivity and dexterity factors is measured by the Towse touch test. Metal ball-bearings are of two sizes and two textures. The bearings are mixed in a center section of a form board and the subject's task is to separate them into two trays in the upper section of the form board. Subjects are blind-folded during the test and the score is the number of correctly sorted balls minus errors. This test has been used principally by Bennett and his co-workers in chamber environments for diagnosing inert gas narcosis (Bennett and Towse 1971).

2. Biomechanical Forces

The forces that a diver can exert underwater have been studied principally as phenomena potentially degraded by the relatively tractionless water environment and by the cooling of the muscle groups involved in force production.

a. Handgrip Strength

The most frequently measured aspect of diver strength has been the handgrip, using a hand dynamometer as the test apparatus. Clarke *et al.* (1958) related handgrip strength to exposure temperature, muscle temperature, and blood flow through the forearm. Optimal performance (measured in both duration of sustained contraction and maximum tension) corresponded to muscle temperatures in the range 25–29°C. Performance fell off at muscle temperature values outside this range. Figure VI-5 illustrates the basic apparatus.

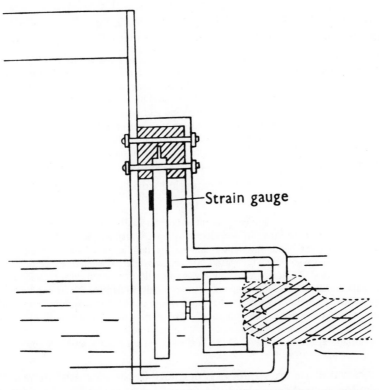

Figure VI-5. Diagrammatic representation of the handgrip strain-gauge dynamometer. [From Clark *et al.* (1958) by permission of the American Physiological Society.]

Handgrip-strength tests have been used underwater to compare bare vs. gloved hand strength (Egstrom *et al.* 1973), to examine the effect of a 20-min exposure to water temperature of 8°C (Bowen 1968), and to assess reduction in strength following a 3-hr exposure of 15.5 and 4.5°C water (Vaughan and Andersen 1973).

b. Hand-Torqueing Strength

Underwater tasks require the application of hand torque as well as grip (e.g., turning screwdrivers, nut runners, knobs, valves, etc.). Apparatus for this test consists of a 0.75-in. diameter brass cylinder instrumented to measure the amount of angular force applied in inch-pounds. Apparatus was designed by U. S. Army Natick Laboratories to assess effects of various hardware under cold, wet conditions (McGinnis *et al.* 1972). Average torque under control conditions was 60 in.-lb as compared to 53 in.-lb after a 2-min immersion in 35°F water. Bare-hand torqueing force was 54.5 in.-lb, which was significantly better than when wearing leather gloves, 46.5 in.-lb, but less effective than when wearing a variety of impermeable gloves, 64–68 in.-lb.

c. Rotary and Linear Forces

Considerable research has been conducted by Streimer and his associates into the physiological costs and work outputs associated with two basic force-production tasks: a rotary work task and a flexion/extension work task. Figure VI-6 illustrates the rotary-task equipment and is described by Streimer (1969*b*) as follows:

> This task required the test subject to grasp the handle of a 6-inch crank with his right hand and rotate this handle in a counterclockwise fashion for a 15-minute period. This rotary effort was done in a self-paced mode against a 9-pound resistance level. The total number of revolutions produced by the subject was recorded at one-minute intervals throughout the 15-minute work period and were used in deriving minute by minute and total work output levels. During task performance, the subject braced himself with his left hand by grasping a fixed horizontal bar located at the level of the crank handle axis of rotation and lying in the plane of handle rotation. All subjects grasped the stabilizing bar at a point 12–18 inches from the crank handle.

Figure VI-7 illustrates the flexion/extension work task equipment; the procedure is described by Streimer (1969*b*) as follows:

> This task required the test subject to exert forces in a flexion/extension mode against a handle constrained to move in a horizontal plane. Grasping the vertically oriented handle in his right hand, the subject moved the handle back and forth against a 9-pound resistance level, at a self-paced rate, and self-selected excursion distance, for a 15-minute period. Typically, the handle was pushed out to a full arm extension and then pulled back to an elbow angle of approximately 30 degrees. Total handle travel distance was registered via a calibrated counter and recorder at one-minute intervals. These data were used in the computation of work output values and efficiency. Bracing by the left hand was effected with the hand positioned at full arm extension on a horizontal bar at the level of the flexion/extension handle, and at a separation distance of about 10 inches.

The rotary task is more significantly degraded by the water environment than the linear task as compared to dry-land baselines: 20–37% reduction in output of rotary task as compared to 16–18% reduction in output of linear task. However, the rotary-task output underwater is more productive relative to the linear task: 0.016–0.045 hp

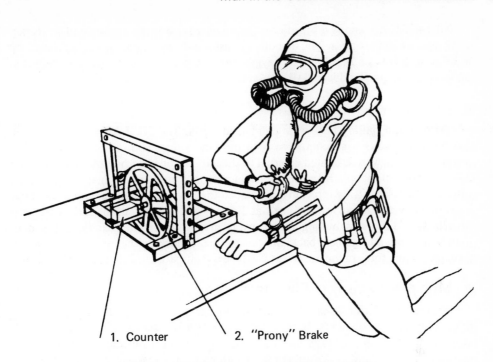

1. Counter   2. "Prony" Brake

Figure VI-6. Underwater rotary work (Streimer 1969).

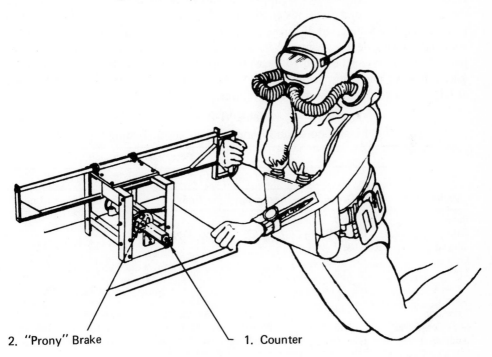

2. "Prony" Brake   1. Counter

Figure VI-7. Underwater linear work (Streimer 1969).

vs. 0.016–0.022 hp, and is more efficient when energy costs are accounted for. Rotary work underwater using a 9-in. handle against a 9-lb resistance produced 0.045 hp at an energy cost of 24.8 liters of oxygen per min per hp. Linear work underwater, on the other hand, produced 0.020 hp at more than twice the physiological cost.

d. Lifting and Pulling Forces

Lifting and pulling tests were conducted during the SEALAB II experiment in 205 fsw and results were compared with dry-land values (Bowen *et al.* 1966). Two torque wrenches were mounted external to the habitat, one in a horizontal orientation 30 in. above a stable platform, the other in a vertical orientation against the habitat exterior. To assess lifting force, divers grasped the horizontal handle with both hands and lifted upward. Mean lifting force was 600 ft-lb as compared to the surface baseline of 628 ft-lb. To assess pulling force, divers grasped a bracing handle in the non-preferred hand and pulled back on the torque wrench with the preferred hand. Average pulling force in the water was 200 ft-lb as compared to 238 ft-lb on the surface.

e. Multiple-Force Test Platforms

Several test platforms have been devised to measure a variety of force applications in underwater environments. Barrett and Quirk (1969) installed a test stand in 50 ft of ocean water at Point Magu to study one- and two-arm pushing strength against vertical, overhead, and deck surfaces both with and without waist restraints. Norman (1969) used the underwater test apparatus shown in Figure VI-8 to gather basic force-exertion data under various conditions of personnel restraints, worksite geometry, and type and direction of forces to be exerted. Egstrom *et al.* (1973) constructed a strength-testing device (Figures VI-9 and VI-10) for installation in the UCLA test pool. The device can be used to measure grip strength, shoulder-girdle adduction, knee extension, and elbow flexion. Table VI-2 presents results of recent tests conducted at UCLA at two different water temperatures.

Table VI-2
Mean Scores and Ranges for Strength Testing [a]

		Force exerted, lb			
		Warm (79°F)		Cold (43°F)	
Strength test	Dry	Beginning	End	Beginning	End
Bare-handed	138.7	131.9	130.4	—	—
Grip	(100–184)	(93–169)	(99–179)	—	—
Gloved	108.7	—	—	104.8	102.1
Grip	(84–134)	—	—	(79–135)	(73–154)
Elbow	73.0	73.5	74.9	73.7	73.0
Flexion	(58–106)	(53–131)	(55–113)	(51–115)	(49–103)
Knee	117.6	120.7	118.3	117.8	118.2
Extension	(77–185)	(75–181)	(61–185)	(79–185)	(74–185)
Shoulder-girdle	94.3	95.6	96.3	91.5	87.4
Adduction	(47–149)	(60–133)	(69–151)	(65–133)	(53–124)

[a] From Egstrom *et al.* (1973). Used by permission of the School of Engineering and Applied Sciences, University of California, Los Angeles.

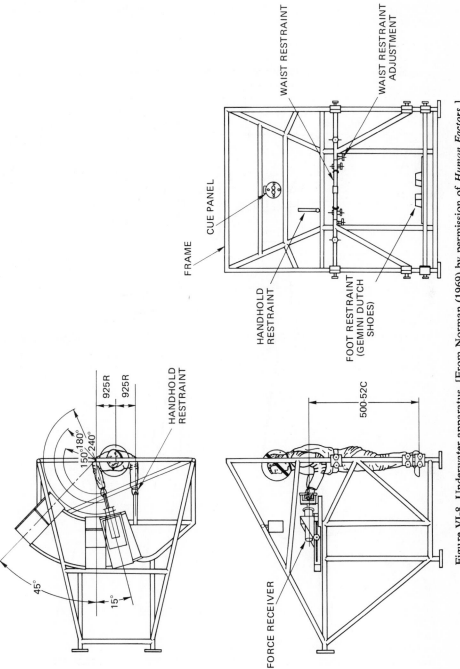

Figure VI-8. Underwater apparatus. [From Norman (1969) by permission of *Human Factors*.]

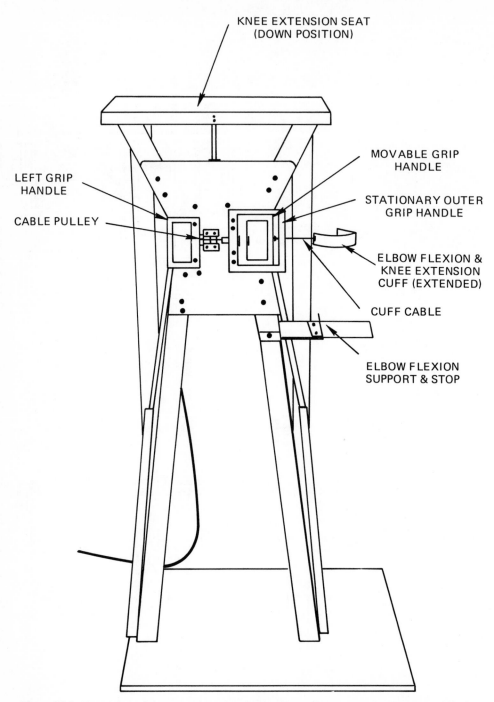

Figure VI-9. Front view of the strength testing device. [From Egstrom *et al.* (1973) by permission of the School of Engineering and Applied Sciences, University of California, Los Angeles.]

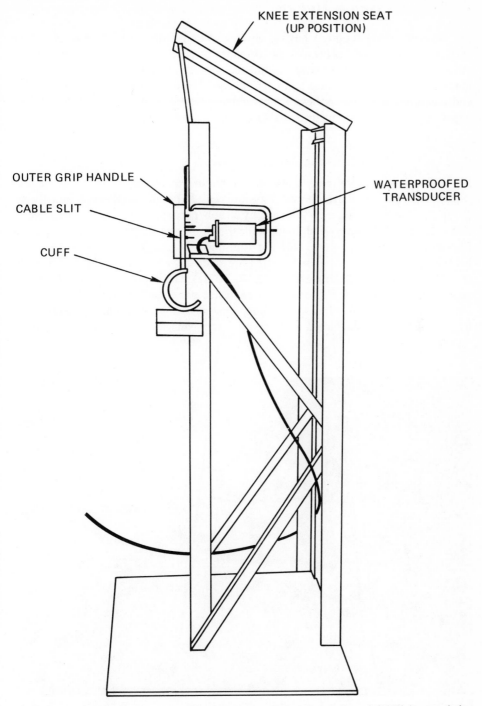

Figure VI-10. Side view of the strength testing device. [From Egstrom *et al.* (1973) by permission of the School of Engineering and Applied Sciences, University of California, Los Angeles.]

3. *Steadiness and Tremor*

Bennett (1966) reported having observed "trembling of the hands, arms, and even the whole body" as behavioral correlates of exposure to a heliox environment at 19–25 ATA. These observations stimulated effort to more precisely specify reactions in the areas of hand steadiness and postural equilibrium by means of a variety of tests and measures. Main categories of measurement devices include tremor transducers, paper/pencil and apparatus tests, balance-rail, and statometer tests.

a. Tremor Transducers

A device to measure velocity of postural tremor was developed for use in chamber experiments by the Royal Naval Physiological Laboratory (Bennett and Towse 1971). A velocity transducer is attached by a rubber sheath to the finger. The subject's task is to extend the finger and hand, while the elbow and forearm rest on his leg (see Figure VI-11). Velocity output of the transducer is analyzed by frequency bands between 2 and 30 Hz. The bandwidths are delta (2–4 Hz), theta (4–8 Hz), alpha (8–13 Hz), beta 1 (13–20 Hz), and beta 2 (20–30 Hz). The importance of the frequency analysis is to enable comparisons with established patterns. For example, normal tremor tends to show a large frequency component in the 8–12-Hz range, while abnormal tremor, such as that associated with nervous system aberrations, occurs in the 3–8-Hz range.

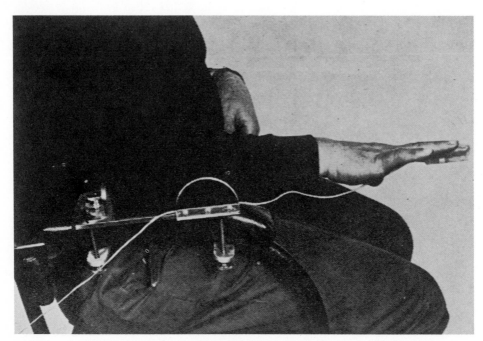

Figure VI-11. Postural tremor device. [From Bachrach and Bennett (1973) by permission of the Aerospace Medical Association.]

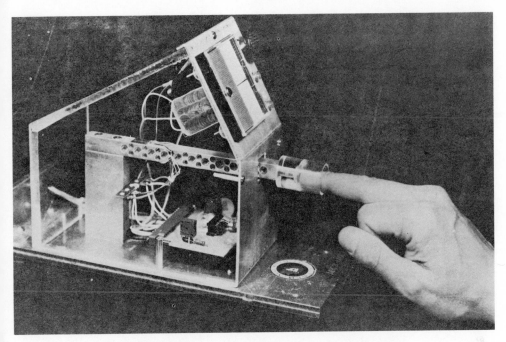

Figure VI-12. Mark III Mod 1 tremor device (Tresansky 1973).

Intentional-tremor devices have been developed at the Naval Medical Research Institute, the most recent of which is illustrated in Figure VI-12. The device consists of a horizontal bar, 7 in. long and ½-in. square, secured at one end to an aluminum baseplate. Mounted near the fixed end of the bar are four strain gauges in a Wheatstone bridge pattern. A plastic tube is fitted over the open end of the bar to guide the subject's finger to the bar end. The subject maintains a constant downward pressure with the tip of the middle finger upon the lever, which is preset to a specific force. Tremor is defined as variations in the subject's ability to maintain a zero reading on a graduated display. The device has been used in several chamber tests under hyperbaric conditions to 49.5 ATA. For a thorough review of applications, see Bachrach and Bennett (1973).

b. Paper/Pencil and Apparatus Tests

Three tests of hand steadiness have been used which appear sensitive to changes in atmospheric pressure, involve a minimum of materials, and are easy to administer and score.

(i) Marked Points Test. This test uses a sheet of paper containing a 10 × 10 cm square, divided into 100 squares of 1 × 1 cm. The test diver is given a pencil and required to make a single dot in as many squares as he can in a 10-sec interval. Score is the number of squares marked less errors (marks placed on lines or more than one dot per square). The test was used by Bühlmann *et al.* (1970) in a chamber test to 31 ATA with encouraging results. Although the data were not tested for statistical

significance, decrements in performance were noted when the test was administered within 10 min of the compression phase of the dive.

(ii) Maze Tracking Test. In this test several parallel lines are drawn the length of a paper. Distance between the parallel lines is $\frac{1}{2}$-in. and the test diver's task is to mark a continuous line between each pair without touching either side. A trial consists of 90 sec and two scores are recorded: number of errors and length of line segment traveled. The test, attributed to Albano by Bennett (1967), is not to be confused with a standardized test, Maze Tracing, an index of spatial scanning (French *et al.* 1963).

(iii) Ball-Bearing Test. Apparatus consists of a tray full of small ball-bearings, tweezers, and a vertical tube of a diameter just slightly larger than the bearings. Figure VI-13 illustrates the apparatus. The test diver's task is to place the ball-bearings in the tube using the tweezers. Score is the number of balls transferred from the tray to the tube in a 1-min interval. The test has been used primarily by Bennett and his associates in chamber experiments involving hyperbaric heliox. Test scores appear to decrease significantly at compression and recover with time at depth. Degree of performance degradation appears related to both rate and magnitude of compression.

Figure VI-13. Ball-bearing test apparatus. [From Bennett (1966) by permission of Pergamon Press, London.]

c. Postural-Equilibrium Tests

(i) Balance-Rail. Apparatus consists of a wide ($2\frac{1}{4}$-in. or 5.7-cm) rail and a narrow rail ($\frac{3}{4}$-in. or 1.9-cm). The test diver stands with his feet aligned with the rail in a heel-to-toe position, arms folded across his chest. The test is administered with eyes open or closed; score is time on the rail. Three tests have been found of value in discriminating performance changes as a function of pressure (Braithwaite *et al. in press*). Test conditions, wide-rail/eyes-closed and narrow-rail/eyes-open, are sensitive to pressure changes between 1 and 40 ATA; beyond this pressure level, performance is at a minimum and these test configurations do not reflect additional pressure changes. At pressures between 40 and 49.5 ATA, performance differences can be detected by a wide-rail/eyes-open test configuration.

(ii) Statometer. A balance board, designed at the Karolinska Institute in Stockholm, Sweden, has been used to measure standing steadiness (Adolfson *et al.* 1972). Apparatus is a small board with three strain gauges for sensing frequency and magnitude of body movements, both fore-aft and side-to-side. The subject stands on the board with feet together and attempts to stand steady. Measurements can be taken with eyes open or with eyes shut. This test has been used as an index of the effects of nitrogen narcosis in high-pressure air environments between 1 and 10 atm. Sensitivity of the test to pressure changes is greatest for the eyes-closed condition and the lateral-sway component of standing steadiness (Figure VI-14). Test–retest reliability of the measures varies with eye condition and sway component (Table VI-3).

A statometer device was used to obtain frequency and amplitude data characteristics of postural disturbances in hyperbaric heliox to 49.5 ATA (Braithwaite *et al. in press*). Performance on the statometer was particularly affected in the pressure range between 31 and 49.5 ATA.

4. Dexterity and Assembly

Finger dexterity, manual dexterity, and assembly skills are recognized as important constituents of many diver tasks and a variety of tests have been used to assess

Table VI-3

Test–Retest Reliability : Correlation between Body Sway in First and Second Recordings at Various Pressure Levels (Depth) in Ten Experienced Divers while Breathing Air[a]

	Lateral movement		Sagittal movement	
Depth	Closed eyes	Open eyes	Closed eyes	Open eyes
1	0.80	0.92	0.83	0.92
2.2	0.83	0.62	0.89	0.68
4	0.59	0.71	0.84	0.91
7	0.95	0.77	0.83	0.85
10	0.67	0.69	0.96	0.75
1	0.74	0.79	0.58	0.57

[a]Reprinted from Adolfson *et al.* (1972) with permission of the Aerospace Medical Association.

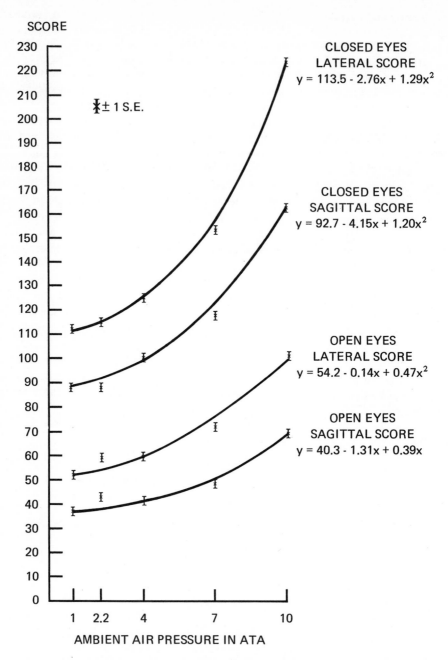

Figure VI-14. Relation between standing steadiness (body sway) in statometer units (score) and pressure level (depth). Means of ten experienced divers while breathing air. Curves are best-fit curves significant at the *p* < 0.001 level, implying a quadratic relationship. [From Adolfson *et al.* (1972) by permission of the Aerospace Medical Association.]

performance in these areas. Several of the dexterity tests are well-established, standardized tests of known technical characteristics and these tend to be used in dry-chamber or cold-air studies. Tests administered in the water, on the other hand, tend to be simplified adaptations of the standardized tests and their characteristics are uncertain.

a. Finger-Dexterity Tests

Factor analytic studies of dexterity tests consistently isolate finger dexterity or fine dexterity from the more general factor, manual dexterity. Finger dexterity involves the coordination of finger movements in performing fine manipulations. Tests that are most heavily loaded on the finger-dexterity factor all require grasping, placing, and assembling small objects where the finger tips are used in the manipulations.

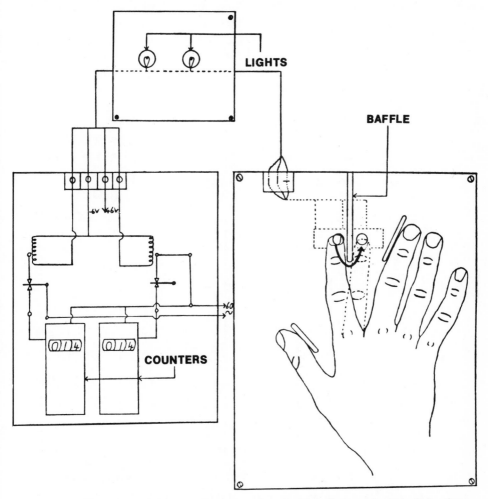

Figure VI-15. Diagram to illustrate apparatus used to study flexion speed of the index finger. [From Hunter *et al.* (1952) by permission of *The Canadian Journal of Medical Science.*]

(i) Hunter Finger-Flexion Test. A simple but sensitive test of the effects of cold on finger stiffness was devised by Hunter *et al.* (1952) and requires rapid, full flexion of the index finger. Apparatus consists of two pushbuttons mounted on a flat surface and separated by a vertical baffle. The hand is positioned on the board so that the tip of the index finger can reach either button; this hand position is fixed by wedges against the thumb and second finger. Figure VI-15 illustrates the apparatus. The task is simply to press alternate buttons as rapidly as possible in a 10-sec interval and the score is number of contacts. Finger movement is significantly reduced as finger-skin temperature falls below 12°C. Results of this test have been compared with results of a tapping test; the flexion test is significantly more sensitive to the effect of cooling (see Figure VI-16).

(ii) O'Connor Finger-Dexterity Test. The O'Connor test apparatus consists of a hinged box constructed so that, when it is lying open, one-half is a surface with 100 small holes and the other half is a tray holding 300 small pins. The subject's task is to pick up three pins simultaneously, using the preferred hand, and to place each in one of the holes. Score is number of pins placed in a 5-min trial (see Figure VI-17). The O'Connor test rates high on factorial purity in that it tends to show high factor loadings on finger dexterity and very low loading on other factors. The test has high test–retest reliability; correlations are reported in the range of 0.60–0.80 (Fleishman and Hempel 1954).

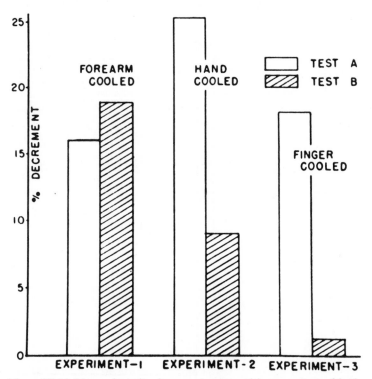

Figure VI-16. Finger dexterity decrement expressed in percentage of both tests A and B when finger, hand, and forearm (with exclusion of the hand) are cooled (LeBlanc 1955).

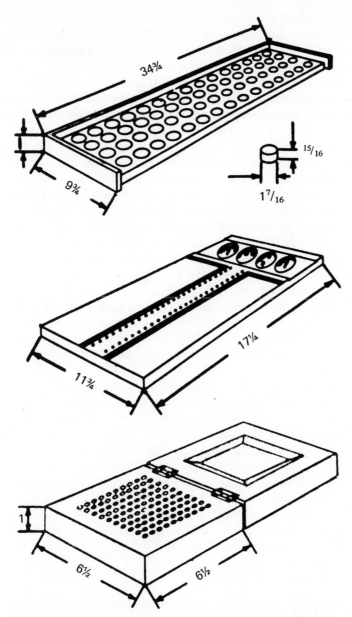

Figure VI-17. Finger dexterity test apparatus. [From Fleishman (1954) by permission of the American Psychological Association.]

(iii) Purdue Pegboard Test. Apparatus consists of a pegboard with two columns of 25 holes and four bins at the top containing either pegs, washers, or collars (Figure VI-17). The subject is required to perform a variety of tasks depending on which form of the test is being administered: Right Hand, Left Hand, Both Hands, or Assembly. The most commonly used form is the Assembly task. In this version, the test subject's task is to assemble in each hole a peg, a washer, a collar, and another washer. Score

is the number of assemblies completed in a 60-sec interval. In the one- and two-hand versions, the subject simply places as many pegs in the holes as possible in a 30-sec time interval using the left, right, or both hands. As with the O'Connor test, the Purdue Pegboard has the characteristic of high factorial purity in that factor analyses show high factor loading on the finger-dexterity factors and insignificant loadings on all others. Test–retest reliabilities tend to vary according to the form of the test used: Assembly has the highest, on the order of 0.75–0.90, and the one-handed versions have the lowest, 0.40–0.60 (Fleishman and Hempel 1954).

(iv) Minnesota Rate of Manipulation Test. Apparatus for the Minnesota test consists of a form board containing 60 holes (four columns of 15) and 60 cylindrical blocks. The blocks are 1.5 in. in diameter and are fitted into the holes in the form board so as to protrude 7/16 in. above the surface (see Figure VI-17). The test is administered in two forms: Turning and Placing. In the Turning test, the blocks are in the holes and the subject's task is to remove them, turn them over, and replace them as rapidly as possible. Score is number of blocks turned in two 35-sec trials. In the Placing task, the blocks are arranged along the sides of the form board and the subject's task is to place blocks in the board as rapidly as possible. Score is the number of blocks placed in two 40-sec trials.

The Minnesota tests are of moderate factorial purity, tending in factor analytic studies to load equally on finger dexterity and manual dexterity. Both versions of the test have very high test–retest reliability. Correlation coefficients of successive administrations are reported in the range of 0.80–0.90 (Fleishman and Hempel 1954).

Finger-dexterity tests have been used to assess effects of nitrogen narcosis: Purdue Pegboard Assembly scores, for example, were degraded by 8% between 1- and 4-atm air pressure (Kiessling and Maag 1962). More often they are used to detect effects of local cooling of the fingers and hands. As compared to the O'Connor and Minnesota tests, the Purdue Pegboard Assembly Test appears to be the most sensitive to differences in exposure temperature (Dusek 1957). Figure VI-18 shows the relative sensitivity to differences in cold exposure of 18 subjects who performed each test under three cold-air conditions and in a control temperature of 75°F.

Bowen (1968) devised an adaptation of the Purdue test for an underwater application. The test, Pegs and Rings, did not significantly discriminate between performance at exposure temperatures between 72 and 47°F.

b. Manual Dexterity Tests

Manual dexterity is defined by tests in which the whole hand (rather than just the finger) manipulates objects; generally they test the ability to make skillful arm–hand movements.

(i) Screwplate Tests. By far the most frequently used apparatus for assessing manual dexterity are the screwplate tests. These tests are nonstandardized; each investigator tends to create a unique test in terms of number of nut and bolt assemblies, their dimensions and arrangements on the plate, the criterion for task completion, and the scoring dimensions. In general the apparatus consists of a vertically mounted brass or aluminium plate with a pattern of holes punched through it, the

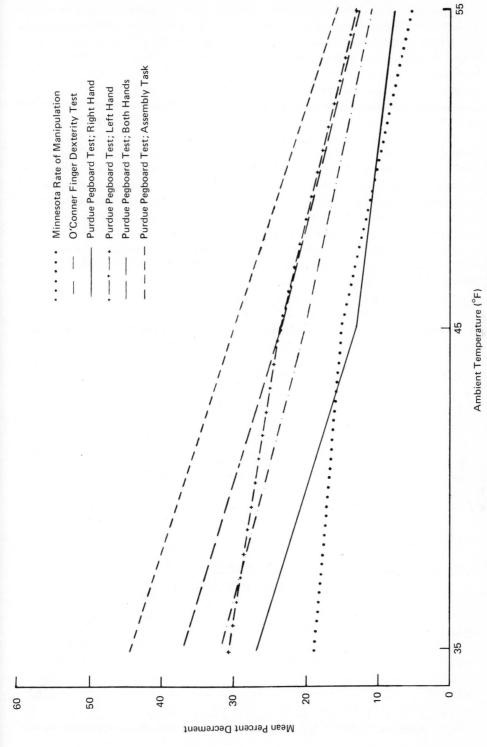

Minnesota Rate of Manipulation
O'Conner Finger Dexterity Test
Purdue Pegboard Test; Right Hand
Purdue Pegboard Test; Left Hand
Purdue Pegboard Test; Both Hands
Purdue Pegboard Test; Assembly Task

Ambient Temperature (°F)

Mean Percent Decrement

Figure VI-18. Percent decrement in performance as a function of ambient temperature (Dusek 1957).

arrangement of holes enabling easy identification of a left and right half. Nut–bolt–washer assemblies are placed in one-half of the plate and the subject's task is to transfer the assemblies to the other side. Score is time to complete task or number of assemblies transferred in a fixed time period. A secondary index is often number of nuts that are judged, by some criterion, to be loose. Very little is known about the factorial composition of screwplate tests and the only report of test–retest reliability indicates a low mark on this test-selection criterion. Obtained coefficients for 18 test divers were 0.46 at shallow depth and 0.35 at 100 ft ocean depth (Baddeley 1966*a*).

Screwplate tests have been used to assess effects of nitrogen narcosis when test personnel breathe compressed air at increasing pressure, to assess the effects of cold exposure, and as a test of the differential effects of chamber vs. ocean depth-equivalent pressures. A screwplate test of manual dexterity used to assess nitrogen narcosis appears less sensitive a measure than an arithmetic test (Adolfson and Muren 1965). Figure VI-19 presents a comparison of decrement-percentage curves for screwplate and arithmetic tests as functions of breathing hyperbaric air.

In cold water, a screwplate test failed to discriminate between short-duration exposures to 72, 62, and 47°F water (Bowen 1968); in another application (Stang and Weiner 1970) a screwplate test successfully distinguished 50°F from 60 and 70°F water over a 90-min exposure (see Figure VI-20).

(ii) Bennett Hand-Tool Test. A standardized test designed to test for proficiency in the use of wrenches and screwdrivers is the Bennett Hand-Tool Test. Apparatus consists of a 9 × 18 in. board with vertical end-pieces 9 in. high. Each end-piece is

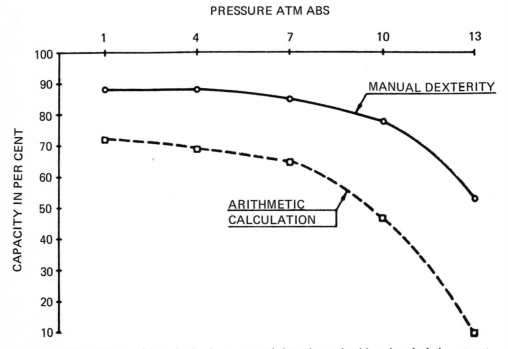

Figure VI-19. Effects of hyperbaric air on manual dexterity and arithmetic calculation at rest. [From Adolfson and Muren (1965) by permission of Sartryck ur Forsvarsmedicin, Stockholm.]

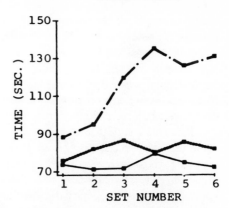

Figure VI-20. The screwplate test. Broken line, 50°F. Heavy solid line, 60°F. Light line, 70°F. Each test had a duration of 15 min. [From Stang and Weiner (1970) by permission of *Human Factors*.]

drilled with 12 holes (three rows of four) and one end is fitted with a nut–bolt–washer combination. Each row contains bolts of a common size, which are different in size from the other two rows. Appropriate size wrenches and screwdriver tools are provided (see Figure VI-21). The subject's task is to unfasten each of the 12 assemblies and transfer them to the other side of the test board. Score is time required to transfer the 12 assemblies. Median time in air is approximately 7 min. Test–retest reliability coefficients between 0.81 and 0.91 are reported in the manual of directions (Bennett, G. K., 1965). Also, percentile norms for various industrial and educational groups are presented.

The Bennett test has been used to assess various kinds of gloves for use in cold, wet conditions, and has been found to be less sensitive than the Minnesota Rate-of-Manipulation/Turning test in differentiating performance under wet/cold vs. dry conditions (McGinnis *et al.* 1972).

(iii) Knot-Tieing Test. Three coils of $\frac{1}{8}$-in.-diameter, braided cotton cord in 8-ft lengths are the only necessary materials. The subject's task is to tie "overhand knot and bight" knots as rapidly as possible for a 30-sec interval. Score is number of knots tied. This task has not been included in factor analytic studies and although its users refer to it as a test of manual dexterity, it would appear to include a large element of finger dexterity.

(iv) Block-Stringing/Packing Tests. In block stringing, materials consist of several 1-in. cubes of wood with 3/16-in. holes drilled through the center of each face. The stringer is a blunted needle 1/16-in. in diameter and $2\frac{1}{2}$ in. in length. A length of string is passed through the eye of the needle. The subject's task is to string blocks as rapidly as possible for 30 sec. Score is number of blocks strung. In the packing task, the subject holds a box in his nonpreferred hand and transfers blocks from a surface to the box with his preferred hand. Score is number of blocks packed in a 30-sec interval.

The knot-tieing, block-stringing, and block-packing tasks have been used to assess the effects of hand cooling on manual dexterity. No information is available regarding factor loadings or test–retest reliability. Sensitivity of the tests to changes in mean weighted skin temperature (MWST) appears high and rectilinear (see Figure VI-22). The tests have been used to separate the effects of local cooling of the hands

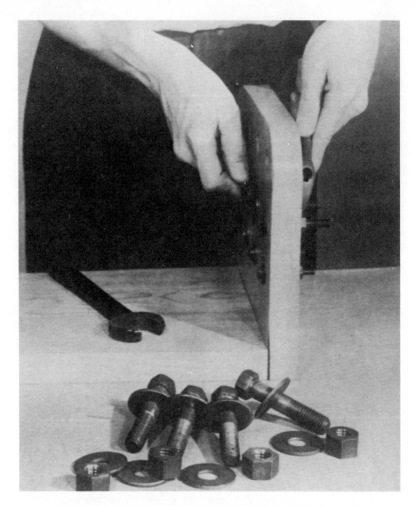

Figure VI-21. Hand-tool dexterity test. [From P. B. Bennett (1965) by permission
of the Royal Naval Personnel Research Committee, London.]

from whole-body cooling. As long as the hands are kept warm, performance in these
tasks is maintained at criterion levels (Gaydos 1958).

c. Assembly Tests

Assembly tests are distinguished from the dexterity tests by the inclusion of task
components beyond simple manipulation of small objects. There are correct and in-
correct ways of fitting pieces together. Assembly tasks sacrifice factor purity in order
to gain the advantage of increased task realism.

(i) Triangle Test. The Triangle Test was developed by Bowen *et al.* (1966) as
one of the performance tests for SEALAB II. The test apparatus includes three 1-ft

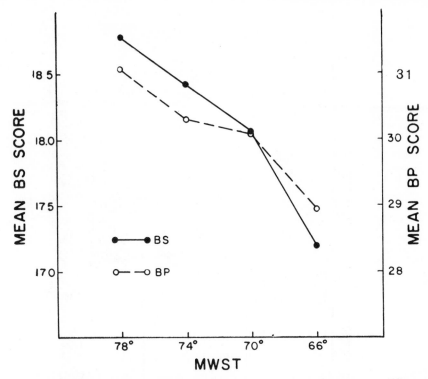

Figure VI-22. Block-stringing (BS) and block-packing (BP) performance at different levels of body cooling. The score for each MWST represents the mean across Ss of the mean BS and BP scores over four 30-sec trials. [From Lockhart (1968) by permission of *Ergonomics*.]

lengths of pipe which can be joined at their ends by nut–bolt–washer assemblies. In one version of the test, "different corners," holes in the pipe ends were placed asymmetrically so that the triangle could be formed in only one arrangement of the three lengths of pipe. Performance with this version of the test could then be compared with the performance using the simpler, symmetric version, which involved only the dexterity component. Score is time to assemble.

(ii) UCLA Pipe Puzzle. A flanged-pipe construction task was developed by UCLA Biotechnology Laboratory personnel for assessing performance in underwater assembly tasks where two-man coordination was required (Weltman *et al.* 1968). Thirteen flanged sections were assembled onto a 4 in. × 5 in. base to form a structure 7 ft high. The sections were fabricated from 2-in. galvanized pipe, valves, connectors, and threaded flanges. The flanges formed the connector points for task assembly and each flange required two $\frac{5}{8}$ in. × 2 in. bolts. Later a pressure test assembly was added to the apparatus with air supply, control valve, and pressure gauge. Subjects work in pairs and use an adjustable wrench and a torque wrench. Score is construction completion time (see Figure VI-23).

The UCLA Pipe Puzzle Assembly Task has been used to compare experienced

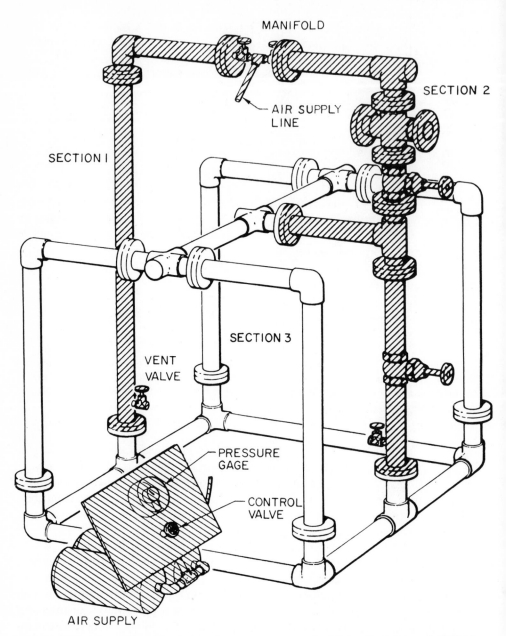

Figure VI-23. Pipe construction task and pressure test console. [From Egstrom *et al.* (1972) by permission of the School of Engineering and Applied Science, University of California, Los Angeles.]

vs. inexperienced divers, the effects of tank vs. open-ocean environments, and water-temperature differences. The task has successfully discriminated performance of experienced from novice divers, but was not sensitive to ocean vs. test-tank environments when experienced divers were tested (Weltman *et al.* 1970). Assembly/disassembly times in water at various temperatures suggest that the pipe-puzzle task is not a sensitive instrument for testing the effects of cold. Mean performance times for different sets of test divers were 26.0, 38.3, and 40.6 min in water temperatures of 26.6, 15.5, and 6.6°C (Weltman *et al.* 1971*a*).

5. Tracking and Coordination

This section presents tests for a complex of ability components which have been called by a variety of labels (such as psychomotor coordination, multilimb coordination, system equalization, and control precision). The abilities presumably tapped by these tests include: the ability to make sensitive, highly controlled positioning adjustments; the ability to use hand controls to control one or more axes of motion in systems with position, velocity, or acceleration dynamics; and the ability to coordinate two hands, two feet, or any combination of hands and feet.

a. Rotary Pursuit Test

Apparatus consists of a turntable driven at a high rate of speed, in which is set a small metal disk. The subject uses a prod-type stylus and attempts to hold the stylus in contact with the target disk as the turntable revolves (see Figure VI-24).

Factor purity of this test is high. In a factor analytic study with 200 subjects the Rotary Pursuit Test loaded primarily on the factor defined as fine control sensitivity and no other factor loadings were significant (Fleishman 1958). Test–retest reliability coefficient was 0.81 over repeated administrations. The test has been used in studies of the effects of prolonged cold exposure on skilled performance. Results revealed that cold reduces the upper limit of performance that can be achieved in a psychomotor task. Subjects practiced the task for 41 successive days—16 days in warm air, 12 days in cold air, and 13 days in warm air. During the cold period, skilled performance fell significantly and remained depressed relative to levels achieved when air temperature was returned to normal (Teichner and Kobrick 1955).

b. Two-Hand Coordination Test

An adaptation of a standardized test of two-hand tracking was devised by Bowen *et al.* (1966) for use on the ocean floor during SEALAB II. Apparatus consisted of a box containing a vertical peg which could be positioned in *xy* coordinates by two controls. Turning the right-hand knob moved the peg along the *x* axis; turning the left-hand knob moved the peg along the *y* axis. Maximum distances were 9 in. along the *x* axis and 12 in. along the *y* axis. The peg extended 1 in. above the box and therefore protruded above a template placed as a lid on the box. An irregular track was cut in the template and the diver's task was to cause the peg to trace through the track

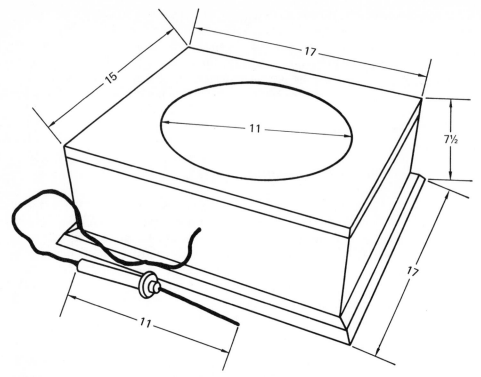

Figure VI-24. Rotary pursuit. [From Fleishman (1954) by permission of the American Psychological Association.]

in the template by properly coordinated movements of the controls. In the ocean test environment, the apparatus suffered corrosion and timing equipment malfunctioned, so that no reliable performance data were obtained. Essentially identical apparatus was later used in fresh water (Bowen 1968). The test was sensitive to differences in exposure temperature; average time to successfully track the template increased significantly as temperature decreased.

c. Multidimensional Pursuit Test

The USAF School of Aviation Medicine developed an apparatus to simulate the multilimb coordination required of aircraft pilots. The tasks required in this test are similar to those performed by pilots of wet submersibles. The apparatus requires the subject to monitor the drift of four eccentrically driven instrument pointers and to keep them within prescribed scale areas by coordinated adjustments of hand and foot controls. Controls include stick, throttle, and rudder pedals; displays are labeled RPM, airspeed, turn, and bank. Electric timers record the total time that all four pointers are within criterion limits. The apparatus essentially presents to the subject four simultaneous compensatory tracking tasks (see Figure VI-25).

Reliability of the test is very high. Coefficients of 0.93 and 0.89 are reported (Kleitman and Jackson 1950, Fleishman 1958). It has been used to assess effects of

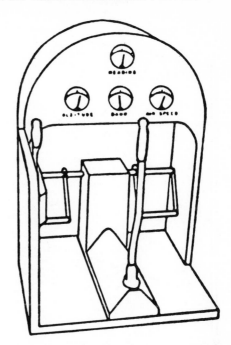

Figure VI-25. Multidimensional pursuit test. [From Fleishman (1958) by permission of the American Psychological Association.]

long-duration cold exposure on multiple compensatory tracking proficiency. With personnel in light clothing and exposed for 3 hr to 70, 55, and 40°F air temperatures, the test significantly discriminated the 40°F exposure from the other two (Payne 1959). Furthermore, qualitative differences in performance on the test were noted in the 40°F air as personnel experienced intervals of time when their concentration was focused on subsets of the overall task.

6. Reaction Time and Vigilance

Tests of simple and choice reaction time and of vigilance are described in this section. Simple visual reaction time typically emerges as a unique performance dimension in factor analytic studies of perceptual–motor tests. It is defined as the speed with which an individual is able to respond to a signal when it occurs. Choice reaction or discrimination reaction time involves an additional process. Associations between pairs of signals and responses must be learned prior to testing; then, when a given signal occurs, a judgmental process intervenes to select the appropriate response from a set of response alternatives. Vigilance monitoring performance is defined by tasks that require the diver to attend to an information source over lengthy time intervals in order to detect signals that occur infrequently and randomly. The monitoring of an obstacle-avoidance display in a wet submersible is a task of this type.

a. Simple Visual Reaction-Time Test

Apparatus for simple visual reaction-time testing consists of a single light stimulus and a single response key, switch, or pushbutton. A warning signal precedes the

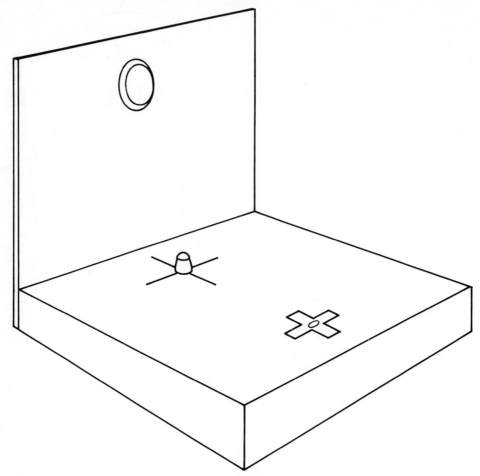

Figure VI-26. Reaction-time test. [From Fleishman (1958) by permission of the American
Psychological Association.]

occurrence of the visual signal by a random interval between 0.5 and 1.5 sec. The
subject sits with his hand on a fixed spot 6 in. from the response key. The warning
signal occurs, followed by the stimulus light, to which the subject responds as quickly
as possible (see Figure VI-26). Standardized test administration is two sets of 20 signals
and score is the accumulated time between signal onset and response. The test is of
very high factor purity, and is highly reliable; test–retest coefficients of 0.82 and 0.86
are reported (Fleishman 1958).

The Simple Visual Reaction-Time Test has been used as an index of nitrogen
narcosis at depth equivalent pressures between 1 and 300 ft. Small but consistent differences were found in mean times between 0.214 sec at the surface and 0.257 sec at 300
ft (Shilling and Willgrube 1937). Performance on this test has also been found to be
sensitive to environmental distractors such as high wind velocity/cold air combinations, but not to differences in mean skin temperature (Teichner 1958).

b. Choice Reaction-Time Test

The simple reaction-time task is made complex by requiring additional processes of the subject to intervene between stimulus presentation and response. The choice or discrimination reaction-time test includes two or more visual signals and two or more responses. Responses are usually associated with signals on a one-to-one basis. In a two-choice test there are two lights and two switches; the subject presses the left switch when the left light is activated and the right switch when the right light is activated. In a four-by-four test there are four lights and four switches or a single

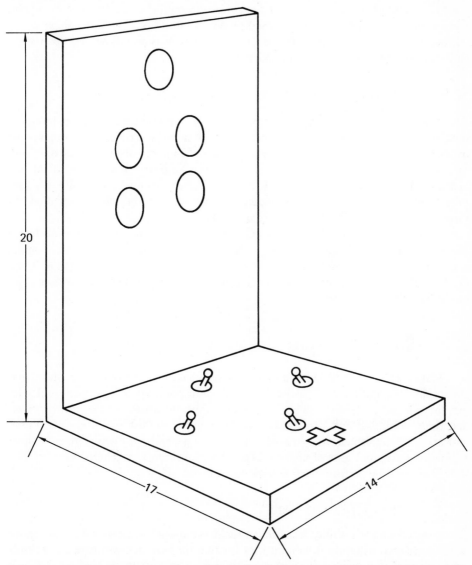

Figure VI-27. Discrimination reaction-time test. [From Fleishman (1958) by permission of the School of Engineering and Applied Sciences, University of California, Los Angeles.]

switch that can be thrown in one of four directions (see Figure VI-27). Choice reaction tests require the subject to first learn the correct associations, then react to the presented stimuli according to the learned rule for selecting among alternative responses.

A variety of such tests have been used in applications relevant to diver behavior. With a two-light, two-switch test, choice reaction speed was found to lengthen by 21% as a function of nitrogen narcosis from breathing air at 1 vs. 4 atm (Kiessling and Maag 1962). Another form, left–right responses to odd–even number stimuli, was used to test for narcotic effects of helium at a depth-equivalent chamber pressure of 1000 ft; no difference in choice reaction time was found to be associated with this condition relative to performance at the surface (Summitt *et al.* 1969). A two-light, two-switch version of the test was used to detect effects of cold water exposure at 70, 60, and 50°F for 90 min. Choice reaction time was significantly impaired by exposure to 50°F water (Stang and Wiener 1970).

c. Peripheral Light-Monitoring Test

A $\frac{1}{4}$-in. piece of Plexiglas tubing backlit by a miniature incandescent bulb was mounted in the periphery of a diver's facemask, 60° into the periphery at a polar angle of 300°. Signals were presented randomly at a rate of 75 per hour for a 10-sec duration. During the vigilance test, the subject's main tasks were dial monitoring, arithmetic tasks, or detecting gaps in Landolt Rings. The subject was provided with a waterproof switch, which served both to indicate detection of the peripheral signal and to turn off the light (Weltman and Egstrom 1966).

Performance on this task was significantly reduced when subjects were tested under simulated depth conditions in a chamber where no actual pressure changes occurred. The test also discriminated between novice and experienced divers (Weltman *et al.* 1971*b*). The test is proposed as a means of assessing generalized stresses associated with diving activities.

7. *Perceptual Processes*

Main categories of perceptual processes included in this section are postural and geographic orientation, time estimation, and perceptual speed—postural and geographic orientation because of their obvious implications to the safety of the diver, and time estimation and perceptual speed because their distortions are often-used indicators of impairment due to cold, narcosis, or generalized diving stress. A small sample of perceptual tests of uncertain factor composition are included because they have been used by performance-oriented researchers.

a. Postural-Orientation Tests

Tests of a diver's ability to orient to the gravitational vertical under water have been conducted primarily in laboratories for the purpose of investigating the sensitivity of vestibular cues to orientation when visual and kinesthetic cues are reduced. The results of this work have practical application since, in the water, visual cues are often severely reduced and postural cues from the kinesthetic receptors are partially nullified. Test apparatus consists of a one- or two-axis tilt table or chair mounted in a test pool (Diefenbach 1961, Nelson 1967). The test diver is strapped in the chair or to the table in ways that minimize pressure cues to orientation. The diver's facemask is blackened and scuba exhaust is led away via a long hose in order to reduce cues from air bubbles. Test procedure is to rotate the diver for a specified number of turns, so that he comes to rest at an angle of tilt from the vertical according to the experimenter's schedule of trials for counterbalancing-order effects. Then the test diver indicates the direction of the gravitational vertical either by adjusting a pointer mechanism or by resetting the chair (table) to the judged vertical position.

In the ocean, tests of orientation have been conducted having a test diver turn a number of somersaults in the water, then both orient his body toward the surface and point to the surface (Ross *et al.* 1969). A second diver held a weighted rope out of sight of the test diver and underwater photographs were taken of the test diver and vertically oriented rope. Photographs were simultaneously taken from front and side views. Later, angular errors were measured from the photographs. Angular measurements were taken from the trunk, the forearm, and the finger; angular error from the vertical was consistently larger for the forearm measures than for the body trunk or finger angles.

Results of all studies support the conclusion that extent of judgmental error depends on the orientation of the diver when tested. Error is small when the diver is in a head-up orientation, and large when he is head down in the water.

b. Geographic-Orientation Tests

Ross *et al.* (1970) have devised a series of ingenious tests to compare geographic orientation underwater with orientation on land. The Triangle Completion Test was used to compare direction and distance judgments on land vs. ocean floor. Rope triangles were staked out on land and on the seabed; the three sides of the triangle measured $20 \times 18 \times 18$ ft. Then one of the sides was removed. Test divers walked (swam) blindfolded along the two sides by holding on to the rope, then attempted to complete the missing third leg of the rope triangle by walking (swimming) the appropriate direction and distance. On land, test divers tended to walk the correct distance and made only small errors of direction, mainly to the right. Underwater, they tended to swim too great a distance and to veer off, tending to swim in a circle right.

The Distance Traveled Test was composed of straight-line courses staked out on land and on the sea floor. Divers were instructed to walk (swim) specified distances between 10 and 75 ft. At each distance specified, divers swam further than they walked.

In the Circle Completion Test center poles were set on land and on the sea floor. A 12-ft. piece of line was attached to the pole at a height of 6 ft and subjects directed to walk (swim) a complete circle. On land they walked too far, but underwater they stopped short.

c. Perceptual-Speed Tests

Perceptual speed is a commonly identified performance dimension accounting for variation in scores of perceptual–motor test batteries. This factor is usually defined as involving ability to make rapid identification of forms or figures and rapid comparisons between similar forms or figures. Tests of perceptual speed require simple test materials and a minimum of training of the divers to be tested, and they are easily administered. They have been most often used as potential indicators of impaired perceptual processes in tests of narcosis in hyperbaric test environments.

(i) Cross Out Tests. These tests use a page full of randomly occurring numbers or letters. The test diver is given a pencil and the instruction to cross out a specific letter (number) whenever it occurs. Score is the number of correct crossouts in a specified time interval, usually on the order of 60 sec. This test has been used to assess effects of nitrogen narcosis in pressure chambers at 1–10 ATA (Shilling and Willgrube 1937, Bennett *et al.* 1967) and to assess potential narcotic effects of heliox at chamber pressures of 16 and 31 ATA (O'Reilly 1973, Parker 1969).

(ii) Color-Naming Test. In this test 100 $\frac{1}{2}$-in. squares painted in ten different colors were arranged in random order in ten rows and ten columns. The test diver was required to name the color sequence, moving from row to row. Score was time to complete the naming task. The test has been correlated with oral temperature over the normal range of diurnal variation. As temperatures were higher, naming speed was faster. Correlation coefficient between temperature and naming time was -0.88 (Kleitman and Jackson 1950).

(iii) Card-Sorting Test. In this test decks of playing cards were made available to test divers in a pressure chamber experiment and they were instructed to sort the cards by suit. Score was the number of correct sortings completed in a 10-min time period. The test discriminated between surface and 4-ATA pressure when air was the breathing gas (Bennett *et al.* 1967).

d. Time-Estimation Test

A measure of the constancy with which a standard time interval can be estimated has been used as an index of response to diving stress (Deppe 1969). At a depth of 33 ft in a diving tank, 13 inexperienced and 8 experienced scuba divers performed a search task with blackened facemasks. In addition they were to tap on the floor at 1-min intervals. Differences in both means and variances of these estimates were significant. Inexperienced divers overestimated the time interval by one-third and their variability was double that of the experienced group.

e. Hidden-Patterns Test

In a test proposed as a test of resistance to distraction, test items consist of a simple geometric pattern followed by four complex patterns in which the simple pattern may or may not be imbedded. The test diver's task is to indicate which test patterns contain the simple figure. The test was used to diagnose effects of breathing high-pressure heliox in pressure chamber tests between 1 and 31 ATA. There were no significant differences in subject performance at any test pressure (Biersner and Cameron 1970*b*).

f. Stroop Word—Color Test

This test has been proposed as a measure of stress sensitivity (Biersner and Cameron 1970b). It examines the ability to persist in a given response set in the presence of a competing response set. The original development of test materials was for the purpose of quantitatively comparing the strength of association between words and color names (Stroop 1935).

The test diver is presented with a series of 100 chips painted either red, green, blue, brown, or purple and his task is to name the colors as rapidly as possible. Next, he is presented a list of 100 words. The words are the names of the five colors, and the words are printed in colored ink using the same five colors. The color of ink and the color name printed are always different, however. In no case, for example, is the word "Blue" printed in blue ink. The task is to name the color of the ink, ignoring the word. The difference in color-naming time between naming colored chips and naming colors (where word associations are a source of interference) is used as a measure of ability to resist the effect of a competing perceptual set.

g. Wechsler Bellevue Digit—Symbol Test

The Wechsler Bellevue test or some variant of it is frequently used in hyperbaric studies, although its factorial composition is ambiguous. Factor analytic studies have coupled the test to memory, perceptual speed, and number facility (Wechsler 1958). A number–symbol code is given to the test diver—numbers zero through nine and a simple symbol associate for each number. On the test sheet are rows of numbers and blanks. The task is to fill the blanks with the correct symbols via reference to the code at the top of the test sheet or via memory. Score is number of correct number–symbol associates completed in one 60-sec trial.

8. Cognitive Processes

Behnke et al. (1935) first called attention to the effects of exposure to hyperbaric environments on diver cognitive processes. They observed nine technically trained researchers performing their regular tasks in animal research under standard laboratory conditions except that the pressure of their breathing gas (air) was raised from 1 to 4 atm. Their qualitatively described observations appear to have guided the development of tests in the cognitive area; they listed the following as cognitive correlates of narcosis:

1. There were frequent errors in arithmetic calculations.
2. Recollection required greater effort.
3. Concentration was comparatively difficult.
4. Powers of concentration were limited.
5. There was a tendency toward fixation of ideas.
6. There was a slowing up of mental activity.

a. Arithmetic-Computation Tests

Simple arithmetic computation is by far the most frequent kind of test used to examine potential decrements in cognitive function as a consequence of exposure to the various conditions of diving. The tests employed have been more or less homemade, with no standardized format. Sample formats include:

1. Perform a set of four problems, one each of addition, subtraction, multiplication and division.
2. Multiply two one-digit numbers, then add (subtract) their product from a two-digit number.
3. Add sets of five two-digit numbers.
4. Multiply a two-digit by a one-digit number.
5. Add two three-digit numbers, then subtract a three-digit number from the sum.
6. Perform arithmetic computations imbedded in map problems and repetitive dive problems.

Whatever the form, arithmetic tests have been found to be highly sensitive to the narcotic effects of high partial pressures of nitrogen. As indices of cold exposure, the arithmetic tests have been less sensitive. Neither accuracy nor time to complete a set of items appears to suffer in response to differences in exposure temperatures. However, if the kind of arithmetic problem selected is more difficult than the simple forms, then number of problems omitted appears to be a correlate of cold exposure. Arithmetic tests have been used in high-pressure heliox environments to 45 ATA, revealing no decrement from levels of performance obtained at the surface.

b. Memory Tests

Standardized memory tests are of two main types: tests of immediate memory span (e.g., Visual Digit Span) and tests of associative memory (French *et al.* 1963). A few applications of the latter test category can be found in research related to man in the sea; the majority, however, are innovations of unknown technical characteristics. Modifications in memory capability have been sought as an effect of cold and as an effect of hyperbaric helium-breathing environments.

(i) Word–Number Test. In a standardized test of associative memory the subject is given 3 min to examine 20 word–number pairs. Following a specified interval, the words are presented in a different order and the task is to supply the correct number to complete each pair. A shortened version of this test was used by Biersner and Cameron (1970*b*) in a dry-chamber test at 31 ATA where heliox was the breathing gas. In a previous study by these same authors (Biersner and Cameron 1970*a*), an operational adaptation of the word–number test was devised, substituting ten quantitative characteristics of a hypothetical salvage dive for the common words and 2-digit numbers of the standardized test. Sample paired-associate items were as follows: dive number—403; depth of dive—257 ft; surface interval—184 min; bottom time—295 min.

(ii) Picture–Number Test. A standardized test of associative memory essentially substitutes pictures for words in the word–number test. Standard procedure is to

allow 4 min for studying the 21 pairs prior to testing with the pictures alone. This test was used by O'Reilly (1973) in a dry heliox chamber environment at 16 ATA.

(iii) Clock Test. This test was developed for use in the water by Bowen (1968). Its characteristics are more akin to tests of immediate memory span than to associative memory; it assesses capability to recall previously studied configurations. Test apparatus consists of a panel display of eight clock faces in two rows of four. On each trial, the experimenter sets each clock face to an hour–minute combination (minute settings are restricted to quarter-hours). The test diver studies the panel for 1 min; then, after a 30-sec wait, he attempts to reproduce the settings on a formatted response sheet. Number of recalls attempted and percent correct are scored. The test was administered to five subjects in air and again in water after a short (16-min) exposure to water temperatures of 16.6 and 8.4°C. Temperature differences did not affect the recall scores, although a significant 22% reduction in accuracy did occur between the in-air and in-water test administrations.

(iv) UCLA Memory Tests. Three types of test have been used at UCLA to explore recall and recognition performances as functions of water temperature. The first test consists of six prose passages about hypothetical shipwrecks, each described by five characteristics: name of ship, type of ship, depth of water, type of bottom, and type of danger to the salvage diver. The test diver studies the six passages for 5 min, then recall is tested by asking the diver to reproduce as many of the 30 shipwreck-characteristic "facts" as he can. Later, recognition is tested by 24 true/false items about the facts in the passages, half of which are true.

A second test uses five sheets of paper on each of which five items are pictured: a tool, an animal, a geometric shape, a pair of numbers, and a pair of letters, for a total of 25 items. The test diver studies the five sheets for 5 min; then a 45-min interval is permitted to elapse prior to testing. Both the prose passage and the pictorial tests have been useful in detecting performance decrement as a function of water temperature.

A third type of memory test was constructed by elaborating upon the UCLA Pipe Puzzle Assembly Task. Five bolt–gasket–washer combinations are required to be inserted into the pipe assembly at specified joints. The combinations are defined by bolt color, washer shape, and gasket material. In addition, five special items are to be built into the construction at specified places: a gauge, a nipple, a valve, a plug, and an elbow. Test divers study the blueprint (see Figure VI-28) for "several minutes" at poolside, then each one enters the water and places the assemblies according to his ability to reecall the blueprint. Performance on this test did not reflect differences in exposure temperature (Egstrom *et al.* 1973).

c. Free-Association Test

The subject is presented a series of common items as stimulus words (ball, tube, hole, etc.) and instructed to respond to each word with related words for 60 sec. Score is average number of related words given per stimulus word. Also scored is the average delay time between responses. The two indices are highly correlated and neither measure appears particularly sensitive to the effects of nitrogen narcosis (see Figures VI-29 and VI-30).

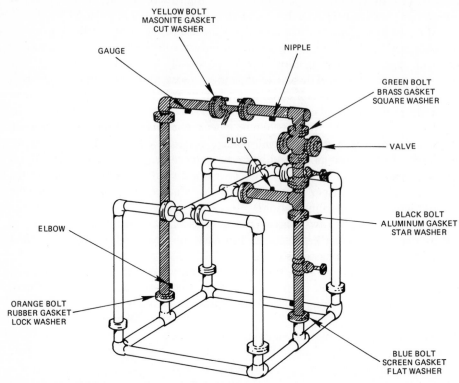

Figure VI-28. The pipe puzzle construction, memory task. [From Egstrom *et al.* (1973) by permission of the School of Engineering and Applied Sciences, University of California, Los Angeles.]

d. Sentence-Comprehension Test

A test was developed on the basis of evidence that the time required to understand a sentence is closely dependent on its structure (Baddeley *et al.* 1968). For example, positive statements are more readily comprehended than negative ones, and active statements more quickly than passive ones. It is assumed that comprehension of linguistic material requires grammatical transformation to simpler forms, and that these transformations provide an intellectual task sufficiently demanding to be sensitive to a reduction in intellectual capacity. The test is comprised of a series of sentences claiming to describe the order of two letters A and B. The sentences do this in a variety of ways, describing order relationships using combinations of six binary conditions:

> positive or negative
> active or passive
> true or false
> precedes or follows
> A or B mentioned first
> letter pair AB or BA

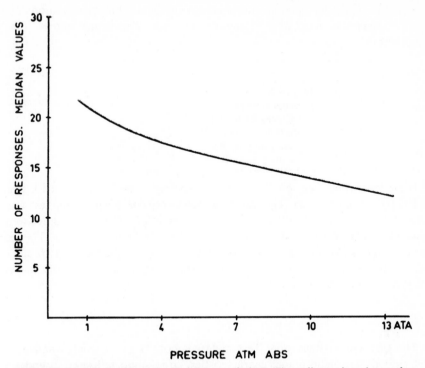

Figure VI-29. Effects of hyperbaric air on associations. The ordinate gives the number of responses per stimulus work. [From Adolfson and Muren (1965) by permission of Sartrych ur Forsvarsmedicin, Stockholm.]

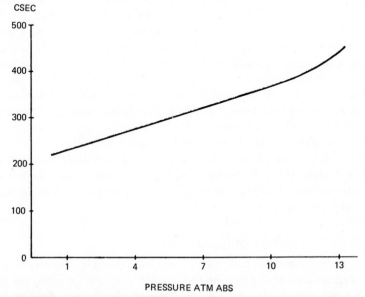

Figure VI-30. Effects of hyperbaric air on associative reaction time. The ordinate gives the median value in centiseconds per response. [From Adolfson and Muren (1965) by permission of Sartrych ur Forvarsmedicin, Stockholm.]

Test consists of the 64 possible combinations of the six binary conditions and the subject completes as many items as possible in 3 min. Test format and sample items are as follows:

Test item	True	False
1. A follows B: BA	×	
2. B precedes A: AB		×
3. A is followed by B: AB		
4. B is not followed by A: BA		
5. B is preceded by A: BA		
6. A does not precede B: BA		

The Sentence-Comprehension Test was administered twice at each of two depths (5 and 100 ft) in the open sea. The test–retest reliability coefficient of the test was 0.78 at the shallow depth and 0.81 at 100 ft (Baddeley *et al.* 1968). Test scores were not sensitive to pressure or narcotic differences at the two depths, however. The test has also been used with experienced and novice divers—in a diving tank at 15 ft and in the ocean at 20 ft. No differences were found in test scores as a function of these conditions (Weltman *et al.* 1970).

e. Conceptual-Reasoning Test

The Conceptual-Reasoning Test materials consist of 32 small wooden blocks which embody five dichotomous characteristics: large–small, tall–short, round–square, hollow–solid, and numbered–lettered. None of the blocks are identical, but 16 have two characteristics in common, 8 have two or more characteristics in common, and 4 blocks are common in at least three ways. The subject is presented a dual-classification diagnostic problem by the experimenter deciding upon the basis of the classification and showing the subject one of the eight blocks that has the chosen characteristics. The subject's task is to determine which two of the five characteristics of the model block are being used by the experimenter as the basis of classification. He does this by a trial and error sequence of selecting blocks similar to the model block on two characteristics, getting positive or negative feedback from the experimenter, then testing another possibility, until he discovers the correct classification basis. The Conceptual-Reasoning Test is essentially a test of inductive reasoning. Score is amount of time to find the correct characteristics.

A 16-block version of the Conceptual-Reasoning Test was used to detect effects of breathing compressed air at 4 atm as compared to a surface-control condition (Kiessling and Maag 1962). Performance at 4 atm was significantly degraded as compared to performance at surface pressure. Average length of time to find correct solutions was increased by 33%.

f. Navigation Problem-Solving Test

Apparatus and materials have been developed to test the effects of long-duration cold water exposure on ability to construct and solve navigation problems. Apparatus consists of a waterproof display for use in the water by the test diver and a control and monitoring display for use by the experimenter. The navigator's display presents intended track of a submersible, yards along track, yards across track (left or right), and time on leg. The controller's console enables the experimenter to establish rates of change for both along- and across-track displays at the navigator's console. Plotting successive pairs of values through time provides a vector representing the submersible's actual track over the bottom. The subject is provided a laminated plastic plotting board and response form for each problem. Other materials included are a 1/20 scale engineer's rule for measuring yards and knots and a protractor for measuring compass headings. Information provided to the test diver by means of the navigator's display enables him to construct a vector triangle composed of the submersible's intended track, actual track, and current vectors. Using the rule and protractor the subject can determine five values from the vector triangle: speed and course of the submersible over the ground, set and drift of the current, and an adjusted heading for the submersible that compensates for the effects of the current. Score is given by time to complete the problem and the accuracy of solutions.

The test has been administered to eight subjects over several hours of exposure to moderate and cold water (15.5 vs. 4.5°C). Test scores were significantly degraded over time of exposure but no consistent differences in time or accuracy were reported as a function of water-temperature differences. Problem omissions occurred in the cold water but not in the moderate water (Vaughan and Andersen 1973).

9. Mood and Adjustment

The development of saturation diving techniques and the increasing interest in the potential of habitats as undersea work stations impose another set of environmental stressors on the diver: isolation and confinement. Ability to cope with the psychological concomitants of these environmental stress factors becomes yet another performance dimension for which tests and measures are required. Information about the ability of divers to adjust to habitat-related stress environments has come from research conducted in the context of two principal habitat efforts: SEALAB and TEKTITE. In both cases the initial efforts, SEALAB I and TEKTITE I, were conducted for the purpose of testing the feasibility of habitat operation; and the second of the series, SEALAB II and TEKTITE II, were more fully planned to incorporate systematic data collection efforts focusing on the behavioral effects of habitat living.

Researchers who conducted the behavioral/habitability aspects of these programs credit Gunderson and Nelson (1966) with providing their programs a basic structure for examining adjustment. From their studies of adjustment to wintering-over in Antarctica, Gunderson and Nelson abstracted three main factors accounting for the variety of measures taken, questionnaires used, and rating forms they administered: task orientation, emotional stability, and social compatibility.

a. Task Orientation

The most consistently used measure of task orientation is an empirical measure of actual time spent in work as opposed to a variety of other activities. Since SEALAB II was designed to test salvage techniques in the water, the principal measures were number of sorties and total time in the water. TEKTITE II was oriented toward the support of a variety of marine sciences activities, which included both in-water data collection and in-habitat analyses. Percentage of time allocated to marine sciences, therefore, was the main index of productivity or work-orientation.

Ratings of the crew members by team leaders was tried in SEALAB II. Each man received a rating on two aspects of his performance: "overall competence and performance as a diver," and "willingness to do his share or more of the common work." Ratings were made on a four-point scale, labeled fair, good, very good, and excellent. The ratings failed to discriminate among divers due to a clustering of ratings at the high end of the scales.

b. Emotional Stability

Mood scales were used in both SEALAB and TEKTITE programs with good success. These are self-rating paper and pencil tests consisting of a list of adjectives and scales for the subject to check the degree to which each adjective applies to him at that moment. Various mood scales are derived from the responses. In SEALAB II a 67-item list was used from which six mood scales were derived: anger, happiness, fear, depression, lethargy, and psychological well-being. In TEKTITE II a 33-item list was used from which six negative and five positive moods were derived. Positive moods included concentration, activation, social affection, pleasantness, and nonchalance; negative mood scales were anxiety, depression, deactivation, aggression, egotism, and skepticism. Positive mood scale scores tended to correlate with objective measures of productivity. In SEALAB II, selected mood scores correlated on the order of 0.50 with dive frequency and total dive time (see Table VI-4); in TEKTITE, positive mood scale scores correlated 0.35 with percent of time spent in work activity and 0.43 with ratings of habitat acceptance. In addition to their correlations with objective measures of work, six of the scales used in TEKTITE II were sensitive to amount of time in the habitat. According to the names of the scales, subjects appeared to become less pleasant, active and social as mission duration increased from 14 to 20 or 30 days. Again these mood shifts tend to correlate with objectve observations, as the participants did less work, slept more, and interacted less as a function of time.

Mood checklists have been used in pressure-chamber experiments. The Multiple Affect Adjective Checklist (anxiety, depression, and hostility scales) was used as a test for anxiety in novice divers exposed to simulated hyperbaric conditions (Weltman et al. 1970). Anxiety scores were significantly higher for subjects tested in the simulated chamber than for subjects tested in a normal surface condition. The Clyde Mood Scale, from which elation/depression scales can be derived, was used to test for possible mood shifts as a consequence of exposure to high hydrostatic pressure during a 1000-ft chamber dive (Summitt et al. 1969).

c. Social Compatibility

In SEALAB II, partial tv monitoring was the main source of behavioral data, while in TEKTITE II both tv and open-microphone monitoring were complete. Therefore, the index of an individual's social characteristics in SEALAB II was based on amount of time spent in the presence of others, while in TEKTITE the index was

Table VI-4

Intercorrelations of Items Related to Performance and Mood[a]

	(1)	(2)	(3)	(4)	(5)	(6)	(7)	(8)	(9)	(10)	(11)	(12)	(13)	(14)
1. Age	1.00													
2. Diving experience	0.67^c	1.00												
3. Ordinal position	0.32	0.25	1.00											
4. Size of hometown	0.05	0.00	-0.15	1.00										
5. Fear	-0.11	0.01	-0.49^c	0.26	1.00									
6. Anger	-0.07	-0.06	0.05	-0.22	0.34	1.00								
7. Happiness	-0.26	-0.23	-0.53^c	0.46^b	0.53^c	-0.33	1.00							
8. Psychological well-being	0.20	0.17	0.57^c	-0.19	-0.56^c	0.22	-0.87^c	1.00						
9. Communication outside SEALAB	-0.39^b	-0.04	-0.39^b	0.23	0.40^b	0.06	0.55^c	-0.46^b	1.00					
10. Gregariousness	-0.09	0.06	0.32	-0.48^b	-0.24	0.20	-0.55^c	0.51^c	-0.18	1.00				
11. Participation in meal preparation	0.16	-0.01	0.00	0.36	-0.16	-0.21	0.22	-0.13	-0.16	-0.38^b	1.00			
12. N performance tests	0.00	-0.21	0.49^c	0.04	-0.25	0.08	-0.09	0.14	-0.33	-0.07	0.64^c	1.00		
13. Diving time	-0.04	0.01	0.52^c	-0.48^c	-0.50^c	0.20	-0.52^c	0.52^c	-0.50^c	0.50^c	-0.10	0.22	1.00	
14. N Sorties	0.13	0.14	0.56^b	-0.37	-0.44^b	0.23	-0.48^c	0.47^b	-0.49^b	0.46^b	-0.14	0.09	0.89^c	1.00
15. Choice as a peer	0.28	0.14	0.25	-0.21	-0.22	0.21	-0.44^b	0.47^b	-0.41^b	0.14	0.19	-0.02	0.28	0.33

[a] From Radloff and Helmreich (1968). Used by permission of the Naiburg Publ. Corp.
[b] $p < 0.05$.
[c] $p < 0.01$.

based on amount of time spent in communication with others. In both programs a gregariousness index was developed. This index correlated well with work productivity measures. In SEALAB II, the gregariousness index correlated with number of dives, $r = 0.46$, and with total diving time, $r = 0.50$; in TEKTITE, the gregariousness index correlated with percent of time spent in marine science work activity $r = 0.59$ (Radloff and Helmreich 1968).

10. *SINDBAD Test Battery*

System for Investigation of Diver Behavior at Depth (SINDBAD) is a perceptual–motor and cognitive test battery designed into a hardware system for administration in wet or dry environments at pressures to 445 lb/in.^{2}. Tests are administered from outside the chamber via an experimenter's console and visual stimuli are presented to the test diver at a porthole. The response apparatus is waterproofed and pressure-proofed. Figure VI-31 illustrates the main components of the SINDBAD equipment system.

Figure VI-31. SINDBAD test battery apparatus (Reilly and Cameron 1968). (A) Artist's rendering of operational SINDBAD I system. (B) Experimenter's console. (C) Subject's response apparatus.

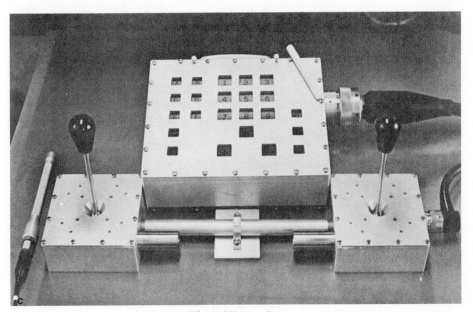

Figure VI-31—*Cont*.

Table VI-5

SINDBAD I Factors and Tests
(Cameron and Every 1970)

Factor	Test	Factor	Test
1. Arm–hand steadiness	Arm tremor	14. Time interval estimation	Interval reproduction
2. Control precision	Position control	15. Vigilance	Visual signal detection
3. Finger dexterity	Key insertion	16. Wrist–finger speed	Tapping
4. Flexibility of closure	Hidden patterns	17. Associative memory	Word–number
5. Length estimation	Shortest road	18. Flexibility of set	Anagrams
6. Manual dexterity	Wrench and cylinder	19. Induction	Letter sets
7. Multilimb coordination	Two-hand tracking	20. Memory span	Visual digit span
8. Perceptual speed	Number comparison	21. Number facility	Addition, subtraction, multiplication, division
9. Reaction time	Visual reaction time		
10. Response orientation	Choice reaction time	22. Spatial scanning	Choosing a path
11. Spatial orientation	Card rotations	23. Time sharing	Track and monitor
12. Speed of arm movement	Horizontal arc	24. Visualization	Surface development
13. System equalization	Acceleration control/rate control	25. Visual monitoring	Terminal digits

Table VI-6

Summary Table of SINDBAD Results (Bain and Berghage 1974)

1. Hidden patterns	28.3	14.7	9–66	23.1	9
2. Shortest road	7.15	3.77	1–16	6.2	7
3. Number comparison	12.6	5.25	7–27	11.8	8
4. Card rotations	15.1	10.3	1–36	13.5	11
5. Key insertion	17.6	3.09	12–23	17.3	18
6. Wrench and cylinder	10.3	2.11	8–14	9.5	8
7. Visual reaction time	0.285	0.036	0.242–0.394	0.275	0.258, 0.272
8. Interval reproduction	16.3	10.2	3.6–32.7	13.2	None
9. Tapping	106	20.3	78–174	104	95, 106
10. Word–number	8.93	4.73	0–18	8.6	9
11. Letter sets	6.18	1.96	2–10	5.9	6
12. Arithmetic	35.1	12.5	10–57	36.5	34, 40, 42, 47
13. Choose a path	3.93	3.09	0–14	3.1	6
14. Surface development	11.3	5.35	2–23	11.2	14
15.[a] Position control	43.8	16.7	28.3–99.7	40.0	None
16.[a] Two-hand tracking	97.1	17.3	63.8–132.9	102	None
17.[a] Choice reaction time	1.09	0.242	0.65–1.54	1.08	None
18.[a] Rate control	99.8	26.7	57.3–141.4	99.5	None
19.[a] Visual signal detection	80.0	8.94	60–90	75.7	80
20.[a] Visual digit span	6.19	0.981	5–8	6.25	7
21.[a] Track and monitor	0.178	0.508	−0.355–−2.00	0.0835	None
22.[a] Terminal digits	89.4	13.4	60–100	95.0	100

[a] Based on $N = 16$, all others based on $N = 27$.

Table
Intercorrelation Matrix

Factor name	1	2	3	4	5	6	7	8	9
1. Flexibility of closure	1.0								
2. Length estimation	−0.17	1.0							
3. Perceptual speed	0.11	0.33	1.0						
4. Spatial orientation	0.58	0.14	0.03	1.0					
5. Finger dexterity	0.17	−0.05	−0.11	0.31	1.0				
6. Manual dexterity	−0.08	0.08	0.00	0.09	0.33	1.0			
7. Reaction time	−0.33	−0.04	−0.11	−0.28	0.16	0.14	1.0		
8. Time interval estimation	−0.11	−0.11	−0.21	−0.16	−0.40	−0.23	0.22	1.0	
9. Wrist–finger speed	0.05	0.04	0.34	−0.06	0.33	0.40	−0.16	−0.21	1.0
10. Associative memory	0.28	−0.28	0.14	0.21	0.10	−0.29	−0.35	0.03	0.21
11. Induction	−0.18	0.03	0.00	0.02	0.13	−0.08	0.17	−0.07	0.04
12. Number facility	0.02	0.09	0.29	0.09	0.12	−0.18	−0.15	0.10	0.10
13. Spatial scanning	−0.30	0.18	−0.03	−0.15	0.26	0.09	0.21	−0.28	0.28
14. Visualization	0.13	−0.48	−0.28	0.17	0.60	0.19	0.16	0.16	0.20
15. Control precision	0.28	−0.11	−0.27	0.07	0.33	−0.22	−0.06	−0.09	0.04
16. Multilimb coordination	−0.06	−0.45	−0.29	−0.56	0.01	−0.13	0.13	−0.01	0.20
17. Response orientation	−0.52	−0.06	0.14	−0.48	0.13	−0.06	0.42	−0.26	−0.06
18. Systems equalization	−0.22	0.01	−0.25	−0.41	−0.18	−0.34	−0.03	−0.22	0.11
19. Vigilance	0.51	−0.24	0.21	0.61	0.03	−0.10	−0.35	0.08	−0.23
20. Memory span	0.33	−0.26	−0.23	0.17	0.02	0.05	−0.02	−0.39	−0.29
21. Time sharing	0.02	0.55	0.20	0.34	0.22	0.45	0.18	−0.42	−0.15
22. Visual monitoring	0.19	0.49	0.58	0.48	−0.10	−0.03	−0.20	0.10	0.05

[a] Correlations between variables 1–14 based on $N = 27$. All others based on $N = 16$.

SINDBAD is capable of administering 25 tests of the kinds reviewed in this section. Factors were selected to represent independent characteristics of man's inventory of basic abilities as estimated from factor analytic studies, and specific tests were selected on the basis of technical criteria such as factor purity, test–retest reliability, and sensitivity. Table VI-5 presents the 25 factors and associated tests which the SINDBAD system is currently programmed to administer.

Studies are in progress at the U. S. Navy Experimental Diving Unit to accumulate normative data descriptive of the population of Navy divers, and to determine the technical characteristics of the test battery in its current configuration. Table VI-6 presents recent descriptive data from a group of 27 Navy divers for 22 of the 25 tests. Table VI-7 illustrates the low intercorrelations among the several tests and Table VI-8 demonstrates the extremely high factorial purity of the test battery. Each test appears to be measuring an independent aspect of performance.

C. Summaries of Performance Research in the Diving Environment

1. Performance in Hyperbaric Air

In 1935, Behnke *et al.* wrote of the behavioral changes that occurred in nine persons performing physiological research in a laboratory where barometric pressure was raised from 1 to 4 atm. They noted mood shifts toward euphoria, various im-

VI-7
(Bain and Berghage 1974)

10	11	12	13	14	15	16	17	18	19	20	21	22
1.0												
0.48	1.0											
0.53	0.53	1.0										
0.16	0.26	0.22	1.0									
0.17	0.26	0.22	0.13	1.0								
0.32	0.19	−0.04	0.09	0.14	1.0							
−0.10	−0.38	−0.17	0.22	−0.14	0.19	1.0						
−0.34	0.17	−0.08	0.26	0.11	−0.13	0.18	1.0					
0.17	−0.15	−0.47	−0.08	−0.49	0.37	0.32	0.10	1.0				
0.17	−0.18	0.22	−0.19	0.35	0.07	−0.35	−0.31	−0.56	1.0			
0.02	0.00	0.03	−0.08	0.04	0.38	0.24	−0.12	−0.20	0.46	1.0		
−0.13	0.44	0.09	−0.19	0.01	−0.13	−0.52	−0.02	−0.40	0.08	0.31	1.0	
0.49	0.43	0.46	−0.42	−0.18	−0.07	0.34	0.16	−0.03	0.06	−0.24	0.27	1.0

pairments of the higher mental processes, and reduced neuromuscular control; they attributed the changes to the narcotic effect of increased partial pressure of nitrogen in air. Efforts to quantify the degree of change associated with increased pressure and to identify variables which interact with the main effect have since occupied a number of researchers and produced a large literature. In general, performance tends to fall off with increased pressure according to a positively decelerating function; that is, performance deteriorates gradually as the pressure increases, then more rapidly as pressure continues to increase. Performance in the cognitive task categories tends to fall further and more sharply than does performance in the perceptual and motor skill areas; however, the specific relationship between air pressure and performance depends on the experience of the man being tested, how practiced he is in the task, how commonplace the test environment is to him, and the aspect of task performance being measured.

a. Cognitive-Task Performance

Table VI-9 presents a sample of studies conducted to define the effects of exposure to hyperbaric air on performance in tasks that involve the higher mental processes. Each of the tasks is different, but those in the first four studies are homogeneous to the extent that each involves the solution of simple arithmetic problems. The measures of task performance in this set include time to complete a fixed number of problems, number of problems attempted in a fixed time period, and either number correct or percent correct. Using the baseline score at 1 atm as a reference value, other scores can be expressed as a percentage change from the baseline, so that different measures

Table VI-8

SINDBAD Factor Analysis Results—Factor Names (Bain and Berghage 1974)

Variable	Spatial orientation	Reaction time	Induction	Length estimation	Wrist-finger speed	Time interval estimation	Spatial scanning	Perceptual speed	Finger dexterity	Number facility	Flexibility of closure	Associative memory	Manual dexterity	Visualization
1. Hidden patterns	0.30	-0.16	-0.10	-0.11	0.03	-0.07	-0.15	0.06	0.09	0.01	-0.90	-0.07	0.06	-0.01
2. Shortest road	0.09	-0.02	0.02	0.96	0.01	-0.04	0.09	0.16	-0.07	0.04	0.09	0.11	-0.04	0.01
3. Number comparison	0.01	-0.03	-0.01	0.18	0.17	-0.12	-0.04	0.94	-0.11	0.15	-0.06	-0.04	0.01	0.02
4. Card rotations	0.91	-0.14	0.03	0.10	-0.07	-0.08	-0.07	0.00	0.17	0.03	-0.29	-0.06	-0.06	-0.01
5. Key insertion	0.14	0.07	0.02	0.00	0.14	-0.24	0.11	-0.07	0.92	0.05	-0.07	-0.05	-0.15	0.06
6. Wrench and cylinder	0.05	0.06	-0.04	0.04	0.20	-0.10	0.01	-0.01	0.15	-0.08	0.05	0.10	-0.95	-0.01
7. Visual reaction time	-0.13	0.94	0.09	-0.04	-0.10	-0.11	0.08	-0.03	0.10	-0.07	0.14	0.12	-0.06	-0.01
8. Interval reproduction	-0.07	-0.11	-0.03	-0.05	-0.07	0.95	-0.14	-0.11	-0.13	0.07	0.06	0.00	0.10	-0.02
9. Tapping	-0.07	-0.12	0.01	0.00	0.90	-0.08	0.15	0.20	0.18	0.02	-0.03	-0.10	-0.23	-0.01
10. Word–number	0.11	-0.23	0.34	-0.22	0.16	-0.01	-0.15	0.07	0.05	0.31	-0.12	-0.75	0.20	0.01
11. Letter sets	0.02	0.10	0.93	0.01	0.01	-0.03	0.12	-0.02	0.07	0.25	0.10	-0.16	0.03	-0.02
12. Arithmetic	0.03	-0.08	0.28	0.04	0.01	0.08	0.13	0.17	0.09	0.90	-0.01	-0.17	0.09	-0.02
13. Choose a path	-0.07	0.08	0.11	0.09	0.14	-0.14	0.94	-0.04	0.12	0.11	0.13	0.08	-0.01	-0.01
14. Surface development	0.12	0.12	0.19	-0.45	0.13	0.26	0.08	-0.14	0.64	0.15	-0.07	0.05	-0.11	-0.41

Variable	Time sharing	Vigilance	Control precision	Response orientation	Visual monitoring	Multilimb coordination	Memory span	Systems equalization
15. Position control	0.06	-0.04	0.96	-0.07	-0.02	0.06	0.17	-0.18
16. Two-hand tracking	0.30	0.21	0.08	0.11	-0.18	0.87	0.19	-0.11
17. Choice reaction time	0.01	0.12	-0.06	0.98	0.08	0.08	-0.03	-0.03
18. Rate control	0.20	0.28	0.22	0.04	0.00	0.10	-0.08	-0.90
19. Visual signal detection	0.05	-0.87	0.05	-0.17	-0.02	-0.21	0.26	0.32
20. Visual digit span	-0.24	-0.27	0.24	-0.04	-0.15	0.20	0.86	0.09
21. Track and monitor	-0.91	0.04	-0.06	0.00	0.15	-0.27	0.21	0.19
22. Terminal digits	-0.12	0.02	-0.02	0.09	0.97	-0.14	-0.10	0.00

Table VI-9

Research on Cognitive Task Performance in Hyperbaric Air

Reference	Pressure levels, ATA	Test environment	Subjects	Task	Measures recorded	Percent decrement relative to performance at 1 ATA				
						ATA 4	5½	7	8½	10
1. Shilling and Willgrube 1937	1–10	Dry chamber	Forty-six student divers and staff	Four problems: one each of addition, subtraction, multiplication, and division	a. Time to complete:	a. 12%	16%	23%	44%	53%
					b. Number correct	b. 3%	4%	6%	12%	16%
2. Adolfson 1965, 1967 Adolfson and Muren 1965	1–13	Dry chamber	Fifteen trained divers	Sets of problems of the following form: 48 + (2 × 4) = ?	Number correct:	ATA 4 2%		7 6%	10 25%	13 60%
3. Baddeley and Flemming 1967	1 and 7	Ocean	Eight experienced amateurs, members of a university diving club	Add sets of five two-digit numbers for 5 min	(at 7 ATA) Number completed: 19% Percent correct: 15%					
4. Bennett et al. 1969	1–10	Dry chamber	Five experienced divers	Multiply a two-digit by a one-digit number for 2 min	a. Number completed:	ATA 4 a. 5%	5½ 10%	7 12%	8½ 18%	10 21%
					b. Number correct:	b. 2%	11%	13%	21%	32%
5. Kiessling and Maag 1962	1 and 4	Dry chamber	Ten subjects: eight experienced, two inexperienced	Conceptual reasoning test	Time to complete (at 4 ATA)	33%				
6. Adolfson 1967	1–13	Dry chamber	Eight experienced divers	Continuous free association	Associative speed (number of associations in 60 sec):	ATA 4 4%		7 20%	10 3%	13 47%
7. Baddeley et al. 1968	1 and 4	Ocean	Eighteen experienced amateur divers	Sentence comprehension test	(at 4 ATA) Number completed: 13% Number correct: 15%					

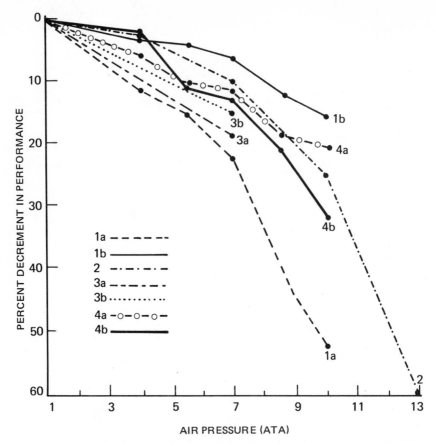

Figure VI-32. Effects of hyperbaric air on arithmetic task performance (see Table VI-9).

can be compared. Figure VI-32 presents graphically the results of arithmetic task performance as reported in studies 1–4 of Table VI-9. Depending on the specific conditions of the testing and the kind of measure used, performance can be seen generally to fall off in the area between 4 and 7 ATA, more sharply between 7 and 10 ATA, and reach a maximum percentage decrement of 60% at 13 ATA, the limit of the variable as studied.

Other kinds of tasks have been used to represent the applications of cognitive processes, but only a few have been used in conjunction with hyperbaric air. Research cited as entries 5–7 of Table VI-9 describes these studies and Figure VI-33 graphically presents their results. Performance in free associating to familiar words and comprehending simple, logical identities appears to generally follow the trend for arithmetic task performance. The decrement in performance on the conceptual reasoning task, however, is markedly more steep at 4 ATA than is that for the previous set of functions. The test involves complex cognitive operations of inductive hypothesis generation and deductive testing of the hypothesis, evaluation of results, and continued iteration of this procedure until a correct solution is discovered. The suggestion is that more

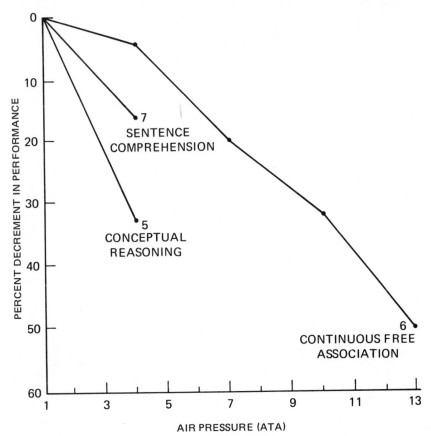

Figure VI-33. Effects of hyperbaric air on other cognitive task performance (see Table VI-9).

complicated tasks will result in a greater rate of deterioration with increased pressure than is currently depicted by the aggregated results of arithmetic task performance.

b. Reaction Time and Perceptual Speed

The Behnke *et al.* (1935) observation of "slowed mental activity" has led a few researchers to explore reaction time and perceptual speed as performance dimensions vulnerable to the narcotic effects of hyperbaric air. Table VI-10 is a characterization of representative research in these areas.

Simple visual reaction time was studied by Shilling and Willgrube (1937) and found to be a linear function of pressure between 1 and 10 ATA. No further study of this performance dimension has been made in the context of hyperbaric air, although 25 years later Kiessling and Maag (1962) used a two-choice reaction-time task as a measure of nitrogen narcosis. These results are presented in Figure VI-34; they show a 20% reduction in simple visual reaction time at 10 ATA, with intervening data

Table VI-10

Research on Reaction Time and Perceptual Speed in Hyperbaric Air

Reference	Pressure levels, ATA	Test environment	Subjects	Task	Measures recorded	Results
1. Shilling and Willgrube 1937	1–10	Dry chamber	Forty-six U. S. Navy student divers and diving staff	Perceptual speed (number crossout)	Number of crossouts	Fewer crossouts 1–10 ATA
					Number of errors	4 ATA −1; 7 ATA −2.5; 10 ATA −8.7 — More frequent errors as compared with 1 ATA
				Simple visual reaction time	Time delay, sec	0.7 1.8 2.6 ATA 1 5½ 7 8½ 10 RT 0.214 0.237 0.242 0.248 0.257 Decrement re 1 ATA — 10% 13% 16% 20%
2. Bennett *et al.* 1967	1, 2, and 4	Dry chamber	Eighty Royal Navy personnel; 19 very experienced, 14 no diving experience	Perceptual speed (card sorting)	Seconds per card	No change 1–4 ATA
					Errors per trial	Increase 1–4 ATA in average errors 1.1 to 2.4
3. Kiessling and Maag 1962	1 and 4	Dry chamber	Ten subjects: eight experienced divers and two medical students	Choice reaction time test: 2 stimuli, 2 responses	Time delay, sec	ATA 1 4 RT 0.237 0.287 Decrement — 21% re 1 ATA

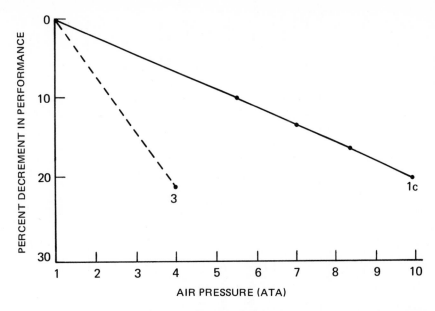

Figure VI-34. Effects of hyperbaric air on simple and choice reaction time (see Table VI-10).

points forming a remarkably precise rectilinear relationship with pressure. Choice reaction time achieves a comparable decline (21%) at only 4 ATA.

Perceptual speed has been examined by a number crossout task and a card-sorting task, with mixed results. Shilling and Willgrube (1937) reported data relative to a baseline at 1 ATA but did not report the baseline values. They reported progressively fewer correct crossouts at 4, 7, and 10 ATA, but the values are small (-1, -2.5, and -8.7) and the baseline unknown. Bennett et al. (1967), using a card-sorting task, found no significant change in sorting speed at 4 ATA relative to a baseline at 1 ATA. Both sets of authors reported very small incremental increase in errors but this appears to be an insensitive measure, since at 10 ATA, test personnel made only 2.6 more errors in the crossout task than they did at 1 ATA. Although the number of studies is small and the amount of data thin, the reaction-time work suggests decrement in performance with increasing pressure. This decrement, however, is less pronounced than is the case with cognitive tasks.

c. Dexterity-Task Performance

In 1935 Behnke et al. observed an "impaired neuromuscular coordination" in their laboratory workers at atmospheric pressures of 3 and 4 ATA. Many years later this aspect of performance became the subject of quantitative research to more precisely describe the values associated with the effects of breathing hyperbaric air. Table VI-11 presents descriptive details of representative search and Figure VI-35 graphically depicts their results. The negatively decelerating function is similar to that describing arithmetic task performance except for the location of the turning point.

Table VI-11

Research on Dexterity-Task Performance in Hyperbaric Air

Reference	Pressure levels, ATA	Test environment	Subjects	Task	Measures recorded	Results
1. Kiessling and Maag 1962	1 and 4	Dry chamber	Ten subjects: eight divers, two medical students	Purdue Pegboard Assembly	Number of assemblies	ATA 1, 4 Number 28.1, 25.9 % Decrement —, 8%
2. Adolfson and Muren 1965	1–13	Dry chamber	Fifteen experienced divers	Screwplate	Time to complete	ATA 4, 7, 10, 13 % Decrement 0, 4%, 10%, 35%
					Number of bolts mounted	ATA 1, 4, 7, 10, 13 N 16.5, 15.6, 15.1, 14.2, 10.0 % Decrement 3%, 6%, 12%, 38%
3. Baddeley 1966	1 and 4	Dry chamber	Eighteen subjects: Sixteen Army divers (Royal Engineers), two amateur divers	Screwplate	Time to complete	ATA 1, 4 Time, sec 196, 206 % Decrement 5%

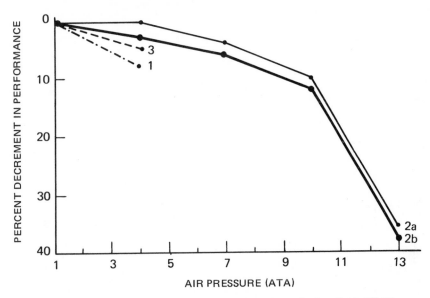

Figure VI-35. Effects of hyperbaric air on dexterity tasks (see Table VI-11).

While the decrement function for the arithmetic task turned sharply downward at 7 ATA, the function for dexterity tasks breaks at 10 ATA. Also, the extent of the maximum decrement is shown to be less than for the perceptual tasks and considerably less than for the cognitive task set.

d. Modifying Factors

Research results as reported are intended to portray the effects of breathing hyperbaric air on three categories of performance: cognitive, perceptual, and dexterity. Main effects, however, are known to be modified by a variety of factors, such as experience of the test diver, extent of practice at the task, rest–work condition of the test diver, and whether the testing was conducted in a dry chamber or in the water. A few samples of reported research are included to illustrate this point and to suggest the difficulty of direct application of research results to operational diving situations.

(i) Diver Experience. Test subjects who are more experienced in diving and more habituated to the conditions of diving will perform better than inexperienced divers exposed to the same conditions. Shilling and Willgrube (1937) presented data, for example, that permit the performance of experienced divers to be isolated from the total test group. Figure VI-36 shows this comparison; experienced divers were consistently less affected by the pressure at each level between 1 and 10 ATA. The task in this instance was the solution of arithmetic problems and the measure was time to complete four problems.

(ii) Resting vs. Working Diver. Susceptibility to the narcotic effects of breathing hyperbaric air is modified by the condition of the diver: resting vs. exercising. Adolfson

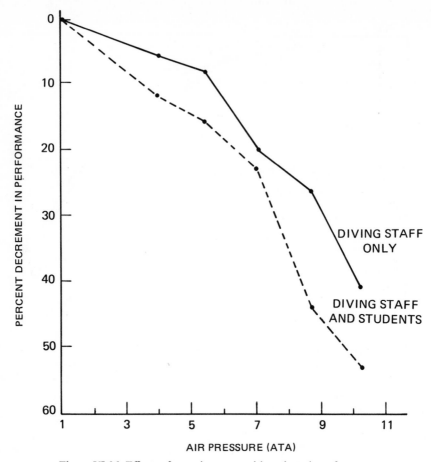

Figure VI-36. Effects of experience on arithmetic task performance.

(1965) has shown this effect for performance in both dexterity and arithmetic tasks (Figures VI-37 and VI-38). Significant decrement in dexterity task performance occurred at 10 ATA when test subjects were at rest, and at 4 ATA when they had been exercising for 5 min prior to testing. On the arithmetic task, performance was significantly reduced at 10 ATA at rest and at 7 ATA following exercise.

(iii) *Practice in the Task.* A persistent problem in the use of both apparatus and pencil/paper tests is the tendency of test personnel to improve with practice in the task. In tests designed to assess performance good experimental procedure calls for a period of task training to a criterion level so that changes in performance can be attributed to the effects of hyperbaric air rather than to the effects of practice. From an operational point of view, the significance of practice in offsetting the deleterious effects of exposure to hyperbaric-air breathing environments is an important consideration. Research data are most likely to be applied to a well-trained diver. That practice in the task does ameliorate the effects of exposure can be seen from data presented by Baddeley *et al.* (1968). Figure VI-39 shows the effect of practice as a modifier of per-

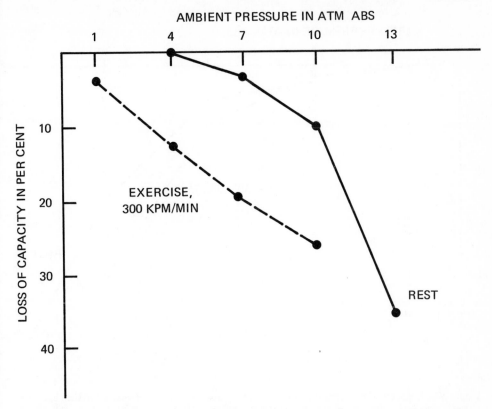

AMBIENT PRESSURE IN ATM ABS

Figure VI-37. Effects of hyperbaric air on manual dexterity at rest and during physical exercise. [From Adolfson (1965) by permission of the *Scandinavian Journal of Psychology*, Stockholm.]

formance decrement in a cognitive task. Test subjects consistently completed more test items and did more problems correct on the second day of testing as compared to the first day for both 1 and 4 ATA pressure. Figure VI-40 shows a similar effect for a dexterity task. Transfer of bolts from one side of a plate to another was accomplished in less time on the second day of testing independent of the difference in pressure (1 vs. 4 ATA).

(iv) Characteristics of the Test Environment. The majority of research into the effects of hyperbaric air on performance has been conducted in a dry-chamber test environment. The main justification for this is the relative ease with which other variables can be excluded or controlled so that results can be attributed to the effect of breathing hyperbaric air rather than to a variety of confounded variables. But the practical diving community is more concerned with predicting diver performance in the water and, more precisely, in the ocean. Testing in a water environment introduces a large number of variables which affect performance: facemask restrictions, visual distortions, suit and equipment encumbrances, air bubbles and distracting noises of breathing, facemask leaks, reduced tactile sensitivity and consequent loss of feedback cues, lack of a stable footing unless heavily weighted, etc. Beyond these variables, an ocean test environment imposes additional conditions which affect performance: the

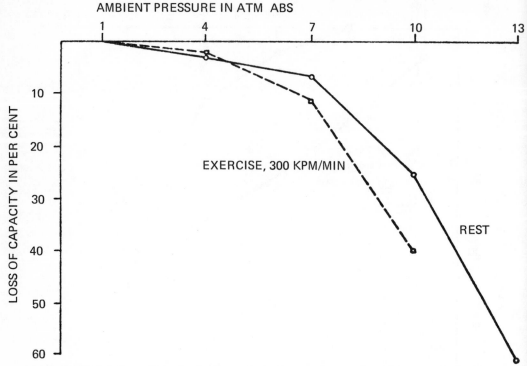

Figure VI-38. Effects of hyperbaric air on calculation capacity at rest and during physical exercise. [From Adolfson (1965) by permission of the *Scandinavian Journal of Psychology*, Stockholm.]

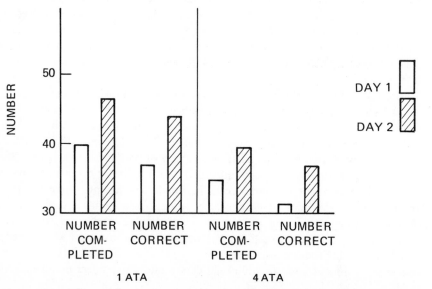

Figure VI-39. Effect of practice in a sentence comprehension task. [Based on data from Baddeley *et al.* (1968); used by permission of *Ergonomics*.]

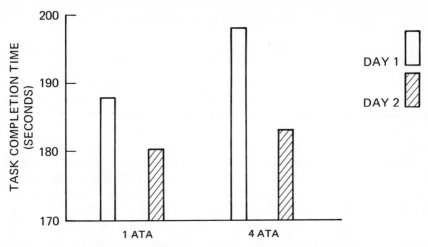

Figure VI-40. Effect of practice in a screwplate dexterity task. [Based on data from Baddeley *et al.* (1968); used by permission of *Ergonomics*.]

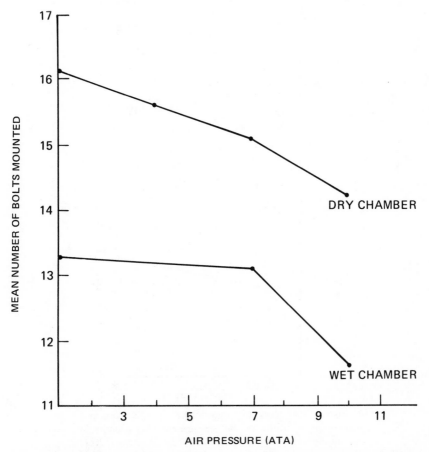

Figure VI-41. Effects of a water environment on performance in a dexterity task. [Based on data from Adolfson (1967); used by permission of *Acta Psychologica Gothoburgensia*, Stockholm.]

surge of a surf zone, the press of a tide or current, reduced visibility, anxiety about self-maintenance, etc.

A suggestion of the effects of these characteristics of the task environment can be seen in two studies. Adolfson (1967) compared performance on a dexterity task in a dry chamber and in a wet chamber at several levels of pressure. Figure VI-41 shows that the number of screw bolts transferred from one side of a plate to the other was consistently less in the wet than in the dry. The percentage reduction was 12% and 13%, respectively, from 1 to 10 ATA, but the absolute difference in level of performance was, on the average, 16% in favor of the dry-chamber condition.

Baddeley (1966a) has shown the effects of testing in the ocean as compared with a dry chamber at 1 and 4 ATA. The task was a screwplate test of manual dexterity similar to that used by Adolfson (1967), but the measure was time to complete the transfer of a fixed number of bolts, rather than number transferred. The results of this study are shown in Figure VI-42. From 1 to 4 ATA in the dry-chamber test environment, performance fell 5%, while the difference between 1 and 4 ATA in the ocean was 49%. Another aspect on Baddeley's (1966a) data is that task performance in the dry chamber at 4 ATA was reduced by 34% in the ocean at 4 ATA.

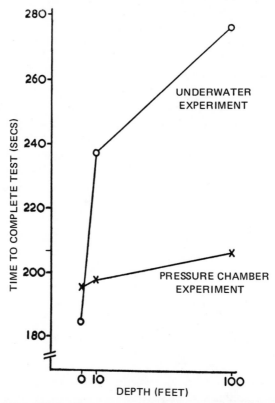

Figure VI-42. Time to complete the screwplate test as a function of depth for the pressure chamber experiment and the underwater experiment. [From Baddeley (1966a) by permission of the American Psychological Association.]

2. *Performance in Hyperbaric Heliox*

The general character of research in high-pressure, helium-based breathing environments is exploratory, probing for signs of physiological and performance dysfunction at ever greater pressure levels. The experiments are conducted in dry chambers with small numbers of highly select test divers; the measurement emphasis is on physiological characteristics; performance measures tend to be included as a secondary aspect of the overall measurement effort. The performance-oriented studies include the full range of psychomotor, perceptual, and cognitive performance dimensions using about the same set of tests and measures as were used in hyperbaric air studies.

The steadiness, tremor, and dexterity performance areas appear to offer most promise as sensitive and reliable indicators of pressure-related dysfunction in helium-breathing environments. This is an interesting contrast to the findings in hyperbaric air, where the cognitive performance dimensions gave the earliest and most reliable signs of degradation. Tests of postural equilibrium, postural and intentional tremor, and fine dexterity tend to show significant impairment under pressure conditions of 30–40 ATA, although several questions remain unanswered—e.g., whether the performance change is a function of inert gas narcosis, absolute pressure level, or compression rate, and whether the effects are adapted to with time at pressure. Other performance dimensions tend to show no change attributable to differences in pressure. Two phenomena are frequently encountered: First, performance on tests administered periodically throughout a compression/decompression cycle tends either to improve or deteriorate as a function of time but independent of the level of pressure; second, performance decrements measured just before or just following the end of the compression phase of the test dive recover with time at the constant-pressure level, particularly in those instances where compression has been acccomplished at high rates. The first phenomenon, behavior change with time of dive but independent of pressure, is usually accounted for as a practice effect when performance improves, and as a motivational effect when performance deteriorates. The second phenomena, performance decrement at compression recovering with continued exposure, is variously attributed to anxiety, temperature–humidity changes accompanying pressure changes in the chamber, and rate of change in the hydrostatic pressure per se.

a. Tremor, Steadiness, and Dexterity Tasks

Table VI-12 presents the findings of representative work relating exposure to hyperbaric heliox-breathing environments and performance in motor-control tasks. Performance tests include direct measures of postural tremor taken via velocity transducer, indirect measures of tremor via tests of hand steadiness or dexterity, and measures of postural equilibrium via balance rails and boards.

On the whole, the evidence tends to show a lack of significant effect except in cases of high rates of compression when divers were tested just prior to or just following the compression phase of the dive. In these instances, significant performance

Table VI-12

Research on Tremor, Steadiness, and Dexterity Task Performance in Hyperbaric Heliox

Reference	Pressure level and compression rate	Test environment	Subjects	Task	Measures recorded	Results
1. O'Reilly 1973	1 vs. 16 ATA two-stage compression 43 hr at depth 15 ft/min compression	Dry chamber	Six subjects: three Makai Range divers, three scuba divers from U. of Hawaii	Minnesota rate of manipulation Screwplate Hand maze	Dexterity: completion time Dexterity: completion time Two-hand coordination: number completed	No decrement Performance improved with practice Performance improved with practice
2. Bachrach *et al.* 1971	1–16 ATA two-stage compression 17 ft/min to 5 ATA 0.67 ft/min to 16 ATA	Dry habitat Aegir	Three subjects	Intentional tremor	Tremor amplitude	No change in magnitude or in magnitudes within frequency bands; intraindividual consistency
3. Bennett 1966	1–25 ATA 100 ft/min compression	Dry chamber	Two to six subjects	Ball-bearing maze tracking	Hand steadiness: number of ball bearings placed in tube; number of parallel line touches in 90 sec	Impaired performance 10–30 min following compression, then recovery
4. Bühlmann *et al.* 1970	1–31 ATA Compression at 30–50 ft/min 75 hr at 31 ATA	Dry chamber	Six subjects: experienced divers of Royal Navy Deep Trials Unit	Drawn points Marked points	Hand steadiness: number of dots placed Hand steadiness: number of dots placed number of balls placed	No change Decrement within 10 min of compression, then recovery

Reference	Conditions	Chamber	Subjects	Tests	Measures	Results
5. Bennett and Towse 1971	1–46 ATA Stage compression 17 ft/min 10 hr at 46 ATA	Dry chamber	Two subjects: one experienced recreational diver, one inexperiencd	Purdue Pegboard Assembly Ball-bearing test Postural tremor	Dexterity: number of assemblies number of balls placed Amplitude per frequency bands	Impairment after compression, then recovery One subject showed increase, other did not
6. Braithwaite et al. in press 7. Berghage et al. 1974	1–49.5 ATA Stage compression 0–14 ft at 14 ft/min 14–400 ft at 5 ft/min 400–1000 ft at 0.67 ft/min 1000–1300 ft at 0.50 ft/min 1300–1600 ft at 0.33 ft/min 7 days at 49.5 ATA	Dry chamber	Six subjects: experienced U.S. Navy diving personnel including two diving medical officers	Balance rail statometer Intentional tremor	Standing steadiness: time on rail, lateral axis amplitude and frequency Tremor amplitude and frequency	Gross disturbance of postural equilibrium at pressures >40 ATA Increase in amplitude at pressures between 31 and 49.5 ATA; also shift in frequency of the tremor signal

impairments as compared to baseline values were noted. Given time at depth, however, the effect apparently diminished, as tests administered 30–60 min after compression yielded performance levels approximating the baseline values. The most recent chamber test to 49.5 ATA, however, presents a strong case for impairment of performance in these task areas. Balance-rail and balance-board tests of postural equilibrium showed consistent and progressive deterioration in performance at 31, 40, and 49.5 ATA test administrations on both compression and decompression sides of the maximum pressure, 49.5 ATA (Braithwaite *et al. in press*). Furthermore, test divers remained seven days at 49.5 ATA with no apparent adaptation to the conditions underlying performance decrement in this task. Also during this chamber test, the NMRI MK3 intentional-tremor device revealed significant changes in both amplitude and frequency as a function of pressure. No consistent evidence was found for or against adaptation of the effect of pressure with time. On successive test administrations while at 49.5 ATA, one version of the test (maintaining a 50-g force) showed improved performance while the other (maintaining a 500-g force) showed continued deterioration (Berghage *et al. in press*).

b. Reaction Time and Perceptual Speed

Tables VI-13 and VI-14 show sets of research studies conducted in the range of 1–46 ATA in heliox environments where reaction-time and perceptual-speed tests were used. Examination of Table VI-13 reveals reaction time studied by simple visual two-choice, and four-choice reaction time with no indication of decrement. Table VI-14 presents essentially an identical picture for perceptual speed: no change in perceptual speed, as evidenced by a variety of tests, to 46 ATA.

c. Cognitive Tasks

Cognitive-task performance as a correlate of hyperbaric helium environments has been studied in two areas: arithmetic computation and associative memory. Since tests of simple arithmetic computation had proven to be sensitive and reliable indicators of nitrogen narcosis in hyperbaric air, they were used again in hyperbaric heliox environments as indicators of helium narcosis. The majority of the evidence shows performance on these tests to be unaffected by helium-based breathing environments to 46 ATA, although P. B. Bennett (1965) reported significant decrements in accuracy scores of 18% at 19 ATA and 42% at 25 ATA. Table VI-15 presents a summary of the relevant studies.

Performance in tests of associative memory, on the other hand, present a mixed picture (Table VI-16). O'Reilly (1973) used a standardized picture–number test of associative memory and found a 25% decrement in performance at 16 ATA as compared to performance at the surface. Time interval between learning and recall was not specified. Biersner and Cameron (1970a) administered word–number tests of associative memory at 19 and 31 ATA. No decrement in performance occurred at either pressure level when the recall interval was short, 1–5 min; however a significant 23% decrement was found at 19 ATA when the recall interval was 60 min. Biersner and Cameron (1970a) attributed this latter result to observed differences among the

Table VI-13

Research on Reaction Time in Hyperbaric Heliox

Reference	Pressure level, ATA	Test environment	Subjects	Task	Measures recorded	Result
1. O'Reilly 1973	1 and 16	Dry chamber	Six divers	Simple visual reaction time	Response latency	No decrement
				Two-choice visual and auditory reaction	Response latency and errors	No decrement in speed; progressive increase in errors over days in the chamber, but not related to pressure
2. Summitt et al. 1969	1–31	Dry chamber	Five trained divers	Two-choice visual reaction	Response latency and errors	No decrement in speed or accuracy
3. Parker 1969	1–34	Dry chamber	Two experienced commercial divers	Four-choice visual reaction	Response latency and errors	No decrement in speed or accuracy

Table VI-14

Research on Perceptual Speed in Hyperbaric Heliox

Reference	Pressure level, ATA	Test environment	Subject	Task	Measures recorded	Result
1. Bennett et al. 1967	1 and 4	Dry chamber	Eighty Royal Navy volunteers of varied diving experience	Card sorting	Number of cards sorted in 10 min; number of errors	No decrement in speed or accuracy
2. O'Reilly 1973	1 and 16	Dry chamber	Six scuba divers	Number comparison test Identical pictures test	Number correct	Equivocal results; one test suggested decrement, the other did not
3. Bühlmann et al. 1970	1–31	Dry chamber	Two to six experienced deep divers	Deciphered letters test[a]	Number correct	No decrement in accuracy
4. Parker 1969	1–34	Dry chamber	Two experienced commercial divers	Letter cancellation test	Number of cancellations; number of errors	No decrement in speed or accuracy; significant individual differences and intraindividual consistency
5. Bennett and Towse 1971	1–46	Dry chamber	Two laboratory staff personnel: one inexperienced and one experienced scuba diver	Wechsler Bellevue digit symbol test	Number correct	No decrement in accuracy

[a] An adaptation of the Wechsler Bellevue digit–symbol test.

Table VI-15

Research on Arithmetic Task Performance in Hyperbaric Heliox

Reference	Pressure level, ATA	Test environment	Subject	Task	Measures recorded	Result
1. Bowen et al. 1966	1 and 7	Dry habitat: SEALAB II	Twenty-seven SEALAB II divers	Two-digit by one-digit multiplication	Number attempted; number correct	No decrement in number of items attempted, number correct, or ratio of correct to attempted
2. O'Reilly 1973	1–16	Dry chamber	Six scuba divers	Repetitive dive table problems	Completion time; number of errors	Improved with practice
3. Bennett, P. B. 1965	1–25	Dry chamber	Six subjects at 19 ATA; four subjects at 25 ATA	Two-digit by one-digit multiplication	Number attempted; number correct	No decrement in number attempted; decrements of 18% (19 ATA) and 42% (25 ATA) in accuracy
4. Bühlmann et al. 1970	1–31	Dry chamber	Two to six experienced deep divers	Two-digit by one-digit multiplication	Number attempted; number correct	No decrements
5. Parker 1969	1–34	Dry chamber	Two experienced commercial divers	Add sets of 11 one-digit numbers	Number attempted; number correct	No decrements; intraindividual consistency in performance level
6. Bennett and Towse 1971	1–46	Dry chamber	Two laboratory staff personnel: one inexperienced and one experienced scuba diver	Two-digit by one-digit multiplication	Number attempted; number correct	No decrements

Table VI-16

Research on Memory Task Performance in Hyperbaric Heliox

Reference	Pressure level, ATA	Test environment	Subject	Task	Measures recorded	Result
1. O'Reilly 1973	1–16	Dry habitat (Aegir habitat moored on surface at Makai Range pier)	Six divers with no experience in saturation diving	Picture–number test of associative memory	Number of correct associations recalled	25% decrement
2. Biersner and Cameron 1970a	1–19	Dry chamber	Twenty U. S. Navy divers	Nonstandardized word–number test of associative memory	Number of correct recalls after 5-min interval	No decrement
					Number of correct recalls after 60-min interval	23% decrement
3. Biersner and Cameron 1970b	1–31	Dry chamber	Ten divers: five in experimental, five in control group; Navy divers and civilian diving technicians	Word–number paired associates test	Number of correct recalls after 1-min interval	No decrement

test divers in anxiety level, and recommended that tests of anxiety be used to screen potential participants in experimental dives.

3. *Work and Maneuver in a Tractionless, Viscous Medium*

Man's potential as an undersea worker, the limitations of his roles, and his work-support requirements are specified by information about his performance characteristics in work and maneuvering tasks. Three areas of performance are included in this section: basic force production, swimming and load carrying, and orientation and maneuvering. Research programs contributing knowledge to these areas of performance have been conducted primarily in the ocean with reasonably well-trained divers. Performance measures are not factor-based tests but job samples approximating operational tasks.

a. Basic Force Production

Because the water offers resistance to movements due to its density, force production undersea is not an approximation of values obtained in air; and, since the diver is usually neutrally buoyant in the water, he must be given restraining aids in order to apply any significant force. Dimensions of force production and conditions which affect force production have been researched principally by Streimer and his associates (Streimer 1969a, Streimer 1969b, Streimer *et al.* 1971, Streimer *et al.* 1972), Barrett and Quirk (1969), and Norman (1969).

(i) Land vs. Water Work Environments. Streimer and his associates conducted a 5-yr program of research under Office of Naval Research sponsorship to describe man's underwater work characteristics for two basic forms of manual work: rotary and linear. The rotary work task was to continuously rotate a crankhandle in a counterclockwise fashion for a 15-min interval. The test diver knelt on the pool or ocean bottom and braced himself with his left hand while performing the rotary work task with his right. The linear task was a push–pull reciprocating work task similar to hacksawing and the time interval and restraint system were as described for the rotary work task (Streimer 1969a). These basic tasks were used in a variety of configurations (resistance levels for both tasks and crankhandle radii for the rotary task) in a series of studies to examine the effects of land vs. water work environments on the power-output/energy-cost relationship.

Table VI-17 presents power-output data for the linear and rotary tasks in a variety of configurations as performed in air and in shallow water. Over all configurations, the average power output for the rotary task on land was 0.045 hp and in water 0.032 hp, a reduction of 29%. For the reciprocating linear task these values were 0.023 and 0.019 hp, a reduction of 17%. The rotary tasks suffered greater percentage performance decrement than did the linear task in the land vs. water comparison, but the rotary task was superior in terms of power output in the water, 0.032 hp for the rotary task vs. 0.019 hp for the linear. Of further interest, it can be seen from Table VI-17 that the most effective configuration for the rotary task was the 9-in handle combined with a 9-lb resistance. This combination yielded the greatest power output,

Table VI-17

Land vs. Underwater Power Production Comparisons, by Work Mode and Configuration (Streimer et al. 1971)

Resistance level, lb	Continuous rotary effort								
	6-in. radius			9-in. radius			12-in. radius		
	Land	Water	% Degradation	Land	Water	% Degradation	Land	Water	% Degradation
3	0.025	0.016	36.0	0.031	0.020	35.4	0.033	0.021	36.3
6	0.041	0.029	29.2	0.047	0.036	23.4	0.052	0.036	30.7
9	0.048	0.038	20.8	0.058	0.045	22.4	0.068	0.045	33.8

Continuous reciprocating linear effort

Resistance level, lb	Land	Water	% Degradation
6	0.019	0.016	15.7
9	0.024	0.020	16.6
12	0.027	0.022	18.5

0.045 hp, and the next-to-least reduction in output vis-à-vis land values, 22%. Furthermore, the 9-lb resistance level, 9-in. cranking radius was determined to be the most efficient from the standpoint of power output per energy cost. Energy cost was defined as liters of oxygen used by the diver per minute of work, and efficiency was defined as liters of O_2 per minute per horsepower. Table VI-18 presents the efficiency characteristics of the range of configurations for both rotary and linear tasks. Again, it is clear from Table VI-18 that rotary work is not only more productive, but is more efficient by a factor of over 100%. Average efficiency of the reciprocating task was 59.6 liters of O_2 per min per hp, while the average efficiency of the rotary task over all of the configurations was 28.2 liters of O_2 per min per hp.

(ii) Depth Effects. Streimer and colleagues conducted a series of tests to determine the effects of increased depth of water on the power-output/cost relationship, and again, both rotary and linear tasks were used as work elements. Tests were conducted at 5, 33, and 66 ft (1.15, 2, and 3 ATA); foot-pounds per minute was used as the unit of output; and oxygen uptake and breathing-gas consumption rate were used as measures of costs. Table VI-19 presents the results of the study series; several important conclusions can be drawn from them. First, power output is independent of water depth, and the relative productivity of linear vs. rotary tasks is not affected by changes in water depth. Linear task output is constant at 660 ft-lb/min (0.020 hp), while rotary task output averages 1320 ft-lb/min (0.040 hp), twice the level of productivity of the linear task. Second, the energy cost of work remains constant across task types and depths. Test divers worked at a level of energy expenditure defined as 1.20 liters of oxygen uptake per minute. Third, the cost factor clearly affected by depth is breathing-gas consumption rate. The number of liters of air used per minute at 33 and 66 ft is two and three times the rate at 5 ft. As determined by this and other studies, gas consumption is a direct, one-to-one linear function of pressure as measured by atmospheres absolute (ATA).

From these data Streimer has derived a generalized "constancy of input" hypothesis. He suggests that, given conditions of self-paced work, divers will select for themselves a relatively low level of energy input permitting them to work in an aerobic (oxygen debt-free) condition over sustained time intervals. This low level of energy expenditure will be constant over variations in task difficulty. Output will be adjusted according to the ease of the task, enabling input to remain constant at a comfortable level.

(iii) Work-Load Effects. Using the self-paced work mode as a definition of baseline power output, test divers were made to work at 10, 20, and 30% greater levels of output (Streimer *et al.* 1972). Baseline power output was 1100 ft-lb/min for the sample of divers who participated in the test using the rotary task, a 10-min work period, at a 33-ft water depth (2 ATA). Test divers were then "forced" to work at 1236, 1352, and 1463 ft-lb/min, approximately 10, 20, and 30% increases over the baseline or self-paced level. Changes in three cost-oriented variables were compared to baseline values: breathing-gas consumption rate, oxygen uptake, and pulmonary efficiency as measured by percent of oxygen removed from the breathing gas (air).

Table VI-20 presents the results of this test series and Figure VI-43 presents the data as percentages of baseline values. Examination of Figure VI-43 reveals an intriguing set of relationships. When men are forced to work at levels of production in

Table VI-18

Comparisons of Productivity and Energy Cost in Underwater Work as a Function of Work Mode for Representative Task Configurations (Streimer *et al.* 1971)

Configuration (resistance level, crank radius)	Continuous rotary						Continuous linear reciprocating	
	6 lb, 6 in.	6 lb, 9 in.	6 lb, 12 in.	9 lb, 6 in.	9 lb, 9 in.	9 lb, 12 in.	6 lb	9 lb
Horsepower	0.029	0.036	0.036	0.038	0.045	0.045	0.016	0.020
Liters of O_2 min^{-1} hp^{-1}	34.4	29.3	28.9	26.3	24.8	25.2	65.7	53.6

Table VI-19

Breathing-Gas Consumption, Oxygen Uptake, and Power Output–Depth Comparisons (Streimer *et al.* 1972)

	5 Foot Depth	33 Foot Depth	66 Foot Depth
Breathing Gas Consumption Rate at Depth, (Liters air/min)	20.5 / Not Available	40.4 / 39.5	59.2 / 61.1
Oxygen Uptake (Liters O_2/min, STPD)	1.03 / 1.15	1.20 / 1.18	1.18 / 1.22
Power Output, (Ft.-lbs/min)	1485 / 660	1287 / 660	1320 / 660

Legend   ☐ - Linear Task   ▨ - Rotary Task

excess of their self-paced level, cost variables change in near-perfect linear correspondence to the percentage increase in power output. Breathing-gas consumption increases on a $1:1\frac{1}{2}$ basis, oxygen uptake increases on a $1:1$ basis, and pulmonary efficiency decreases on a $1:\frac{1}{2}$ basis.

(iv) Direction of Force Application. Norman (1969) determined average magnitude of force application against a lever using underwater test apparatus and six experienced scuba divers in good physical condition. One variable in the test program

Table VI-20

Gas Consumption, Oxygen Uptake, and Pulmonary Efficiency for Work Loads 10, 20, and 30% beyond the Self-Paced Level (Streimer *et al.* 1972)

	Self-paced work load (1110 ft-lb/min)	+10% (1236 ft-lb/min)	+20% (1352 ft-lb/min)	+30% (1463 ft-lb/min)
Breathing-gas consumption, liters/min	40.81	46.83	53.27	62.28
Oxygen uptake, liters/min	1.31	1.43	1.53	1.66
Pulmonary efficiency, % O_2 removed from air	3.24	3.05	2.87	2.68

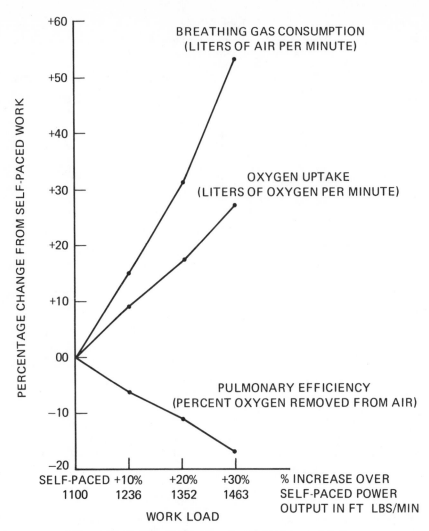

Figure VI-43. Effects of increased work load on cost indices.

was direction of force application against the handle. Table VI-21 presents the average force in pounds exerted against the lever for both impulse and sustained levels of effort. Impulse force was maximum effort sustained over a 1-sec interval; sustained effort was for 4 sec. The entries in Table VI-21 are averages over a large series of runs that included a range of diver restraint systems to be described shortly. In general, it can be seen that push–pull directions are approximately twice as productive as either up–down or left–right directions of force application; and forces applied over a 1-sec interval are approximately twice as great as can be applied over a 4-sec interval.

Barrett and Quirk (1969) determined average force applications against vertical, overhead, and deck work surfaces for 1-, 3-, and 6-min time intervals. While Norman's data were averaged over a variety of restraint systems, Barrett and Quirk's data were obtained using a waist-restraint system for two-handed tasks and a left-hand restraint

Table VI-21

Mean Sustained and Impulse Forces (lb)
as a Function of Direction of Application[a]

Direction	Sustained	Impulse
Push	20	52
Pull	21	51
Up	12	25
Down	16	29
Right	12	23
Left	14	26
Mean	16	34

[a] Reprinted from Norman (1969) with permission
of *Human Factors*.

system for one-handed tasks. Test were conducted in the ocean at a depth of 50 ft
and results are presented in Figure VI-44. Differences in force application were significant for one vs. two hands and for time intervals; differences according to orientation of the work surface were negligible and statistically not significant. The 6-min
interval was judged by the majority of test divers to be a very tedious effort, at or
very near the limiting condition for diver work tasks.

(v) *Restraint Systems.* A major variable in the determination of how much force
the average diver can exert in underwater work tasks is the nature of the restraining
systems. Norman (1969) conducted a detailed and thorough study of restraint systems
for applications to the space program. The study was conducted under water as a
simulation of weightlessness, and except for the condition that test personnel wore
pressurized Apollo space suits rather than diving apparel, the results should have
direct application to the working diver. Results are presented in Table VI-22 for
both sustained (4-sec) and impulse (1-sec) time intervals, and for six directions of
force application. The hand-and-shoes restraint condition is most like the condition
under which the Streimer group conducted their tests, and it is interesting to note

Table VI-22

Mean Sustained and Impulse Forces (lb) as a Function of Restraints[a]

	Restraints														
	None		Hand		Waist		Shoes		Hand and waist		Hand and shoes		Waist and shoes		Hand, waist, and shoes	
Direction	S	I	S	I	S	I	S	I	S	I	S	I	S	I	S	I
Push	0	35	1	41	15	43	4	46	29	57	30	62	35	58	43	69
Pull	0	43	2	43	22	46	4	48	31	51	33	61	37	57	38	61
Up	0	19	5	21	10	23	17	28	14	23	18	31	17	28	17	29
Down	2	23	9	26	10	23	21	33	16	26	26	37	19	30	21	32
Right	0	18	10	22	12	22	9	22	15	25	16	28	14	23	16	27
Left	0	18	17	29	12	22	8	23	17	28	21	34	15	25	19	30

[a] Reprinted from Norman (1969) with permission of *Human Factors*.

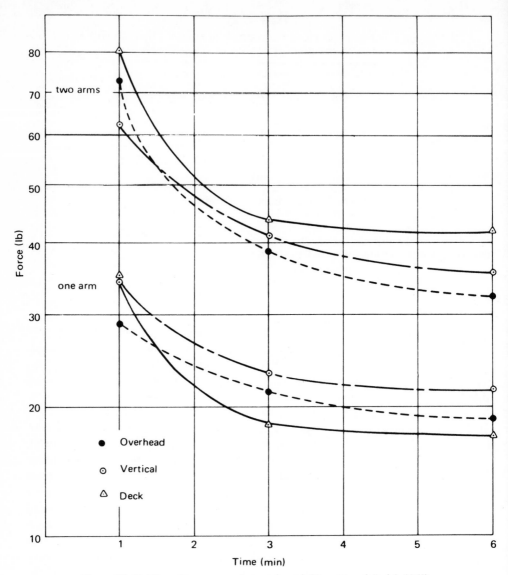

Figure VI-44. Diver force curves (approximate) (Barrett and Quirk 1969).

from Table VI-22 that this restraint condition is optimal for up–down and left–right force applications, and roughly the equivalent of any other two- or three-variable restraint system for push–pull.

b. Swimming and Load Carrying

(i) Swim Speed and Energy Costs. Although estimates of the range of swim speeds accomplished in underwater, finned swimming have been as high as 1.0–2.4 knots (Christianson *et al.* 1965), this result was obtained for a 30-sec time interval

Table VI-23

Oxygen Consumption and Respiratory Minute Volume (RMV) at
Different Work Rates (U. S. Navy 1970)

Activity		Oxygen consumption,[b] liters/min (STPD)	RMV,[c] liters (BTPS)
Rest	Bed rest (basal)	0.25	6
	Sitting quietly	0.30	7
	Standing still	0.40	9
Light work	Slow walking on hard bottom[a]	0.6	13
	Walking, 2 mph	0.7	16
	Swimming, 0.5 knot (slow)[a]	0.8	18
Moderate work	Slow walking on mud bottom[a]	1.1	23
	Walking, 4 mph	1.2	27
	Swimming, 0.85 knot (average speed)[a]	1.4	30
Heavy work	Maximum walking speed, hard bottom[a]	1.5	34
	Swimming, 1.0 knot[a]	1.8	40
	Maximum walking speed, mud bottom[a]	1.8	40
Severe work	Running, 8 mph	2.0	50
	Swimming, 1.2 knots[a]	2.5	60
	Uphill running	4.0	95

[a] Underwater activity.
[b] All figures are average values. There is considerable variation between individuals. STPD means standard conditions.
[c] The RMV values are approximate for the corresponding oxygen consumption. Individual variations are large. BTPS means body temperature, existing barometric pressure, saturated with water vapor at body temperature.

at a maximum kicking effort. Considering sustained swimming on the order of 1 hr, a more reasonable estimate of swimming speeds is 0.5–1.2 knots (U. S. Navy 1970). Table VI-23 compares various values along the swim-speed range with standard air environment activity in terms of work load. Work load in this table can be viewed by category from rest to severe work or in terms of both oxygen uptake in liters per minute and respiratory minute volume. Slow swimming at 0.5 knot is comparable to walking at 2 mph; swimming at 1.2 knots is comparable to running at 8 mph in terms of oxygen consumption rates. Figure VI-45 presents average oxygen consumption rates for selected speeds within the range 0.5–1.2 knots for a sample of U. S. Navy Underwater Demolition Team swimmers and a sample of "good" swimmers from the U. S. Navy Experimental Diving Unit. These men are among the most highly select and well-conditioned underwater swimmers in the national population and O_2 consumption rates in Figure VI-45 may be considerably underestimated if applied to men in less than excellent physical condition.

Morrison (1973) measured oxygen consumption rates for a sample of Royal Navy Harbor Clearance divers instructed to swim at maximum sustainable effort for 10 min. Tests were conducted at water depths of 6, 78, and 176 ft and oxygen uptake was determined to average 2.8 liters/min and to be invariant as a function of depth. This finding substantiates the results of Streimer's work with rotary and linear tasks and extends the range within which O_2 uptake is known to be independent of depth

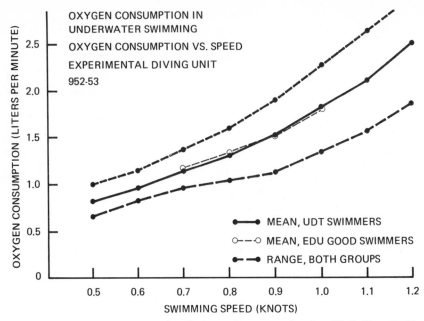

Figure VI-45. Oxygen consumption during underwater swimming (U. S. Navy 1970).

from 66 to 176 ft. Morrison's work has been used to establish a design guide for semi-closed underwater breathing apparatus: Oxygen flow rates must provide 3.0 liters of O_2/min in order to accommodate the short-term, high-work-load, energy-expenditure requirements of the working diver/swimmer.

(ii) Load Carrying. Andersen (1968) and Andersen *et al.* (1969) studied diver characteristics as load carriers. An initial test series examined swim speed as a function of drag in transporting neutrally buoyant loads. The basic load consisted of a canvas field pack and a flotation bladder. The field pack was weighted to 23.5 lb as in an operational condition, then made neutrally buoyant by inflation of the flotation bladders. Inflated, each outfit was sized at 11 × 12 × 9 in. (1188 in.²). Test divers were students and staff scuba divers from Scripps Institute of Oceanography. They carried no packs, one, two, and three packs in tests of drag effects on swimming speed over a 780-ft distance and at 30-ft ocean depth. Figure VI-46 shows the percentage decrement in swim speed from a baseline defined by the no-pack condition. This figure shows an essentially linear relationship between step increases in drag and percentage reduction in swim speed; over the 780-ft swim distance, speed was reduced by 15, 22, and 29%.

In a second series of field tests, the packs were made 3, 6, and 9 lb negative and test swims were made to examine the effects of negative weight in addition to drag effects. Test results showed that swim speed was not significantly reduced beyond the effects of drag alone by the addition of up to 9 lb weight.

A third test series was conducted to expand the range of weight carried. Packs were made 25 lb negative at 15 ft (pool bottom) and the diver's task was to transport one, two, and three packs (25, 50, and 75 lb) a distance of 180 ft. Only in the 25-lb

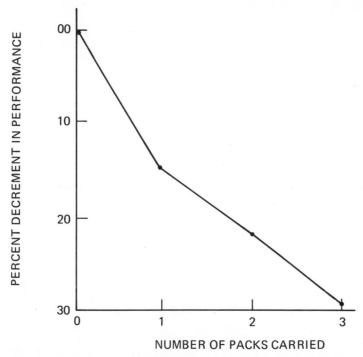

Figure VI-46. Swimming speed percentage decrement as a function of increased drag in carrying neutrally buoyant packs. [Based on data from Andersen *et al.* (1969).]

negative weight condition could the test divers swim with the pack; at 50 and 75 lb (two and three packs) the divers walked along the bottom carrying or dragging the packs. Apparently the limiting condition of the use of swimmers vs. underwater walkers as load carriers lies between 25 and 50 lb negative.

c. Orientation and Maneuvering

(i) Postural and Geographic Orientation. The characteristics of the underwater environment tend to reduce visual and kinesthetic cues to the direction of the gravitational vertical; the diver can become uncertain about which direction is up. Both laboratory experiments (Mann and Passey 1951, Margaria 1958, Brown 1961, Diefenbach 1961, and Nelson 1967) and field tests (Ross *et al.* 1969) support the general conclusion that the magnitude of error in estimation of the vertical is a function of the estimator's angle of tilt from the vertical. When the diver is in a head-down orientation his mean error estimation is 30°; when he is in a head-up orientation, mean error is only 7°. Mean error in estimating the gravitational vertical appears to approximate a normal distribution around 180°, the head-down position (see Figure VI-47). Furthermore, the magnitude of the errors of estimation of the vertical is a function of the amount of time the subject has been exposed to a given tilt angle. The longer

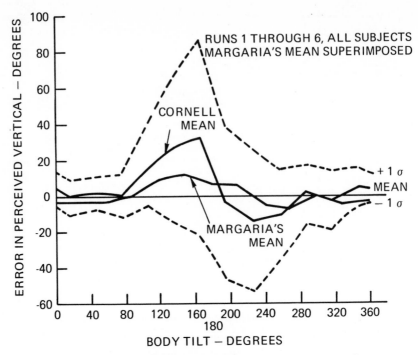

Figure VI-47. Average blindfolded directional error corrected for setting error vs. body tilt. [From Diefenbach (1961).]

the diver is tilted with reference to the vertical, the greater will be his error in estimating the direction of the vertical (see Figures VI-48 and VI-49).

Margaria (1958) reports an important learning effect in judging the gravitational vertical. Errors as great as 180° were made by naive subjects during their first four or five experiences with tilt.

Ross *et al.* (1970), conducted a series of ocean tests to compare performance involving various aspects of direction and distance judgments in land vs. underwater travel. The purpose of these tests was to define the capability of an unaided diver to geographically orient himself on the sea floor. In the Triangle Completion Test, divers attempted to swim a direction and distance which would close an isosceles triangle. As compared to their performance on a comparable land task, divers swam too great a distance and made larger angular errors (Figure VI-50). Of particular interest was that three of the divers made veering swims to the right, tending to swim in a right-handed circle.

In the Distance Traveled Test the task was to swim (walk) a series of distances: 15, 30, 45, 60, and 75 ft. Divers tended to swim further underwater than they walked on land, and the shorter distances were more poorly estimated than the longer ones.

Finally the diver's ability to estimate circles and sectors of circles was tested and compared to identical trials on land. A 12-ft line was attached to a center pole at a 6-ft height off the ocean floor. Divers were to swim the full 12-ft-radius circle and various segments of the full circle. Compared to performance on land, divers tended

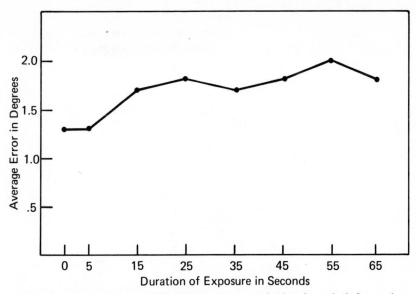

Figure VI-48. Average error of adjustment to gravitational vertical for various durations of exposure to body tilt. [From Mann and Passey (1951) by permission of the American Psychological Association.]

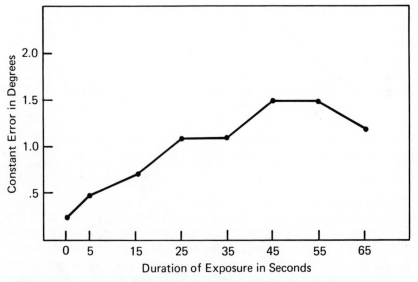

Figure VI-49. Mean constant error in direction of initial tilt for adjustment to the gravitational vertical for various durations of exposure to body tilt. [From Mann and Passey (1951) by permission of the American Psychological Association.]

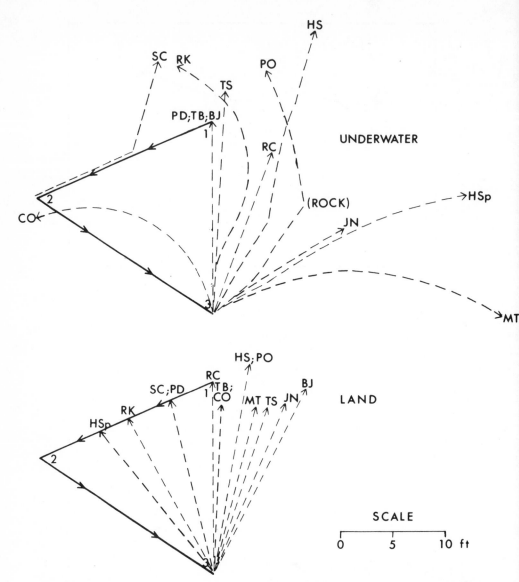

Figure VI-50. Routes taken by blindfolded subjects under water and then on land when attempting to complete the third side of the triangle. [From Ross *et al.* (1970) by permission of *Human Factors.*]

to stop short. Mean angular error on land was +9° compared to −48° underwater. The perimeter of the circle was 75 ft and the mean stopping point was 15 ft short. Ross *et al.* (1970) call attention to the discrepancy of underestimating linear distances and overestimating circular distances under water. They offer two alternate explanations for further tests. The first is that divers were attempting to estimate angles rather than distances in the circle-completion task. The second is that more effort is required to swim in a circle than in a straight line, and that distance judgments are biased by estimates of effort expended.

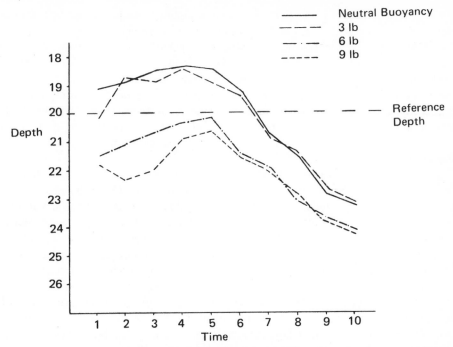

Figure VI-51. Average depth profile for four weight conditions over time increments from beginning to end of test run (Anderson 1968).

(ii) Course and Depth Keeping. Compass-aided swimmers have been tested for course-keeping performance on an underwater test range (Andersen 1968). The test range consisted of an anchored buoy as a starting marker and a 250-ft baseline staked out on the ocean floor perpendicular to a line 780 ft from the starting buoy. Average heading error for the compass-aided, 780-ft swim was 5.2°.

Depth-keeping performance of swimmers using depth gauges has been studied as a function of buoyancy condition (Andersen 1968). Instructed to swim at 20-ft, overall depth keeping was found to be plus or minus 1 ft. Profiles over time of swim, however, revealed a constant error in depth as time progresses. The profiles were also significantly affected by diver buoyancy and weight conditions: neutral and 3 lb negative vs. 6 or 9 lb negative (see Figure VI-51).

4. Performance in the Cold

Among the several environmental factors affecting man's performance under water, cold has been one of the most extensively studied, and the literature covers the range of performance categories. Although research on the performance of submerged, wet-suited divers is a relatively recent activity, a considerable knowledge about the effects of hand and body cooling exists in the literature concerning cold air. This literature is relevant to the extent that some physiological consequences of cold-air exposure are reported so that performance changes can be related to a common denominator such as finger, hand, or mean skin temperature.

Unfortunately, the research literature has several weaknesses which limit the strength of statements that can be made with confidence about cold-water exposure/ performance-consequences relationships. The first is the difficulty in linking the *cause* of performance decrement to temperature changes or body heat loss as separate from the decrement occasioned by a variety of other conditions, such as: psychological distraction effects of initial exposure to a novel environmental feature, motivational effects of long-duration exposures, and lack of adequate familiarity with the task requirements in the water.

A second problem is the uncertainty concerning measures to use as a definition of cold in a physiological sense. Performance-oriented research on the effects of cold tends to use the more readily available measurements such as skin surface and core temperatures to describe a test diver's state of coldness at the time his performance is assessed. Webb (1973), however, advocates body heat loss as the key physiological variable, and reports a lack of significant correlation between caloric heat loss and various combinations of skin and rectal temperature measurements. Teichner (1968) has stressed the measurement of compensatory processes or controlling events rather than controlled events in assessing the physiological effects of exposures to cold. Rectal temperature, for example, is a controlled event in Teichner's conceptualization, while changes in blood flow rates and volumes and metabolic changes are the controlling events.

A third problem is the tendency of researchers to relate performance changes directly to the temperature of the ambient environment, ignoring altogether the intervening physiological concomitants of exposure.

A fourth major source of confusion in determining performance decrement lies in the tests used to describe performance. Although numerous standardized apparatus and paper/pencil tests exist which have known properties, such as internal consistency, test–retest reliability, factor loadings, and other technical characteristics, there is a widespread tendency to pass these tests by in favor of homemade tests which have the advantage of being tailored to the specific circumstances of the dive but the disadvantage of having unknown characteristics in all the technical aspects of psychometric testing.

A particularly challenging aspect of performance measurement in cold-water exposures is to unravel the confounding of perceptual and cognitive task performance by changes in hand and finger dexterity. Paper/pencil tests of these higher order functions show decrement upon long-duration cold exposure in part because the hands manipulate the pencil less skillfully.

Finally, the literature of cold effects, far more than in other performance areas, is afflicted by graphic presentations of functional relationships between measures of cold and measures of performance for the purpose of showing trends, when statistical tests of the performance data fail to show a significant effect of the temperature conditions.

In spite of these hazards, a synopsis of the main relationships can be made, but the operational application-oriented reader should be cautious of the state of the science in this area.

a. Tactile Sensitivity

Perhaps the most well-substantiated effect of exposure to cold is the reduction in sensitivity of the skin. The generalized experimental paradigm is to vary the temperature of the skin surface of the hand or finger and test for threshold sensitivity. Threshold changes with skin temperature have been determined for two-point and two-edge discrimination, absolute pressure, vibratory sensitivity, and pressure-reproduction accuracy. The majority of research results reveals some common characteristics: There are wide interindividual differences in sensitivity, high intraindividual consistency, and the nature of the relationship with skin temperature tends to be L-shaped. Sensitivity thresholds change only slightly over a wide range of skin temperatures until a skin temperature of approximately 8°C is reached. From 8 to 0°C the function accelerates sharply. Figure VI-52 is typical of skin-sensitivity/skin-temperature relationships. The average range of two-edge discrimination, for example, is 1 mm at a hand-skin temperature of 30°C and is 5 mm at 0°C. The loss in discrimination is slow at first but becomes increasingly rapid as the freezing point is approached (Mills 1956). Using an identical measure, Mackworth (1953) reported the range of individual differences as 0.5–2.0 mm at a finger-skin temperature of 33°C. These basic findings were reproduced in a study by Morton and Provins (1960) and further verified in recent field studies by Bowen (1968) and Stang and Weiner (1970).

b. Grip Strength

Grip strength has been shown to be related to muscle temperature. Clarke *et al.* (1958) investigated hand grip in four test subjects on two dimensions of performance: duration of sustained contraction and maximum exerted force. The test subject's forearm was immersed 30 min in water at seven temperatures, 2, 10, 14, 18, 26, 34, and 42°C, and hand grip was tested under water using a strain gauge dynamometer. Short-duration, maximum-force applications were not affected by 30-min immersions

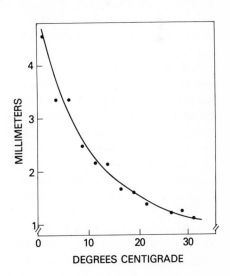

Figure VI-52. Relation between the two-edge threshold and the skin temperature. [From Mills (1956) by permission of the American Physiological Society.]

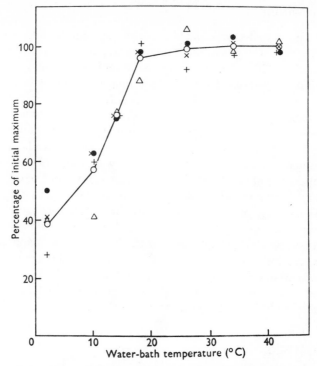

Figure VI-53. Individual and mean maximal tensions, expressed as a percentage of initial maximum tension, recorded after the forearm had been immersed in water at each of seven temperatures for 30 min. [From Clarke *et al.* (1958) by permission of the *Journal of Physiology*, London.]

in water at temperatures above 18°C. In cooler water, maximum force declined sharply from baseline values. Figure VI-53 shows the percent decrement in maximum hand-grip strength for 30-min exposures to water temperature between 2 and 42°C. In 2°C water, maximum hand grip was reduced by 40% of preimmersion levels. The second aspect of hand-grip strength, duration of sustained contraction, revealed a different relationship with variations in exposed temperature. Optimal performance occurred in water temperatures of 18°C and declined in both warmer and cooler water. Individual differences in forearm diameter suggested that an optimal muscle temperature accounted for the difference and that the subjects with the thinner arms reached optimum value in slightly warmer water than did the subjects with thicker arms (see Figure VI-54).

Temperature of the brachioradialis muscle near the elbow was measured between 1 and 4 cm beneath the skin. Figure VI-55 shows the muscle temperature gradients for 30-min exposures to water at 2, 10, 18, and 26°C. Duration of sustained contraction in hand-grip performance as a function of muscle temperature is shown in Figure VI-56 for successive trials. The capacity for sustained contraction was optimal at a muscle temperature of approximately 27°C, some 6–10°C cooler than *normal* muscle

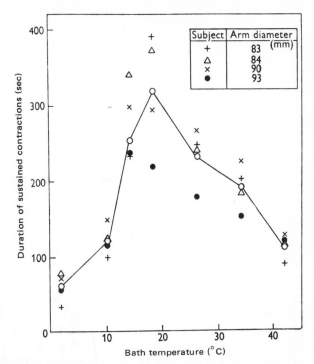

Figure VI-54. Individual and mean durations of sustained contractions in water at seven temperatures. [From Clarke *et al.* (1958) by permission of the *Journal of Physiology*, London.]

temperature. At muscle temperatures both higher and lower than 27°C, capacity for sustained hand grip was reduced.

Operationally oriented research on man in the sea typically has used short-duration, maximum-effort kinds of tests for hand grip, and test divers are protected by wet suits in the areas of the muscle groups related to hand grip. Therefore, it is not surprising that tests conducted under these conditions have yielded negative results

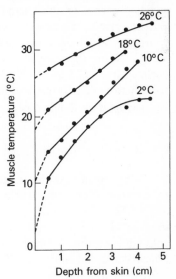

Figure VI-55. Temperature gradients measured vertically from the skin through the forearm after it had been immersed in water at 2, 10, 18, and 26°C for 30 min. [From Clarke *et al.* (1958) by permission of the *Journal of Physiology*, London.]

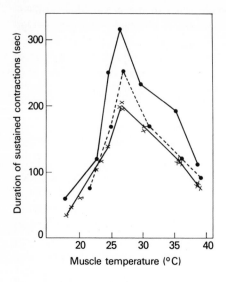

Figure VI-56. The average duration of five sustained contractions at recorded muscle temperatures. The upper curve represents the first contraction, while the third, fourth, and fifth contractions are all represented by the lower curve. [From Clarke *et al.* (1958) by permission of the *Journal of Physiology*, London.]

concerning hand-grip/water-temperature relationships or hand-grip/hand-skin-temperature relationships. For example, Bowen (1968) tested wet-suited subjects in the water approximately 24 min at water temperatures of 16.5 and 8.5°C and found no significant differences in hand-grip strength either as a function of water temperature or as compared to dry-land baseline values. Similarly, Egstrom *et al.* (1973) found no differences in bare-hand grip strength of wet-suited subjects after 1 hr exposure to 26°C water. Egstrom's findings suggest that the use of five-fingered Neoprene rubber gloves may degrade grip strength in moderate water temperatures as compared to bare-hand performance. In air, mean bare-handed grip strength (short-term maximum force) was 63 kg; gloved grip strength was 49 kg.

Vaughan and Andersen (1973) appear to have gotten muscle temperatures sufficiently cold to affect hand grip in spite of the protection afforded by wet suits. Baseline grip strengths were determined for eight U. S. Navy combat swimmers, then compared with test values taken immediately following a 3-hr exposure to 4.5°C water. During this exposure, average skin temperature measured at the upper arm declined from 32.1 to 21.6°C. Average baseline grip strength was significantly reduced from 56 to 47 kg of force. Within 1 hr of water exit, however, grip strength had returned to baseline level in all cases.

c. Steadiness and Tremor

While steadiness and tremor have not been tested in cold water, there is evidence from two studies conducted in cold-air environments that cold exposure significantly affects these performance dimensions. Peacock (1956) used a test of postural tremor to reveal significant decrement in aiming steadiness of 14 riflemen as a consequence of cold-air exposure. Tremor signals were analyzed into vertical and horizontal components and the more significant increase over baseline tremor values was in the horizontal component.

Lockhart (1968) cooled subjects in air to two levels of mean weighted skin temperature, 25.5 and 21.0°C, and used a test of steadiness aiming to compare performance under these conditions of whole-body cooling. Steadiness performance was significantly poorer at the lower skin temperature, and the steadiness test was a more sensitive indicator of the changed physiological condition than were two dexterity tests, block stringing and Purdue Pegboard Assembly.

d. Dexterity and Assembly

Finger dexterity is reduced by exposure to cold apparently by two mechanisms: the increased viscosity of the synovial fluid in the joints, and the cooling of the muscle groups that control finger movement. Hunter *et al.* (1952) demonstrated a significant reduction in speed of finger flexion when skin temperature of the finger was approximately 10°C; and LeBlanc (1955) used Hunter's test of finger flexion speed to show that performance decrement increased when the hand or the hand and forearm were cooled in addition to cooling just the fingers. Decrement in speed of finger movement was 20% when the finger alone was cooled, 25% when hand and fingers were cooled, and 50% when fingers, hand, and forearm were cooled.

The general finding that tactile sensitivity of the fingers is substantially reduced at about 8°C finger-skin temperature, combined with the indication of significant finger stiffness at 10°C skin temperature, would suggest general decrement in a wide range of manual-dexterity related tasks in this skin temperature range. This conclusion has been borne out by the results of a series of studies conducted at U. S. Army

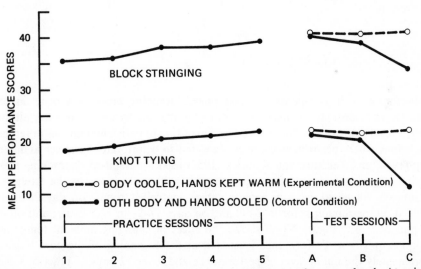

Figure VI-57. Comparison of mean performance scores on a complex manual task. At points *A*, *B*, and *C*, hand-skin temperatures were above 80°F in the experimental condition and above 80, 60–65, and 50–55°F, respectively, in the control condition. Mean body-surface temperatures were 85, 82, and 78°F, respectively, in both conditions. [From Gaydos (1958) by permission of the American Physiological Society.]

Quartermaster Research and Engineering Center, Natick, Massachusetts. Gaydos and Dusek (1958), for example, trained subjects in two tasks involving finger dexterity (knot tying and block stringing) in order to account for practice effects, then tested them in two conditions (one in which hands and overall mean skin temperatures were the same and another in which the hands were kept warm while the mean skin temperature of the body was cooled). Figure VI-57 shows that as long as the hands were kept warm, performance was maintained at criterion levels, and only when the hands were made approximately 10°C did performance deteriorate.

Clark (1961) also used the knot-tying task to assess effects of maintaining a given level of hand-skin temperature for progressively longer time intervals. Significant differences in knot-tying performance were found between 12.5 and 15.5°C hand-skin temperature conditions. Clark concluded that a hand-skin temperature of 15.5°C is required in order to maintain criterion levels of dexterity performance (Figure VI-58).

If the hands can be kept warm, 32°C, performance in a range of dexterity tasks is not affected by whole-body cooling except at very low values of mean weighted skin temperature. Lockhart (1966, 1968) reported decrement in some dexterity tasks at mean weighted skin temperatures in the range 19–21°C. Even at this level of overall body cooling, screwplate and Purdue Pegboard Assembly task performances were not affected.

In two studies of whole-body immersion of wet-suited test subjects, Bowen (1968) found no statistically significant differences in a screwplate task and a modified version of the Purdue Pegboard Assembly task in response to a 25-min exposure to water at 8.5°C; Stang and Weiner (1970), however, found significant performance differences on a range of nonstandardized dexterity and assembly tasks between 90-min exposures to 10°C water as compared to 15.5 and 21.0°C exposures. Figure VI-59 presents the latter results.

e. Tracking and Coordination

Performance in compensatory and pursuit tracking tasks as well as in other visual–motor coordination tasks appears to be affected by cold exposure, although reported results do not enable the specification of skin or core temperatures or temperature regions where performance can be expected to deteriorate. Teichner and Wehrkamp (1954) and Teichner and Kobrick (1955) used a pursuit-rotor tracking task to examine effects of ambient air temperature on skill acquisition. In the latter study five subjects stayed 41 days in a temperature-controlled room; on days 1–16 they practiced on the pursuit rotor in 24°C air; on days 17–28 in 12.5°C air; and on days 29–41 in 24°C air. Figure VI-60 shows the effect of changing ambient temperature on skill acquisition. Performance levels fell at the beginning of the cold testing to approximately the initial level of performance and, over 12 days of practice sessions in the cold, improved at a reduced rate. When ambient temperature was increased, performance effectiveness increased markedly.

Payne (1959) used a Link trainer type of apparatus to test the effects of cold exposure on a four-variable compensatory tracking task. Seventy-two USAF person-

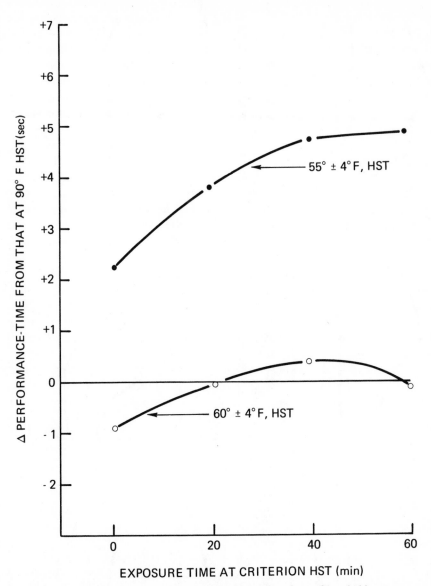

Figure VI-58. Changes (Δ) in manual performance as functions of hand-skin temperatures and duration of cold exposure (positive changes are decrements). [From Clark (1961) by permission of the American Psychological Association.]

nel dressed in light clothing of an insulative capability of 1 clo* made test runs of 160 min in air temperatures of 21, 12.5, and 4.5°C. Performance in the task at 4.5°C was significantly poorer than performance in the warmer temperatures. Figure VI-61 shows mean performance levels over 20 8-min cycles of work on the tracking task for

* 1 clo = 0.18°C per kcal per m² per hr, or that amount of insulation that will transfer 5.56 kcal per m² per hr per °C.

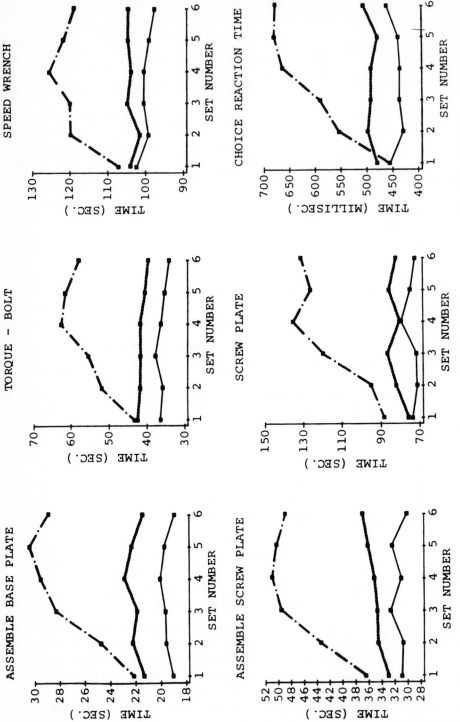

Figure VI-59. Performance measures versus set number for all tasks at 50°F (broken line), 60°F (heavy solid line), and 70°F (light line). Each task (or measure) was performed once during each set. Each set had a duration of 15.25 min; therefore, the six sets represent a total exposure of 91.5 min.

[From Stang and Wiener (1970) by permission of *Human Factors*.]

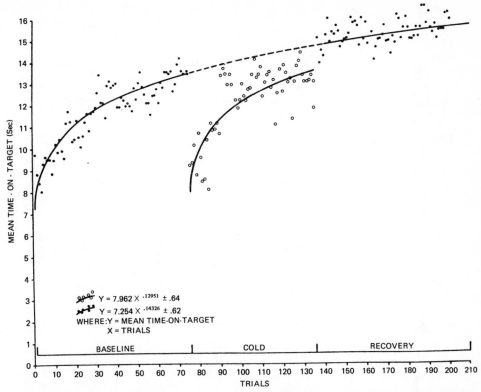

Figure VI-60. Pursuit rotor performance as a function of practice under two different ambient temperatures. [From Teichner and Kobrick (1955) by permission of the American Psychological Association.]

three temperature levels. Payne reports that not only was the tracking performance quantitatively poorer in the cold air, but qualitative differences were also noted. Test subjects became increasingly irritable and experienced periods when they attended to only one or two components of the four-variable task.

Research with wet-suited subjects working in the water has yielded mixed results. Bowen (1968) used a two-hand coordination task in air, and in shallow-water, short-duration (20 min) exposures at 16.6 and 8.4°C. The effect of temperature was significant in increasing the length of time required to complete the test. Using dry-land performance as a baseline, completion time increased by 25% and 56% for the two cold-water conditions.

Vaughan and Mavor (1972) and Vaughan and Swider (1972) report results of studies conducted with U. S. Navy combat swimmers controlling a wet submersible in operational environments. Variability in depth-holding performance was used as a measure of tracking capability of men in cold water, and changes in depth control were related to rectal and skin temperatures over a 4-hr period of exposure to water at 16.5°C. Average hourly skin temperature decreased from 30.2 to 27.5°C between the first and fourth hour, and average rectal temperature fell from 37.9 to 36.7°C;

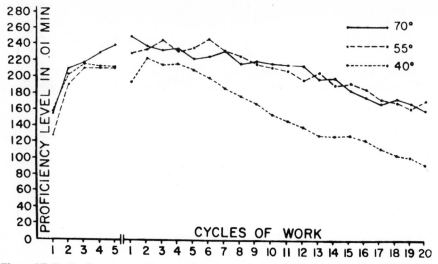

Figure VI-61. Decline in tracking proficiency as a function of ambient temperature. Acquisition curves prior to 30-min rest period represent training trials conducted in an ambient temperature of 70°F. [From Payne (1959) by permission of the American Physiological Society.]

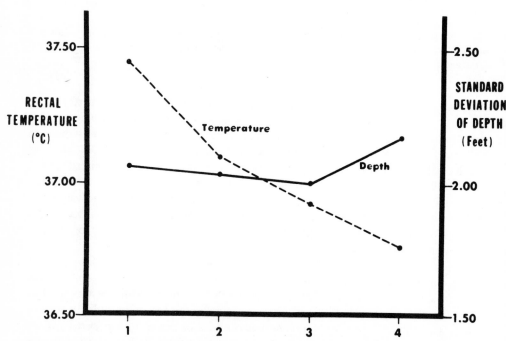

Figure VI-62. Pilot average hourly rectal temperature vs. depth control variability as a function of exposure duration ($N = 6$) (Vaughan and Mavor 1972).

but no significant differences were found in depth-control performance (see Figure VI-62). Depth control, as estimated by the standard deviation, was approximately ±2 ft throughout the 4-hr exposure interval and was invariant with skin and core temperature.

In a later test series, both depth and heading deviations were measured during 4-hr exposures to colder water, 6°C. As a consequence of the colder ambient water, average hourly skin temperature fell from 26.1 to 23.4°C between the first and fourth hours, while average core temperature fell from 37.7 to 36.5°C. Again no effect of these changes was noted in the depth-control data, but heading control was significantly degraded (see Figure VI-63).

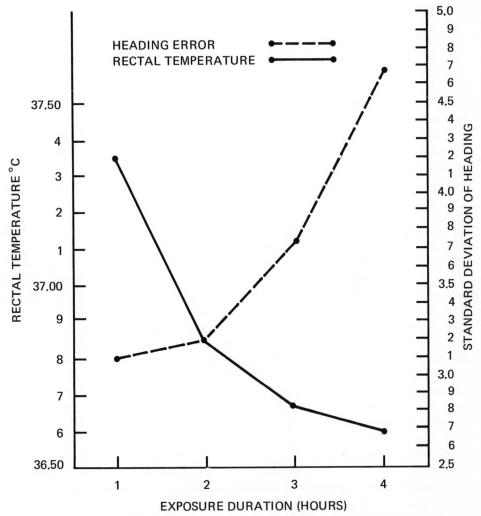

Figure VI-63. Pilot average hourly rectal temperature vs. heading control error for 4-hr exposure in 6°C water (N = 6) (Vaughan and Swider 1972).

f. Reaction Time and Vigilance

Simple visual reaction time has been studied as a function of exposure to high-velocity, low-temperature winds (Teichner 1958). Mean skin temperature varied in the range 27–32.5°C but no significant relationship with reaction time was found. Reaction time was significantly affected by high-velocity, low-temperature winds independently of changes in skin temperature. Teichner interpreted these findings by postulating a "distraction effect," a psychological, attentional phenomenon which operates to degrade attention-dependent performance in the presence of unusual environmental conditions and in the absence of any physiological change. The "distraction hypothesis" has been used by Bowen (1968), Stang and Wiener (1970), and Vaughan and Andersen (1973) to account for similar findings: performance degradation in the presence of extreme environmental conditions with no measurable change in physiological state.

Vigilance performance as a function of cold has been studied in three test series with wet-suited, fully immersed divers. Stang and Wiener (1970) exposed divers to 10, 15.5, and 21°C water for 90 min at shallow depth during which time they were tested six times. Vigilance in this study was defined as time to respond to a two-light stimulus while solving arithmetic problems. Response time increased significantly over the six testings in water at 10°C but not at 15.5 or 21°C. Figure VI-64 shows plots of mean skin temperature and response delay for the three exposure conditions. Of interest is the seeming lack of a "distraction effect." Response delay progressively increases as a function of exposure time and decreasing mean skin temperature in water of 10°C.

Vaughan and Andersen (1973), on the other hand, report on research which clearly supports the distraction hypothesis. Test personnel performed a vehicle-control task (depth and heading control) as the primary task and monitored a CRT display for simulated obstacles. A simple button-press response was made to the signal and both percent of targets detected and response time for detected signals were used as indices of vigilance performance. Well-conditioned, well-trained U. S. Navy UDT personnel were test divers and the tests were conducted at two water temperatures, 4.5 and 15.5°C. Test divers were in the water for 3 hr, out of the water for the fourth hour, then back in the water for a fifth and sixth hour. Figures VI-65 and VI-66 reveal both a distraction effect and a long-term fatigue effect, part of which may be attributable to cold discomfort. Percent of targets detected (Figure VI-65) was 100% in the first hour of exposure to the 15.5°C water and 92% in 4.5°C water, a statistically significant difference. In the second hour, however, performances under the two exposure conditions were not different even though physiological changes in skin and core temperatures were apparent (Figures VI-67 and VI-68). Performance in the third hour continued to decline, although not differentially, as a function of exposure temperature or differences in skin temperature. After 1 hr out of the water, performance on the vigilance task declined between the fifth and sixth hours and, again, the exposure conditions did not have a differential effect. A similar description fits the latency performance data presented in Figure VI-66.

Egstrom *et al.* (1972) conducted a vigilance study in the same paradigm (primary task with ancillary detection task), which showed no significant effect of water-

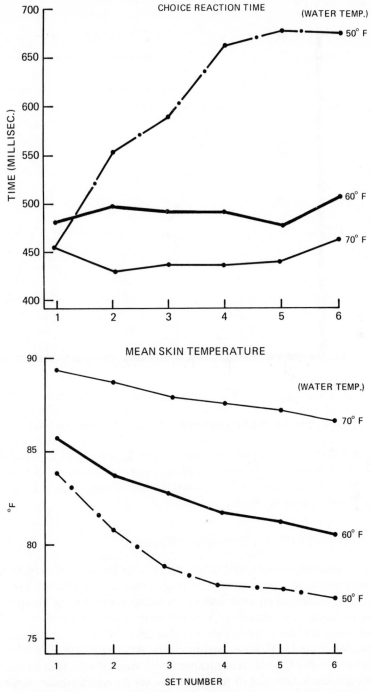

Figure VI-64. Response time delay and mean skin temperature for three exposure conditions in a vigilance task. [From Stang and Wiener (1970) by permission of *Human Factors*.]

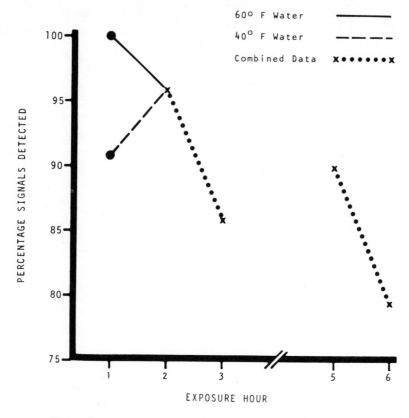

Figure VI-65. Detection percentage (Vaughan and Andersen 1973).

temperature differences. Test personnel were exposed to water at 25.5 and 4.5°C for 45 min. During this time they worked on an assembly task and responded to a peripheral light signal as a secondary task. Detection percentage was 77% in the warm water and 71% in the cooler water.

g. Memory and Other Cognitive Processes

Memory, problem-solving, and other *higher order* cognitive processes have only recently been tested for effects of cold-water exposure, and the testing has been conducted with nonstandardized tests and with tailor-made tasks not comparable with those used in earlier and previously reviewed studies concerning the effects of inert gas narcosis. Bowen (1968) used a clock test of short-term recall to compare the effects of dry-land, water, and cold-water environments. After approximately 17 min in the water at 16.6 and 8.4°C, relatively inexperienced scuba divers studied for 1 min a display of eight clock faces set to random times on the quarter-hour. After a 30-sec delay, the subjects attempted to reproduce the eight clock settings. The ratio of number correct to number attempted was used as a score. There was a significant reduction in recall percentage as performed in the water as compared to dry land (62% vs. 79%),

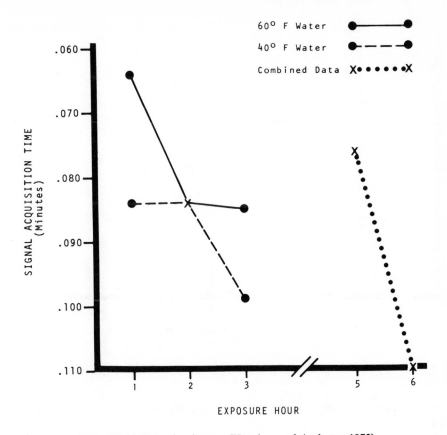

Figure VI-66. Detection latency (Vaughan and Andersen 1973).

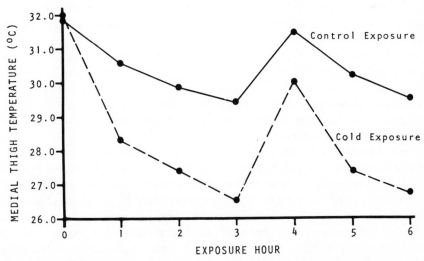

Figure VI-67. Mean hourly skin temperature: medial thigh (Vaughan and Andersen 1973).

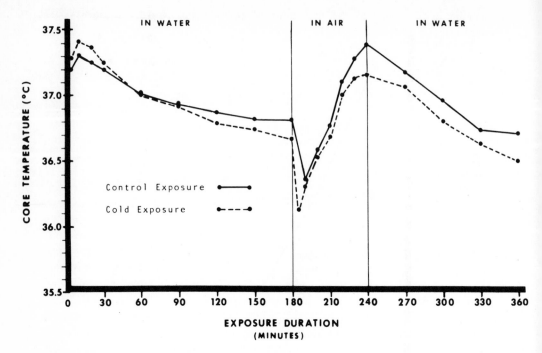

Figure VI-68. Mean core temperature profiles during 6-hr exposure phases (Vaughan and Andersen 1973).

but no difference as a function of water temperature: 62% in 16.6°C water vs. 61% in 8.4°C water.

Egstrom *et al.* (1972, 1973) have examined the effects of warm vs. cold water exposure on learning material in air and remembering it under water; and the reverse, learning under water and remembering in air. In one case a prose passage containing 30 facts was read and studied for 5 min after the subjects had been submerged 40 min. Immediately upon water exit they were tested for recall and for recognition of the learned material. In the second case, the material consisted of pages of picture–number associations, which the subjects studied for 5 min in the air, then submerged for 40 min, and were tested. In both cases recall was significantly reduced for the cold-water condition as compared to the warm-water condition. Also in both tests, recognition performance represented a significant improvement over recall performance for the cold-water exposure condition only. In the warm-water tests, recall and recognition scores were not different.

Bowen (1968) and Vaughan and Andersen (1973) have used a variety of ad hoc tasks representing the exercise of higher order cognitive functioning (such as symbol-processing, set exceptions, map problem-solving, and navigation problem-solving) with a common effect: significant decrement in performance (accuracy or time) when the in-water performance is compared with a dry-land performance, but no decrement as a function of exposure to widely different temperatures. Both authors have noted the tendency of test subjects to omit test items as a significant feature of performance in cold water, and Vaughan and Andersen (1973) again report a "distraction effect" influencing performance in the first hour of performance in very cold exposure temperature, 4.5 as compared to 15.5°C.

5. *Adjustment to Isolation and Confinement in Habitats*

Concern over the effects of habitat living initially focused on biomedical and behavioral feasibility, gradually shifted toward issues of productivity and habitability, and currently lie in the areas of shallow-water saturation and independence from surface support systems. Among U. S. sponsored habitat programs, two were principally feasibility tests: the Navy's SEALAB I project, in which four men spent 11 days at 193 ft near Bermuda, and the Department of Interior's TEKTITE I project, wherein four men spent 60 days at 49 ft in Great Lameshur Bay, St. John. Those two projects concentrated on biomedical measurement and assessment, with relatively minor interest in behavioral research, mostly observational and based on impressions gained from post-mission interviews with the divers. Both follow-on programs, SEALAB II and TEKTITE II, included major behavioral research projects, and most of the operational data about adjustment to conditions of submerged habitat living come from these two projects. Recent habitat programs, such as FLARE, PRINUL, and HYDRO-LAB sponsored by U. S. National Oceanographic and Atmospheric Administration, are principally shallow-water, marine-science, *in situ* laboratory facilities used by diving scientists for training and research projects of relatively short duration.

a. Performance and Adjustment in SEALAB II

During August-October 1965, three teams of 10 men spent 15 days in the SEALAB habitat at 205 ft ocean depth. The men were highly select, well-trained, and highly motivated toward successful achievement of their missions: to survive and do productive work on the ocean floor. The habitat was a 12-ft diameter cylinder 57 ft in length partitioned into three areas: work/recreation, galley, and bunkroom. The breathing gas was 78% helium, 18% nitrogen, and 4% oxygen. In general, the living conditions were primitive: crowded, with a minimum of privacy and storage space; noisy, due to the operation of air compressors and other machinery; uncomfortable, due to lack of humidity control and poor air circulation; and with communication made difficult, due to helium-distorted speech.

(i) Criterion Measures. Performance and adjustment measurement followed a three-factor criterion model developed by Gunderson and Nelson (1966) from factor analytic studies of adjustment in groups of men wintering-over in Antarctica. Successful accommodation to conditions of isolation and confinement there was accounted for by three clusters of attributes: task orientation, emotional stability, and social compatibility. In SEALAB II, task orientation was assessed by analysis of dive records and by team leader ratings on work-related scales. Emotional stability was assessed by the periodic administration of a Mood Adjective Checklist and by records of attendance at meals. Social compatibility was assessed principally by a sociometric choice questionnaire, from which an index of "cohesiveness" was derived, and by

Table VI-24.

Intercorrelations of Criteria[a]

	1	2	3	4	5	6	7	8	9
1. Diving time	1.00[b]								
2. Leader rating	0.40[b]	1.00							
3. Fear	−0.50[c]	−0.47[b]	1.00						
4. Meals missed	−0.24	−0.22	0.35	1.00					
5. Gregariousness	0.50[c]	0.10	−0.24	−0.06	1.00				
6. Time in work area	−0.37[b]	−0.47[b]	0.28	0.00	0.36	1.00			
7. Sociometric peer choice	0.27	0.59[c]	−0.22	−0.40[b]	0.14	−0.28	1.00		
8. Outside calls	−0.50[c]	−0.44[b]	0.40[b]	−0.37[b]	−0.18	−0.29	−0.41[b]	1.00	
9. Meal preparation	−0.10	0.04	−0.16	−0.44[b]	−0.38[b]	−0.11	0.19	−0.16	1.00

[a] From Radloff and Helmreich (1968). By permission of the Naiburg Publ. Corp.
[b] $p < 0.05$.
[c] $p < 0.01$.

observations of amount of time spent in social interaction with other team members, from which a "gregariousness" index was derived. Secondary sociability indicators were number of social contacts directed away from the team (topside phone calls), and number of meals prepared for the others. Table VI-24 shows the intercorrelations among these several measurements and indices of adjustment to the conditions of habitat living. Perhaps of most interest are the correlations with diving time, the principal measure of productivity used in the SEALAB II behavioral research program. Individuals who spent the most time in the water also received higher ratings from the team leader, expressed less fear, spent more time in social communication with teammates, spent less time in the habitat work area, and made fewer phone calls outside the habitat than their less productive peers.

(ii) Temporal Trends. Time in the habitat was divided into two halves and various measures of performance and adjustment were compared for suggestions of trends. Two principal indicators of adjustment showed positive trends. First, all three teams spent more time in the water during the second half of the mission; both the frequency and the duration of dives increased significantly. Averaging across the three teams, SEALAB II divers increased their time in the water by 24% during the second week. Second, cohesiveness of each team increased over the period of the mission. Teams were assembled on the basis of considerations other than sociometric choice, and, indeed, team composition was random according to sociometric choices made prior to the missions. The results of the post-mission administration of the sociometric choice questionnaire showed a significant shift toward the selection of teammates, particularly in team I. Table VI-25 presents the choice data.

Although the men reported feelings of fatigue and sleep loss, there was no shift either in time of arising or in mood.

(iii) Habitability Problems. Assessment of habitat features and conditions were made from post-mission interview material. The most frequently mentioned habitat design problem was the inadequate space around the hatch area for diving gear and apparel. The divers needed a larger wet room in which to prepare for dives, dress, and undress. A second common complaint was the excessive time required to maintain the habitat and its life-support systems. Other frequently mentioned factors were as follows: high humidity, poor air circulation, lack of storage space, high noise level, helium-speech distortion, and difficulty in sleeping.

Table VI-25

Sociometric Choices of SEALAB Aquanauts of Own-Team Members before and after Test (Miller *et al.* 1967)

Time of choice	Total all teams	Team 1	Team 2	Team 3
Before test	147	47	27	73
After test	250	102	47	101
Change before to after	+103	+55	+20	+28

b. Performance and Adjustment in TEKTITE

The TEKTITE program made good use of several findings of the SEALAB program. The design of the TEKTITE habitat, for example, included a wet room with more adequate space and facilities for pre- and post-dive activities. Humidity and temperature controls were provided. The teams were smaller by half, providing more living space and privacy. Based on the SEALAB complaint of the high cost of maintenance, the TEKTITE team composition concept included a habitat maintenance engineer and four diving scientists.

The TEKTITE habitat was in two main sections connected by a passageway. Each section was further subdivided into upper and lower halves. One section included a wet room and an equipment room; the other section included a control room and crew quarters. The habitat was installed in 49 ft of water in Lameshur Bay, St. John, where the water temperature and visibility were both pleasantly high. The breathing gas was 92% nitrogen and 8% oxygen.

In addition to the advances in habitat design and site selection, the TEKTITE behavioral research program was more sophisticated than in SEALAB II. Since the principal investigators in TEKTITE were the same as in SEALAB, the behavioral research programs were based on common concepts and data collection methods were expanded and formalized vis-à-vis SEALAB. Observational data collection was expanded by more complete tv coverage of the habitat and more frequent sampling of activities, locations, and communications of the teams; also, a formalized habitat environment methodology was developed.

(i) Time Allocation. In TEKTITE I, four marine scientists spent 60 days in the habitat and its more immediate environs (the range of swim distance was 1800 ft from the habitat), and detailed records of time as spent in main categories of activity were compared to those of average American men and women. Figure VI-69 portrays the comparison. The TEKTITE habitat dwellers can be seen to be basically similar to the average land dweller in their time allocations. They tended to sleep slightly more and play slightly less than their land-based counterparts; their maintenance activities, while about twice as time-consuming as the average man, were half of the average woman. As the 60-day mission wore on, time spent in habitat maintenance tended to decline, as did time spent sleeping, while work and recreation tended to absorb increasing proportions of time. Figure VI-70 presents the 60-day history of time allocation among the major categories of activity.

In TEKTITE II, 40 scientists and eight habitat engineers made up ten teams of five members each. Each team spent two or three weeks in the habitat and the overall allocation of time to main categories of activity is shown in Figure VI-71. Of particular interest was the finding that approximately 20% of the time was spent in leisure. Table VI-26 presents a breakdown of specific activities by order of priority. In contrast to records of their actual leisure-time activities, when the divers were asked to give their opinion as to how they spent their leisure time, reading books topped the list.

(ii) Criterion Measures. Three criteria of productivity and adjustment were used in TEKTITE II: time spent in marine science work, time spent in communication with teammates and surface control, and self-administered mood scales. Overall,

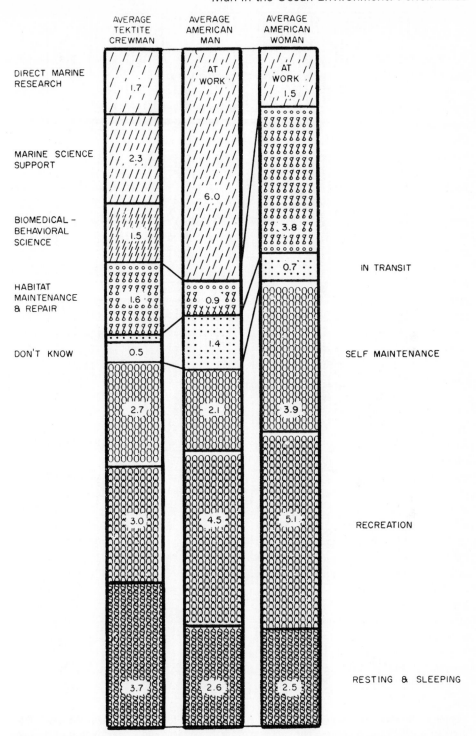

Figure VI-69. Comparison of the average day's time use in hours (from 6:15 a.m. to 11:45 p.m.) by the TEKTITE I crew and the average American man and woman (Radloff *et al.* 1970).

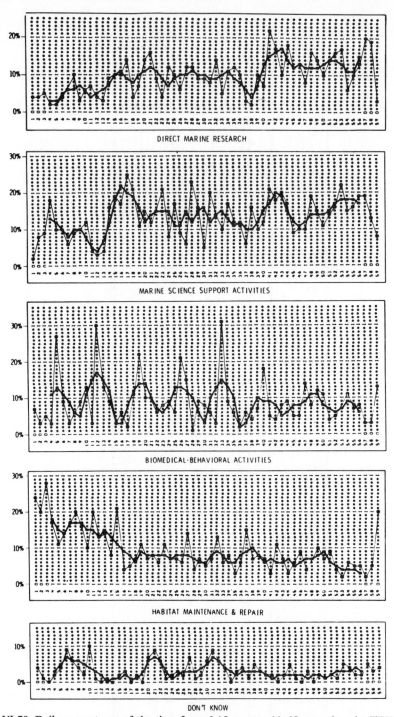

Figure VI-70. Daily percentages of the time from 6:15 a.m. to 11:45 p.m. that the TEKTITE I crew spent in various activities. The curves show daily averages and seven-day moving averages. (Radloff *et al.* 1970.)

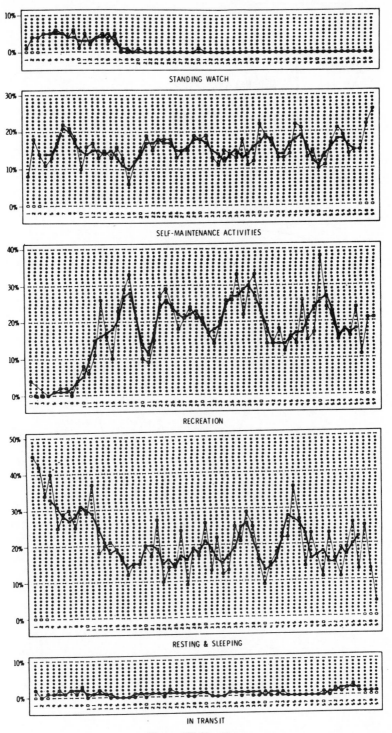

STANDING WATCH

SELF-MAINTENANCE ACTIVITIES

RECREATION

RESTING & SLEEPING

IN TRANSIT

Figure VI-70—*Cont.*

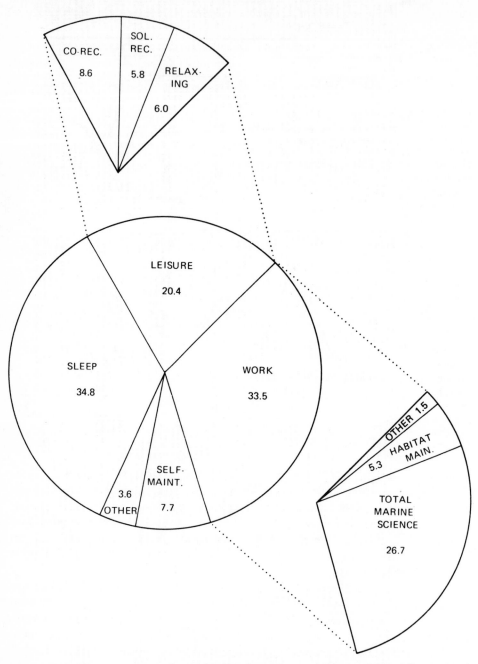

Figure VI-71. Mean percentage of total mission time spent on habitat activities (all aquanauts) (Helmreich 1971a).

Table VI-26

Distribution of Various Leisure Activities in Terms of Frequency, Length of Duration, and Total Duration (Mission A) (Nowlis *et al.* 1971)

Leisure activity	Number of times observed	Average duration, min	Total duration, min
Nonmission-relevant conversation between crew	701	5.98	4195
Listening to audio cassettes	167	12.62	2108
Watching T.V. films	71	27.45	1949
Snacking	126	4.90	617
Reading a leisure book	37	15.98	591
Leafing through a book or reading a magazine	60	8.52	511
Writing a letter	19	21.26	404
Napping	6	61.83	371
Looking out view ports	105	2.90	305
Resting quietly	24	12.17	292
Spontaneous interpersonal actions	38	6.68	254
Taking camera pictures	61	3.79	231
Watching topside on T.V.	73	2.56	187
Tinkering with personal gear	36	5.14	185
Nonmission-relevant conversation with topside	46	3.65	168
Quietly watching or listening to other crew members	53	3.06	162
Playing musical instrument	5	17.40	87
Wandering	18	3.72	67
Fixing audio or T.V. tapes	7	8.86	62
Physical exercise for fun	3	13.67	41
Reading letters	4	10.00	40
Looking through leisure facilities for something to do	7	3.71	26
Games	1	21.00	21
Waiting	2	9.00	18
Reading newspaper	2	8.50	17
Sketching	2	6.50	13
Listening to radio	3	3.33	10

productivity was correlated with gregariousness, $r = 0.59$, as in SEALAB II. Those who spent the most time in marine science activities also spent the most time in social communication with their teammates. In contrast to SEALAB II results, however, there was a positive correlation, $r = 0.35$, between productivity and communication with topside personnel. This may be explained by the differences in communication content: In SEALAB II, topside contacts were social, while in TEKTITE, topside contacts were operational.

The Mood Adjective Checklist was scored for five positive and six negative moods. Overall, the positive moods dominated and surprisingly low scores were recorded for anxiety (Table VI-27). Mood scores were correlated with productivity. Divers who expressed more highly positive moods spent more time in work activities, $r = 0.35$; while those who scored high on negative scales, such as depression, spent more time in solitary recreation, $r = 0.38$.

(iii) Temporal Trends. Activity patterns changed in consistent and significant ways between the first and second halves of a mission. The ten missions were separated into long (three-week) and short (two-week) missions and shifts in activity patterns

Table VI-27

Mean Scores on the Mood Adjective Check List
(Self-Administered Daily in the Habitat)
(Nowlis *et al.* 1971)

Mood	Mean score[a]	Mood	Mean score
Concentration	3.81	Deactivation	1.81
Activation	3.81	Aggression	0.51
Social affection	2.80	Egotism	0.51
Pleasantness	2.78	Skepticism	0.44
Nonchalance	2.10	Depression	0.33
		Anxiety	0.25

[a] The lower the score, the less common was the feeling in day-to-day life in the habitat.

were analyzed; the shorter missions revealed more significant shifts than did the longer missions. Table VI-28 presents percentage time distributions for first vs. second halves of the mission for two-week and three-week missions. Both long and short missions show a decrease in total work and habitat maintenance, and an increase in sleep between the first and second halves.

Shifts in mood scale scores as a function of days in the habitat are shown in Figure VI-72. The positive moods, such as pleasantness, social affection, concentration, and activation, can be seen to fall with time, while negative moods, anxiety and depression, change only slightly.

(iv) Habitability Problems. Habitability problems were diagnosed by means of three procedures: two rating forms and a post-mission debriefing. The rating forms included a Habitability Assessment Rating Scale (HARS), which listed 63 specific items in and features of the habitat to be rated by the divers. Each item was assessed on six criterion dimensions: performance of function, comfort in use, convenience of location, ease of maintenance, aesthetic value, and safety. Ratings were made five days prior to mission completion. Figure VI-73 presents means of obtained ratings in order of least well-designed aspects of the habitat. The divers tended to rate most positively those items designed for recreation, and to rate negatively those items intended to provide support for scientific and engineering tasks, and the provision of information. Access to news was the most negatively rated item. Correlations of HARS ratings with mood and performance data revealed a tendency for those divers with high HARS ratings to score higher on the pleasantness mood scale, $r = 0.43$, and to do more total work, $r = 0.35$.

A second rating form was the TEKTITE Environmental Assessment form (TEA). This form was designed around a diver activity-by-support feature matrix. Its purpose was to assess the degree to which various features of the habitat supported the scientist's or engineer's activities. This form was administered four days prior to the completion of each mission. Table VI-29 presents mean ratings for general habitat environmental features (rows) in terms of activity categories (columns). The lowest ratings were accorded habitat characteristics intended to support inside-the-habitat scientific and maintenance work.

Post-mission interviews were conducted with each diver and complaints were

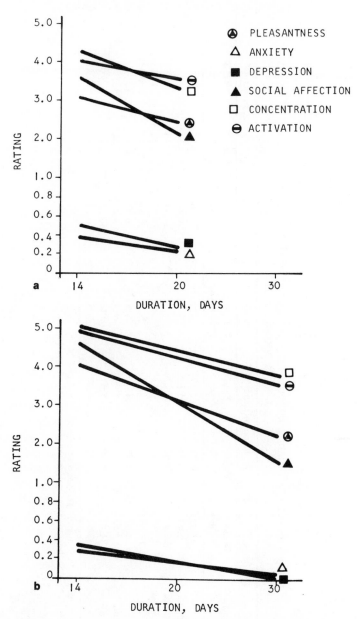

Figure VI-72. Effect of mission duration on mood (Nowlis *et al*. 1971).

Table VI-28

Changes in Activity Variables between First and Second Halves of Long and Short Missions
(Helmreich 1971*b*)

Variable	Long missions[a]				Short missions[b]			
	Mean % 1st half	Mean % 2nd half	Trial F-ratio	Probability	Mean % 1st half	Mean % 2nd half	Trial F-ratio	Probability
Total work	31.49	29.34	5.59	0.025	40.00	32.75	26.79	0.0002
Habitat maintenance	5.45	4.23	15.82	0.0008	6.39	4.60	11.55	0.0039
Total marine science	26.93	26.03	<1	N.S.	31.54	26.33	20.93	0.0005
Direct marine science	12.88	14.35	4.41	0.044	15.95	13.42	11.03	0.0045
Total leisure	19.59	19.93	<1	N.S.	17.86	23.47	30.64	0.0001
Co-recreation	8.47	8.36	<1	N.S.	6.37	10.67	101.46	0.0000
Solitary recreation	5.57	6.17	1.80	N.S.	5.41	5.64	<1	N.S.
Sleep	34.49	36.30	15.87	0.0008	31.35	33.44	5.99	0.025
Gregariousness	32.53	31.47	2.82	N.S.	30.71	32.81	12.82	0.0028

[a] Six missions, $N = 30$. For long mission analyses of variance, $df = 1$ and 24.
[b] Four missions, $N = 20$. For short mission analyses of variance, $df = 1$ and 16.

Table VI-29

Mean Environmental Assessment Scores[a] (Nowlis et al. 1971)

| | Sleep | Food | | Recreation | | | Work | | | | Hygiene | | | Average |
		Eating	Preparation	Exercise & active rec.	Games, books, entertainment	Social interaction	Science inside	Maintenance inside	Access to outside	Work outside	Waste elim.	Washing, shower, ing	Overall	
Is there enough room?	3.12	2.76	2.46	2.32	2.83	2.80	1.69	2.15	2.51	×	2.61	3.12	2.98	2.62
Is the lighting of the area satisfactory?	3.27	3.41	3.20	3.26	3.34	3.16	2.72	3.08	3.08	2.77	3.30	3.32	3.18	3.16
Is the location of the area satisfactory?	3.12	2.98	2.90	2.67	2.89	3.00	1.95	2.31	2.71	3.06	2.79	3.18	×	2.82
Is the layout of the area satisfactory?	3.03	2.78	2.63	2.50	2.77	2.89	1.86	2.07	2.49	×	2.81	2.97	2.87	2.66
Is it quiet enough?	2.22	2.59	2.63	2.38	2.37	2.38	2.20	2.34	2.24	×	2.47	2.62	2.31	2.40
Is there a lack of odor?	3.15	3.00	2.83	3.18	3.19	3.11	2.73	3.14	2.68	×	2.38	2.83	2.90	2.92
Is the temperature satisfactory?	3.48	3.59	3.35	3.37	3.45	3.63	3.37	3.45	3.34	3.56	3.57	3.63	3.59	3.49
Is the humidity satisfactory?	3.60	3.66	3.62	3.47	3.57	3.66	3.51	3.61	3.47	×	3.61	3.62	3.68	3.59
Is enough time allowed?	2.80	3.32	3.24	2.86	2.86	3.25	2.88	2.64	3.19	3.26	3.21	3.38	×	3.07
Are the times available OK?	3.36	3.24	3.24	3.21	3.00	3.06	3.21	2.84	3.57	3.60	3.37	3.42	×	3.26
Is there good selection and variety?	×	2.75	×	2.30	2.67	×	2.48	×	×	×	×	×	×	2.58
How does the habitat affect the activity in general?	3.16	3.10	2.79	2.47	2.75	2.97	2.32	2.55	3.20	3.46	2.66	3.27	×	2.90
Average	3.12	3.08	2.95	2.82	2.98	3.07	2.54	2.74	2.91	3.26	2.94	3.18	3.07	2.96

[a] N = 41. Key: 1, Poor. 2, Fair. 3, Very good. 4, Excellent. ×, Not applicable.

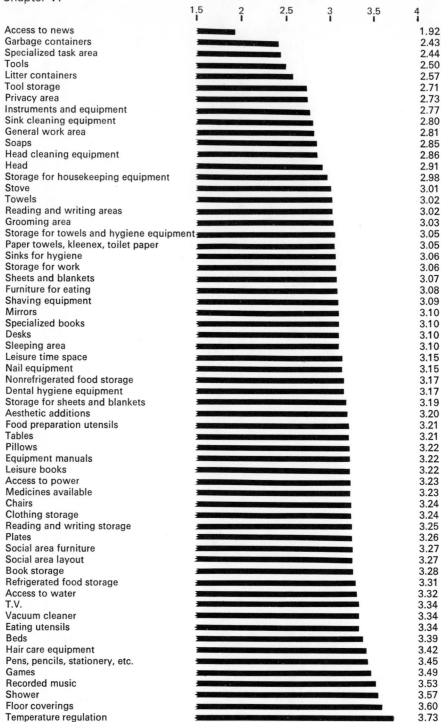

Access to news	1.92
Garbage containers	2.43
Specialized task area	2.44
Tools	2.50
Litter containers	2.57
Tool storage	2.71
Privacy area	2.73
Instruments and equipment	2.77
Sink cleaning equipment	2.80
General work area	2.81
Soaps	2.85
Head cleaning equipment	2.86
Head	2.91
Storage for housekeeping equipment	2.98
Stove	3.01
Towels	3.02
Reading and writing areas	3.02
Grooming area	3.03
Storage for towels and hygiene equipment	3.05
Paper towels, kleenex, toilet paper	3.05
Sinks for hygiene	3.06
Storage for work	3.06
Sheets and blankets	3.07
Furniture for eating	3.08
Shaving equipment	3.09
Mirrors	3.10
Specialized books	3.10
Desks	3.10
Sleeping area	3.10
Leisure time space	3.15
Nail equipment	3.15
Nonrefrigerated food storage	3.17
Dental hygiene equipment	3.17
Storage for sheets and blankets	3.19
Aesthetic additions	3.20
Food preparation utensils	3.21
Tables	3.21
Pillows	3.22
Equipment manuals	3.22
Leisure books	3.22
Access to power	3.23
Medicines available	3.23
Chairs	3.24
Clothing storage	3.24
Reading and writing storage	3.25
Plates	3.26
Social area furniture	3.27
Social area layout	3.27
Book storage	3.28
Refrigerated food storage	3.31
Access to water	3.32
T.V.	3.34
Vacuum cleaner	3.34
Eating utensils	3.34
Beds	3.39
Hair care equipment	3.42
Pens, pencils, stationery, etc.	3.45
Games	3.49
Recorded music	3.53
Shower	3.57
Floor coverings	3.60
Temperature regulation	3.73

Figure VI-73. Habitability assessment rating scale means ($N = 43$) (Nowlis *et al.* 1972). Scoring system: 5 = superlative, 4 = very good, 3 = ordinary, 2 = poor, 1 = very poor.

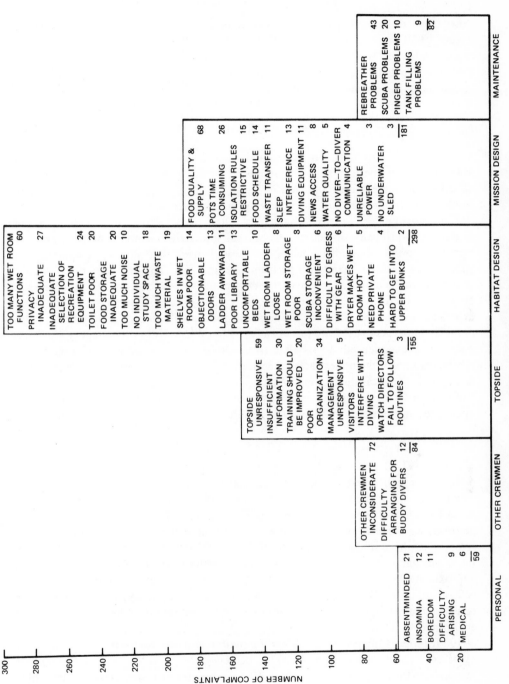

Figure VI-74. Number of complaints (from debriefing tapes) (Nowlis *et al.* 1971).

recorded as part of these sessions. A total of 859 complaints were noted, approximately 18 per diver. These complaints were categorized according to their referent: personal, other crewmen, topside personnel, habitat design, mission design, and maintenance. Figure VI-74 presents a summary of these findings. By far the most frequent source of complaint was the habitat design. The complaint data are in correspondence with the TEA ratings in calling out the divers' dissatisfaction with design features related to the accomplishment of their work.

6. *Reactions to Generalized Stress in the Diving Experience*

In previous sections, environmental factors have been examined as isolated influences on the performance characteristics of the diver. Nitrogen and helium gas narcosis, tractionlessness, cold, and isolation/confinement have been treated as independent environmental factors modifying man's performance under water. In this section two kinds of research studies are summarized: those that treat the diving experience holistically, including the several interactions among multiple environmental factors as they affect performance; and those that attempt to quantify the effects of generalized anxiety on performance undersea. The first category of studies is usually conducted in open-ocean test environments where all the multiple stressors operate interactively. The second category contrasts the performance of novice vs. experienced divers as means of illuminating the contribution of generalized anxiety in the diving experience. Care has been taken to exclude from this category studies of novice and experienced divers that focus on learning or accommodating to the special features of the underwater environment. These studies tend to be concentrated in the area of visual distortions under water and the effect of diving experience in reducing performance or judgmental error in estimating size and distance of objects, orientation in the water, judging swim distances, hand–eye coordination, etc.

a. Dexterity and Assembly Tasks

Table VI-30 presents a set of research projects; it incorporates two features: performance concerned manual-dexterity/assembly types of tasks, and testing was conducted both in the ocean and in a less stressful environment so that generalized stress effects could be assessed. The degree of decrement in dexterity/assembly task performance depends on the baseline condition used for comparison. Figure VI-75 summarizes those studies in Table VI-30 that used dry-land performance as a baseline. From this figure it can be seen that shifting from dry land to shallow water produces approximately a 30% decrement and that the percentage decrement increases practically as a linear function of ocean depth. Previous review of hyperbaric air research revealed a range of 3–8% reduction in performance between 1 and 4 ATA in chamber environments. Decrement in performance of dexterity/assembly tasks in the ocean at 100 ft (approximately 50%) is clearly a matter of more than can be accounted for by narcotic effects of nitrogen. At 200 ft, decrement vis-à-vis dry-land performance is seen to be in the range of approximately 60–65% even though a helium-based breathing gas was used and the test subjects were very experienced SEALAB II divers. As the diver works deeper, the visibility is less, it is usually colder, and there is greater risk of

Table VI-30

Research on Dexterity Task Performance under Variations of General Environmental Stress

Reference	Test environment	Test divers	Task	Measures	Results		
1. Bowen et al. 1966	Dry land, 15-ft lake depth, 205-ft ocean depth (heliox)	Three to fourteen U. S. Navy SEALAB II divers	Triangle test a. Simple: same corners, large bolts b. Complex: different corners, small bolts	Completion time, sec	Surface a. 60 b. 85	15 ft 80 110	205 ft 100 135
2. Baddeley 1966	Dry land, 10-ft ocean depth, 100-ft ocean depth	Eighteen divers: Sixteen Royal Engineer divers; two experienced amateur divers	Screwplate	Completion time, sec	Surface 185	10 ft 237	100 ft 276
3. Baddeley and Flemming 1967	10-ft ocean depth, 200-ft ocean depth (heliox)	Eight experienced amateur divers	Screwplate	Completion time, sec	10 ft 206	200 ft 268	
4. Weltman et al. 1970	15-ft pool depth, 20-ft ocean depth	a. Five experienced divers b. Fifteen novice divers	UCLA Pipe Puzzle Assembly	Completion time, min	Pool a. 22.4 b. 35.9	Ocean 23.1 45.2	

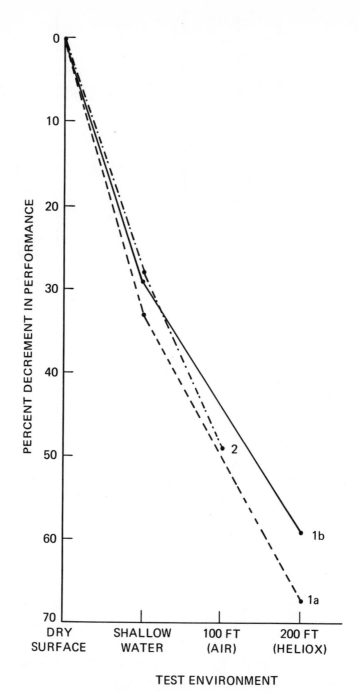

Figure VI-75. Decrement in dexterity/assembly task performance from dry-land baseline.

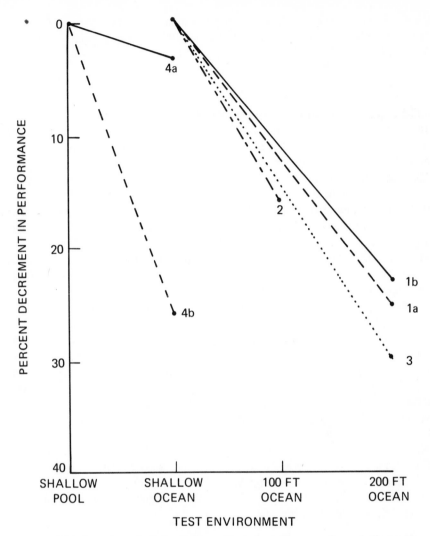

Figure VI-76. Decrement in dexterity/assembly task performance from shallow-water baselines.

losing buoyancy control or becoming lost. All of these variables interact to degrade performance in the ocean far beyond the simple effects of breathing-gas pressure as previously reviewed. Figure VI-76 presents the same data except that shallow-water performance is used as a baseline for assessing performance decrement at greater depths. Of special interest is the study by Weltman *et al.* (1970), which shows the comparative response of experienced and novice divers in performing an assembly task in a shallow pool vs. in the ocean at a comparable depth. Experienced divers were not affected by the difference in test environments, while the novice divers suffered a 26% decrement. The remaining studies (Baddeley 1966a, Bowen *et al.* 1966, Baddeley and Fleming 1967) were conducted with reasonably experienced divers and yielded highly consistent results: When shallow-water performance is a baseline, performance of dexterity/assembly tasks in the ocean at 100 and 200 ft can be expected to degrade by approximately 15 and 30%, respectively.

b. Vigilance Monitoring

The effects of generalized stress on vigilance task performance have been demonstrated in studies of novice scuba divers. In an initial study (Weltman and Egstrom 1966) 15 scuba trainees were tested on the surface, at 15-ft pool depth, and at 25-ft ocean depth. The main task was either to monitor a meter readout or to do arithmetic problems, and the secondary task was to respond to a randomly occurring light mounted in the periphery of the facemask. Speed of response to the light was used as the measure of vigilance performance. Average response delay was 1.06 sec on the surface, 1.43 sec in the pool, and 1.87 sec in the ocean. Using dry land as a baseline, the ocean condition produced a 76% increase in response delay. When the shallow-water test condition is used as a baseline, performance decrement in the ocean becomes 31%. The latter figure appears to be accountable as a generalized stress effect, based on a second study of the phenomenon (Weltman et al. 1970). In this study two groups of 15 diver trainees were tested with the peripheral light monitoring procedure: one group in standard dry-land conditions and the second in a simulated hyperbaric chamber. Illumination, temperature, and ambient noise levels were approximately the same in the two test conditions, but the experimental group were led to believe that they were being tested under pressure. The simulation was accomplished by testing in chamber with simulated sounds of hissing air and progressive movement of a pressure-gauge needle indicating hyperbaric conditions. This group averaged 29% fewer signal detections as compared to the control group. Support for the suggestion that a generalized anxiety about the diving situation accounts for the decrement is provided by two measures: heart rate and self-ratings on an anxiety scale. Weltman et al. (1970) reported an average increase of ten beats per minute in the heart rates of the chamber vs. the control group. Also, the test personnel in the chamber group scored significantly higher on the anxiety scale in a post-test administration of the Multiple Affect Adjective Checklist.

c. Time Estimation

Deppe (1969) compared experienced and novice scuba divers on ability to estimate time underwater. Divers were tested on the bottom of a 33-ft tank. The divers' facemasks were blacked out and they were to swim about on the bottom groping for coins. Additionally, they were to tap on the bottom at intervals of 1 min over a total test trial of 10 min. The average estimate of the experienced divers was exactly 1.00 min with a standard deviation of 0.19; while the average estimate of the novices was 1.37 min with a standard deviation of 0.37. Deppe attributed the difference to the effect of stress producing an overload on coping behavior.

In terms of average performance decrements, comparisons of novice and experienced divers have yielded remarkably consistent results across dexterity/assembly and perceptual-task categories, as summarized in Table VI-31.

Table VI-31

Effects of Generalized Stress
on Performance

Task category	Percent decrement attributable to anxiety
Assembly	26
Vigilance	
Detection time	31
Percent detection	29
Time estimation	37

References

ADOLFSON, J. Deterioration of mental and motor functions in hyperbaric air. *Scand. J. Psychol.* 6(1):26–32 (1965).

ADOLFSON, J. Human performance and behavior in hyperbaric environments. *Acta. Psychol. Gothoburgensia* 6:1–74 (1967).

ADOLFSON, J., and A. MUREN. Air breathing at 13 atmospheres. Psychological and physiological observations. *Forsvarsmedicin* 1:31–37 (Jan. 1965).

ADOLFSON, J., L. GOLDBERG, and T. BERGHAGE. Effects of increased ambient air pressures on standing steadiness in man. *Aerosp. Med.* 43:520–524 (May 1972).

ANDERSEN, B. G. Diver performance measurement: underwater navigation and weight carrying capabilities. Groton, Conn., General Dynamics, Electric Boat Division, Tech. Rep. on Contract ONR N00014-67-CO447 (1968).

ANDERSEN, B. G., F. L. ALLEN, and J. C. LAMB. Diver performance measurement: Transporting neutrally buoyant objects manual movement by heavy objects. Groton, Conn., General Dynamics, Electric Boat Division (1969).

BACHRACH, A. J., and P. B. BENNETT. Tremor in diving. *Aerosp. Med.* 44:613–623 (June 1973).

BACHRACH, A. J., D. R. THORNE, and K. J. CONDA. Measurement of tremor in the Makai Range 520-foot saturation dive. *Aerosp. Med.* 42:856–860 (Aug. 1971).

BADDELEY, A. D. Influence of depth on the manual dexterity of free divers: a comparison between open sea and pressure chamber testing. *J. Appl. Psychol.* 50:81–85 (Feb. 1966a).

BADDELEY, A. D. Diver performance and the interaction of stresses. In: Lythgoe, J. N., and J. D. Woods, eds. *Underwater Association Report 1966–1967*, pp. 35–46. Carshalton, England, T.G.W. Industrial and Research Promotions, for the Underwater Association of Malta (1966b).

BADDELEY, A. D., and N. C. FLEMMING. The efficiency of divers breathing oxygen–helium. *Ergonomics* 10:311–319 (May 1967).

BADDELEY, A. D., J. W. DeFIGUEREDO, J. W. H. CURTIS, and A. N. WILLIAMS. Nitrogen narcosis and performance under water. *Ergonomics* 11:157–164 (Mar. 1968).

BAIN, E. C., and T. E. BERGHAGE. Preliminary evaluation of SINDBAD tests. U. S. Navy Exp. Diving Unit (Oct. 1973).

BARRETT, F. B., and J. T. QUIRK. Diver performance using handtools and hand-held pneumatic tools. U. S. Nav. Civ. Eng. Lab., Rep. NCEL-R-653 (Dec. 1969).

BEHNKE, A. R., R. M. THOMSON, and E. P. MOTLEY. The psychologic effects from breathing air at 4 atmospheres pressure. *Amer. J. Physiol.* 112:554–558 (July 1935).

BENNETT, G. K. *Hand Tool Dexterity Test, Manual of Directions*, revised edition. New York, The Psychological Corporation (1965).

BENNETT, P. B. Psychometric impairment in men breathing oxygen–helium at increased pressures. Alverstoke, U. K., Roy. Nav. Physiol. Lab., Med. Res. Counc., Rep. 251 (1965).

BENNETT, P. B. *The Aetiology of Compressed Air Intoxication and Inert Gas Narcosis.* Oxford, Pergamon Press (1966).

BENNETT, P. B. Performance impairment in deep diving due to nitrogen, helium, neon and oxygen. In: Lambertsen, C. J., ed. *Underwater Physiology. Proceedings of the Third Symposium on Underwater Physiology, March 1966, Washington, D. C.,* pp. 327–340. Baltimore, Williams and Wilkins (1967).

BENNETT, P. B., and E. J. TOWSE. Performance efficiency of men breathing oxygen–helium at depths between 100 feet and 1500 feet. *Aerosp. Med.* 42:1147–1156 (Nov. 1971).

BENNETT, P. B., E. C. POULTON, A. CARPENTER, and M. J. CATTON. Efficiency at sorting cards in air and a 20 percent oxygen–helium mixture at depths down to 100 feet and in enriched air. *Ergonomics* 10:53–62 (Jan. 1967).

BENNETT, P. B., K. N. ACKLES, and V. J. CRIPPS. Effects of hyperbaric nitrogen and oxygen on auditory evoked responses in man. *Aerosp. Med.* 40:521–525 (May 1969).

BERGHAGE, T. E., L. E. LASH, W. R. BRAITHWAITE, and E. D. THALMANN. Intentional tremor on a helium–oxygen chamber dive to 49.5 ATA, *in preparation.*

BIERSNER, R. J., and B. J. CAMERON. Cognitive performance during a 1000-foot helium dive. *Aerosp. Med.* 41:918–920 (Aug. 1970a).

BIERSNER, R. J., and B. J. CAMERON. Memory impairment during a deep helium dive. *Aerosp. Med.* 41:658–661 (June 1970b).

BOWEN, H. M. Diver performance and the effects of cold. *Hum. Factors* 10:445–464 (June 1968).

BOWEN, H. M., B. ANDERSEN, and D. PROMISEL. Studies of divers' performance during the SEALAB II project. *Hum. Factors* 8:183–199 (June 1966).

BRAITHWAITE, W. R., T. E. BERGHAGE, and J. H. CROTHERS. Postural equilibrium and vestibular response at 49.5 ATA. *Undersea Biomed. Res., in press.*

BROWN, J. L. Orientation to the vertical during water immersion. *Aerosp. Med.* 32:209–217 (Mar. 1961).

BÜHLMANN, A. A., M. MATTHYS, G. OVERRATH, P. B. BENNETT, D. H. ELLIOTT, and S. P. GRAY. Saturation exposures at 31 ata in an oxygen–helium atmosphere with excursions to 36 ata. *Aerosp. Med.* 41:394–402 (Apr. 1970).

CAMERON, B. J., and M. G. EVERY. An integrated measurement system for the study of human performance in the underwater environment. *Operator's Handbook.* Falls Church, Va., BioTechnology, Inc. (1970).

CHRISTIANSON, R. A., G. WELTMAN, and G. H. EGSTROM. Thrust forces in underwater swimming. *Hum. Factors* 7:561–568 (Dec. 1965).

CLARK, R. E. The limiting hand skin temperature for unaffected manual performance in the cold. *J. Appl. Psychol.* 45:193–194 (June 1961).

CLARKE, R. S. J., R. F. HELLON, and A. R. LIND. The duration of sustained contractions of the human forearm at different muscle temperatures. *J. Physiol. (London)* 143:454–473 (Oct. 1958).

DEPPE, A. H. Overload and sensory deprivation: time estimation in novice divers. *Percept. Mot. Skills* 29:481–482 (Oct. 1969).

DIEFENBACH, W. S. The ability of submerged subjects to sense the gravitational vertical. Buffalo, N. Y., Cornell Aeronautical Laboratory (1961).

DUNN, J. M. Psychomotor functioning while breathing varying partial pressures of oxygen–nitrogen. Brooks Air Force Base, Tex., Sch. Aerosp. Med., Rep. 62-82 (1962).

DUSEK, E. R. Manual performance and finger temperatures as a function of ambient temperature. U. S. Army, Natick Lab., Environ. Protection Res. Div., Tech. Rep. EP-68 (1957).

EGSTROM, G. H., G. WELTMAN, A. D. BADDELEY, W. J. CUCCARO, and M. A. WILLIS. Underwater work performance and work tolerance. Los Angeles, Univ. Cal., Sch. Eng. Appl. Sci., Rep. UCLA-ENG-7243 (July 1972).

EGSTROM, G. H., G. WELTMAN, W. J. CUCCARO, and M. A. WILLIS. Underwater work performance and work tolerance. Los Angeles, Univ. Cal., Sch. Eng. Appl. Sci., Rep. UCLA-ENG-7318 (1973).

FLEISHMAN, E. A. Dimensional analysis of psychomotor abilities. *J. Exp. Psychol.* 48:437–454 (Dec. 1954).

FLEISHMAN, E. A. Dimensional analysis of movement reactions. *J. Exp. Psychol.* **55**:438–453 (May 1958).

FLEISHMAN, E. A., and W. E. HEMPEL. A factor analysis of dexterity tests. *Personnel Psychol.* **7**:15–32 (1954).

FOWLER, B., and K. N. ACKLES. Narcotic effects in man of breathing 80–20 argon–oxygen and air under hyperbaric conditions. *Aerosp. Med.* **43**:1219–1224 (Nov. 1972).

FRENCH, J. W., R. B. EKSTROM, and L. A. PRICE. *Manual for Kit of Reference Tests for Cognitive Factors.* Princeton, N. J., Educational Testing Service (June 1963).

GAYDOS, H. F. Effect on complex manual performance of cooling the body while maintaining the hands at normal temperatures. *J. Appl. Physiol.* **12**:373–376 (May 1958).

GAYDOS, H. F., and E. R. DUSEK. Effects of localized hand cooling versus total body cooling on manual performance. *J. Appl. Physiol.* **12**:377–380 (May 1958).

GUNDERSON, E. K. E., and P. D. NELSON. Criterion measures for extremely isolated groups. *Personnel Psychol.* **19**:67–80 (Spring 1966).

HELMREICH, R. Patterns of aquanaut behavior. In: Miller, J. W., J. G. VanDerwalker, and R. A. Waller, eds. *TEKTITE II. Scientists-in-the-Sea*, pp. VIII-30–VIII-45. Washington, D. C., U. S. Department of the Interior (1971a).

HELMREICH, R. The TEKTITE II human behavior program. Austin, Tex., Univ. Tex., Dept. Psychol., Social Psychol. Lab., Rep. TR-14 (Mar. 15, 1971b).

HORVATH, S. M., and A. FREEDMAN. The influence of cold upon the efficiency of man. *J. Aviation Med.* **18**:158–164 (Feb. 1947).

HUNTER, J., E. H. KERR, and M. G. WHILANS. The relation between joint stiffness upon exposure to cold and the characteristics of synovial fluid. *Can. J. Med. Sci.* **30**:367–377 (Oct. 1952).

KIESSLING, R. J., and C. H. MAAG. Performance impairment as a function of nitrogen narcosis. *J. Appl. Psychol.* **46**:91–95 (Apr. 1962).

KLEITMAN, N., and D. P. JACKSON. Body temperature and performance under different routines. *J. Appl. Physiol.* **3**:309–328 (Dec. 1950).

LeBLANC, J. S. Impairment of manual dexterity in the cold. Manitoba, Can., Def. Res. Northern Lab., Rep. DRNL 4/55 (1955).

LOCKHART, J. M. Effects of body and hand cooling on complex manual performance. *J. Appl. Psychol.* **50**:57–59 (Feb. 1966).

LOCKHART, J. M. Extreme body cooling and psychomotor performance. *Ergonomics* **11**:249–260 (May 1968).

MACKWORTH, N. H. Finger numbness in very cold winds. *J. Appl. Physiol.* **5**:533–543 (Mar. 1953).

MANN, C. W., and G. E. PASSEY. The perception of the vertical: V. Adjustment of the postural vertical as a function of the magnitude of postural tilt and duration of exposure. *J. Exp. Psychol.* **41**:108–113 (Feb. 1951).

MARGARIA, R. Wide range investigations of acceleration in man and animals. *J. Aviation Med.* **29**:855–871 (Dec. 1958).

McGINNIS, J. M., J. M. LOCKHART, and C. K. BENSEL. A human factors evaluation of cold-wet handwear. U. S. Army, Natick Lab., Tech. Rep. 73-22-PP (Apr. 1972).

MILLER, J. W., R. RADLOFF, H. M. BOWEN, and R. L. HELMREICH. The SEALAB II human behavior program. In: Pauli, D. C., and G. P. Clapper, eds. Project SEALAB report. An experimental 45-day undersea saturation dive at 205 feet, pp. 245–271. Off. Nav. Res. Rep. ACR-124 (Mar. 8, 1967).

MILLS, A. W. Finger numbness and skin temperature. *J. Appl. Physiol.* **9**:447–450 (Nov. 1956).

MILLS, A. W. Tactile sensitivity in the cold. In: Fisher, F. R., ed. *Protection and Functioning of the Hands in Cold Climates*, pp. 76–86. Washington, D. C., National Academy of Sciences–National Research Council (1957).

MORRISON, J. B. Oxygen uptake studies of divers when fin swimming with maximum effort at depths of 6–176 feet. *Aerosp. Med.* **44**:1120–1129 (Oct. 1973).

MORTON, R., and K. A. PROVINS. Finger numbness after acute local exposure to cold. *J. Appl. Physiol.* **15**:149–154 (Jan. 1960).

NELSON, J. G. The effect of water immersion and body position upon perception of the gravitational vertical. U. S. Nav. Air Devel. Cent., Rep. NADC-MR-6709 (1967).

NORMAN, D. G. Force application in simulated zero gravity. *Hum. Factors* 11:489–505 (Oct. 1969).

NOWLIS, D., H. H. WATTERS, and E. C. WORTZ. Habitability assessment program. In: *TEKTITE II. Scientists-in-the-Sea*, pp. VIII-68–VIII-89. Washington, D. C., U. S. Department of the Interior (1971).

NOWLIS, D. P., E. C. WORTZ, and H. WATTERS. TEKTITE II habitability program. Los Angeles, Cal., Air Research Manufacturing Co., Rep. 71-6192 (Jan. 14, 1972).

O'REILLY, J. P. Behavioral effectiveness at 16 ata. Honolulu, Univ. Hawaii, Sea Grant Tech. Rep. UNIHI-SEAGRANT-TR-73-01 (Apr. 1973).

PARKER, J. W. Performance effects of increased ambient pressure. II. Helium–oxygen saturation and excursion dive to a simulated depth of 110 feet. U. S. Nav. Submar. Med. Cent. Rep. SMRL 596 (Sept. 1969).

PAYNE, R. B. Tracking proficiency as a function of thermal balance. *J. Appl. Physiol.* 14:387–389 (May 1959).

PEACOCK, L. J. A field study of rifle aiming steadiness and serial reaction performance as affected by thermal stress and activity. U. S. Army Med. Res. Lab., Rep. 231 (1956).

RADLOFF, R., and R. HELMREICH. *Groups under Stress: Psychological Research in SEALAB II.* New York, Appleton-Century-Crofts (1968).

RADLOFF, R., R. MACH, and N. ZILL. Behavioral program. In: Pauli, D. C., and H. A. Cole, eds. *Project TEKTITE I*, pp. A-13–A-47. Off. Nav. Res., Rep. DR-153 (Jan. 1970).

REILLY, R. E., and B. J. CAMERON. An integrated measurement system of the study of human performance in the underwater environment. Falls Church, Va., Bio-Technology, Inc. (1968).

ROSS, H. E., S. D. CRICKMAR, N. V. SILLS, and E. P. OWEN. Orientation to the vertical in free divers. *Aerosp. Med.* 40:728–732 (July 1969).

ROSS, H. E., D. J. DICKINSON, and B. J. JUPP. Geographical orientation under water. *Hum. Factors* 12:13–23 (Feb. 1970).

SHILLING, C. W., and W. W. WILLGRUBE. Quantitative study of mental and neuromuscular reactions as influenced by increased air pressure. *U. S. Nav. Med. Bull.* 35:373–380 (Oct. 1937).

STANG, P. R., and E. L. WIENER. Diver performance in cold water. *Hum. Factors* 12:391–399 (Aug. 1970).

STREIMER, I. A study of work-producing characteristics of underwater operations. Downey, Cal., North American Rockwell Corp., Space Division, Rep. SD 68-347 (1969*a*).

STREIMER, I. A study of work-producing characteristics of underwater operations as a function of depth. Downey, Cal., North American Rockwell, Space Division, Rep. SD 69-712 (Nov. 1969*b*).

STREIMER, I., D. P. W. TURNER, K. VOLKMER, and P. PRYOR. Experimental study of diver performance in manual and mental tasks at 66 feet. San Diego, Cal., Man Factors, Inc., Rep. MFI 71-115 (Sept. 1971).

STREIMER, I., D. P. W. TURNER, K. VOLKMER, and P. PRYOR. A study of forced-pace work characteristics at a 33-foot working depth. San Diego, Cal., Man Factors, Inc., Rep. MFI 72-123 (Sept. 1972).

STROOP, J. R. Interference in serial verbal reactions. *J. Exp. Psychol.* 18:643–661 (Dec. 1935).

SUMMITT, J. K., J. S. KELLEY, J. M. HERRON, and H. A. SALTZMAN. Joint U. S. Navy–Duke University 1000-foot saturation dive. U. S. Navy Exp. Diving Unit, Rep. NEDU 3-69 (1969).

TEICHNER, W. H. Manual dexterity in the cold. *J. Appl. Physiol.* 11:333–338 (Nov. 1957).

TEICHNER, W. H. Reaction time in the cold. *J. Appl. Psychol.* 42:54–59 (Feb. 1958).

TEICHNER, W. H. Interaction of behavioral and physiological stress reactions. *Psychol. Rev.* 75:271–291 (July 1968).

TEICHNER, W. H., and J. L. KOBRICK. Effects of prolonged exposure to low temperature on visual–motor performance. *J. Exp. Psychol.* 49:122–126 (Feb. 1955).

TEICHNER, W. H., and R. F. WEHRKAMP. Visual-motor performance as a function of short duration ambient temperature. *J. Exp. Psychol.* 47:447–450 (June 1954).

TRESANSKY, G. J. The NMRI Mark III Mod. I tremor device. U. S. Nav. Med. Res. Inst., Rep. 2 (May 1973).

U. S. NAVY. *U. S. Navy Diving Manual*. Washington, D. C., U. S. Navy Department (Mar. 1970) (NAVSHIPS 0994-001-9010).

VAUGHAN, W. S., JR. and B. G. ANDERSEN. Effects of long-duration cold exposure on performance of tasks in naval inshore warfare operations. Landover, Md., Oceanautics, Inc., Rep. on Contract ONR-N00014-73-C-0309 (Nov. 1973).

VAUGHAN, W. S., JR. and A. S. MAVOR. Diver performance in controlling a wet submersible during four-hour exposures to cold water. *Hum. Factors* **14**:173–180 (Apr. 1972).

VAUGHAN, W. S., JR. and J. S. SWIDER. Crew performance in swimmer delivery vehicle operations (U). Landover, Md., Whittenburg, Vaughan Associates, Inc. (May 1972)

WEBB, P. Rewarming after diving in cold water *Aerosp. Med.* **44**:1152–1157 (Oct. 1973).

WECHSLER, D. *The Measurement and Appraisal of Adult Intelligence*. Baltimore, Williams and Wilkins (1958).

WEITZ, J. Vibratory sensitivity as a function of skin temperature. *J. Exp. Psychol.* **28**:21–36 (Jan. 1941).

WELTMAN, G., and G. H. EGSTROM. Perceptual narrowing in novice divers. *Hum. Factors* **8**:499–506 (Dec. 1966).

WELTMAN, G., G. H. EGSTROM, R. E. ELLIOTT, and H. S. STEVENSON. Underwater work measurement techniques: initial studies. Los Angeles, Univ. Cal., Sch. Eng. Appl. Sci., Rep. on Contract ONR N00014-67-A-0111-0007 (1968).

WELTMAN, G., R. A. CHRISTIANSON, and G. H. EGSTROM. Effects of environment and experience on underwater work performance. *Hum. Factors* **12**:587–598 (Dec. 1970).

WELTMAN, G., G. H. EGSTROM, M. A. WILLIS, and W. CUCCARO. Underwater work measurement techniques: final report. Los Angeles, Univ. Cal., Sch. Eng. Appl. Sci., Rep. on Contract ONR-N00014-67-0111-0007 (1971*a*).

WELTMAN, G., J. E. SMITH, and G. H. EGSTROM. Perceptual narrowing during simulated pressure-chamber exposure. *Hum. Factors* **13**:99–107 (Apr. 1971*b*).

VII

Decompression Sickness

A. *Introduction*

Man first began to spend substantial amounts of time under increased pressure soon after the development of the air compressor, early in the nineteenth century. The majority of such exposures occurred in European tunneling and caisson work. The process of decompressing took some time because of the cumbersome air locks and the general need to climb many steps back to the surface. During and after these decompressions men were noted to exhibit symptoms of distress but this was generally attributed to the unhealthy gas environment that resulted from poor ventilation, candle-burning, and tunneling under marshes where trapped sulfurous gases were often found.

As machinery and methods improved, the tunnels became longer and were often much deeper. The decompression was shortened because improved air locks permitted rapid decompression. These changes in work habits resulted in a major increase in the incidence of crippling and fatal sequelae. Since air quality in the work areas had been substantially improved, attending physicians had to look elsewhere for the causative factor; they soon decided that the decompression was too short for the men to become reacclimated to the lower pressure. Indeed, prolonged decompression did substantially reduce the incidence of severe symptoms.

The salient features of decompression sickness (DCS) were established during the period from 1870 to 1910. Hoppe's early suggestion that sudden death in compressed air workers was due to liberation of gas from blood and tissues was confirmed by the work of Paul Bert. In 1878 Bert performed the first definitive studies oriented toward understanding the processes involved in reacclimation. He observed that bubbles formed within animal tissues following rapid decompression and that bubbles, if present in adequate numbers, could kill or paralyze the animals. The higher the pressure and the longer the exposure, the greater was the ensuing volume of bubbles. Slow decompression suppressed bubble formation. These findings have been repeatedly confirmed and form the basis of our understanding of the malady that we now call *decompression sickness.**

---

* Other names used in recent times include dysbarism and "bends."

467

The requirement of a slow decompression represents a major limitation on man's ability to dive safely to great depths. This limitation is always contrary to the goal of the planned activity. Traumatic injuries at diving depths represent a major hazard because treatment must await the completion of a slow decompression. Thus, enormous effort has been expended to develop the most rapid and safe decompression methods.

J. S. Haldane and his associates (Boycott *et al.* 1908) attempted to generalize the required decompression schedule in terms of the exposure depth and bottom time* so as to be able to calculate the so-called "decompression schedule" for any set of diving circumstances. Using goats, they found that so long as the pressure was less than 2 ATA, the goat could return to the surface directly without any ill effects.

The Haldane generalization postulated that the body absorbs excess nitrogen when exposed to increased pressure and that this is the source of gas needed to generate bubbles during decompression. A slow decompression permits the exhalation of the excess nitrogen, thereby inhibiting bubble formation. Nitrogen is distributed throughout body tissues according to the relative solubility and blood flow to that tissue.

Rapid decompression slows nitrogen elimination and thus allows more nitrogen to remain in the body for bubble formation. Neither Haldane nor anyone since has successfully validated or repudiated the model. The concept has been extremely useful in guiding research, and the model has been extensively used to correlate empirical diving results.

The military and commercial demand for shorter decompression times has been met with the development of procedures that result in an acceptably low incidence of decompression sickness. In order to achieve this low incidence, diving medical officers have also evolved an extensive set of limitations on the divers and the equipment they use. The evaluation of these procedures and limitations has involved empirical data based on extensive human experimentation.

The recent availability of scuba equipment has led to an enormous amount of diving for sport or science that is not subject to the same controls. It is a considerable testimony to the existing procedures that they have worked so well for scuba divers. Table VII-1 lists a series of hyperbaric exposures that result in demands for decompression procedures that pose special problems in the use of existing methods. It would be fortuitous if the empirical techniques that are now available could be extrapolated to all diving situations; but it is highly improbable.

In view of this situation, this chapter is organized to provide the basic information needed to understand existing decompression procedures and their limits. Diagnosis and treatment of decompression accidents are detailed and sources of advice and assistance are identified. This chapter is outlined to help in planning a program of hyperbaric exposure so as to avoid a decompression accident. Such planning inevitably leads to a restriction on the amount of exposure one would like to have in order to accomplish a job. Yet experience has shown that a philosophical acceptance of this limitation is highly preferable to the risk of a severe decompression catastrophe.

This chapter will deal only with decompression sickness. Other forms of barotrauma, such as air embolism, are discussed in Chapter III.

It will become evident that equipment design and performance play a key role in

---

* See the section on prevention, this chapter, for a detailed definition of these terms.

Table VII-1

Hyperbolic Exposures for Which Decompression Schedules Have Not Been Expressly Developed

Hyperbaric exposure	Special features not tested in available schedules
Hyperbaric medicine	a. Patients with altered circulation and no experience in hyperbaric exposure and excessive O_2 exposure b. Physicians and professional staff who may be older and in poorer physical condition than normal divers and therefore not able to tolerate the same risks
Scuba sport diving	a. Age and physical condition of divers b. Lack of appropriate treatment facilities (chambers, physicians, extra gas, medicine) c. Variable pressure exposure d. Increased hazard of underwater accident
Scientific diving	a. Lack of appropriate treatment facilities due to (generally) lower financial support b. Variable pressure exposure c. Increased bottom time d. Increased hazard of underwater accident due to (1) use of scientific equipment; (2) need to observe dangerous life forms in more dangerous environments (caves, canyons, etc.)
Deep water diving	a. Depths are so great that pressure, per se, may alter biological response to decompression b. Need to use habitats (gas-phase decompression) c. Appearance of decompression sickness at depths greater than the normal treatment depths d. Existence of pressure-related symptoms that complicate diagnosis of decompression sickness (compression arthralgia, HPNS, narcosis)

the prevention of decompression sickness. The provision of a diver with a consistent life support system (respiratory, thermal, guidance, propulsive, etc.) is an excellent way to minimize the risk in any given diving operation.

B. Factors Relevant to the Pathogenesis of Decompression Sickness

Decompression sickness (DCS) occurs when man is subjected to reduced environmental pressure that causes bubbles of inert gas to form and grow within tissues or the vascular space. Physical factors relevant to the nucleation and growth of bubbles are of importance, as they permit one to assess the environmental conditions that would lead to the greatest risk of DCS (pressure, temperature, noise). Biological factors that influence the transport of inert gas throughout the body can be characterized. Responses of the body to the insult of bubble formation are less well known.

This section will discuss the underlying physical and biological factors that have been shown to be relevant to the process of bubble formation, growth, and dissolution as a result of hyperbaric exposures. Biological factors relevant to the pathology of DCS are well described in the medical literature.

Note that the extensive studies on decompression sickness that have been made in association with altitude (hypobaric) exposure are not always relevant to hyperbaric exposures. They will be cited only where the factors have been shown to be of importance in hyperbaric environments.

1. *Physical Factors*

a. Inert Gas Solubility

During hyperbaric exposure, the partial pressure of respired inert gases is generally different than it is at atmospheric pressure with air ($P_{N_2} = 0.79$ ATA). As shown in Chapter III, the partial pressure of an inert gas in a breathing mixture is related to the environmental pressure P_{abs} by

$$P_x = F_x P_{abs} \tag{1}$$

where F_x is the fraction of the inert gas in the gas mixture. It is generally assumed that the ideal gas law $PV = nRT$ is applicable to inert gases in diving environments, validating Eq. (1). See Chapter IV for further details on the properties of inert gases in hyperbaric breathing mixtures.

All the inert gases (He, H_2, N_2, Ne, Ar) dissolve in blood and tissue according to Henry's law, viz.

$$C_x = \alpha_x P_x \tag{2}$$

where C_x is the dissolved concentration [generally expressed as cm^3 (STP)/cm^3 fluid] and α_x is the absorption constant [cm^3(STP)/cm^3 fluid atm] for the inert gas denoted by the subscript x. Thus, when an individual is exposed to an elevated partial pressure of inert gas, the inert gas tends to be absorbed into the body through the lungs and to dissolve into the tissues of the body until they are equilibrated with the respired gas.

After a period of time, equilibration occurs [as per Eq. (2)] and no more gas may dissolve. Thus, at any elevated pressure there is a maximum of gas that may dissolve in the body tissues.

The Bunsen solubility coefficient α_x depends on the temperature, type of tissue or fluid, and type of inert gas. Over the normal range of diving, it is *not* a function of absolute pressure (Poynting effect), partial pressure of the inert gas, or the presence of other inert gases.*

The temperature dependence of the solubility coefficient α_x is given by

$$\alpha_x = \alpha_{x,0} e^{-\Delta H/RT} \tag{3}$$

where ΔH is the heat of solution, R is the gas constant, and T is the absolute temperature in degrees Kelvin ($T\,^\circ K = 273 + T\,^\circ C$). Table VII-2 presents the preexponential

---

* While these statements have been experimentally validated for water at depths as deep as 50 ATA, they have not been fully tested for biological fluids; therefore, at the recently attained depths of 50 ATA there may be some influence of pressure on solubility. Its significance in developing decompression procedures is probably negligible.

Table VII-2

Solubility Parameters for Various Gases in Water

$$\alpha_x = \alpha_{x,0}e^{-\Delta H/RT}$$

Gas x	$\alpha_{x,0}$, cm³ (STD)/cm³ ATA	ΔH, kcal/g mole	$\alpha_x(37.0°C)$	Reference
N_2	2.210×10^{-4}	2,465	12.30×10^{-3}	Douglas 1964
He	8.193×10^{-9}	8,520	8.66×10^{-3}	Weiss 1971
H_2	1.830×10^{-8}	8,422	16.70×10^{-3}	Morrison 1952
Ne	5.317×10^{-10}	10,256	9.61×10^{-3}	Douglas 1964
Ar	2.721×10^{-4}	2,777	25.90×10^{-3}	Douglas 1964
O_2	2.645×10^{-4}	2,757	23.60×10^{-3}	Douglas 1964

factor $\alpha_{x,0}$ and ΔH for the various inert gases in water. Since ΔH is a positive number, the solubility decreases at elevated temperatures.

The solubility of inert gases is greater in fat tissue than in lean tissue, due to the tendency for greater solubility in lipoidal fluids. Table VII-3 tabulates inert gas solubility for blood, lean tissue, fat tissue, and for a lipoidal fluid (olive oil) at 37°C. Tissues behave very similarly to fluids in their solubility characteristics. The solubility increases with increasing molecular weight of the gas, except for H_2. Similarly, the partition coefficient between olive oil and water ($\lambda = \alpha_{0.0.}/\alpha_{water}$) also increases with molecular weight of the gas.

Thus, fat tissues represent large reservoirs for all the inert gases and the absolute amount held increases with the molecular weight of the gas being inhaled. On the other hand, the partitioning of the gas from the gas phases into the liquid requires that few molecules go into the liquid in order to saturate it. For example, the volume of bubbles that could form if one saturates 100 cm³ of fat tissue with air is 4.2 cm³. Haldane and others (Boycott *et al.* 1908) have established that a dive involving a 1-atm change in pressure is apparently safe for man. We may thus surmise that: (a) this volume loading of gas is tolerable, (b) that bubbles do not form, or (c) there is some variant in between these extremes. The question of bubble nucleation and growth is presented in the next subsection.

Table VII-3

Solubility Coefficients for Inert Gases in Biological Fluids at 37°C

Gas	Blood[a] α	Lean tissue α	Fat α	Olive oil α
H_2	—	—	—	0.0484
He	0.0159	—	—	0.0159
N_2	0.0130+	0.012+	0.062+	0.067
Ar	—	—	—	0.14
Ne	—	—	0.020	0.019
O_2	0.0223	0.023	—	0.112
CO_2	0.488	—	—	1.25

[a] 150 g Hb/100 ml.

b. Bubble Nucleation and Growth

In order to transpose dissolved gas within a pure liquid into a bubble within that liquid, the physical environment of the liquid must be changed so that the bubble-containing state can form spontaneously. Thermodynamic arguments guide this process. A few examples serve to introduce the problem.

In Figure VII-1(a) a liquid is in equilibrium with N_2 gas at 2 ATA within a cylinder that is isolated by a diaphragm that transmits ambient pressure. The valve is closed in (b); the concentration of dissolved gas in the liquid is then equal to 2α and the system is at equilibrium.

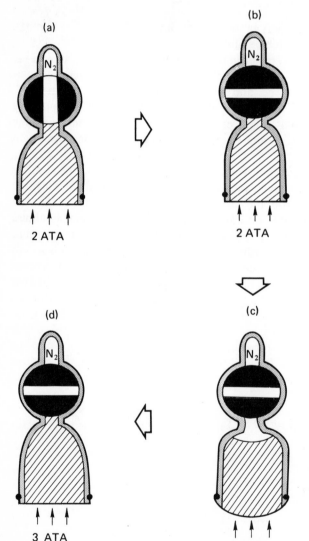

Figure VII-1. The equilibrium between an inert gas and water in a closed system. The two-phase system is isolated from the exterior environment by the container. Pressure is transmitted to the lower end across an impermeable diaphragm. The valve may be rotated to isolate the liquid from the gas phase.

If the ambient pressure is reduced to 1 ATA [Figure VII-1(c)], bubbles form and grow by the process of N_2 diffusion from the liquid. The bubble will grow until the concentration of dissolved N_2 is α; at that point equilibrium is reestablished. The diaphragm distends to compensate for the increased volume below the valve.

If the ambient pressure is then raised to 3 ATA [Figure VII-1(d)], the bubbles will dissolve in the liquid. When totally dissolved, the concentration of N_2 is again 2α. The system is less than saturated but at pressure equilibrium so long as the valve remains closed. If the valve is now opened so that the liquid is in contact with the gas, gas will dissolve in the liquid until it is saturated (i.e., the concentration is 3α).

Thus, Figure VII-1 demonstrates the principle that when the absolute pressure on a fluid is decreased below the partial pressure of the dissolved gas, there is a tendency for the gas to leave the solution going into the gas phase which exists at a lower pressure. It will do this until the partial pressure in the gas phase equals the dissolved partial pressure (1 atm, in this case). It also shows that bubbles may be dissolved by raising the ambient pressure to a value where the equilibrium state results in subsaturation.

There are other ways to produce supersaturation. If we equilibrated a liquid with pure N_2 at 1 atm at 25°C in the equipment shown in Figure VII-1(a), the dissolved gas concentration would be

$$C_{N_2} = \alpha_{N_2}{}^{25°} = 0.02 \text{ cm}^3(\text{STP})/\text{cm}^3$$

If the valve were then closed and the temperature raised to 37.0°C, the partial pressure of N_2 in the liquid would now be

$$P_{N_2}(37°) = C_{N_2}/\alpha_{N_2}{}^{37°} = 0.02/0.0127 = 1.5 \text{ ATA}$$

but the absolute pressure is still only 1 ATA. So this solution is also supersaturated [in the same sense as supersaturation was produced in Figure VII-1(c)]; a gas phase would again tend to form so that enough N_2 would diffuse out of the liquid to reduce its partial pressure to 1 ATA. So nucleation of a gas phase tends to occur whenever the N_2 partial pressure in the water exceeds the ambient pressure, regardless of how we attained this state of affairs.

Yet, nucleation and growth are kinetic processes (i.e., they proceed at a finite rate). Thus, through equilibrium considerations we can describe the final state of a process that involves nucleation and growth. But we must use kinetic equations to describe the rate of these processes. The body is only transiently supersaturated, so it is essential to estimate nucleation and growth rates if we are to be able to assess the potential hazard of a decompression procedure.

Since all kinetic processes are described by a general equation, we may seek to characterize nucleation and growth of bubbles in this way. The general kinetic equation is

$$\text{rate of a process} = \frac{\text{driving force for the process}}{\text{resistance to the process}} \tag{4}$$

or

$$J_i = \frac{\Delta E_i}{R_i} \tag{5}$$

where i denotes the process.

In bubble nucleation processes the driving force is represented by the difference between the dissolved-gas partial pressure and the absolute pressure $P_x - P_{abs}$. By convention, whenever the difference is positive, the rate of bubble nucleation is positive and bubbles will continue to form.

The resistance to nucleation RN cannot be theoretically predicted, because of our imprecise knowledge of the thermodynamic states of water. However, it is known empirically that several factors do influence the process; it occurs much more rapidly at elevated temperatures, and it occurs much more easily at a liquid–surface interface. This is especially true if the interface is hydrophobic.*

De novo bubble formation rates have been estimated. Under diving conditions the rates would appear to be so slow as to preclude bubble formation. The process may be greatly accelerated by ultrasonic energy, as it acts to transiently produce substantial reductions in the local fluid pressure. This occurs because the driving force for the nucleation rate is $P_x - \bar{P}_{abs}$, and $\bar{P}_{abs}$, the local fluid pressure, is now given by

$$\bar{P}_{abs} = P_{abs} + P_u \sin(\omega t) \qquad (6)$$

where P_u is the amplitude of the ultrasonically produced pressure wave of frequency ω^{-1}. Whenever P_u is large compared to P_{abs}, the nucleation rate is enhanced. This occurs because the negative segment of the acoustic wave produces bubble growth at a greater rate than that of the bubble dissolution produced by the positive segment.

Sources of ultrasonic energy are abundantly available in the diving environment; noise and vibration are predominant external sources. The acoustic inhomogeneity of the body makes it an excellent site for acoustic focusing effects that would tend to amplify a normally weak ultrasonic wave.

As the dissolved gas concentration increases, the energy needed for cavitation decreases. Figure VII-2 shows the cavitation threshold for water and benzine; 100% saturated fluid at 1 ATA total pressure requires only 1 atm of ultrasonic energy to induce cavitation.

As the hydrostatic pressure is increased on a fluid saturated at 1 ATA, the ultrasonic energy required for cavitation increases proportionately (Figure VII-3). Cavitation is *not* strongly temperature dependent. As the ultrasonic frequency increases, the required cavitation energy increases exponentially. Thus, one is generally concerned with sound sources below 100 kHz; higher frequency sources are generally too weak to cause cavitation.

Transient ultrasonic pulses can induce nucleation, but generally more energy is required. Figure VII-4 shows that a 0.1-sec pulse will cavitate a liquid, but it requires about ten times the power required under constant-irradiation conditions.

Viscous fluids and tissues (*gels*) require more ultrasonic energy for bubble formation. Figure VII-5 shows that the power requirements increase as the logarithm of the fluid viscosity.

Some investigators have suggested that bubble nucleation rates would never be a rate-limiting step in bubble formation because there are always micronuclei that exist. Thus, bubble appearance merely depends on the growth of the micronuclei in the

---

* It also appears the RN is a function of the degree of supersaturation. Thus, at low degrees of supersaturation the process is exceedingly slow; at higher values, it increases rapidly and then the process is substantially proportional to the degree of supersaturation.

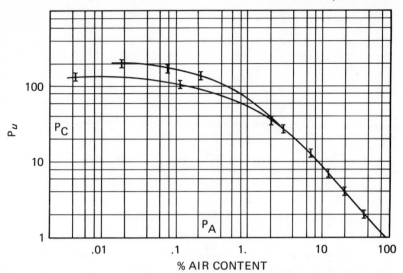

Figure VII-2. Cavitation threshold of water (upper curve) and benzine (petroleum ether) (lower curve) as a function of percentage of gas concentration at a hydrostatic pressure of 1 atm and temperature of 22°C. P_u is the cavitation threshold in bars.

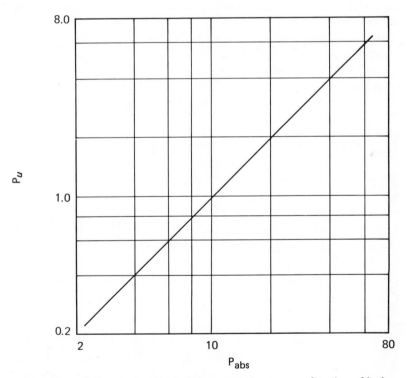

Figure VII-3. Cavitation threshold of air-saturated water as a function of hydrostatic pressure. P_u is the cavitation threshold in bars; P_{abs} is the atmospheric pressure in bars. $T = 22°C$.

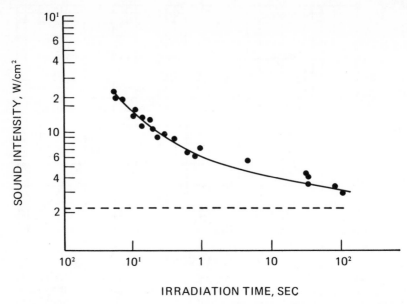

Figure VII-4. Average sound intensity as a function of time between switching instrument on and start of cavitation. Excitation frequency is 365 kHz in tap water. The intensity below which no cavitation appears in 15 min of irradiation is shown as a dashed line.

presence of a supersaturated tissue. Harvey (1951) demonstrated that such micronuclei do not exist in human blood. Their presence in peripheral tissue has recently been inferred but not yet unequivocally demonstrated.

Once the bubble forms and reaches a size of about 0.1 μm, it is viewed as a stable bubble (i.e., $2\gamma/r < P_x - P_{\text{abs}}$). At this point the pressure inside the bubble P_B is given

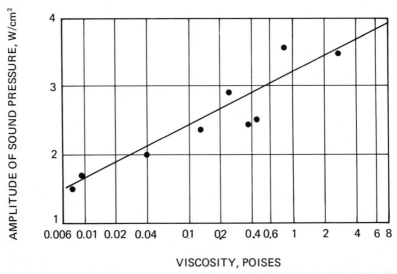

Figure VII-5. Amplitude of cavitation sound pressure against viscosity of liquid.

by $P_{abs} + (2\gamma/r)$, where r is the bubble radius and γ is the surface tension. If $P_B < P_x$, there is a gradient of partial pressure from the liquid to the bubble and the bubble will continue to grow by absorbing gas from the surrounding liquid. This process of growth will cease when the partial pressure in the liquid is reduced to the level of P_B (by the process of gas transport into the bubbles). Now $P_B \simeq P_{abs}$ when $r \geq 20\ \mu m$ (i.e., in water, $\gamma \simeq 72\ dyn/cm$ and $P_{abs} - P_B = 0.07\ ATA$ when $r = 20\ \mu m$). So, surface tension effects are only important for very small bubbles.

The rate of the bubble growth is given by

$$J_G = (P_x - P_B)/R_G \tag{7}$$

Normally, the resistance term R_G is governed by diffusion of the gas molecules in the liquid phase and is inversely proportional to the Bunsen solubility coefficient

$$R_G = \frac{K_G}{\alpha_x D_x} \tag{8}$$

where D_x is the molecular diffusivity of the gas in the liquid phase, expressed as cm^2/sec. The constant of proportionality K_G is dependent on the bubble geometry, the presence of other bubbles, and the location of the bubble in the tissue. Table VII-4 gives the diffusivity D_x of various gases in fluids and tissues.*

It is not possible to predict diffusivities for liquid solvents based on molecular considerations. Tissue diffusivities are unavailable. However, one can estimate diffusivities based on the well-known Stokes–Einstein equation, $D = kT/6\pi\mu r$, where k is the Boltzmann gas constant, T is the absolute temperature, μ is the viscosity of the solvent, and r is the molecular radius of the solute molecule.

The temperature dependence of D is based on the ratio of T/μ. Since the logarithm of viscosity is generally inversely dependent on absolute temperature (i.e., $\ln \mu = B/T + A$), diffusivity will increase with absolute temperature, but in a complex manner.

The diffusivity of gases in tissue is normally determined by measuring the steady-state flux through a flat piece of tissue of known thickness. If the gas solubility is known, one can estimate diffusivity by use of the linear diffusion form of Eq. (7):

$$J_G = \frac{DA\alpha}{t} \Delta P \tag{9}$$

where A and t are the area and thickness of the tissue section that is exposed to a gas pressure gradient of ΔP. The diffusivity determined from Eq. (9) is a bulk average value. The relative contribution of diffusion through cells or between cells is unknown.

The product $D\alpha$ is often called the "permeability" of the tissue.† This is a useful term for heterogeneous media because of the uncertainties in estimating α.

---

* There are some reports in the literature that D_x in tissue may be much less than in plasma. The argument revolves around the methods whereby the diffusion coefficient is measured (steady-state vs dynamic methods). Presentation of the data in Table VII-4 is offered to show the influence of gas molecular weight and tissue type, using the same measurement method.

† Many early investigators referred to permeability as the ratio $J/(A\ \Delta P)$. This is really the "specific flux density" and is of little use because the tissue thickness is seldom reported in these studies.

Table VII-4

Diffusion Coefficients of Gases in
Fluids and Tissues $T = 37°C$

Gas	Fluid/tissue	$D \times 10^5$ cm²/sec
H_2	Water	3.04
H_2	Serum	2.68
N_2	Water	1.32
N_2	Serum	1.15
O_2	Water	1.82
O_2	Lung	1.38
O_2	Connective tissue	0.58

Notice that if nucleation must occur to create bubbles, at least one dissolved gas must be present in a supersaturated form. Recent results at great depth where pressure was not altered, but He and N_2 were exchanged in the breathing mix, resulted in symptoms normally associated with subcutaneous bubbles (see the section on diagnosis, this chapter). The gases were switched in such a way that the total dissolved gas partial pressure ($P_{N_2} + P_{He}$) would have been substantially elevated in subcutaneous tissue. If micronuclei existed, bubbles could have resulted by increased growth rates. No such anomalous appearance of symptoms has ever been reported in subsaturation, low-pressure diving where only one inert gas was involved.

In conclusion, bubble nucleation or growth may be important in the production of *in vivo* bubbles during decompression procedures. No generalizations can be made that will assist in determining which factor is the dominant one for a given type of dive. Yet dive procedures are often selected that would be safe based on the dominance of nucleation as the rate-limiting step, yet would be unsafe for a growth-dominated process of *in vivo* bubble production.

2. Biological Factors

The whole process of inert gas dissolution into body tissues begins at the interface with the environment. Inert gases enter the body mainly through the lungs; transdermal transport, while finite, is probably only important when one is concerned with subcutaneous bubbles. Absorption of gas from trapped gas (e.g., intestinal) does occur, but is generally not a factor because of the limited amount of gas (see Chapter III for complications of trapped-gas pockets during decompression).

Dissolved gases are carried by blood flow from the lungs to the peripheral tissues, where they diffuse into tissue cells. During decompression, the process is reversed.

Ambient pressure is transmitted throughout the body. During compression and decompression certain (relatively) isolated anatomical zones may not equilibrate instantly, but the time lag for equilibration is probably minimal. One may assume, then, that by relating a gas partial pressure in a tissue to the environmental pressure one has the correct gradient for bubble nucleation and growth.

The biologically important factors are the dynamic processes of (a) pulmonary exchange; (b) blood flow and distribution between and within tissues; and (c) diffusion and convection in peripheral tissues.

a. Pulmonary Exchange

Inert gases are absorbed in the lungs by transalveolar diffusion into blood. The amount of inert gas needed to fully equilibrate the blood with the gas in the alveolar space is a small fraction of that present in the lungs under normal inflation. For example, the maximum rate of N_2 transport across the lung would be about 160 cm^3 (STD)/min. This is a small fraction of the 8 liters/min air flow in the lungs under normal conditions (see Chapter IV for a detailed discussion of respiratory exchange). Thus, inert gas absorption occurs without influencing pulmonary gas volumes; that is, the process of O_2 and CO_2 exchange is considered to be independent of inert gas exchange:

$$PA_x = P_{abs} - PA_{O_2} - PA_{CO_2} - P_{H_2O}^{37°} \tag{10}$$

One can estimate the inert-gas partial pressure in the alveoli from a knowledge of the respired gas mixture. The maximum contribution of PA_{CO_2} is 60 mm Hg, and this is partly compensated by PA_{O_2} being somewhat less than $F_{O_2} \times P_{abs}$. A reasonable approximation of Eq. (10) would be to neglect all factors but those due to the inert gas and O_2:

$$PA_x \simeq F_x \times P_{abs} \tag{11}$$

where F_x is the inert gas fraction in the breathing gas. Equation (11) becomes even more valid at greater depths.

For the purposes of developing safe decompression schedules, we need to estimate the absorption of inert gases using assumptions that yield a degree of absorption that is equal to or greater than the actual amount. Equation (9) provides such a conservative estimation.

The oscillatory nature of respiration has little or no effect on decompression estimates. This is a valid approximation because the breathing frequency is 16 min^{-1} or more, while the highest frequency associated with inert gas transport within the body is about 0.2 min^{-1}; therefore, the respiratory oscillations are thoroughly damped by the much slower transport frequencies of the cardiovascular system.

The diffusional resistance of alveolar epithelium is so small that pulmonary venous blood is in essential equilibrium with alveolar gas (which, as we stated above, is in essential equilibrium with inspired gas). This seems to be a valid approximation for inert gas uptake; again, it represents a conservative estimation of gas absorption.

b. Transdermal Exchange

Inert gases are exchanged across the skin by diffusion based on the following equation:

$$J_x = \frac{D_x \alpha_x A}{t} (F_x P_{abs} - P_x) \tag{12}$$

where A is the whole body area, t is the average skin thickness, and P_x is the mixed venous inert gas partial pressure. When a diver is submerged, the surrounding fluid has a low inert gas pressure (for N_2, 0.79 atm; for other inert gases, ~ 0 atm). In these environments inert gas is lost from the diver.

Chamber decompressions that involve breathing a gas mixture from a mask (e.g., pure O_2) provide the setting where transdermal exchange may play a dominant role. The appearance of local skin reactions in divers who were breathing one inert gas while living in a habitat of another inert gas has been explained as being due to the additive effect of the two inert gases diffusing in opposite directions.

There is enough gas transport that Behnke (1969) was able to measure transcutaneous N_2 transport by measuring N_2 washout from the lungs with only a 1-ATA gradient in N_2 partial pressure.

During decompression in a chamber (where the skin is surrounded by high inert gas partial pressure), O_2 breathing may not be as useful as it would be in the water (where $P_i \sim 0$), because of transcutaneous gas transport. In general, experience shows that safe decompressions in such a chamber environment will often lead to unsafe schedules when subsequently tested in the cold water (whether effects are due to reduced local blood flow or gas diffusion is not known).

In most other diving environments transdermal exchange is neglected in the development of decompression procedures.

c. Blood Flow and Distribution between and within Tissues

During hyperbaric exposure the arterial blood carries inert gas to all the tissues of the body. Each tissue is presented with a limited amount of gas per minute, which is (in cm^3 of gas delivered to tissue i per minute) $J_i = C_a Q_i$, where Q_i is the blood flow rate into tissue i.

In the previous section we showed that $C_a = F_x P_{abs} \alpha_x$; thus

$$J_i = F_x \alpha_x P_{abs} \cdot f_i Q \tag{13}$$

where f_i is the fraction of the cardiac output $Q(cm^3/min)$ that is delivered to tissue i. The tissue cannot saturate any faster than inert gas is supplied by blood flow. If a tissue's rate of saturation is controlled by Eq. (13), it is called a perfusion-limited tissue.

The process of achieving a maximum of tissue saturation is shown schematically in Figure VII-6. The entering dissolved gas totally distributes between blood and tissue so the partial pressure of venous blood is identical to the average value of the tissue $P_{i,x}$. The ratio of $\alpha_x/\alpha_{i,x}$ is the partition coefficient for blood and tissue, λ_i, and is generally less than 1.0. A fatty tissue, where $\lambda_i \sim 0.2$, would saturate more slowly than a lean tissue with the same degree of blood supply because of the need to deliver more moles of gas to achieve equilibrium. Figure VII-6 also includes the dynamic mass balance model for inert gas within the tissue.

For the case where absorption occurs while the external pressure is constant over a long period of time, the tissue absorbs gas according to the following general equation (see Figure VII-7):

$$\left[\frac{P_a - P_i(t)}{P_a - P_i(0)} \right]_x = 1 - e^{-t/\tau_x} \tag{14}$$

where

$$\tau_x = \left(\frac{\alpha_i}{\alpha} \right)_x \left(\frac{\rho_i W_i}{1000 f_i Q} \right)$$

is the time constant for the tissue.

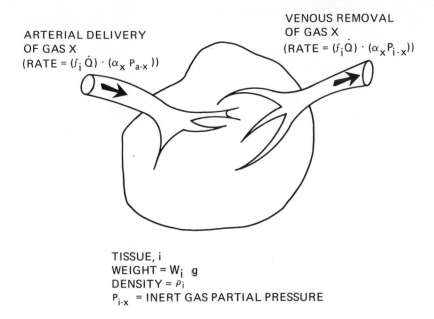

ARTERIAL DELIVERY
OF GAS X
$(RATE = (f_i \dot{Q}) \cdot (\alpha_x P_{a-x}))$

VENOUS REMOVAL
OF GAS X
$(RATE = (f_i \dot{Q}) \cdot (\alpha_x P_{i-x}))$

TISSUE, i
WEIGHT = W_i g
DENSITY = ρ_i
P_{i-x} = INERT GAS PARTIAL PRESSURE

RATE EQUATION FOR UPTAKE & WASHOUT:

$$\frac{dP_{i-x}}{dt} = \left(\frac{1000 f_i \dot{Q}}{\rho_i W_i}\right)\left(\frac{\alpha_x}{\alpha_{i-x}}\right)\left(P_{a-x} - P_{i-x}\right)$$

Figure VII-6. Inert gas exchange in a perfusion-limited tissue.

In subsequent sections the time constant will not be directly expressed. Rather, we will use the more traditional factor, called the half-time of the tissue. It is defined as

$$t_{\frac{1}{2}x} \equiv (\ln 2) \times \tau_x \tag{15}$$

Any tissue in the body will absorb inert gas at either the rate defined for the perfusion-limited tissue or at a slower rate. The blood flow term $(\rho_i W_i/1000 f_i Q$, expressed as cm^3 of tissue $\cdot$ min^{-1}/cm^3 of blood flow) has been measured for many tissues in the body by a variety of techniques. Table VII-5 is a summary of such flow rates based on the compilation of Jones (1951).

Tissue perfusion rates depend on exercise, inspired P_{O_2}, temperature, conditioning, nutritional state, subject size, and even (perhaps) the rate of change of environmental pressure. These influences have been generalized as follows:

1. Exercise increases muscle perfusion proportional to muscular work but causes a concomitant decrease in skin blood flow.
2. Elevated arterial O_2 partial pressures depress blood flow in almost all tissues that have been measured.
3. Decreased ambient temperature has little effect on central tissue blood flow but drastically reduces peripheral perfusion rates.
4. An aquanaut in good condition performs work at lower than normal levels of cardiac output.

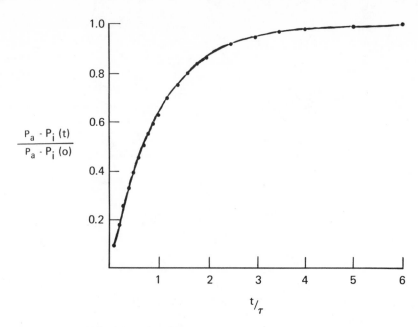

Figure VII-7. Generalized plot of inert gas absorption by a perfusion-limited tissue [see Eq. (14) in the text].

5. Overly obese and slender people exceed the normal range of cardiac output per kg BW and therefore probably have altered tissue perfusion rates (in addition, the obese subject has a higher fat content, which is reflected in an altered partition coefficient).

6. Rapid compression has been shown to drastically reduce local blood flow rates (bone and peripheral tissue), although this may not influence overall tissue perfusion rates.

Table VII-5

Tissue Perfusion Rates for a 70-kg Man (1.85 m²) under Rest Conditions with an Expected Cardiac Output of 5.8 liters/min

Tissue	Blood flow rate, cm^3 of blood/cm^3 of tissue·min^{-1}	Tissue volume cm^3
Lungs	4.8	1,200
Kidney	4.9	270
Heart	0.5	300
Adrenals, testes, prostate	1.0	80
Brain	0.54	1,400
Marrow	0.21	1,400
Hepato-portal	2.1	3,145
Muscle connective tissue, skin	0.022–0.0089	45,000–55,000
Fat	0.013–0.039	15,000–5,000

While there is great temptation to consider the process of inert gas exchange to be influenced by only these factors, the situation is really far more complex. Ultimately, we evaluate the adequacy of decompression schedules in terms of the existence of symptoms which are really only partially related to blood flow. Because of this uncertainty in predicting the influence that a hyperbaric exposure will exert on blood flow, decompression procedures should be tested under conditions closely simulating the actual environment. Also, schedules should be used in only those circumstances where one is sure that hemodynamic factors will not lead to lower than normal tissue perfusion. (For example, it is hazardous for an obese person to dive with U. S. Navy procedures that were tested on slender men.)

d. Site of Bubble Nucleation and Growth

The other role of the cardiovascular system in decompression is as a possible site for bubble nucleation or growth. The vascular tree is a compliant space (more so than tissue), so a rapid volume expansion due to bubble nucleation would result in little mechanical resistance. In addition, turbulence at heart valves or around rigid arterial bifurcations could result in the formation of short-lived micronuclei. These would normally dissolve. If they were carried into tissues where gas depots could provide a supersaturated environment, they could grow to a size greater than the critical radius. With these nuclei lodged in venules, they would be in the most supersaturated tissue regions. Bubbles so positioned would not represent a serious resistance to venous flow. Thus, they could persist and grow to considerable size before being dislodged into the the central venous circulation. Harvey and his associates (Harvey 1951) demonstrated that such micronuclei do not persist in circulating venous blood. This does not preclude their existence in peripheral vascular beds or in tissue spaces.

The severe cases of decompression sickness involve circulating bubbles that somehow bypass the lungs. When this condition occurs, the normal blood flow is drastically altered. The ability to transport inert gas from peripheral tissues is greatly reduced because the induced state of circulatory shock will result in an almost complete cessation of blood flow in the peripheral tissues. Even after therapeutic recompression completely dissolves the bubbles, the damage to the circulatory system (release of clotting factors, lipoproteins, alteration in platelet function, edema, etc.) may require days to recover, during which time the blood distribution will be abnormal. Thus, divers are normally kept from diving for at least a week after experiencing DCS.

e. Diffusion and Convection in Tissues

The concept of perfusion-limited tissues has been reasonably well established for the central tissues of the body, such as the CNS, heart, liver, kidney, and muscles. But in the peripheral tissues the assumptions are less well founded. In these tissues the cells have evolved so as to endure a more variable blood flow and one that is regulated by other functionalities, such as heat balance, response to trauma, and response to external stimuli (thermal and mechanical). In these tissues, the capillary morphology is irregular and, often, large avascular regions exist (e.g., cartilage, synovial fluid). Lymph flow in these regions may be a dominant means of nutrient transport. The distance between capillaries may be so great that diffusion will play a dominant role in the process of saturating the tissue with an inert gas.

Since oxygen must somehow reach these isolated cells, one can assume that inert gases do likewise (having similar solubilities and diffusivities). But if the process of diffusion dominates the delivery of inert gas to these regions, then there must be substantial gradients in partial pressure between these cells and the nearest vascular sections. During decompression these peripheral tissues would have inert gas partial pressures substantially greater than the average vascular levels, resulting in the peripheral tissues being a site of maximum supersaturation.

During a hyperbaric exposure, these tissues would all exhibit saturation times greater than those of the perfusion-limited tissues. The estimation of saturation times for these tissues is dependent on both the gas diffusivity (Table VII-4) in these tissues and the geometry of the tissue. Based on realistic anatomical considerations, saturation time constants of 300 min have been estimated, where

$$\tau = \delta^2/D \qquad (16)$$

where δ is some characteristic diffusion distance (e.g., mean intercapillary separation).

It is important to note that shifting from one gas to another will change a diffusion-limited time constant in a different manner than it will for a perfusion-limited tissue. A change in breathing gases implies a change in the rate of tissue saturation in a perfusion-limited tissue based on the relative gas solubilities, while it depends on the relative gas diffusivities in a diffusion-limited tissue.

One must be careful not to assume that these long saturation times necessarily correlate with empirical whole-body saturation times for developing decompression schedules. The *slow* tissue defined by Eq. (14) may not be a suitable site for bubble nucleation, due to its low compliance to distention (e.g., a gelled liquid has a greater resistance to cavitation than does the ungelled liquid). These tissues do, however, represent a reservoir from which gas slowly emerges such that it could contribute to local vascular bubble growth long after decompression is complete. [Unfortunately, there are tissues that are neither perfusion-limited nor diffusion-limited, but are partially controlled by both processes. In these tissues a 30% change in blood flow might result in only a 10% change in the rate of inert gas absorption.]

Whole-body washout studies. Measurement of gas exchange within the body yields evidence for the simultaneous existence of many rates of gas exchange. If a subject breathes a gas mixture devoid of N_2 and the excreted nitrogen is measured, a multi-exponential curve of N_2 excretion vs. time is obtained. Analysis of this curve yields evidence for at least four tissue domains characterized by a volume and a time constant. Other than providing direct experimental evidence for the spectrum of time constants required to adequately characterize man, these measurements are of little utility in defining the rate-controlling processes in inert gas exchange at tissue level, where bubbles presumably originate.

The fast time constants will correlate with the perfusion-limited time constants, but the slow time constants will not. Measurement of the time constant for a diffusion-limited tissue is complicated by the small rate of gas evolution from the tissue.

3. Summary

1. Gas solution and dissolution *in vivo* follow expected physicochemical laws.

2. Bubble nucleation and growth rates can be characterized by physical models; but the rates of these processes are unpredictable from first principles and depend strongly on the details of the local site for determining the controlling factors.

3. Inert gas absorption into body tissues requires anywhere from a few minutes to over 24 hr to reach a state of equilibrium with the gas environment.

4. While the laws of physics predict that bubbles will form at the site of maximum supersaturation, experimental conditions do not permit the assessment of whether this occurs *in vivo*.

5. The site of symptoms may not correlate with the site of bubble formation or the site of bubble growth, due to the involvement of the circulatory system.

C. Diagnosis

The basis for diagnosing decompression sickness is an interactive physical examination of the subject. The symptoms are qualitative impressions of discomfort and can often be confused with minor accidents that occur during the dive. While one could be tempted to wait for confirmation of the severity of these symptoms, it is far wiser to treat the first symptoms, if at all possible.

1. Incidence of Decompression Sickness and Appearance of Symptoms

Because of the limited testing of decompression schedules, there remains a finite incidence of the disease even when all instructions have been followed. There is no way to determine which exposures give the greatest risk. One always has to be aware of the possibility of suffering this disease.

The symptoms of decompression sickness are generally classed as mild (Type 1) or serious (Type II). They may occur any time during the decompression or soon thereafter. A review of U. S. Navy records of subsaturation showed that:

50% of the symptoms had occurred within 30 min of surfacing
85% of the symptoms had occurred within 1 hr of surfacing
95% of the symptoms had occurred within 3 hr of surfacing

Similarly, Figure VII-8 shows the time of onset of both Type I and Type II symptoms for 127 cases of decompression sickness recorded from a Canadian diving experience. Notice that 42 cases (33%) occurred prior to surfacing. Saturation diving results in a higher incidence (86%) of bends at depth. (This fact may only be due to the experimental nature of virtually all decompressions from saturation.)

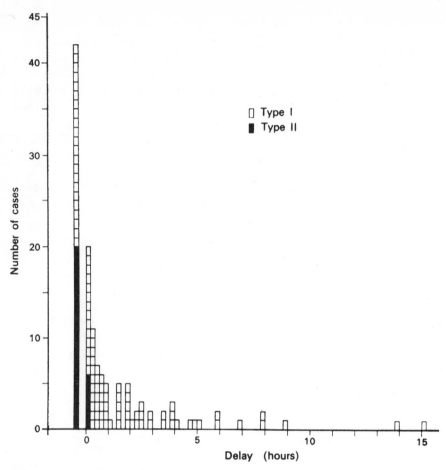

Figure VII-8. Decompression sickness—time of onset of cases in relation to the time of reaching surface. A total of 127 cases from the Canadian Forces Institute of Aviation Medicine and Royal Canadian Navy Diving Establishments, 1962–1967. The 42 cases that occurred during decompression are shown together before the time of surfacing. [From Kidd and Elliott (1969) by permission of Bailliere and Tindall, London.]

Military dives deeper than 10 ATA result in substantially more bends at depth (46%) than do dives at less than 10 ATA (10%). There are no comparable figures available for civilian diving. Tunnelers experience symptoms both at depth and after surfacing, with the mean time for symptoms being 2 hr.

2. Major Symptoms

The major symptoms of decompression sickness are (Type I) pain, dizziness, paralysis, shortness of breath, extreme fatigue, and collapse. These symptoms are best described for laymen (who must often make the first diagnosis) in the *U. S. Navy Diving Manual* (1970):

Occasionally the skin may show a blotchy and mottled rash. There may be small red spots that vary in size from a pinhead to the size of a dime. Sometimes mottling is so pronounced that the skin takes on an appearance like that of pink marble, and the term of "marbeling" is applied.

A typical case of decompression sickness may begin with itching or burning of a localized area of the body. This symptom may spread and then finally become localized again. There may be a feeling of tingling or numbness of the skin. In rare cases, the man may have a sensation of ants crawling over him.

Pain, which is the most frequent and predominating symptom is of a deep and boring character. Divers describe it as being felt in the bone or in the joint. Usually the pain is slight when first noticed and then becomes progressively worse until it is unbearable. The pain usually is not affected by movement of the area, but it may be temporarily relieved by vigorous rubbing or hot applications. The most frequently confused situation is that of a diver suffering a muscle strain or a joint sprain during a dive. However, this condition can usually be distinguished by the fact that strains and sprains are painful to touch and motion, while pain in a joint from decompression sickness is generally not. Swelling and discoloration usually occur with a sprain but are rare in uncomplicated cases of decompression sickness. A diver who has pain that might be a symptom of decompression sickness should have *treatment* by recompression, even though the pain may turn out to have been from a strain or sprain. WHEN IN DOUBT, TREAT BY RECOMPRESSION. Failure to treat doubtful cases is the most frequent cause of lasting injury.

Abdominal pain after a dive has frequently been followed by symptoms and signs of spinal cord involvement (i.e. weakness or paralysis of legs; burning, tingling, or numbness of legs or feet). It is important to recognize that the onset of abdominal pain in decompression sickness is a serious symptom requiring immediate recompression *treatment* as such. Careful examination of the diver upon his arrival at depth in the chamber should determine the extent of spinal cord involvement and the time necessary for complete resolution of the involvement in order to select the proper table for treatment.

When dizziness occurs, the diver feels that the world is revolving about him and that he is falling to one side. Frequently, he will have ringing in the ears at the same time that dizziness occurs. History and physical examination become important when these symptoms occur because they also can follow middle-ear damage, as from squeeze.

Serious symptoms are those caused by bubbles in the brain, spinal cord, or lungs. These symptoms require longer *treatment* than the pain-only type, and it is very important not to overlook them when they are present. Many of the serious symptoms are so well defined that the diver is certain to notice and report the symptoms, or the signs are so obvious that his tenders cannot miss them. However, it is quite possible to miss some of the less obvious signs and symptoms or to fail to recognize the milder disorders such as simple weakness, partial paralysis, or defective vision. Do not let a serious case be treated inadequately simply because no one bothered to check! For example, occasionally a diver who complains only of pain in an arm or a leg will also be found to have weakness or partial paralysis when he is examined thoroughly. It is also important to know *all* that is wrong with the patient so that you can be sure he is relieved of all his symptoms during treatment.

The *U. S. Navy Diving Manual* (1970) also provides guidance on performing the diagnosis:

The following are the most important things to check when examining a man prior to treatment or when trying to determine whether all symptoms have been relieved:

(a) *How does he feel:* (Ask him.)

   1. Pain—where and how severe? Changed by motion? Sore to touch or pressure? Bruise marks in the area?
   2. Mentally clear?
   3. Weakness, numbness, or peculiar sensations anywhere?
   4. Can he see and hear clearly?
   5. Can he walk, talk, and use his hands normally?
   6. Any dizziness?
   (b) *Does he look and act normal:* (Don't merely take his word if he says that he is all right.)
   1. Can he walk normally? Any limping or staggering?
   2. Is his speech clear and sensible?
   3. Is he clumsy or does he seem to be having difficulty with any act of movement?
   4. Can he keep his balance when standing with his eyes closed?
   (c) *Does he have normal strength:* (Check his strength against your own and compare his right side with his left.)
   1. Normal handgrip?
   2. Able to push and pull strongly with both arms and legs?
   3. Able to do deep knee bends and other exercises?
   (d) *Are his sensations normal:*
   1. Can he hear clearly?
   2. Can he see clearly both close (reading) and distant objects? Normal vision in all directions?
   3. Can he feel pinpricks and light touches with a wisp of cotton all over his body? (Note that some areas are normally less sensitive than others—compare with yourself if in doubt.)
   (e) *Look at his eyes:*
   1. Are the pupils normal size and equal?
   2. Do they close down when you shine a light in his eyes?
   3. Can he follow an object around normally with his eyes?
   (f) *Check his reflexes* if you know how.
   Note that it should not take a great deal of time to examine a man reasonably well. Especially when you are under pressure in the chamber, there is seldom time to waste, but do not shortchange the patient. If there is real need for haste, having him walk and do a few exercises will usually show (or call to his attention) the more serious defects. In all cases where there is any doubt, treat the diver as though he is suffering from decompression sickness. If you are not sure that he is completely free from *serious* symptoms, use the longer table. Remember that time and air are much cheaper than joints and brain tissue.

Tables VII-6 and VII-6B present the diagnosis process in such a manner that one can differentiate among possible causative factors. (Gas embolism, pneumothorax, and mediastinal emphysema are discussed in Chapter III.)

These symptoms occur with the following frequency, as summarized by the U. S. Navy (1970): pain, 89% (leg, 30%/arm, 70%); dizziness, 5.3%; paralysis, 2.3%; and shortness of breath, 1.6%. The Canadians have found pain associated with 75% of their cases. Reports on tunnelers indicate a higher incidence of pain in the legs than in the arms.

The skin is often involved in Type I symptoms (20% of the cases reported). The visual symptoms are always preceded by intense itching, generally about the trunk of the body. Local vasodilation yields a red spot with a cyanotic, blue center, indicative of local stasis. These symptoms most often appear after long, shallow dives.

Table VII-6A

Diagnosis of Decompression Sickness[a]

Signs and symptoms	Skin bends	Pain only	CNS bends	Chokes	Brain damage	Spinal cord damage	Pneumothorax	Mediastinal emphysema
Pain: Head					× ×			
Back			×					× ×
Neck								
Chest			×	× ×		×	× ×	×
Stomach			× ×			×		
Arms/legs		× ×				×		
Shoulders		× ×				×		
Hips		× ×				×		
Unconsciousness			× ×	×	× ×	×	×	
Shock			× ×	×	× ×	×	×	
Vertigo			× ×		×			
Visual difficulty			× ×		× ×			
Nausea/vomiting			× ×		× ×			
Hearing difficulty			× ×		× ×			
Speech difficulty			× ×		× ×			
Balance lack			× ×		× ×			
Numbness	×		× ×		× ×	×		×
Weakness		×	× ×		× ×	×		
Strange sensations	×		× ×		× ×	×		
Swollen neck								× ×
Short of breath			×	×	×	×	×	×
Cyanosis				×	×	×	×	×
Skin changes	× ×							
Eye tracking			× ×		× ×			

[a] × × Probable; × possible cause.

Table VII-6B

Diagnosis of Decompression Sickness: Confirming Information

Diving history	Yes	No	Patient examination	Yes	No
Decompression obligation?	——	——	Does diver feel well?	——	——
Decompression adequate?	——	——	Does diver look and act normal?	——	——
Blow-up?	——	——	Does diver have normal strength?	——	——
Breath-hold?	——	——	Are diver's sensations normal?	——	——
Nonpressure cause?	——	——	Are diver's eyes normal?	——	——
Previous exposure?	——	——	Are diver's reflexes normal?	——	——
			Is diver's pulse rate normal?	——	——
			Is diver's gait normal?	——	——
			Is diver's hearing normal?	——	——
			Is diver's coordination normal?	——	——
			Is diver's balance normal?	——	——
			Does the diver feel nauseous?	——	——

For tunnelers and divers decompressing in chambers, "skin bends" are often encountered. They are not useful as predictors of subsequent symptoms and are transient in nature. Treatment is seldom provided. Behnke (1970) reported that 25% of the Type II cases and 12.5% of the Type I cases of decompression sickness in the BART project exhibited skin symptoms. Lymphatic occlusion occurs in a small percentage of cases. It is manifested by local swelling and soreness, with the skin over the affected area having a *pigskin* appearance due to pitting edema.

The symptoms associated with Type II decompression sickness are more random in occurrence, probably because they reflect severe systemic involvement caused by bubbles that have been discharged into the central circulation. The bubbles (or their hematologic sequelae) may lodge in any of the central organs.

Table VII-7 reports the incidence of Type II symptoms reported for Canadian divers. Note that the severe pulmonary and CNS symptoms may be due to pulmonary barotrauma (see Chapter III) as well as decompression sickness. Diagnosis of the source of the symptoms should wait until treatment has been initiated. This differential diagnosis will be necessary in carrying through the treatment and planning for the use of auxiliary measures (e.g., with severe decompression sickness, one may wish to use plasma expanders and heparin, which are not indicated for acute treatment of pulmonary gas emboli).

3. *New Methods of Diagnosis*

In recent years there has been substantial research into the development of new methods of diagnosing decompression sickness. These methods center around direct detection of the bubbles or the biochemical concomitants to the bubbles. These

Table VII-7

The Distribution of Type II Symptoms[a]

Symptomatology	Number	Percentage
Dizziness	9 ⎫	
Confusion	2 ⎬ 31	
Disorientation	2	31
Nystagmus	1 ⎭	
Eye signs: diplopia	2 ⎫	7
tunnel vision	1 ⎭	7
Sensory impairment: hyperaesthesine	4 ⎫	25
paraesthesiae	7 ⎭	25
Motor weakness or loss	9 ⎫	25
Dysarthria	2 ⎭	25
Nausea	2 ⎫	4
Respiratory: dyspnoea	3 ⎬	
coughing	1 ⎭	9
Total	45	

[a] From Bennett and Elliott (1969) by permission of Bailliere and Tindall, London. Type II manifestations presenting symptoms may occur in more than one category.

methods have been developed to assist in the testing of experimental decompression schedules by providing a more quantitative end point to the disease. A general finding with these newer methods has been the realization (long suspected) that many processes occur well before the subject experiences the conventional symptoms. In many cases the symptoms never occur, even though the new method reveals a valid alteration in the measured parameter, indicating a disturbed homeostasis.

a. Physical Methods

Major advances in ultrasonic instrumentation have resulted in several *real-time* systems for bubble detection. Almost all the methods use frequencies greater than 1 MHz; at these frequencies they utilize the fact that ultrasound cannot penetrate air bubbles, but will penetrate tissues. The method involves purposely introducing ultrasonic energy and measuring the energy that either penetrates the tissue (*transmission* mode) or is reflected back off the bubbles (*reflected* mode). Some methods can detect stagnant bubbles lodged in peripheral tissues (the presumed source of bubbles) and some can only detect the circulating bubbles.

The ultrasonic energies utilized are presumed to be safe, based on information published concerning ultrasonic cavitation in supersaturated liquids. Figure VII-9 shows that the energy needed to induce cavitations rises sharply with frequency, so

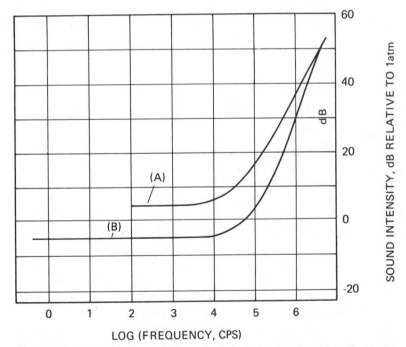

Figure VII-9. Cavitation threshold as a function of ultrasonic frequency. Frequency dependence of onset of cavitation in a liquid. (A) Distilled, filtered, and degassed water; (B) tap water containing 20% air. [From Galloway (1954) by permission of The American Institute of Physics.]

that the estimated sound pressure requirements at 1 MHz are over 1000 atm relating to 1 ATA. Although this power intensity is substantially higher than anyone has used to date, little work has been done to account for the possible focusing or resonance-producing possibilities that could occur in the inhomogeneous environments in the body (both conditions are known to occur *in vivo* and could result in local sound energies well in excess of the energy source). Until such time as the safety has been well demonstrated, these instruments should only be utilized by trained investigators.

The Doppler Ultrasonic Flowmeter (DUF) was first used to detect intravascular bubbles as an air emboli monitor in open-heart surgery. The device was subsequently used to detect bubbles in peripheral veins of divers after surfacing from dives in which no other symptoms were seen.

The DUF detects moving objects by the principle of detecting the Doppler shift of ultrasound that is produced. The Doppler shift is related to the velocity of reflecting particles v by

$$\Delta f = \frac{2fv \cos \theta}{c}$$

where f is the ultrasonic frequency, c is the speed of sound in the fluid, and θ is the angle between the incident ultrasonic beam and the vector of the velocity v.

This frequency shift is normally monitored audibly because it falls in the audible range for all but the highest particle velocities.

The Transcutaneous DUF (TDUF) permits one to measure qualitative blood-flow changes within vessels located at a depth below the skin; the depth is defined by the geometry of the probe. Most TDUF's operate by continuous emission; reception occurs at a separate crystal. Since the angle between the probe and the blood vessel is unknown, the absolute velocity cannot be measured.

When a bubble moves through the field of the TDUF a transient drop in Doppler frequency is detected. While other vascular *debri* (thrombi, platelet, aggregates, etc.) can produce somewhat similar sounds, the bubbles do generate a unique signal that can be detected with a high degree of certainty (probably because of the reflection off the air–blood interface). The bubble size cannot be measured by this method without an *in vivo* reference.

A special transducer has been developed to permit monitoring of the pulmonary artery in man. With this device bubbles have been detected prior to the occurrence of symptoms. Using a second transducer, it has been determined that these "silent" bubbles [so-named by Behnke (1955) long before they were detectable] come from a single extremity. Since all such circulating bubbles must pass through the pulmonary artery, this seems like the best monitoring location for detection of moving bubbles.

Bubbles have been detected in humans after a hyperbaric exposure in which they did not experience any of the classic symptoms of decompression sickness. They have been detected in repeat breath-hold dives to 15 m. The locus of most bubbles that have been detected with pulmonary artery monitoring can be determined by selected motion of each extremity until one is found that causes an increased bubble incidence in the pulmonary artery. Confirmation is then made by use of a peripheral vein TDUF monitor placed over the appropriate vein.

Rubissow (Rubissow and MacKay 1971) has reported on a scanning ultrasonic system that has a 2–5 μm detection limit for bubbles. He has shown that the amplitude of the imaged reflected bubble may be confused with reflection from a multiplicity of smaller bubbles. With the use of a step recompression he was able to measure the actual bubble size. This system could be developed for field use in much the same way as a portable X-ray machine.

Martin *et al.* (1973) have reported detecting bubbles in human trials of a novel transmission ultrasonic system. The unit utilizes the fact that ultrasonic reflection off a bubble is nonlinear and tends to generate higher harmonics of the incident ultrasound. They used a pair of 115-kHz transducers mounted on each side of the thigh so the beam transected the great saphenous and femoral veins, as well as the skin and musculature. By filtering the received signal at the second harmonic (230 kHz) they identified a characteristic *signature* of bubbles. Their studies indicate that the unit could be used in a hyperbaric environment. However, the subjects need to be motionless, and the identification of the bubble signature is subject to interpretation. Often, the subjects experienced symptoms before any bubbles were detected, unlike the results of the DUF studies.

Acoustic-optical imaging is able to detect stagnant bubbles (Buckles and Knox 1969). This method produces a direct optical image of the acoustic energy either reflected from or transmitted through the tissue. It is at a very early stage of development, and would generally only be available to large research or treatment centers.

b. Biochemical Methods

Measurements of hematological changes during decompression are based on one of two concepts: (a) that bubbles will initiate measurable changes; or (b) that the stress of inadequate decompression will induce alterations that would predispose the blood to bubble formation. Bubbles could induce hematological changes because the gas–blood interface is highly active and could lead to protein denaturation or alterations in clotting factors. Extravascular bubbles can disrupt cell membranes, releasing a wide variety of enzymes that are indicative of tissue trauma. Alternatively, bubbles growing in vessels could block blood flow and result in the release of enzymes or other indicators of cellular hypoxia.

Clearly, such biochemical studies are not practicable for real-time monitoring. However, they do represent a measure of decompression *insult* that is more sensitive than clinical symptoms. As such they are useful as determinants in the development of new decompression schedules that should have much lower incidence of clinical symptoms.

Studies have shown marked elevations of CPK and haptoglobins, which are measures of tissue damage, in divers exposed to saturation conditions and subsequently decompressed. No reports of decompression sickness were made by these subjects. Others have shown that there is an increase in circulating lipids.

D. *Prevention*

Decompression sickness is prevented by proper preparation for a dive and use of decompression procedures. Some divers still suffer decompression sickness even though they adhere to schedules. This is because the tables were developed to permit a finite incidence of decompression sickness (if made 100% safe, they would be prohibitively long). While this is acceptable to divers who have good treatment facilities nearby, this risk may be excessive for others who are less well equipped. For example, a successful decompression schedule for compressed air workers in the U. K. is one that reduces the incidence of decompression sickness to below 2% for exposures over 4 hr between depths of 40 and 90 ft (2.2 and 3.7 ATA).

In this section the development and use of decompression procedures will be presented. When the requirements of a dive exceed the range of available tables, the change in procedures will be indicated, where known. The decompression schedules discussed here are presented together at the end of this chapter.

1. Preparation

Diver selection (see Chapter XI) is necessary so as to prohibit from diving those participants who will probably have difficulty in decompressing or be especially prone to decompression sickness. The decompression schedules have generally been developed with experienced, conditioned subjects who range from 19 to 45 yr in age and meet military standards of physical fitness. Increasing age and body weight increases one's susceptibility to decompression sickness. Subjects who have suffered diseases that impaired their circulatory or respiratory system are also high-risk subjects.

Tunneling schedules are somewhat different in order to take into account the generally greater variance in subject physical characteristics. Behnke (1970) has reported that adequate selection criteria are necessary to maintain an acceptably low incidence of decompression sickness and avoid aseptic bone necrosis. The selection physical examination must be given by a physician trained in this field of environmental medicine; the examination should include: age, weight (correlated against general body structure), confirmation of auditory tube patency and activity, pulmonary pathology, radiographic survey of long bones, equilibrium disturbances, and response to a suitable cardiovascular stress test (e.g., Step-Test Exercise). See Chapter XI for further details. Schedules have been validated only on men, even though many women have been safely decompressed using the U. S. Navy air decompression tables.

Comparison of *safe* decompression schedules developed in different countries suggests that there may be national and racial differences in susceptibility to DCS. This factor has never been tested. The observed differences could also reflect variations in aquanauts' or tunnelers' response to pain or the diving officers' selection of subjects (e.g., in order to validate a schedule some investigators will use a subject who is known to be highly susceptible to DCS while others would reject such a person from the protocol).

Most subjects in preparation for testing decompression schedules are *worked-up*; that is, they make several exposures to safe pressures near the one to be tested, within a period of days before the test dive. Worked-up divers are more resistant to DCS; in fact, in tunneling work it is well known that most of the occurrences of decompression sickness are limited to men who are returning to work after a layoff.

The majority of accidents occur to the relatively inexperienced subjects. Proper equipment selection and maintenance are also critical to the prevention of decompression sickness. Equipment failures at depth (especially breathing equipment) are a major cause of accelerated decompression. Each set of diving equipment is designed for use in a certain environment in order to maximize safety. For example, a standard scuba tank, used at depths shallower than 60 ft (the recommended limit) will safely limit one's bottom time so as to never require decompression stops.

In summary, careful preparation involves several general rules:

1. Plan a dive carefully to be sure the proper decompression schedule is selected. Consider possible changes in plans that would necessitate alternative decompression procedures.
2. Determine the estimated amount of gas required for the mission. This will depend on the type of equipment utilized and the anticipated work load while on the bottom. The longer the exposure (and therefore, the longer the decompression), the more important it is to have reserve gas available for emergencies.
3. There are limits to the use of certain equipment and gas mixtures. These limits should never be exceeded!
4. ANY TIME ANYONE SKIPS A SCHEDULED DECOMPRESSION STOP, PRESUME HE WILL SUFFER DECOMPRESSION SICKNESS AND SEEK TREATMENT IMMEDIATELY.

2. Operational Factors that Influence Decompression Procedures

During a hyperbaric exposure, one exercises at a certain work level. Then, during decompression, one normally minimizes exercise. So decompression procedures are normally developed with subjects performing a certain level of work consistent with the anticipated use of the schedules. Schibli and Bühlmann (1972) have shown that the amount of decompression one needs is directly related to the level of work performed at depth. This undoubtedly relates to the increased inert gas absorption that would occur during exercise. Figure VII-10 shows this increased decompression time for a subsaturation $He-O_2$ dive profile. This increased decompression need will be greatest for short dives, where the perfusion-limited tissues dominate and are not saturated. No simple rule is available to adjust decompression schedules for extreme exercise, mainly because of the empirical nature of the models. This variability is nicely avoided in saturation diving.

On the other hand, moderate exercise during decompression, especially to counter the influence of a cold water environment, should enhance inert gas washout and lower the risk of decompression sickness. There is, however, no conclusive proof that this is true. Exercise should be avoided after surfacing from a satisfactory decompression.

Oxygen breathing is often used during decompression to accelerate inert gas washout. While this does increase the inert gas partial pressure gradient from tissues to gas, it also tends to reduce blood flow rate. Decompressions using oxygen are generally faster than air decompressions; however, there are no general rules on how

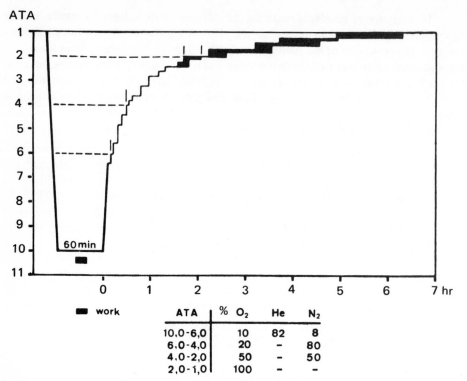

Figure VII-10. The influence of exercise on safe decompression time following subsaturation diving. Decompression schedules from a 60-min, 10-ATA (297-fsw) dive on an 82–8–10 mixture of He–N₂–O₂. The divers performed 15 min of exercise at an 80-W rate; there were 15 dives without work and 39 dives with work included. The darkened area indicates the extension in decompression time required to compensate for the exercise. [From Schibli and Bühlmann (1972) by permission of *Helvetica Medica Acta*, Schwabe and Co., Basel.]

to alter a standard air decompression schedule for oxygen breathing. In addition, there is the risk of oxygen toxicity (see Chapter IV). Studies on O_2 breathing during decompression have utilized resting divers and are therefore not applicable to scuba diving. In tunneling, O_2 breathing has been used, but one has to educate the workers as to its explosive hazards.

Similarly, there are many vasoactive drugs that one might consider using to enhance gas washout during decompression. But the influence of drugs on decompression schedules has not been evaluated. (The use of drugs in the treatment of decompression sickness is discussed later in this chapter under treatment.)

3. Decompression Schedules

Decompression schedules are tables and instructions that indicate how to safely decompress a diver from a hyperbaric exposure utilizing specific gas mixtures and life support equipment. The schedules are generally presented as a function of three variables: (1) depth of the dive, (2) time spent at depth (bottom time), and (3) available breathing gas.

These schedules are developed for use with specific equipment and generally have been developed for specific commercial or military needs. They carry medical–legal validity only within the context for which they are developed; most carry disclaimers of liability from use of the tables by anyone other than the designated user.

The schedules are generally developed in response to a specific need, such as when new equipment becomes available or an operational task must be carried out for which existing schedules are inadequate. Guided by the principles put forth earlier, it is possible to develop a model based on existing tables for calculating new schedules.* Most new tables are so complex that computers are utilized.

A matrix of test dives is calculated with the model that spans the desired depth and bottom time required. Then the new schedules must be validated experimentally using human volunteers and (generally) a test chamber capable of simulating the operational environment. If the schedules are found to be unsafe, they are modified empirically so as to achieve a certain number of symptom-free dives. Once such a matrix has been validated, the model is adjusted to reflect the necessary empirical modifications that occurred during testing. Then the modified model is used to calculate other schedules within the matrix limits of depth and bottom time.

The diving schedules are finally tested in the operational environment. This often requires further refinement of the schedules and the model.

a. Models Utilized to Compute Decompression Schedules

Models used to develop and correlate schedules must be viewed as empirical tools for curve-fitting. Ever since Haldane suggested that the parameters of the model had physiological and anatomic relevance, there has been an ongoing attempt to validate that concept. While such a goal is laudable, it is hardly necessary for the model to be useful for the purpose described above. As will be seen below, any model that is useful for schedule correlation is so complex as to make it exceedingly difficult to ever prove that it is an accurate reflection of actual *in vivo* processes. The literature thoroughly documents the failures of models that are used to extrapolate to new environments.

All models must have certain general characteristics. First, the process of inert gas absorption into body tissues during the hyperbaric exposure must be characterized, both for the projected dive profile and for the to-be-determined decompression exposure. During decompression, the tissue partial pressures must be compared to the environmental pressure in order to assess the state of supersaturation. Then, the process of bubble formation/growth must be described as a function of this supersaturation and some criteria of acceptability utilized. That is, one must decide what degree of supersaturation or bubble size should be viewed as hazardous. The model permits computation of a decompression depth–time profile that never exceeds this limit, yet minimizes the decompression time. Since most of the diving in this century has utilized the stage method of decompression (i.e., decrease the pressure rapidly to the next stage and hold at that pressure until the next step), the empirical values for permissible supersaturation include the safety inherent in this procedure. The results are *not* directly applicable to continuous decompressions, as discussed below.

---

* The general goal is to develop one model capable of handling all diving environments; less ambitious models can be more empirical, resulting in great mathematical simplifications.

During staged decompression, one remains at the depth until the tissue tensions reduce to a level such that the next step in pressure may be accomplished. That is, before the pressure is reduced, the tissue tension must be at the level acceptable after the pressure reduction; this implies that the tissue tension during most of the holding period is substantially less than the tolerable limit.

Models that have been most successfully utilized in the development of decompression schedules are introduced in the following subsections.

(i) Inert Gas Transport. Because the decompression schedule is dependent on so many factors, the model must be written in order to account for the important practical factors* encountered in the diving environment.

All models must be capable of representing inert gas exchange throughout the time domain represented by the dive. Earlier in the chapter the dynamics of inert gas exchange were shown to fall into three general time domains: (a) *early stages*, dominated by perfusion-limited tissues; (b) *late stages*, where peripheral tissues with long time constants (~ 300 min) contribute most of the expired gas; and (c) *intermediate stages*, where all tissues are substantial contributors to the expired inert gas. A single model that covers all three zones is the most general, but also the most complex to manipulate.

A generalized way to handle this mass transfer process is shown in Figure VII-11. Blood flow through the lungs results in equilibration of arterial blood with inspired gas (i.e., $PA_x = P_x = F_x P_{abs}$). This blood is carried to a variety of capillary beds, where exchange with tissues occurs. Tissues 1 through n are perfusion-limited. Tissue r is perfusion-limited; it is broken into several zones, chosen such that r is a perfusion-limited tissue and r', r'', etc., are diffusion-limited tissues.

It is well established that one must compute the inert gas partial pressure in body tissues. However, British investigators and others have adhered to the older Haldane simplification of equating the inert gas absorption in a tissue to the total ambient pressure. Thus, for a diver on air at 33 ft of sea water (2 ATA) Haldane said a tissue could absorb 33 ft of excess nitrogen, when in fact it can only absorb 25.4 ft of excess nitrogen ($0.8 \times 33 = F_{N_2} \times P_{abs}$).

The Haldanian approach is adequate for dives where a single gas mixture is utilized (e.g., air). The resulting conclusions about permissible supersaturation need only be systematically modified to use either description. But for diving with inert gas mixtures that change during the dive, it is preferable to utilize real inert gas partial pressures. Gas partial pressures will be utilized throughout this chapter.

Most empirical models are simplifications of the general model in Figure VII-11. The Workman model (Workman 1965) considers only perfusion-limited tissues; the spectrum of time constants is much greater than for the previous Haldane model, upon which it is based. This is necessary because the Haldane model was based on short dives of shallow depths; Workman's model had to encompass experience with much longer dives, including saturation exposures.

---

* A more restricted use of modeling has been proposed by Barnard (1967) to assist in the interpolation between decompression profiles from a series of experimentally successful dives where all variables were constant except for depth (saturation dives). The technique is perfectly valid and will be discussed further.

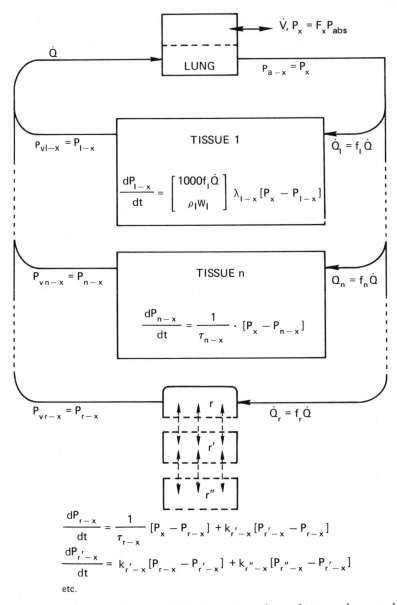

Figure VII-11. Generalized model for inert gas exchange between tissues and respired gases.

Table VII-8 lists the half-times ($t_i/1.44$) used by Workman (1965) to characterize inert gas transport. Values are given for both N_2 and He, the two common inert gases used in modern diving. Each tissue was considered to exchange gas with arterial blood independent of other tissues; during the dive, the arterial inert gas partial pressure is specified as a function of time and each tissue achieves a certain gas loading based on this imposed forcing function $P_x(t)$.

Table VII-8

Table of Maximum Allowable Tissue Tensions (*M* Values) of Nitrogen and Helium
for Tissues of Various Half-Times (Workman 1965) [a]

$t_{1/2}$ min	Depth of decompression stop, ft									
	10	20	30	40	50	60	70	80	90	100
Nitrogen										
5	104	122	140	158	176	194	212	230	248	266
10	88	104	120	136	152	168	184	200	216	232
20	72	87	102	117	132	147	162	177	192	207
40	56	70	84	98	112	126	140	154	168	182
80	54	67	80	93	106	119	132	145	158	171
120	52	64	76	88	100	112	124	136	148	160
160	51	63	74	86	97	109	120	132	143	155
200	51	62	73	84	95	106	117	128	139	150
240	50	61	72	83	94	105	116	127	138	149
Helium										
5	86	101	116	131	146	161	176	191	206	221
10	74	88	102	116	130	144	158	172	186	200
20	66	79	92	105	118	131	144	157	170	183
40	60	72	84	96	108	120	132	144	156	168
80	56	68	80	92	104	116	128	140	152	164
120	54	66	78	90	102	114	126	138	150	162
160	54	65	76	87	98	109	120	131	142	153
200	53	63	73	83	93	103	113	123	133	143
240	53	63	73	83	93	103	113	123	133	143

[a] *M* values are in feet of sea water equivalent. $t_{1/2}$ is the tissue half-time.

Decompression schedules in the U. K. are based on a model that implies only one tissue is involved and that inert gas absorption in that tissue is diffusion-controlled. Hempleman (1967) has been the major proponent for this model. Since he was working with short dives (less than 100 min) he could approximate the gas absorption dynamics by a simple square root of time factor (a well-known approximation of an infinite series of exponential terms taken at times that are short compared to the saturation time). This approach does not require the direct calculation of tissue inert gas partial pressures. Rather, a factor $\phi(t)$ is computed that relates to the quantity of inert gas absorbed after a tissue is exposed to a unit increase in external partial pressure.

$$\phi(t) = 1 - \frac{8}{\pi^2}\left(e - Kt + \frac{1}{9}e^{-9Kt} + \frac{1}{25}e^{-25Kt} + \cdots\right) \qquad (17)$$

The empirical value of K (based on evaluating experimental data) proposed by Hempleman (1969) is

$$K = 0.007928$$

when t is expressed as minutes. The total gas absorbed during a dive is computed by dividing the dive into a series of step changes in depth and using the superposition principle to sum the results. When decompressing, the value K is increased by 50% because Hempleman suggests that inert gas transport is impeded during decompression. Decompression tables for both caisson and tunnel workers and deep sea divers have been developed with this model and are in use.*

Exercise during a dive has a profound influence on gas exchange and presumably on decompression schedules. Bühlmann (1969) found that exercising divers required more decompression time than when they rested during the same dive. His model (based on a modified Haldane concept) was modified by increasing the half-time of the fast tissues during the exercise period. Workman's (1965) model was based on dives in which hard work was performed; so the empirical constants that are used implicitly contain the influence of exercise.

When oxygen breathing is used, it is customary to make the following assumptions: (a) Neglect the reduction in blood flow to most tissues (i.e., no change in half-time spectrum of tissues) and (b) include a 20% leak of inert gas around the mask (i.e., the fraction of inert gas is 0.20, not zero as would occur if perfect seals could be made).

There is no fully satisfactory way to account for breathing inert gas mixtures (so-called tri-mix gas) or for switching between inert gases during the dive. One may choose to account for them separately, but ultimately they must be combined so as to estimate the total inert gas loading of each tissue. There are no decompression procedures that permit one to randomly change inert gas during the dive.

In recent years *saturation diving* has gained great favor because it prolongs bottom time without increasing the decompression requirements. The concept is based on the assumption that if a diver stays at a certain depth he will ultimately become saturated with the inert gas he is breathing. From then on he does not incur any more decompression obligation. Clearly, the diver must be provided extensive life support means to perform a saturation dive (food, shelter, heat, etc.).

The time required to reach saturation can be inferred from the inert gas models described above. Workman utilizes a slow tissue with a half-time of 240 min. This tissue will be saturated (to within 1%) after six half-times or 24 hr. Recent data on air saturation diving indicate that there may be slow tissues with 500-min half-times. These tissues would require 48 hr to saturate.

(ii) Safety Criteria. Haldane proposed that the body could tolerate dissolved inert gas so long as the decompression is carried out by reducing the pressure in successive steps where the ratio of computed tissue inert gas tension to final pressure never exceeds a value of 2.0. This conclusion was based on subsaturation studies where the final pressure was always 1 ATA; Haldane (Boycott *et al.* 1908) said this ratio should not be used at depths in excess of 165 ft.

Hawkins *et al.* (1935) were the first to point out that the Haldane ratio was too conservative for the fast-half-time tissues and not conservative enough for slow tissues.

---

* Hempleman's model is presented as an example of a simplified model that has little potential for extrapolation to diving environments outside the limits within which it has been tested (e.g., greater depths or mixed inert gases). The fact that the model has empirical utility is testimony to the hazards of applying the criterion of rigorous anatomical relevance to a model such as Workman's.

These concepts were utilized in developing the revised U. S. Navy Standard Air Decompression Tables (U. S. Navy 1943).

Studies on deep diving carried out during the late 1940's clearly indicated that the Haldane ratio must be reduced at elevated pressures. This fact is consistent with the notion that the ratio is a measure of the tendency for bubbles to form under the influence of supersaturation (see this chapter, physical factors). It is also consistent with the alternative theory that bubble growth is controlled by minimizing the degree of supersaturation. That is, the concept of a permissible degree of supersaturation is perfectly sound even though we have not proven whether bubble nucleation or growth is the process that controls decompression.

It has been determined empirically that the Haldane ratio decreases as absolute pressure increases in a parabolic fashion. Figure VII-12 shows this dependence as it has been determined by Bühlmann (1969). Two curves are presented; the shorter half-time tissues can utilize a higher supersaturation ratio at all depths.

Schedules were computed using the data from Figure VII-12 such that the permissible ratio is determined at each stop depth according to the computed values of tissue inert gas tension.

The U. S. Navy air decompression schedules incorporate this depth dependence of supersaturation. The schedules were the first ever computed on a digital computer. They utilize a tissue half-time spectrum of 5, 10, 20, 40, 80, and 120 min. The permissible supersaturation was selected for each tissue as a unique function of the tissue's nitrogen tension at the end of the dive. This permissible supersaturation value was used throughout all subsequent stops up to 10 fsw. Surfacing from this stop was controlled by the surfacing ratios originally proposed by Van der Aue et al. (1951). Figure VII-13 depicts both the surfacing values and the ratios used at depth for the U. S. Navy air schedules. These same values have been used to compute the U. S. Navy repetitive dive schedules.

Early studies with He as a breathing gas indicated that the helium Haldane ratio could be increased over the value used for N_2, permitting shorter decompressions from

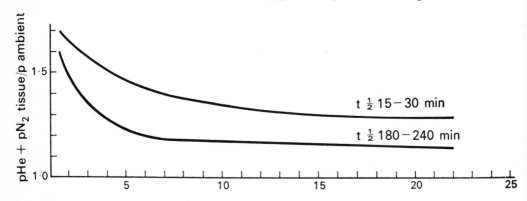

Figure VII-12. The Haldane ratio as a function of depth (N_2–He–O_2 mixtures). The permissible supersaturation of N_2 and/or He for the half-time spectrum used by Bühlmann in the calculation of human decompression schedules. The tables used stage decompression (see Figure VII-10) and can be used for any gas mixture at sea level or altitude. [From Schibli and Bühlmann (1972) by permission of *Helvetica Medica Acta*, Schwabe and Co., Basel.]

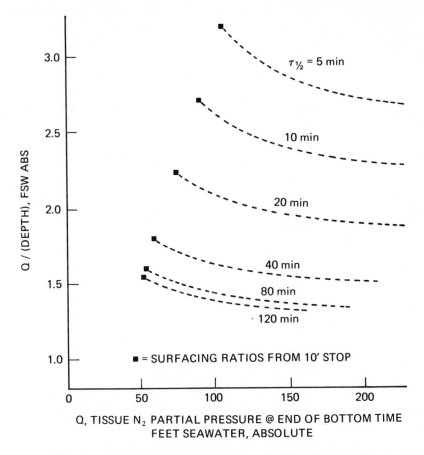

Figure VII-13. Permissible supersaturation ratios utilized in the U. S. Navy's air decompression schedules (des Granges 1956).

equivalent dives. Ever since, investigators have continuously been modifying the values as helium dives have progressed to deeper depths for longer exposure times. The data of Bühlmann are probably the most comprehensive available because of his extensive experience with very deep dives. Figure VII-12 shows how the Haldane ratio decreases at great depth. Bühlmann (1969) used tissue half-times that cover the range of 15–240 min, in essential agreement with Workman (1965).

Workman suggested that the Haldane tables were also compatible with the concept that there is a permissible partial pressure difference which governs decompression rates. He proposed that each tissue can sustain a certain amount of inert gas partial pressure and that this quantity M increased linearly with depth; at all depths, $M > P$. A review of published data for both air and He–O_2 dives yielded initial estimates of these M values for each of the presumed tissue constants.*

---

* Both Rashbass (1955) and Duffner et al. (1959) had used this "pressure head" concept for He–O_2 dives where the difference was independent of depth and only one tissue was considered. Their experimental data were subsequently used by Workman for establishment of the M values.

Workman's method requires that the permissible supersaturation difference be determined at each stop depth. His method deviates from the earlier methods in that the M value is not a function of the inert gas content of the tissue; rather, it is uniquely dependent on the depth of the dive. This presumption greatly simplifies the computation of schedules. Workman (and others who now utilize this method) has never presented a theoretical justification for this method of dealing with the depth dependence of permissible supersaturation of staged or continuous decompressions ranging in time from a few minutes to saturation exposures.

Table VII-8 summarizes the empirical parameters of the Workman (1965) model. A major benefit of this model is its generality and mathematical simplicity. It can be easily programmed on analog and digital computers. It is expressed in terms of inert gas partial pressures, permitting the use of any type of gas mixture; in principle, the composition of the mixtures can be changed at any point in the dive. It permits the calculation of staged or continuous decompression, ranging in time from a few minutes to saturation exposures.

The values presented in Table VII-8 are the so-called Workman M values. The values in the first column ($P = 10$ ft) are the tissue partial pressures that permit one to come directly to the surface ($P = 0$ ft). If one wishes to decompress in stages other than 10-ft steps (e.g., 0.5-ft steps, or *quasicontinuous* ascent), the allowable tissue partial pressure is given by

$$M_P = M_0 + (\Delta M/\Delta P)P \tag{18}$$

when the values of M_0 and $\Delta M/\Delta P$ are given in Table VII-8.

Hempleman's model (1967) utilizes the safety criteria urged by Rashbass (1955), i.e., that a constant excess of 30 ft of N_2 partial pressure in tissue over the ambient pressure is acceptably safe. The present Royal Navy diving tables are calculated on this basis.

The use of this constant allowable supersaturation is too conservative for long dives of medium depth, especially as it applies to compressed air workers. Hempleman (1969) has adopted a modified Haldane ratio that varies with depth and is given by

$$r = 400/P_{abs} + 180 \tag{19}$$

where P_{abs} is the absolute pressure in pounds per square inch. Tables based on this ratio and the diffusion-controlled inert gas transport (equal uptake and washout rates) have yielded a 1% incidence of decompression sickness when used between 79 and 95 ft (over 40,000 exposures) by U. K. compressed air workers.

Clearly the development of safety criteria for use in the calculation of decompression tables involves medical, ethical, and legal considerations. The limited experimental data, combined with the wide range of operational environments encountered in undersea activities, make scientific judgment only a partial contribution to the selection of such criteria. Extrapolation of any criterion beyond the bounds within which it has been tested is exceedingly hazardous. The models described in this section are presented because they are the basis for most of the decompression procedures used in the noncommunist world.

The following section is a guide to decompression procedures that can be used for a wide variety of undersea exposures. The actual tables are given at the end of this chapter.

b. Decompression Schedules Available for Human Use

A decompression schedule consists of: (a) a table of depth–time relations; (b) a set of instructions for the selection and use of (a); and (c) a set of limitations that define the diving environment in which the schedule has been tested and is intended for use. In this section the schedules will be detailed that have been developed, tested, and reported in the published literature. Where schedules have been developed in several countries for the same environment (e.g., air diving with hard-hat gear), the American schedules will be reported.

Decompression schedules that have been developed by commerical diving firms are often proprietary and not available for public disclosure. In emergency situations, however, these firms are often contacted for assistance and will freely offer advice on suitable decompression procedures based on their experience.

The decompression schedules are classified by the type of inert gas (air, mixed gases) and by the length of dive exposure (subsaturation, saturation).

c. Air Decompression Schedules

(i) Subsaturation Exposures. Subsaturation diving with air is the most common form of hyperbaric exposure. Literally all tunneling is performed under these conditions; most scuba and hard-hat diving is done with air. The preponderance of non-military diving tasks require only air (harbor salvage and repair, sports diving). This type of diving is subdivided into three environmental zones (see Figure VII-14): (I) *shallow* diving with short bottom times, requiring no decompression, (II) *normal* dives requiring decompression, and (III) *exceptional* or *emergency* dives requiring the use of relatively untested decompression schedules. These subsaturation air schedules are the most complex type of available tables because the diver must decompress in the water where many other risk factors are encountered. Consequently, there is great motivation to minimize the decompression schedule. In addition, there are a variety of emergency procedures available that permit the dive master to cope with special problems that may occur during decompression.

Many subsaturation air diving operations call for repetitive dives within a 12-hr period in order to optimize the use of personnel. There is always the risk that residual excess N_2 will persist in the diver from the first dive, enhancing his risk of decompression sickness following the second dive. Procedures have been developed to account for this excess risk, based on a knowledge of the first dive (depth and bottom time) and the time between dives (surface interval).

The U. S. Navy decompression schedules are the procedures of choice for all such air diving (Schedules 1.1–1.7 at the end of this chapter). The no-decompression limits are defined in Schedules 1.1 and 1.2. Repetitive dives require the assignment of a group designator that reduces the available bottom time in the second dive; this group designator is obtained by the use of Schedules 1.2 and 1.3.

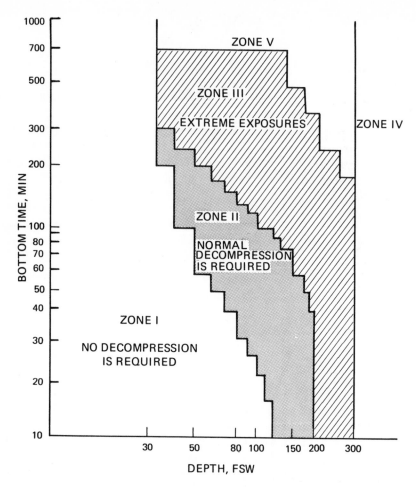

Figure VII-14. This figure permits one to determine the decompression risk associated with a dive that consists of a selected depth and bottom time. Such a dive
may require: no decompression (I), use of normal decompression (II), or use of
extreme exposure decompressions (III); dives in zone IV ARE NOT RECOM-
MENDED. Decompressions from zone V are covered under saturation decompression (Figure VII-15) (U. S. Navy 1971).

If one elects to minimize in-the-water decompression, one can have the diver
breathe oxygen, directly surface, and transfer to a decompression chamber where he is
recompressed to a suitable depth and subsequently decompressed. Schedule 1.6 is used
for this procedure; if either oxygen breathing is not available or the diver suffers O_2
toxicity, use Schedule 1.5. These schedules have, for the most part, never been actually
tested. They were prepared using the same parameters that were used to compute the
normal tables.

Figure VII-15 outlines these air decompression tables and the sequence in which
they are utilized for repetitive dives. Table VII-9 lists the general rules governing the

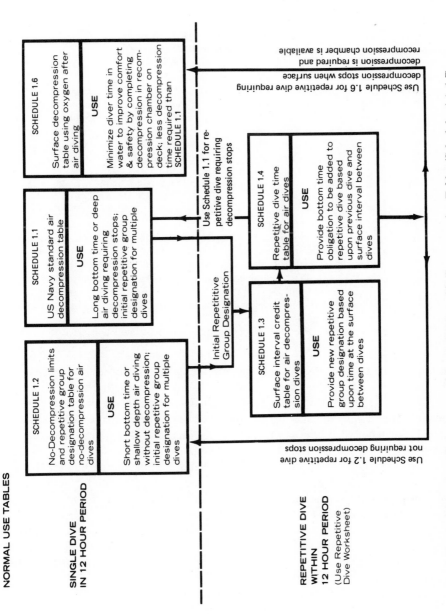

Figure VII-15. A guide to the use of the U. S. Navy Air Decompression Tables (Schedules 1.1–1.7).

Table VII-9

Limitations to U. S. Navy Air Decompression Schedules

Relevant terminology: see Decompression Schedules 1.1–1.7 at the end of this chapter

Equipment suitable for diving:

Open-circuit scuba (normal: 60 ft/60 min; exceptional: 130 ft/20 min)

Deep-sea gear (normal: 190 ft/40 min; exceptional: 250 ft/40 min)

Lightweight gear (normal: 60 ft/60 min; exceptional: 90 ft/60 min)

Variations in rate of ascent

1. Ascend from all dives at 60 ft/min unless otherwise noted
2. If unable to maintain the ascent rate of 60 ft/min:
    A. If the delay was at a depth greater than 50 ft, increase the bottom time of the dive by the difference between the time used in ascent and the time that should have been used at a rate of 60 ft/min; decompress according to the requirements of the new total bottom time
    B. If the delay was at a depth less than 50 ft, increase the first stop by the difference between the time used in ascent and the time that should have been used at the rate of 60 ft/min

Specific instructions for use of decompression tables

1. All dives which are not separately listed are covered in the tables by the next deeper and next longer schedule; do not interpolate
2. Enter the tables at the listed depth that is exactly equal to, or is the next greater than, the maximum depth attained during the dive
3. Select the bottom time of those listed for the selected depth that is exactly equal to or is next greater than the bottom time of the dive
4. Use the decompression stops listed on the line for the selected bottom time
5. Ensure that the diver's chest is maintained as close as possible to each decompression depth for the number of minutes listed
6. Commence timing each stop on arrival and resume ascent when specified time has elapsed
7. Observe all special table instructions
8. Always fill out Repetitive Dive Worksheet or similar systematic guideline (see Record Keeping Section)

selection of the appropriate schedule and the rules governing ascent rate, missed stops, etc.

The U. S. Navy Air Subsaturation Schedules have been extensively used by commercial and government divers. A recent review of fleet experience indicated that a bends incidence of 0.69% had resulted from their use. This figure is substantially better than that achieved with earlier schedules.

Tunneling and caisson work requires special decompression schedules. The workers cover a broader spectrum of the population; in addition, the freedom from immersion permits more mobility and diversity of action during decompression. The exposure time (analogous to bottom time) is much longer. Since this type of work leads to five exposures per week over sustained periods, the chronic effects of repeated decompressions must be considered. First, there is the development of resistance to decompression sickness; this is particularly apparent at the start of a job, or after a layoff of greater than ten days.

The more insidious manifestations of chronic decompressions is the delayed appearance of aseptic bone necrosis. Although the exact etiology of this crippling disease is still uncertain, definitive evidence exists that it arises due to decompressions from the especially long exposures that characterize compressed air work. Thus, a major impetus for the development of better decompression schedules for compressed air workers is the reduction of the incidence of aseptic necrosis.

The wide variability in age, weight, and training of compressed air workers demands that they be subjected to a stringent physical examination as an employment condition. Behnke (1970) reported that 18.5% of the examinees for work on the San Francisco tunnels were disqualified, "chiefly because of excess weight and pulmonary pathology." All men found with juxta-articular lesions were disqualified.

The incidence of decompression sickness with modern decompression procedures is generally less than 2% in compressed air work. Behnke (1970) reported an incidence of only 135 cases out of the 80,360 man-decompressions in the BART project; this is an incidence of 0.17%. In spite of the earlier indications that obese older men should be more susceptible to decompression sickness, no such distinctions could be drawn in the San Francisco experience.

More striking is the reduction in incidence of bone lesions that accompanied the adoption of the so-called "Washington State Tables" (Schedules 2.1–2.5). No bone lesions have been found in either the Washington State or San Francisco tunneling. This low incidence is consistent with the studies that have been carried out in the United Kingdom using the newly devised Blackpool Tables.

Decompression from compressed air work (tunnels and caissons) occurs in an air-filled chamber. Traditionally, U. S. workers decompress by a series of constant venting maneuvers, changing the decompression rate at preassigned pressures. The pressure is expressed in pounds per square inch (gauge), psig. The British procedure differs in that the Haldanian staging method is utilized. Figure VII-16 is the comparative decompression (U. K. and U. S.) following a 4-hr work shift at 24 psig (2.63 ATA).

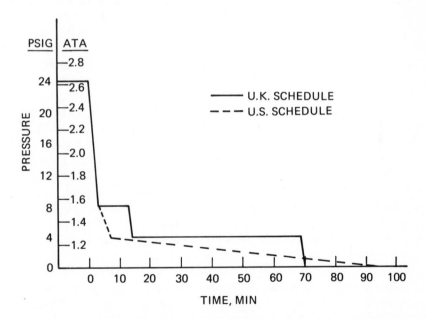

Figure VII-16. Decompression schedules for compressed air workers, resulting from a 4.0-h shift at 24 psig (2.63 ATA). The breathing gas is air. [Wisconsin (1971) and Decompression Sickness Panel (1973). Reprinted by permission from the Department of Industry, Labor and Human Relations, State of Wisconsin.]

Decompression schedules for this environment are generally developed by contractors for the insurance companies. The recent enactment of the Occupational Safety and Health Act (OSHA) resulted in the adoption of a national standard tunneling decompression procedure [the Washington State Tables (Schedules 2.1–2.5)]. These schedules were originally calculated by Duffner (1962), who utilized a three-tissue, perfusion-limited model (30, 60, 120 min half-times).* The use of a *split shift* (decompress to 1 ATA for lunch) was abolished on the grounds that it doubled the decompression hazard to the worker.

The permissible supersaturation in the Washington State Tables was based on U. S. Navy experience at that time; the 30- and 60-min tissues had surfacing N_2 partial pressures of 21 psig and that for the 120-min tissue had to be less than 18.6 psig N_2 pressure. The first decompression *stage* was at a rate of 5 psi/min (0.33 atm/min) and could not exceed a 16-psi drop in pressure or approach closer to the surface than 4 psi. The rate is then adjusted for the next stage based on a permissible supersaturation for the mean depth of the stage. This is repeated until reaching the 4-psi stop, where the final stage is commenced at the slowest rate of ascent consistent with arriving at the surface with safe supersaturation criteria.

Figure VII-17 indicates the total decompression time required for compressed air work at varying pressures and over the normal working time. The shaded area indicates the zones where the schedules have been extensively utilized. The dotted lines indicate the extreme exposures. The heavy bisecting lines separate the region where these tables are more conservative than the U. S. Navy air schedules from the region where the Extreme Exposure U. S. Navy air table is more conservative. This lack of conservatism in the compressed air tables at long bottom times is due to the use of a single value for permissible supersaturation. The U. S. Navy schedules utilize a permissible supersaturation that decreases at greater depth (see Figure VII-13). These decompression schedules have yielded less than 2% DCS incidence in three different states (Washington, California, and Wisconsin) and there have been no reported cases of aseptic bone necrosis.

The long decompression times required by these tables provides a strong economic incentive for finding alternative decompression procedures. The use of O_2 breathing has been proposed, but it has never been utilized in the U. S. because of the fire hazards and the uncertain effects of chronic HPO exposure. German compressed air workers have successfully utilized O_2 during decompressions in over 15,000 man-decompressions. The O_2 is limited to 2 ATA and is breathed by mask. The chambers are vented to keep the O_2 concentration to less than 30%. The saving in decompression time is generally greater than 50% [e.g., a 6-hr exposure to 35 psig (3.3 ATA) is reported to yield an O_2 decompression of 91 min (1.94% incidence) as compared to the 233 min required on the Washington State Tables].

(ii) Saturation Exposures. With the emergence of scientific interest in shallow water exploration, the inadequacies of subsaturation air diving became apparent. There emerged a strong need to prolong diving time in the range of 30–150 fsw and permit greater flexibility in diving profile. Saturation diving offers the opportunity to

---

* Duffner used absolute pressures in his model; for consistency they have been converted here to N_2 partial pressures, since air is used throughout the tables.

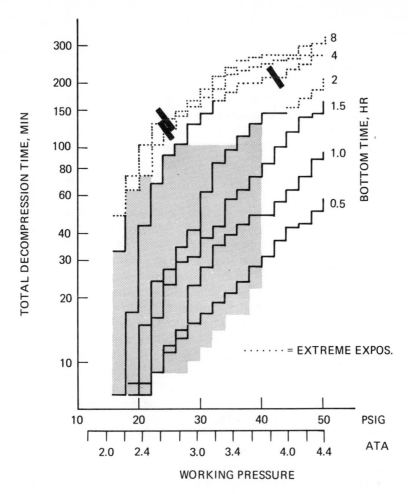

Figure VII-17. Total decompression time for compressed air work as a function of working pressure and bottom time (Schedules 2.1–2.5). The shaded zone indicates the recommended operating zone wherein the schedules have been most thoroughly tested. The U. S. Navy Extreme Exposure Tables (see Schedules 1.1–1.7) are considered safer than these schedules for 4- and 8-h bottom times when pressures exceed 42 and 22 psig, respectively. These zones are indicated with solid slash marks. See Schedules 2.1–2.5 for the detailed tables.

extend the no-decompression limits to these depths, making it a particularly attractive diving technique. By utilizing an underwater habitat, one could also remain at depth for greater periods of time and thus accomplish more useful work.*

Also, using air in such an underwater habitat results in great simplification of the diver-support equipment (see Chapter IX). All that is needed is an air compressor that can ventilate the habitat so as to keep the P_{CO_2} below tolerable limits. So long as the oxygen partial pressure is maintained less than 0.4 atm, oxygen toxicity is unlikely.

* Saturation diving was first developed to permit deeper diving on He–O₂ gas mixtures.

This limit of $P_{O_2} \leq 0.4$ atm presumably limits air saturations to 2 ATA (33 fsw). However, air-saturation diving has been carried out at depths as great as 99 fsw with no apparent ill effects from O_2 toxicity. Unfortunately, decompression schedules from such exposures have not been extensively tested; so it is preferable to retain a 2-ATA limit.

The most extensive shallow-water saturation-decompression experience has utilized mixtures of N_2 and O_2, keeping the O_2 fraction below that in air (generally, it is adjusted so that $P_{O_2} \simeq 0.2$ ATA). This results in a higher tissue inert-gas partial pressure at saturation for any habitat depth than would be achieved if air were the breathing mixture. These decompression schedules for such dives require more decompression time than would be needed for an air-saturated dive to the same depth. Thus, in the absence of air-saturation schedules, one should use the schedules available for N_2–O_2 saturation ($F_{O_2} < 0.21$) selected at the same saturation depth.

1. Surface decompression from air saturation at depths up to 50 fsw. In the event an emergency evacuation from an air-saturated habitat becomes necessary, one may surface for a brief interval as long as a recompression chamber is available on-site (a normal safety requirement for all saturation diving).

The recommended procedure is to use the tables developed for the 50-ft TEKTITE mission (Schedule 3.1). These tables were developed after experiments demonstrated that one could tolerate a 15-min surface interval after surfacing from a 40-fsw habitat where the breathing gas is a N_2–O_2 mixture (91/9) (equivalent to 51 fsw breathing air). Schedule 3.1 requires a recompression to 60 fsw, regardless of the saturation depth. Oxygen breathing is used intermittently. (Note: This schedule was also used for treatment of decompression sickness that occurred during the TEKTITE missions.)

When direct ascent to the surface is not possible, the decompression procedure is much more complicated. The method requires a transfer capsule-deck decompression (DDC) facility; with such equipment the divers may be decompressed along a continuous ascent profile after being transferred to the DDC and raised to the surface.

2. Decompression schedules utilizing air. The saturation decompression schedules developed by Hamilton *et al.* (1973) are based on the same values of allowable supersaturation as will be recommended below for excursion diving; therefore, they are recommended for use up to the depths to which they have been tested, 60 fsw. The general procedure involves decompressing at a constant rate of 1/6 ft/min to a given depth; from that depth to the surface a slower and continually changing rate of ascent is used. The only variable factor in the decompression is the depth at which one switches from 1/6 ft/min to the slower rate. Figure VII-18 shows the decompression from 60 fsw. The numbers along the curve indicate the saturation depth that results in an intersection with the continuous curve. Always use the saturation depth value at the 5-ft increment deeper than the saturation depth. Table VII-10 indicates the rate of ascent in 2- and 3-ft increments from 30 fsw to the surface, as is shown in Figure VII-18.

Decompression from saturation diving in the open sea poses several technical problems. If the saturation depth is less than 33 fsw (2 ATA), one can theoretically ascend directly to the surface at a reasonable rate. Because of the population variability in response to this decompression, the use of oxygen breathing in the habitat prior

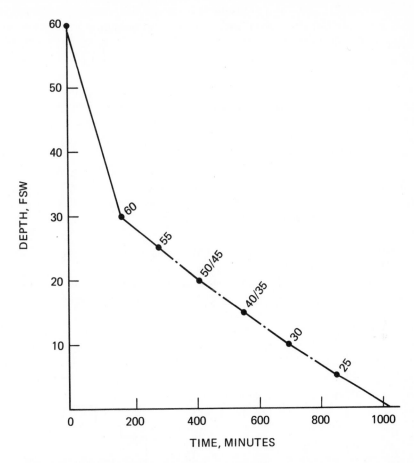

Figure VII-18. Decompression profile for saturation diving with air at depths up to 60 fsw. The numbers next to the curve indicate the point on this profile where decompressions from the depths below 60 fsw occur. See Table VII-10 for further details. (Hamilton *et al.* 1973.)

to ascent is probably warranted. Saturation diving should never be undertaken without the advice of a trained physician.

Figure VII-19 permits one to estimate the decompression time required from air saturation at any depth up to 60 fsw.

3. Saturation-excursion diving. Seldom is the ocean floor level around a habitat. In most habitat operations it has been necessary to undertake excursions from the saturation depth to accomplish one's goals. Generally, these excursions are short enough so that no decompression is necessary upon returning to the habitat. With increased exposure time, however, decompression stops are required. Excursions to shallower depths must be restricted so as to avoid inducing decompression sickness while at the excursion depth.

Table VII-10

Air Saturation Decompression Procedures [a]

Saturation depth, fsw	Procedure	Initial depth	Final depth	Rate, ft/min
60	Ascend at 1/6 ft/min to:	30	28	1/22
		28	25	1/24
50	Ascend at 1/6 ft/min to:	25	23	1/26
		23	20	1/27
50/45	Ascend at 1/6 ft/min to:	20	18	1/27
		18	15	1/28
40/35	Ascend at 1/6 ft/min to:	15	13	1/29
		13	10	1/30
30	Ascend at 1/6 ft/min to:	10	8	1/31
		8	5	1/32
25	Ascend at 1/6 ft/min to:	5	3	1/33
		3	0	1/34

[a] Notes: 1. Do not ascend in any aircraft for 24 h following end decompression, or in an unpressurized aircraft for 48 h.
2. Do not begin decompression less than 3 h after a descending excursion. If any excursion is performed between 3 and 12 hr before beginning ascent, add 5 fsw to the depth where ascent slows from 1/6 ft/min to the slower rate.
3. See Figures VII-18 and VII-19 for graphic details of the decompression profile.

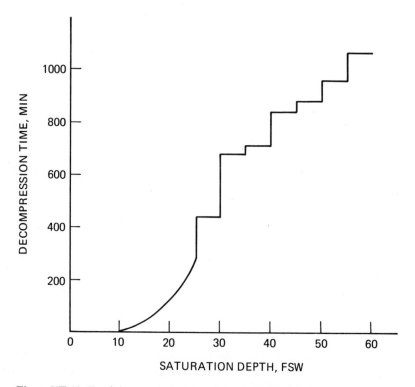

Figure VII-19. Total decompression time for air saturation diving at depths up to 60 fsw. See Table VII-10 for further details.

Descending excursion dives from saturation are entirely analogous to dives from the surface, where we are initially saturated at 0 fsw (1 ATA). The experience of Hamilton *et al.* (1973) has shown that saturation at a pressure in excess of 1 ATA results in extension of the no-decompression limits, due to the use of an allowable supersaturation matrix that is most conservative at 1 ATA. Thus, while a man can dive 80 ft from the surface with a 40-min bottom time without any decompression, a similar 80-ft excursion from a 30-fsw saturation platform results in a 68-min bottom time.

Since one is primarily interested in no-decompression excursion diving from habitats, detailed tables worked out by Hamilton are included at the end of this chapter (Schedules 4.1 and 4.2). Figure VII-20 is a crossplot of these data that permits one to determine the no-decompression limits for any value of saturation depth and excursion distance to a saturation depth of 70 fsw. By entering Figure VII-20 with a desired working depth, it is possible to choose a series of no-decompression excursion times and saturation depths. The tradeoff between excursion time and excessive saturation depth is often dependent on the goals of the diving mission and the available surface support equipment.

Repetitive excursion diving from air saturation diving platforms has not been simply worked out. The full spectrum of tissues can influence the decompression, depending on the types of dive combinations that are required. It is possible to arrive at conservative estimates of the bottom time for a repetitive dive by utilizing the no-decompression limits at each saturation level (Figure VII-20) and the repetitive dive Schedules 1.2–1.4 for normal air diving from the surface (these schedules appear at the end of this chapter). The method is as follows:

1. When the first dive utilizes the maximum no-decompression bottom time, obtain a group designator from Schedule 1.2 by entering the excursion depth (working depth − saturation depth) and adopting the maximum designator. Then use the surface interval, Schedule 1.3, to reduce this designator according to the time spent in the habitat at the saturation pressure. Then obtain the residual nitrogen time from Schedule 1.4, using the excursion depth; this time must be subtracted from the no-decompression time available for the second dive (obtain this no-decompression time from Figure VII-20).

2. When the bottom time in the first dive is less than the no-decompression limit, a *surface equivalent* excursion time (SEET) must be calculated for entry into Schedule 1.2. This SEET is obtained by multiplying the real excursion time (ET) by the ratio of maximum ET at the surface to maximum ET at the saturation depth D_{SAT}; i.e.,

$$\text{SEET} = \text{ET} \times \frac{\text{ET}_{MAX} \text{ at } D_{SAT} = 0}{\text{ET}_{MAX} \text{ at } D_{SAT}}$$

Use this SEET and the excursion depth to obtain a group designator from Schedule 1.2. After correcting for the surface interval, obtain the residual nitrogen time from Schedule 1.4 and subtract it from the maximum no-decompression ET at the excursion depth of the repetitive dive.

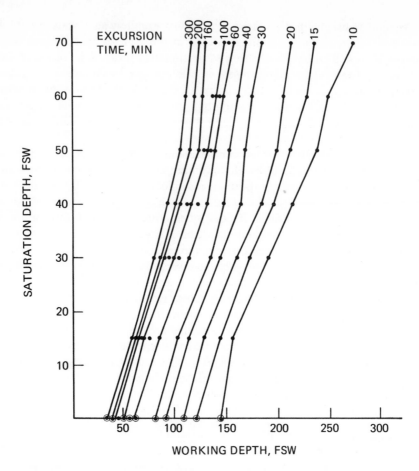

Figure VII-20. The air saturation depth as a function of the desired working depth and excursion time. No decompression is required for the excursion times. (Hamilton *et al.* 1973.)

Decompression schedules for repetitive diving beyond the no-decompression limits have not been developed. *Such procedures should be avoided!* In an emergency, use the U. S. Navy Decompression Tables, substituting excursion depth (working depth − saturation depth) for depth in the tables.

Excursion to depths shallower than the saturation depth carry the attendant risk of inducing decompression sickness while the diver is in the water. There was an incident in the TEKTITE dive of symptoms occurring when a diver made an excursion shallower than the habitat *after* having made a descending excursion earlier in the day. The NOAA instructions (Hamilton *et al.* 1973) accompanying the ascending excursion table (see Schedule 4.2) are worth quoting:

Timing: Begin timing on arrival at excursion depth (bottom time does not include transit times). Rates: Ascend at 10 to 30 fpm. Descend at 60 fpm or faster if desired. If bends symptoms are noted, descend immediately to habitat. If descent to habitat is held up by ear problems, it is preferable to discontinue descent (or even momentarily ascend a bit) and clear the ear, rather than incur ear damage in order to adhere strictly to the table.

The ascent excursion schedules are based on a limited set of data, except in the region where there is no time limit. This latter region is based on the use of the same matrix of allowable supersaturations that was used for the descending excursion and decompression tables mentioned earlier. Notice that the tables suggest that one may decompress directly to the surface and safely remain there 36 min before returning to the habitat.

(iii) Air Decompression Schedules for Fresh Water Diving. Scientific and sport diving is often carried out in fresh water lakes and quarries. One may use the normal schedules given in the preceding sections, but corrected for depth. That is, 1 atm of pressure is equivalent to 33 ft of salt water and to 30 ft of fresh water. So a dive in fresh water to x feet is equivalent to a salt water dive of $(33/30)x$ feet. By entering the diving tables at this hypothetical depth, the appropriate decompression schedule can be selected. Each stop depth must be converted to the equivalent fresh water depth [i.e., stop depth in ffw = stop depth in fsw $\times$ (33/30)].

(iv) Air Decompression Schedules for Personnel and Patients within a Hyperbaric Chamber. Patients who are treated with high-pressure oxygen (HPO) are generally not at risk during decompression. The pressures are generally limited to less than 6 ATA and much of the exposures are spent breathing pure O_2. It is the attendants who are at risk; they are usually exercising, breathing air, and are most anxious to leave the chamber when the treatment is complete.

Modifications of the U. S. Navy air tables (Schedules 1.1–1.7) are probably the best decompression schedules for these purposes. Decompression rates between stops should be slowed to 15 ft/min (the usual 60 ft/min is appropriate for a diver anxious to leave the water). The last 10 ft (0.3 atm) should take 5 min. Finally, most centers utilize Behnke's (1967) recommendation that personnel breathe pure oxygen (readily available!) for 15 min during the decompression, while resting, and preferably at or about 60 fsw (2.82 ATA).

d. Mixed-Gas Decompression Schedules

Mixed-gas decompression schedules refer to diving environments in which gases other than air or O_2 are being utilized. Historically, the first deviation from air breathing was the use of helium to replace nitrogen in an attempt to reduce the decompression time required with air. Helium–oxygen mixtures were successfully utilized in the rescue and subsequent first U. S. Navy salvage of a sunken submarine, the SQUALUS. At present, there is a variety of gas mixtures that have been successfully used in operational diving.

The use of gas mixtures substantially complicates safe decompression. Since no simple procedures have been developed to correlate N_2 and He dives that are performed at the same depth for the same bottom times, He–O_2 decompression schedules must be developed from diving experience. However, there are substantially fewer controlled He–O_2 dives than there are air dives and the problem is further complicated by changing gas mixtures during the dive (e.g., from He–O_2 to air) or in utilizing a third inert gas, such as neon (tri-mix).

This section will, therefore, report only on fully tested schedules and refrain from speculation on the schedules to be utilized with novel gas mixtures. Any gas mixture complicates the logistics of a dive and should only be utilized by qualified professionals who have successfully tested their decompression schedules in chambers. The recent observation of bends occurring with a change in inert gases, *without a decrease in pressure*, is an ominous comment on the state of knowledge regarding mixed-gas decompressions. It is always desirable to have a physician and treatment chamber available when utilizing mixed gases.

Mixed-gas decompression procedures utilize the concept that inert gas partial pressures are the determinant variables in computing decompression schedules. The relevant supersaturation parameters are the inert gas partial pressures in each tissue and the local absolute hydrostatic pressure.

One of the major reasons for using gas mixtures is to keep the P_{O_2} within tolerable limits (see Chapter IV for a discussion of O_2 toxicity). At a P_{O_2} in excess of 0.4 atm, one is limited in the time he may breathe O_2; this time decreases as the diver increases his work effort. The toxic results may be pulmonary or CNS abberrations. The generally accepted P_{O_2} time limits for working divers are given in Chapter IV.

Working with mixed gases greatly complicates life-support equipment. No longer can a simple air compressor be utilized. Extensive backup procedures and equipment must assure an adequate supply of the mixed gas. Emergency procedures should be available in case the mixed-gas source is inadequate and decompression must be completed using air. Everything possible must be done to conserve the gas. Toward this goal, many sophisticated systems recirculate the gas to the diver after passing it through a CO_2 scrubber. This recycled gas is mixed with fresh gas from the gas supply (see Chapter IV, discussion of breathing mixtures, for details of such equipment). Because the recycled gas has less O_2 than the fresh supply, the diver breathes a gas mixture that has a lower P_{O_2} than the supply. In computing decompression schedules, one must use the actual P_{O_2} (or an approximation) that the diver is breathing.

The following paragraphs divide mixed-gas decompression procedures into those utilized for He–O_2, N_2–O_2, and tri-gas mixtures.

(i) Helium–Oxygen Gas Mixtures. The use of helium as a respired inert gas was first proposed by the chemist Hildebrand (Sayers *et al.* 1925) because it is so much less lipid-soluble than is nitrogen. He reasoned that the long-half-time tissues in the body are really fat depots that store considerable quantitites of nitrogen. Helium, being much less soluble, ought to be rapidly removed; in fact, the half-time of a perfusion-limited tissue (see biological factors earlier in this chapter) ought to be reduced by the ratio of fat/blood partition coefficient between the two inert gases. In fact, this concept has been experimentally validated for other inert gases of higher molecular weight, so it surely does apply to helium and nitrogen.

Unfortunately, diving experience with helium has not reduced decompression time as predicted by the He–N_2 partition coefficient. The probable explanation is that the really slow tissues are not purely perfusion-limited, but also include some diffusion dependence. Alternatively, the slow tissues could be aqueous, rather than lipoidal, where the partition coefficients are not so different between He and N_2.

Helium diving has two major advantages: *The diver does not suffer inert gas narcosis*, so he may safely dive deeper than on air (the lower density of He–O_2 also makes breathing easier at depth), and *the decompression time is shorter* than the equivalent air decompression time (Table VII-11). The major disadvantages (other

Table VII-11

Comparative No-Decompression Limits for
Scuba Diving with Air as He–O$_2$

Depth, fsw	No-decompression time limit	
	He–O$_2$[a]	Air
40	260	—
60	130	60
80	60	40
100	35	25
120	25	15
140	15	10
160	10	5
180	5	5

[a] U. S. Navy Schedules for Mk VI Semiclosed scuba with 68% He, 32% O$_2$.

than logistic complications) are that *the high thermal conductivity of helium* results in heating requirements and/or elevated respiratory heat loss (see Chapter IV, discussions of breathing mixtures and of cold) and *the alteration in speech* that accompanies He–O$_2$ breathing makes communications more difficult (see Chapter X). Oxygen toxicity is an ever-present hazard in He–O$_2$ diving because of the longer and deeper exposures. Table VII-12 lists the ten rules of safety for He–O$_2$ diving.

1. *Hard-hat schedules.* The first extensive use of He–O$_2$ was carried out in the late 1930's at the U. S. Navy Experimental Diving Unit. The schedules for hard-hat diving were then extensively tested during the salvage of the U. S. S. SQUALUS. These schedules were subsequently recalculated and retested by Molumphy (1950), resulting in the publication of the schedules that are still in use today for hard-hat diving (see Schedules 5.1–5.4 and 7.1 at the end of this chapter). These schedules were calculated with a Haldanian model that included the following assumptions:

Breathing gases
1. Constant $F_{O_2}(0.14–0.21)$.
2. Oxygen breathing at 60 and 50 ft (assume $F_{O_2} = 0.8$).
3. Express depth in terms of helium partial pressure, so that

$$P_{He} = (d + 33)[1.0 - (F_{O_2} - 0.02)]$$

where d is the depth and F_{O_2} is the oxygen fraction.
Helium uptake and washout:
1. Assume a tissue spectrum where $t_{\frac{1}{2}} = 5, 10, 20, 30, 40, 50, 60, 70$ min.
2. Use twice the actual bottom time to calculate the helium loadings in each tissue (to account for exercise at depth).
3. Treat the nitrogen that is equilibrated at surface (26 ft) as helium.
Permissible supersaturation:
A constant ratio of tissue partial pressure to local barometric pressure was selected as (1.7:1.0). This was not depth dependent, nor did it vary throughout the tissue spectrum.

Table VII-12

The Ten Rules for He–O$_2$ Breathing, to Avoid Oxygen Toxicity Problems

---

Loss of He–O$_2$ *Supply*

Deeper than 50 ft

1. Shift to air, come all the way out in accordance with Emergency Air Table (Schedule 5.4); no surface decompression

Loss at 50-ft stop

2. Shift to air (or He–O$_2$); complete stop in accordance with Emergency Air Table (Schedule 5.4) (or He–O$_2$, Schedule 5.3); can surface decompress after 30 ft stop; O$_2$ time is good time

Loss at 40-ft stop

3. Not within surface decompression or emergency surface decompression limits: same treatment as above

4. Within emergency surface decompression limits: Surface-decompress diver; double missed time of required water stop for surface decompression and add to chamber stop

5. Within normal surface decompression limits: Surface-decompress normally

O$_2$ *Toxicity Symptoms*

Symptoms at 50-ft stop

6. Ascend to 40-ft stop; shift to air (or He–O$_2$); can surface-decompress after 30-ft stop; disregard missed time at 50 ft

Symptoms at 40-ft stop

7. Not within surface decompression or emergency surface decompression limits: Ascend to 30 ft stop; shift to air (or He–O$_2$); can surface-decompress after 30 ft stop; disregard missed time at 40 ft

8. Within emergency surface decompression limits: Surface-decompress diver; double missed time of required water stop for surface decompression and add to chamber stop

9. Within normal surface decompression limits: Surface-decompress normally

10. Symptoms during chamber stop: Remove mask; complete decompression in accordance with Emergency Air Table (Schedule 5.4); O$_2$ time is good time

---

The decompression tables available for hard-hat diving (Schedules 5.1–5.4 and 7.1) are summarized in Figure VII-21. As with air diving, there are no-decompression limits, normal diving (limited by O$_2$ toxicity hazards), exceptional exposure tables, emergency tables (with and without He–O$_2$), and surface decompression procedures. Repetitive diving is NOT permitted within a 12-hr period.

The schedules have been used extensively by the U. S. Navy and many commercial firms. The U. S. Navy incidence of decompression sickness has been reported to be 0.83% when using the schedules.

2. *Mixed-gas scuba schedules.* The U. S. Navy developed He–O$_2$ decompression procedures for modified scuba equipment. They were developed in order to reduce the risk of CO$_2$ retention, oxygen toxicity, nitrogen narcosis, and the overexertion of breathing denser air. The equipment involves a rebreathing circuit; Figure VII-22 is a schematic of the U. S. Navy Mk. VI semiclosed diving apparatus. (Commerical diving systems are available that utilize the same decompression procedures.) The gas mixture has an elevated F_{O_2} (0.32 is normal and 0.40 is permitted) into the system at a rate that is dependent on the depth and exercise level of the dive. After mixing with the recycled gas in the inhalation bag, the F_{O_2} is lowered to a predictable level. By carrying a separate bottle of pure oxygen and a crossover switch, divers can decompress at 30 and 20 fsw on pure oxygen.

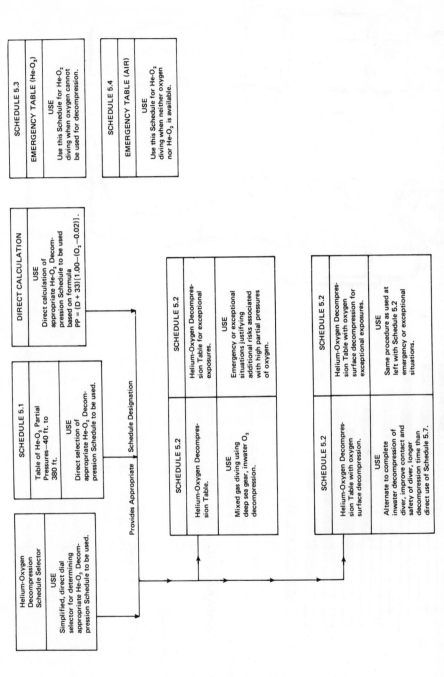

Figure VII-21. A summary of U. S. Navy Decompression Tables for the use of He–O₂ breathing while diving with deep sea diving equipment. REPETITIVE DIVING IS NOT ALLOWED WITH LESS THAN A 12-HR SURFACE INTERVAL (see Schedules 5.1–5.9). (U. S. Navy 1971.)

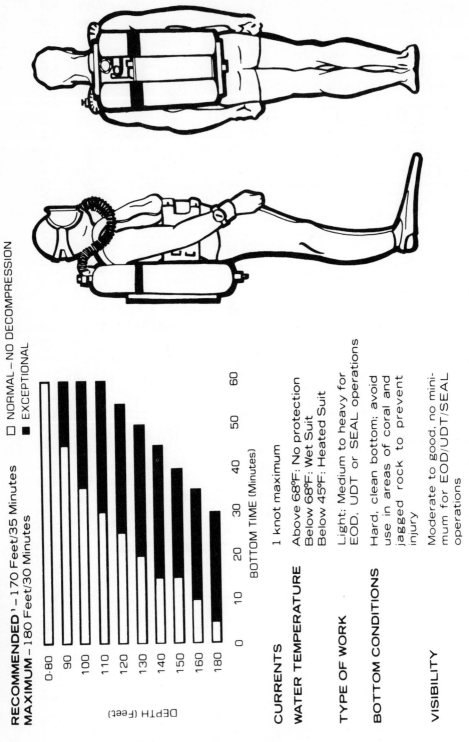

Figure VII-22. U. S. Navy Mk. VI semiclosed diving apparatus (U. S. Navy 1971).

DEPTH/DURATION LIMITS

RECOMMENDED¹ – 170 Feet/35 Minutes

MAXIMUM – 180 Feet/30 Minutes

☐ NORMAL – NO DECOMPRESSION

■ EXCEPTIONAL

CURRENTS   1 knot maximum

WATER TEMPERATURE   Above 68°F: No protection
                      Below 68°F: Wet Suit
                      Below 45°F: Heated Suit

TYPE OF WORK   Light: Medium to heavy for
                      EOD, UDT or SEAL operations

BOTTOM CONDITIONS   Hard, clean bottom; avoid
                      use in areas of coral and
                      jagged rock to prevent
                      injury

VISIBILITY   Moderate to good, no mini-
                      mum for EOD/UDT/SEAL
                      operations

The mixed-gas scuba decompression schedules (Schedules 5.5–5.9 at the end of this chapter) were developed by several U. S. Navy investigators. They were completed (including the repetitive schedules) by Workman and Reynolds (1965). The model utilized a tissue half-time spectrum of 5–120 min. The permissible supersaturation was expressed as a constant pressure difference, varying for each tissue (Table VII-13).

The procedures for using these tables are shown in Figure VII-23. Initial tables are selected based on the projected dive profile and equipment available. Repetitive diving schedules are selected and computed on the same basis as was used for air diving. Note that emergency procedures are available for decompressing with air and for surface decompression. The schedules have been extensively tested in both chambers and the open sea environment.

3. *Decompression procedures from saturation diving.* Saturation diving while breathing a He–O$_2$ mixture has become an operational reality during the last 10 yr. Chamber tests as deep as 1700 fsw have been carried out and operational diving procedures exist for depths of 1000 fsw. Saturation is accomplished by maintaining the oxygen at a constant P_{O_2} throughout the dive and decompression. Since decompression occurs inside chambers, oxygen breathing from masks is often used at shallow depths.

The development of decompression procedures for great depths has shown the general inadequacy of all the models in extrapolating previous experience into new domains of greater depths and bottom times. Schreiner and Kelly (1967) articulate the general Haldanian model for various types of saturation decompression procedures. In such calculations one is only concerned with the longest half-time tissue of the model. They showed that the decompression consists of an initial rapid ascent, or "pull," followed by a slow rate of ascent, given by

$$\frac{dP}{dt} = -\frac{\ln 2}{t_{\frac{1}{2}}}(P_{O_2} + \Delta P) \tag{20}$$

where ΔP is the permissible supersaturation of the slowest body tissue. The initial pull is done to establish a gradient in helium partial pressure from the tissue to the ambient air. The pull is a constant amount so long as no excursion diving has occurred in the 12 hr prior to starting the decompression.

The selection of $t_{\frac{1}{2}}$ and ΔP has undergone steady revision. Workman (1965) presented a matrix of permissible supersaturation values that included a 240-min tissue. The M values [i.e., ΔP in Eq. (20)] (see Figure VII-24) exhibit two general

Table VII-13

Safe Helium Partial Pressure
upon Surfacing

$t_{\frac{1}{2}}$, min	$P_{He} - 33$, fsw
5	83
10	59
20	41
40	29
80	23
120	20

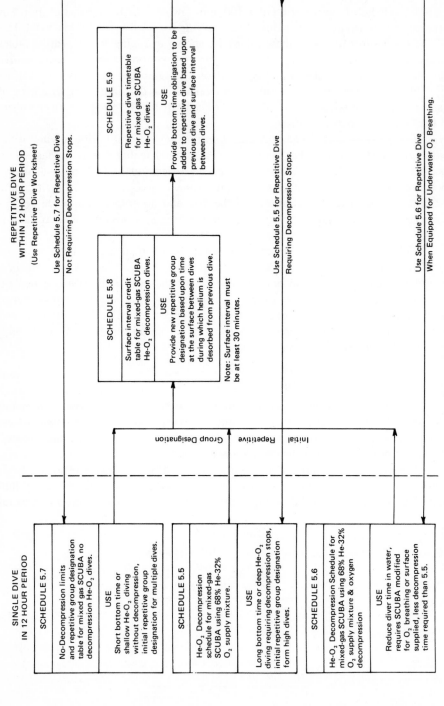

MIXED GAS SCUBA

SINGLE DIVE
IN 12 HOUR PERIOD

SCHEDULE 5.7

No-Decompression limits
and repetitive group designation
table for mixed gas SCUBA no
decompression He-O₂ dives.

USE

Short bottom time or
shallow He-O₂ diving
without decompression,
initial repetitive group
designation for multiple dives.

SCHEDULE 5.5

He-O₂ Decompression
schedule for mixed-gas
SCUBA using 68% He-32%
O₂ supply mixture.

USE

Long bottom time or deep He-O₂
diving requiring decompression stops,
initial repetitive group designation
form high dives.

SCHEDULE 5.6

He-O₂ Decompression Schedule for
mixed-gas SCUBA using 68% He-32%
O₂ supply mixture & oxygen
decompression

USE

Reduce diver time in water,
requires SCUBA modified
for O₂ breathing or surface
supplied, less decompression
time required than 5.5.

REPETITIVE DIVE
WITHIN 12 HOUR PERIOD
(Use Repetitive Dive Worksheet)

Use Schedule 5.7 for Repetitive Dive
Not Requiring Decompression Stops.

SCHEDULE 5.9

Repetitive dive timetable
for mixed gas SCUBA
He-O₂ dives.

USE

Provide bottom time obligation to be
added to repetitive dive based upon
previous dive and surface interval
between dives.

SCHEDULE 5.8

Surface interval credit
table for mixed-gas SCUBA
He-O₂ decompression dives.

USE

Provide new repetitive group
designation based upon time
at the surface between dives
during which helium is
desorbed from previous dive.

Note: Surface interval must
be at least 30 minutes.

Use Schedule 5.5 for Repetitive Dive
Requiring Decompression Stops.

Use Schedule 5.6 for Repetitive Dive
When Equipped for Underwater O₂ Breathing.

Initial   Repetitive   Group Designation

Figure VII-23. A summary of U. S. Navy Decompression Schedules for use with mixed-gas scuba (U. S. Navy 1971).

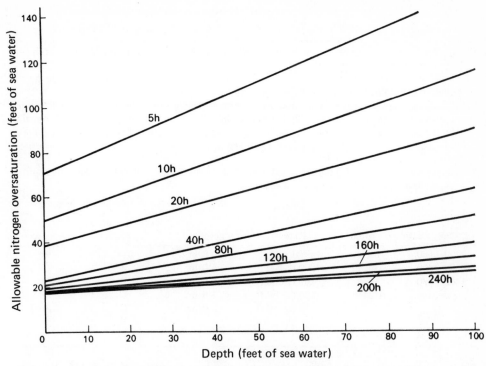

Figure VII-24. The permissible supersaturation of helium partial pressure as a function of depth and tissue half-time [data of Workman and Reynolds (1965)].

characteristics that have been extensively validated: (1) As $t_{\frac{1}{2}}$ increases, M decreases, (2) as depth increases, M increases linearly (except for the M value of the slowest tissue, which is independent of depth). Thus, one would project from the above equation for dP/dt that a linear rate of ascent would yield a safe decompression schedule.

During testing of these schedules, several compromises were made that ultimately yield a much slower decompression than these considerations suggest is necessary. Overnight stops were instituted so as to permit the divers time to rest. Because the divers are relatively secure in chambers, the need to achieve a speedy decompression was deemed secondary to a safe procedure. So these slower schedules have never really been tested against the model and should not be interpreted as implying that a much longer tissue half-time is required.

The U. S. Navy procedures utilize a decompression curve that is a series of ascent rates, gradually slowing down as the surface is approached. The decompression begins with a 30-fsw pull. Table VII-14 summarizes the procedure.

4. *Excursion diving from saturation.* One may ascend up to 30 fsw from a saturation pressure once within a 24-hr period. Descending excursions depend on the saturation depth and the bottom time that is desired. Schedules 6.1–6.3 (at the end of this chapter) indicate permissible bottom times and a repetitive group designation used for repetitive excursion diving. Schedules 6.3 may be used to compensate for time in the

Table VII-14

General Instructions for Saturation Diving Decompression Schedules

Divers should normally remain at saturation depth for 24 h subsequent to the last excursion for the purpose of reequilibrium of tissues

Chamber oxygen environment should be maintained at 0.30–0.32 atm; carbon dioxide levels should be maintained less than 0.5% surface equivalent

"Initial ascent" represents the initial rapid ascent rate at the commencement of decompression to establish a gradient for inert gas elimination; to find the initial ascent distance, find from the habitat internal credit table the repetitive group designation of the diver who has the most residual helium; the initial ascent distance is given in the table below, opposite the appropriate repetitive group designation letter

To adequately decompress from a saturated condition the diver must follow both the daily routine schedule and rate of ascent schedule given below

Daily routine schedule		Rate of ascent schedule		Initial ascent distance	
				Repetitive group designation	Feet of initial ascent
2400–0600	Stop	Initial ascent	10 ft/hr	—	30
0600–1400	Ascend	1000 ft–200 ft	6 ft/hr	A	25
1400–1600	Stop	200 ft–100 ft	5 ft/hr	B	20
1600–2400	Ascend	100 ft– 50 ft	4 ft/hr	C	15
		50 ft– 0 ft	3 ft/hr	D	10
				E	5
				F	0

habitat between dives. These excursion tables have been extensively tested by the U. S. Navy.

5. *Emergency abort schedules from saturation habitats.* Abort schedules were calculated using the matrix from Figure VII-24. They have not been extensively tested. The schedules are included at the end of this chapter (Schedule 7.1).

(ii) N_2–O_2 Gas Mixtures. Generally nitrogen–oxygen mixtures are only used for saturation diving in shallow depths. The mixture is used to reduce the P_{O_2} to avoid toxic responses. Decompression procedures have been developed for both excursion diving from saturation and decompression to sea level. The schedules are based on extensive compilations of Hamilton *et al.* (1973) (Schedules 4.1 and 4.2). The matrix of tissue half-times and permissible supersaturations was previously discussed under air saturation diving.

E. *Treatment*

In all diagnosed cases of decompression sickness, the patient must be moved rapidly to a recompression chamber where he may be treated. All planning for diving activities should include the identification of the nearest treatment facility and a plan for being able to get to the facility rapidly in the case of an accident.

There are two phases in the treatment of DCS: *diagnosis* and *emergency first aid* during transportation to a facility and *treatment* in a recompression chamber. The first phase is generally supervised by nonmedical personnel. The second phase is generally directed by a medically trained chamber operator with a local physician being available for consultation. The purpose of this section is to identify the procedures and necessary equipment required of nonmedical personnel.

DCS treatment procedures have emerged from the need to provide medical care during the development of decompression tables. These procedures generally pre-supposed the rapid availability of a treatment chamber and a high degree of experience on the part of all participants. Because diagnosis under these circumstances occurs soon after the dive, it was satisfactory to merely recompress the diver in order to dissolve the bubbles that were causing the symptoms.

The emergence of sport diving introduced new complications to the treatment of decompression sickness. Unsophisticated divers were often struck with severe DCS because they grossly disobeyed the appropriate decompression schedule. In addition, they were often afflicted for long periods before reaching treatment facilities. These patients present the physician with disseminated gas space and severe tissue damage that occurs because of mechanical distention of tissues or the resulting long-term hypoxia. The symptoms are caused by massive intravascular bubbles, or their sequelae.

Tunnel workers and commercial diving fishermen often present the same kind of complex disease picture when presented for treatment because they are reluctant to miss work, or are long distances from a treatment center. When DCS occurs in divers undergoing a saturation decompression, special procedures are required because they are often struck at depths below the prescribed treatment depths.

1. *Diagnosis and Emergency Action during Transportation to a Recompression Chamber*

Diagnosis of decompression sickness has been already described in this chapter. The goal of the diagnosis is to decide, if at all possible, whether the diver suffers from pain-only bends, severe bends, or other form of pressure-induced trauma (air embolism, barotrauma, bruise, broken bones, etc.). This is not only important in deciding which kind of emergency care to apply, but may be decisive in deciding where to go for treatment. During diagnosis the patient should lie down with the legs higher than the head. Any loss of heart beat or respiration should be immediately dealt with by standard first aid methods. THESE PROCEDURES SHOULD ALWAYS BE CONTINUED, UNLESS A PHYSICIAN CERTIFIES THAT THE PATIENT IS DEAD!

Whenever the patient is moved, the head should be kept lower than the feet to minimize the chance of a brain air embolism occurring as a sequelae to severe DCS. Avoid flying the patient at any substantial altitude, as this lower pressure can adversely affect the patient's condition. In all cases of DCS, recompression therapy is necessary no matter how long it takes to reach a chamber since adverse sequelae may occur from residual tissue damage.

Recompression of a patient by reimmersion in the sea should be avoided. The most minimal treatment schedule available requires 6 hr under pressure and very few divers can stand this exposure.

If oxygen is available, it should be used on the patient (at the very least, it enhances inert gas washout).

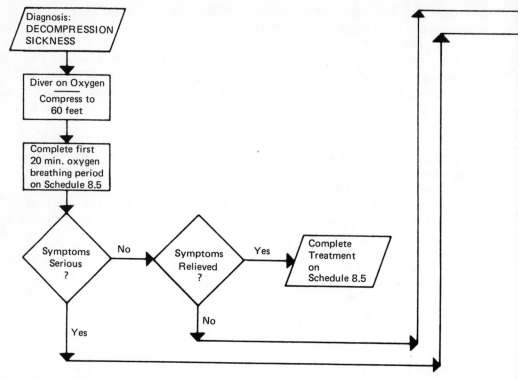

Figure VII-25. Logic chart for patients who are suffering "pain-only" Type I decompression sickness (U. S. Navy 1973).

2. Recompression Treatment

While most treatment facilities have their own trained personnel, this section will provide details on the treatment procedures as an aid to designing such facilities and to planning a dive, and in the event sufficient trained personnel are not available.

A family of treatment tables has been designed to accommodate different treatment facilities and the different severities of DCS. All treatments start by compressing the patient (and, if there is room, a tender) to one of two fixed pressures (60 or 165 fsw). If oxygen is available from surface-supplied bottles, tables should be selected to utilize O_2 breathing. Treatment success has improved markedly when oxygen breathing was used. The major benefit of recommended treatment tables is that the subsequent decompression phase has been thoroughly tested.

Figure VII-25 is a logic chart for the treatment of patients suffering from pain-only bends, where a facility for oxygen breathing is available. The treatment schedules noted (Schedules 8.4–8.6) are included at the end of this chapter. Table VII-15 lists the general rules for carrying out such treatments.

If there is a recurrence of symptoms during the decompression, use Figure VII-26 to guide in the subsequent recompression and treatment. If symptoms recur

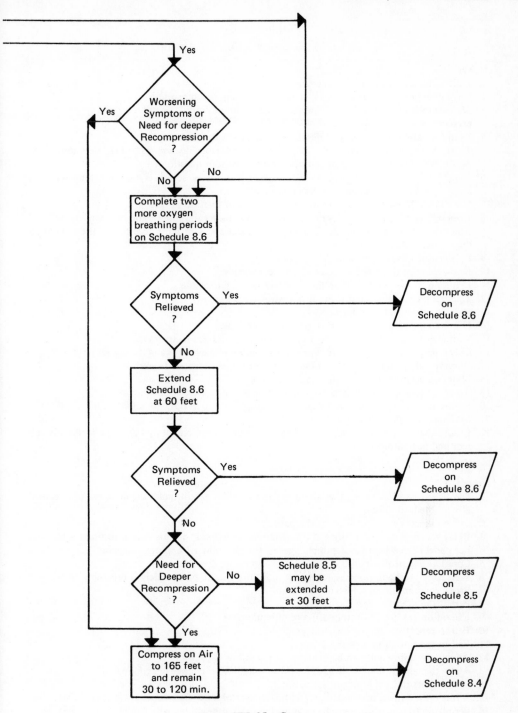

Figure VII–25—*Cont.*

Table VII-15

Rules for Recompression Treatment

Always
1. Follow the treatment tables accurately
2. Have qualified tender in chamber at all times during recompression
3. Maintain the normal descent and ascent rates; maintain rapid descent if serious symptoms are present
4. Examine patient thoroughly at depth of relief or treatment depth
5. Treat an unconscious patient for air embolism or serious decompression sickness unless the possibility of such a condition can be ruled out without question
6. Consider the use of 80% helium–20% oxygen in cases of serious symptoms, recurrence of symptoms, or when patient has difficulty breathing
7. Use oxygen if available; ensure that the patient can tolerate oxygen.
8. Use Oxygen Treatment Tables (Schedules 8.5–8.8) under the supervision of a Medical Officer, either directly or by telephone; a qualified medical assistant must accompany the patient in the chamber during treatment
9. Be alert for oxygen poisoning if oxygen is used
10. Know what to do in the event of oxygen convulsion
11. Maintain oxygen usage within the time and depth limitations
12. Take all precautions against fire if oxygen is used
13. Provide water and sand buckets
14. Use fire-retardant paint and materials in the chamber
15. Ventilate the chamber according to specified rates and gas mixtures
16. Check patient's condition before and after coming to each stop and during long stops
17. Observe patient for at least 6 h after treatment for recurrence of symptoms
18. Maintain accurate time-keeping and recording
19. Assure proper decompression of all personnel entering the chamber
20. Ensure that the chamber and its auxiliary equipment are in operational condition at all times
21. Maintain a well-stocked medical kit at hand
22. Ensure that all personnel are trained in operation of equipment and are able to do any job required in treatment

Never
1. Permit any shortening or other alteration to the tables except under the direction of a trained Diving Medical Officer
2. Exceed descent rate tolerated by the patient
3. Delay recompression for physical examination or first aid; do this during recompression
4. Let patient sleep between depth changes or for more than 1 h at any one stop
5. Continue ascent if patient's condition worsens
6. Wait for a mechanical resuscitator; use mouth-to-mouth immediately if breathing ceases
7. Break rhythm during artificial respiration
8. Permit the use of oxygen below 60 ft
9. Use oil on any oxygen fitting or piece of equipment
10. Fail to report symptoms early (diver)
11. Fail to treat doubtful cases
12. Allow gas supply tanks to be depleted or reach low capacity
13. Allow damage to door seals and dogs; use minimum force in "dogging down"
14. Leave doors dogged after pressurization
15. Allow open flames, matches, cigarette lighters, or pipes to be carried into the chamber
16. Permit electrical appliances to be used during oxygen-breathing or when chamber atmosphere is compressed air

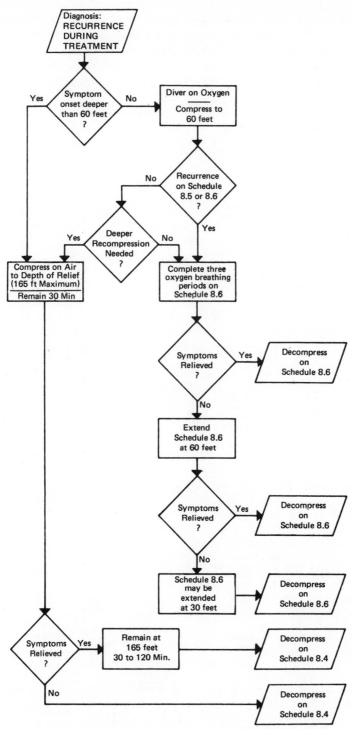

Figure VII-26. Recurrence during treatment (U. S. Navy 1973).

after the treatment is complete, use Figure VII-27 for guiding the subsequent treatment. After treatment, always keep the patient under observation for at least 6 hr.

In the event oxygen is not available for treatment of pain-only decompression sickness, use Schedule 8.1 if symptoms are relieved by the time the patient is at 66 ft. Schedule 8.2 is to be used for such patients when symptoms are relieved at depths

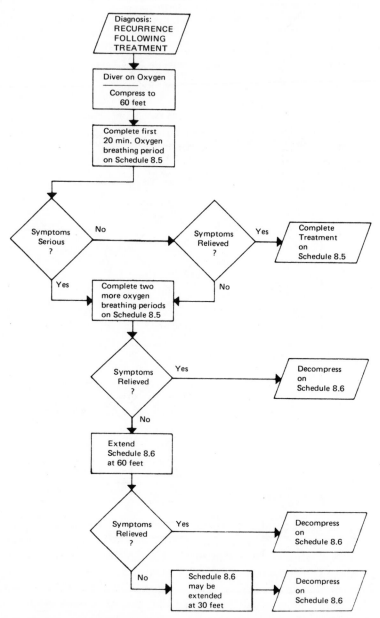

Figure VII-27. Recurrence following treatment (U. S. Navy 1973).

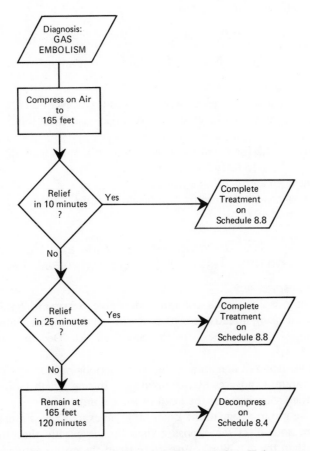

Figure VII-28. Treatment of gas embolism and Type II decompression
sickness (U. S. Navy 1973).

greater than 66 ft. Schedule 8.3 is useful when symptoms are relieved within 30 min
at 165 ft.

For patients who are suffering severe decompression sickness or are suspected of
having an air embolus, the preferred treatment procedure is outlined in Figure VII-28.
These schedules all require oxygen breathing at depth. If oxygen is not available, use
Schedule 8.4. Any recurrence of symptoms is treated as in Figure VII-26 and VII-27.

Patients suffering severe decompression sickness often experience residual symptoms due to tissue damage and hypoxia. These patients should be attended by a
physician.

F. *Decompression Schedules*

1. *U. S. Navy Decompression Schedules for Subsaturation Air Diving*

(See Table VII-9 in text for a guide to the use of these schedules.)

a. Schedule 1.1: U. S. Navy Standard Air Decompression Table

(i) Special Instructions
1. Rate of ascent is not critical between stops for stops of 50 ft or less.
2. If dive was particularly cold or strenuous, use next longer bottom time schedule.
3. See Schedule 1.2 for repetitive groups in no-decompression dives.
(ii) Example. Dive to 82 ft for 36 min.
1. Select next greater depth, i.e., 90 ft.
2. Select next greater bottom time opposite 90 ft, i.e., 40 min.
3. Stop 7 min at 10 ft.

b. Schedule 1.2: No-Decompression Limits and Repetitive Group Designation Table for No-Decompression Air Dives

(i) Special Instructions
1. No-decompression limits column: Allowable maximum bottom time that permits surfacing directly at 60 ft/min with no decompression stops.
2. For longer bottom times use the Standard Air Decompression Schedule (Schedule 1.1).
3. Repetitive group designation table: Time periods in each vertical column are the maximum exposures at various depths during which a diver will remain within the group listed at the head of the column.
4. Repetitive group designation: Enter table on exact or next greater depth than exposure and select the exposure time that is exactly the same as or next greater than the actual exposure time. Read the group designation (letter) at the top of the column for the next dive.
5. Exposure times beyond 5 hr and less than 40-ft depth are beyond field requirements of this table.
(ii) Example. A dive to 32 ft for 45 min.
1. Select next greater depth, i.e., 35 ft.
2. Select next greater exposure time than 45 min, i.e., 50 min.
3. Read designation at top of column, i.e., group E.

c. Schedule 1.3: Surface Interval Credit Table for Air Decompression Dives

(i) Special Instructions.
1. Surface interval time in the schedule is in hours and minutes (7:59 = 7 hr and 59 min).

2. Surface interval must be at least 10 min.
3. Repetitive group designation after surface interval: Enter the schedule on the diagonal slope using the group designation from previous dive. Read horizontally until the actual surface interval is equal to or between the interval shown in the schedule. Read the new group designation at the top of the column.
4. Dives following surface intervals of more than 12 hr are not repetitive dives. Use actual bottom times in the Standard Air Decompression Schedules to compute decompression for such dives.

(ii) Example. Find new group designation after dive to 110 ft for 30 min and a time on the surface of 1 hr and 30 min.

1. The previous repetitive group from the last column of the 110/30 schedule of the Standard Air Decompression Table is J.
2. Locate "J" in the diagonal column.
3. Follow the schedule across horizontally.
4. The 1 hr, 30 min interval lies between the times 1:20 and 1:47.
5. Diver has lost sufficient inert gas to place him in the group at the top of vertical column G.
6. Use this new group designation to determine residual nitrogen time to be credited toward repetitive dive.

d. Schedule 1.4: Repetitive Dive Timetable for Air Dives

(i) Special Instructions
1. Bottom times listed in this schedule are called "residual nitrogen times."
2. Residual nitrogen time is the time a diver is to consider he has already spent on bottom when a repetitive dive is started to a specific depth.
3. Residual nitrogen time: Enter the schedule horizontally with the repetitive group from the surface interval credit table. Read directly the bottom time to be added to the repetitive dive in the depth column for that dive.

(ii) Example. The group designation from the surface interval credit table from a previous dive is H. How much bottom time must be added (residual nitrogen time) for a repetitive dive to 110 ft?

1. Enter the schedule horizontally at H.
2. Read in the 110-ft depth column the residual nitrogen to be added: 27 min.
3. The schedule shows that one must start a dive to 110 ft as though he had already been on the bottom 27 min.
4. Use the Standard Air Decompression Schedule or the No-Decompression Schedule to determine dive schedule for repetitive dive.

e. Schedule 1.5: U. S. Navy Standard Air Decompression Table for Exceptional Exposures

Special Instructions. The schedule includes only schedules of decompression for exceptional or emergency cases. The great demands placed on a diver's endurance by emergencies necessitating the use of this schedule are such that complete assurance of success of the decompression schedules is impossible.

Never follow a dive covered by this schedule with a repetitive dive. The Diving Officer must weigh the need for any dive in this schedule for exceptional exposures against the increased danger and demands on the diver's physical endurance.

f. Schedule 1.6: Surface Decompression Table Using Oxygen

Special Instructions. The use of surface decompression provides the advantages of added comfort and security for the diver. Routine use of this technique requires a recompression chamber equipped with proper oxygen-breathing equipment. Use of this schedule may be indicated in certain emergency situations where a surface interval must come between the dive and the major part of decompression. Although it is possible for the decompression period following the surface interval to be in the water, recompression in a chamber is always preferable.

In the event of oxygen toxicity symptoms or failure of the oxygen supply, decompress according to Schedule 1.8 disregarding time spent on oxygen. Use of this technique exposes the driver to a brief surface interval between his leaving the water and his attaining the scheduled decompression stop depth in the recompression chamber. The interval must be as short as possible.

When surface decompression is to be used, this schedule is employed in place of the standard Air Decompression Schedule (Schedule 1.1).

Column 3: time of ascent to the first stop or to the surface at a rate of 25 ft/min. Column 4 (water stops): time spent at tabulated stops using air. If no stops are required, ascend to surface at 25 ft/min. When stops are required, use a 25 ft/min ascent rate to first stop. Take an additional minute between stops. Ascend from 30-ft stop to surface at 30 ft/min.
Column 5 (surface interval): Surface interval shall not exceed 5 min and includes 1 min ascent from 30-ft stop, 3 min 30 sec for landing the diver on deck and undressing, and time of descent from surface to 40 ft in recompression chamber (30 secs).
Column 6: During the period of oxygen breathing the chamber should be ventilated unless an oxygen elimination system is used.

g. Schedule 1.7: Surface Decompression Table Using Air for Air Diving

(i) Special Instructions. This schedule may be used for surface decompression from an air dive in the event that oxygen toxicity or failure of the oxygen supply prevents the use of Schedule 1.6. When surface decompression on air is to be used, this schedule is employed in place of the Standard Air Decompression Schedule (Schedule 1.1). If this schedule is used as a result of oxygen toxicity problems with Schedule 1.6, disregard previous time spent on oxygen when decompressing according to this schedule. There is no surface decompression schedule for use following a dive on the Standard Air Decompression Schedule for Exceptional Exposures (Schedule 1.5).

1. All ascent and descent rates are 60 ft/min.

Schedule 1.1

U. S. Navy Standard Decompression Table (U. S. Navy 1971)

Depth	Bottom time, min	Time to first stop, min:sec	Decompression stops, ft					Total ascent, min:sec	Repetitive group
			50	40	30	20	10		
40	200	0:00					0	0:40	(*)
	210	0:30					2	2:40	N
	230	0:30					7	7:40	N
	250	0:30					11	11:40	O
	270	0:30					15	15:40	O
	300	0:30					19	19:40	Z
50	100	0:00					0	0:50	(*)
	110	0:40					3	3:50	L
	120	0:40					5	5:50	M
	140	0:40					10	10:50	M
	160	0:40					21	21:50	N
	180	0:40					29	29:50	O
	200	0:40					35	35:50	O
	220	0:40					40	40:50	Z
	240	0:40					47	47:50	Z
60	60	0:50					0	1:00	(*)
	70	0:50					2	3:00	K
	80	0:50					7	8:00	L
	100	0:50					14	15:00	M
	120	0:50					26	27:00	N
	140	0:50					39	40:00	O
	160	0:50					48	49:00	Z
	180	0:50					56	57:00	Z
	200	0:40				1	69	71:00	Z
70	50	1:00					0	1:10	(*)
	60	1:00					8	9:10	K
	70	1:00					14	15:10	L
	80	1:00					18	19:10	M

Schedule 1.1—Cont.

Depth	Bottom time, min	Time to first stop, min:sec	| Decompression stops, ft 50	40	30	20	10	Total ascent, min:sec	Repetitive group
80	90	1:00					23	24:10	N
	100	1:00					33	34:10	N
	110	0:50				2	41	44:10	O
	120	0:50				4	47	52:10	O
	130	0:50				6	52	59:10	O
	140	0:50				8	56	65:10	N
	150	0:50				9	61	71:10	N
	160	0:50				13	72	86:10	N
	170	0:50				19	79	99:10	N
80	40	1:10					0	1:20	(*)
	50	1:10					10	11:20	K
	60	1:10					17	18:20	L
	70	1:10					23	24:20	M
	80	1:00				2	31	34:20	N
	90	1:00				7	39	47:20	N
	100	1:00				11	46	58:20	O
	110	1:00				13	53	67:20	O
	120	1:00				17	56	74:20	N
	130	1:00				19	63	83:20	N
	140	1:00				26	69	96:20	N
	150	1:00				32	77	110:20	N
90	30	1:20					0	1:30	(*)
	40	1:20					7	8:30	J
	50	1:20					18	19:30	L
	60	1:20					25	26:30	M
	70	1:10				7	30	38:30	N
	80	1:10				13	40	54:30	N
	90	1:10				18	48	67:30	O
	100	1:10				21	54	76:30	N
	110	1:10				24	61	86:30	N

Depth (ft)	Bottom time (min)	Time to first stop	30	20	10	Total ascent time	Repetitive group
	120	1:10	5	32	68	101:30	N
	130	1:00		36	74	116:30	N
100	25				0	1:40	(*)
	30	1:30			3	4:40	L
	40	1:30			15	16:40	K
	50	1:20		2	24	27:40	L
	60	1:20		9	28	38:40	N
	70	1:20		17	39	57:40	O
	80	1:20		23	48	72:40	O
	90	1:10	3	23	57	84:40	N
	100	1:10	7	23	66	97:40	N
	110	1:10	10	34	72	117:40	N
	120	1:10	12	41	78	132:40	N
110	20				0	1:50	(*)
	25	1:40			3	4:50	H
	30	1:40			7	8:50	J
	40	1:30		2	21	24:50	L
	50	1:30		8	26	35:50	M
	60	1:30		18	36	55:50	N
	70	1:20	1	23	48	73:50	O
	80	1:20	7	23	57	88:50	N
	90	1:20	12	30	64	107:50	N
	100	1:20	15	37	72	125:50	N
120	15				0	2:00	(*)
	20	1:50			2	4:00	H
	25	1:50			6	8:00	I
	30	1:50			14	16:00	J
	40	1:40		5	25	32:00	L
	50	1:40		15	31	48:00	N
	60	1:30	2	22	45	71:00	O
	70	1:30	9	23	55	89:00	O
	80	1:30	15	27	63	107:00	N
	90	1:30	19	37	74	132:00	N
	100	1:30	23	45	80	150:00	N
130	10				0	2:10	(*)

Schedule 1.1—*Cont.*

Depth	Bottom time, min	Time to first stop, min:sec	Decompression stops, ft 50	40	30	20	10	Total ascent, min:sec	Repetitive group
	15	2:00					1	3:10	F
	20	2:00					4	6:10	H
	25	2:00					10	12:10	J
	30	1:50				3	18	23:10	M
	40	1:50				10	25	37:10	N
	50	1:40			3	21	37	63:10	O
	60	1:40			9	23	52	86:10	Z
	70	1:40			16	24	61	103:10	Z
	80	1:30		3	19	35	72	131:10	Z
	90	1:30		8	19	45	80	154:10	Z
140	10	2:10					0	2:20	(*)
	15	2:10					2	4:20	G
	20	2:00					6	8:20	I
	25	2:00				2	14	18:20	J
	30	2:00				5	21	28:20	K
	40	1:50			2	16	26	46:20	N
	50	1:50			6	24	44	76:20	O
	60	1:50			16	23	56	97:20	Z
	70	1:40		4	19	32	68	125:20	Z
	80	1:40		10	23	41	79	155:20	Z
150	5	2:20					0	2:30	C
	10	2:20					1	3:30	E
	15	2:10					3	5:30	G
	20	2:10				2	7	11:30	H
	25	2:10				4	17	23:30	K
	30	2:00			5	8	24	34:30	L
	40	2:00			12	19	33	59:30	N
	50	2:00			19	23	51	88:30	O
	60	1:50		3	19	26	62	112:30	Z
	70	1:50		11	19	39	75	146:30	Z

Depth (feet)	Bottom time (min)	Time to first stop	| 50	40	30	20	10	Total ascent time	Repetitive group
160	80	1:40		9	19	33	84	173:30	Z
	5						0	2:40	D
	10	2:30					1	3:40	F
	15	2:20					4	7:40	H
	20	2:20				1	11	16:40	J
	25	2:20				3	20	29:40	K
	30	2:10				7	25	40:40	M
	40	2:10			2	11	39	71:40	N
	50	2:00			7	23	55	98:40	N
	60	2:00		2	16	23	69	132:40	N
170	5						0	2:50	D
	10	2:40					2	4:50	F
	15	2:30				2	5	9:50	H
	20	2:30				4	15	21:50	J
	25	2:20			2	7	23	34:50	L
	30	2:20			4	13	26	45:50	M
	40	2:10		1	10	23	45	81:50	O
	50	2:10		5	18	23	61	109:50	N
	60	2:00	2	15	22	37	74	152:50	N
180	5						0	3:00	D
	10	2:50					3	6:00	F
	15	2:40				3	6	12:00	I
	20	2:30			1	5	17	26:00	K
	25	2:30			3	10	24	40:00	L
	30	2:30			6	17	27	53:00	N
	40	2:20		3	14	23	50	93:00	O
	50	2:10	2	9	19	30	65	128:00	N
190	5						0	3:10	D
	10	2:50				1	3	7:10	G
	15	2:50				4	7	14:10	I
	20	2:40			2	6	20	31:10	K
	25	2:40			5	11	25	44:10	M
	30	2:30		1	8	19	32	63:10	N
	40	2:30		8	14	23	55	103:10	O

Schedule 1.2

No-Decompression Limits and Repetitive Group Designation Table for
No-Decompression Air Dives (U. S. Navy 1971)

Depth, ft	No-decompression limit, min	A	B	C	D	E	F	G	H	I	J	K	L	M	N	O
10		60	120	210	300											
15		35	70	110	160	225	350									
20		25	50	75	100	135	180	240	325							
25		20	35	55	75	100	125	160	195	245	315					
30		15	30	45	60	75	95	120	145	170	205	250	310			
35	310	5	15	25	40	50	60	80	100	120	140	160	190	220	270	310
40	200	5	15	25	30	40	50	70	80	100	110	130	150	170	200	
50	100		10	15	25	30	40	50	60	70	80	90	100			
60	60		10	15	20	25	30	40	50	55	60					
70	50		5	10	15	20	30	35	40	45	50					
80	40		5	10	15	20	25	30	35	40						
90	30		5	10	12	15	20	25	30							
100	25		5	7	10	15	20	22	25							
110	20			5	10	13	15	20								
120	15			5	10	12	15									
130	10			5	8	10										
140	10			5	7	10										
150	5			5												
160	5				5											
170	5				5											
180	5				5											
190	5				5											

Schedule 1.3

Surface Interval Credit Table for Air Decompression Dives[a] (U. S. Navy 1971)

Group	Z	O	N	M	L	K	J	I	H	G	F	E	D	C	B	A
Z	0:10–0:22	0:23–0:34	0:35–0:48	0:49–1:02	1:03–1:18	1:19–1:36	1:37–1:55	1:56–2:17	2:18–2:42	2:43–3:10	3:11–3:45	3:46–4:29	4:30–5:27	5:28–6:56	6:57–10:05	10:00–12:00[b]
O		0:10–0:23	0:24–0:36	0:37–0:51	0:52–1:07	1:08–1:24	1:25–1:43	1:44–2:04	2:05–2:29	2:30–2:59	3:00–3:33	3:34–4:17	4:18–5:16	5:17–6:44	6:45–9:54	9:55–12:00[b]
N			0:10–0:24	0:25–0:39	0:40–0:54	0:55–1:11	1:12–1:30	1:31–1:53	1:54–2:18	2:19–2:47	2:48–3:22	3:23–4:04	4:05–5:03	5:04–6:32	6:33–9:43	9:44–12:00[b]
M				0:10–0:25	0:26–0:42	0:43–0:59	1:00–1:18	1:19–1:39	1:40–2:05	2:06–2:34	2:35–3:08	3:09–3:52	3:53–4:49	4:50–6:18	6:19–9:28	9:29–12:00[b]
L					0:10–0:26	0:27–0:45	0:46–1:04	1:05–1:25	1:26–1:49	1:50–2:19	2:20–2:53	2:54–3:36	3:37–4:35	4:36–6:02	6:03–9:12	9:13–12:00[b]
K						0:10–0:28	0:29–0:49	0:50–1:11	1:12–1:35	1:36–2:03	2:04–2:38	2:39–3:21	3:22–4:19	4:20–5:48	5:49–8:58	8:59–12:00[b]
J							0:10–0:31	0:32–0:54	0:55–1:19	1:20–1:47	1:48–2:20	2:21–3:04	3:05–4:02	4:03–5:40	5:41–8:40	8:41–12:00[b]
I								0:10–0:33	0:34–0:59	1:00–1:29	1:30–2:02	2:03–2:44	2:45–3:43	3:44–5:12	5:13–8:21	8:22–12:00[b]
H									0:10–0:36	0:37–1:06	1:07–1:41	1:42–2:23	2:24–3:20	3:21–4:49	4:50–7:59	8:00–12:00[b]
G										0:10–0:40	0:41–1:15	1:16–1:59	2:00–2:58	2:59–4:25	4:26–7:35	7:36–12:00[b]
F											0:10–0:45	0:46–1:29	1:30–2:28	2:29–3:57	3:58–7:05	7:06–12:00[b]
E												0:10–0:54	0:55–1:57	1:58–3:22	3:23–6:32	6:33–12:00[b]
D													0:10–1:09	1:10–2:38	2:39–5:48	5:49–12:00[b]
C														0:10–1:39	1:40–2:49	2:50–12:00[b]
B															0:10–2:10	2:11–12:00[b]
A																0:10–12:00[b]

[a] The upper set of repetitive groups indicates the group at the end of the surface interval (He–O_2 dives). The diagonal set of repetitive groups indicates the group at the beginning of the surface interval from previous dive.

[b] Dives following surface intervals of more than 12 hr are not repetitive dives. Use actual bottom times in the helium–oxygen decompression tables to compute decompression for such dives.

Schedule 1.4

Repetitive Dive Timetable for Air Dives (U. S. Navy 1971)

Repetitive group	Repetitive dive depth (air dives), ft															
	40	50	60	70	80	90	100	110	120	130	140	150	160	170	180	190
A	7	6	5	4	4	3	3	3	3	3	2	2	2	2	2	2
B	17	13	11	9	8	7	7	6	6	6	5	5	4	4	4	4
C	25	21	17	15	13	11	10	10	9	8	7	7	6	6	6	6
D	37	29	24	20	18	16	14	13	12	11	10	9	9	8	8	8
E	49	38	30	26	23	20	18	16	15	13	12	12	11	10	10	10
F	61	47	36	31	28	24	22	20	18	16	15	14	13	13	12	11
G	73	56	44	37	32	29	26	24	21	19	18	17	16	15	14	13
H	87	66	52	43	38	33	30	27	25	22	20	19	18	17	16	15
I	101	76	61	50	43	38	34	31	28	25	23	22	20	19	18	17
J	116	87	70	57	48	43	38	34	32	28	26	24	23	22	20	19
K	138	99	79	64	54	47	43	38	35	31	29	27	26	24	22	21
L	161	111	88	72	61	53	48	42	39	35	32	30	28	26	25	24
M	187	124	97	80	68	58	52	47	43	38	35	32	31	29	27	26
N	213	142	107	87	73	64	57	51	46	40	38	35	33	31	29	28
O	241	160	117	96	80	70	62	55	50	44	40	38	36	34	31	30
Z	257	169	122	100	84	73	64	57	52	46	42	40	37	35	32	31

Schedule 1.5

U. S. Navy Standard Air Decompression Table for Exceptional Exposures (U. S. Navy 1971)

Depth, ft	Bottom time, min	Time to first stop, min:sec	Decompression stops, ft													Total ascent time, min:sec
			130	120	110	100	90	80	70	60	50	40	30	20	10	
40	360	0:30													23	23:40
	480	0:30													41	41:40
	720	0:30													69	69:40
60	240	0:40												2	79	82:00
	360	0:40												20	119	140:00
	480	0:40												44	148	193:00
	720	0:40												78	187	266:00
80	180	1:00												35	85	121:20
	240	0:50											6	52	120	179:20
	360	0:50											29	90	160	280:20
	480	0:50											59	107	187	354:20
	720	0:40										17	108	142	187	455:20
100	180	1:00										1	29	53	118	202:40
	240	1:00										14	42	84	142	283:40
	360	0:50									2	42	73	111	187	416:40
	480	0:50									21	61	91	142	187	503:40
	720	0:50									55	106	122	142	187	613:40
120	120	1:20										10	19	47	98	176:00
	180	1:10									5	27	37	76	137	284:00
	240	1:10									23	35	60	97	179	396:00
	360	1:00								18	45	64	93	142	187	551:00
	480	0:50							3	41	64	93	122	142	187	654:00
	720	0:50							32	74	100	114	122	142	187	773:00
140	90	1:30									2	14	18	42	88	166:20
	120	1:30									12	14	36	56	120	240:20
	180	1:20								10	26	32	54	94	168	386:20
	240	1:10							8	28	34	50	78	124	187	511:20

Schedule 1.5—*Cont.*

Depth, ft	Bottom time, min	Time to first stop, min:sec	130	120	110	100	90	80	70	60	50	40	30	20	10	Total ascent time, min:sec
160	360	1:00						9	32	42	64	84	122	142	187	684:20
	480	1:00						31	44	59	100	114	122	142	187	801:20
	720	0:50					16	56	88	97	100	114	122	142	187	924:20
170	70	1:50									1	17	22	44	80	166:40
	80	1:30									8	17	19	51	98	183:50
	90	1:20						2	10	12	12	14	34	52	120	246:50
	120	1:20					4	10	22	28	18	32	42	82	156	356:50
	180	1:10					18	24	30	42	34	50	78	120	187	535:50
	240	1:00				22	34	40	52	60	50	70	116	142	187	681:50
	360	1:00			14	22	42	56	91	97	98	114	122	142	187	873:50
	480	0:50			14	40	42	56	91	97	100	114	122	142	187	1007:50
180	60	2:10									5	16	19	44	81	168:00
190	50	2:20									4	13	22	33	72	147:10
	60	2:20									10	17	19	50	84	183:10
200	5	3:10													1	4:20
	10	3:00												1	4	8:20
	15	2:50											1	4	10	18:20
	20	2:50											3	7	27	40:20
	25	2:50										2	7	14	25	49:20
	30	2:40									2	8	9	22	37	73:20
	40	2:30								2	6	16	17	23	59	112:20
	50	2:30								12	13	30	24	39	75	161:20
	60	2:20							10	24	28	40	38	51	89	199:20
	90	1:50					1	10	10	42	48	70	64	74	134	324:20
	120	1:40				6	10	10	24	98	100	114	106	98	180	473:20
	180	1:20		1	10	10	18	24	24	42	100	114	122	142	187	685:20
	240	1:20	12	22	36	40	44	56	82	98	100	114	122	142	187	1058:20
210	5	3:20													1	4:30
	10	3:10												2	4	9:30

Depth	Time	(to 1st stop)							Total
220	15	3:00				1	5	13	22:30
	20	3:00				4	10	23	40:30
	25	2:50			2	7	17	27	56:30
	30	2:50			4	9	24	41	81:30
	40	2:40		4	9	19	26	63	124:30
	50	2:30	1	9	17	19	45	80	174:30
230	5	3:30						2	5:40
	10	3:20					2	5	10:40
	15	3:10				2	5	16	26:40
	20	3:00			1	3	11	24	42:40
	25	3:00			3	8	19	33	66:40
	30	2:50		1	7	10	23	47	91:40
	40	2:50		6	12	22	29	68	140:40
	50	2:40	3	12	17	18	51	86	190:40
240	5	3:40						2	5:50
	10	3:20				1	2	6	12:50
	15	3:20				3	6	18	30:50
	20	3:10			2	5	12	26	48:50
	25	3:10			4	8	22	37	74:50
	30	3:00		2	8	12	23	51	99:50
	40	2:50	1	7	15	22	34	74	156:50
	50	2:50	5	14	16	24	51	86	202:50
250	5	3:50						2	6:00
	10	3:30				1	3	6	14:00
	15	3:30				4	6	21	35:00
	20	3:20			3	6	15	25	53:00
	25	3:10		1	4	9	24	40	82:00
	30	3:10		4	8	15	22	56	109:00
	40	3:00	3	7	17	22	39	75	167:00
	50	2:50	8	15	16	29	51	94	218:00
	5	3:50					1	2	7:10
	10	3:40				1	4	7	16:10
	15	3:30			1	4	7	22	38:10
	20	3:30			4	7	17	27	59:10
	25	3:20	1	2	7	10	24	45	92:10

Schedule 1.5—Cont.

Depth, ft	Bottom time, min	Time to first stop, min:sec	130	120	110	100	90	80	70	60	50	40	30	20	10	Total ascent time, min:sec
	30	3:20									6	7	17	23	59	116:10
	40	3:10								5	9	17	19	45	79	178:10
	60	2:40					4	10	10	10	12	22	36	64	126	298:10
	90	2:10		8	10	10	10	10	10	28	28	44	68	98	186	514:10
260	5	4:00												1	2	7:20
	10	3:50											2	4	9	19:20
	15	3:40										2	4	10	22	42:20
	20	3:30									1	4	7	20	31	67:20
	25	3:30									3	8	11	23	50	99:20
	30	3:20								2	6	8	19	26	61	126:20
	40	3:10							1	6	11	16	19	49	84	190:20
270	5	4:10												1	3	8:30
	10	4:00											2	5	11	22:30
	15	3:50										3	4	11	24	46:30
	20	3:40									2	3	9	21	35	74:30
	25	3:30								2	3	8	13	23	53	106:30
	30	3:30								3	6	12	22	27	64	138:30
	40	3:20							5	6	11	17	22	51	88	204:30
280	5	4:20												2	2	8:40
	10	4:00										1	2	5	13	25:40
	15	3:50									1	3	4	11	26	49:40
	20	3:50									3	4	8	23	39	81:40
	25	3:40								2	5	7	16	23	56	113:40
	30	3:30							1	3	7	13	22	30	70	150:40
	40	3:20						1	6	6	13	17	27	51	93	218:40
290	5	4:30												2	3	9:50
	10	4:10										1	3	5	16	29:50
	15	4:00									1	3	6	12	26	52:50
	20	4:00									3	7	9	23	43	89:50

Decompression stops, ft

Depth, ft	Bottom time, min	Time to first stop, min:sec	120	110	100	90	80	70	60	50	40	30	20	10	Total ascent time, min:sec
	25	3:50							3	5	8	17	23	60	120:50
	30	3:40							5	6	16	22	36	72	162:50
	40	3:30							7	15	16	32	51	95	228:50
300	5	4:40											3	3	11:00
	10	4:20									1	3	6	17	32:00
	15	4:10								2	3	6	15	26	57:00
	20	4:00							2	3	7	10	23	47	97:00
	25	3:50						2	3	6	8	19	26	61	129:00
	30	3:50						2	5	7	17	22	39	75	172:00
	40	3:40					3	6	9	15	17	34	51	90	231:00
	60	3:00	4	4	6	8	10	14	14	28	32	50	90	187	460:00

Extreme exposures: 250 and 300 ft

Depth, ft	Bottom time, min	Time to first stop, min:sec	Decompression stops, ft																			Total ascent time, min:sec
			200	190	180	170	160	150	140	130	120	110	100	90	80	70	60	50	40	30	20	10	
250	120	1:50							5	10	10	10	10	16	24	24	36	48	64	94	142	187	684:10
	180	1:30					4	8	8	10	22	24	24	32	42	44	60	84	114	122	142	187	931:10
	240	1:30					9	14	21	22	22	40	40	42	56	76	98	100	114	122	142	187	1109:10
300	90	2:20					3	8	8	10	10	10	10	16	24	24	34	48	64	90	142	187	693:00
	120	2:00			4	8	8	8	8	10	14	24	24	24	34	42	58	66	102	122	142	187	890:00
	180	1:40	6	8	8	8	14	20	21	21	28	40	40	48	56	82	98	100	114	122	142	187	1168:00

Schedule 1.6
Surface Decompression Table Using Oxygen (U. S. Navy 1971)

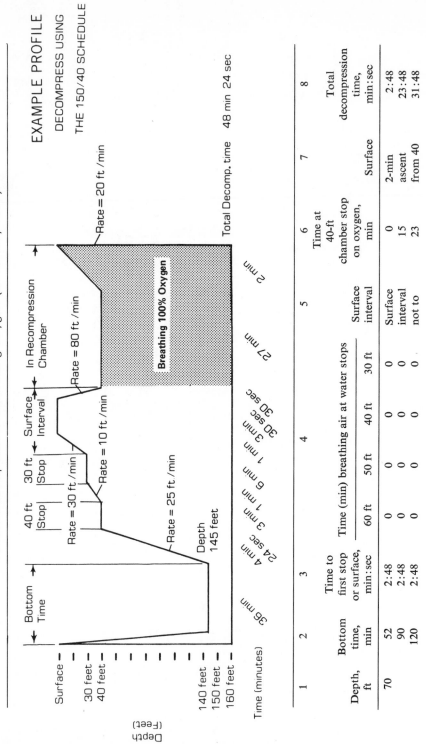

EXAMPLE PROFILE

DECOMPRESS USING

THE 150/40 SCHEDULE

			Time (min) breathing air at water stops							
	Bottom time, min	Time to first stop or surface, min:sec	60 ft	50 ft	40 ft	30 ft	Surface interval	Time at 40-ft chamber stop on oxygen, min	Surface	Total decompression time, min:sec
Depth, ft	2	3	4				5	6	7	8
70	52	2:48	0	0	0	0	Surface interval not to	0	2-min ascent from 40	2:48
	90	2:48	0	0	0	0		15		23:48
	120	2:48	0	0	0	0		23		31:48

Depth	Bottom time				Time			exceed 5 min	ft in chamber to surface while breathing oxygen	total
80	150	0	0	0	2:28	0	0	0	31	39:48
	180	0	0	0	2:48	0	0	0	39	47:48
	40	0	0	0	3:12	0	0	0	0	3:12
	70	0	0	0	3:12	0	0	0	14	23:12
	85	0	0	0	3:12	0	0	0	20	29:12
	100	0	0	0	3:12	0	0	0	26	35:12
	115	0	0	0	3:12	0	0	0	31	40:12
	130	0	0	0	3:12	0	0	0	37	46:12
	150	0	0	0	3:12	0	0	0	44	53:12
90	32	0	0	0	3:36	0	0	0	0	3:36
	60	0	0	0	3:36	0	0	0	14	23:36
	70	0	0	0	3:36	0	0	0	20	29:36
	80	0	0	0	3:36	0	0	0	25	34:36
	90	0	0	0	3:36	0	0	0	30	39:36
	100	0	0	0	3:36	0	0	0	34	43:36
	110	0	0	0	3:36	0	0	0	39	48:36
	120	0	0	0	3:36	0	0	0	43	52:36
	130	0	0	0	3:36	0	0	0	48	57:36
100	26	0	0	0	4:00	0	0	0	0	4:00
	50	0	0	0	4:00	0	0	0	14	24:00
	60	0	0	0	4:00	0	0	0	20	30:00
	70	0	0	0	4:00	0	0	0	26	36:00
	80	0	0	0	4:00	0	0	0	32	42:00
	90	0	0	0	4:00	0	0	0	38	48:00
	100	0	0	0	4:00	0	0	0	44	54:00
	110	0	0	0	4:00	0	0	0	49	59:00
	120	0	0	0	4:00	0	0	0	53	63:00
110	22	0	0	0	4:24	0	0	0	0	4:24
	40	0	0	0	4:24	0	0	0	12	22:24
	50	0	0	0	4:24	0	0	0	19	29:24
	60	0	0	0	4:24	0	0	0	26	36:24
	70	0	0	0	4:24	0	0	0	33	43:24
	80	0	0	0	3:12	0	0	1	40	51:12
	90	0	0	0	3:12	0	0	2	46	58:12

Schedule 1.6—*Cont.*

1	2	3	4				5	6	7	8
		Time to	Time (min) breathing air at water stops					Time at 40-ft		Total
Depth, ft	Bottom time, min	first stop or surface, min:sec	60 ft	50 ft	40 ft	30 ft	Surface interval	chamber stop on oxygen, min	Surface	decompression time, min:sec
	100	3:12	0	0	0	5		51		66:12
	110	3:12	0	0	0	12		54		76:12
120	18	4:48	0	0	0	0		0		4:48
	30	4:48	0	0	0	0		9		19:48
	40	4:48	0	0	0	0		16		26:48
	50	4:48	0	0	0	0		24		34:48
	60	3:36	0	0	0	2		32		44:36
	70	3:36	0	0	0	4		39		53:36
	80	3:36	0	0	0	5		46		61:36
	90	3:12	0	0	3	7		51		72:12
	100	3:12	0	0	6	15		54		86:12
130	15	5:12	0	0	0	0		0		5:12
	30	5:12	0	0	0	0		12		23:12
	40	5:12	0	0	0	0		21		32:12
	50	4:00	0	0	0	3		29		43:00
	60	4:00	0	0	0	5		37		53:00
	70	4:00	0	0	0	7		45		63:00
	80	3:36	0	0	6	7		51		75:36
	90	3:36	0	0	10	12		56		89:36
140	13	5:36	0	0	0	0		0		5:36
	25	5:36	0	0	0	0		11		22:36
	30	5:36	0	0	0	0		15		26:36
	35	5:36	0	0	0	0		20		31:36
	40	4:24	0	0	0	2		24		37:24
	45	4:24	0	0	0	4		29		44:24
	50	4:24	0	0	0	6		33		50:24

Depth	Bottom time	Time to first stop	50	40	30	20	10	Total ascent time
	55	4:24				7	38	56:24
	60	4:24				8	43	62:24
	65	4:00			3	7	48	70:00
	70	3:36		2	7	7	51	79:36
150	11	6:00					0	6:00
	25	6:00				0	13	25:00
	30	6:00				0	18	30:00
	35	4:48			0	4	23	38:48
	40	4:24		0	3	6	27	48:24
	45	4:24		0	5	7	33	57:24
	50	4:00	0	2	5	8	38	66:00
	55	3:36	2	5	9	4	44	77:36
160	9	6:24					0	6:24
	20	6:24				0	11	23:24
	25	6:24			0	0	16	28:24
	30	5:12		0	0	2	21	35:12
	35	4:48		0	4	6	26	48:48
	40	4:24		3	5	8	32	61:24
	45	4:00	3	4	8	6	38	73:00
170	7	6:48					0	6:48
	20	6:48				0	13	25:48
	25	6:48			0	0	19	31:48
	30	5:12		0	3	5	23	44:12
	35	4:48		4	4	7	29	57:48
	40	4:24	4	4	8	6	36	72:24

Schedule 1.7

Surface Decompression Table Using Air for Air Diving (U. S. Navy 1971)

All Ascent Rate = 60 ft/min

Total Ascent Time = 82 min /40 sec

Decompress Using 150/40 Schedule

Depth, ft	Bottom time, min	Time to first stop, min:sec	Time at water stops, min			Surface	Chamber stops (air), min		Total ascent time, min:sec
			30	20	10		20	10	
40	230	0:30			3	Time on		7	14:30
	250	0:30			3	surface		11	18:30
	270	0:30			3	not to		15	22:30
	300	0:30			3	exceed 3		19	26:30
50	120	0:40			3	min		5	12:40
	140	0:40			3	and 30 sec		10	17:40

Depth (ft)	Bottom time (min)	Time to first stop	30 ft	20 ft	10 ft	Total time of ascent
	160	0:40		3	21	28:40
	180	0:40		3	29	36:40
	200	0:40		3	35	42:40
	220	0:40		3	40	47:40
	240	0:40		3	47	54:40
60	80	0:50		3	7	14:50
	100	0:50		3	14	21:50
	120	0:50		3	26	33:50
	140	0:50		3	39	46:50
	160	0:50		3	48	55:50
	180	0:50		3	56	63:50
	200	0:40		3	69	80:10
70	60	1:00		3	8	16:00
	70	1:00		3	14	22:00
	80	1:00		3	18	26:00
	90	1:00		3	23	31:00
	100	1:00		3	33	41:00
	110	0:50		3	41	52:20
	120	0:50		4	47	59:20
	130	0:50		6	52	66:20
	140	0:50		8	56	72:20
	150	0:50		9	61	78:20
	160	0:50		13	72	93:20
	170	0:50		19	79	106:20
80	50	1:10		3	10	18:10
	60	1:10		3	17	25:10
	70	1:10		3	23	31:10
	80	1:00		3	31	42:30
	90	1:00		7	39	54:30
	100	1:00		11	46	65:30
	110	1:00		13	53	74:30
	120	1:00		17	56	81:30
	130	1:00		19	63	90:30
	140	1:00	26	26	69	126:30
	150	1:00	32	32	77	146:30

Schedule 1.7—Cont.

Depth, ft	Bottom time, min	Time to first stop, min:sec	50	40	30	20	10	Surface	Chamber 20	Chamber 10	Total ascent time, min:sec
90	40	1:20					3			7	15:20
	50	1:20					3			18	26:20
	60	1:20					3			25	33:20
	70	1:10				3			7	30	45:40
	80	1:10				13			13	40	71:40
	90	1:10				18			18	48	89:40
	100	1:10				21			21	54	101:40
	110	1:10				24			24	61	114:40
	120	1:10				32			32	68	137:40
	130	1:00			5	36			36	74	156:40

Depth, ft	Bottom time, min	Time to first stop, min:sec	50	40	30	20	10	Surface	Chamber 20	Chamber 10	Total ascent time, min:sec
100	40	1:30					3	Time on surface not to exceed 3 min and 30 sec	3	15	23:30
	50	1:20				3			9	24	35:50
	60	1:20				3			17	28	45:50
	70	1:20				3			23	39	64:50
	80	1:20				23			23	48	99:50
	90	1:10			3	23			23	57	111:50
	100	1:10			7	23			34	66	124:50
	110	1:10			10	34			41	72	155:50
	120	1:10			12	41				78	177:50
110	30	1:40					3		3	7	15:40
	40	1:30				3			3	21	33:00
	50	1:30				3			8	26	43:00
	60	1:30				18			18	36	78:00
	70	1:20			1	23			23	48	101:00

Depth (ft)	Bottom time (min)	Time to first stop	50 ft	40 ft	30 ft	20 ft	10 ft	Total ascent time
120	80	1:20			7	23	57	116:00
	90	1:20		3	12	30	64	142:00
	100	1:20		3	15	37	72	167:00
130	25	1:50					6	14:50
	30	1:50					14	22:50
	40	1:40			3	5	25	39:10
	50	1:40			9	15	31	67:10
	60	1:30			15	22	45	97:10
	70	1:30			16	23	55	116:10
	80	1:30			19	27	63	138:10
	90	1:30		3	19	37	74	173:10
	100	1:30		8	23	45	80	189:10
140	25	2:00					10	19:00
	30	1:50			3	3	18	30:20
	40	1:40			9	10	25	51:20
	50	1:40			16	21	37	88:20
	60	1:40			19	23	52	113:20
	70	1:30			19	24	61	131:20
	80	1:30			23	35	72	170:20
	90	1:30		3	23	45	80	203:20
150	20	2:10					6	15:10
	25	2:00			3	3	14	26:30
	30	2:00			5	5	21	37:30
	40	1:50			16	16	26	66:30
	50	1:50			24	24	44	104:30
	60	1:50		4	23	23	56	124:30
	70	1:40		10	32	32	68	161:30
	80	1:40			41	41	79	200:30
160	20	2:10				3	7	19:40
	25	2:10				4	17	31:40
	30	2:10			5	8	24	46:40
	40	2:00			12	19	33	82:40
	50	2:00		3	19	23	51	115:40
	60	1:50		11	19	26	62	142:40
	70	1:50	3	19	19	39	75	189:40

Schedule 1.7—*Cont.*

Depth, ft	Bottom time, min	Time to first stop, min:sec	Time at water stops, min					Surface	Chamber stops (air), min		Total ascent time, min:sec
			50	40	30	20	10		20	10	
160	80	1:40	1	17	19	50			50	84	227:40
	20	2:20				3			3	11	23:50
	25	2:20				7			7	20	40:50
	30	2:10			2	11			11	25	55:50
	40	2:10			7	23			23	39	98:50
	50	2:00		2	16	23			23	55	125:50
	60	2:00		9	19	33			33	69	169:50
	70	1:50	1	17	22	44			44	80	214:50
170	15	2:30				3			3	5	18:00
	20	2:30				4			4	15	30:00
	25	2:20			2	7			7	23	46:00
	30	2:20			4	13			13	26	63:00
	40	2:10		1	10	23			23	45	109:00
	50	2:10		5	18	23			23	61	137:00
	60	2:00	2	15	22	37			37	74	194:00
	70	2:00	8	17	19	51			51	86	239:00
180	15	2:40				3			3	6	19:10
	20	2:30			1	5			5	17	35:10
	25	2:30			3	10			10	24	54:10
	30	2:30			6	17			17	27	74:10
	40	2:20		3	14	23			23	50	120:10
	50	2:10	2	9	19	30			30	65	162:10
	60	2:10	5	16	19	44			44	81	216:10
190	15	2:50				4			4	7	22:20
	20	2:40			2	6			6	20	41:20
	25	2:40			5	11			11	25	59:20
	30	2:30		1	8	19			19	32	86:20
	40	2:30		8	14	23			23	55	130:20
	50	2:20	4	13	22	33			33	72	184:20
	60	2:20	10	17	19	50			50	84	237:20

2. Do not exceed the 3 min, 30 sec time limit on the surface.
3. No time saving result from use of this schedule in place of the Standard Air Decompression Schedule; comfort and security of the diver are the only advantages.

(ii) Example. Using the surface decompression technique with air, determine the dive profile for an air dive to 145 ft and a bottom time of 36 min.

2. *Decompression Schedules for Compressed Air Work*

a. Decompression Tables

Explanation. The decompression schedules are computed for working chamber pressures from 0 to 14 lb, and from 14 to 50 psig inclusive by 2-lb increments and for exposure times for each pressure extending from $\frac{1}{2}$ to over 8 hr inclusive. Decompressions will be conducted by two or more stages with a maximum of four stages, the latter for a working chamber pressure of 40 psig or greater.

Stage 1 consists of a reduction in ambient pressure ranging from 10 to a maximum of 16 lb/in.² but in no instance will the pressure be reduced below 4 lb at the end of stage 1. This reduction in pressure in stage 1 will always take place at a rate not greater than 5 lb/min.

Further reduction in pressure will take place during stage 2 and subsequent stages as required at a slower rate, but in no event at a rate greater than 1 lb/min.

Decompression Schedule 2.1 indicates in the body of the schedule the total decompression time in minutes for various combinations of working chamber pressure and exposure time.

Decompression Schedule 2.2 indicates for the same various combinations of working chamber pressure and exposure time the following: (a) the number of stages required; (b) the reduction in pressure and the terminal pressure for each required stage; (c) the time in minutes through which the reduction in pressure is accomplished for each required stage; and (d) the pressure reduction rate in minutes per pound for each required stage.

Example 1. Four-hour working period at 20 lb gauge.
Decompression Schedule 2.1: Twenty pounds for 4 hr. Total decompression time: 43 min.
Decompression Schedule 2.2: Stage 1. Reduce from 20 to 4 lb at the uniform rate of 5 lb/min. Elapsed time, stage 1: 3 min.
Stage 2 (final stage). Reduce pressure at a uniform rate from 4 to 0 lb gauge over a period of 40 min. Rate: 0.10 lb/min or 10.00 min/lb. Stage 2 (final) elapsed time: 40 min.
Total time: 43 min.

Example 2. Five-hour working period at 24 lb gauge.
Decompression Schedule 2.1: 24 lb for 5 hr. Total decompression time: 117 min.
Decompression Schedule 2.2: Stage 1. Reduce pressure from 24 to 8 lb at uniform rate of 5 lb/min. Elapsed time stage 1: 3 min.

Stage 2. Reduce pressure at a uniform rate from 8 to 4 lb over a period of 4 min. Rate: 1 lb/min. Elapsed time, stage 2: 4 min.

Transfer men to Special Decompression Chamber maintaining the 4-lb pressure during the transfer operation.

Stage 3 (final stage). In the Special Decompression Chamber, reduce the pressure at a uniform rate from 4 to 0 lb gauge over a period of 110 min. Rate: 0.037 lb/min or 27.5 min/lb. Stage 3 (final) elapsed time: 110 min.

Total time: 117 min.

b. Repetitive Diving Procedures

The information contained in the following pages is adapted from the U. S. Navy Diving Tables and is to be used when an individual will enter a compressed air environment more than once within a 12-hr period.

The Department of Industry, Labor and Human Relations may accept alternate methods of decompression for repetitive exposures provided the licensed physician submits his proposed procedures to the Department of Industry, Labor and Human Relations for its review and approval.

The Department of the Navy is in no way liable for the use or misuse of Schedules 2.3–2.5.

c. Suggestions for the Guidance of Compressed Air Workers

1. Eat moderately before going on shift.

2. Be temperate. Avoid excessive alcoholic beverages the night before or within 8 hr of going on shift.

3. Sleep at least 7 hr daily.

4. Take extra outer clothing into the tunnel when going on shift and wear it during decompression to avoid chilling during that period.

5. Do not sit or rest in a cramped position during decompression.

6. Do not exercise during decompression. This does not mean you cannot move around to avoid sitting in one position. Decompress according to schedule, for this means safety and freedom from compressed air illness or air pains. It also safeguards against damage to the bones.

7. Do not do hard exercise immediately after decompression.

8. Do not take a hot bath or shower within 6 hr of decompressing. Moderately warm bath or shower is permissible.

9. Do not go to sleep in a cramped position after decompressing.

10. Do not allow yourself to become chilled within 6 hr after decompression.

11. Report at once to the physician in charge if you suspect you are suffering from air pains or decompression sickness. Men suffering from compressed air illness should not be given any intoxicating liquors.

12. IF AFTER DECOMPRESSING YOU DEVELOP "NIGGLES" OR AIR PAINS THAT PERSIST LONGER THAN A HALF-HOUR, CALL THE MEDICAL LOCK *AT ONCE*.

13. If you become ill away from the job site, communicate at once with the physician in charge.

14. Wear your identification bracelet so it will be known what to do with you in an emergency.

15. Stay within a 30-mile radius of the recompression facility for at least 1 hr after locking out.

16. Do not reenter the man lock if suffering from air pains or decompression sickness.

17. Do not engage in scuba diving at depths greater than 33 ft within 12 hr of coming off shift. Do not engage in any scuba diving within 12 hr of going on shift.

18. Do not fly in any aircraft for at least 12 hr after coming off shift.

19. See that you are reexamined as required by the Department of Industry, Labor and Human Relations.

Schedule 2.1
Decompression Schedule for Compressed Air Work (Wisconsin 1971)

Important note: The pressure reduction in each stage is accomplished at a uniform rate. Do not interpolate between values shown on the schedules. Use the next higher value of working chamber pressure or exposure time should the actual working chamber pressure or the actual exposure time, respectively, fall between those for which calculated values are shown in the body of the schedules.

Work pressure, psig	Working period, hr									
	$\frac{1}{2}$	1	$1\frac{1}{2}$	2	3	4	5	6	7	8
	Total decompression time, min									
0–12	3	3	3	3	3	3	3	3	3	3
14	6	6	6	6	6	6	6	6	16	16
16	7	7	7	7	7	7	17	33	48	48
18	7	7	7	8	11	17	48	63	63	73
20	7	7	8	15	15	43	63	73	83	103
22	9	9	16	24	38	68	93	103	113	128
24	11	12	23	27	52	92	117	122	127	137
26	13	14	29	34	69	104	126	141	142	142
28	15	23	31	41	98	127	143	153	153	165
30	17	28	38	62	105	143	165	168	178	188
32	19	35	43	85	126	163	178	193	203	213
34	21	39	58	98	151	178	195	218	223	233
36	24	44	63	113	170	198	223	233	243	253
38	28	49	73	128	178	203	223	238	253	263
40	31	49	84	143	183	213	233	248	258	268
42	37	56	102	144	189	215	245	260	263	268
44	43	64	118	154	199	234	254	264	269	269
46	44	74	139	171	214	244	269	274	289	299
48	51	89	144	189	229	269	299	309	319	319
50	58	94	164	209	249	279	309	329	—	—

Schedule 2.2
Decompression Schedule for Compressed Air Work (Wisconsin 1971)
Do not interpolate. Use next higher value for conditions not computed.

Working chamber pressure, psig	Working period, hr	Stage No.	Pressure reduction, psig		Time in stage, min	Pressure reduction rate, min/lb	Total time decompress, min
			From	To			
14	½	1	14	4	2	0.20	
		2	4	0	4	1.00	6
	1	1	14	4	2	0.20	
		2	4	0	4	1.00	6
	1½	1	14	4	2	0.20	
		2	4	0	4	1.00	6
	2	1	14	4	2	0.20	
		2	4	0	4	1.00	6
	3	1	14	4	2	0.20	
		2	4	0	4	1.00	6
	4	1	14	0	2	0.20	
		2	4	0	4	1.00	6
	5	1	14	4	2	0.20	
		2	4	0	4	1.00	6
	6	1	14	4	2	0.20	
		2	4	0	4	1.00	6
	7	1	14	4	2	0.20	
		2	4	0	14	3.50	16
	8	1	14	4	2	0.20	
		2	4	0	14	3.50	16
	Over 8	1	14	4	2	0.20	
		2	4	0	30	7.50	32
16	½	1	16	4	3	0.20	
		2	4	0	4	1.00	7
	1	1	16	4	3	0.20	
		2	4	0	4	1.00	7
	1½	1	16	4	3	0.20	
		2	4	0	4	1.00	7
	2	1	16	4	3	0.20	
		2	4	0	4	1.00	7
	3	1	16	4	3	0.20	
		2	4	0	4	1.00	7
	4	1	16	4	3	0.20	
		2	4	0	4	1.00	7
	5	1	16	4	3	0.20	
		2	4	0	4	3.50	17
	6	1	16	4	3	0.20	
		2	4	0	30	7.50	33
	7	1	16	4	3	0.20	
		2	4	0	45	11.25	48
	8	1	16	4	3	0.20	
		2	4	0	45	11.25	48
	Over 8	1	16	4	3	0.20	
		2	4	0	60	15.00	63
18	½	1	18	4	3	0.20	
		2	4	0	4	1.00	7

Schedule 2.2—*Cont.*

Working chamber pressure, psig	Working period, hr	Stage No.	Pressure reduction, psig From	To	Time in stage, min	Pressure reduction rate, min/lb	Total time decompress, min
	1	1	18	4	3	0.20	
		2	4	0	4	1.00	7
	1½	1	18	4	3	0.20	
		2	4	0	4	1.00	7
	2	1	18	4	3	0.20	
		2	4	0	5	1.25	8
	3	1	18	4	3	0.20	
		2	4	0	8	2.00	11
	4	1	18	4	3	0.20	
		2	4	0	14	3.50	17
	5	1	18	4	3	0.20	
		2	4	0	45	11.25	48
	6	1	18	4	3	0.20	
		2	4	0	60	15.00	63
	7	1	18	4	3	0.20	
		2	4	0	60	15.00	63
	8	1	18	4	3	0.20	
		2	4	0	70	17.50	73
	Over 8	1	18	4	3	0.20	
		2	4	0	84	21.00	87
20	½	1	20	4	3	0.20	
		2	4	0	4	1.00	7
	1	1	20	4	3	0.20	
		2	4	0	4	1.00	7
	1½	1	20	4	3	0.20	
		2	4	0	5	1.25	8
	2	1	20	4	3	0.20	
		2	4	0	12	3.00	15
	3	1	20	4	3	0.20	
		2	4	0	12	3.00	15
	4	1	20	4	3	0.20	
		2	4	0	40	10.00	43
	5	1	20	4	—	0.20	
		2	4	0	60	15.00	63
	6	1	20	4	3	0.20	
		2	4	0	70	17.50	73
	7	1	20	4	3	0.20	
		2	4	0	80	20.00	83
	8	1	20	4	3	0.20	
		2	4	0	100	25.00	103
	Over 8	1	20	4	3	0.20	
		2	4	0	110	27.50	113
22	½	1	22	6	3	0.20	
		2	6	0	6	1.00	9
	1	1	22	6	3	0.20	
		2	6	0	6	1.00	9
	1½	1	22	6	3	0.20	
		2	6	0	13	2.20	16

Schedule 2.2—*Cont.*

Working chamber pressure, psig	Working period, hr	Stage No.	Pressure reduction, psig		Time in stage, min	Pressure reduction rate, min/lb	Total time decompress, min
			From	To			
	2	1	22	6	3	0.20	
		2	6	0	21	3.50	24
	3	1	22	6	3	0.20	
		2	6	0	35	5.85	38
	4	1	22	6	3	0.20	
		2	6	0	65	10.83	68
	5	1	22	6	3	0.20	
		2	6	0	90	15.00	93
	6	1	22	6	3	0.20	
		2	6	0	100	16.67	103
	7	1	22	6	3	0.20	
		2	6	0	110	18.35	113
	8	1	22	6	3	0.20	
		2	6	0	125	20.80	128
	Over 8	1	22	6	3	0.20	
		2	6	0	130	21.70	133
24	½	1	24	8	3	0.20	
		2	8	4	4	1.00	
		3	4	0	4	1.00	11
	1	1	24	8	3	0.20	
		2	8	4	4	1.00	
		3	4	0	5	1.25	12
	1½	1	24	8	3	0.20	
		2	8	4	4	1.00	
		3	4	0	16	4.00	23
	2	1	24	8	3	0.20	
		2	8	4	4	1.00	
		3	4	0	20	5.00	27
	3	1	24	8	3	0.20	
		2	8	4	4	1.00	
		3	4	0	45	11.25	52
	4	1	24	8	3	0.20	
		2	8	4	4	1.00	
		3	4	0	85	21.25	92
	5	1	24	8	3	0.20	
		2	8	4	4	1.00	
		3	4	0	110	27.50	117
	6	1	24	8	3	0.20	
		2	8	4	4	1.00	
		3	4	0	115	28.80	122
	7	1	24	8	3	0.20	
		2	8	4	4	1.00	
		3	4	0	120	30.00	127
	8	1	24	8	3	0.20	
		2	8	4	4	1.00	
		3	4	0	130	32.50	137
	Over 8	1	24	8	3	0.20	
		2	8	4	8	2.00	

Schedule 2.2—*Cont.*

Working chamber pressure, psig	Working period, hr	Stage No.	Pressure reduction, psig		Time in stage, min	Pressure reduction rate, min/lb	Total time decompress, min
			From	To			
		3	4	0	140	35.00	151
26	½	1	26	10	3	0.20	
		2	10	4	6	1.00	
		3	4	0	4	1.00	13
	1	1	26	10	3	0.20	
		2	10	4	6	1.00	
		3	4	0	5	1.25	14
	1½	1	26	10	3	0.20	
		2	10	4	6	1.00	
		3	4	0	20	5.00	29
	2	1	26	10	3	0.20	
		2	10	4	6	1.00	
		3	4	0	25	6.25	34
	3	1	26	10	3	0.20	
		2	10	4	6	1.00	
		3	4	0	60	15.00	69
	4	1	26	10	3	0.20	
		2	10	4	6	1.00	
		3	4	0	95	23.75	104
	5	1	26	10	3	0.20	
		2	10	4	8	1.33	
		3	4	0	115	28.80	126
	6	1	26	10	3	0.20	
		2	10	4	8	1.33	
		3	4	0	130	32.50	141
	7	1	26	10	3	0.20	
		2	10	4	9	1.50	
		3	4	0	130	32.50	142
	8	1	26	10	3	0.20	
		2	10	4	9	1.50	
		3	4	0	130	32.50	142
	Over 8	1	26	10	3	0.20	
		2	10	4	30	5.00	
		3	4	0	130	32.50	163
28	½	1	28	12	3	0.20	
		2	12	4	8	1.00	
		3	4	0	4	1.00	15
	1	1	28	12	3	0.20	
		2	12	4	8	1.00	
		3	4	0	12	3.00	23
	1½	1	28	12	3	0.20	
		2	12	4	8	1.00	
		3	4	0	20	5.00	31
	2	1	28	12	3	0.20	
		2	12	4	8	1.00	
		3	4	0	30	7.50	41
	3	1	28	12	3	0.20	
		2	12	4	10	1.25	

Schedule 2.2—*Cont.*

Working chamber pressure, psig	Working period, hr		Decompression data				
		Stage No.	Pressure reduction, psig		Time in stage, min	Pressure reduction rate, min/lb	Total time decompress, min
			From	To			
		3	4	0	85	21.20	98
	4	1	28	12	3	0.20	
		2	12	4	14	1.75	
		3	4	0	110	27.50	127
	5	1	28	12	3	0.20	
		2	12	4	20	2.50	
		3	4	0	120	30.00	143
	6	1	28	12	3	0.20	
		2	12	4	20	2.50	
		3	4	0	130	32.50	153
	7	1	28	12	3	0.20	
		2	12	4	20	2.50	
		3	4	0	130	32.50	153
	8	1	28	12	3	0.20	
		2	12	4	32	4.00	
		3	4	0	130	32.50	165
	Over 8	1	28	12	3	0.20	
		2	12	4	50	6.25	
		3	4	0	130	32.50	183
30	$\frac{1}{2}$	1	30	14	3	0.20	
		2	14	4	10	1.00	
		3	4	0	4	1.00	17
	1	1	30	14	3	0.20	
		2	14	4	10	1.00	
		3	4	0	15	3.75	28
	$1\frac{1}{2}$	1	30	14	3	0.20	
		2	14	4	10	1.00	
		3	4	0	25	6.25	38
	2	1	30	14	3	0.20	
		2	14	4	14	1.40	
		3	4	0	45	11.25	62
	3	1	30	14	3	0.20	
		2	14	4	17	1.70	
		3	4	0	85	21.20	105
	4	1	30	14	3	0.20	
		2	14	4	30	3.00	
		3	4	0	110	27.50	143
	5	1	30	14	3	0.20	
		2	14	4	35	3.50	
		3	4	0	130	32.50	165
	6	1	30	14	3	0.20	
		2	14	4	35	3.50	
		3	4	0	130	32.50	168
	7	1	30	14	3	0.20	
		2	14	4	45	4.50	
		3	4	0	130	32.50	178
	8	1	30	14	3	0.20	
		2	14	4	55	5.50	

Schedule 2.2—*Cont.*

Working chamber pressure, psig	Working period, hr	Decompression data						
		Stage No.	Pressure reduction, psig		Time in stage, min	Pressure reduction rate, min/lb	Total time decompress, min
			From	To			
		3	4	0	130	32.50	188
	Over 8	1	30	14	3	0.20	
		2	14	4	71	7.10	
		3	4	0	130	32.50	204
32	½	1	32	16	3	0.20	
		2	16	4	12	1.00	
		3	4	0	4	1.00	19
	1	1	32	16	3	0.20	
		2	16	4	12	1.00	
		3	4	0	20	5.00	35
	1½	1	32	16	3	0.20	
		2	16	4	15	1.25	
		3	4	0	25	6.25	43
	2	1	32	16	3	0.20	
		2	16	4	22	1.83	
		3	4	0	60	15.00	85
	3	1	32	16	3	0.20	
		2	16	4	28	2.33	
		3	4	0	95	23.75	126
	4	1	32	16	3	0.20	
		2	16	4	40	3.33	
		3	4	0	120	30.00	163
	5	1	32	16	3	0.20	
		2	16	4	45	3.75	
		3	4	0	130	32.50	178
	6	1	32	16	3	0.20	
		2	16	4	60	5.00	
		3	4	0	130	32.50	193
	7	1	32	16	3	0.20	
		2	16	4	70	5.83	
		3	4	0	130	32.50	203
	8	1	32	16	3	0.20	
		2	16	4	80	6.67	
		3	4	0	130	32.50	213
	Over 8	1	32	16	3	0.20	
		2	16	4	93	7.75	
		3	4	0	130	32.50	226
34	½	1	34	18	3	0.20	
		2	18	4	14	1.00	
		3	4	0	4	1.00	21
	1	1	34	18	3	0.20	
		2	18	4	14	1.00	
		3	4	0	22	5.50	39
	1½	1	34	18	3	0.20	
		2	18	4	25	1.80	
		3	4	0	30	7.50	58
	2	1	34	18	3	0.20	
		2	18	4	35	2.50	

Schedule 2.2—*Cont.*

Working chamber pressure, psig	Working period, hr	Stage No.	Pressure reduction, psig		Time in stage, min	Pressure reduction rate, min/lb	Total time decompress, min
			From	To			
		3	4	0	60	15.00	98
	3	1	34	18	3	0.20	
		2	18	4	43	3.10	
		3	4	0	105	26.25	151
	4	1	34	18	3	0.20	
		2	18	4	55	3.93	
		3	4	0	120	30.00	178
	5	1	34	18	3	0.20	
		2	18	4	62	4.43	
		3	4	0	130	32.50	195
	6	1	34	18	3	0.20	
		2	18	4	85	6.07	
		3	4	0	130	32.50	218
	7	1	34	18	3	0.20	
		2	18	4	90	6.43	
		3	4	0	130	32.50	223
	8	1	34	18	3	0.20	
		2	18	4	100	7.15	
		3	4	0	130	32.50	233
	Over 8	1	34	18	3	0.20	
		2	18	4	115	8.23	
		3	4	0	130	32.50	248
36	½	1	36	20	3	0.20	
		2	20	4	16	1.00	
		3	4	0	5	1.25	24
	1	1	36	20	3	0.20	
		2	20	4	16	1.00	
		3	4	0	25	6.25	44
	1½	1	36	20	3	0.20	
		2	20	4	30	1.88	
		3	4	0	30	7.50	63
	2	1	36	20	3	0.20	
		2	20	4	40	2.50	
		3	4	0	70	17.50	113
	3	1	36	20	3	0.20	
		2	20	4	52	3.25	
		3	4	0	115	28.75	170
	4	1	36	20	3	0.20	
		2	20	4	65	4.06	
		3	4	0	130	32.50	198
	5	1	36	20	3	0.20	
		2	20	4	90	5.63	
		3	4	0	130	32.50	223
	6	1	36	20	3	0.20	
		2	20	4	100	6.25	
		3	4	0	130	32.50	233
	7	1	36	20	3	0.20	
		2	20	4	110	6.88	

Schedule 2.2—*Cont.*

Working chamber pressure, psig	Working period, hr	Stage No.	Pressure reduction, psig From	To	Time in stage, min	Pressure reduction rate, min/lb	Total time decompress, min
		3	4	0	130	32.50	243
	8	1	36	20	3	0.20	
		2	20	4	120	7.50	
		3	4	0	130	32.50	253
	Over 8	1	36	20	3	0.20	
		2	20	4	140	8.75	
		3	4	0	130	32.50	273
38	½	1	38	22	3	0.20	
		2	22	6	16	1.00	
		3	6	0	9	1.50	28
	1	1	38	22	3	0.20	
		2	22	6	16	1.00	
		3	6	0	30	5.00	40
	1½	1	38	22	3	0.20	
		2	22	6	20	1.25	
		3	6	0	50	8.34	73
	2	1	38	22	3	0.20	
		2	22	6	30	1.88	
		3	6	0	95	15.83	128
	3	1	38	22	3	0.20	
		2	22	6	35	2.19	
		3	6	0	140	23.35	178
	4	1	38	22	3	0.20	
		2	22	6	50	3.12	
		3	6	0	150	25.00	203
	5	1	38	22	3	0.20	
		2	22	6	55	3.44	
		3	6	0	165	27.50	223
	6	1	38	22	3	0.20	
		2	22	6	70	4.38	
		3	6	0	165	27.50	238
	7	1	38	22	3	0.20	
		2	22	6	85	5.32	
		3	6	0	165	27.50	253
	8	1	38	22	3	0.20	
		2	22	6	95	5.93	
		3	6	0	165	27.50	263
	Over 8	1	38	22	3	0.20	
		2	22	6	110	6.88	
		3	6	0	165	27.50	278
40	½	1	40	24	3	0.20	
		2	24	8	16	1.00	
		3	8	4	4	1.00	
		4	4	0	8	2.00	31
	1	1	40	24	3	0.20	
		2	24	8	16	1.00	
		3	8	4	5	1.25	
		4	4	0	25	6.25	49

Schedule 2.2—*Cont.*

Working chamber pressure, psig	Working period, hr	Decompression data					
		Stage No.	Pressure reduction, psig		Time in stage, min	Pressure reduction rate, min/lb	Total time decompress, min
			From	To			
	$1\frac{1}{2}$	1	40	24	3	0.20	
		2	24	8	16	1.00	
		3	8	4	20	5.00	
		4	4	0	45	11.25	84
	2	1	40	24	3	0.20	
		2	24	8	25	1.56	
		3	8	4	20	5.00	
		4	4	0	95	23.75	143
	3	1	40	24	3	0.20	
		2	24	8	30	1.88	
		3	8	4	30	7.50	
		4	4	0	120	30.00	183
	4	1	40	24	3	0.20	
		2	24	8	45	2.81	
		3	8	4	35	8.75	
		4	4	0	130	32.50	213
	5	1	40	24	3	0.20	
		2	24	8	47	2.94	
		3	8	4	53	13.25	
		4	4	0	130	32.50	233
	6	1	40	24	3	0.20	
		2	24	8	55	3.44	
		3	8	4	60	15.00	
		4	4	0.	130	32.50	248
	7	1	40	24	3	0.20	
		2	24	8	65	4.06	
		3	8	4	60	15.00	
		4	4	0	130	32.50	258
	8	1	40	24	3	0.20	
		2	24	8	75	4.70	
		3	8	4	60	15.00	
		4	4	0	130	32.50	268
	Over 8	1	40	24	3	0.20	
		2	24	8	95	5.93	
		3	8	4	60	15.00	
		4	4	0	130	32.50	288
42	$\frac{1}{2}$	1	42	26	3	0.20	
		2	26	10	16	1.00	
		3	10	4	6	1.00	
		4	4	0	12	3.00	37
	1	1	42	26	3	0.20	
		2	26	10	16	1.00	
		3	10	4	12	2.00	
		4	4	0	25	6.25	56
	$1\frac{1}{2}$	1	42	26	3	0.20	
		2	26	10	16	1.00	
		3	10	4	23	3.83	
		4	4	0	60	15.00	102

Schedule 2.2—*Cont.*

Working chamber pressure, psig	Working period, hr	Stage No.	Pressure reduction, psig		Time in stage, min	Pressure reduction rate, min/lb	Total time decompress, min
			From	To			
	2	1	42	26	3	0.20	
		2	26	10	16	1.00	
		3	10	4	30	5.00	
		4	4	0	95	23.75	144
	3	1	42	26	3	0.20	
		2	26	10	16	1.00	
		3	10	4	50	8.34	
		4	4	0	120	30.00	189
	4	1	42	26	3	0.20	
		2	26	10	17	1.06	
		3	10	4	65	10.83	
		4	4	0	130	32.50	215
	5	1	42	26	3	0.20	
		2	26	10	27	1.69	
		3	10	4	85	14.18	
		4	4	0	130	32.50	245
	6	1	42	26	3	0.20	
		2	26	10	27	1.69	
		3	10	4	100	16.67	
		4	4	0	130	32.50	260
	7	1	42	26	3	0.20	
		2	26	10	30	1.88	
		3	10	4	100	16.67	
		4	4	0	130	32.50	263
	8	1	42	26	3	0.20	
		2	26	10	35	2.19	
		3	10	4	100	16.67	
		4	4	0	130	32.50	268
	Over 8	1	42	26	3	0.20	
		2	26	10	60	3.75	
		3	10	4	100	16.67	
		4	4	0	130	32.50	293
44	½	1	44	28	3	0.20	
		2	28	12	16	1.00	
		3	12	4	8	1.00	
		4	4	0	16	4.00	43
	1	1	44	28	3	0.20	
		2	28	12	16	1.00	
		3	12	4	20	2.50	
		4	4	0	25	6.25	64
	1½	1	44	28	3	0.20	
		2	28	12	16	1.00	
		3	12	4	27	3.38	
		4	4	0	72	18.00	118
	2	1	44	28	3	0.20	
		2	28	12	16	1.00	
		3	12	4	40	5.00	
		4	4	0	95	23.75	154

Schedule 2.2—*Cont.*

Working chamber pressure, psig	Working period, hr	Stage No.	Pressure reduction, psig		Time in stage, min	Pressure reduction rate, min/lb	Total time decompress, min
			From	To			
	3	1	44	28	3	0.20	
		2	28	12	16	1.00	
		3	12	4	60	7.50	
		4	4	0	120	30.00	199
	4	1	44	28	3	0.20	
		2	28	12	16	1.00	
		3	12	4	85	10.62	
		4	4	0	130	32.50	234
	5	1	44	28	3	0.20	
		2	28	12	16	1.00	
		3	12	4	105	13.13	
		4	4	0	130	32.50	254
	6	1	44	28	3	0.20	
		2	28	12	16	1.00	
		3	12	4	115	14.38	
		4	4	0	130	32.50	264
	7	1	44	28	3	0.20	
		2	28	12	16	1.00	
		3	12	4	120	15.00	
		4	4	0	130	32.50	269
	8	1	44	28	3	0.20	
		2	28	12	16	1.00	
		3	12	4	120	15.00	
		4	4	0	130	32.50	269
	Over 8	1	44	28	3	0.20	
		3	28	12	40	2.50	
		3	12	4	120	15.00	
		4	4	0	130	32.50	293
46	½	1	46	30	3	0.20	
		2	30	14	16	1.00	
		3	14	4	10	1.00	
		4	4	0	15	3.75	44
	1	1	46	30	3	0.20	
		2	30	14	16	1.00	
		3	14	4	25	2.50	
		4	4	0	30	7.50	74
	1½	1	46	30	3	0.20	
		2	30	14	16	1.00	
		3	14	4	35	3.50	
		4	4	0	85	21.20	139
	2	1	46	30	3	0.20	
		2	30	14	16	1.00	
		3	14	4	47	4.70	
		4	4	0	105	26.25	171
	3	1	46	30	3	0.20	
		2	30	14	16	1.00	
		3	14	4	65	6.50	
		4	4	0	130	32.50	214

Schedule 2.2—*Cont.*

Working chamber pressure, psig	Working period, hr	Decompression data					
		Stage No.	Pressure reduction, psig		Time in stage, min	Pressure reduction rate, min/lb	Total time decompress, min
			From	To			
	4	1	46	30	3	0.20	
		2	30	14	16	1.00	
		3	14	4	95	9.50	
		4	4	0	130	32.50	244
	5	1	46	30	3	0.20	
		2	30	14	16	1.00	
		3	14	4	120	12.00	
		4	4	0	130	32.50	269
	6	1	46	30	3	0.20	
		2	30	14	16	1.00	
		3	14	4	125	12.50	
		4	4	0	130	32.50	274
	7	1	46	30	3	0.20	
		2	30	14	16	1.00	
		3	14	4	140	14.00	
		4	4	0	130	32.50	289
	8	1	46	30	3	0.20	
		2	30	14	16	1.00	
		3	14	4	150	15.00	
		4	4	0	130	32.50	299
	Over 8	1	46	30	3	0.20	
		2	30	14	25	1.56	
		3	14	4	160	16.00	
		4	4	0	130	32.50	318
48	½	1	48	32	3	0.20	
		2	32	16	16	1.00	
		3	16	4	12	1.00	
		4	4	0	20	5.00	51
	1	1	48	32	3	0.20	
		2	32	16	16	1.00	
		3	16	4	35	2.92	
		4	4	0	35	8.75	89
	1½	1	48	32	3	0.20	
		2	32	16	16	1.00	
		3	16	4	45	3.75	
		4	4	0	80	20.00	144
	2	1	48	32	3	0.20	
		2	32	16	16	1.00	
		3	16	4	60	5.00	
		4	4	0	110	27.50	189
	3	1	48	32	3	0.20	
		2	32	16	16	1.00	
		3	16	4	90	7.50	
		4	4	0	120	30.00	229
	4	1	48	32	3	0.20	
		2	32	16	16	1.00	
		3	16	4	120	10.00	
		4	4	0	130	32.50	269

Schedule 2.2—*Cont.*

Working chamber pressure, psig	Working period, hr	Stage No.	Pressure reduction, psig		Time in stage, min/lb	Pressure reduction rate, min/lb	Total time decompress, min
			From	To			
	5	1	48	32	3	0.20	
		2	32	16	16	1.00	
		3	16	4	140	11.67	
		4	4	0	130	32.50	299
	6	1	48	32	3	0.20	
		2	32	16	16	1.00	
		3	16	4	160	13.33	
		4	4	0	130	32.50	309
	7	1	48	32	3	0.20	
		2	32	16	16	1.00	
		3	16	4	170	14.17	
		4	4	0	130	32.50	319
	8	1	48	32	3	0.20	
		2	32	16	16	1.00	
		3	16	4	170	14.17	
		4	4	0	130	32.50	319
50	$\frac{1}{2}$	1	50	34	3	0.20	
		2	34	18	16	1.00	
		3	18	4	14	1.00	
		4	4	0	25	6.25	58
	1	1	50	34	3	0.20	
		2	34	18	16	1.00	
		3	18	4	40	2.86	
		4	4	0	35	8.75	94
	$1\frac{1}{2}$	1	50	34	3	0.20	
		2	34	18	16	1.00	
		3	18	4	55	3.93	
		4	4	0	90	22.50	164
	2	1	50	34	3	0.20	
		2	34	18	16	1.00	
		3	18	4	70	5.00	
		4	4	0	120	30.00	209
	3	1	50	34	3	0.20	
		2	34	18	16	1.00	
		3	18	4	100	7.15	
		4	4	0	130	32.50	249
	4	1	50	34	3	0.20	
		2	34	18	16	1.00	
		3	18	4	130	8.58	
		4	4	0	130	32.50	279
	5	1	50	34	3	0.20	
		2	34	18	16	1.00	
		3	18	4	160	11.42	
		4	4	0	130	32.50	309
	6	1	50	34	3	0.20	
		2	34	18	16	1.00	
		3	18	4	180	12.85	
		4	4	0	130	32.50	329

Schedule 2.3

Repetitive Diving Procedures[a] (Wisconsin 1971)

Pressure, psig	Repetitive group															
	A	B	C	D	E	F	G	H	I	J	K	L	M	N	O	Z
4	60	120	210	300												
7	35	70	110	160	225	350										
9	25	50	75	100	135	180	240	325								
11	20	35	55	75	100	125	160	195	245	315						
13	15	30	45	60	75	95	120	145	170	205	250	310				
16	5	15	25	40	50	60	80	100	120	140	160	190	220	270	310	
18	5	15	25	30	40	50	70	80	100	110	130	150	170	230	270	300
22	—	10	15	25	30	40	50	60	70	80	90	110	140	160	200	240
27	—	10	15	20	25	30	40	50	55	60	70	80	100	120	140	200
31	—	5	10	15	20	25	35	45	45	50	60	70	80	100	130	170
36	—	5	10	15	20	25	30	35	40	—	50	60	70	90	110	150
40	—	5	10	12	15	20	25	30	—	40	—	50	60	80	90	130
45	—	5	7	10	15	20	22	25	30	—	40	50	—	60	80	120
49	—	—	5	10	13	15	20	25	—	30	—	40	50	70	70	100

[a] Instructions for use: The tabulated compressed air exposure times are in minutes. The times at the various pressures in each vertical column are the maximum exposures during which a compressed air worker will remain within the group listed at the head of the column. To find the repetitive group designation enter the table on the exact or next greater working pressure than that to which exposed and select the listed exposure time exact or next greater than the actual exposure time. The repetitive group designation is indicated by the letter at the head of the vertical column where the selected exposure time is listed. For example: An exposure in compressed air was for 45 min at 26 psig. To determine the repetitive group enter the table at 27 psig (the next higher pressure, as 26 psig, is not listed) and move horizontally until 50 min (the next greater tabulated exposure time, as 45 min, is not listed), then move vertically to the top of the column where H is shown as the repetitive group.

Schedule 2.4

Repetitive Diving Procedures: Open Air Interval Credit Table[a,b] (Wisconsin 1971)

Repetitive group at end of open air interval

Beginning group	Z	O	N	M	L	K	J	I	H	G	F	E	D	C	B	A
Z	0:10–0:22	0:23–0:34	0:35–0:48	0:49–1:02	1:03–1:18	1:19–1:36	1:37–1:55	1:56–2:17	2:18–2:42	2:43–3:10	3:11–3:45	3:46–4:29	4:30–5:27	5:28–6:56	6:57–10:05	10:00–12:00*
O		0:10–0:23	0:24–0:36	0:37–0:51	0:52–1:07	1:08–1:24	1:25–1:43	1:44–2:04	2:05–2:29	2:30–2:59	3:00–3:33	3:34–4:17	4:18–5:16	5:17–6:44	6:45–9:54	9:55–12:00*
N			0:10–0:24	0:25–0:39	0:40–0:54	0:55–1:11	1:12–1:30	1:31–1:53	1:54–2:18	2:19–2:47	2:48–3:22	3:23–4:04	4:05–5:03	5:04–6:32	6:33–9:43	9:44–12:00*
M				0:10–0:25	0:26–0:42	0:43–0:59	1:00–1:18	1:19–1:39	1:40–2:05	2:06–2:34	2:35–3:08	3:09–3:52	3:53–4:49	4:50–6:18	6:19–9:28	9:29–12:00*
L					0:10–0:26	0:27–0:45	0:46–1:04	1:05–1:25	1:26–1:49	1:50–2:19	2:20–2:53	2:54–3:36	3:37–4:35	4:36–6:02	6:03–9:12	9:13–12:00*
K						0:10–0:28	0:29–0:49	0:50–1:11	1:12–1:35	1:36–2:03	2:04–2:38	2:39–3:21	3:22–4:19	4:20–5:48	5:49–8:58	8:59–12:00*
J							0:10–0:31	0:32–0:54	0:55–1:19	1:20–1:47	1:48–2:20	2:21–3:04	3:05–4:02	4:03–5:40	5:41–8:40	8:41–12:00*
I								0:10–0:33	0:34–0:59	1:00–1:29	1:30–2:02	2:03–2:44	2:45–3:43	3:44–5:12	5:13–8:21	8:22–12:00*
H									0:10–0:36	0:37–1:06	1:07–1:41	1:42–2:23	2:24–3:20	3:21–4:49	4:50–7:59	8:00–12:00*
G										0:10–0:40	0:41–1:15	1:16–1:59	2:00–2:58	2:59–4:25	4:26–7:35	7:36–12:00*
F											0:10–0:45	0:46–1:29	1:30–2:28	2:29–3:57	3:58–7:05	7:06–12:00*
E												0:10–0:54	0:55–1:57	1:58–3:22	3:23–6:32	6:33–12:00*
D													0:10–1:09	1:10–2:38	2:39–5:48	5:49–12:00*
C														0:10–1:39	1:40–2:49	2:50–12:00*
B															0:10–2:10	2:11–12:00*
A																0:10–12:00*

[a] Instructions for use: Open air interval time in the table is in hours and minutes (7:59 means 7 hours and 59 minutes). The open air interval must be at least 10 minutes. An open air interval of more than 12 hours does not require additional decompression or the use of this table.

Find the repetitive group designation letter (from the previous shift) on the diagonal slope. Enter the table horizontally to select the open air interval time that in exactly between the actual open air times shown. The repetitive group designation for the end of the open air interval is the head of the vertical column where the selected open air interval time is listed.

For example: A previous shift was at 35 psi for 1 hour. The worker remains in open air for 1 hour and 30 minutes and wishes to find the new repetitive group designation. The repetitive group from the Repetitive Group Designation Table is L. Enter the open air interval credit table along the diagonal line labeled L. The 1 hour and 30 minute open air interval lies between the times 1:26 and 1:49. Therefore the worker is placed in group H (at the head of the vertical column selected).

[b] The asterisks indicate that compressed air exposures following open air intervals of more than 12 hr are not considered multiple exposures. Actual compressed air exposure time will be used for the determination of decompression time for open air intervals greater than 12 hr.

Schedule 2.5

Repetitive Diving Procedure[a] (Wisconsin 1971)

Repetitive group	Repetitive exposure pressure, psig							
	18	22	27	31	36	40	45	49
A	7	6	5	4	4	3	3	3
B	17	13	11	9	8	7	7	6
C	25	21	17	15	13	11	10	10
D	37	29	24	20	18	16	14	13
E	49	38	30	26	23	20	18	16
F	61	47	36	31	28	24	22	20
G	73	56	44	37	32	29	26	24
H	87	66	52	43	38	33	30	27
I	101	76	61	50	43	38	34	31
J	116	87	70	57	48	43	38	34
K	138	99	79	64	54	47	43	38
L	161	111	88	72	61	53	48	42
M	187	124	97	80	68	58	52	47
N	213	142	107	87	73	64	57	51
O	241	160	117	96	80	70	62	55
Z	257	169	122	100	84	73	64	57

[a] Instructions for use: The compressed air exposure times listed in the schedule are called "residual nitrogen times" and are the times a compressed air worker is to consider he has already spent in compressed air when he starts a repetitive exposure to a specific pressure. They are in minutes. Enter the table horizontally with the repetitive group designation from the Open Air Interval Credit Table (Schedule 2.4). The time in each vertical column is the number of minutes that would be required (at a pressure listed at the head of the columns) to saturate to the particular group. For example, the final group designation from the Open Air Interval Credit Table (Schedule 2.4) on the basis of a previous exposure and open air interval is H. It is planned to reenter compressed air at a pressure of 42 psig. What time must be added to the actual time spent in compressed air? Enter Schedule 2.5 on row H. Since 42 psig is greater than 40 psig but less than 45 psig, use the longer time of 33 min. This means that the compressed air worker enters the compressed air environment as though he had already been at 42 psig for 33 min. The exposure time listed in Schedule 2.5 is added to the actual time spent in compressed air. Decompression is carried out based on the sum of the actual exposure time and the time for Schedule 2.5 for the pressure encountered.

3. *Surface Decompression Table for Saturation Diving with Air (Schedule 3.1)*

a. Limits

1. Air: 50 fsw.
2. N_2–O_2: Depths equivalent to 66 ft N_2 tension.

b. Surfacing Instructions

1. The maximum time permitted to surface transfer to the chamber and begin recompression is 15 min. Recompress to 60 fsw, irrespective of the saturation depth.
2. Ascent to the surface at rates up to 20 ft/min is acceptable.

Schedule 3.1

Surface Decompression Table for Saturation Diving with Air (Beckman and Smith 1972)

Depth, fsw	Time, min	Total decompression time, min	Breathing medium
60	20	20	Oxygen
↓	5	25	Oxygen
55	20	45	Air
↓	5	50	Air
50	20	70	Oxygen
↓	5	75	Oxygen
45	20	95	Air
↓	5	100	Air
40	20	120	Oxygen
↓	15	135	Air
25	60	195	Air
↓	5	200	Air
20	90	290	Air
20	30	320	Oxygen
↓	5	325	Oxygen
15	90	415	Air
15	60	475	Oxygen
↓	5	480	Air
10	120	600	Air
10	60	660	Oxygen
↓	5	665	Oxygen
5	150	815	Air
5	60	875	Oxygen
↓	5	880[a]	Air
Surface			

[a] Total decompression time: 880 min, or 14 hr and 40 min. Total 100% oxygen inhalation: 4 hr and 50 min.

4. *Air Excursion Tables for Saturation Diving on Air or N₂–O₂ Mixtures*

NOAA OPS Excursion Instructions. Schedules 4.1 and 4.2.

(*a*) *Criteria.* Habitat gas: oxygen 0.2 atm or greater, balance nitrogen. Excursion gas: air or N_2–O_2 mix having greater than 0.2 atm oxygen.

(*b*) *Descending excursions.* Timing: Begin timing on departure from habitat depth (bottom time includes descent time).

Rates: Descend as fast as desired. A slow descent is preferred, but reduces bottom time.

Schedule

NOAA OPS Table for Descending Excursions : No-Stop

Habitat depth, fsw	Excursion																
	80	85	90	95	100	105	110	115	120	125	130	135	140	145	150	155	160
30	350	267	156	113	91	78	68	60	55	50	45	40	36	32	28	24	22
35	*	*	283	229	143	108	89	77	68	61	54	46	41	37	34	31	28
40	*	*	*	301	240	202	147	112	92	80	70	59	50	44	39	35	32
45	*	*	*	*	323	253	210	181	137	108	91	69	56	48	42	38	34
50	*	*	*	*	*	350	267	219	187	164	140	86	64	53	45	40	36
55	*	*	*	*	*	*	314	245	203	174	153	137	86	63	52	45	40
60	*	*	*	*	*	*	*	284	224	187	161	142	127	85	63	52	45
65	*	*	*	*	*	*	*	315	236	191	162	145	128	111	85	63	51
70	*	*	*	*	*	*	*	*	279	213	174	148	129	114	103	84	62
75	*	*	*	*	*	*	*	*	*	*	*	288	228	191	165	145	95
80		*	*	*	*	*	*	*	*	*	*	*	*	*	317	225	215
85			*	*	*	*	*	*	*	*	*	*	*	*	*	*	328
90				*	*	*	*	*	*	*	*	*	*	*	*	*	*
95				*	*	*	*	*	*	*	*	*	*	*	*	*	*
100					*	*	*	*	*	*	*	*	*	*	*	*	*
105						*	*	*	*	*	*	*	*	*	*	*	*
110							*	*	*	*	*	*	*	*	*	*	*
115								*	*	*	*	*	*	*	*	*	*
120									*	*	*	*	*	*	*	*	*

ª Asterisk indicates up to 6 hr (360 min). Read instructions before using.

Ascend at any desired rate up to 30 fpm.

(*c*) *Ascending excursions*. Begin timing on arrival at excursion depth ("bottom" time does not include transit times).

Rates: Ascend at 10–30 ft/min.

Descend at 60 ft/min or faster if desired.

If bends symptoms are noted, descend immediately to habitat. If descent to habitat is held up by ear problems, it is preferable to discontinue descent (or even momentarily ascend a bit) and clear the ear, rather than incur ear damage in order to adhere strictly to the schedule.

4.1

Decompression Time (min) [a] (Hamilton *et al*. 1973)

depth, fsw

165	170	175	180	185	190	195	200	205	210	215	220	225	230	235	240	245	250
18	15	13	12	11	10	9	8	8	7	7	6	6	5	5	5	5	—
25	22	20	16	14	13	11	10	9	9	8	7	7	6	6	6	5	5
30	28	25	23	21	17	15	13	12	11	10	9	8	8	7	7	6	6
31	29	27	25	23	22	21	18	16	14	12	11	10	9	9	8	8	7
33	30	28	26	24	22	21	20	19	18	16	14	13	12	11	10	9	8
36	32	30	27	25	24	22	21	20	19	18	17	15	13	12	11	10	9
39	35	32	29	27	25	23	22	21	19	18	17	17	15	13	12	11	10
44	39	35	32	29	27	25	23	22	20	19	18	17	16	16	14	12	11
51	44	39	35	31	29	26	25	23	21	20	19	18	17	16	15	14	13
66	53	45	40	35	32	29	27	25	23	22	20	19	18	17	16	16	15
122	70	55	47	41	36	32	29	27	25	23	22	20	19	18	17	16	15
265	225	95	66	54	46	40	36	32	29	27	25	23	22	22	19	18	17
*	339	275	168	97	63	55	47	41	37	33	30	28	26	24	23	21	20
*	*	*	306	227	143	113	80	62	52	46	40	37	33	31	28	26	25
*	*	*	*	341	281	193	135	109	93	72	59	50	44	40	36	33	31
*	*	*	*	*	354	308	262	174	129	107	77	62	53	46	41	38	35
*	*	*	*	*	*	*	334	292	257	176	132	83	65	55	48	43	39
*	*	*	*	*	*	*	*	347	303	270	243	163	91	68	57	49	44
*	*	*	*	*	*	*	*	*	*	329	291	261	237	101	72	59	51

Schedule 4.2

NOAA OPS Table for Ascending Excursions (Hamilton et al. 1973)

Habitat depth, fsw	Excursion depth, fsw																		
	0	5	10	15	20	25	30	35	40	45	50	55	60	65	70	75	80	85	90
30	36	48	60	*	*	*	*												
35	30	37	48	60	*	*	*	*											
40	24	31	40	52	60	*	*	*	*										
45	17	24	31	40	52	60	*	*	*	*									
50	12	18	25	32	42	60	*	*	*	*	*								
55	7	13	18	25	32	42	60	*	*	*	*	*							
60	—	7	13	18	25	32	42	60	*	*	*	*	*						
65	—	—	8	14	20	27	34	44	60	*	*	*	*	*					
70	—	—	—	8	14	20	27	34	44	60	*	*	*	*	*				
75	—	—	—	—	9	15	21	28	36	47	60	*	*	*	*	*			
80	—	—	—	—	—	9	15	21	28	36	47	60	*	*	*	*	*		
85	—	—	—	—	—	5	10	16	23	30	37	48	60	*	*	*	*	*	
90	—	—	—	—	—	—	5	10	16	23	30	37	48	60	*	*	*	*	*
95	—	—	—	—	—	—	—	6	12	18	24	31	40	52	60	*	*	*	*
100	—	—	—	—	—	—	—	—	6	12	18	24	31	40	52	60	*	*	*
105	—	—	—	—	—	—	—	—	—	7	13	18	25	32	42	60	*	*	*
110	—	—	—	—	—	—	—	—	—	—	7	13	18	25	32	42	60	*	*
115	—	—	—	—	—	—	—	—	—	—	—	7	13	18	25	32	42	60	*
120	—	—	—	—	—	—	—	—	—	—	—	—	7	13	18	25	32	42	60

[a] Asterisk indicates no time limit. Read instructions before using.

5. *Decompression Schedules for Subsaturation He–O₂ Diving*

a. Schedule 5.1: He–O₂ Partial Pressures—40–380 ft

b. Schedule 5.2: Helium–Oxygen Decompression Schedule (Normal and Exceptional Exposures)

Special Instructions on Table Selection. Select proper partial pressure schedules by using one of the following procedures.

(*a*) *In-water decompression.*
1. Rate of ascent to first stop

$$= \frac{\text{bottom depth} - \text{depth of first stop}}{\text{time to first stop}}$$

2. Rate of ascent between stops is 60 ft/min.
3. Switch to O_2 and ventilate 25 scf at 50 ft, or at 40 ft if there is no 50-ft stop. Ventilation time is included in stop time.
4. Remain on O_2 for time of 50- and 40-ft stops.
5. Ascend on O_2 to surface at a uniform rate during the last minute of the 40-ft stop.

(*b*) *Surface decompression.* For schedules in which the first stop is 40 ft:

1. Ventilate with O_2 as above.
2. Remain on O_2 for 10 min at 40 ft.
3. Surface diver in 1 min.
4. Repressurize diver to 40 ft in recompression chamber with air.
5. Diver breathes O_2 by mask for full time of 40-ft stop.
6. During last 5 min of decompression time surface diver at uniform rate while breathing O_2.
7. Maximum allowable time for steps 4 and 5 is 5 min.

For schedules in which the first stop is 50 ft:

1. Ventilate with O_2 and stay at 50 ft for full stop time.
2. Ascend to 40-ft stop and stay at 40 ft for time of 50-ft stop.
3. Surface in 1 min and follow procedure above.

c. Schedule 5.3: Emergency Schedule (He–O₂)

Use this schedule in an emergency when oxygen cannot be used for decompression, owing to failure of the oxygen supply or to symptoms of oxygen poisoning. Use a He–O₂ mix containing a minimum of 16% oxygen. If the impossibility of using oxygen is known in advance, use the regular schedule up to the first oxygen stop, then shift to the emergency schedule (He–O$_2$).

d. Schedule 5.4: Emergency Schedule (Air)

Use this schedule in an emergency when neither oxygen nor He–O$_2$ can be used during decompression. Using this schedule, rate of ascent to the first stop should be the same as listed in the partial pressure tables, but should not exceed 60 ft/min. Rate of ascent on subsequent stops is not critical as long as full decompression is received at each stop.

e. Schedule 5.5: Helium–Oxygen Decompression Schedule for Mixed-Gas Scuba Using 68% Helium–32% Oxygen Supply Mixture

Special Instructions. This schedule is for use only with semiclosed-circuit apparatus. The Standard Supply Mixture (68% He–32% O$_2$) can be used to a maximum depth of 200 ft for 30 min. The Alternate Gas Mixture (60% He–40% O$_2$) is limited to a maximum depth of 80 ft and is intended mainly for use at depths less than 50 ft. Rate of descent should not exceed 75 ft/min. Rate of ascent from bottom and between stops should be 60 ft/min.

f. Schedule 5.6: Helium–Oxygen Decompression Table for Mixed-Gas Scuba Using 68% Helium–32% Oxygen Supply Mixture and Oxygen Decompression

Special Information. This schedule allows for oxygen decompression from helium–oxygen mixed-gas dives and provides significant savings in decompression time as compared with Schedule 5.5. This schedule can only be used with semiclosed-circuit apparatus modified with oxygen cylinder and injection system or by supplying the diver on a descending line with surface-supplied oxygen delivered to a demand regulator at the required decompression depth.

1. The first oxygen stop is at 20 or 30 ft.
2. Two minutes are provided in the schedule to secure the helium–oxygen injection system and purge the breathing bag thoroughly three times with oxygen at the first oxygen stop.
3. Decompression time at the first oxygen stop does not start until after the required 2 min for oxygen purging has elapsed.

g. Schedule 5.7: No-Decompression Limits and Repetitive
    Group Designation Mixed-Gas Scuba No-Decompression
    Table for Helium–Oxygen Dives

(i) Special Instructions. This schedule for use only with Mixed Gas Scuba
(68% He–32% O_2).

1. No-decompression limits column: Allowable maximum bottom time that
   permits surfacing directly at 60 ft/min with no decompression stops.
2. For longer bottom times use the He–O_2 decompression table for mixed-gas
   scuba, Schedule 5.5.
3. Repetitive group designation schedule: Time periods in each vertical column
   are the maximum exposures at various depths during which a diver will remain
   within the group listed at the head of the column.
4. Repetitive group designation: Enter the schedule on the exact or next greater
   depth than exposure and select the exposure time that is exactly the same as
   or next greater than the actual exposure time. Read the group designation
   (letter) at the top of the column for the next dive.
5. Exposure times for depths less than 40 ft are listed up to 12 hr although this
   is considered beyond field requirements for this schedule.

(ii) Example. A dive to 42 ft for 45 min.

1. Select next greater depth, i.e., 50 ft.
2. Select next greater exposure time than 45 min, i.e., 55 min.
3. Read the repetitive group designation at the top of the column, i.e., group E.

h. Schedule 5.8: Surface Interval Credit Table for Mixed-Gas
    Scuba Helium–Oxygen Decompression Dives

(i) Special Instructions. Surface interval time in the schedule is in hours and
minutes (1:30 = 1 hr and 30 min).

1. Surface interval must be at least 30 min.
2. Repetitive group designation after surface interval: Enter the table on the diagonal slope using the designation from previous dive. Read horizontally until
   the actual surface interval is exactly between the intervals shown in schedule
   5.5. Read the new group designation at the top of the column.
3. Dives following surface intervals of more than 12 hr are not repetitive dives.
   Use actual bottom times in the helium–oxygen decompression schedules to
   compute decompression for such dives.

(ii) Example. Find new group designation after dive to 110 ft for 30 min and a time on the surface of 1 hr and 30 min.

1. The previous repetitive group from the last column of the 110/30 schedule of the helium–oxygen decompression schedules is G.
2. Locate G in the diagonal column.
3. Follow across horizontally.
4. The 1 hr 30 min interval lies between the times 1:01 and 1:40.
5. Diver has lost sufficient inert gas to place him in the group at the top of vertical column E.

i. Schedule 5.9: Repetitive Dive Timetable for Mixed-Gas Scuba Helium–Oxygen Dives

(i) Special Instructions. Bottom times listed in this schedule are called residual helium times. Residual helium time is the time a diver is to consider he has already spent on the bottom when a repetitive dive is started to a specific depth.

To find bottom time to be added to schedule for repetitive dive enter the schedule horizontally with the repetitive group from the Surface Interval Credit Table (Schedule 5.8). Read directly the bottom time to be added to the repetitive dive in the depth column for that dive.

(ii) Example. The group designation from the Surface Interval Credit Table from a previous dive is H. How much bottom time must be added (residual helium time) for a repetitive dive to 110 ft?

1. Enter the schedule horizontally at H.
2. Read in the 110-ft depth column the residual helium time to be added: 39 min.
3. The schedules show that one must start a dive to 110 ft as though he had already been on the bottom 39 min.
4. Use Helium–Oxygen Decompression Schedule 5.5 or 5.6 or the No-Decompression Schedule to determine dive schedule for repetitive dive.

Schedule 5.1

Helium–Oxygen Partial Pressures—40–380 ft (U. S. Navy 1971)

Depth, ft	Oxygen, percent												
	15	16	17	19	21	23	25	30	35	40	45	50	55
40	64	63	63	61	60	58	57	b	b	b	b	b	b
50	73	72	71	69	68	66	64	60	56	b	b	b	b
60	81	80	80	78	76	74	72	67	63	58	54	b	b
70	90	89	88	86	84	82	80	75	70	64	59	54	54
80	99	98	97	94	92	90	88	82	76	71	65	59	
90	108	106	105	103	100	98	95	89	83	77	71	64	
100	116	115	114	111	108	106	103	96	90	83	76		
110	125	123	122	119	116	113	111	103	96	89	82		
120	134	132	131	127	124	121	118	111	103	95			
130	142	141	139	136	133	129	126	118	110	102			
140	151	149	148	144	141	137	134	125	116				
150	160	158	156	152	149	145	141	132	123				
160	168	166	165	161	157	155	149	139					
170	177	175	173	169	165	161	157	147					
180	186	184	182	177	173	169	165	154					
190	195	192	190	186	181	177	172						
200	203	201	199	194	189	185	180						
210	212	209	207	202	197	192	188						
220	221	218	216	210	205	200	195						
230	229	227	224	219	214	203	203						
240	238	235	233	227	222	216							
250	247	244	241	235	230	224							
260	255	252	250	244	238								
270	264	261	258	252	246								
280	273	270	267	260	254								
290	282	278	275	269									
300	290	287	284	277									
310	299	295	292	285									
320	308	304	301										
330	316 [a]	313	309										
340	325	321	318										
350	334	330	326										
360	342	338											
370	351	347											
380	360	356											

[a] Boxed numbers indicate exposures which exceed the limit for a 30-min exposure at 1.6 atm P_{O_2}.

[b] No-decompression cutoff depth for oxygen = {(52.8/oxygen percent) − 33. [52.8/(D + 33)] × 100 = maximum oxygen percent. (D + 33 × oxygen percent)/33 = effective atmospheres of oxygen. Partial pressure = (D + 33) × [1.00 − (decimal equivalent of O_2%−0.02)]. 16% is considered the minimum oxygen percentage for surface breathing. The partial pressure tables are not designed for shifting to lower percentages of oxygen during the dive. Therefore, those partial pressure tables requiring a lower percent of oxygen are for emergency use only.

Schedule 5.2

Helium–Oxygen Decompression Table (Normal and Exceptional Exposures) (U. S. Navy 1971)

Partial pressure	Bottom time, min	Time to first stop, min	Decompression stops, ft																Total ascent time, min
			180	170	160	150	140	130	120	110	100	90	80	70	60	50	40	
60	10	4															0	4
	20	4															0	4
	30	4															0	4
	40	4															0	4
	60	4															0	4
	80	4															6	8
	100	2															7	9
	120	2															9	11
	240	2															13	15
70	10																6	9
	20																7	10
	30																9	12
	40																10	13
	60																15	18
	80																17	20
	100																22	25
	120																25	28
	140																27	30
	160																29	32
	180																31	34
	200																31	34
	220																33	36
	240																33	36
80	10																6	9
	20																10	13
	30																13	16
	40																17	20

90	60	24	27
	80	32	35
	100	40	43
	120	42	45
	140	45	48
	160	47	50
	180	48	51
	200	48	51
	220	48	51
	240	50	53
	10	8	11
	20	15	18
	30	18	21
	40	23	26
	60	35	38
	80	45	48
	100	50	53
	120	55	58
	140	58	61
	160	60	63
	180	60	63
	200	62	65
	220	62	65
	240	63	66
100	10	3 / 10	13
	20	17	20
	30	24	27
	40	31	34
	60	47	50
	80	56	59
	100	63	66
	120	67	70
	140	70	73
	160	72	75

Schedule 5.2—*Cont.*

Partial pressure	Bottom time, min	Time to first stop, min	180	170	160	150	140	130	120	110	100	90	80	70	60	50	40	Total ascent time, min
	180																73	76
	200																73	76
	220																73	76
	240																75	78
110	10	3															12	15
	20																21	24
	30																31	34
	40																39	42
	60																56	59
	80																67	70
	100																75	78
	120																78	81
	140																81	84
	160																83	86
	180																84	87
	200																84	87
	220																85	88
	240																86	89
120	10	3															14	17
	20																25	28
	30																36	39
	40																47	50
	60																66	69
	80																77	80
	100																84	87
	120																87	90
	140																90	93
	160																92	95
	180																93	96

Depth (ft)	Bottom time (min)				
130	200		0	93	96
	220		0	95	98
	240	3	0	97	100
140	10		0	16	19
	20		0	29	32
	30		0	42	45
	40		0	53	56
	60		0	73	76
	80		0	86	89
	100		0	92	95
	120		10	96	99
	140		10	99	102
	160		10	92	105
	180		10	93	106
	200		10	94	107
	220		0	95	108
	240	3	0	96	109
150	10		0	19	22
	20		0	34	37
	30		0	49	52
	40		0	62	65
	60		0	82	85
	80		10	94	97
	100		10	99	102
	120		10	97	110
	140		12	98	111
	160		13	99	112
	180		14	99	114
	200		15	99	115
	220		10	99	116
	240	3	10	99	117
	10		10	11	24	0
	20		10	28	41	0
	30		10	45	58	0
	40			59	79	7

Schedule 5.2—*Cont.*

Partial pressure	Bottom time, min	Time to first stop, min	180	170	160	150	140	130	120	110	100	90	80	70	60	50	40	Total ascent time, min
	60														7	10	78	98
	80														7	10	90	110
	100														7	10	96	116
	120														7	11	98	119
	140														7	13	99	122
	160														8	15	99	125
	180														9	15	99	126
	200														10	16	99	128
	220														11	16	99	129
	240														12	16	99	130
160	10	3												0	0	10	12	25
	20													0	7	10	33	53
	30													0	7	10	50	70
	40													0	7	10	65	85
	60													0	7	10	84	104
	80													0	7	10	96	116
	100													0	7	13	99	122
	120													0	9	16	99	127
	140													0	15	16	99	133
	160													0	18	16	99	136
	180													0	20	16	99	138
	200													0	22	16	99	140
	220													0	23	16	99	141
	240													7	19	16	99	144
170	10	3												0	7	10	15	35
	20													0	7	10	36	56
	30													0	7	10	55	75
	40													0	7	10	70	90
	60													7	6	10	83	109

Decompression stops, ft

Depth	Time						
	80	127	98	10	9	7	
	100	135	98	14	13	7	
	120	142	99	16	17	7	
	140	147	99	16	21	8	
	160	151	99	16	22	11	
	180	152	99	16	23	11	
	200	153	99	16	23	12	
	220	155	99	16	23	14	
	240	157	99	16	23	16	
180 (3)	10	37	17	10	0	7	0
	20	61	41	10	0	7	0
	30	83	62	10	1	7	0
	40	101	77	10	4	7	0
	60	122	92	10	10	7	0
	80	137	98	13	14	7	0
	100	147	99	15	18	9	7
	120	155	99	16	21	5	7
	140	158	99	16	22	9	7
	160	163	99	16	23	11	7
	180	165	99	16	23	15	7
	200	167	99	16	23	17	7
	220	169	99	16	23	19	7
	240	171	99	16	23	21	7
190 (4)	10	41	20	10	0	7	0
	20	65	44	10	0	7	0
	30	92	67	10	4	7	0
	40	110	81	10	8	0	7
	60	133	96	15	11	5	7
	80	149	99	16	15	9	7
	100	158	99	16	19	13	7
	120	166	99	16	23	17	7
	140	170	99	16	23	19	9
	160	173	99	16	23	20	11
	180	176	99	16	23	21	13
	200	178	99	16	23	22	14

Schedule 5.2—Cont.

Partial pressure	Bottom time, min	Time to first stop, min	\multicolumn{15}{c}{Decompression stops, ft}	Total ascent time, min
			180	170	160	150	140	130	120	110	100	90	80	70	60	50	40	
	220												15	23	23	16	99	180
	240												17	23	23	16	99	182
200	10	4											0	7	0	10	22	43
	20												7	0	2	10	50	73
	30												7	0	7	10	69	97
	40												7	4	9	10	84	118
	60												7	9	13	12	93	138
	80											7	3	13	18	15	99	159
	100											7	6	16	21	16	99	169
	120											7	8	20	23	16	99	177
	140											7	11	21	23	16	99	181
	160											7	15	23	23	16	99	187
	180											7	17	23	23	16	99	189
	200											7	18	23	23	16	99	190
	220											7	20	23	23	16	99	192
	240											8	20	23	23	16	99	193
210	10	4										0	7	0	0	10	25	46
	20											0	7	0	4	10	53	78
	30											7	0	3	7	10	74	105
	40											7	0	7	10	10	86	124
	60											7	4	10	14	13	98	150
	80											7	8	14	18	16	99	166
	100											7	12	17	23	16	99	178
	120											8	15	21	23	16	99	186
	140											10	17	21	23	16	99	190
	160											12	17	22	23	16	99	193
	180											14	18	22	23	16	99	196
	200											16	18	23	23	16	99	199
	220											17	19	23	23	16	99	201

U.S. Navy Standard Air Decompression Table (continued)

Depth (ft)	Bottom time (min)	Decompression stops (ft) 80	70	60	50	40	30	20	10	Total
210	240			18	20	23	23	16	99	203
220	10		0	0	7	0	0	10	28	49
220	20		0	7	0	1	6	10	57	85
220	30		0	7	0	6	7	10	79	113
220	40		7	7	3	9	10	13	90	133
220	60		7	0	9	11	17	13	98	159
220	80		7	3	11	15	20	16	99	172
220	100		7	6	14	19	23	16	99	188
220	120		7	8	18	23	23	16	99	198
220	140		7	11	18	23	23	16	99	201
220	160		7	14	19	23	23	16	99	205
220	180		7	15	20	23	23	16	99	207
220	200		8	16	20	23	23	16	99	208
220	220		9	17	20	23	23	16	99	210
220	240		0	19	20	23	23	16	99	213
230	10		0	0	7	0	2	10	30	53
230	20		7	7	0	3	7	10	61	92
230	30		7	7	2	6	9	10	81	119
230	40		7	0	6	9	11	14	93	140
230	60	0	7	4	9	12	18	16	99	167
230	80	0	8	8	12	17	21	16	99	184
230	100	0	10	12	15	20	23	16	99	196
230	120	0	6	14	19	23	23	16	99	206
230	140	7	9	16	20	23	23	16	99	211
230	160	7	11	18	20	23	23	16	99	216
230	180	7	13	19	20	23	23	16	99	218
230	200	7	7	19	20	23	23	16	99	220
230	220	7	0	19	20	23	23	16	99	222
230	240	7	7	19	20	23	23	16	99	224
240	10	0	7	7	0	0	3	10	33	57
240	20	0	0	0	1	4	7	10	65	98
240	30	0	0	0	5	7	10	11	85	128
240	40	7	7	3	7	9	13	15	95	149
240	60	7	0	8	10	14	18	16	99	175
240	80	7	3	10	14	18	23	16	99	194

Note: A value "4" appears adjacent to each of the depth-group labels (220, 230, 240).

Schedule 5.2—*Cont.*

Partial pressure	Bottom time, min	Time to first stop, min	180	170	160	150	140	130	120	110	100	90	80	70	60	50	40	Total ascent time, min
	100									7	6	12	17	23	23	16	99	207
	120									7	7	16	19	23	23	16	99	214
	140									7	11	16	20	23	23	16	99	219
	160									7	13	19	20	23	23	16	99	224
	180									8	15	19	20	23	23	16	99	227
	200									8	17	19	20	23	23	16	99	229
	220									9	17	19	20	23	23	16	99	230
	240									11	17	19	20	23	23	16	99	232
250	10	4								0	7	0	0	2	4	10	35	62
	20									0	7	0	2	5	7	10	68	103
	30									7	0	2	6	7	10	10	87	133
	40									7	0	5	8	9	14	12	96	155
	60								0	7	4	8	11	14	19	16	99	182
	80								0	7	7	11	16	18	23	16	99	201
	100								0	7	10	14	19	23	23	16	99	215
	120								7	3	12	17	19	23	23	16	99	223
	140								7	4	15	18	19	23	23	16	99	228
	160								7	7	16	19	19	23	23	16	99	233
	180								7	9	17	19	20	23	23	16	99	237
	200								7	11	17	19	20	23	23	16	99	239
	220								7	12	17	19	20	23	23	16	99	240
	240								7	13	17	19	20	23	23	16	99	241
260	10	4								0	7	0	0	2	4	10	37	64
	20									7	0	0	3	7	7	10	70	108
	30									7	0	4	6	8	10	10	89	138
	40									7	2	5	9	9	14	13	96	159
	60								7	0	7	9	12	16	21	16	99	191
	80								7	3	9	13	15	21	23	16	99	210
	100								7	6	11	14	19	23	23	16	99	222

	120			7	8	13	19	20	23	23	16	99	232
	140			7	11	15	19	20	23	23	16	99	237
	160			8	13	17	19	20	23	23	16	99	242
	180			9	14	17	19	20	23	23	16	99	244
	200			10	16	17	19	20	23	23	16	99	247
	220			11	16	17	19	20	23	23	16	99	248
	240			13	16	17	19	20	23	23	16	99	250
270	10	4		0	7	0	0	0	4	4	10	40	69
	20			0	7	0	2	4	6	7	10	74	114
	30			7	0	2	5	6	9	10	10	92	145
	40			7	0	3	8	9	10	15	14	96	166
	60		0	7	3	7	10	14	16	21	16	99	197
	80		0	7	6	10	13	17	23	23	16	99	218
	100		7	2	9	13	16	20	23	23	16	99	232
	120		7	4	11	14	19	20	23	23	16	99	240
	140		7	5	14	15	19	20	23	23	16	99	245
	160		7	7	15	17	19	20	23	23	16	99	250
	180		7	9	16	17	19	20	23	23	16	99	253
	200		7	11	16	17	19	20	23	23	16	99	255
	220		7	13	16	17	19	20	23	23	16	99	257
	240		7	15	16	17	19	20	23	23	16	99	259
280	10	4		0	7	0	0	2	3	4	10	42	72
	20			7	0	0	2	6	6	8	10	78	121
	30			7	0	3	6	6	9	13	10	93	151
	40		7	0	2	5	8	8	12	16	13	98	173
	60		7	0	6	8	10	14	19	23	16	99	206
	80		7	3	8	11	14	17	23	23	16	99	225
	100		7	5	11	13	16	20	23	23	16	99	237
	120		7	8	12	16	19	20	23	23	16	99	247
	140		7	10	16	17	19	20	23	23	16	99	254
	160		8	13	16	17	19	20	23	23	16	99	258
	180		9	14	16	17	19	20	23	23	16	99	260
	200		10	15	16	17	19	20	23	23	16	99	262
	220		12	15	16	17	19	20	23	23	16	99	264
	240		14	15	16	17	19	20	23	23	16	99	266

Schedule 5.2—Cont.

Partial pressure	Bottom time, min	Time to first stop, min	180	170	160	150	140	130	120	110	100	90	80	70	60	50	40	Total ascent time, min
290	10	4						0	0	7	0	0	3	3	4	10	46	77
	20							0	7	0	0	4	6	7	7	10	81	126
	30							7	0	1	5	5	9	9	12	10	96	158
	40						0	7	0	4	6	8	9	12	17	15	98	180
	60						0	7	4	6	8	12	15	18	23	16	99	212
	80						7	0	7	9	11	15	17	23	23	16	99	231
	100						7	2	9	11	15	17	20	23	23	16	99	246
	120						7	4	11	13	16	19	20	23	23	16	99	255
	140						7	5	13	16	17	19	20	23	23	16	99	262
	160						7	8	14	16	17	19	20	23	23	16	99	266
	180						7	10	15	16	17	19	20	23	23	16	99	269
	200						7	12	15	16	17	19	20	23	23	16	99	271
	220						7	13	15	16	17	19	20	23	23	16	99	272
	240						7	14	15	16	17	19	20	23	23	16	99	273
300	10	5				0	0	0	7	0	0	0	4	3	4	10	49	82
	20					0	0	7	0	0	2	6	6	6	9	10	83	134
	30					0	0	7	0	2	5	5	9	9	14	12	94	162
	40					0	0	7	0	5	7	8	11	13	17	15	98	186
	60					0	7	0	6	7	9	12	15	20	23	16	99	219
	80					0	7	2	8	10	12	16	19	23	23	16	99	240
	100					0	7	5	10	12	15	19	20	23	23	16	99	254
	120					0	7	8	11	16	17	19	20	23	23	16	99	264
	140					0	8	9	14	16	17	19	20	23	23	16	99	269
	160					0	8	13	15	16	17	19	20	23	23	16	99	274
	180					7	3	13	15	16	17	19	20	23	23	16	99	276
	200					7	5	14	15	16	17	19	20	23	23	16	99	279
	220					7	6	14	15	16	17	19	20	23	23	16	99	280
	240					7	9	14	15	16	17	19	20	23	23	16	99	283
310	10	5				0	0	0	7	0	0	2	3	3	5	10	52	87

Depth	Time															
320	20		0	0	0	7	0	0	4	5	6	6	11	10	84	138
	30		0	0	7	0	0	5	5	7	8	9	14	12	96	168
	40		0	0	7	0	3	5	8	8	11	13	18	15	99	192
	60		0	0	7	3	6	7	10	12	18	22	23	16	99	228
	80		0	7	0	6	9	11	12	16	19	23	23	16	99	246
	100		0	7	1	9	10	14	17	19	20	23	23	16	99	263
	120		0	7	4	11	12	14	17	19	20	23	23	16	99	270
	140		0	7	5	12	15	16	17	19	20	23	23	16	99	277
	160		7	7	8	14	15	16	17	19	20	23	23	16	99	282
	180		7	7	10	14	15	16	17	19	20	23	23	16	99	284
	200		7	7	12	14	15	16	17	19	20	23	23	16	99	286
	220		7	8	13	14	15	16	17	19	20	23	23	16	99	288
	240		7	9	13	14	15	16	17	19	20	23	23	16	99	289
	10	5	0	0	0	7	0	0	0	3	3	3	7	10	54	92
	20		0	0	7	0	0	2	4	5	6	7	10	10	85	141
	30		0	0	7	0	2	4	5	7	8	11	15	13	98	175
	40		0	7	0	1	4	6	7	8	12	15	19	16	99	199
	60		0	7	0	5	6	9	11	13	17	20	23	16	99	231
	80		0	7	3	7	9	11	13	17	20	23	23	16	99	253
	100		0	7	5	9	11	13	17	19	20	23	23	16	99	267
	120		7	7	7	12	13	16	17	19	20	23	23	16	99	277
	140		7	9	9	12	15	16	17	19	20	23	23	16	99	283
	160		7	11	11	14	15	16	17	19	20	23	23	16	99	288
	180		7	13	13	14	15	16	17	19	20	23	23	16	99	290
	200		7	13	13	14	15	16	17	19	20	23	23	16	99	293
	220		7	13	13	14	15	16	17	19	20	23	23	16	99	294
	240		7	13	13	14	15	16	17	19	20	23	23	16	99	296
330	10	5	0	0	0	7	0	0	0	4	3	3	7	10	56	95
	20		0	0	7	0	0	3	5	5	6	8	10	10	88	147
	30		0	7	0	0	4	4	6	7	9	11	17	13	98	181
	40		0	7	0	4	4	6	7	9	12	16	20	16	99	205
	60		0	9	2	6	8	9	11	14	17	23	23	16	99	240
	80		7	11	6	8	8	13	14	19	20	23	23	16	99	261
	100		7	11	7	10	13	16	17	19	20	23	23	16	99	277
	120		7	13	9	12	13	16	17	19	20	23	23	16	99	283

Schedule 5.2—Cont.

Partial pressure	Bottom time, min	Time to first stop, min	Decompression stops, ft															Total ascent time, min
			180	170	160	150	140	130	120	110	100	90	80	70	60	50	40	
	140				7	6	11	13	15	16	17	19	20	23	23	16	99	290
	160				7	8	13	14	15	16	17	19	20	23	23	16	99	295
	180				7	10	13	14	15	16	17	19	20	23	23	16	99	297
	200				7	12	13	14	15	16	17	19	20	23	23	16	99	299
	220				9	12	13	14	15	16	17	19	20	23	23	16	99	301
	240				10	12	13	14	15	16	17	19	20	23	23	16	99	302
340	10	5		0	0	0	7	0	0	0	2	3	3	4	7	10	59	100
	20			0	0	7	0	0	2	3	4	6	5	10	10	10	90	152
	30			0	0	7	0	1	4	5	6	8	8	13	17	14	98	186
	40			0	7	0	1	4	5	7	7	10	12	17	22	16	99	212
	60			0	7	0	5	6	8	9	11	15	20	23	23	16	99	247
	80			0	7	2	7	8	10	13	15	19	20	23	23	16	99	267
	100			0	7	5	9	9	13	16	17	19	20	23	23	16	99	281
	120			7	1	7	10	13	15	16	17	19	20	23	23	16	99	291
	140			7	2	9	12	14	15	16	17	19	20	23	23	16	99	297
	160			7	4	10	13	14	15	16	17	19	20	23	23	16	99	301
	180			7	5	12	13	14	15	16	17	19	20	23	23	16	99	304
	200			7	6	12	13	14	15	16	17	19	20	23	23	16	99	305
	220			7	8	12	13	14	15	16	17	19	20	23	23	16	99	307
	240			7	10	12	13	14	15	16	17	19	20	23	23	16	99	309
350	10	5		0	0	0	7	0	0	0	3	3	3	4	7	10	61	103
	20			0	0	7	0	0	2	4	5	7	8	9	10	10	90	157
	30			0	7	0	0	3	5	5	6	8	9	13	18	14	98	191
	40			0	7	0	2	4	6	7	8	10	13	16	22	16	99	215
	60			7	0	3	5	6	9	10	13	16	18	21	23	16	99	251
	80			7	0	7	7	8	11	13	15	19	20	23	23	16	99	273
	100			7	2	8	8	12	13	16	17	19	20	23	23	16	99	288
	120			7	4	9	11	13	15	16	17	19	20	23	23	16	99	297
	140			7	6	11	13	14	15	16	17	19	20	23	23	16	99	304

The following table is printed rotated on the page. Its columns are labelled by bottom time (minutes); the far-left entries read **360** and **5**.

160	180	200	220	240	10	20	30	40	60	80	100	120	140	160	180	200	220	240
307	309	311	313	314	108	163	196	222	257	279	294	303	310	313	315	317	319	321
99	99	99	99	99	64	94	99	99	99	99	99	99	99	99	99	99	99	99
16	16	16	16	16	10	10	14	16	16	16	16	16	16	16	16	16	16	16
23	23	23	23	23	7	13	18	23	23	23	23	23	23	23	23	23	23	23
23	23	23	23	23	5	9	13	17	23	23	23	23	23	23	23	23	23	23
20	20	20	20	20	3	7	11	14	19	20	20	20	20	20	20	20	20	20
19	19	19	19	19	3	5	8	11	16	19	19	19	19	19	19	19	19	19
17	17	17	17	17	2	5	7	8	12	17	17	17	17	17	17	17	17	17
16	16	16	16	16	2	4	5	7	11	13	16	16	16	16	16	16	16	16
15	15	15	15	15	0	4	4	6	8	11	15	15	15	15	15	15	15	15
14	14	14	14	14	0	0	4	5	8	10	11	14	14	14	14	14	14	14
13	13	13	13	13	0	1	3	5	7	9	12	13	13	13	13	13	13	13
11	12	12	12	12	7	0	0	1	5	7	8	9	11	12	12	12	12	12
9	9	11	11	11	0	7	7	0	0	2	6	7	9	11	11	11	11	11
7	8	8	10	11	0	0	0	7	7	7	0	1	3	4	5	7	9	11
					0	0	0	0	0	0	7	7	7	7	7	7	7	7

Schedule 5.3

Emergency Table for Use in Helium–Oxygen
Diving—Emergency Table (He–O$_2$)
(U. S. Navy 1971)

Decompression stop depth, ft	Decompression stop time, min
50	26
40	30
30	35
20	42
10	55

Schedule 5.4

Emergency Table for Use in Helium–Oxygen
Diving—Emergency Table (Air) (U. S. Navy 1971)

Stops, ft	Depth to:						
	100 ft	150 ft	200 ft	250 ft	300 ft	350 ft	400 ft
190							3
180							11
170							12
160						9	12
150						13	13
140					4	13	14
130					14	15	15
120					16	16	16
110				13	16	17	17
100				18	18	18	18
90			7	19	19	20	20
80			22	22	22	22	22
70			24	24	24	24	24
60		22	26	26	26	27	27
50		30	30	30	30	30	30
40	14	35	35	35	35	35	35
30	42	42	42	42	42	42	42
20	52	52	52	52	52	52	52
10	68	68	68	68	68	68	68

Schedule 5.5

Helium–Oxygen Decompression Table for Mixed-Gas Scuba Using 68% Helium—32% Oxygen Supply Mixture (U. S. Navy 1971)

Depth, ft	Bottom time, min	Time to first stop, min:sec	Decompression stops, ft					Total ascent time, min:sec	Repetitive group
			50	40	30	20	10		
40	260						0	0:40	L
50	180						0	0:50	L
	200	0:40					20	20:50	L
60	130						0	1:00	L
	150	0:50					20	21:00	L
	170	0:50					35	36:00	L
70	85						0	1:10	J
	100	1:00					15	16:10	K
	115	1:00					25	26:10	L
	130	1:00					40	41:10	L
80	60						0	1:20	I
	70	1:00				5	10	16:20	J
	80	1:00				10	15	26:20	K
	90	1:00				10	25	36:20	K
	100	1:00				10	35	46:20	K
90	45						0	1:30	H
	60	1:10				5	10	16:30	J
	70	1:10				5	20	26:30	K
	85	1:10				10	30	41:30	L
100	35						0	1:40	G
	50	1:20				5	15	21:40	J
	60	1:20				10	20	31:40	K
	70	1:10			5	15	25	46:40	K
110	30						0	1:50	G
	40	1:30				5	10	16:50	H
	50	1:30				10	20	31:50	J
	65	1:20			5	15	25	46:50	L
120	25						0	2:00	G

Schedule 5.5—*Cont.*

Depth, ft	Bottom time, min	Time to first stop, min:sec	Decompression stops, ft					Total ascent time, min:sec	Repetitive group
			50	40	30	20	10		
	35	1:40				5	10	17:00	I
	45	1:30			5	10	15	32:00	K
	55	1:30			10	15	20	47:00	L
130	20						0	2:10	F
	30	1:50				5	10	17:10	I
	40	1:40			5	10	15	32:10	J
	50	1:30		5	5	15	20	47:10	L
140	15						0	2:20	E
	25	2:00				5	10	17:20	G
	35	1:50			5	10	20	37:20	J
	45	1:40		5	10	15	20	52:20	K
150	15						0	2:30	E
	20	2:10				5	10	17:30	G
	30	2:00			5	10	15	32:30	J
	40	1:50		5	10	15	20	52:30	K
160	10						0	2:40	E
	20	2:10			5	5	10	22:40	G
	35	2:00		5	10	10	20	47:40	K
170	10						0	2:50	E
	20	2:20			5	5	10	22:50	H
	35	2:10		5	10	15	20	52:50	K
180	5						0	3:00	C
	10	2:40				5	10	18:00	E
	20	2:20		5	5	10	10	33:00	H
	30	2:20		5	10	15	20	53:00	K
190	10	2:50				5	10	18:10	E
	20	2:30		5	5	10	20	43:10	H
	30	2:20	5	5	10	15	25	63:10	K
200	10	3:00				5	15	23:20	F
	20	2:40		5	5	10	20	43:20	I
	30	2:30	5	5	10	15	35	73:20	K

Schedule 5.6

Helium–Oxygen Decompression Table for Mixed-Gas Scuba Using 68% Helium–32% Oxygen Supply Mixture and Oxygen Decompression (U.S. Navy 1971)

Depth, ft	Time, min	He–O$_2$ 50 ft	He–O$_2$ 40 ft	Oxygen 30 ft	Oxygen 20 ft	Repetitive group
60	170				20	L
70	115				15	L
	130				25	L
80	80				15	K
	90				20	K
	100				25	K
90	70				15	K
	85				25	L
100	50				15	J
	60				20	K
	70			5	20	K
110	50				15	J
	65			5	20	L
120	45			5	15	K
	55			10	20	L
130	40			5	15	J
	50		5	5	20	L
140	35			5	15	J
	45		5	5	20	K
150	30			5	15	J
	40		5	10	20	K
160	20			5	10	G
	35		5	10	20	K
170	20			5	10	H
	35		5	10	20	K
180	20		5	5	10	H
	30		5	10	20	K
190	20		5	5	15	H
	30	5	5	10	20	K
200	20		5	5	20	I
	30	5	5	10	25	K

[a]Allow 2 min to complete bag purge to oxygen.

Schedule 5.7

No-Decompression Limits and Repetitive Group Designation Mixed-Gas Scuba
No-Decompression Table for Helium–Oxygen Dives (U. S. Navy 1971)

Depth, ft	No-decompression limit, min	Repetitive group (He–O$_2$ dives)											
		A	B	C	D	E	F	G	H	I	J	K	L
10		70	190	720									
20		25	60	95	145	215	335	720					
30		15	35	60	80	110	145	185	245	335	525	720	
40	260	10	25	40	55	70	90	110	140	165	200	245	260
50	180	10	20	30	40	55	70	85	100	120	140	160	180
60	130	5	15	25	35	45	55	65	75	90	105	120	130
70	85	5	10	20	30	35	45	55	65	75	85		
80	60	5	10	15	25	30	40	45	55	60			
90	45	5	10	15	20	30	35	40	45				
100	35	5	9	15	20	25	30	35					
110	30	5	8	12	15	20	25	30					
120	25		5	10	15	20	22	25					
130	20		5	10	15	17	20						
140	15		5	10	12	15							
150	15		5	10	12	15							
160	10		5	6	8	10							
170	10			6	8	10							
180	5			5									

Schedule 5.8

Surface Interval Credit Table for Mixed-Gas Scuba Helium–Oxygen Decompression Dives[a] (U. S. Navy 1971)

L	K	J	I	H	G	F	E	D	C	B	A
L 0:00	0:31	0:41	0:51	1:21	1:41	2:01	2:31	3:11	4:01	5:11	7:11
0:30	0:40	0:50	1:20	1:40	2:00	2:30	3:10	4:00	5:10	7:10	12:00[b]
	K 0:00	0:31	0:41	1:01	1:21	1:51	2:21	3:01	3:51	5:01	7:01
	0:30	0:40	1:00	1:20	1:50	2:20	3:00	3:50	5:00	7:00	12:00[b]
		J 0:00	0:31	0:41	1:01	1:31	2:01	2:41	3:31	4:41	6:41
		0:30	0:40	1:00	1:30	2:00	2:40	3:30	4:40	6:40	12:00[b]
			I 0:00	0:31	0:51	1:21	1:51	2:21	3:11	4:21	6:21
			0:30	0:50	1:20	1:50	2:20	3:10	4:20	6:20	12:00[b]
				H 0:00	0:31	0:51	1:31	2:01	2:51	4:01	6:01
				0:30	0:50	1:30	2:00	2:50	4:00	6:00	12:00[b]
					G 0:00	0:31	1:01	1:41	2:31	3:41	5:41
					0:30	1:00	1:40	2:30	3:40	5:40	12:00[b]
						F 0:00	0:36	1:11	2:01	3:11	5:11
						0:35	1:10	2:00	3:10	5:10	12:00[b]
							E 0:00	0:41	1:31	2:41	4:41
							0:40	1:30	2:40	4:40	12:00[b]
								D 0:00	0:51	2:01	4:01
								0:50	2:00	4:00	12:00[b]
									C 0:00	1:21	3:11
									1:20	3:10	12:00[b]
										B 0:00	2:01
										2:00	12:00[b]
											A 0:00
											12:00[b]

[a] The upper set of repetitive groups indicates the group at the end of the surface interval (He–O₂ dives). The diagonal set of repetitive groups indicates the group at the beginning of the surface interval from previous dive.

[b] Dives following surface intervals of more than 12 hr are not repetitive dives. Use actual bottom times in the helium–oxygen decompression tables to compute decompression for such dives.

Schedule 5.9

Repetitive Dive Timetable for Mixed-Gas Scuba Helium–Oxygen Dives (U. S. Navy 1971)

Repetitive group	Repetitive dive depth (He–O₂ dives), ft																
	40	50	60	70	80	90	100	110	120	130	140	150	160	170	180	190	200
A	13	10	8	7	6	6	5	5	4	4	4	4	3	3	3	3	3
B	26	21	17	14	13	11	10	9	8	8	7	7	6	6	6	5	5
C	40	32	26	22	19	17	15	14	13	12	11	10	9	9	8	8	8
D	56	44	35	30	26	23	20	19	17	15	14	13	13	12	11	11	10
E	74	57	45	38	33	29	26	23	21	19	18	17	16	15	14	13	13
F	93	71	56	47	40	35	31	28	26	24	22	20	19	18	17	16	15
G	115	86	67	56	47	42	37	33	30	28	26	24	22	21	20	19	18
H	139	102	79	66	55	48	43	39	35	32	29	28	26	24	22	21	20
I	168	120	92	76	64	56	49	44	40	37	33	31	29	27	26	24	23
J	203	141	105	87	72	63	55	49	45	41	37	35	32	31	28	27	25
K	248	165	120	98	80	71	61	55	49	45	42	39	36	34	32	30	28
L	305	191	137	111	91	79	68	61	55	50	46	43	40	37	35	33	31

6. *Decompression Procedures for Excursion Diving from Saturation, While Breathing He–O₂ Gas Mixtures*

If the ascent rate exceeds 60 ft/min, the diver should pause at a pressure 10 ft greater than this saturation depth for the time that should have been taken in an ascent at 60 ft/min.

If the ascent rate is slower than 60 ft/min:

1. If the additional time used in ascent does not take the diver beyond the no-decompression limit for his excursion, then it is only necessary to consider the additional delay as part of the excursion time for subsequent excursions.
2. If the additional time used in ascent takes the diver beyond the no-decompression limit for his excursion, pressurize the habitat to the depth at which the delay occurs. Thereafter, he may safely return to the habitat.

a. Depth Limitations

Schedule 6.1 should be used when the excursion dives are made from a saturation exposure at any depth in the range from 150 to 300 ft of sea water. Schedule 6.2 should be used when the excursion dives are made from a saturation exposure at any depth in the range from 300 to 600 ft of sea water. No schedules are presently approved for excursion diving from saturation exposures less than 150 ft of sea water.

b. Habitat Pressure

The saturation pressure of the habitat will normally be kept constant while the excursion diver is absent. The pressure of the habitat may be increased in an emergency and the excursion diver safely brought back to a saturation depth greater than the one from which he left. *Under no circumstances* should the excursion diver be brought back to a habitat saturation pressure less than the pressure from which he departed.

c. Habitat Atmosphere Control

At any time during the saturation exposure in the habitat deeper than 14 ft sea water the oxygen partial pressure should be maintained at 0.3 ATA (220 mm Hg), the carbon dioxide partial pressure should not be greater than 0.0025 ATA (2 mm Hg), the nitrogen partial pressure should not be greater than 1.2 ATA (900 mm Hg), and the balance of the total pressure should be that of helium.

d. UBA Gas Mixture

The gas mixture supply to the underwater breathing apparatus (UBA) of the diver during the excursion should be helium–oxygen with the proportion and the flow rate adjusted so that the oxygen partial pressure in the diver's inhaled breath (bag level) should generally be between 0.8 and 1.0 ATA (610–760 mm Hg). Fluctuations of the bag level in the range between 0.5 and 1.5 ATA (380–1140 mm Hg) are acceptable.

Positive measures should be taken to prevent the oxygen partial pressure falling below 0.4 ATA (305 mm Hg) or rising above 1.6 ATA (1215 mm Hg).

e. Description of Schedules

1. No-Decompression Schedule. Gives the number of minutes permitted at any excursion depth for a no-decompression excursion.

2. Repetitive Group Designation Schedule. Gives the repetitive group designation for any excursions that may have preceded the repetitive excursion.

3. Repetitive Excursion Timetable. Gives the number of minutes of residual helium time that must be subtracted from the no-decompression limits of the repetitive excursion to give the maximum actual excursion time that will still permit a no-decompression return to the saturation depth.

5. Repetitive Group Designation Schedule. Gives the repetitive group designation for the sum of the residual helium time and the actual excursion time of the repetitive excursion.

5. Habitat Interval Credit Schedule. Gives credit for the release of residual helium from the diver's body during the interval of the saturation pressure of the habitat between excursions.

f. Instructions for Use of Schedules

(i) Schedules 6.1 and 6.2. Schedule 6.1 should be used when the saturation gauge depth is between 150 and 300 fsw. Schedule 6.2 should be used when the saturation gauge depth is between 300 and 600 fsw.

1. Read the exact, or next greater, depth to which the excursion dive is to be made in the depth of excursion column.
2. Read in the no-decompression limits column the maximum allowable excursion time that is permitted without requiring decompression stops.

Example: A 40-ft excursion dive is to be conducted from a saturated depth of 220 ft. What is the allowable excursion time for which no decompression is required?
Use: Schedule 6.1, 150–300 ft.
Enter: Depth of excursion column at "plus 50 ft" (next greater than 40 ft).
Read: 270 min in the no-decompression limits column.

1. The diver may spend up to 270 min at the 40 ft excursion depth before requiring decompression.
2. Read in the repetitive group designation column the repetitive letter group that corresponds to the actual, or next greater than actual, excursion time.

Example: In the above dive, the diver actually spends 70 min at the 40-ft excursion depth. What is his repetitive group designation?
Enter: Depth of excursion column at "plus 50 ft."
Read: Horizontally to the 100-min column (next greater than 70 min).
Read: Vertically to the repetitive group designation C.

(ii) Schedule 6.3. Schedule 6.3 should be used only when the gauge depth of the saturation exposure is between 150 ft and 600 ft of sea water. Habitat interval time in the schedule is in hours and minutes: 5:30 means 5 hr and 30 min. With the repetitive group designation from the previous excursion (from either Schedule 6.1 or 6.2) find that letter on the diagonal slope of the schedule. Enter the schedule horizontally to select the listed habitat interval that is equal to or next greater than the actual habitat interval time. The repetitive group designation for the end of the habitat interval is at the head of the vertical column where the selected habitat interval is listed.

Example: After the above dive, the diver spends 6 hr in the habitat. What is his new repetitive group designation?

Enter: Schedule 6.3 on the diagonal slope at the letter C (group designation from previous dive).

Read: Horizontally to the 6:30 column (next greater than 6 hr).

Read: Vertically to the new repetitive group designation, B.

The diver has lost sufficient inert gas to place him in repetitive group B.

If a repetitive excursion dive is to be conducted, enter the repetitive group designation column (Schedule 6.1 or 6.2) corresponding to the letter group from the habitat interval credit table. Enter the depth of excursion column corresponding to the exact or next greater depth of the repetitive dive. Read the residual helium time at the intersection of these columns.

Example: Following the above dive, a repetitive excursion dive to 60 ft below the saturation depth is planned for the same diver. How much residual helium time does the diver have? How long may the excursion dive be before decompression is required?

Enter: Schedule 6.1, repetitive group designation column under the letter B (group designation from habitat interval credit schedule after 6 hr in habitat).

Enter: Depth of excursion column at the "plus 75 ft" depth (next greater than 60 ft).

Read: At intersections of these two columns, residual helium time is 40 min.

Read: No-decompression limits column corresponding to the "plus 75 ft" depth is 150 min.

Subtract: 150 − 40 (residual helium time) = 110 min.

The diver may remain at the 60-ft excursion depth for 110 min before decompression is required.

Schedule 6.1
Excursion Timetable for Saturation at a Gauge Depth between 150 and 300 fsw (U. S. Navy 1971)

Depth of excursion from saturation exposure	No-decompression limits, min	Repetitive group designation					
		A	B	C	D	E	F
Plus 25 ft	—	60	150	300	600	—	—
50	270	30	60	100	150	210	270
75	150	20	40	65	90	120	150
100	60	10	20	30	40	50	60

Schedule 6.2

Excursion Timetable for Saturation at a Gauge Depth between
300 and 600 fsw (U. S. Navy 1971)

Depth of excursion from saturation exposure	No-decompression limit, min	Repetitive group designation					
		A	B	C	D	E	F
Plus 25 ft	—	60	150	300	600	—	—
50	270	30	60	100	150	210	270
75	150	20	40	65	90	120	150
100	100	15	30	45	60	80	100
125	75	10	20	30	45	60	75
150	60	10	20	30	40	50	60

Schedule 6.3

Chamber Interval Credit Table for Saturation Exposure
at a Gauge Depth between 150 and 600 fsw[a]

	F	E	D	C	B	A
F	To 1:00	2:30	4:00	6:30	12:00	24:00
	E	1:30	3:00	5:30	10:00	24:00
		D	2:00	4:00	8:00	24:00
			C	2:30	6:30	24:00
				B	4:00	24:00
					A	24:00

[a] Upper set of repetitive groups indicates repetitive group at the end of the chamber interval (before repetitive excursion). Diagonal set of repetitive groups indicates group at the beginning of the chamber interval.

7. Emergency Abort Schedule (Schedule 7.1)

In the event that a scheduled saturation dive must be aborted prior to reaching saturation conditions, a decompression profile must be run. The decompression rates shown have been calculated for use in the event of an emergency occurring under circumstances beyond the scope of the exceptional exposure tables (Schedule 5.2). The decompression rates used were derived by computer using the Workman calculation method. Fifty-foot intervals were used as convenient levels for emergency stops. Each depth interval is further subdivided into three exposure conditions, namely 120, 240, and 360 min. Use the next greater depth and time exposure if the dive is terminated at a level not stated in the schedule. For example, if a 40-ft saturation dive is aborted after 150 min, the abort schedule to be used would be the 50/240 schedule.

Schedule 7.1
Emergency Abort Schedules (U. S. Navy 1971)

50-ft dive		100-ft dive		150-ft dive	
Depth	Rate	Depth	Rate	Depth	Rate
50/120		100/120		150/120	
50–8	10 ft/min	100–48	10 ft/min	150–90	10 ft/min
8–3	2 min/ft	48–35	1.5 min/ft	90–74	1 min/ft
3–0	4 min/ft	35–22	4 min/ft	74–53	3 min/ft
		22–12	6 min/ft	53–49	5 min/ft
		12–0	9 min/ft	49–25	10 min/ft
				25–0	20 min/ft
50/240		100/240		150/240	
50–10	10 ft/min	100–52	10 ft/min	150–95	10 ft/min
10–9	6 min/ft	52–48	1 min/ft	95–90	1 min/ft
9–5	4 min/ft	48–39	4 min/ft	90–74	3 min/ft
5–0	7 min/ft	39–20	6 min/ft	74–44	10 min/ft
		10–0	20 min/ft	44–0	20 min/ft
50/360		100/360		150/360	
50–15	10 ft/min	100–54	10 ft/min	150–95	10 ft/min
15–12	2 min/ft	54–47	3 min/ft	96–90	3 min/ft
12–4	7 min/ft	47–25	10 min/ft	90–60	10 min/ft
4–0	9 min/ft	25–0	20 min/ft	60–0	[a]

200-ft dive		250-ft dive		300-ft dive	
Depth	Rate	Depth	Rate	Depth	Rate
200/120		250/120		300/120	
200–130	10 ft/min	250–170	10 ft/min	300–210	10 ft/min
130–100	1 min/ft	170–155	1 min/ft	210–190	1 min/ft
100–90	3 min/ft	155–135	2 min/ft	190–170	2 min/ft
90–56	10 min/ft	135–130	3 min/ft	170–130	10 min/ft
56–0	[a]	130–90	10 min/ft	130–0	[a]
		90–0	[a]		
200/240		250/240		300/240	
200–135	10 ft/min	250–175	10 ft/min	300–215	10 ft/min
135–130	1 min/ft	175–170	1 min/ft	215–200	2 min/ft
130–115	3 min/ft	170–160	2 min/ft	200–160	10 min/ft
115–80	10 min/ft	160–120	10 min/ft	160–0	[a]
80–0	[a]	120–0	[a]		
200/360		250/360		300/360	
200–136	10 ft/min	250–182	10 ft/min	300–230	10 ft/min
136–134	8 min/ft	182–140	10 min/ft	230–185	10 min/ft
134–98	10 min/ft	140–0	[a]	185–0	[a]
98–0	[a]				

Schedule 7.1—*Cont.*

350-ft dive		400-ft dive		450-ft dive	
Depth	Rate	Depth	Rate	Depth	Rate
350/120		400/120		450/120	
350–250	10 ft/min	400–280	5 ft/min	450–310	5 ft/min
250–220	1 min/ft	280–260	2 min/ft	310–280	12 min/ft
220–170	11 min/ft	260–210	12 min/ft	280–0	*a*
170–0	*a*	210–0	*a*		
350/240		400/240		450/240	
350–250	2 ft/min	400–300	10 ft/min	450–340	1 ft/min
250–200	12 min/ft	300–250	12 min/ft	340–300	12 min/ft
200–0	*a*	250–0	*a*	300–0	*a*
350/360		400/360		450/360	
350–270	1 ft/min	400–330	10 ft/min	450–370	3 ft/min
270–230	12 min/ft	330–280	10 min/ft	370–320	12 min/ft
230–0	*a*	280–0	*a*	320–0	*a*

500-ft dive		550-ft dive		600-ft dive	
Depth	Rate	Depth	Rate	Depth	Rate
500/120		550/120		600/120	
500–360	10 ft/min	550–410	10 ft/min	600–460	30 ft/min
360–300	12 min/ft	410–350	12 min/ft	460–400	12 min/ft
300–0	*a*	350–0	*a*	400–0	*a*
500/240		550/240		600/240	
500–390	2 ft/min	550–440	10 ft/min	600–500	30 ft/min
390–340	12 min/ft	440–390	12 min/ft	500–440	12 min/ft
340–0	*a*	390–0	*a*	440–0	*a*
500/360		550/360		600/360	
500–420	30 ft/min	550–470	30 ft/min	600–520	30 ft/min
420–370	12 min/ft	470–420	12 min/ft	520–470	12 min/ft
370–0	*a*	420–0	*a*	470–0	*a*

Schedule 7.1—*Cont.*

650-ft dive		700-ft dive		750-ft dive	
Depth	Rate	Depth	Rate	Depth	Rate
650/120		700/120		750/120	
650–510	30 ft/min	700–550	30 ft/min	750–600	30 ft/min
510–440	12 min/ft	550–500	12 min/ft	600–540	12 min/ft
440–0	[a]	500–0	[a]	540–0	[a]
650/240		700/240		750/240	
650–550	30 ft/min	700–590	30 ft/min	750–640	30 ft/min
550–490	12 min/ft	590–540	12 min/ft	640–590	12 min/ft
490–0	[a]	540–0	[a]	590–0	[a]
650/360		700/360		750/360	
650–570	30 ft/min	700–620	30 ft/min	750–670	30 ft/min
570–520	12 min/ft	620–570	12 min/ft	670–610	12 min/ft
520–0	[a]	570–0	[a]	610–0	[a]

800-ft dive		850-ft dive	
Depth	Rate	Depth	Rate
800/120		850/120	
800–660	30 ft/min	850–700	30 ft/min
660–590	12 min/ft	700–640	12 min/ft
590–0	[a]	640–0	[a]
800/240		850/240	
800–690	30 ft/min	850–740	30 ft/min
690–640	12 min/ft	740–690	12 min/ft
640–0	[a]	690–0	[a]
800/360		850/360	
800–720	30 ft/min	850–770	30 ft/min
720–670	12 min/ft	770–720	12 min/ft
670–0	[a]	720–0	[a]

[a] Assume standard saturation dive decompression schedule.

8. *U. S. Navy Treatment Schedules for Decompression Sickness*

a. Schedule 8.1: Air Treatment

1. Use. Treatment of pain-only decompression sickness when oxygen cannot be used and pain is relieved at a depth less than 66 ft.
2. Descent rate. 25 ft/min.
3. Ascent rate. One minute between stops.
4. Time at 100 ft. Includes time from the surface.

b. Schedule 8.2: Air Treatment

1. Use. Treatment of pain-only decompression sickness when oxygen cannot be used and pain is relieved at a depth greater than 66 ft.
2. Descent rate. 25 ft/min.
3. Ascent rate. One minute between stops.
4. Time at 165 ft. Includes time from the surface.

c. Schedule 8.3: Air Treatment

1. Use. Treatment of serious symptoms when oxygen cannot be used and symptoms are relieved within 30 min at 165 ft.
2. Descent rate. 25 ft/min.
3. Ascent rate. One minute between stops.
4. Time at 165 ft. Includes time from surface.

d. Schedule 8.4: Air Treatment

1. Use. Treatment of serious symptoms or gas embolism when oxygen cannot be used and when symptoms are not relieved within 30 min at 165 ft.
2. Descent rate. 25 ft/min.
3. Ascent rate. One minute between stops.
4. Time at 165 ft. Includes time from the surface.

e. Schedule 8.5: Oxygen Treatment

1. Use. Treatment of pain-only decompression sickness when oxygen can be used and symptoms are relieved within 10 min at 60 ft. Patient breathes oxygen from the surface.
2. Descent rate. 25 ft/min.
3. Ascent rate. 1 ft/min. Do not compensate for slower ascent rates. Compensate for faster rates by halting the ascent.
4. Time at 60 ft. Begins on arrival at 60 ft.
5. If oxygen breathing must be interrupted, allow 15 min after the reaction has entirely subsided and resume schedule at point of interruption.
6. If oxygen breathing must be interrupted at 60 ft, switch to Schedule 8.6 upon arrival at the 30-ft stop.

7. Tender breathes air throughout. If treatment is a repetitive dive for the tender or the table is lengthened, the tender should breathe oxygen during the last 30 min of ascent to the surface.

f. Schedule 8.6: Oxygen Treatment

1. Use. Treatment of decompression sickness when oxygen can be used and symptoms are not relieved within 10 min at 60 ft. Patient breathes oxygen from the surface.
2. Descent rate. 25 ft/min.
3. Ascent rate. 1 ft/min. Do not compensate for slower ascent rates. Compensate for faster rates by halting the ascent.
4. Time at 60 ft. Begins on arrival at 60 ft.
5. If oxygen breathing must be interrupted, allow 15 min after the reaction has entirely subsided and resume schedule at point of interruption.
6. Tender breathes air throughout. If treatment is a repetitive dive for the tender or the table is lengthened, the tender should breathe oxygen during the last 30 min of ascent to the surface.

g. Schedule 8.7: Oxygen Treatment

1. Use. Treatment of gas embolism when oxygen can be used and symptoms are relieved within 15 min at 165 ft.
2. Descent rate. As fast as possible.
3. Ascent rate. 1 ft/min. Do not compensate for slower ascent rates. Compensate for faster rates by halting the ascent.
4. Time at 165 ft. Includes time from the surface.
5. If oxygen breathing must be interrupted, allow 15 min after the reaction has entirely subsided and resume schedule at point of interruption.
6. Tender breathes air throughout. If treatment is a repetitive dive for the tender or the table is lengthened, the tender should breathe oxygen during the last 30 min of ascent to the surface.

h. Schedule 8.8: Oxygen Treatment

1. Use. Treatment of gas embolism when oxygen can be used and symptoms moderate to a major extent within 30 min at 165 ft.
2. Descent rate. As fast as possible.
3. Ascent rate. 1 ft/min. Do not compensate for slower ascent rates. Compensate for faster ascent rates by halting the ascent.
4. Time at 165 ft. Includes time from the surface.
5. If oxygen breathing must be interrupted, allow 15 min after the reaction has entirely subsided and resume schedule at point of interruption.
6. Tender breathes air throughout. If treatment is a repetitive dive for the tender or the table is lengthened, the tender should breathe oxygen during the last 30 min of ascent to the surface.

Schedule 8.1
Recompression Treatment Table (Air Treatment)
(U. S. Navy 1973)

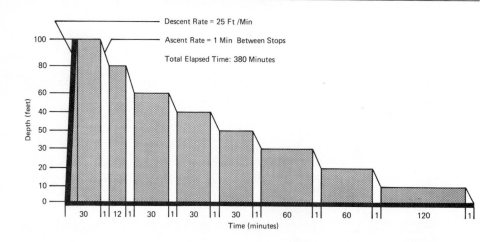

Depth, ft	Time, min	Breathing medium	Total elapsed time, min
100	30	Air	30
80	12	Air	43
60	30	Air	74
50	30	Air	105
40	30	Air	136
30	60	Air	197
20	60	Air	258
10	120	Air	379
0	1	Air	380

Schedule 8.2
Recompression Treatment Table (Air Treatment)
(U. S. Navy 1973)

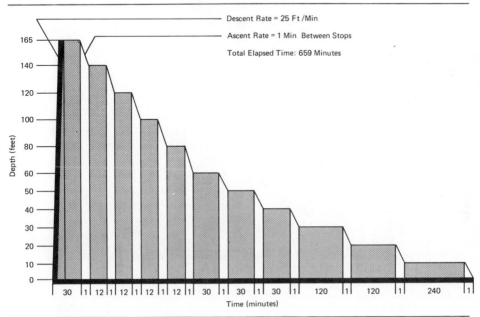

Descent Rate = 25 Ft /Min

Ascent Rate = 1 Min Between Stops

Total Elapsed Time: 659 Minutes

Depth, ft	Time, min	Breathing medium	Total elapsed time, min
165	30	Air	30
140	12	Air	43
120	12	Air	56
100	12	Air	69
80	12	Air	82
60	30	Air	113
50	30	Air	144
40	30	Air	175
30	120	Air	296
20	120	Air	417
10	240	Air	658
10	240	Air	658
0	1	Air	659

Schedule 8.3
Recompression Treatment Table (Air Treatment)
(U. S. Navy 1973)

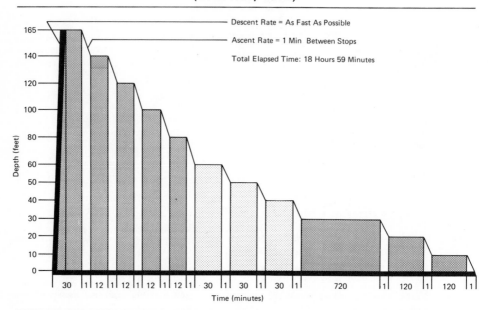

Depth, ft	Time	Breathing medium	Total elapsed time, hr:min
165	30 min	Air	0:30
140	12 min	Air	0:43
120	12 min	Air	0:56
100	12 min	Air	1:09
80	12 min	Air	1:22
60	30 min	Oxygen (or air)	1:53
50	30 min	Oxygen (or air)	2:24
40	30 min	Oxygen (or air)	2:55
30	12 hr	Air	14:56
20	2 hr	Air	16:57
10	2 hr	Air	18:58
0	1 min	Air	18:59

Schedule 8.4
Recompression Treatment Table (Air Treatment)
(U. S. Navy 1973)

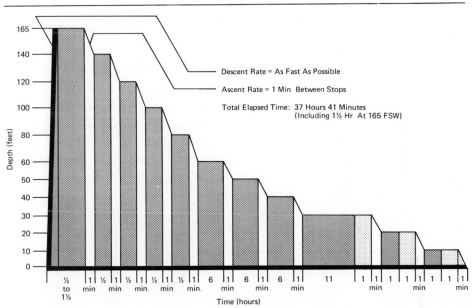

Descent Rate = As Fast As Possible

Ascent Rate = 1 Min Between Stops

Total Elapsed Time: 37 Hours 41 Minutes
(Including 1½ Hr At 165 FSW)

Depth, ft	Time	Breathing medium	Total elapsed time, hr:min
165	½–1½ hr	Air	1:30
140	½ hr	Air	2:01
120	½ hr	Air	2:32
100	½ hr	Air	3:03
80	½ hr	Air	3:34
60	6 hr	Air	9:35
50	6 hr	Air	15:36
40	6 hr	Air	21:37
30	11 hr	Air	32:38
30	1 hr	Oxygen (or air)	33:38
20	1 hr	Air	34:39
20	1 hr	Oxygen (or air)	35:39
10	1 hr	Air	36:40
10	1 hr	Oxygen (or air)	37:40
0	1 min	Oxygen	37:41

Schedule 8.5
Recompression Treatment Table (Oxygen Treatment)
(U. S. Navy 1973)

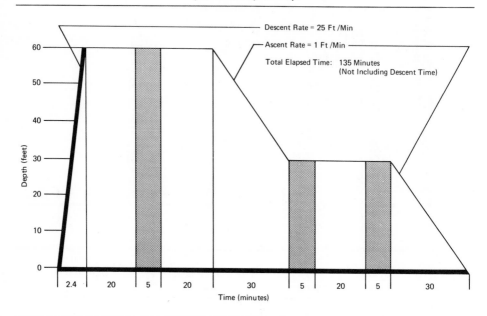

Depth, ft	Time, min	Breathing medium	Total elapsed time,[a] min
60	20	Oxygen	20
60	5	Air	25
60	20	Oxygen	45
60–30	30	Oxygen	75
30	5	Air	80
30	20	Oxygen	100
30	5	Air	105
30–0	30	Oxygen	135

[a] Does not include descent time.

Schedule 8.6
Recompression Treatment Table (Oxygen Treatment)
(U. S. Navy 1973)

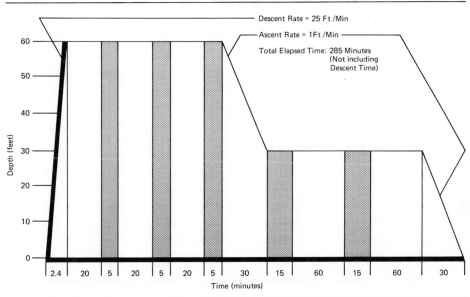

Depth, ft	Time, min	Breathing medium	Total elapsed time,[a] min
60	20	Oxygen	20
60	5	Air	25
60	20	Oxygen	45
60	5	Air	50
60	20	Oxygen	70
60	5	Air	75
60–30	30	Oxygen	105
30	15	Air	120
30	60	Oxygen	180
30	15	Air	195
30	60	Oxygen	255
30–0	30	Oxygen	285

[a] Does not include descent time.

Schedule 8.7
Recompression Treatment Table (Oxygen Treatment)
(U. S. Navy 1973)

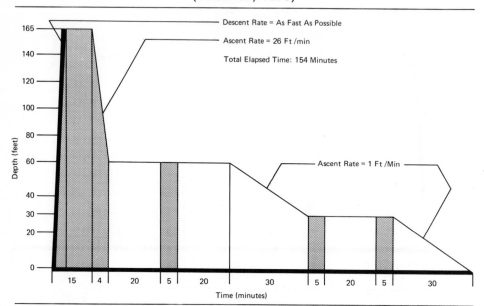

Depth, ft	Time, min	Breathing medium	Total elapsed time, min
165	15	Air	15
165–60	4	Air	19
60	20	Oxygen	29
60	5	Air	44
60	20	Oxygen	64
60–30	30	Oxygen	94
30	5	Air	99
30	20	Oxygen	119
30	5	Air	124
30–0	30	Oxygen	154

Schedule 8.8
Recompression Treatment Table (Oxygen Treatment)
(U. S. Navy 1973)

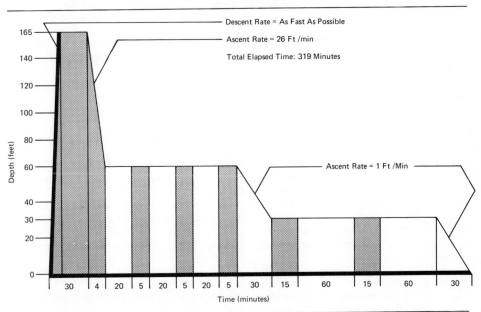

Depth, ft	Time, min	Breathing medium	Total elapsed time, min
165	30	Air	30
165–60	4	Air	34
60	20	Oxygen	54
60	5	Air	50
60	20	Oxygen	79
60	5	Air	84
60	20	Oxygen	104
60	5	Air	109
60–30	30	Oxygen	139
30	15	Air	154
30	60	Oxygen	214
30	15	Air	229
30	60	Oxygen	289
30–0	30	Oxygen	319

References

BARNARD, E. E. P. The treatment of decompression sickness developing at extreme pressures. In: Lambertsen, C. J., ed. *Underwater Physiology. Proceedings of the Third Symposium on Underwater Physiology*, pp. 156–164. Baltimore, Williams and Wilkins (1967).

BECKMAN, E. L., and E. M. SMITH. TEKTITE II. Medical supervision of the Scientists-in-the-Sea. *Texas Rep. Biol. Med.* **30**:1–204 (Fall 1972).

BEHNKE, A. R. Investigations concerned with problems of high altitude flying and deep diving—applications of certain findings pertaining to physical fitness to the general military service. *Milit. Surg.* **90**:9–28 (Jan. 1942).

BEHNKE, A. R. Some early studies on decompression. In: Bennett, P. B., and D. H. Elliott, eds. *The Physiology and Medicine of Diving and Compressed Air Work*, pp. 226–251. Baltimore, Williams and Wilkins (1969).

BEHNKE, A. R. Medical aspects of pressurized tunnel operations. *J. Occup. Med.* **12**:101–112 (Apr. 1970).

BOYCOTT, A. E., G. C. C. DAMANT, and J. S. HALDANE. The prevention of compressed air illness. *J. Hyg.* (*London*) **8**:342–443 (1908).

BUCKLES, R. G., and C. KNOX. In vivo bubble detection by acoustic-optical imaging techniques. *Nature* **222**:771–772 (May 24, 1969).

BÜHLMANN, A. A. The use of multiple inert gas mixtures in decompression. In: Bennett, P. B., and D. H. Elliott, eds. *The Physiology and Medicine of Diving and Compressed Air Work*, pp. 357–385. Baltimore, Williams and Wilkins (1969).

DECOMPRESSION SICKNESS PANEL, MEDICAL RESEARCH COUNCIL. A medical code of practice for work in compressed air. London, England, Construction Industry Research and Information Association (Feb. 1973) (Rep. CIRIA 44).

DES GRANGES, M. Standard air decompression table. U. S. Navy Exp. Diving Unit, Rep. 5-57 (Dec. 3, 1956).

DOUGLAS, E. Solubilities of oxygen, argon, and nitrogen in distilled water. *J. Phys. Chem.* **68**:169–180 (Jan. 1964).

DUFFNER, G. J. Decompression sickness and its prevention among compressed air workers. City of Seattle, Washington (Dec. 20, 1962).

DUFFNER, G. J., J. F. SNYDER, and L. L. SMITH. Adaptation of helium–oxygen to mixed-gas scuba. U. S. Navy Exp. Diving Unit, Rep. 3-59 (1959).

GALLOWAY, W. J. An experimental study of acoustically induced cavitation in liquids. *J. Acoust. Soc. Amer.* **26**:849–857 (Sept. 1954).

HAMILTON, R. W., JR., D. J. KENYON, M. FREITAG, and H. R. SCHREINER. NOAA OPS I and II: Formulation of excursion procedures for shallow undersea habitats. Tarrytown, N. Y., Union Carbide Tech. Cent., Environ. Physiol. Lab., Rep. UCRI 731 (July 31, 1973).

HARVEY, E. N. Physical factors in bubble formation. In: Fulton, J. F., ed. *Decompression Sickness*, pp. 90–114. Philadelphia, W. B. Saunders and Co. (1951).

HAWKINS, J. A., C. W. SHILLING, and R. A. HANSEN. A suggested change in calculating decompression tables for diving. *U. S. Nav. Med. Bull.* **33**:327–338 (July 1935).

HEMPLEMAN, H. V. Decompression procedures for deep, open sea operations. In: Lambertsen, C. J., ed. *Underwater Physiology. Proceedings of the Third Symposium on Underwater Physiology*, pp. 255–266. Baltimore, Williams and Wilkins (1967).

HEMPLEMAN, H. V. British decompression theory and practice. In: Bennett, P. B., and D. H. Elliott, eds. *The Physiology and Medicine of Diving and Compressed Air Work*, pp. 291–318. Baltimore, Williams and Wilkins (1969).

JONES, H. B. Gas exchange and blood-tissue perfusion factors in various body tissues. In: Fulton, J. F., ed. *Decompression Sickness*, pp. 278–321. Philadelphia, W. B. Saunders Co. (1951).

KIDD, D. H., and D. H. ELLIOTT. Clinical manifestations and treatment of decompression sickness in divers. In: Bennett, P. B., and D. H. Elliott, eds. *The Physiology and Medicine of Diving and Compressed Air Work*, pp. 464–490. Baltimore, Williams and Wilkins Co. (1969).

MARTIN, F. E., J. E. HUDGENS, and J. W. WONN. Manned hyperbaric demonstration of incipient bubble detection using nonlinear ultrasonic propagation. Annapolis, Md., Westinghouse Electric Corp., Ocean Res. Eng. Cent., Rep. OER-73-16 (May 31, 1973).

MOLUMPHY, G. G. He–O_2 decompression tables. U. S. Navy Exp. Diving Unit, Rep. 8-50 (Sept. 26, 1950).

MORRISON, J. J. The salting-out of non-electrolytes. Part I. The effect of ionic size, ionic charge, and temperature. *J. Chem. Soc.* (*London*) (Pt. III):3814–3818 (Oct. 1952).

RASHBASS, C. Investigation into the decompression tables. Alverstoke, England, Roy. Nav. Physiol. Lab., Med. Res. Counc., Roy. Nav. Pers. Res. Comm., Rep. UPS 151 (Oct. 1955).

RUBISSOW, G. J., and R. S. MACKAY. Ultrasonic imaging of in vivo bubbles in decompression sickness. *Ultrasonics* 9:225–234 (Oct. 1971).

SAYERS, R. R., W. P. YANT, and J. H. HILDEBRAND. Possibilities in the use of helium–oxygen mixtures as a mitigation of caisson disease. U. S. Dept. Interior, Bur. Mines, Rep. 2670 (Feb. 1925).

SCHIBLI, R. A., and A. A. BÜHLMANN. The influence of physical work upon decompression time after simulated oxygen–helium dives. *Helv. Med. Acta* 36:327–342 (Oct. 1972).

SCHREINER, H. R., and P. L. KELLY. Computation methods for decompression from deep dives. In: Lambertsen, C. J., ed. *Underwater Physiology. Proceedings of the Third Symposium on Underwater Physiology*, pp. 275–299. Baltimore, Williams and Wilkins (1967).

SPENCER, M. P., S. D. CAMPBELL, J. L. SEALEY, F. C. HENRY, and J. LINDBERGH. Experiments on decompression bubbles in the circulation using ultrasonic and electromagnetic flowmeters. *J. Occup. Med.* 11:238–244 (May 1969).

U. S. NAVY. *U. S. Navy Diving Manual.* Washington, D. C., U. S. Navy Department (Mar. 1970) (NAVSHIPS 0994-001-9010).

U. S. NAVY SUPERVISOR OF DIVING. *Handbook: U. S. Navy Diving Operations.* Washington, D. C., U. S. Navy Department, Naval Ships Systems Command (1971) (NAVSHIPS 0994-009-6010).

U. S. NAVY SUPERVISOR OF DIVING. *U. S. Navy Recompression Chamber Operators Handbook.* Washington, D. C., U. S. Navy Department, Naval Ship's System Command (1973) (NAVSHIPS 0994-014-5010).

VAN DER AUE, O. E., R. J. KELLAR, E. S. BRINTON, G. DARRON, H. D. GILLIAM, and R. J. JONES. Calculation and testing of decompression tables for air dives employing the procedure of surface decompression and the use of oxygen. U. S. Navy Exp. Diving Unit, Unit Rep. 1 on Proj. NM002007 (Nov. 1951).

WEBSTER, E. Cavitation. *Ultrasonics* 1:39–48 (Jan./Mar. 1963).

WEISS, R. F. The solubility of nitrogen, oxygen and argon in water and seawater. *Deep Sea Res.* 17:721–735 (1970).

WISCONSIN STATE DEPARTMENT OF INDUSTRY, LABOR AND HUMAN RELATIONS. Work under compressed air, Appendix A. In: *Wisconsin Administrative Code of Rules of the Department of Industry, Labor and Human Relations.* Madison, Wisconsin (Aug. 1971). (No. 188).

WORKMAN, R. D. Calculation of decompression schedules for nitrogen–oxygen and helium–oxygen dives. U. S. Navy Exp. Diving Unit, Rep. 6-65 (May 26, 1965).

WORKMAN, R. D., and J. L. REYNOLDS. Adaptation of helium–oxygen to mixed gas scuba. U. S. Navy Exp. Diving Unit, Rep. 1-65 (Mar. 1, 1965).

VIII

Operational Safety Considerations

A. *Introduction*

Underwater activity confronts man with forces and resulting physiological effects that are not encountered in his normal terrestrial environment. "These forces impose definite limits and can cause serious accidents. The diver's safety depends upon his knowledge of these factors and his ability to recognize and handle them" (U. S. Navy 1970).

Many potential dangers can be almost completely avoided if they are recognized. Ignorance is the diver's worst enemy.

In a detailed study of 38 skilled scuba divers Taylor (1959) found that: "In the opinion of the divers, the greatest hazards to life resulting from diving are man-made; namely, panic, ignorance, foolhardiness, other careless divers, poor training, air embolism and bends." The number of divers citing each hazard is listed here:

Panic	23
Ignorance	4
Foolhardiness	3
Other careless divers	3
Poor training	2
Air embolism	2
Bends	1

Figure VIII-1 presents a graphic view of accidents, illnesses, hazards, and environmental difficulties which might occur under different exposures to different divers. In the center is the diver who has been carefully selected and adequately trained, and whose predive condition is satisfactory. In spite of all of this he is acutely aware of the altered sensory input associated with his underwater environment, as detailed in the first concentric circle. Circle number 2 shows the physiological and psychological problems inherent in underwater work. The diver may not be aware of most of these effects until the danger point is reached. The outer circle, number 3,

629

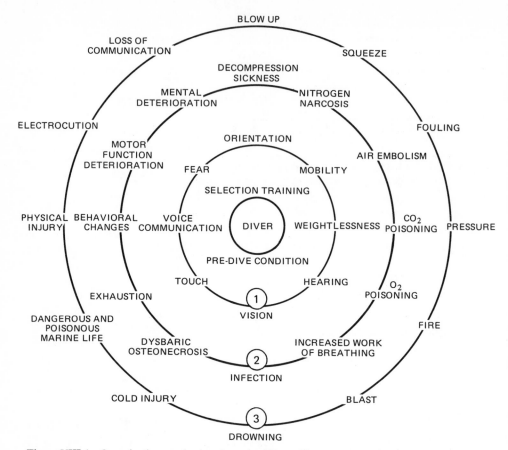

Figure VIII-1. Organization and planning. Accidents, illnesses, hazards, and environmental difficulties affecting performance. The diver—selected, trained, and in good condition. (1) The immediate circle of sensory awareness. (2) Physiological and psychological problems inherent in underwater work. (3) Possible environmental accidents.

lists the possible environmental accidents. It should be noted that there is a good deal of interrelation between the conditions listed in the three circles and in some cases a cause-and-effect relationship.

It is the purpose of this chapter to discuss the chief hazards confronting the diver and the ways in which they can be most successfully guarded against, from organizational planning on through escape and rescue.

B. *Organization and Planning*

It is imperative that in all underwater activity, whether military or civilian, a knowledgeable person be "in charge." Someone must be responsible for the well-being and safety of the divers, and for the efficiency of the operation. Regardless of how much fun it may be, it is well to remember that all diving is associated with hazards not encountered on dry land. A number of the hazards peculiar to diving will be considered in this chapter.

Planning must precede any diving operation; the diver must be assured of proper equipment, proper assistance from a buddy or assistant, and adequate management from the surface crew. In any operation bottom time is at a premium and time spent in careful planning should result in greater efficiency as well as greater safety.

In planning, the first consideration is to determine and define the objective. What is to be accomplished, is it feasible—and necessary—and what is the best way to proceed? Can the objective be accomplished without exposing a diver? A plan should be outlined and all phases considered in detail.

Underwater conditions must be considered. What is the water depth at the work-site, and what is the type of bottom? Will tides and currents be a problem? Is everything ready on the surface, including personnel and equipment, and are weather conditions satisfactory? One of the most common errors is to plan inadequately the supporting effort.

The diver should be given step-by-step instructions for the task he is to accomplish, and also told of the organization of the supporting group who will make sure that the lines, the tools, and the other equipment respond to his every need.

When diving becomes routine there is a tendency to relax and lay aside the planning and preparation for emergencies; this is when accidents occur. Be sure all necessary equipment is available and in good working order, and particularly be sure that personnel are ready for any emergency. This includes knowing the location of the nearest recompression chamber and the nearest medical facilities, and having appropriate transportation available.

It is imperative that no person should participate in any underwater activity who has not been selected in accordance with standard criteria (see Chapter XI, discussion of *selection*) and who has not received training appropriate to the task at hand (see Chapter XI, discussion of *training*).

1. *General Safety Precautions*

Safety instructions for special types of diving will be considered in the next section. In this section we present a *check list* common to all diving activities. It is taken from the *U. S. Navy Diving Manual*, Section 1.7:

1. Have you exhausted all other plans of action before putting a diver down?
2. Have you carefully planned the total operation?
3. Have the divers been thoroughly briefed and do they understand what is to be accomplished and how?
4. Have you notified all interested activities that diving operations are in progress?
5. Is the type of gear you have chosen adequate and safe for the job?
6. Have competent supervisory personnel been designated?

7. Have arrangements been made to obtain medical assistance in case of emergency?
8. Is the recompression chamber ready for use, or have you determined the availability of the nearest chamber?
9. Is a copy of the appropriate decompression tables available?
10. Have you taken into consideration all environmental factors: weather, sea state, depth, type of bottom, etc.?

a. Personnel

It is important that all divers have been properly selected (Chapter XI, *selection*) and trained (Chapter XI, *training*) and are in condition to dive (see this chapter, discussion of *predive conditions*). In addition to the above, the following checkoff list is important:

1. Are the divers qualified for the type of job contemplated, and are they qualified for the depth anticipated?
2. Have all the divers been trained in the use of the equipment planned for use?
3. Have you checked the predive condition of your divers?

b. Equipment

1. Has the equipment been tested to determine if it is in working order and suitable for the job?
2. Do you have any spare equipment in case of need?
3. Have you an adequate supply of compressed gas available?

c. Safety during Diving Operations

1. Is the diving boat moored in the most advantageous position to minimize effort by the diver to reach the work?
2. Are you displaying the proper warning signals? Proper flags are laid down by international and national authorities. Proper lights at night?
3. Have all efforts been made to prevent the divers from becoming fouled on the bottom?
4. Have the divers been instructed not to cut any lines until they have ascertained the purpose for which they are being used?
5. If work is being performed inside a wreck, have you made arrangements for one diver to tend the lines of the diver working inside from the point of entry?
6. If explosives are being used, have you taken measures to prevent a charge being set off when a diver is in the water?
7. If electric power for underwater welding or cutting is being used, is the diver properly dressed?
8. Have you made provisions for decompressing the divers should this be necessary?

Remember: In all cases, the depth of the water and the condition of the diver, especially with regard to fatigue, rather than the amount of work to be done shall determine the amount of time the diver is to spend on the bottom.

d. Recompression Chambers

See Chapter VII, on *decompression sickness*.

e. Safe Handling of Compressed-Gas Cylinders

See Chapter IV, on *the physiology of breathing mixtures*.

2. *Special Situations*

The above general considerations cannot be expected to cover all situations. The following are some diving problems requiring special plans and arrangements (CIRIA 1972):

Diving in shark-infested waters
Diving in wrecks
Diving on or near ship's bottom
Diving in the vicinity of ship's propellers
Diving in locks, basins, and culverts
Diving to place underwater explosives
Diving in sewers, and in natural and artificial underground waters
Diving from drilling rigs and other high platforms
Diving from a pipe laybarge
Diving from a basket slung from a crane
Diving from a diving bell operated from a crane
Divers operating a wet submersible

3. *Diving at Altitude*

The decompression schedules normally used by sports and professional divers at sea level must have adjustments made to compensate for lowered atmospheric pressure, e.g., in a mountain lake. Extensive research was conducted by Bühlmann *et al.* (1973) on decompression following diving in mountain lakes. They point out that, "the major factor that causes a decompression accident is the ratio of nitrogen pressure in the tissue to ambient pressure; therefore, for the same depth and the same bottom time but for a lower ambient pressure, longer decompression times are necessary."

According to CIRIA (1972), the adjustments should be as follows:

1. Altitudes of less than 100 m ($\sim$300 ft): No adjustment is required.
2. Dives between altitudes of 100 and 300 m ($\sim$300–1000 ft): Add one-fourth of the depth to give the depth of the dive.

3. Dives between altitudes of 300 and 2000 m ($\sim$1000–6500 ft): Add one-third of the depth to give the depth of the dive.
4. Dives between altitudes of 2000 and 3000 m ($\sim$6500–10,000 ft): Add one-half to the depth to give the depth of the dive.

4. *Polar or Ice Diving*

Additional hazards must be overcome when diving in very cold environment and under the ice. However, if proper diving equipment, support equipment and protective measures are used, it is possible to overcome the problems encountered as a result of the cold. A complete guide to working under polar conditions has been written by Jenkins (1973) and should be carefully studied before undertaking diving under the ice.

All accidents and the resulting emergencies are much more dangerous and difficult to handle in frigid temperatures. For example, falling into the water in the tropics can be a lark but falling into the water in the arctic can be rapidly fatal. Do not try to swim unless forced to by lack of a life jacket or other flotation equipment. Float with the life jacket and await rescue. Do not exercise in the cold water in an attempt to keep warm, for it will have the reverse effect. Stay clear of entry holes when not required to assist the diver, for the edge is invariably slippery. Never work around a hole by yourself.

"In the event of a breathing-system failure, the diver should: switch to his backup system if available; notify his partner; exit to the surface with buddy; or exit to the surface without buddy if necessary and have him recalled by surface personnel. In the event of a tear or flooding of the exposure suit, the diver should surface immediately for the extreme chilling effect of polar water will render the diver in thermal stress within minutes" (Jenkins 1973).

If uncontrolled ascent occurs due to a suit blowup or lost weight belt, the diver should: exhale during ascent; relax against the ice; if suit blowup, release pressure with exhaust valve; signal tender to haul in the tether; and wait for assistance from buddy and/or tender. A lost diver should ascend to the overhead ice cover immediately, relax as much as possible, and await rescue. Under polar environmental conditions it is particularly important to have a well-planned and rehearsed set of emergency procedures.

C. *Predive Conditions*

The importance of an individual's physical and psychological condition in coping with a stressful situation is well known and should be emphasized here. This is not only a general requirement for selection, but is equally or even more important for each dive. The following is a brief description of conditions that should be checked before a diver begins the day's underwater activity. Particular care must be exercised if the planned task is difficult or hazardous.

1. *Age*

Although no definitive research can be found dealing with the relationship of age to diving, there is a generally accepted dictum that men over 45 should not be called upon for strenuous deep-diving missions (see Chapter XI, on *selection*).

2. *Drugs*

Some drugs can produce significant temporary adverse effects on the diver's mental processes, motor coordination, physical stamina, and tolerance to cold. Even with marijuana, "in warm water the diver may become ultrarelaxed, sleepy, unaware, lazy, and his work ability may be reduced significantly" (Somers 1972). Thus anyone taking drugs should be most carefully evaluated and certainly no one who is drug-addicted should be allowed to dive.

3. *Alcohol*

The immediate apparent effects of consumption of alcoholic beverages are mental disorientation, impaired physical coordination, vertigo, poor judgment, and general physical weakness, all of which are incompatible with safe diving. Thus, at least 24 hr should elapse between heavy drinking and diving. Individuals should not dive for at least 4 hr after drinking any alcoholic beverage, and a chronic alcoholic should never be allowed to dive (see Chapter XI, on *selection*).

4. *Cigarette Smoking*

A heavy smoker (1 + packs per day) loses lung capacity and thus ability to adjust quickly to a changing pressure environment. A worker under compressed air either as a tunnel laborer or a diver would be wise not to smoke at all. The long-time heavy smoker should not be selected for more arduous underwater tasks.

5. *Diet*

Evidence points to the desirability of a low-fat, but otherwise normally balanced diet (see Chapter IV, discussion of *diet and metabolism*) for an active diver. Diving should not be undertaken until 2 hr after the last meal.

6. *Obesity*

The evidence here is incontrovertible—a fat rat or a fat man is more susceptible to decompression sickness than a lean one. Some authorities say that for regular underwater or compressed-air work the weight of the individual should not be more than 10% over that given in standard table (see Chapter IV, discussion of *diet and metabolism*, and Chapter XI, on *selection*).

7. Fatigue

Physical fatigue predisposes the individual to decompression sickness and to a decline in both quality and quantity of production for a given amount of effort and is, therefore, to be avoided whenever possible. Psychological fatigue is closely related to physiological fatigue and the resultant feelings of strain, irritation, and general discomfort lead to a disturbed physiological state with its deleterious effects. It must be remembered that motivation and job satisfaction are especially significant aspects of hazardous, physically and psychologically demanding occupations such as deep diving. The signs, symptoms, and associated behavior of acute and chronic fatigue have been prepared by Behnke (1970) and are given in Table VIII-1. Fatigue is discussed in more detail in this chapter in the section on *other hazards*.

8. Exercise

A diver should be in the best physical condition. Evidence shows that regular exercise builds up the cardiorespiratory system as well as the musculature. Exercise on the bottom or under pressure leads to early and more complete saturation of the body tissues with the gas being breathed and, thus, predisposes the diver to decompression sickness; exercise during the later stages of decompression helps to eliminate the dissolved gases more quickly and is, therefore, desirable.

9. Emotional Instability

The stable or *normal* individual is usually able to adapt to stress because of a large measure of compensatory reserve, but there is a true saying that everyone has his breaking point. The supervisor must be constantly aware of the possibility of emotional instability. Behnke (1970) has prepared a checkoff list of danger signs and symptoms, which is presented in Table VIII-1.

10. Respiratory Infections

Infections such as colds, sore throat, and sinusitis are important to consider because pressure equalization may be impaired or the infection may be made worse; also, a diver with a respiratory infection is not working at peak efficiency. Chest colds and bronchitis add other complications, since coughing is a problem for anyone who must hold a mouthpiece in place, and breathing resistance is increased during such infections. Except in an emergency, never let a man dive with fever or other signs of respiratory infection.

Table VIII-1

Manifestations of Acute and Chronic Fatigue : Impairments Which
Preclude an Individual on any Given Day from Diving—A
Checkoff List for the Supervisor Responsible[a]

| Intellectual functions | Impairment referable to somatic systems | | | Affective behavior |
	System or region	Overt	Covert	
Difficulty in thinking and in concentration, impaired memory, insight, and judgment	Cephalic	Drowsiness	Heavy head, headache	Anxiety, tension, irritability, exaggerated fears
Fixation of ideas	Respiratory	Shortness of breath, shallow, rapid breathing	Feeling of suffocation	Depression, lethargy
	Vasomotor	Sweating, pallor, rapid pulse	Weakness	
	Cardiac	Decreased ability to work; rise in diastolic blood pressure	Precordial distress	Euphoria, excitement, hilarity, pugnacity
	Gastric	—	Loss of appetite, distress, nausea	
	Intestinal	Diarrhea	Cramps	
	Neuromuscular	Impaired coordination, tremors, speech disorders	Fatigue	

[a] Axiom: The actions of the emotionally aroused individual are unpredictable. From Behnke (1970) with permission of the Athletic Institute, Chicago.

D. *Underwater Blast*

Underwater blast is an important problem for all underwater workers, and the hazard has increased with the greater use of swimmers and divers in handling both underwater demolition and blasting work. The factors which determine the degree of injury sustained by personnel in the water at the time of an underwater blast are: proximity to the source of the blast; size and character of the explosive; the medium through which the force is transmitted; the depth and/or degree of submersion of the swimmer or diver; and the protection worn.

The medium through which the force is transmitted, i.e., air or water, makes an enormous difference. World War II demonstrated that for detonations yielding the same energy, underwater blast was more deadly than air blast at the same standoff range (Hoff and Greenbaum 1954). This is due to the fact that the pressure of the shock waves decays much more rapidly in air than in water. A man who would be unharmed by an air explosion of a hand grenade at 15 ft distance (providing he is out of the direct line of shrapnel) would undoubtedly be killed if a similar charge were exploded at the same distance from him underwater (Committee on Amphibious Operations 1952).

1. *Physical Aspects of the Explosion*

a. The Initial Compression Wave

In an underwater detonation the explosive force is exerted in all directions as compression waves. The steep-fronted or first wave is known as the "shock wave" (Cole 1965).

The pressure level in the spherical wave depreciates more rapidly with distance than the inverse first power law, but eventually approaches this limit at large distances (Cole 1965). This concept is illustrated in Figure VIII-2. Pressure level behind the maximum pressure falls off exponentially. The front of the pulse remains steep and the time of decay of the pulse tends to increase as the high-pressure front of the wave gets farther and farther ahead of the low-pressure tail.

b. Reflection of the Primary Pressure Pulse

Additional pressure pulses are experienced at some distance from the explosion due to the primary pressure pulse being reflected off the bottom and from the surface of the water. In both cases, the angle of incidence is equal to the angle of reflection. In general, the pressure wave reflected from the bottom tends to add to the initial pressure wave, while that reflected from the water surface reduces the effect.

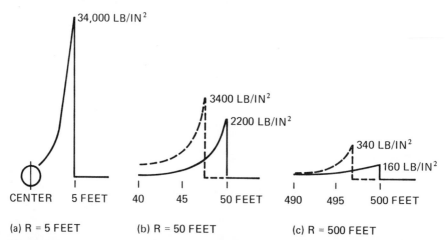

Figure VIII-2. The pressure distribution around a 300-lb TNT charge at three different times (distances) after completion of detonation. For comparison, the pressure waves which would exist if the earlier state (condition *a*) were propagated as an acoustic wave are indicated by the dashed curve. [From Cole (1965) by permission of Princeton University Press.]

The reflection of the pulse off the bottom is always somewhat distorted, even if the reflector is hard rock. With soft mud the distortion is very obvious, and the reflection may sometimes seem to come from a reflecting layer some way below the actual surface of the mud. For very soft mud, there is no reflection at all. If the charge is actually fired on the bottom there is no reflected pulse, but the primary pulse may be stronger than it would have been for the charge in mid-water, if the bottom material is hard.

The reflection of the pressure pulse at the air–water boundary results in a direct pressure pulse (pressure) and a reflected pulse (tension) separated by a time interval corresponding to the difference between the two acoustic paths. This is illustrated in Figure VIII-3. The reflected pulse, coming from the surface, subtracts from the initial pulse. For practical purposes, the effect is to suppress the latter part of the initial pressure pulse, and thus to diminish the time-integral of the pressure impulse, without affecting the maximum pressure.

According to a recent report of the Naval Ordnance Laboratory (Christian and Gaspin 1974), "the impulse of the shock wave is greatly influenced by the duration of the pulse." When "the pulse duration is determined by the time at which the surface-reflected wave arrives and cuts off the tail of the shock wave," this time t_s depends on the depth of the charge D, the depth of the swimmer y, and the slant range R from the charge. The authors give the following equation for t_s:

$$t_s \,(\text{m sec}) = 0.2[(R^2 + 4yD)^{1/2} - R] \tag{1}$$

This is assuming that the constant sound velocity in water is 5 f/m sec. The impulse I of the shock wave up to time t_s is

$$I \,(\text{psi-m sec}) = 1.3 \times 10^3 W^{0.64} R^{-0.91}[1 - e^0] \tag{2}$$

where $Q = -16.67 t_s W^{-0.26} R^{-0.18}$.

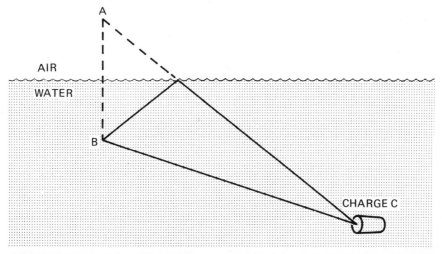

Figure VIII-3 Reflection of the initial pressure pulse from the water surface. The path difference between the positive (pressure) and negative (tension) pulses = CA − CB. The tension wave arrives later by the interval (AC − CB)/a, where a is the velocity of sound in water (5000 ft/sec). [From Wakeley (1945) by permission of Lancet, Ltd., London.]

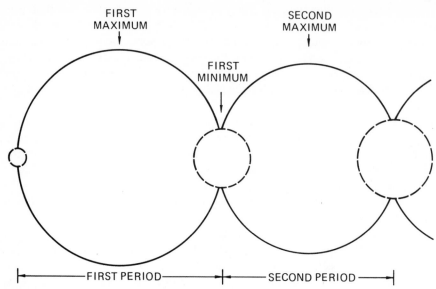

Figure VIII-4. Oscillation of the gas products from an underwater explosion. The period of oscillation, in the absence of boundary effects, is related to the internal energy of the gas and the hydrostatic pressure, being proportional to the cube root of energy and inverse five-sixths power of the pressure. [From Cole (1965) by permission of Princeton University Press.]

c. The Gas Sphere and Subsidiary Pulses

The initial shock wave is succeeded by other pressure waves resulting from oscillations of the gas bubble produced by the explosion (Cole 1965). The inertia of the water and the elastic properties of the gas provide the conditions for an oscillating system, and the bubble undergoes repeated cycles of expansion and contraction. This process is illustrated in Figure VIII-4. These oscillations may persist for ten or more cycles. When the bubble is at minimum radius the pressure is at maximum. They are relatively small compared to the initial shock wave, and will not be considered further.

d. Surface Effects

The features visible above the surface depend considerably on the initial depth of the charge. They are quite spectacular for shallow explosions, but virtually undetectable for great depths. Where the sum of the direct and reflected pressure pulses surpasses the critical value that the water surface can sustain (500 psi), the water surface rises, producing a characteristic *dome* (Wolf 1970). The breakthrough of the bubble occurs sometime later. In some cases, plumes of water may rise hundreds of feet into the air. These effects become less pronounced for deeper depths of explosion and may be represented only by an emulsion of explosion products, or *slick*, spreading across the surface of the water (Cole 1965).

2. *Mechanisms of Injury*

Damage or injury from underwater blast is caused by the shock wave. The under-water swimmer is vulnerable at many fluid-gas interfaces in his body that are subject to pressure–tension phenomena (Committee on Amphibious Operations 1952). The lungs and intestines are the principal sites of injury. Because the head also contains air spaces, damage to these spaces and to the central nervous system can occur if the diver's head is underwater (U. S. Navy 1970). The parts of the body that do not con-tain air or gas, such as the muscles and bones, transmit the shock wave practically without disruption and are usually not damaged (Committee Amphibious Operations 1952).

a. Peak Pressure

Shock-wave peak pressure and shock-wave impulse are of equal importance with relation to the degree of damage produced. The *U. S. Navy Diving Manual* (1970) uses the peak pressure as the damage variable, assuming that a pressure greater than 500 psi will cause injury to the lungs and intestinal tract, and that one greater than 2000 psi will cause certain death.

The Committee on Amphibious Operations (1952) takes a more conservative view of the maximum sustainable pressure (200 psi for deterrent and 300–400 psi for lethal pressure).

According to Christian and Gaspin (1974), the safety level is at about 125 psi (provided the impulse does not exceed 5.5 psi-m sec). It is pointed out, however, that no internal injuries were found to occur in animal experiments at psi of several hundred as long as the impulse did not exceed 2 psi-m sec. They state that the "com-putation of safe standoff ranges involves finding the locations in the water at which the peak shock wave pressure is 50 psi at the same time that the shock-wave impulse is 2 psi-m sec." The equation for peak pressure P is given as

$$P = A(W^{1/3}/R)^\alpha \tag{3}$$

where W is explosive charge weight, R is the slant range between explosion and swim-mer, and A and α are constants that vary with the explosive material. The equation given for TNT, when charge weight is given in pounds and range is given in yards, is (see Figure VIII-5)

$$P\,(\text{psi}) = 6242(W^{1/3}/R)^{1.13} \tag{4}$$

b. Momentum or Impulse

As has been noted above, the impulse of the shock wave is greatly influenced by the direction of the pulse (Figure VIII-3). Initially, the impulse is the same in all directions from the charge but, as has been noted, the surface-reflected tension wave decreases the initial impulse by combining with the outgoing shock wave and cutting off the decaying tail of the wave (see Figure VIII-6). At shallower depths, the reflected wave arrives closer behind the shock wave, causing a shorter pressure pulse, and

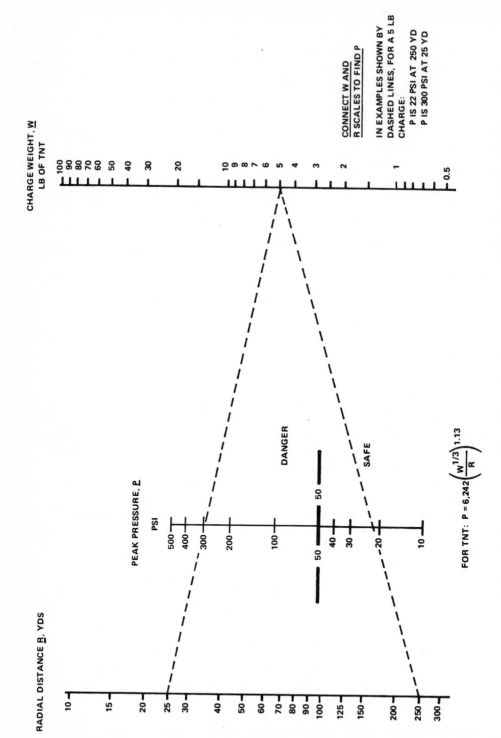

Figure VIII-5. Determination of peak pressure for given weight of TNT and slant range (Christian and Gaspin 1974).

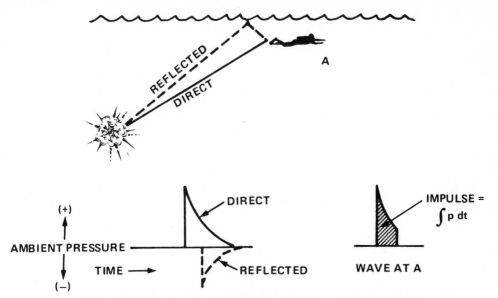

Figure VIII-6. Shock wave characteristics (Christian and Gaspin 1974).

reducing the impulse. Thus, the shallow swimmer will receive a smaller impulse than the deep swimmer, provided the slant range from the charge is the same in both cases. This effect of depth also applies to the charge itself: A swimmer at constant depth will receive less impulse from a shallow charge than from a deep one (see Figures VIII-7 and VIII-8).

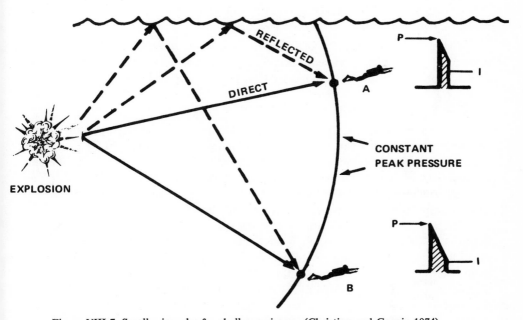

Figure VIII-7. Smaller impulse for shallow swimmer (Christian and Gaspin 1974).

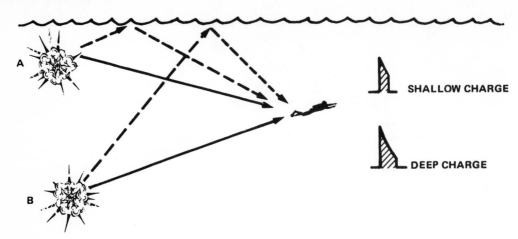

Figure VIII-8. Smaller impulse for shallow charge (Christian and Gaspin 1974).

c. Degree of Immersion

The organs containing gas are most likely to sustain underwater blast damage only if they are immersed in the water. In partial immersion, all the wetted area will be subject to the maximum pressure, and the pressure will terminate latest for those parts that are immersed most deeply, due to the reflection of the pressure wave from the surface of the water. Thus, the time-integral of pressure will be greatest for the deepest parts, dropping to practically nothing at the surface itself. This nonuniform application of pressure may lead to a greater tendency for the cells to burst, and a tendency for all the tissues to be squeezed upward (Wakeley 1945).

3. *Clinical Aspects and Pathology*

The most consistent areas of injury in underwater blast are the lungs and the abdominal air-containing viscera. Occasionally, central nervous system lesions are found, probably secondary to blood shifts or air emboli (Wolf 1970).

a. Effect on the Lungs and Thoracic Cavity

Massive hemorrhages, due to rupture of alveolar walls and tearing of the lungs, are the most prominent feature of severe blast injury (Clemedson 1956). These are the results of pressure differences between the lungs and the pleural space from different accelerations of the structural thoracic elements (Clemedson and Granstom 1950). Arterial air embolism, induced by rupture of boundaries between alveolar spaces and alveolar capillaries, has also been demonstrated (Benzinger 1951).

Patients suffering from blast injury often display extreme difficulty in exhaling. Respiration may be slow and shallow, accompanied by pain in the chest (Barrow and Rhoades 1944). In severely injured animals, cessation of breathing lasting from a few seconds to more than 1 min may occur and, in the more severely injured, lasts until death. Respiratory changes are related to the degree of lung damage and are due primarily to vagal nerve reflexes elicited in the damaged lungs (Clemedson 1949). Benzinger (1950) regarded this reflex action as a protective mechanism that served to restrict lung activity to the minimum necessary to maintain life, sparing the damaged lung tissue and preventing further bleeding.

Severe lung damage and the accompanying complications also have an effect upon the heart, circulation, and nervous system.

(i) Effect upon the Heart. Animal experiments have demonstrated an instantaneous bradycardia, which may range from a slight slowing to an excessive reduction of the heart rate. Sometimes the first postdetonation heart beat does not occur for 30 sec or more, which has been attributed to vagal nerve reflexes elicited in the damaged lungs (Clemedson 1949, Benzinger 1950).

(ii) Effect upon Circulation. The pathologic–anatomic changes in the lungs cause a contraction or obstruction of the pulmonary capillary bed, and an increase in pulmonary pressure, which will, in turn, cause a drop in aortic and systemic arterial pressure (Clemedson 1956).

(iii) Effect upon the Nervous System. Pathological changes in the nervous system from severe blast injury have been reported, including headache, depressed reflexes, and sensorial changes (Hamlin 1943).

(iv) Terminal Effects. In severe lung damage with massive pulmonary hemorrhages and obstruction of the airway, suffocation may result. In most cases, however, lung damage is not sufficient to be the direct cause of death, death being due to circulatory failure, air embolism, or complications such as bronchopneumonia (Clemedson 1956). Mechanisms involved in circulatory failure are circulatory shock due to closing of the respiratory passages (Zuckerman 1941) and cardiac insufficiency and pulmonary edema caused by obstruction of the pulmonary circulation (Clemedson 1949).

The occurrence of air embolism in blast injury appears to be significant where death occurs soon after the time of injury (Benzinger 1951). Air is most commonly found in the coronary arteries and in the left side of the brain. In some cases, the basal brain vessels may be more or less virtually filled with air.

b. Effect on the Abdomen and Air-Filled Viscera

The vulnerability of the low abdominal viscera to blast injury is probably secondary to their containing relatively large amounts of gas and having relatively nonmuscular walls (Wolf 1970). As a rule damage is restricted to the lower abdomen (Wakeley 1945) and there is notable absence of injury to the liver, spleen, kidneys, or bladder (Hoff and Greenbaum 1943). The characteristic finding is a distended and silent abdomen. There may be rupture of intestinal vessels leading to blood clots and obstruction of the lower part of the small intestine, or there may be perforation of the intestine leading to generalized infection the abdominal cavity (McMullin 1943). The abdominal X-ray (Gates 1943) may reveal gaseous distention of the intestines and free gas may be seen in the abdominal cavity.

4. Treatment

A person who is exposed to an underwater blast of any severity should receive medical attention whether or not he appears to be seriously wounded. The patient may be asymptomatic immediately after the injury but should be admitted for observation in case complications develop. Appropriate examinations and tests for lung and abdominal injury should be made. These may include a complete blood count and differential, chest X-ray, and plain film and upright of the abdomen (Wolf 1970). Treatment is similar to that for one who is suspected of having total body trauma and should include counteracting shock, maintenance of fluid intake by transfusion, and avoidance of medication or nourishment by mouth (McMullin 1943). The major decision involves the need for surgical intervention in the abdomen. Oxygen administration has been unreservedly recommended for the treatment of lung-blast injury (Clemedson 1956).

5. Protective Measures

Extent of blast injury may be reduced if a diver wears a kapok garment surrounding his chest and abdomen (Medical Research Council 1945). Protection is afforded by a layer of air held within the kapok, but efficiency is reduced if the kapok gets wet. The possible use of fiberglass suits has also been investigated (House et al. 1955). However, the amount of material required for significant protection is too bulky and buoyant to be practicable. Use of rigid material in conjunction with a foam substance may increase the effectiveness of protection (U. S. Navy 1970).

Preliminary work (Committee on Amphibious Operations 1952) indicates that a unicellular plastic suit would considerably reduce the peak pressure and impulse received by a swimmer. Approximate calculations have been made by acoustical theory, and give the ratio of transmitted to incident pressure in terms of the impedance Z_1 of water and the impedance Z_2 of the protective layer, such that

$$\frac{\text{transmitted pressure}}{\text{incident pressure}} = \frac{4Z_2{}^2}{Z_1{}^2} \tag{5}$$

If the protective layer is foam plastic such as Neoprene, the pressure attentuation factor should be somewhere between 80 and 200; however, there is no experimental data to support this.

The most important and probably only sure way to avoid injury is to be out of the water at the time of the blast. The swimmer or diver should attempt to reach the surface and get as much of his body out of the water as possible, taking advantage of anything that he can climb up on or use for flotation. If he must remain in the water, he should float or swim face up to put the thicker tissues of the back between the vulnerable organs and the explosion. Interposing an air-containing appliance—such as a life vest—between the vulnerable organs and the surface of the water will tend to reflect some of the shock wave at its air–water interface and absorb some of the remainder (Wolf 1970).

E. Fire Safety

1. Introduction

An oxygen-enriched atmosphere (OEA) is an atmosphere which contains greater than 21 mole % oxygen and/or contains a partial pressure of oxygen greater than 0.21

atm. Oxygen-enriched atmospheres are encountered with the use of compressed air and in many decompression procedures. Compressed air at a pressure equivalent to 200 fsw has an oxygen partial pressure of 1.5 atm; at 300 fsw, 2 atm. After a dive has been completed, the diver must be slowly decompressed. Toward the end the decompression process is often hastened by enriching the gas mixture with oxygen. Treatment of decompression sicknesses may require the inhalation of 100% oxygen at depths of 60 fsw. This is done by mask, but exhalation of oxygen into the chamber will rapidly enrich the oxygen content of the chamber atmosphere and under these conditions it is exceedingly difficult, even with frequent flushing, to maintain oxygen concentration below 25% (Dorr 1971a).

In the event of fire, immediate escape is impractical because of the physiological danger of a sudden decrease in pressure. Even when the atmosphere is only air at ordinary pressure, the risk of injury or death from fire in an enclosed space from which there is no immediate escape is extremely severe.

The presence of oxygen-enriched environments in hyperbaric chambers requires a knowledge of fire hazards associated with these environments and strict adherence to safety procedures. Although fires in hyperbaric chambers have been rare, they occur with astonishing suddenness and often fatal results (Harter 1967a, 1967b; Swan 1967; NASA 1967).

Fires must therefore be prevented if possible; or, if this is impossible, they must be held to burning rates and locations at which they can be easily and quickly extinguished.

2. History

On March 22, 1945, a fire occurred in a recompression chamber aboard a Navy diving ship. Two men subsequently died and another was severely burned. The chamber was at a depth of 40 ft and the subjects were breathing from oxygen inhalators. This fire led to a number of improvements in the outfitting of Navy diving chambers (Harter 1967b).

Nevertheless, on February 16, 1965, a fire in a hyperbaric research chamber at the U. S. Experimental Diving Unit, Washington, D. C., resulted in the deaths of two Navy divers. At the time of the fire the divers were at a pressure of 41.4 psi (60 fsw), in an atmosphere with the approximate composition 27% (by volume) oxygen, 36.5% nitrogen, and 36.5% helium. A portable carbon dioxide scrubber was in operation, serving the dual purpose of removing carbon dioxide and of obtaining homogeneity of the gas mixture in the chamber. One minute and 10 sec before the fire 2.5 ft^3 of oxygen was admitted to the chamber over a 5-min period. The point of introduction of the oxygen was near the scrubber to provide rapid mixing of the oxygen with the existing atmosphere. After the accident, the Investigation Board determined that the fire was initiated by a localized heat source in the scrubber motor. Overheating of the motor, caused by faulty operation of the centrifugal throw-out switch, resulted in the motor running on the starter windings. The filter, downstream of the motor, was one normally used for filtering jet fuel, and following manufacture had been tested in a kerosene mixture. Residual kerosene was probably the primary fuel for this fire (Harter 1967b).

Then on January 31, 1967, a fire broke out in a hypobaric chamber at Brooks Air Force Base. Two airmen had gone into a chamber of pure oxygen at $\frac{1}{2}$ atm to tend some laboratory animals, and during that time a fire broke out. Although only a little over 30 sec elapsed while the fire was recognized, emergency procedures were performed, the chamber was back at ground level, two medical officers were on hand, and the doors were opened, both airmen were killed. It was calculated that the temperature inside the chamber reached 800°F within 14 sec. It was believed that the fire was initiated by the fractured insulation of a lamp cord allowing a short to the aluminum floor plate (Swan 1967).

This incident was followed by the Apollo spacecraft fire in which three astronauts were killed (NASA 1967).

3. Fire Aspects

To reduce the possibility of fire in an oxygen-enriched environment it is necessary to eliminate one of three fundamental prerequisites for combustion: (a) an atmosphere capable of supporting combustion, (b) fuel, or (c) an ignition source. In practice, it is often difficult or impossible to remove any one of these requirements with absolute certainty. Attempts should therefore be made to reduce the contributing effects of all three essential factors; i.e., electrical items that may serve as potential sources of ignition should be avoided, the quantity of flammable materials in the chamber should be reduced, and the absolute oxygen concentration and the oxygen partial pressure within the chamber should be minimized as much as possible.

a. Ignition

(i) Sources of Ignition. Combustion involves strongly exothermic reactions between gases and vapors resulting in the generation of high temperatures. The initiation mechanisms involved in the flame reaction are complex and not too well understood. For a particular reaction system the minimum energy which the molecules must possess to permit chemical interaction is usually referred to as the "activation energy." For many fuel–oxygen combinations at room temperature the activation energy is much greater than the average energy of the molecules. An increase in temperature increases the number of molecules in the activated state and the reaction rate increases. As the temperature is further increased, eventually enough fuel and oxygen molecules react and sufficient additional thermal energy is released to enable the combustion reaction to become self-sustaining until one or the other, or both, of the reactants have essentially been consumed. Regardless of whether the combustible material is a solid, liquid, or gas, initiation of the flame reaction occurs in the gas or vapor phase. Depending on (a) the type of ignition source, (b) the specific chemical nature and physical character of the combustible, and (c) the composition and pressure of the atmosphere, the minimum initiation energy for flame will vary (Botieri 1967).

In general, at a given pressure the minimum ignition energy varies inversely with the change in volume percent oxygen. For a fixed-percent oxygen content, the minimum ignition varies inversely with the change in pressure. Depending on the particular "combustible," there exists a minimum oxygen concentration and a minimum pressure below which ignition, from a practical standpoint, is not possible. An increase in environmental temperature reduces the minimum ignition energy requirements and enhances the possibility of autoignition (Botieri 1967).

General sources of fire ignition are summarized in Table VIII-2. The most probable source of ignition in a hyperbaric chamber is from its electrical system, assuming the elementary *no smoking* rule is observed. Electrical systems may be as complex as those found in research chambers or as simple as those used in the field. Some chambers have no internal electrical systems at all and a very low probability of accidental fire. But more often than not one sees internal electrical lighting, blowers, scrubbers, and communication systems. Malfunctions in such systems are often able to supply the heat necessary to ignite flammable materials, as evidenced by chamber fires at the Experimental Diving Unit and Brooks Air Force Base (Dorr 1971b).

Investigation of the potential hazard of a circuit interruption (a quick disconnection of a male plug from a female socket under a 115-V, 20-amp load) reveals that in the absence of flammable vapors this is a fairly low-probability ignition source. In recent tests, a spark jumping between such contacts failed to ignite an adjacent piece of paper, even in an oxygen environment (Dorr 1971a). Where vapors are present, as in a surgical operating room containing such products as ether or cyclopropane, these and even cooler electrostatic sparks represent a very real hazard.

Electric motors, such as the type that may be used in scrubbers or fans, pose a potentially more hazardous threat than that of a short-lived spark. Particular concern should be given to a locked or jammed motor which can cause enough overheat to start a chamber fire.

In general, static sparks likely to be generated by personnel can possess sufficient energy to ignite combustible gases and powders in both air and oxygen. These sparks, however, provide a very low probability for ignition of solid materials such as cotton or nylon. Appropriate procedures should be followed, nevertheless, to minimize the likelihood of static spark generation from personnel and equipment (Botieri 1967). Generation of electrostatic sparks may be avoided by using materials that conduct electricity and by keeping the humidity high (Roth 1964).

Table VIII-2
Possible Sources of Ignition

Sparks	Hot surfaces	Hot gases
Electrostatic sparks	Heating by shear rates	Gas flames
Frictional or impact sparks	Heating of wires by electricity	Shock waves
Sparks or arcs caused by interrupting the flow of electric currents	Incadescent carbon wear particles from electric motor brushes	Adiabatic compression
	Friction	

(ii) Effect of Inert Gas on Ignition. The likelihood of ignition of a combustible is primarily influenced by the oxygen content of the environment, and less by the nature of the inert gas. An inert gas such as nitrogen or helium essentially provides a physical obstacle to the effective interaction of a fuel and oxygen molecule in at least three ways: (a) by promoting reaction chain termination, (b) by promoting the formation of free radicals which do not lead to branching chain-reaction propagation, and (c) by cooling the reaction and reactants.

For inert gases to have a pronounced effect on ignition energy they must be present in relatively high concentrations. The specific effect on ignition energy requirements by the typical ignition sources will vary with the particular inert gas involved. These effects in certain instances correlate with the heat capacity and thermal conductivity properties of the inert gases (Botieri 1967).

b. Atmosphere

Oxygen concentration and partial pressure are highly variable parameters in most diving situations. Even though oxygen is always adjusted within the relatively narrow limits required for proper respiration, the range of potential fire hazard touches each extreme. Very rapid fire propagation has been observed in chambers containing a physiologically tolerable, oxygen-enriched atmosphere, although certain atmospheres in which man can function normally will not support combustion at all.

Since both partial pressure and percentage of oxygen (or inert gas) have an effect on combustion, the combustion situation found in diving chambers is complex. In general, a constant-percentage mixture of oxygen and inert gas (such as air) supports combustion better as pressure is increased, and even a slight enrichment of oxygen in a mixture will increase the fire risk. In a helium–oxygen environment ignition is more difficult but flame propagates more rapidly than in nitrogen–oxygen mixtures.

(i) Combustion in Compressed Air. Combustion of paper at various pressures of compressed air has been investigated (Cook *et al.* 1967*a*). As shown in Figure VIII-9, the temperature required to ignite paper in compressed air falls off rapidly as the pressure is increased.

When one speaks about the effect of pressure on burn rate, it is necessary to specify the angle of the burning material. In the horizontal position there is little or no increase in burning rate as the pressure is increased. At greater angles the burning rate increases appreciably as the pressure becomes greater, at least up to a pressure of 300 fsw, and this rate increases as the angle becomes larger, up to 45°; at steeper angles the rate of increase with increasing pressure starts to fall off. In the vertical position, the burning rate becomes erratic. This effect is illustrated in Figure VIII-10.

In any position except horizontal, the burning rate of filter paper increases with pressure. At first, however, the rate of increase becomes progressively less until, at pressures somewhere above 300 fsw, the rate levels off. This effect is shown in Figure VIII-11, for filter paper at an angle of 45°.

The fire hazard in a chamber filled with compressed air increases as the air pressure increases (to about 300 fsw) for two reasons: (a) the higher the pressure, the greater the ease of ignition of combustible material, and (b) the higher the pressure, the greater (except in the horizontal position) the rate of combustion. The hazard

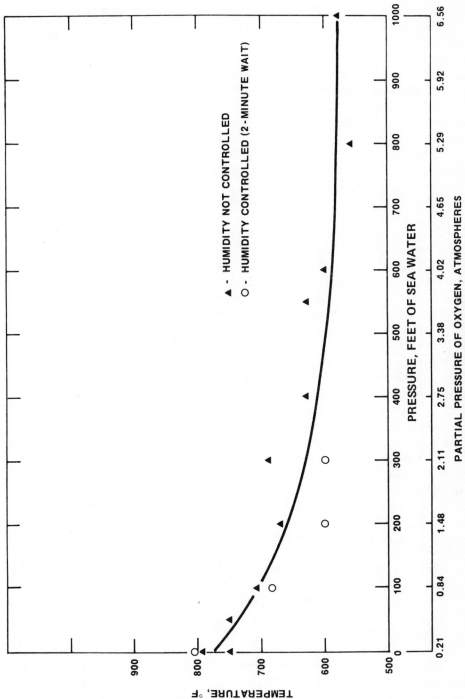

Figure VIII-9. Variation with pressure of the ignition temperature of filter paper strips at an angle of 45° in compressed air (Cook *et al.* 1967*a*).

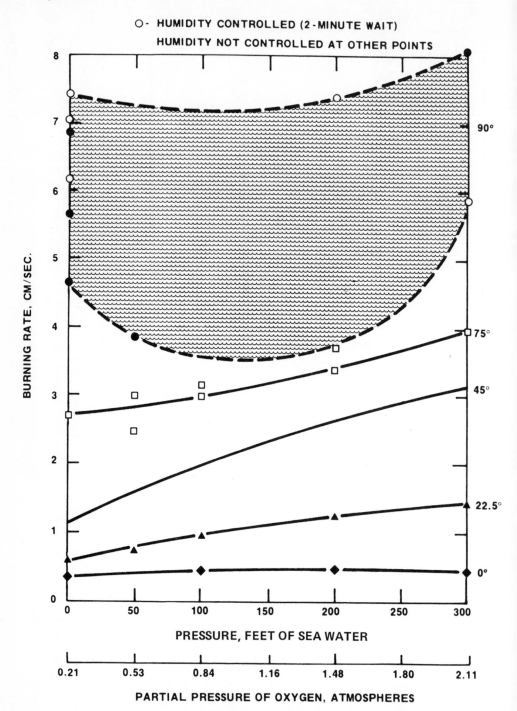

Figure VIII-10. Variation with pressure of burning rate of filter paper strips held at various angles in air, 0–300 ft (Cook *et al.* 1967*a*).

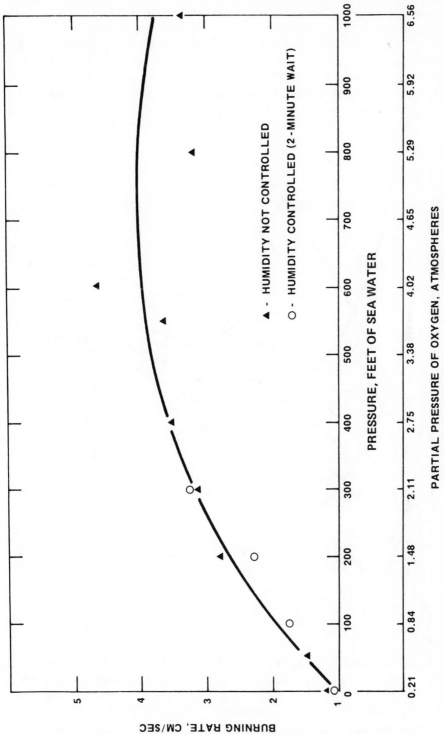

Figure VIII-11. Experimental results for the burning of filter paper strips at an angle of 45° in compressed air (Cook *et al.* 1967a).

is greatest when flammable material is so located within the chamber that combustion is propagated vertically.

(ii) Effect of Inert Gas Composition on Combustion. Burning rates of filter paper positioned at a 45° angle have been measured in oxygen–nitrogen and oxygen–helium mixtures containing 15–100% oxygen over the pressure range of 1–10 ATA (Cook *et al.* 1967*a*, 1967*c*). When the burning rate is compared with the total pressure

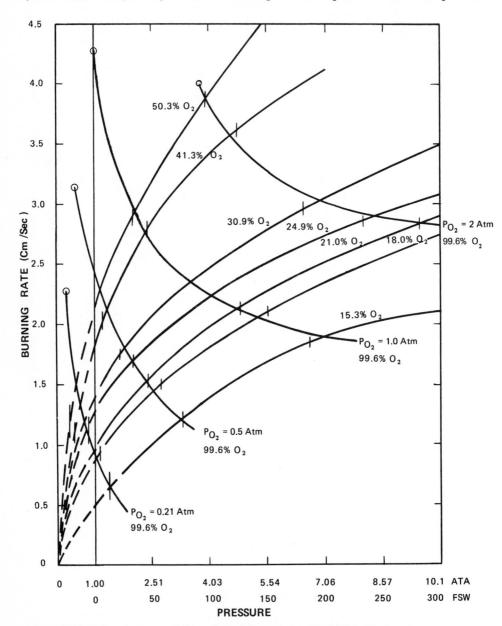

Figure VIII-12. Burning rates of filter paper strips at an angle of 45° in N_2–O_2 mixtures.

in oxygen–nitrogen mixtures (Figure VIII-12), the burning rate is found to accelerate in all gas mixtures as total pressure is increased; when the burning rate is compared with percentage of oxygen content in the gas mixture, the burning rate is found to accelerate at all pressures as the oxygen content is increased.

Burning rate is affected in the same way when the breathing mixture is oxygen–helium (Figure VIII-13); it is accelerated by either an increase in the total pressure

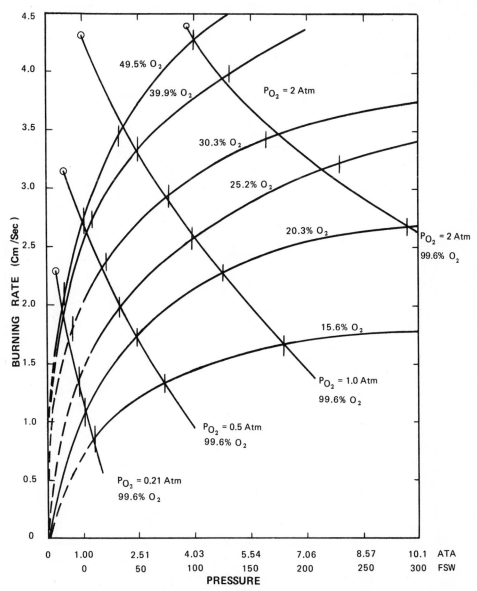

Figure VIII-13. Burning rates of filter paper strips at an angle of 45° in He–O₂ mixtures (Dorr 1971b).

of any given gas mixture or an increase in the oxygen concentration at any given total pressure.

Although it is more difficult to overheat materials in a helium than in a nitrogen environment (since the thermal conductivity of helium is almost six times that of nitrogen), once ignited, rate of flame spread is greater in the helium atmosphere (Cook 1967). In Figure VIII-14 (Dorr 1971b) the ratio of the burning rate in helium R_{He} to the burning rate in nitrogen R_{N_2} is plotted relative to oxygen percentage (in the range of 15–50%) at five pressures between 0 and 200 fsw. On the whole, the ratio is greater than one; that is, combustion proceeds faster in a helium environment than in nitrogen (oxygen concentration and total pressure remaining the same). The difference in burning rate is most pronounced at atmospheric pressure in a 30%

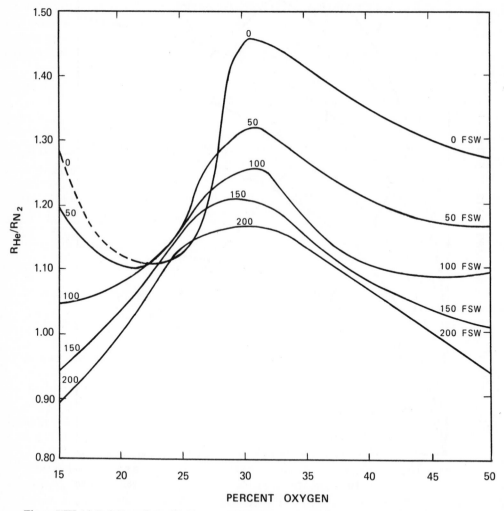

Figure VIII-14. Relative effect of helium and nitrogen as oxygen diluents on the rate of combustion of filter paper strips at an angle of 45° (Dorr 1971b).

oxygen atmosphere, and tends to diminish as the total pressure of the mixture is increased. A reduction of the difference in densities of oxygen–helium and oxygen–nitrogen mixtures at higher pressures may account for this (Dorr 1970).

Huggett et al. (1966) derived a simplified mathematical model of the burn-rate mechanism. In this model diffusion and heat conduction are considered to be the rate-controlling processes, which eliminates chemical reaction rates from the model. The model cannot, however, be used to predict either the absolute burning rates or the effect of gas composition. From the model it would appear that polyatomic gases would be more effective in reducing rates of flame spread than monatomic or diatomic gases because of their higher volumetric heat capacities. A gas such as CF_4 should be highly effective in preventing flame propagation, but physiological activity and high density prevent its use in life-support systems. In spite of the good correlation obtained between burn rates and heat capacities, Huggett et al. conclude that "the much larger flame spread-rate in helium–oxygen atmospheres as compared to nitrogen–oxygen atmospheres is attributable primarily to the higher thermal conductivities and presumably also higher diffusivities of the helium mixtures, and only secondarily to the higher flame temperatures associated with the lower heat capacity of helium." Thus it would appear that the heat capacity correlation may be of interest only in defining the region of noncombustion (Cook et al. 1967a).

An example of the practical significance of the effect of gas composition on burn rates is the situation illustrated in Figure VIII-15 by the position of the points, which represent various oxygen concentrations at sea level pressure (1 ATA). An oxygen increase from 20 to 25% may seem small at first, but it is enough to increase the burning rate by 25%. This degree of change may be extremely important during a long chamber decompression when oxygen is breathed by mask. Oxgyen exhaled into the chamber will quickly raise the level to values well above 25%, and it is almost impossible to reduce this significantly by purging the chamber with air. About the only way this situation can be managed safely is to provide for overboard dump of exhaled oxygen. In any case, flammable materials and sources of ignition must be minimized.

(iii) Region of Noncombustion. The risk of fire is not always increased in hyperbaric environments. At high pressures, the oxygen percentage in a mixture of oxygen and inert gas is normally reduced in order to avoid the danger of oxygen toxicity. This decrease in the oxygen percentage may also render the atmosphere incapable of supporting combustion under certain carefully defined conditions (Cook et al. 1968, Dorr and Schreiner 1969). Three zones representing complete combustion, incomplete combustion, and no combustion for filter paper in oxygen–nitrogen mixtures are shown in Figure VIII-15. Isobars showing oxygen partial pressures representing minimum and maximum partial pressures normally used for respirable environment are also shown. It is seen that the area denoted ABC lies both within the region of noncombustion and between the two oxygen isobars; thus it will maintain life but will not support the combustion of ordinary flammable materials such as paper and cotton cloth. The narcotic effect of nitrogen at high pressure, however, makes its use as a diluent gas inadvisable at partial pressures exceeding approximately 5.5 atm. Therefore, only the area denoted ADE is both physiologically acceptable and free from all risk of accidental fire. A somewhat larger area lies within the zone of incomplete combustion. The absence of narcosis in oxygen–helium mixtures

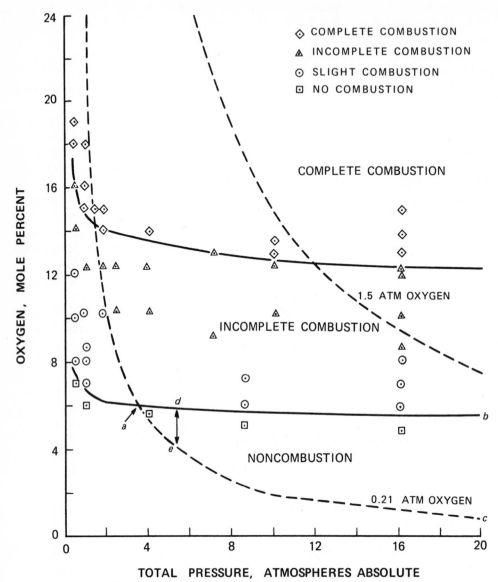

Figure VIII-15. Three combustion zones for vertical paper strips in N_2–O_2 mixtures. [From Cook *et al.* (1968). by permission of the American Chemical Society, Washington, D. C.]

produces a much larger and more practical range of gas composition and pressure (area ABC in Figure VIII-16) that is physiologically satisfactory and still devoid of combustion hazards.

Such atmospheric conditions, denoted the region of noncombustion, can be successfully applied to increase the safety of dry-chamber underwater welding. In such an operation, the torch flame or electric arc provides a continuous and highly energetic ignition source. Whether the atmosphere is capable of supporting combus-

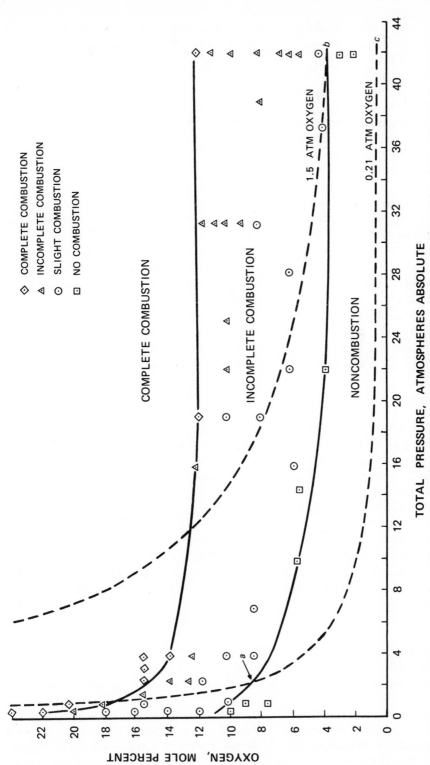

Figure VIII-16. Three combustion zones for vertical paper strips in He–O₂ mixtures. [From Cook *et al.* (1968) by permission of the American Chemical Society, Washington, D. C.]

tion, the probability of accidental fire is invariably increased, since the welders' clothing, breathing hose and mask, electrical insulation, hydrocarbons, etc. are flammable. Operating within the region of noncombustion reduces this fire risk while still providing a life-supporting atmosphere (Cook 1967, Hamilton 1970).

(iv) Flammability Limits of Hydrogen–Oxygen Mixtures. At depths greater than 1000 fsw, helium–oxygen mixtures may become too dense for comfortable breathing. It has been suggested that hydrogen be partially or entirely substituted for helium at these pressures, as the density of hydrogen is only about half that of helium.

Two problems arise when hydrogen is used as a diluent for oxygen: (a) how to produce the mixture without risk of explosion, and (b) how to prevent fire or explosion while the diver is in the hydrogen–oxygen environment. In each case it is essential to keep the oxygen concentration below a certain critical minimum (the upper flammability limit) for a significant hydrogen–oxygen reaction to occur. It has been determined that the upper flammability limit is 5.3% oxygen (Dorr 1971*b*); no significant variation has been observed within a pressure range of 0–1350 fsw. Incorporating a safety factor of 1.5, it appears that a 3.5% oxygen–96.5% hydrogen mixture could be used in complete safety, provided the proper precautions are taken in preparation of the mixtures and safeguards are observed against leakage to the atmosphere (and ignition sources). A 3.5% oxygen mixture would become respirable at 175 fsw (oxygen partial pressure of 0.21 atm). At the greater depths where hydrogen would most likely be used as a diving gas, the oxygen concentration could be considerably reduced and the safety factor significantly increased.

c. Materials

When it is impractical to maintain the hyperbaric environment within the region of noncombustion and if all possible sources of ignition within the chamber cannot be precluded with absolute certainty, great care should be exercised in the selection of materials to be placed in the chamber. The results of standard flammability tests in the ambient atmosphere are not valid indicators of the burning behavior the same material would exhibit under hyperbaric conditions. Studies have been made testing materials for combustion in air at atmospheric pressure in compressed air and in oxygen–nitrogen mixtures containing increasing amounts of oxygen at a total pressure of 1 ATA (Dorr 1971*b*, Dorr and Schreiner 1969, Cook *et al.* 1967*b*). The materials were rated according to ten classifications shown in Table VIII-3. Recommended applications for various materials in oxygen-enriched hyperbaric chambers are given in Table VIII-4 (Dorr 1971*b*).

4. *Methods of Fire Prevention and Control*

a. Fire Detection

Because fire spreads so rapidly in oxygen-enriched atmospheres, reliable detection techniques capable of initiating extinguishing or other emergency action are a mandatory protection requirement. Detectors for this application should provide

Table VIII-3
Scale of Fire Resistance[a]

Class 0. Burns readily in air at atmospheric pressure

Class 1. Has an appreciable higher ignition temperature and/or burns at an appreciably lower rate in air at 1 ATA pressure than cotton cloth or paper; an example of a Class 1 material is wool

Class 2. Nonflammable or self-extinguishing in air at 1 ATA pressure

Class 3. Self-extinguishing or burns slowly in air at a pressure of 100 ft of sea water (4.03 ATA)

Class 4. Self-extinguishing or burns slowly in air at a pressure of 200 fsw (7.06 ATA)

Class 5. Self-extinguishing or burns slowly in a mixture of 25% oxygen and 75% nitrogen at a pressure of 1 ATA

Class 6. Self-extinguishing or burns slowly in a mixture of 30% oxygen and 70% nitrogen at a pressure of 1 ATA

Class 7. Self-extinguishing or burns slowly in a mixture of 40% oxygen and 60% nitrogen at a pressure of 1 ATA

Class 8. Self-extinguishing or burns slowly in a mixture of 50% oxygen and 50% nitrogen at a pressure of 1 ATA

Class 9. Nonflammable in 100% oxygen at a pressure of 1 ATA

[a] From Cook *et al.* (1967*b*) by permission of the Textile Institute, Princeton, N. J.

Table VIII-4
Fire-Resistant Materials (Dorr 1971*b*)

Comments	Material	Class	Comments
Fabric	Teflon-Coated beta Fiberglas (4484)	9	Teflon coating increases comfort and durability
Fabric	Beta Fiberglas (4190B)	9	Completely nonflammable under all conditions; chief disadvantages are possible skin irritation and low abrasion resistance (resulting in poor durability)
Fabric	Teflon	8	Heavy, low moisture regain, lack of wearer comfort
Fabric	PBI	7	Suitably flame resistant in compressed air; comfortable as clothing and bedding; however, is not commercially available at present
Fabric	Durette	6	Available (at relatively high cost) and suitable for use in compressed air in terms of flame resistance and comfort; appears to be the best compromise choice of fabrics at this time
Elastomer	Fluorel 1071	8	Significantly increases fire safety of breathing masks and hoses
Paper	Nonflammable paper (Dynatech)	9	Writing quality inferior to Scheufelen paper
Paper	Scheufelen Paper	9	Good choice for all paper articles that are used in chamber
Electrical insulation	Kapton	9	Generally unavailable as pre-insulated wire
Electrical insulation	Teflon	8	Teflon-insulated wire commercially available; wires should always be enclosed in additional mechanical protection and should be protected from overheating

Table VIII-5

Available Test-Chamber Fire-Detection Systems

(Adapted from Botieri 1967)

Type of System	Specific to fire or ignition	Approximate response times, sec	Applicability	Volume coverage
Smoke detectors, large particles[a]	No	2–5	No	Yes
Smoke detectors, small particles[a]	No	2–5	No	Yes
Infrared solid-state detectors	Yes[b]	0.1[b]	Yes	Yes
Thermocouples	No	0.05	No	No
Continuous-element overheat detectors	No	5	No	No
Eutectic-alloy-type	No	5	No	No
Ultraviolet detectors	Yes	0.010	Yes	Yes

[a] Excellent for incipient fire detection. May detect visible fog also. As during decompression.
[b] Specificity, fast response, and sensitivity are not available in one system.

volume surveillance and permit early and rapid detection of incipient combustion as well as flame. Table VIII-5 indicates various types of detectors currently available for fire detection. In general, these detectors rely upon either the temperature rise or radiation emission; some use a combustible product with the flame process for activation. Certain of these detectors, such as the overheat or rate-of-temperature-rise detectors, are not acceptable for oxygen-enriched environments because of their inherent slow response time and limited volume coverage. Flame radiation (UV and IR) sensors and smoke detectors in combination appear to be ideally suited for this application (Botieri 1967).

b. Fire Extinguishers

Fire extinguishment is accomplished by four basic physical or physical–chemical actions: (1) cooling the combustible material to a temperature below that required for ignition or the evolution of flammable vapors; (2) smothering the fire by reducing the oxygen or fuel concentration to a level that will not support combustion; (3) separating fuel and oxidizer by removing one of them or by mechanically separating the two (the major mechanism of mechanical protein foam on jet fuel fires, often referred to as "blanketing" action); and (4) chemical interference or inhibition of the reactions occurring in the flame front or just before the flame front. The mode of action and effectiveness of several extinguishing agents are shown in Table VIII-6 (Botieri 1967).

Agents such as nitrogen and carbon dioxide depend primarily on dilution of the oxygen content to a noncombustible level. In the absence of a special breathing system to protect against carbon dioxide toxicity, carbon dioxide extinguishment cannot safely be employed in a hyperbaric chamber. In an attempt to replace carbon dioxide as a gaseous fire-quenching agent, both nitrogen and helium have been considered and tested. It appears that neither of these inert gases is of value (except by the rapid

Table VIII-6

Candidate Fire Extinguishing Agents for Oxygen-Enriched
Atmospheres (adapted from Botieri 1967)

		Compatibility	
Agent	Mode of action[a]	Personnel	O_2 fires
Water	1, 2, 5	Excellent	Good
Foam	1, 3, 5	Good	Unknown
Dry chemical (NaHCO$_3$, ABC)	3, 4, 5	Good	Unknown
CO_2	1, 2	Fair	Poor
N_2	2	Poor (anoxia)	Poor
CF_3Br	1, 2, 4	Good[b]	Very good

[a] Mode of action: 1. Quenching (cooling). 2. Inerting (O_2 dilution). 3. Blanketing. 4. Chemical inhibition. 5. Radiation shielding.
[b] May decompose in heat to yield toxic products.

dilution method in which the entire chamber atmosphere is rapidly diluted), as it is not possible to maintain a significant concentration of the gas in a given location. Due to their narcotic properties, more research is needed before gases of heavier molecular weight can be considered for such use under pressure (Bond 1966).

Although dry chemical agents should provide rapid suppression of flame and excellent radiation shielding when initially discharged, the permanency of the fire extinguishment is doubtful (Botieri 1967).

Although monobromotrifluoromethane (CBrF$_3$, Freon 1301) and chlorobromomethane (Freon 1011) have been shown to be effective extinguishing agents, a delayed application of the extinguishing agent could produce toxic pyrolysis products if the fire had a head start (Thomas 1967, Vernot 1967, Carter 1967).

High-expansion foam has proved an effective means of extinguishing fires that have been allowed to build up to their full intensity (Dorr and Schreiner 1969), but there is presently little knowledge as to the physiological danger of such agents due to pyrolysis products or otherwise.

At the present state of the art, due to safety considerations, the best extinguishing agent for use in hyperbaric chambers is water. Water extinguishment operates primarily by cooling. It works best if it strikes the flame, or wets the fire, but wetting most substances will retard or prevent their burning, even in oxygen. In spray form, water may not put the fire out immediately, but spread is halted and from this point on extinguishment is almost certain. The spray can be continued indefinitely, assuring safety of chamber occupants (Eggleston 1967). Water at a spray density of 5 ml/cm^2 min ($1\frac{1}{4}$ gal/ft^2 min) applied for 2 min is required to extinguish cloth burning in 100% oxygen at atmospheric pressure (NFPA 1969). Water spray systems require special design for hyperbaric chamber applications. The pressure at the spray nozzles must be about 50 psi above chamber pressure to produce the desired degree of atomization and droplet velocities. Spray pattern of nozzles might be affected by chamber pressures (Eggleston 1967). To compensate for the reduced coverage at elevated pressures, design of the system must provide an adequate number of strategically located nozzles, to wet all possible exposed areas in the chamber no matter what the chamber pressure may be (Ault and Carter 1967). Pressurization is best obtained from a compressed gas source, since pumps have a startup time. Simultaneously with dis-charge of the water, all electrical power to the chamber should be discontinued to prevent shorting and electrical shocks to personnel within the chamber (Carter 1967).

A manually directable fire hose might permit occupants of a chamber to control localized small or incipient fires, and because its use would be less catastrophic than activation of a deluge system, it might be used more quickly and with less damage to chamber apparatus.

It is imperative that water-deluge systems be thoroughly tested at all applicable operating pressures and conditions. Numerous installations have been disappointing when given realistic tests, and redesign has been necessary.

c. General Safety Procedures

General rules for providing fire safety in a hyperbaric chamber may be sum-marized as follows:

1. Oxygen concentration and/or partial pressure should be as low as possible, preferably within the region of noncombustion. It is necessary to use an overboard dump system where pure oxygen is breathed by mask in a chamber.
2. Ignition sources must be eliminated.
3. Combustibles should be minimized, with the complete exclusion of flam-mable liquids and gases.
4. If combustible materials must be employed, the type, quantity, and arrange-ment in the chamber must be carefully controlled.
5. Fire walls and other containment techniques should be utilized to isolate potential high-risk fire zones.
6. A fixed fire-extinguishing system should incorporate smoke detectors and both manual and automatic swtiches; it should provide rapid and sufficient agent discharge.

d. Areas in Need of Further Study

To provide fire safety under hyperbaric conditions, the major area in need of further study is that of fire extinguishment. Although high-expansion foam is an effective means of extinguishing fires, there is no present knowledge about the physio-logical danger of such agents due to pyrolysis or otherwise, and under hyperbaric conditions. Likewise with Freon—which also has been shown to be an effective ex-tinguishing agent—toxic high-temperature pyrolysis products are created, but the severity of their toxicity on humans has not been adequately determined.

Finally, since inert gases are theoretically excellent extinguishing agents, the possibility of using the nontoxic, heavy, inert gas krypton should be investigated. Although krypton is expensive and would cause narcosis at high partial pressures, it might well fill a specialized requirement. The problem with using the inert gases helium and nitrogen as extinguishing agents is the difficulty in maintaining a significant concentration of gas in a given location. The use of a much denser inert gas such as krypton might prove more effective.

F. *Electrical Safety*

Using electric power in operational and research diving chambers poses two serious hazards—fire and electrocution. The problem of chamber fires and the role of electrical sources of ignition has been covered under fire safety. Here we will discuss electrocution.

In a dry or *wet* diving chamber, electrocution is a potential hazard when instrumentation and other current-consuming devices are in operation. Sea water and perspiration increase the shock hazard. In cases of severe electrical shock requiring immediate treatment, rescue is difficult or impossible.

1. *Physiology of Electric Shock*

a. General

Physiological effects occur when the body becomes part of an external electrical circuit. The magnitude of the current flow through the body depends on the voltage between the electrical connections and the resistance of the body. The part of the body that is located between the two points of electrical contact forms an *inhomogeneous volume conductor*, in which the distribution of the current flow is determined by the local tissue conductivity (Cromwell 1973). The current has the greatest density at the contact points; it spreads radially outward from the point of entrance, and collects again at the exit point (Finkelstein and Roth 1968). With alternating current, the contact points are both entrance and exit points. In comparison to the bones, the resistance of the trunk (the tissues and organs can be considered a conducting gel bathed in saline) is negligible (Cromwell 1973).

Basically, electric current can affect body tissues in two different ways: by generating heat and by disrupting electrochemical action potentials of motor nerves and muscles. Figure VIII-17 shows the approximate current range and the resulting effects for a 1-sec exposure to various levels of 60-Hz alternating current applied externally to the body.

Electrical energy dissipated in tissue resistance causes a temperature rise, which, if great enough, can cause tissue damage and burns. In fact, burns and blisters are the most common result of electrical accidents. With household current, electrical burns are usually limited to localized damage at or near the contact points, but at higher voltages larger areas of the body may be involved (Cromwell 1973). Blistering or burns destroy the protective resistance of the skin, permitting greater currents to flow. If the current flow continues long enough, the heat may raise the body temperature to a value sufficient to cause death (Dalziel 1956). Electrical burns are slow to heal but seldom become infected.

Impulse transmission through sensory and motor nerves (which involve electrochemical potentials) can be disrupted by extraneous electric currents. Stimulation of motor nerves or muscles can occur, causing muscle fiber contraction. A high-intensity stimulus can cause tetanus of the muscles, in which all the muscle fibers are contracted with the maximum force possible (Cromwell 1973). Very often the body of an electrocuted person is bent so that the head and the heels are bent backward and the body

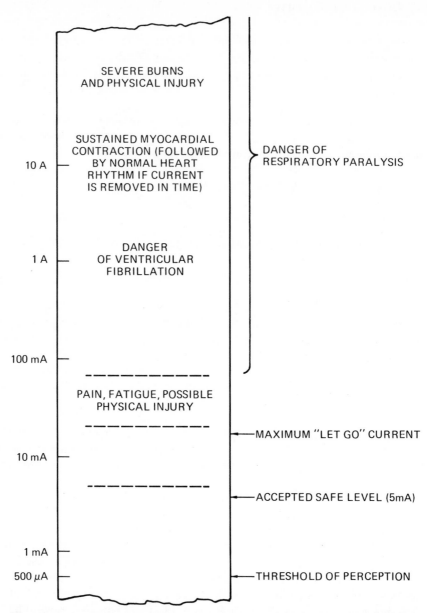

Figure VIII-17. Approximate current range and the resultant physiological effects for a 1-sec exposure to various levels of 60-Hz alternating current applied externally to the body. It is assumed that the path of the current includes the chest region, arm-to-arm or arm-to-leg diagonal. [From Cromwell (1973) by permission of Prentice-Hall International, Englewood Cliffs, New Jersey.]

bowed forward (Finkelstein and Roth 1968). The involuntary muscular contractions often prevent the victim from freeing himself or letting go of a live wire, and in some cases, are great enough to fracture a bone (Dalziel 1956). This tetanizing effect is especially hazardous or fatal when vital organs are affected.

b. The Heart

The organ most susceptible to electric current is the heart (Cromwell 1973). A tetanizing stimulation of the heart results in complete myocardial contraction, which stops the pumping action of the heart and interrupts the blood circulation. If the current is not removed within a few minutes, brain damage, then death results from lack of oxygen to the brain. If this tetanizing current is removed within a short time, heart beat resumes spontaneously.

A current of a lower intensity that excites only a portion of the heart's muscle fibers can be more dangerous than one that completely tetanizes the heart. This partial excitation can change the electrical propagation pattern in the myocardium, de-synchronizing the heart's activity and resulting in random, ineffectual muscle activity. When this condition—called fibrillation—occurs in the ventricles, the heart ceases to pump blood. Ventricular fibrillation is usually not reversible and regular heart rhythm does not resume when the current is removed. To reverse the situation, the heart must be tetanized by an external current source, or defibrillator. Ventricular fibrillation is the most frequent cause of death in fatal electrical accidents.

2. Influencing Factors

Factors which determine the type and extent of physiological damage from electroshock are point of electrical contact and body pathway of conduction, voltage and current, duration of exposure, and type and frequency of the electrical source.

a. Body Pathway of Conduction

Location of the vital organs relative to the current path through the body is of great importance. An arm-to-arm or arm-to-leg diagonal path would be most dangerous to the heart and respiratory system. The cranium is an excellent shield for the brain, since bone is a poor conductor. Electrical energy will pass around the cranium in extracranial soft tissue, if given the opportunity, and the scalp and cranium may be severely burned focally with minimal penetration of the brain (Finkelstein and Roth 1968).

b. Current Intensity

Electrical intensity is described in terms of electrical current. How much voltage is required for a specific amount of current flow depends on the electrical resistance. The greater the voltage applied to a constant resistance, the larger the current will be.

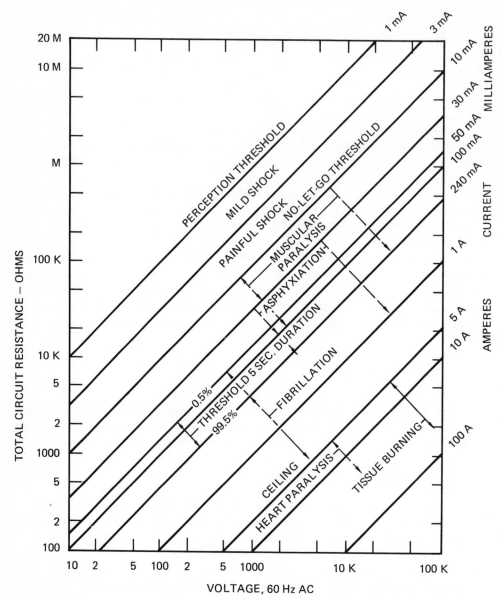

Figure VIII-18. Estimate of physiological thresholds based on current flow from arm to arm or arm to leg, for 150-lb subjects, and fibrillation current for a 5-sec duration. Heavier subjects have lighter threshold currents in direct proportion to weight (Finkelstein and Roth 1968). [After R. H. Lee, Human electrical safety, ISA-Mono-110, Instrument Society of America, Pittsburgh (1965).]

On the other hand, the current diminishes as the resistance is increased. These relationships are represented by Ohm's law,

$$I \, (\text{current}) = \frac{E \, (\text{voltage})}{R \, (\text{resistance})}$$

The range of physiological effects of electric current as a function of total circuit resistance and voltage for a 60-Hz ac current on a 150-lb human is given in Figure VIII-18.

Lethal effects of electrical shock are due to the current flowing through the body and are only indirectly related to voltage and body resistance. However, if the voltage is high enough to break down contact resistance, the current will begin flowing through the body. Dry skin has a relatively high resistance compared to the soft tissues beneath, but it is reduced considerably when the skin is wet.

c. Duration

Time of exposure has a direct relationship to the degree of permanent electroshock damage. There is a rapid decrease in muscular strength from pain and fatigue associated with severe involuntary muscular contractions, and the let-go ability would probably decrease rapidly with duration of contact (Dalziel 1956). Additionally, both skin resistance and total body resistance decrease contact (Finkelstein and Roth 1968).

d. Type and Frequency of Current

Alternating current (ac) is more dangerous than direct current (dc). Besides stimulating sweating, which lowers the skin's resistance, it also causes more severe muscular contractions (Frye 1965). Direct current under 220 V seldom leads to death, whereas even 25 V ac may be dangerous (tetanization of respiratory muscles) if the body is well grounded (Finkelstein and Roth 1968).

Current of 60 Hz is the frequency considered most dangerous to human safety. Above 60 Hz, the degree of muscular contraction decreases and the possibility of respiratory arrest and ventricular fibrillation is reduced. However, as the frequency increases, the heat produced and the possibility of serious burns are increased (Camishion 1966). The effect of frequency on perception and paralysis is shown in Table VIII-7 and the frequency sensitivity of let-go current is shown in Figure VIII-19.

When a steady direct current passes through tissue, the tissue behaves like a simple electrolytic resistance path; in considering alternating currents passing through biological systems, the concept of impedance must be included (Finkelstein and Roth 1968). For dc, impedance and resistance are the same, but for ac, impedance is a complex function of resistance and capacitive reactance. Inductance does not appear to be a factor.

Table VIII-7

Effect of AC Frequency on Perception
and Paralysis [a]

f, Hz	Fraction of 60-Hz effect
0	0.2
10	0.9
60	1.0
300	0.8
1,000	0.6
10,000	0.2
RF	0.01

[a] Finkelstein and Roth (1968). After R. H. Lee, *Human Electrical Safety*, ISA-Mono-110, Instrument Society of America, Pittsburgh, (1965).

3. *Underwater Shock Hazards*

As shown in Figure VIII-20, the electrical shock hazard is much more serious for a person touching a live circuit in a *wet* condition than for one who is completely submerged. However, a swimmer submerged in a conducting medium does not necessarily have to make a direct contact with a live circuit to receive the effects of an electric current. Just his presence in a conducting medium in which an electric field is present is sufficient.

Physiological effects on the swimmer depend upon the magnitude and the direction of current density produced within the swimmer's body (Smoot and Bentel 1971). The body current depends partly on body orientation with regard to direction of the current flow in the water. For two electrodes in a conducting medium the voltage field is similar to that shown in Figure VIII-21. The current flows normal to the equipotential lines, and the current flow in the swimmer will be proportional to the number of voltage gradient lines that are cut. It is important to note, however, that potential to a good ground is high.

An electric field gradient in salt water of 0.2 volts per foot (V/ft) can cause a current flow of 5 milliamperes (mA) through the chest; higher gradients produce proportionally higher currents (Hackman and Glasgow 1968). A 2 V/ft gradient in salt water could cause loss of muscular control in the leg, making it impossible to swim.

4. *Protective Measures*

Electrical power supplied from a power plant uses the earth as the common ground for power line circuits; the difference in electrical potential between the live and ground contacts is delivered when the plug is placed into the receptacle. In most electrical instruments and current-consuming devices, the power supply is connected at some point to the chassis. If the chassis is not properly grounded, it and any metal parts in contact with it will have a higher potential than true ground since there is

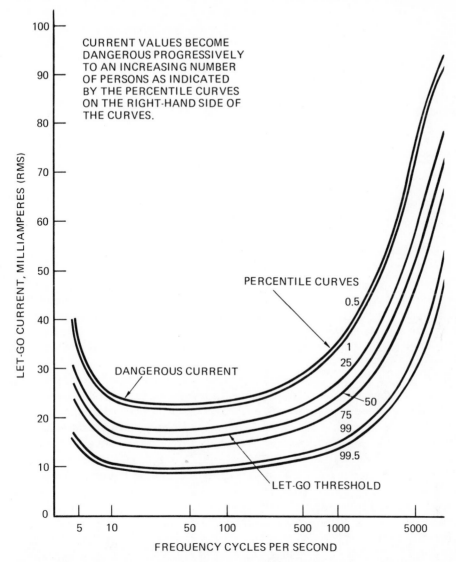

CURRENT VALUES BECOME DANGEROUS PROGRESSIVELY TO AN INCREASING NUMBER OF PERSONS AS INDICATED BY THE PERCENTILE CURVES ON THE RIGHT-HAND SIDE OF THE CURVES.

Figure VIII-19. Effect of frequency on let-go currents for men. Values for women are approximately 66% of those shown. [From Dalziel (1956) by permission of the Institute of Electrical and Electronics Engineers, New York.]

commonly some leakage of current to the chassis (Camishion 1966). This potential constitutes a source of electrical hazard to a grounded person. Protection should consist of adequate insulation, proper grounding of circuits, and the use of safety devices.

All current-consuming devices should be encased in a well-grounded metal container. Metal is recommended so that all electrical conductors are enclosed by a material which is at ground potential. Use of insulating materials, such as the double insulation found on commercially available hand-tool, rather than an outer metal

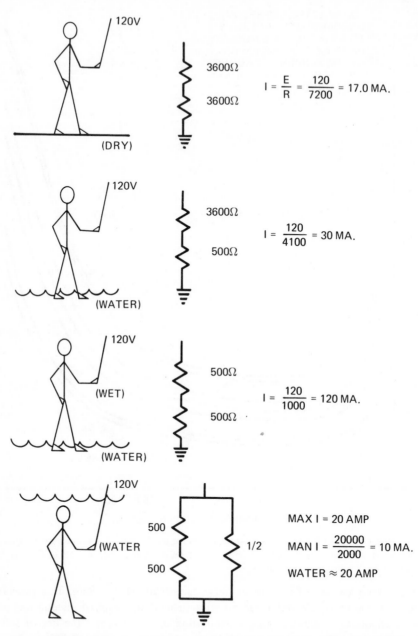

Figure VIII-20. Estimate of severity of electrical shock under different conditions. [From Hackman and Glasgow (1968) by permission of the Marine Technology Society, Washington, D. C.]

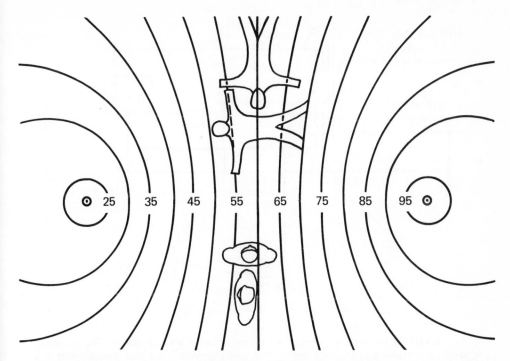

Figure VIII-21. Electrical field in a conducting medium. The current flow in the body will be proportional to the number of gradient lines that are cut (Hackman and Glasgow 1968).

container may be hazardous because these materials depend on an absence of moisture. If such insulation gets wet, particularly with sea water, it provides a steady path for current through the operator. On the other hand, if a metal-covered electrical device should get wet, the operator would be relatively safe since he would constitute a shunt of several kilohms across a path to ground having a resistance of less than 1 ohm to possibly a few ohms (Langley 1969).

Electrical power wires should be enclosed in grounded metal, preferably armored cable (Hamilton *et al.* 1970). Properly armored cable makes electrical shock virtually impossible; it provides protection against mechanical damage and provides an easily inspected ground lead.

As further protection, isolation transformers can be used on all ac power leads (Hamilton *et al.* 1970). Such transformers have grounded electrostatic shields between the primary and secondary windings and eliminate any detectable capacitive coupling effects. Such a transformer is shown in Figure VIII-22. The secondary winding provides a line-voltage supply that is not grounded and that should result in a negligible current through the leakage capacity to ground.

Finally, ground-fault detection systems are available which either sound an alarm or interrupt the power if appreciable (i.e., 5 mA) current leakage to the ground should accidentally occur (Hamilton *et al.* 1970). Although their use is not mandatory if the previously discussed precautions are taken, their incorporation into any chamber or diving system is recommended.

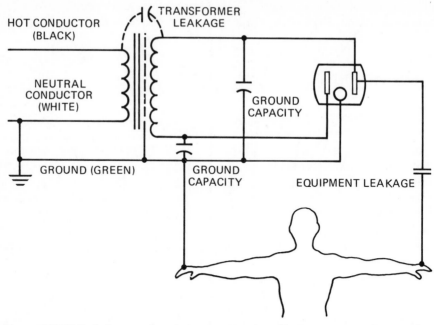

Figure VIII-22. Isolation transformer for the reduction of leakage current. The secondary winding provides a line voltage supply that is not grounded and that should result in a negligible current through the leakage capacity to the ground. [From Cromwell (1973) by permission of Prentice-Hall International, Englewood Cliffs, New Jersey.]

In the case of electrical circuits in the proximity of an underwater swimmer, the following recommendations have been made (Hackman and Glasgow 1968):

1. Exposed area of potentially hot conductors should be minimized.
2. Positions between conductors in the current flow path should be avoided.
3. Potential faults should be either very close together or very far apart.

5. Treatment

The victim must be removed from contact with the circuit as soon as possible. Power should be shut off at its source, if possible, or the wire cut using insulated gloves or insulated cutting instruments. Otherwise, forcible separation of the victim from the wire must be accomplished by means of nonconducting materials.

If the victim does not appear to be breathing, artificial respiration must be continued without interruption until he revives. If the heart has stopped, cardio-pulmonary resuscitation should also include closed-cardiac massage until the heart can be defibrillated. As soon as possible, oxygen should be administered (*The Merck Manual*).

The victim must be kept warm. When conscious, he may be given warm drinks. He should be maintained in a horizontal position until there is no further evidence of peripheral vascular collapse or respiratory embarrassment. Epinephrine and other cardiac stimulants are to be avoided, since they promote ventricular fibrillation of the anoxic myocardium. Hemorrhages and burns should be treated appropriately.

G. *Drowning*

1. *Importance of the Problem*

In any given year drowning accounts for about 7% of the accidental deaths reported in the U. S., or between 7000 and 8000 individuals. In the 1–4-yr age group drowning ranks third as a cause of accidental death, preceded only by motor vehicle accidents and fires and burns. In the 5–15-yr age group, drowning ranks second, with motor vehicle accidents first. In the group 1–24 yr of age no single disease accounts for as many deaths as drowning. In a report from Florida for 1964, of a total of 442 drowning deaths, 17 were skin divers. The U. S. Navy (1970) reports that drowning is the most frequent cause of death in self-contained diving equipment. In fact, most of the fatal casualties among divers and underwater swimmers are due to drowning. The problem is important to all underwater workers since they are frequently called upon to rescue and treat the apparently drowned person, or may themselves be patients.

2. *Causes of Drowning*

With a deep-sea, hard-hat rig, drowning is extremely unlikely, but could occur if the diver finds himself upside-down with the spit-cock open or the chin-button depressed, or in the head-down position with a torn or ruptured suit. With lightweight gear, drowning can follow loss or ditching of the mask (which may be necessary when the air supply is interrupted).

When self-contained breathing equipment is used there are numerous possibilities: loss or flooding of mask or mouthpiece; exhaustion of the gas supply; failure or improper functioning of gear; surface exposure in rough water; and almost any mishap followed by failure of emergency procedures, or by panic. But the most common cause of drowning is physical exhaustion resulting from swimming after surfacing (U. S. Navy 1970). Any accident causing unconsciousness while in any type of diving gear is life threatening.

The skin or breath-hold diver is more likely to drown from loss of consciousness or blackout. This is usually due to hyperventilation just prior to entering the water. "Hyperventilation for a few minutes can increase the alveolar oxygen tension to 140 mm Hg from a normal 100 mm Hg. At the same time the carbon dioxide level may fall very quickly from the normal 40 mm Hg to as low as 15 mm Hg" (Dumitru and Hamilton 1963). "Under these circumstances it is possible for the underwater swimmer to lose consciousness before the impulse to breathe becomes overwhelming." This may, and many times has, led to drowning. Craig (1961) reports eight incidents which indicate "that hyperventilation before breath-holding and exercise may delay the onset of the urge to breathe. Before the partial pressure of CO_2 increases significantly, the O_2 may decrease to a degree incompatible with high-level cerebral function. In five cases of drowning also reported, this chain of events is likely to have occurred."

Cold water may lead to markedly lowered body temperature, or hypothermia, which in turn may lead to death. According to Keatinge (1968), "Approximately 1000 people die from immersion in British coastal and inland waters every year. These deaths are usually ascribed to drowning, but probably include a high proportion of deaths due to hypothermia. Studies of wartime casualities at sea had shown that hypothermia rather than drowning was the main cause of death among people who left sinking ships." He makes two additional points on survival: Wear all clothing possible together with a life-jacket or other flotation gear; and do not exercise in the water, since this only speeds the cooling of the body. Hypothermia may have some beneficial effects in near drowning according to Farthmann and Davidson (1965): "Determined resuscitation efforts should therefore be made in any clinical situation of this kind [near drowning in cold fresh water]. The combined effects of lowered body temperature and fresh water drowning, although aggravating each other in some respects, offer a chance of survival, mainly due to the beneficial action of hypothermia." For further discussion of hypothermia see the section on cold in Chapter 4.

3. Physiology of Drowning

The sequence of events when a person is underwater without air to breathe is first a reflex closure of the glottis and spasm of the bronchi probably occurring at the first threat of the inhalation of water (Anon. 1962). Fear may also potentiate the spasm of the glottis, and certainly cold-water immersion causes the glottis to close. Submergence with no breathing leads to slowing of the heart rate (bradycardia) and may lead rather quickly to cardiac irregularity or to complete cardiac arrest. The bradycardia together with slowing of the peripheral circulation during the breath-hold underwater is nature's attempt to shunt the blood to the brain and heart in a life-saving effort. This period may last for several minutes, until unconsciousness from lack of oxygen results, or the accumulated carbon dioxide stimulates the respiratory center. In the latter case, violent inspiratory efforts result in the inhalation of a considerable volume of water (Fisher 1970). Swallowing of water, coughing, and vomiting occur, followed by loss of consciousness and terminal gasping.

If the person can be extricated before any large amount of fluid has entered the lungs, there is a good chance of recovery, either spontaneous or as the result of artificial respiration. Once the lungs have been flooded, the situation is grave. What happens to the lungs and the circulatory system depends to some extent on whether the victim has inhaled fresh or salt water, but the pathophysiology of drowning is not the province of this handbook. Clinical signs and symptoms would be valuable only for the physician, but a study of the records of 50 Navy cases (Fuller 1963) of non-fatal submersion requiring resuscitation are of sufficient general interest to be included in Table VIII-8.

Table VIII-8

Clinical Signs and Symptoms of the Postimmersion Syndrome

Central nervous system
  Coma—variable duration after rescue
  Convulsions—common during resuscitation
  Jaw muscle spasm
  Abnormal reflexes—usually transient
  Headache
  Visual disturbances—occasional complaint for first 1–2 days
  Loss of memory
  Delirium
Cardiovascular system
  Cardiac arrhythmias—revert with oxygenation
  Cardiac arrest—ventricular fibrillation not recorded
  Slow heart rate
  Low blood pressure—occasional high blood pressure
Blood
  Blood concentration—both fresh and salt water
  Free hemoglobin in blood—rare, transient, fresh water
  Increased number of white blood cells
Respiratory system
  Rapid respiration
  Difficult breathing
  Cough
  Bloody, frothy sputum
  Chest pain
  Laryngospasm
  Pulmonary rales
  Distant breath sounds
  Abnormal lung radiograph—common, usually clears in a few days
  Bluish skin color
Gastrointestinal system
  Vomiting—common during and after resuscitation
  Abdominal distention—swallowed water and air forced into stomach during resuscitation
  Intense thirst—sea water
Genitourinary system
  Albumin in the urine
  Hemoglobin in the urine—rare
General
  Fever
  Rigor
  Acidosis—transient
  Excess of chlorine in the blood (sea water)—transient
  Below normal level of chlorine in the blood (fresh water)—transient
  Increased amount of sugar in the blood
  Urea in the blood

4. Treatment

Drowning is a true emergency where each second counts. As soon as the victim is out of the water he should be held in such a position as to allow water to drain out of the respiratory tract. Then, if he is not breathing spontaneously, immediate and

Table VIII-9
Treatment of Drowning[a]

Treatment	Objective
Emergency	
Expired air resuscitation	Emergency gas exchange
Airway clearance	Emergency gas exchange
Closed-chest cardiac compression	Emergency circulation
Urgent	
Intermittent positive-pressure oxygen	Oxygenation and to prevent pulmonary complications
Epinephrine and/or calcium chloride into the heart	Prepare for defibrillation and treat cardiac arrest
Sodium bicarbonate intravenously	Treat acidosis
Defibrillation	Restore normal circulation
Tracheal intubation	Airway management
Tracheal suction, tracheostomy, and/or bronchoscopy	Airway management
Gastric intubation	Decompress stomach and prevent vomiting
Sedation	Treat restlessness and convulsions
Morphine, digitalis, and/or theophylline	Prevent and treat pulmonary edema
Hypothermia	Protect central nervous system
Chlorpromazine	Support peripheral circulation, and as adjuvant with hypothermia
Transfusion of blood	Replace blood volume and lysed cells
Exchange transfusion	Clear plasma of hemoglobin
Vasoactive drugs	Support blood pressure
Plasma expanders	Support blood volume
Parenteral fluids suitably chosen	Restore normal fluid and electrolyte pattern
Supportive	
Steroids	Prevent and treat aspiration pneumonia
Antibiotics	Prevent and treat bacterial pneumonia
Nebulized detergents	Promote bronchial drainage

[a] From Greene (1965) by permission of the American Physiological Society.

continuous artificial respiration is necessary. If no pulse can be felt or heart sounds heard, closed-chest cardiac massage is indicated and must be continued while the paitient is being taken to the hospital. Regardless of how well the patient responds to emergency treatment, hospitalization is essential because there may well be later complications and the near-drowning victim must be observed carefully. Steps in treatment are presented in Table VIII-9.

5. Prevention

The *U. S. Navy Diving Manual* (1970) lists the following under prevention:

Adequate training and drill in emergency procedures.
Proper equipment in good condition.

Use of lifejacket with scuba; lifeline with lightweight outfit.

Good diving practices; adequate preparations.

Appropriate boats, floats, etc.; readiness for going to aid of diver in distress.

These items of prevention cannot be overemphasized, and a number of sections of this handbook take them up in detail. Here we only emphasize the buddy system as being mandatory for scuba and skin diving; the other forms of diving or underwater work have tenders who can handle emergencies.

H. *Other Hazards*

1. *Marine Life*

"Of the thousands of forms of animal and plant life to be found in the ocean, there are relatively few that constitute a real hazard to the diver. However, some species are dangerous and may in some instances inflict serious wounds, poisoning, or violent death. Most difficulties can be prevented if the diver is made sufficiently aware of potential problems. Marine-life hazards of concern to the diver consist of two general types, those that produce wounds, and those that inflict stings" (U. S. Navy 1970). The *U. S. Navy Diving Manual* (1970) gives a good outline of the problem (see Tables VIII-10 and VIII-11).

The most complete and profusely illustrated study of marine life was reported in three volumes by Halstead (1965) and is an excellent reference.

Table VIII-10

Marine Life, Sharks (U. S. Navy Diving Manual)

Name	Danger[a]	Maximum size, ft	Appearance[b]	Behavior	Where found
White shark	4+	30	Slate-brown to black on back	Savage, aggressive	Oceanic, tropical, subtropical, warm temperate belts, especially in Australian waters
Mako shark	4+	30	Slender form, deep blue-gray on back	Savage	Oceanic, tropical, and warm temperate belts
Porbeagle shark	2+	12	Dark bluish-gray on back	Sluggish except when pursuing prey	Continental waters, northern Atlantic; allied forms in North Pacific, Australia, New Zealand

Table VIII-10—*Cont.*

Name	Danger[a]	Maximum size, ft	Appearance[b]	Behavior	Where found
Tiger shark	2+	30	Short snout, sharply pointed tail	Can be vigorous and powerful	Tropical and subtropical belts of all oceans, inshore and off shore
Lemon shark	2+	11	Yellowish-brown on back, broadly rounded snout	Found in salt-water creeks, bays, and sounds	Inshore western Atlantic, northern Brazil to North Carolina, tropical West Africa
Lake Nicaragua sharks	2+	10	Dark gray on back	Found in shallow water	Fresh-water species of Lake Nicaragua
Dusky shark	1+	14	Bluish or leaden-gray on back	Found in shallow water	Tropical and warm temperate waters on both sides of Atlantic
White-tipped shark	3+	13	Light gray to slate-blue on back	Indifferent, fearless	Tropical and subtropical Atlantic and Mediterranean deep offshore waters
Sand shark	2+	10	Bright gray-brown on back	Stays close to bottom	Indo-Pacific, Mediterranean, tropical West Africa, South Africa, Gulf of Maine to Florida, Brazil, Argentina
Gray nurse shark	3+	10	Pale gray on back	Swift and savage	Australia
Ganges River shark	4+	7	Gray on back	Ferocious, attacks bathers	Indian Ocean to Japan, ascends fresh water rivers
Hammerhead shark	4+	15	Ashy-gray on back; flat, wide head	Powerful swimmers	Warm temperate zone of all oceans, including Mediterranean Sea, out at sea or close inshore

[a] 1+ means minimum danger: 4+ means maximum danger.
[b] All sharks listed are some shade of white on undersides.

Table VIII-11

Marine Life, Other Forms

Name	Danger	Maximum size	Appearance	Behavior	Where found
Great barracuda	4+	6–8 ft	Long and slender; large mouth	Swift, fierce, easily attracted	Tropical and sub-tropical waters, West Indies, Brazil, northern Florida; in the Indo-Pacific from Red Sea to Hawaiian Islands
Groupers	2+	12 ft; 700 lb	Bulky type of body	Curious and bold; voracious feeders	Around rocks, caverns, old wrecks
Moray eels	1+	10 ft	Long, narrow, snakelike	Attack when provoked	Tropical and sub-tropical bottom dwellers
Killer whales	4+	—	Jet black head and back, white under-parts	Ruthless, ferocious	All oceans and seas, tropical to polar. *Caution*: Leave water immediately if sighted
Sea lions	1+	—	Resemble seals but are larger	Curious; fast swimmers	Northern waters
Sea urchins	2+	—	Small, spiny animals	Needle-sharp spines, small venomous pincers	Tropical and temperate zones, ocean floor on rocks and coral reefs
Corals	1+	—	—	Extremely sharp	Tropical and sub-tropical waters
Barnacles, mussels	1+	—	—	Inflict deep cuts	Rocks, pilings, wrecks
Giant clams	2+	Several hundred pounds	—	Trap legs and arms of victim between shells	Abound in tropical waters
Portuguese man-of-war	3+	6-in. diameter	Tentacles up to 50 ft long	Stings with cells on tentacles	Tropical waters
Sea wasp	4+	—	Tentacles up to 50 ft long	Stings with cells on tentacles	Northern Australia, Philippines, Indian Ocean
Octopuses	2+	25 ft	Arms radiating from head	Hold with tentacles; also bite	Underwater caves
Cone shells	2+	—	Colorful shells	Penetrate skin with venom-filled teeth on proboscis (trunk)	Widespread

Table VIII-11—*Cont.*

Name	Danger	Maximum size	Appearance	Behavior	Where found
Horned sharks	1+	—	Spines anterior to back fins		
Stingrays	1+	Several feet	Spine on top of tail, flat body	Drive spine into leg when stepped on	Tropical to temperate waters
Catfish	1+	—	Venomous dorsal and pectoral spines	—	Tropical and temperate, mostly fresh water, some marine
Weeverfish	1+	—	Venomous dorsal and pectoral spines	Toxic to the nervous system and blood	Eastern Atlantic and Mediterranean
Scorpionfish	1+	—	Venomous back, anal, and pelvic spines	Toxic to the nervous system and blood	Tropical and temperate
Sea snakes	3+	9 ft	Resemble snakes; venomous fangs	Boldness varies	Tropical Pacific and Indian Ocean River mouths to far at sea

2. *Fouling*

The diver with lifeline and airhose is more likely to become fouled than the scuba diver, but either may become entangled with various underwater obstructions. Not only is there the possibility of entanglement of lifeline, airhose, cylinders, other equipment, or the body itself, with some underwater obstruction, but there may be a cave-in of a tunnel, or the shifting of heavy objects near the diver or in the route of exit from a wreck.

The surface-supplied diver, if eventually released, usually suffers nothing more than fatigue or exhaustion, but the scuba diver may have to ditch his equipment and come to the surface by free escape with its dangers; or, if trapped, use up the gas supply and become asphyxiated. Recommended action is outlined by the *U. S. Navy Diving Manual* (1970) as follows:

> Whether a diver emerges safely from fouling depends very much on his own actions even though the help of another diver is often required to free him. The diver must observe the following:
>
>   a. Remain calm; think.
>   b. Describe situation to tender or call buddy's attention to situation.
>   c. Carefully and systematically attempt to determine cause of fouling and clear self. Use knife cautiously to avoid cutting airhose or breathing apparatus.
>   d. In fouling with self-contained apparatus, regard ditching and free ascent as a last resort, but prepare to do this in case it proves necessary.
>   e. If efforts to clear prove futile, be quiet and wait for air.

f. Remember that frantic, ill-planned efforts not only usually fail, but can make the situation worse. Futile struggling and panic can result in death from exhaustion and shock.

Fouling can usually be prevented by using proper precautions and common sense. But it is most important to have a buddy if scuba diving and to have a tender and reserve diver on the surface if diving with surface-supplied equipment.

3. Infections

In this section we consider only those infections etiologically associated with the wet, hyperbaric environment. As Hamilton and Schreiner (1968) say:

> . . . if divers choose to live for many days in perennially humid undersea habitats, their skin provides a fertile nesting ground for all sorts of undesirable growths, not all of which are of marine origin. These are in fact the familiar plagues of children's summer camps—fungus infections, athlete's foot, swimmer's ear. Long-duration exposures in undersea habitats will require better medications than presently available, or a more hospitable environment. This can be provided almost perfectly by keeping the atmosphere in the habitat dry.

Sciarli *et al.* (1973) report the condition of four divers in a French program who lived at a depth of 300 m for eight days. Although the atmosphere of the chamber was "disinfected by a device of the Aerovap type" and the divers used a "preventive local disinfectant," they still developed postules, conjunctivitis, skin rash, and had earaches. "The most common problem was painful inflammation and secretion in the auditory canals, without involvement of the tympanum . . ." The 86% humidity was believed to be the cause, for in subsequent dives when the humidity was kept between 40 and 50%, these conditions did not occur.

Oser (1972) reports a similar experience in the German habitat HELGOLAND, where the aquanauts lived under conditions involving a pressure of 2.0 atm. The health of the aquanauts was medically supervised and "infections involving parts of the ear made necessary almost daily medical inspections."

The problem of upper-respiratory infections as a deterrent to hyperbaric activity is presented earlier in this chapter under prediver *conditions*. The important problem of otitis media is detailed in Chapter III under aural barotrauma.

4. Fatigue

Overexertion, fatigue, exhaustion, respiratory embarrassment, panic, and accident is the too often repeated sequence of events leading to a fatality. Webster (1966) reports that in over half the diver fatalities in one study, physical exhaustion probably led to the cause of death. Sawyer (1972) lists "fatigue and exhaustion" as the primary factor leading to injury of scuba divers.

Everyone has had the experience of becoming out of breath from working too hard or running too fast. The respiratory response to overexertion takes time to develop fully, so it is possible to exceed the normal work capacity by quite a margin before realizing it. Under normal conditions on land there is seldom any problem, as shortness of breath passes rapidly when work is slowed or stopped. But underwater, especially at increased depths, the situation may become serious. Breathing resistance both in the diver's own airways and particularly in self-contained diving equipment make the work of breathing difficult and, coupled with anxiety, may lead to real air hunger. "The feeling of impending suffocation is far from pleasant and it has led more than one inexperienced diver into panic and a serious accident" (U. S. Navy 1970). An experienced diver will generally have a good idea of his own capacity and of the limits of his gear. When using unfamiliar equipment, he gradually increases his work rate so that he will not exceed his limits and will not suddenly be faced with severe shortness of breath. If he does find himself in that situation, he is able to avoid panic because experience has shown that the sensation will pass if he reduces or stops his work.

In spite of experience there are conditions which lead to exceeding the known limits. Working against strong currents, or on an unusually muddy bottom; a diving job requiring heavy exertion or unusually prolonged task; wasting effort early in the dive; and perhaps most important, efforts of the diver to free himself when fouled, particularly if the efforts are poorly planned and executed and thus ineffectual, are good examples of such conditions. There are also conditions that can reduce the ability to work hard: excessive breathing resistance in self-contained breathing apparatus; carbon dioxide buildup; a too low oxygen content in the breathing mixture; fouling of air by oil fumes or carbon monoxide; and excessive cold or inadequate protection.

"Overexertion has aspects other than those concerned with breathing. There are definite limits to what a man's muscles can do, and it is not easy to judge how much strength one has in reserve. A diver who wastes energy or tires himself unnecessarily may find that it is difficult to complete his task. A man who neglects his physical condition may discover that what was once a normal work rate now tires him very quickly" (U. S. Navy 1970).

Physical exertion increases the rate of blood circulation as well as the respiratory rate. Consequently, during exertion under pressure, larger amounts of nitrogen (or other inert gases) are transported to the tissues per unit of time. "Consider the circumstances where a diver is working hard under water, e.g. moving heavy objects, swimming against a strong current, etc. This diver's tissues may absorb excessive nitrogen equivalent to 10–20 minutes of extra diving time under normal conditions, and if he was on a dive-schedule of 60 min at 60 feet (the 'no-decompression' limit for that depth), he may suffer decompression sickness if he surfaces without decompression stops" (Somers 1972).

The symptoms of overexertion and exhaustion, as listed by the *U. S. Navy Diving Manual* are: extreme fatigue, increasing weakness, labored breathing, and anxiety and tendency toward panic. Bachrach (1970) presents a table prepared by Behnke which lists manifestations of acute and chronic fatigue: performance impairments, behavioral deterioration, and physiological signs and symptoms (see Table VIII-1).

For treatment the *U. S. Navy Diving Manual* says that: the diver should stop and rest if possible, inform buddy or tender, terminate dive if resting fails to help, and surface when practical, observing proper rate of ascent and decompression stops if required. His buddy should render all possible assistance, particularly in getting him to the surface. The surface personnel should give help in getting the diver aboard and should provide rest, warmth, and nourishment.

Prevention is the most important aspect and should be possible in most cases. The diver should do the following: know his own limits and stay within them; discontinue the dive if it exceeds his powers; use good gear in good condition; concentrate on training and experience to help eliminate panic; employ weights and line when working in strong current; stop to rest and ventilate before becoming overfatigued; and wear adequate cold-water protection.

5. *Flying after Diving*

The first authenticated case to be published of an in-flight incident (Miner 1961) occurred on an intercontinental flight when several members of the crew suffered decompression sickness. The crew members involved had been scuba diving for several hours during the morning and afternoon prior to the evening flight involved. Since that time other cases have been reported, and both animal and human experiments have demonstrated the hazards of flight after diving. For example, in a carefully controlled experiment (Furry *et al.* 1967) the no-bends threshold was determined for a group of dogs, and then they were subjected to an altitude exposure of 10,000 ft (10.1 psia) after surface decompression intervals of 1, 3, 6, and 12 hrs:

> In the first series of 14 animals exposed to 10,000 feet (10.1 psia) pressure altitude after 1 hour surface interval, 13 (92.9%) developed decompression sickness. Eight (61.5%) of these dogs did not recover with recompression to sea level and in addition had to be recompressed in the high-pressure chamber before recovery was complete. In the second series of 10 dogs with a 3-hour surface interval, 3 (30%) developed decompression sickness, but all recovered with a return to sea level pressure. In the third group of 18 animals with a 6-hour surface interval, 5 (27.8%) developed decompression sickness and 3 (60%) of them had to be recompressed in the high-pressure chamber before recovery was complete. In the fourth group of 10 animals with a 12-hour surface interval, none exhibited signs of decompression sickness at altitude.

Based on this work the following policy was recommended: "All personnel who have engaged in compressed-air diving to depths of 25 feet of water or its equivalent should not fly in other than pressurized commercial aircraft or to a cabin altitude greater than 8,000 feet or its equivalent within 12 hours following the termination of a compressed-air exposure" (Furry *et al.* 1967).

Following research with human volunteers (Edel *et al.* 1969) it was concluded: "Scuba divers who stay strictly within the limits (depth-time) of the standard U. S. Navy's 'No-Decompression' limits and repetitive group designation table for no-decompression dives, for a period not exceeding 12 hours will not develop decompression sickness if, after diving, they allow a minimum 2-hour surface interval before flying in a pressurized commercial aircraft. Divers who make dives beyond these 'no decompression' limits should allow a surface interval of 24 hours before decompression to a commercial aircraft's cabin altitude pressure if they are to avoid the risk of the bends." The French (Lavernhe 1970) agree with this dictum but add that any diver who shows symptoms of decompression sickness should observe a minimum period of 24 hr before boarding an aircraft. They also recommend the inhalation of pure

Table VIII-12

Type of dive	Time interval between diving and flying	Maximum altitude (or effective altitude in pressurized aircraft)
Requiring no stops	Up to 1 hr	300 m (~1000 ft) (e.g., helicopter)
	1 to 2 hr	1500 m (~5000 ft)
	Over 2 hr	Unlimited
Requiring stops	Up to 4 hr	300 m (~1000 ft) (e.g., helicopter)
	4–8 hr	1500 m (~5000 ft)
	8–24 hr	5000 m (~16,500 ft)
	Over 24 hr	Unlimited

oxygen as treatment. The CIRIA (1972) group consider it inadvisable to fly above 600 m (approximately 2000 ft) in any aircraft within 24 hr of completing a dive. They further recommend a period of 2 hr between diving operations and flying in a pressurized aircraft, if dives were carried out with no stops; and a period of 24 hr if dives were carried out with stops. In addition they conclude that if flying following diving is essential, the rules given in Table VIII-12 will minimize the risk.

I. *Problems of Escape and Rescue*

1. *Submarines*

a. Escape

Although general submarine activity is not considered in this handbook, the following material is presented because rescue and salvage crews will find it useful and because many of the problems encountered illustrate the difficulties of underwater high-pressure activities.

All the world's navies have had their share of submarine disasters. There have been a number of men who have escaped from shallow depths (less than 200 ft), and a few cases of successful rescue of trapped men. The first recorded escape was from a German submarine in 1851 in Kiel Harbour in 60 ft of water. Wilhelm Brauer, trapped with a number of other men, realized that the only hope of opening the hatches was to raise the internal pressure by flooding. This allowed the hatches to burst open and some of the men shot to the surface (Miles 1969).

The first escape in the British Navy was a dramatic event. In 1916, H. M. Submarine E-41 collided with another submarine and sank in 30 ft of water, with a number of men escaping through the conning tower. Stoker Petty Officer Brown found himself alone behind the watertight bulkhead in the engine room. He began to flood the compartment so as to equalize pressure and enable him to open the hatch. The hatch popped open and a bubble of air was released and then the hatch slammed shut. This was repeated several times until the air in the engine room was almost gone and the hatch remained open, allowing Brown to escape. It is well to remember that no escape is possible unless there is equalization of the air pressure on the outside, since the hatch cannot be opened against pressure. This means either flooding an entire compartment and the men leaving one after the other, or flooding a small compartment (chamber, tube, tower, trunk) that holds two to four men and can be rapidly flooded and drained after each escape.

Since there was obvious need for some way for men to breathe during flooding and escape, there was early development of escape equipment. The German Navy developed the Draeger Breathing Apparatus; the U. S. Navy, the Momsen Lung (Submarine Escape Apparatus); and the British, the Davis Submarine Escape Apparatus (DSEA). These were self-contained equipment, and the DSEA was even used as a diving rig during World War II in the Mediterranean Campaign. But all of these were rather complicated pieces of equipment, and, since all that was needed was an air space for breathing, the U. S. Navy has developed the Steinke Hood. The British also have a hooded system (Elliott 1967). In the British device

> . . . the principles of the hooded ascent are simple. Each man enters the escape chamber (one man at a time) wearing a combined exposure-suit, buoyancy stole, and hood. The chamber is flooded with no increase in pressure until the vent is shut. The pressure then rises rapidly, in about 20 seconds regardless of depth, until the hatch opens. During compression the escaper keeps his stole and lungs inflated via a special valve—the Hood Inflation System (HIS)—which supplies air at a small positive pressure above the chamber pressure. When the hatch opens, the man, being buoyant and standing in a smooth-bored open-ended tube, automatically starts his ascent. During ascent the air in the buoyancy stole expands through special valves to flush fresh air into the hood from which the man may breathe. On arrival at the surface, the hood can be unzipped and the immersion suit blown up to give thermal protection (Barnard 1971).

The feasibility of this method has been worked out and demonstrated both at the submarine-escape training tank, H. M. S. Dolphin, and at sea on H. M. S. Osiris at depths down to 600 ft. Free ascent with no breathing equipment but by simply exhaling on the way to the surface is not only possible but is taught in the U. S. Navy.

However, escape is fraught with a number of dangers and problems: equalization of pressure, nitrogen narcosis, oxygen toxicity, decompression sickness, drowning, the effects of noxious gases under pressure, and fear. For these and other reasons, individual escape is limited to depths not in excess of 600 ft in the British Navy (Elliott 1971), and this only for individual tower escapes, where pressurization time is 20 sec, bottom time is 1.5 sec, and the original pressure in the submarine compartment was not over 2 ATA. During flooding or as the result of an accident there is bound to be some pressure buildup, which must be considered. In the U. S. Navy individual escape is possible—and for shallower depths (under 100 ft), group or compartment escape is feasible—using the new hood or by "free" escape (Alvis 1952). Escapes from greater depths have been made: Steinke, 304 ft; Bond, 307 ft; the British, 600 ft. But these were with highly trained and motivated people, under well-controlled conditions.

b. Rescue

The U. S. Navy has always stressed rescue from the outside and the use of the McCann rescue chamber to remove 33 men from the U.S.S. SQUALUS on the bottom at 240 ft is cited as the classic example. Unfortunately, the "bell" can only be used at a depth at which a diver can function and the last two nuclear submarines lost were at depths where a diver could not have worked.

To provide for rescue at greater depths the U. S. Navy has developed the Deep Submergence Rescue Vehicle (DSRV) (Blair 1969), which has the following minimum characteristics:

Operating depth	1050 m (minimum)
Ascent/descent rate	3 m/min
Speed submerged	5 knots (maximum)
Submerged endurance	3 knots for 12 hrs
Maneuverability	Hover in 1-knot current
Mating attitude	All systems operable at 45° solid angle
Transportable weight	30,500 kg (maximum)
Length	15 m (maximum) transportable by a C-141A
Beam	2.4 m
Personnel	Operational crew: pilot, copilot, and mid-sphere operator
Rescue capability	Up to 24 rescuees per trip
Pressure range	Control sphere 0.8–3.7 ATA; rescue and mid-sphere 0.8–5.0 ATA
Life support	24 hr (minimum for 27 persons); emergency individual life support for 3 hr (minimum)
Pressure hull	Inner hull consists of three interconnecting 2.3-m-diameter spheres of HY-140 steel

Locating the disabled submarine is of first importance, and this may be accomplished by homing in on the submarine indicator buoy which will normally be released, come to the surface, and repeat the code number of the submarine, "SOS," "SUBSUNK," and a steady note of 30 sec for d.f. purposes every 2 min.

Once the submarine is located and contact established, the decision to recommend individual escape or await the arrival of the "bell" or the DSRV must be made, based on depth of the disabled submarine, conditions inside, and availability of rescue equipment.

2. Submersibles and Habitats

There are many different types of submersible vehicles, bells, and habitats and at the present time both escape and rescue are difficult, if not impossible in a large proportion of them (Jones 1969).

A plea was made to begin development of an emergency hyperbaric rescue system for the U. S. (Myrick 1972). One such candidate system consists of a lightweight, portable hyperbaric chamber; a heavy-lift, high-speed rescue helicopter; and a shore-based medical facility for comprehensive care. The U. S. Coast Guard HH-3F helicopter is recommended as the transportation vehicle. A chamber of glass-filament-wound composite material weighing around 1000 lb will provide the necessary hyperbaric environment. Such a system would certainly be a step in the right direction, but its usefulness depends on the patient (diver) being on the surface to be put into the portable hyperbaric chamber. This is not good enough, particularly if the underwater worker is saturated even at a fairly shallow depth (60 ft). What is needed is a way for a diver to move from the habitat to a pressurized vessel that can be kept under pressure for the long period of decompression. The vessel must be capable of mating with a decompression chamber aboard ship, or one that can be transported ashore.

As pointed out by Manuel (1970):

> The capability for submerged dry transfer from one vehicle to another is an absolute requirement for subsea operations. Submerged dry transfer allows passage from one vehicle to another without exposure to currents, cold, marine animals, and, in single atmosphere transfer, without exposure to sea pressure.
>
> Applications will be as varied as the uses of underwater vehicles and structures themselves, but presently evident uses are dry transfer from a delivery vehicle such as a lock-out submarine to a bottom structure, either at sea pressure or atmospheric pressures. These structures might be scientific habitats such as Tektite, or operational, or commercial structures such as SeaLab or subsea production facilities. This capability also has great significance as a rescue and salvage device for small submarines.

A unique rescue method was described for the Helgoland Underwater Laboratory (Burton 1971):

> This is a one-man rescue chamber that can be entered from the decompression chamber without getting one's feet wet, is hermetically sealed, and can be released to float independently to the surface to be picked up by helicopter or lifeboat. For evacuation without surface assistance, a life raft is attached to the top of the pressurized hull and this can be released to pop up to the surface followed by the crew when they have passed through the decompression chamber.

Another chamber has been developed which take into consideration the problem of rescue (Hunley 1968):

> The submersible work chambers have two compartments separated by a pressure-tight hatch. The upper compartment, kept at atmospheric pressure, is occupied by a control technician and a relief diver. The lower compartment is pressurized to ambient sea pressure and is occupied by the working diver, in modified hardhat diving dress, who enters and leaves by the lower hatch. The upper compartment can be pressurized in about 4 minutes to allow the relief diver to go to the aid of the working diver if required. Under normal conditions, neither the relief diver nor the technician would require decompression upon surfacing, since their compartment is maintained at atmospheric pressure.

Ocean Systems, Inc. completed the ADS-IV system in March 1966. Its personnel-transfer capsule mates to the connecting trunk between the two, two-man deck decompression chambers.

The development of diver-lockout capability, which allows use of the best capabilities of divers and submersibles, is gaining acceptance. One principal reason is the safety feature of having the diver on an umbilical under constant observation from the forward compartment of the submersible, with the ability to get to him quickly and thus not leave him stranded on the bottom. After one such diver operation from the SHELF DIVER submersible the two divers reentered the open after-compartment, closed the hatch, and "SHELF DIVER then powered out from under the platform and ascended to the surface; during the ascent, decompression of the divers aft had already started. The submarine, with the divers still under pressure in the aft compartment, was then retrieved on the deck of the supply boat and mated to a sphere and tunnel for transfer under pressure to a deck decompression chamber. The submarine was then ready to take on another set of divers and start below" (Bailey 1972).

A system worthy of detailed study is the U. S. Navy's Deep Diving System MK-2 Mod. 0 (Princevalle 1970). It is a saturation diving complex with a number of components. The one of primary interest in the context of escape and rescue is the personnel transfer capsule (PTC). The PTC may be used during emergency rescue operations by housing four men for the required decompression cycle. This constitutes a situation in which submariners can be rescued from a sunken submarine. In this mode of operation the PTC would normally be mated to the DDC, which can contain 34 survivors, and the total system maintained at pressures up to 5 ATA for up to 48 hr maximum to provide the basic requirements for the maintenance of life during the period of decompression.

The characteristics of the Personnel Transfer Capsules are:

Overall height	9 ft, $10\frac{1}{2}$ in.
Internal diameter	7 ft.
Buoyancy	1000 lb positive
Weight in air	25,000 lb

This system is the one designed to meet the need for a salvage program and the Man-in-the-Sea program.

The Johnson Sea-Link tragedy certainly demonstrated that much more needs to be done to make escape and rescue from submersibles a possibility or even a probability.

References

ALVIS, H. J. Submarine rescue and escape. *Arch. Ind. Hyg.* **6**:293–304 (1952).

ANONYMOUS. *Lancet* (1):468–469 (Mar. 3, 1962).

AULT, W. E., and D. I. CARTER. The influence of hyperbaric chamber pressure on water-spray patterns. *Fire J.* **61**:48 (1967).

BACHRACH, A. J. Diving behavior. In: *Human Performance and Scuba Diving. Proceedings of the Symposium on Underwater Physiology, Scripps Institute of Oceanography, La Jolla, Calif., April 10–11, 1970*, pp. 119–138. Chicago, The Athletic Institute (1970).

BAILEY, V. R., J. LaCERDA, and J. F. MANUEL. Diver lockout and observation submersibles. A perspective of participation in offshore operations. In: *1972 Offshore Technology Conference May 1–3, Houston, Texas. Preprints*, Vol. I, pp. 547–553. Published by the Conference (1972).

BARNARD, E. E. P. Submarine escape from 600 feet (183 metres). *Proc. Roy. Soc. Med.* **64**:1271–1273 (Dec. 1971).

BARROW, D. W., and H. Y. RHOADS. Blast concussion injury. *J.A.M.A.* **125**:900–902 (1944).

BEHNKE, A. R., Jr. Reaction 1 (to paper on Diving Behavior). In: *Human Performance and Scuba Diving. Proceedings of the Symposium on Underwater Physiology, Scripps Institute of Oceanography, La Jolla, Calif., April 1970*, pp. 139–143. Chicago, The Athletic Institute (1970).

BENZINGER, T. In: *German Aviation Medicine. World War II*. Vol II, p. 1225. Washington, D.C. U. S. Government Printing Office (1950).

BENZINGER, T. Causes of death from blast. *Amer. J. Physiol.* **167**:767 (Dec. 1951).

BLAIR, W. C. Human factors in deep submergence vehicles. *Mar. Technol. Soc. J.* **3**:37–46 (Sept/ Oct. 1969).

BOND, G. F. Safety factors in chamber operations. In: *Fundamentals of Hyperbaric Medicine*, pp. 141–143. Washington, D. C., National Academy of Sciences–National Research Council (1966) (Publ. 1298).

BOTIERI, P. B. Fire protection for oxygen enriched atmosphere applications. In: *Proceedings of Fire Hazards and Extinguishment Conference*, pp. 39–69. Brooks AFB, Texas, Air Force Syst. Command, Rep. AMD-TR-67-2 (May 23, 1967).

BÜHLMANN, A. A., R. SCHIBLI, and D. GEHRING. Experimentelle Undersuchungen uber die Dekrompression nach Tauchgangen in Bergseen bie verminder tem Luftdruck. *Schweiz Med. Wochenschr.* **103**:378–383 (Mar. 10, 1973).

BURTON, R. Helgoland underwater laboratory. In: *Sea Frontiers* **17**:335–341 (Nov./Dec. 1971).

CAMISHION, R. C. Electrical hazards in the research laboratory. *J. Surg. Res.* **6**(5):221–227 (May 1966).

CARTER, D. I. Fire extinguishment and protective clothing evaluations. In: *Proceedings of Fire Hazards and Extinguishment Conference*, pp. 69–105. Brooks AFB, Texas, Air Force Syst. Command, AMD-TR-67-2 (May 23, 1967).

CHRISTIAN, E. A., and J. B. GASPIN. Swimmer safe standoffs from underwater explosions. U. S. Nav. Ordn. Lab., Rep. NOLX-80 (1974).

CIRIA UNDERWATER ENGINEERING GROUP. *The Principles of Safe Diving Practice 1972*. London, CIRIA Underwater Engineering Group (1972).

CLEMEDSON, C. J. An experimental study on air blast injuries. *Acta Physiol. Scand.* **18**(Suppl. 61): 1–220 (1949).

CLEMEDSON, C. J. Blast injury. *Physiol. Rev.* **36**(3):336–354 (July 1956).

CLEMEDSON, C. J., and S. A. GRANSTOM. Studies of the genesis of " rib markings " in lung blast injury. *Acta. Physiol. Scand.* **21**(2–3):131–144 (1950).

COLE, R. H. *Underwater Explosions*. New York, Dover Publications (1965).

COMMITTEE ON AMPHIBIOUS OPERATIONS. Effects of underwater blast. In: Revelle, R., chm. Panel on Underwater Swimmers. Washington, D.C. National Academy of Sciences–National Research Council (1952).

COOK, G. A. Combustion safety in diving atmospheres. In: *Proceedings of Fire Hazards and Extinguishment Conference*, pp. 139–147. Brooks AFB, Texas, Air Force Syst. Command, Rep. AMD-TR-67-2 (May 23, 1967).

COOK, G. A., R. E. MEIERER, and B. M. SHIELDS. Screening of flame-resistant materials and comparison of helium and nitrogen for use in diving atmospheres. Tonawanda, N. Y., Union Carbide Corp., First Summary Rep. on Contr. N00014-66-C-0149 (Mar. 21, 1967a).

COOK, G. A., R. E. MEIERER, and B. M. SHIELDS. Combustibility tests on several flame-resistant fabrics in compressed air, oxygen-enriched air, and pure oxygen. *Text. Res. J.* **7**:591–599 (July 1967b).

COOK, G. A., R. E. MEIERER, B. M. SHIELDS, and H. E. NEVINS. Effects of gas composition on burning rates inside decompression chambers up to 300 fsw. In: *Under-Ocean Technology. Proceedings of the 54th Annual Meeting of the Compressed Gas Association, New York, Jan. 17, 1967*, pp. 31–42. Published by the Association (1967c).

COOK, G. A., V. A. DORR, and B. M. SHIELDS. Region of non-combustion in nitrogen–oxygen and helium–oxygen atmospheres. *Ind. Eng. Chem. Processes. Des. Devel.* **7**:308 (1968).

CRAIG, A. B., Jr. Underwater swimming and loss of consciousness. *J.A.M.A.* **176**:255–258 (Apr. 1961).

CROMWELL, L. *Biomedical Instrumentation and Measurements*. Englewood Cliffs, N. J., Prentice Hall (1973).

DALZIEL, C. F. The effects of electric shock on man. Washington, D. C., U. S. Atomic Energy Commission, Office of Health and Public Safety, Safety and Fire Protection Branch. Reprinted from *IRE Trans. Med. Electron.* (May 1956).

DORR, V. A. Fire studies in oxygen-enriched atmospheres. *J. Fire Flammability* **1**:91–106 (1970)

DORR, V. A. Effects of environmental parameters upon combustion of fire resistant materials, potential electrical sources of ignition and analysis of combustion products. Third summary report on combustion safety in diving atmospheres. Contr. N00014-6G-C-0169. Tarrytown, N.Y., Ocean Systems, Inc. (Jan. 31, 1971*a*).

DORR, V. A. Compendium of hyperbaric safety research. Final report on combustion safety in diving atmospheres. Contr. N00014-66-C-0169. Tarrytown, N. Y., Ocean Systems, Inc. (Feb. 28, 1971*b*).

DORR, V. A., and H. A. SCHREINER. Region of non-combustion, flammability limits of hydrogen–oxygen mixtures, full scale combustion and extinguishing tests and screening of flame-resistant materials. Second summary report on combustion safety in diving atmospheres. Contr. N00014-66-C-0149. Tonawanda, N. Y., Ocean Systems, Inc. (May 1, 1969).

DUMITRU, A. P., and F. G. HAMILTON. Underwater blackout—a mechanism of drowning. *GP* 29:123–125 (Apr. 1963).

EDEL, P. O., J. J. CARROLL, R. W. HONAKER, and E. L. BECKMAN. Interval at sea-level pressure required to prevent decompression sickness in humans who fly in commercial aircraft after diving. *Aerosp. Med.* 10:1105–1110 (Oct. 1969).

EGGLESTON, L. A. Evaluation of fire extinguishing systems for use in oxygen rich atmospheres. San Antonio, Texas, Southwest Research Institute, Final Rep. on SwRI Prj. 03-2094 (1967).

ELLIOTT, D. H. Submarine escape from one hundred fathoms. *Proc. Roy. Soc. Med.* 60:617–620 (July 1967).

ELLIOTT, D. H. Submarine escape from 600 feet using rapid compression and buoyant ascent. In: *1971 Offshore Technology Conference, April 19–21, Houston, Texas. Preprints,* Vol. II, pp. 191–194. Published by the Conference (1971).

FARTHMANN, E. H., and A. I. G. DAVIDSON. Fresh water drowning at lowered body temperature. An experimental study. *Amer. J. Surg.* 109:410–415 (Apr. 1965).

FINKELSTEIN, S., and E. M. ROTH. Electric current. In: *Compendiums of Human Responses to the Aerospace Environment,* Vol. 1, Sect. 1–6. NASA CR CA-1205(1). Albuquerque, New Mex., The Lovelace Foundation for Medical Education and Research (Nov. 1968).

FISHER, R. S. Immersion injury and drowning. In: Harrison, J. R., ed. *Principals of Internal Medicine,* pp. 721–723. New York, McGraw-Hill (1970).

FRYE, J. Electric shock. *Electron. World* 1965 (Dec.): 50–51.

FULLER, R. H. Drowning and the postimmersion syndrome, a clinicopathologic study. *Mil. Med.* 128:22–36 (Jan. 1963).

FURRY, D. E., E. REEVES, and E. BECKMAN. Relationship of scuba diving to the development of aviators' decompression sickness. *Aerosp. Med.* 38:825–828 (Aug. 1967). (Also appeared as NMRI Res. Rep. 5 on MF 011.99–1001, 1966.)

GATES, R. Roentgen findings in immersion blast injury. *U. S. Nav. Med. Bull.* 41(1):12–19 (Jan. 1943)

GREENE, D. G. Drowning. In: Fenn, W. O., and H. Rahn, eds. *Handbook of Physiology.* Section 3: *Respiration.* Vol. II, pp. 1195–1204. Washington, D. C., American Physiological Society (1965).

HACKMAN, D. J., and J. S. GLASGOW. Underwater electric shock hazards. *J. Ocean Technol.* 2(3): 49–56 (1968).

HALSTEAD, B. *Poisonous and Venomous Marine Animals.* Vols. I, II, and III. Washington, D. C., U. S. Government Printing Office (1965).

HAMILTON, R. W., Jr. Life support for underwater pipeline welding. In: *Marine Technology 1970,* Vol. I, pp. 159–166. Washington, D. C., Marine Technology Society (1970).

HAMILTON, R. W., Jr., and H. R. SCHREINER. Putting and keeping man in the sea. *Chem. Eng.* 75:263–270 (June 1968).

HAMILTON, R. W., Jr., T. D. LANGLEY, and V. A. DORR. Safe instrumentation for physiological research in the hyperbaric environment. *Trans. N. Y. Acad. Sci.* 32:458–470 (Apr. 1970).

HAMLIN, H. Neurological observations on immersion blast injuries. *U. S. Nav. Med. Bull.* 41(1): 26–31 (Jan. 1943).

HARTER, J. V. A review of the Navy safety program. In: *Proceedings of Fire Hazards and Extinguishment Conference,* pp. 128–138. Brooks AFB, Texas, Air Force Syst. Command, Rep. AMD-TR-67-2 (May 23, 1967*a*).

HARTER, J. V. Fire at high pressure. In: Lambertsen, C. J., ed. *Underwater Physiology, Proceedings of the Third Symposium on Underwater Physiology*, pp. 55–80. Baltimore, Williams and Wilkins (1967b).

HOFF, E. B. C., and L. J. GREENBAUM. *A Bibliographical Sourcebook of Compressed Air, Diving and Submarine Medicine*, Vol. 1. Washington, D. C., Department of the Navy (1943).

HOFF, E. B. C., and L. J. GREENBAUM. *A Bibliographical Sourcebook of Compressed Air, Diving and Submarine Medicine*, Vol. II. Washington, D. C., Department of the Navy (1954).

HOUSE, D. G., P. PRENDERVILLE, B. D. WILSON, S. A. W. WILSON, and A. H. BEBB. Protection against underwater blast. Comparison of steel and fibre glass. U. K., Med. Red. Counc., Roy. Nav. Pers. Res. Comm., Rep. RNP 55/846, UWB 42 (Sept. 1955).

HUGGETT, C., G. von Elbe, and W. HAGGERTY. The combustibility of materials in oxygen–helium and oxygen–nitrogen atmospheres. Brooks AFB, Texas, USAF Sch. Aerosp. Med., Rep. SAM-TR-66-85 (1966).

HUNLEY, W. H. Deep ocean work systems. In: Brahtz, J. F., ed. *Ocean Engineering. Goals, Environment, Technology*, pp. 493–552, New York, Wiley (1968).

JENKINS, W. T. A summary of diving techniques used in polar regions. U. S. Nav. Coastal Syst. Lab., Prelim. Rep. on ONR Res. Proj. RF-51-523-101 (July 1973).

JONES, R. A. Emergency ascent. A deep submersible's last hope for return. *Nav. Eng. J.* **81**:23–27 (Dec. 1969).

KEATINGE, W. R. Immersion hypothermia. *Trans. Soc. Occup. Med.* **18**:73–74 (Apr. 1968).

LANGLEY, T. D. Personal communication (March 1969).

LAVERNHE, J. Plongee sous-marine et voyage aerien. *Presse Med.* **78**(32):1449 (June 27, 1970).

MANUEL, J. F. Submerged dry transfer from a diver lock-out submarine. In: *Progress into the Sea. Transactions of the Symposium, 20–22 Oct. 1969, Washington, D. C.*, pp. 197–199. Washington, D. C., Marine Technology Society (1970).

MCMULLIN, J. J. A. Foreword to symposium on immersion blast injuries. *U. S. Nav. Med. Bull.* **41**:1–2 Jan. (1943).

MEDICAL RESEARCH COUNCIL. Protection of divers against underwater explosions. U. K., Roy. Nav. Pers. Res. Comm., Rep. RNP 47/374, UWB 1 (Nov. 1945).

MERCK, SHARP AND DOME RESEARCH LABORATORIES. *The Merck Manual of Diagnosis and Therapy.* Tenth edition. Rahway, N. J., Merck, Sharp and Dome Research Laboratories (1961).

MILES, S. *Underwater Medicine.* Third edition. Philadelphia, J. B. Lippincott (1969).

MINER, A. D. Scuba hazards to air crew. Business pilots safety bulletin 61–204. New York, Flight Safety Foundation (1961).

MYRICK, J. A. A rescue system for medical diving emergencies. In: *1972 Offshore Technology Conference, May 1–3, Houston, Texas. Preprints*, Vol. I, pp. 557–561. Published by the Conference (1972).

NASA (National Aeronautics and Space Administration). Report of Apollo 204 Review Board to the Administrator. Washington, D. C., National Aeronautics and Space Administration (1967).

NFPA (National Fire Protection Association). *Manual on Fire Hazards in Oxygen-Enriched Atmospheres.* NFPA 53-M, 1969 edition. Boston, Mass., National Fire Protection Association (1969).

OSER, H. Medizinische Erfahrungen beim Einsatz des Unterwasserlaboratoriums ' Helgoland ' in Herbst 1971. DFVLR-Nachrichten, pp. 307–308 (Aug. 1972).

PRINCEVALLE, R. Deep diving system MK-2. In: *Progress into the Sea. Transactions of the Symposium, 20–22 Oct. 1969. Washington, D. C.* pp. 57–70. Washington, D. C., Marine Technology Society (1970).

ROTH, E. M. Space cabin atmospheres. Part II—Fire and blast hazards. Washington, D. C., National Aeronautics and Space Administration (1964) (NASA SP-48).

SAWYER, R. N. Some aspects of scuba in college health. *J. Amer. Coll. Health Ass.* **20**:323–327 (June 1972).

SCIARLI, R., F. SICARDI, C. LEMAIRE, and D. PROSPERI. Mycobacteriologie et plongee a saturation. *Bull. Medsubhyp* **9**:15–21 (Mar. 1973).

SMOOT, A., and C. A. BENTEL. Development of a shock hazard test procedure for underwater swimming pool lighting fixtures. Melville, L. I., N. Y. Underwriters' Elec. Dep. Bull. Res. 60 (Nov. 1971).

SOMERS, L. H. Research diver's manual. Ann Arbor, Mich., Univ. Mich., Sea Grant Tech. Rep. 16, MICHU-SG-71-212 (Aug. 1972).

SWAN, A. G. Two man space environment simulator accident. In: *Proceedings of Fire Hazards and Extinguishment Conference*, pp. 120–127. Brooks AFB, Texas, Air Force Syst. Command, Rep. AMD-TR-67-2 (May 23, 1967).

TAYLOR, J. D. The otolaryngologic aspects of skin and scuba diving. *Laryngoscope* **69**:809–858 (July 1959).

THOMAS, A. A. Pathology report on the toxicity of the pyrolysis products of Freon 1301. In: *Proceedings of Fire Hazards and Extinguishment Conference*, pp. 118–119. Brooks AFB, Texas, Air Force Syst. Command, Rep. AMD-TR-63-2 (May 23, 1967).

U. S. NAVY. *U. S. Navy Diving Manual*. Washington, D. C., U. S. Navy Department (Mar. 1970) (NAVSHIPS 0994-001-9010).

VERNOT, E. H. Inhalation toxicity and chemistry of pyrolysis products of bromotrifluoromethane. In: *Proceedings of Fire Hazards and Extinguishment Conference*, pp. 107–117, Brooks AFB, Texas, Air Force Syst. Command, Rep. AMD-TR-67-2 (May 23, 1967).

WAKELEY, C. P. G. Effect of underwater explosions on the human body. *Lancet* (1):715–718 (June 9, 1945).

WEBSTER, D. P. Skin and scuba diving fatalities in the U. S. In: U. S. Public Health Rep. 81, pp. 703–711 (Aug. 1966).

WOLF, N. M. Underwater blast injury. A review of the literature. U. S. Nav. Submar. Med. Cent., Rep. SMRL 646 (Oct. 26, 1970).

ZUCKERMAN, S. Problems of blast injuries. *Brit. Med. J.* **1**:94 (1941).

Operational Equipment

A. *Hand-Held Tools*

1. *Underwater Tool Design*

Recent advances in life-support systems have increased man's underwater endurance and work capacity. To enable man to perform the more complex underwater work tasks efficiently at greater depths, more mechanical aids are needed to augment man's capability.

A diver's work performance is greatly impaired by the effects of the underwater environment. Distorted senses, disturbed balance, diminished vision, limited communications, cold temperatures, and a need to constantly check life-support equipment affect work efficiency. Buoyancy limits the torque which the worker can apply. Many jobs require tethering the diver, or securing him so that he will have two hands free.

Because of such difficulties, the efficiency of the underwater worker depends to a great extent on the tools he uses. Designers must provide tools that handle easily and function effectively. Unfortunately, tools that operate efficiently in air often become inefficient and sometimes totally useless or extremely hazardous underwater.

Low temperatures of the ocean depths drain the diver's energy, degrading performance already impaired by the heavy suit he wears. Exerting force to perform a task becomes extremely difficult because of limitations of breathing apparatus. In addition, gloves and other protective clothing that must be worn eliminate direct contact between hand and tool. More serious, perhaps, are the limitations on worker vision. Already hampered by a facemask, the diver also finds that the water distorts vision and diminishes available light.

Keeping track of tools is also a problem. The diver must carry or attach to his person whatever supplies and tools he needs because he usually works alone. Silt and lack of light make it easy to lose sight of tools and supplies. Partially offsetting the many problems the diver faces, his ability under water to move in three dimensions aids him in many cases.

While some hand tools will always be needed, experience has revealed that for most tasks power tools are the best means of increasing diver efficiency. In addition to basic utility and usability, two considerations are important in the design of powered tools for marine environments: selection of materials able to withstand the effects of salt water and pressure, and selection of the safest and most efficient power source. Hackman (1970) has commented briefly on materials for underwater tools:

> Tool designers have adequate information for selection of most materials for marine tools. Data are available for marine wasting, pitting, corrosion effects, fouling, velocity effects, cavitation, galvanic effects, and stress-corrosion cracking. Unlike underwater structures, which predominantly use a single material to circumvent galvanic effects, tools require dissimilar materials. Hammers and wrenches must be tough; bearings and gears have to resist wear.
>
> Fabrication methods are often as important as the materials. For example, to prevent crevice corrosion along welds, it is necessary to heat treat, grind, and polish seams. Galvanic action may also be expected where different metals and where some metals and nonmetals join, unless properly isolated.
>
> For some purposes, as in hoses, and cables, rubbers and plastics are superior to metals, yet joints between elastomers and metals frequently present serious problems since the former seldom bond well to corrosion-resistant metals and poor seals can leak.

Many other elements must combine to ensure that a tool is effective, such as minimizing torque, selecting optimum weight and size, providing comfortable handles, and enabling divers to feel the position of the operating tool. Overcoming torque and linear reaction is hardest. The diver can be thrown off balance easily by the force of a tool turning a drill, a tap, or a wrench. Moreover, this instability diminishes the diver's ability to direct enough pressure against the tool to make it operate efficiently.

Both length and weight of tools are considerations. An overly long tool keeps the diver too far from his work, and interferes with his view. Large tools are difficult to carry, handle, and store. Heavy tools affect the diver's stability and waste his energy. The closer it is to neutral buoyancy, the more efficient is the tool.

Underwater switches pose a special problem. One answer is a switch sealed with a diaphragm across the pivot. Another is a magnetic switch, with the external handle incorporating a magnet. As the magnet moves, a small switch inside the sealed motor activates, controlling a larger switch through a relay.

Tools slip easily in a diver's gloved hands, so handles must have positive grips and should not cramp the hand. Hand tools are far more useful when they require only one hand for normal operation.

These design problems are typical of the great number that designers must consider. Each can greatly affect the efficiency of the diver and also the cost of working under the sea. Designs become increasingly complicated as divers descend to greater depths and move farther from the land.

2. Survey of Tool Categories

This section describes hand tools available to underwater workers, summarized from data of Bowen and Hale (1971) and Bayles (1970).

a. Standard Hand-Held Tools

These include pipe wrench, hammers, hack saws, bolt cutters, chisels, crescent wrench, screwdrivers (regular round-blade, flat-head, round-blade Phillips head, Allen head), wrenches (socket with ratchet, box-end, open-end, crescent).

b. Special Hand-Held Tools

1. Screwdriver with ratchet-palm-grip handle. Can be equipped with either standard or Phillips blades; a flat palm-grip provides four times the normal turning power; ratchet adjusts to forward, reverse, or locked positions.

2. Socket driver with palm-grip handle. Similar to (1), but utilizes a socket instead of a screwdriver blade.

3. Selective-socket driver with palm-grip handle. Consists of a series of concentric hexagonal tubes; when pushed onto a bolt head or nut and properly aligned, all tubes smaller than the nut are depressed and the outer tubes encasing the nut become the socket for turning it; each tool accommodates several nut sizes and a number of tool sizes are available.

c. Self-Contained Power Tools

The majority of these tools employ an explosive charge for the necessary working force.

1. Stud driver. Hand-held, explosively activated device for driving a threaded, solid stud into steel plate or concrete; average extraction force or pullout strength varies from 8000 lb for a $\frac{3}{8}$-in.-thick plate to 29,000 lb for a $1\frac{1}{8}$-in. plate; the tool may be reloaded underwater by replacing the used barrel with another sealed preloaded barrel.

2. Cable cutter. Consists of a steel jaw drilled to accommodate a cutter piston with a chisel edge; a torsion spring holds the cutter in place on the cable; the firing pin is locked with a safety pin.

3. Power velocity pad eye. Prefabricated and explosive-powered; provides attachment points for lift forces.

d. Remotely Supplied Power Tools

These are powered electrically, hydraulically, pneumatically, mechanically, or by means of gaseous fuel from the surface.

1. Oxygen arc electrode torch. Used for cutting through steel plates or cables, burning holes, and welding.

2. Oxygen–hydrogen torch. Used infrequently because the oxygen arc electrode torch is more effective.

3. U-slot cable cutter. Uses electric power to fire a special ballistic cartridge, which actuates a cutter punch to sever the cable; it can cut steel cables from $\frac{1}{16}$ to $1\frac{1}{2}$ in. in diameter.

4. Knife-edge hook. This is a 3-ft-long, V-shaped hook used to sever light cables; it uses mechanical power from the surface ship.

5. Cable lift attachments. Assorted hooks and other terminal gear attached by cables to the ship's winches or cranes to provide lifting force to move objects.

6. Chipping tools. Hand-held and usually pneumatically powered, used to clean off marine deposits.

7. Drills and rotary tools. Usually hand-held devices powered electrically, hydraulically, or pneumatically; some have both a steady torque and an impact capability; used to bore holes or to turn bolts or nuts.

8. Tunneling washing nozzles. Nozzles are fitted to fire hoses through which high-pressure water is pumped by the surface ship's pumps; used to clear sediment, or tunnel under heavy objects; the nozzles are designed to substantially reduce the back force applied to the diver.

9. Hydraulic jacks. Used to displace heavy objects; a typical jack can move 60 tons up to 18 in. per stroke.

10. Impact wrenches. Can be used for torqueing, drilling, and tapping.

11. Hydraulic drills. These are not as desirable as the impact wrench because of the possibility of the bit jamming and causing injury to the diver.

12. Hydraulic chain saws. Available for cutting wood under water.

13. Hydraulic disk grinders. Used for cleaning surfaces, grinding nuts, and grinding bevels in metal prior to welding operations.

3. Torqueing Tools

The category of underwater tools dealing with torqueing (screwdriving, but not tightening and bolting) is probably among the most widely used by divers, and will thus be covered in more detail than other types of tools, such as cutting tools, drilling and punching tools, fastening tools, and miscellaneous tools having potential application in underwater work tasks.

Torqueing tools are classified in Table IX-1, based on the functional requirement of the tool system, which, in turn, is dictated by two factors: the work to be accomplished, and the ability of the diver to furnish control and reaction to the tool.

Most underwater equipment and systems that must be dismantled or assembled under water use large, sturdy fasteners that require power wrenches. However, in many situations—for example, where access space is limited—a hand-torque tool has distinct advantages and may be the only type of tool useful for a given work task.

The amount of rotary force that can be applied to a screwdriver handle by a diver in a semiweightless condition is limited, sufficient only to torque small fasteners. Such fasteners have the following disadvantages under water: (1) small screws and bolt-type fasteners are difficult to handle with a gloved hand or in cold water; (2) poor visibility limits a diver's ability to properly position and align small screws and bolts; (3) divers have difficulty keeping the driver head of the tool positioned on the screw or bolt head; and (4) prolonged submersion, resulting in corrosion and/or excessive marine growth, obscures the head of a small fastener, making it difficult to position a screwdriver tip or socket on the fastener for removal.

Although screw-head type does not contribute significantly to performance

Table IX-1
Classification of Torqueing Tools (Anderson 1971)

Power source	Tool item nomenclature	Biomechanical force requirements
Hand torqueing	Screwdrivers and nutrunners	Rotary force (hand torque)
	Hand wrenches	Forearm–wrist strength striking force
Power torqueing	Pneumatic impact wrench	Arm strength with elbow flexion (push lift)
	Hydraulic impact wrench	Backlift; arm strength with elbow flexion (push lift)

efficiency, tests have indicated a slight efficiency advantage for the Phillips and Allen head screws. Also, divers appear to prefer working with the Allen and Phillips screws because both types permit a more positive engagement of the screwdriver, and the Allen screws remain on the screwdriver without falling off.

Knowledge of the absolute force values obtained with various types of handles aids in the development of tool-design criteria. First, the torque values of a specific item can be used to ensure that the tool will meet specific requirements. Second, the comparative capabilities and advantages of design features can be traded off to select design criteria. Hand torque values for three rotary force handle types are given in Table IX-2. Three handle types were tested: a typical screwdriver handle, a globe, and a knob with eight sinusoidal finger-grip indentations. Test results indicated the increased rotary force capability of the $3\frac{1}{4}$-in. knob over the standard screwdriver by a factor of approximately two.

Torqueing tools currently being used in underwater work are, for the most part, common land tools which are either used under water in their off-the-shelf forms or have been modified to meet the constraints imposed by submergence under pressure. In developing human engineering design criteria for torqueing tools used by divers, the general goal has been to simplify the tools in terms of their handling and operating requirements and to develop a modular tool system approach that will reduce the total number of tools that a diver needs at an underwater work site. Specific human engineering recommendations for each of the torqueing tool categories are summarized in Table IX-3. Human engineering factors are discussed later in this chapter.

Table IX-2
Hand Torque Values for Three Rotary Force Handle Types[a]

Test apparel	Hand torque value, in./lb		
	Screwdriver	2-in. Ball	$3\frac{1}{4}$-in. Knob
Without full-pressure suit	63.34	79.58	117.92
With full-pressure suit (unpressurized)	54.17	72.08	129.75
With full-pressure suit (pressurized)	50.16	58.80	105.67

[a] From Pierce (1963) by permission of *Human Factors*.

Table IX-3
Summary of Human Engineering Design Recommendations (Andersen 1971)

Design variables	Human engineering recommendations
Screwdrivers and nutrunners	
General	The number of individual tools carried by a diver can be minimized by the use of a modular screwdriver tool
Size, shape, and weight	Handle diameter: 1.0–1.25 in. Handle length: 4.0–4.25 in. Handle grips: 6-in. longitudinal grooves from 1/4 to 3/8 in. in width, 1/8 in. deep Hand knob attachment (Palm Grip): 2.75–3.25-in. diameter with sinusoidal finger-grip indentations Material: dielectric plastic handle Blade length: 4–6 in., 5/16 in. diameter Blade tip: hex head, Allen head, and Phillips head preferred over flat head tip
Displays and labels	The nature and simplicity of screwdrivers and nutrunners do not warrant extensive labeling; however, where a socket adapter handle is used, individual sockets may be color-coded by size; under conditions of poor visibility, sockets should be notched for ready size identification
Controls	Ratchet drive handgrips are provided with a control to select the direction of ratchet; recommended control type is a two-position lever switch; for underwater use, this control should be of a size that can be operated with a gloved hand; a minimum length of 1/2 in. and a depth of 1/4 in. are recommended
Force application	Rotary force application limits for a standard handle driver are estimated at 50–60 in.-lb; by the addition of a 3.25-in. rotary knob (i.e., Palm Grip handle) on the end of the standard handle, rotary force application can be increased to 105–130 in.-lb
Hand wrenches	
General	Wrench heads used in hand wrench tools fall into four general categories: (1) adjustable open end, (2) fixed open end, (3) box end, and (4) socket head with ratchet drive; the most versatile of these designs is the socket head, which under most conditions can be used in place of the box end or open end wrench; the socket wrench will also decrease the total weight and number of tools carried by a diver
Size, shape, and weight	Material: fabrication of corrosion-resistant material to minimize jamming and maintenance Wrench head: should be fully enclosed, as in the socket and box end wrenches Handle design: wrenches should have the capacity for adding varying lengths of extension handles for increased mechanical advantage Weight: overall tool weight, including sockets, should not exceed 10 lb
Displays and labels	Sockets should be color-coded by size; under poor visibility conditions, additional coding may be supplied by notching sockets for ready size identification
Controls	Wrenches should be furnished with ratchet mechanisms, sealed against sand and other foreign matter; a two-position control

Table IX-3—*Cont.*

Design variables	Human engineering recommendations
Force application	lever should be used, of a size operable with a gloved hand; a minimum 1/2-in. length and 1/4-in. depth are recommended; positive locking mechanisms should be provided for mounting sockets securely to the drive shaft Wrenches should be supplied with torque-limiting mechanisms allowing divers to control the level of torque output and providing a direct display of the amount of torque transferred to a stud-bolt assembly
Power impact wrenches General	Underwater tests and operational work experience with pneumatic and hydraulic impact wrenches clearly indicate a performance advantage over conventional hand wrenches in terms of time savings and work output; however, their high cost, maintenance requirements, logistic support requirements, and size and weight constraints must be considered as limiting factors in their overall acceptability and general utility; hydraulic impact wrenches are generally preferred because they are not depth-limited, have greater power output, and are easier to maintain than pneumatic impact wrenches
Size, shape, and weight	Impact wrenches with a pistol grip configuration are designed primarily for one-hand operation and should not exceed 10 lb in weight; center of gravity should be located toward the handle or directly over the area where the hand grips the pistol handle; overall tool length should not exceed 12 in., in-line design impact wrenches are used for 1-in. or greater socket drives and, because of their weight, require two hands to manipulate; wrenches designed for two-hand operation should not exceed 20 lb in weight and for overall handling ease should not exceed 15 in. in length
Displays and labels	Sockets should be color-coded by size; under conditions of poor visibility, additional coding may be provided by notching sockets for ready size identification; secondary controls should be position-coded; displacement between the control positions should be 30° to allow for positive tactile discrimination
Controls	Handgrips and trigger mechanism: Handgrips should be fully enclosed to protect the operator's hand; handgrip diameter of 3/4 to $1\frac{1}{2}$ in. is recommended; length should be at least $4\frac{1}{4}$ in. to accommodate the full breadth of a gloved hand; handgrip shape may be round or oval in cross section or contour-molded to the shape of the hand; trigger control should provide for four-finger grasping; recommended separation between the trigger and the heel of the hand is $2\frac{1}{2}$ in. with trigger in the open position, and $1\frac{1}{2}$–2 in. with trigger in the closed position; force resistance should be between 5 and 12 lb
Force application	Torque levels generated by power impact wrenches may exceed the limits of that required by a given nut/bolt assembly; impact wrenches should be supplied with torque-limiting mechanisms which allow divers to control the level of torque output and provide direct display of the amount of torque transferred to a stud-bolt assembly

4. Power Tools (Selected)

Due to the utility and high potential for remotely supplied power tools, an evaluation of diver performance, tool performance, and maintenance for several hand-held underwater power tools is provided and summarized in Table IX-4. In general, at depths to 60 ft, hydraulic tools were very effective and practical, while pneumatic tools, although effective, required excessive maintenance. At greater depths, hydraulic tools retain their effectiveness, but pneumatic tools lose effectiveness because of the compressibility of gas. Hydraulic tools generally supply more energy per unit of tool weight than do pneumatic tools; thus, the diver can perform work more rapidly using hydraulic tools (Black and Barrett 1971).

All of the power tools (in particular the chain saw) are potentially dangerous. Thorough procedures must be established for safe tool use, general diving safety, and rescue techniques. The use of scuba gear provides the working diver with greater

Table IX-4

Conclusions and Recommendations for Tools and Accessories
(Black and Barrett 1971)

Item	Conclusions	Recommendations
Pneumatic impact wrench	Depth-limited and requires excessive maintenance, but light and relatively easy to handle	Design for ease in assembly and disassembly to facilitate maintenance; ensure that water cannot enter tool when not in operation to preclude clogging parts with foreign matter
Hydraulic impact wrench	Very effective for drilling, tapping, and bolt and nut impacting	Forward–reverse controls, which should be at rear of motor; on/off control should be actuated by whole hand and designed for rapid emergency release
Hydraulic chain saw	Effective for light use, but lacks adequate power for heavy work	On/off control as above; spikes (dog) below chain to catch into log required to exert force in line with saw and increase leverage
Hydraulic grinder	Very effective for underwater use, but difficult to handle; both hands are required for operation; tethering and/or staging required	On/off control as above
Hoses and connectors	Pneumatic hoses adequate, but hydraulic hoses stiff and lack proper swivel mechanisms at the motor connection	Tests should be conducted to determine if lighter more flexible hoses can be used; the feasibility of using concentric hoses with the supply hose located inside the return hose and a single swivel connector should be investigated; the swivel should be designed to permit rotation in two phases
Tool accessories	Both keyless and key-type are time-consuming to use and are subject to jamming on impact tools; tool handling devices and systems to locate accessories in murky water are required	All tool bits should be provided with adapters that would permit direct mating to impact motor for maximum diver efficiency; special tool-holding devices should be fabricated to permit prearrangement of all tool bits by size; all items should be visible and accessible to the diver

freedom for positioning on work surfaces. With surface-supported diving systems the diver must take added precautions to ensure that his lifeline and air hose do not become entangled with the tool umbilicals or damaged by the tool.

B. *Power Sources and Requirements*

Expansion of diver operations in the sea demands portable diver-support power plants. Divers are becoming increasingly concerned with broad-range activities—underwater salvage and repair, large site preparation and inspection, equipment emplacement, underwater construction, and final assembly and adjustment of massive and complex industrial equipment. The need for propulsion of the diver and his equipment, illumination of relatively large work areas, and operation of powered tools will require underwater power plants of 5–25 kW ratings, or even higher.

Increased use of closed-cycle breathing apparatus and small mobile power plants for support of the diver in a free-swimming mode is expected. Freedom from umbilical entanglements and restrictive connections to the surface will be essential for many divers to be able to work efficiently and safely together on the assembly of large undersea structures. Overall assembly-area support, such as general illumination, can be supplied from a fixed point in the work area, such as the personnel transfer capsule or vehicle. However, the working tools that are to be used directly by the diver to provide close support for electric and hydraulic tools and localized high-intensity illumination may best be powered by a small mobile power plant, compact enough for the diver to transport with him to his work station.

1. *Power Requirements*

For construction and maintenance operations in the sea, effective support of the working diver demands that he be furnished with adequate power to allow his various functional tasks to be accomplished uninhibited by power limitations. Some of the factors involved are illustrated in Figure IX-1. Power requirements can be categorized into three major areas: (1) propulsion and positioning, which enable the diver to rapidly achieve his work station; (2) illumination and working tool systems, which allow him to carry on his functional tasks on station; and (3) instrumentation and communication gear, which provide for his protection, system monitoring, and coordination.

The propulsion requirements for relative motion between the diver and the sea are dependent on the velocity and on the "drag area" of the diver and his equipment. Propulsion requirements for various means of diver transportation are shown in Figure IX-2. Speed limitations of the various methods are indicated by the dashed lines and are as follows: unpropelled diver, 2 knots; diver with a traction unit, 4–5 knots; one-man wet submersible, 6–7 knots; two-man lock-out submarine, 8–9 knots. Speeds of 6 or 7 knots appear to be adequate for most missions. Because drag increases as the square of the velocity, requiring approximately four times as much power to go twice as fast, speed requirements above these levels should be carefully scrutinized.

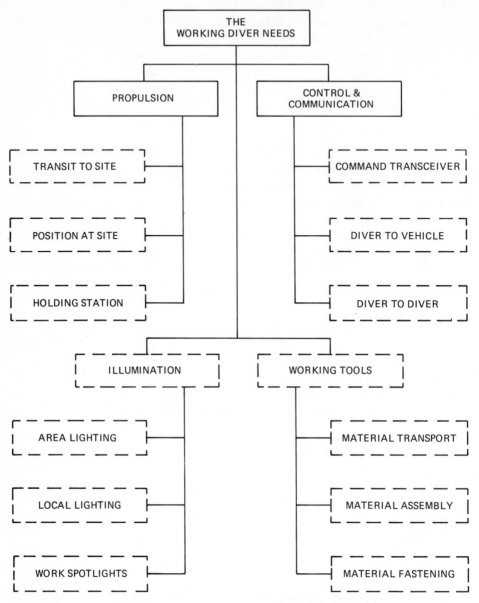

Figure IX-1. Power requirements for the working diver. [From Edwards (1970) by permission of Marine Technology Society, Washington, D. C.]

Underwater construction and maintenance work will require support of the diver by appropriate tools and appliances. Table IX-5 lists some of the typical operations which the diver must carry out in the sea. The table also lists order-of-magnitude figures for hydraulic flow or electric power used in such operations, as well as a typical time per operation. The power requirements for the tool operations in terms of their hydraulic equivalents are shown in Figure IX-3. The indicated power levels are based on overall 50% efficiency for conversion of energy from the electric

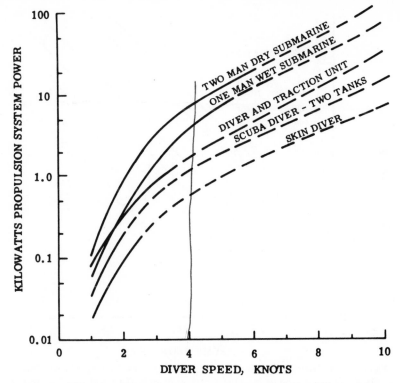

Figure IX-2. Propulsion requirements for diver transportation. [From Edwards (1970) by permission of Marine Technology Society, Washington, D. C.]

motor into hydraulic energy and back again as mechanical energy at the hydraulic motor shaft. (Direct electrically powered systems are not as widely considered, due to the basic shock hazard.) Such a fluid system results in a prime power-supply requirement of somewhat under 1 kW for each GPM of flow at 1000 psi. Small hand tools will, therefore, require a prime power supply of somewhat under 10 kW, while larger hydraulic positioning systems will approach 25 kW in power requirement.

Table IX-5
Tool Operations[a]

	Hydraulic, gpm	Electric, kW	Time per operation
Sawing and cutting	6	1	10–30 min
Punching and drilling	4	0.5	1–3 min
Riveting and bolting	8	2	10–30 sec
Hammering and forming	10	2	30–100 min
Positioning large loads	30	10	10–30 min
Chipping and grinding	4	0.5	30–100 min

[a] From Edwards (1970) by permission of Marine Technology Society, Washington. D. C.

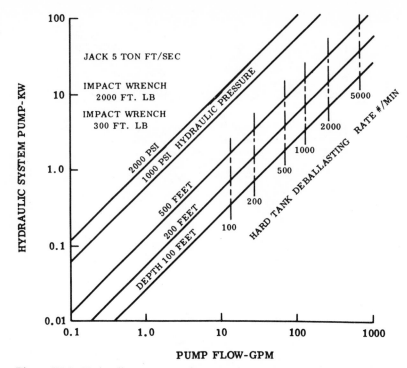

Figure IX-3. Hydraulic power requirements for work system. [From Edwards (1970) by permission of the Marine Technology Society, Washington, D. C.]

Deballasting of hard tanks, for adjustment of the components to be handled to neutral buoyancy, is a representative activity which can be undertaken with a hydraulic pumping system. The lower portion of Figure IX-3 illustrates the pumping power required to evacuate sufficient water from the ballast tanks to produce adjustments of the buoyancy to neutrality at rates of 100–5000 lb/min at working depths from 100 to 500 ft. Here again major working capabilities in the sea will lead to power requirements approaching 25 kW (Edwards 1970).

2. Hydraulic Power Sources

The criteria for selecting a particular hydraulic circuit and the components for an underwater hydraulic power source are different from those used for designing a surface hydraulic power source. A hydraulic pump's tolerance for salt water and reliability are generally more important than the pump's volumetric efficiency. Pressure and flow gauges must be able to withstand severe mechanical vibration, hydraulic fluid shock, and salt-spray corrosion. Hydraulic hose expansion under internal pressure must be minimized, but not at the expense of excessive weight. Hydraulic hose couplings should be easy to couple and uncouple but should not separate when dragged over the side of a boat. Hydraulic fluid must provide adequate lubricity, have a low viscosity in cold water, and retard damage to the hydraulic

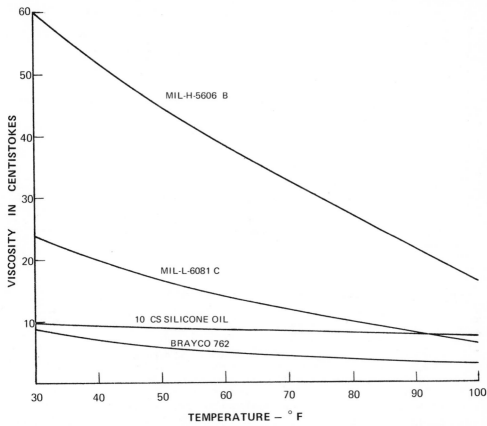

Figure IX-4. Relationship between viscosity and temperature at atmospheric pressure (USNSRDL 1969).

system from intruding salt water. The lower the viscosity, the more efficient is the operation of electric motors, hydraulic equipment, and almost all parts that rotate or translate. Viscosity is increased as pressure increases and as temperature, and pressure ranges found in deep underwater equipment can vary from 20 to 2000 times the surface viscosity at atmosphere pressure (USNSRDL 1969). The relationship of pressure, temperature, and viscosity is shown in Figures IX-4 and IX-5.

The Naval Civil Engineering Laboratory has been working on the development of hydraulic power sources. In a recent report on this ongoing program (Liffick and Black 1972) some of the problems and their solutions are discussed as follows:

> Hydraulic power sources are designed for operation with either open-center or closed-center hydraulic systems. Open-center hydraulic systems continuously cycle hydraulic fluid through hoses to the tool. When the tool is not in operation, the tool motor is bypassed and the hydraulic fluid goes to the reservoir through the return hose. With a closed-center system there is flow through the hydraulic hoses to the tool only if it is operating. [Otherwise, the hydraulic fluid is bypassed to the reservoir at the pump, or pump operation ceases.]

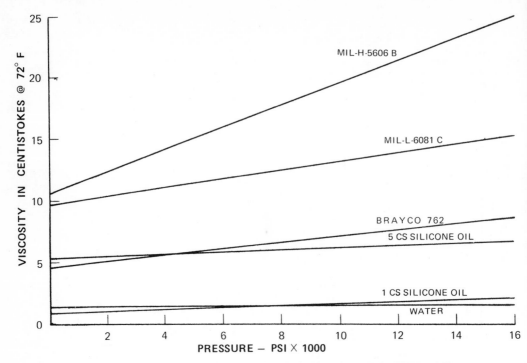

Figure IX-5. Relationship between viscosity and pressure (USNSRDL 1969).

It is preferable to use open-center systems when hydraulic fluid is supplied to a tool through long hoses because the viscosity of the fluid remains lower and more constant. . . . There are, however, certain applications where closed-center systems are used. Most hydraulic cutters and jacks cannot be operated with an open-center system because of their physical configuration. Closed-center tools are used with electro-hydraulic power sources because they can be used to provide multiple tool capability with a minimum of hydraulic valves and piping. . . .

The purpose of a hydraulic power source for diver tools is to convert energy from a prime mover to fluid power for tool operations. The hydraulic pump can be driven by an internal combustion engine or an electrical motor with power from a generator or batteries. . . .

Hydraulic power sources for diving work to depths of several hundred feet are usually topside units driven by a diesel engine. Dual hydraulic hoses run from the surface to the underwater site. The practical maximum depth for surface-supplied hydraulic power results from the accumulator effect of the pressure hose and from the difference in specific gravity between the hydraulic fluid and sea water. The accumulator effect in the pressure hose is caused by an increase in hose volume when pressure increases for tool operation. The internal volume of a 3/4-inch ID, synthetic braid hose might increase as much as 10% when a tool is energized. This results in sluggish tool operation, particularly with impact wrenches and cutters. . . .

A 100-ft column of seawater exerts a pressure of 44.5 psi. Most hydraulic fluids have a specific gravity of approximately 0.86 [see Table IX-6]; therefore, a 100-ft column exerts a pressure of 38.3 psi (0.86 × 44.5 psi). Thus, for every 100 ft of depth there is a differential pressure of 4.2 forcing seawater into the hydraulic tool. This is not a problem when the tool is operating, because the internal fluid pressure is greater than the differential pressure. However, when the tool is not operating, the differential

Table IX-6
Significant Characteristics of Selected Fluids[a]

Property	Brayco 762	MIL-H-5606B	MIL-H-6083L	Silicone Fluid
Specific gravity	0.855	0.867	0.874	0.94
Viscosity at 100°F, centistokes	3.4	14.5	12.0	8.6
Flash point, °F	210	200	220	355
Dielectric breakdown voltage, 0.05-in. gap at 77°F, Vdc	22,000	20,000	25,500	15,000
Bulk modulus at 10,000 psi, 45°F, in psi	2.8×10^5	3.1×10^5	3.1×10^5	21.5×10^5
Cost	High	Low	Low	Very high

[a] From Briggs (1971) by permission of the American Society of Mechanical Engineers, New York.

pressure will force water past the shaft seal into the hydraulic tool. Shaft seals are designed to keep hydraulic fluid in the tool and are often ineffective in the reverse direction. This differential pressure can also force water into hydraulic couplings when they are disconnected under water.

[While the specific gravity of the silicone fluids is high, they are not recommended because of their high price, poor lubricity, high compressibility, and because they do not inhibit sea water corrosion to any extent. A comparison of the compressibility of various fluids is shown in Figure IX-6.]

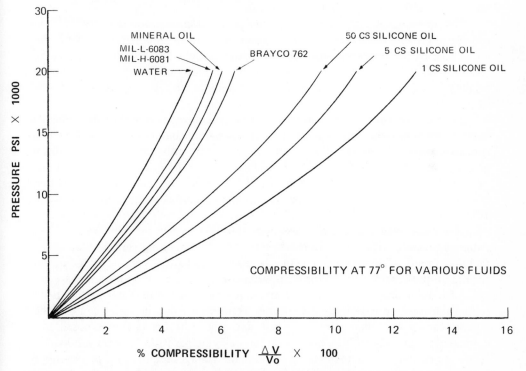

Figure IX-6. Compressibility of various fluids. [From Briggs (1971) by permission of the American Society of Mechanical Engineers, New York.]

For work in water deeper than several hundred feet, an electrohydraulic power source is usually preferred because electrical energy can be transmitted long distance more efficiently than hydraulic power. In applications where surface support is undesirable, batteries can be substituted for electrical power from the surface.

3. Batteries

Batteries are often used as a power source in underwater engineering. One use is for low-power devices, such as communication devices, where portability overrides the high cost of battery power. Also, some electrical hazards associated with wire-delivered power can be avoided by use of batteries so that hand-held tools can be safer to operate.

Conventional lead–acid batteries have been adapted for use in the sea by pressure-compensating their electrolyte chambers to allow direct submersion and by using buoyancy materials to adjust the power plant volume to provide neutral buoyancy. These batteries are most commonly used as secondary batteries for storing energy obtained from some other source to time-average peak power demands and for certain other long-life, low-drain applications. They have historically been used only where weight and size are not serious considerations (Ketler 1971). When the cost of the buoyancy material ($1–2/lb, giving buoyancy of 44 lb/ft^3) is added to the cost of the battery itself ($150–350/kWhr) the initial expense for lead-acid battery systems ranges from $300 to 600/kWhr (Edwards 1970).

Mercury, nickel–cadmium, silver–zinc, silver–cadmium, and alkaline batteries provide improvements in the volume and weight of diver-support power plants, but initial cost of the battery is substantially greater, about $600–700/kWhr for long-lived units (Edwards 1970). These batteries are generally too costly for high-power, primary energy sources because of the great amount of total watt-hours involved (Ketler 1971).

Topping the scale in the high-power, short-life category is the seawater-activated battery commonly used in applications requiring short bursts of energy after long periods of dormant storage. These batteries frequently utilize magnesium anodes in conjunction with a cathode containing silver salts. Since the electrolyte is formed only once, and is not replenished during the life of the system, the useful life of the battery is limited to about a week, with special cases extending as long as 30 days (Ketler 1971).

Hydron seawater batteries have been developed over the past years. These batteries generate continuous electrical power over an operating period ranging from months to years. Using sea water as the electrolyte, the batteries are free-flooding and require no pressure protection for deep water use. Shelf life, an important character-istic affecting overall logistics costs, is unlimited because the seawater battery is dry until submerged into ocean water. Series connection of cells to generate higher voltage cannot be reliably accomplished since the cells are openly exposed to a common seawater electrolyte. Because of low cell voltages (0.25 V under full load), a special ac/dc converter is required to generate the voltages required for power-supply loads such as tape recorders, lights, acoustic devices, and amplifiers. A conversion efficiency of up to 80% from the 0.25-V input has been achieved (Ketler 1971).

Radioisotope thermoelectric generators provide a long-life capability with a wide range of power. Typical costs for these devices range from $10,000 for a 0.1-W unit to about $100,000 for a 25-W unit having a lifetime up to 5 yr. In addition to the very high cost, a disadvantage of the radioisotope thermoelectric generator is the difficulty of assuring the control and safety of the ^{90}Sr radioisotope fuel for satisfying the AEC licensing requirements. There are also [238]Pu-fueled devices available for extremely low (micro- to milliwatt) power levels; however, licensing and costs for this radioisotope may present obstacles to its use (Ketler 1971).

4. Electrical/Electronic Equipment

For some underwater engineering needs it is necessary to use conventional electric, electromechanical, and electrohydraulic power systems. Each of these systems requires the use of relays, circuit breakers, fuses, solenoids, motors, and other common devices. Three methods are commonly used to solve the problem of operating this type of equipment under water. One is placing the equipment in pressure-resistant tanks, external to the manned capsule; but the tanks are expensive, subject to leaks, and require electrical and hydraulic penetrations that can resist high pressure. A second method is to place the equipment inside the manned capsule; however, space is limited and a serious fire hazard is present if relays, circuit breakers, and fuses are in the presence of pure oxygen. Failure or shorts in the external cabling could produce catastrophic failure in the electrical penetrators. Finally, equipment is often placed in sea-ambient-compensated boxes, external to the manned capsule. Such systems require careful selection of the fluid used.

Unique problems are encountered when electrical and electronic equipment is used in a high-pressure, helium–oxygen environment. Figure IX-7 shows the gaseous makeup of the atmosphere in which the equipment must operate at varying depths. For example, at 600 ft the pressure is 267 psig and is approximately, by volume, 92.5% helium, 5.9% nitrogen, and 1.6% oxygen. At 1000 ft the pressure is 444 psig and approximately 95.4% helium, 3.6% nitrogen, and 1.0% oxygen. The density of the environment also increases with depth. At 600 ft the density of the helium–oxygen environment is about 4 times greater than that of atmospheric air and at 1000 ft it is about 5.5 times greater. Because of the greater density, the volumetric heat capacity is greater than that of atmospheric air by about the same ratio as the density. The thermal conductivity of the helium environment is some six times that of air at standard conditions. Also, moisture is always present, and it is usually salt water.

High thermal conductivity affects the operation of heat-activated devices such as thermal circuit breakers or fuses. Electric motors may be operated at loads slightly greater than design capacity because of the high rate of heat dissipation; however, this is not recommended. Fan motors must be oversized because of the greater density of the environment. A fan motor operating at pressures for 1000-ft depth would have to have a rating about 5.5 times greater than a fan circulating the same volume of air at sea level.

Since helium is a very light gas, it permeates most materials and migrates into any void or inadequately sealed container. It is much easier to seal a container to withstand water pressure at 10,000 ft than it is to restrict the penetration of helium at 1000 ft.

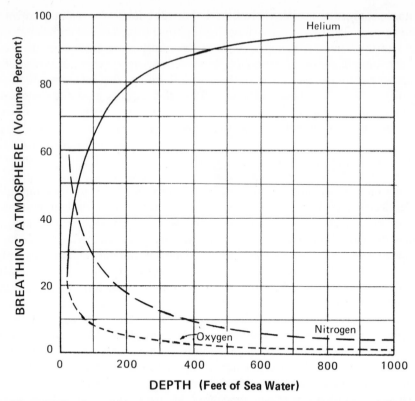

Figure IX-7. Composition of a nitrogen–oxygen environment at various depths. [From Porter and Banks (1970) by permission of the Institute of Electrical and Electronics Engineers, Inc., New York.]

Probably the greatest problem encountered can be related directly to pressure. Cases on electronic components such as transistors, capacitors, filters, and sealed switches have crushed at various depths. Thermal-control sensing elements of the aneroid and liquid-filled type have been found to be pressure sensitive.

Even though the oxygen level is very low (approximately 1% at 1000 ft) all designs must tolerate a 20% oxygen level because in dead air spaces the helium–oxygen mixture can stagnate and the helium tends to migrate out, leaving a high concentration of oxygen. For this reason no open-switch contacts should be permitted. All electrical equipment should either be installed in helium-purged enclosures inside the submerged vehicle or in pressure-proof, watertight containers on the exterior. A helium-purged enclosure should be connected to a helium supply which keeps the interior of the enclosure flushed with dry helium at approximately 1.5 psi greater than the ambient pressure of the vehicle. This prevents buildup of oxygen pockets, and open-switch contacts can be used inside these enclosures. The 1.5-psi differential pressure makes it much easier to seal around switch bushings, potentiometer shafts, and electrical connectors than if the container were required to withstand the full pressure at 1000 ft. The helium-purged enclosure is also much lighter.

Investigations have revealed that cases enclosing transistors can be made of aluminum, copper, or stainless steel, and that the stainless steel cases withstand pressures up to 578 psig or approximately 1300 ft. In commercial equipment where designs are fixed, large capacitors are always a possible problem. One solution for high-capacitance capacitors is to place them on a separate printed circuit, which is then encapsulated in an aluminium can with Epocast 233. This approach gives physical strength to the components and gives a thick, void-free barrier between the components and the environment. Lower capacitance capacitors do not present the same type of problem. To solve the pressure problem with the smaller capacitors, solid tantalum capacitors can be substituted where possible. Where this is not possible, good results have been obtained with Cornell-Dubilier type NLW and Mallory type MTP capacitors. In order to eliminate pressure problems with the large capacitors, paralleling smaller capacitors is one method of obtaining the higher values if space is available (Porter and Banks 1970).

When a piece of electronic equipment cannot be placed in a helium-purged enclosure or pressure-proof container, it has been found that open-relay contacts can be replaced by reed relays; miniature limit switches can be replaced with Micro Switch Number HM1 hermetically sealed miniature switches; and pushbutton switches can be replaced by pushbutton reed switches. These have been successfully tested under pressure equivalent to a depth of 1000 ft. "Environmental sealed military-type toggle switches are used in the helium-purged enclosure since they are provided with a seal around the toggle handle which will withstand the 1.5 psi differential pressure. These switches must be vented to prevent rupture when decompressed after being saturated in a helium–oxygen environment" (Porter and Banks 1970).

Hydraulic–magnetic circuit breakers have been successfully tested for use to 850 ft. They have been used to replace fuses and thermal-type circuit breakers and can serve both as a switch and a circuit protector. The case must be vented to prevent the same failure as described for the sealed switches. "The circuit breaker has a sealed toggle handle and an 'O' ring groove to seal against the panel of the helium-purged enclosure. Silicon-controlled rectifiers can be used to switch large a-c loads replacing open-relay contacts. Large power transistors constructed similar to SCRs can be used to switch large d-c loads. Triacs may also be used to switch a-c loads" (Porter and Banks 1970).

Many ordinary materials—such as adhesives, paints, lubricants, plastics and other synthetics—outgas solvent vapors and may be toxic or noxious. Although these materials may not present problems in normal usage, many of them become intolerable in a closed environment. Man's toxicity limits at increased pressures are not yet known. Further research in this area is very important so that the designer can choose materials that are safe as well as reliable. At present, the careful selection of components coupled with extensive testing is the only solution for the designer (Porter and Banks 1970).

Sea ambient compensation, a compensation system in which chambers are filled with a fluid and pressurized to a few psi above sea ambient pressure, reduces the leak possibility, reduces the cost of the chamber, and permits all power (other than control circuitry) to be kept outside of the manned capsule.

For pressure-compensating fluids in which electric motors, controllers, circuit protection devices, switches, relays, solenoids, electronic equipment, and batteries are immersed, the dielectric properties of the fluid itself should be high enough to permit some degradation of the dielectric strength by the contamination which invariably will occur. The dielectric strength involves the resistance per unit volume, the withstanding voltage, and the dissipation factor (see Table IX-6).

5. Electrohydraulic and Cryogenic Pneumatic Power Sources

As technology increases the capability of the diver to spend longer periods of time at greater and greater depths, problems with surface-support systems become apparent. Longer hoses are cumbersome for divers and for topside handling. Also, they decrease overall system efficiencies because of increased pressure losses, thus requiring increased topside power. Depth inefficiencies with increased depth are more critical with pneumatic tools because of the compressibility of the gas. Surface conditions, such as high winds and inclement weather, cause divers to be burdened with moving supply lines, which may be dangerous or impossible to handle. Self-contained submersible power modules can circumvent many surface-support problems and thus increase the diver's work capability. Submersible power sources may be located either on the bottom near the divers or mounted on habitats or underwater vehicles, depending on the type of diver task.

The cryogenic power module, a gas-generating system with liquid nitrogen as an energy source, is designed for operating conventional pneumatic tools to depths of 120 ft. It is "capable of operating tools requiring up to 20 SCFM [standard cubic feet per minute] of gas for 15 minutes from a module 3 feet high, 1-1/2 feet in diameter, weighing 200 pounds empty and 290 pounds when filled with liquid nitrogen" (Black 1971).

The tool supply circuit consists of a heat exchanger for converting liquid nitrogen to gaseous nitrogen, and pneumatic quick-connects for coupling the system to the tool to be operated. When a tool is operated, the pressure in the dewar forces liquid through the heat exchanger, where it is converted to gas at ambient temperature, which is supplied to the tool at a pressure approximately equal to that of the dewar pressure.

The electrohydraulic power module is a submersible, closed-centered oil hydraulic system. It can supply 2.2 gal/min of hydraulic oil continuously for 3 hr. Submersible lead-acid batteries supply 36 V dc to operate a 3-hp, 1800-rpm electric motor, which is directly coupled to a constant-volume hydraulic pump. The module, designed to operate at depths to 120 ft, and complete with batteries, pressure-compensating oil, and lifting frame, weighs approximately 1500 lb in air.

The expected running times for a tool requiring 20-SCFM flows at depths to 160 ft using liquid nitrogen as the primary power source are shown in Figure IX-8. If the diver uses the tool during 50% of the time at the work site (normal work cycle), the total available time is doubled. When a diver uses scuba gear, time at the work site is normally limited to nondecompression times (which are a function of the working depth at which the diver can make a normal ascent to the surface at 60 ft/min without stopping for decompression). Dashed lines in Figure IX-8 show the maximum allowable time for various depths during nondecompression diving.

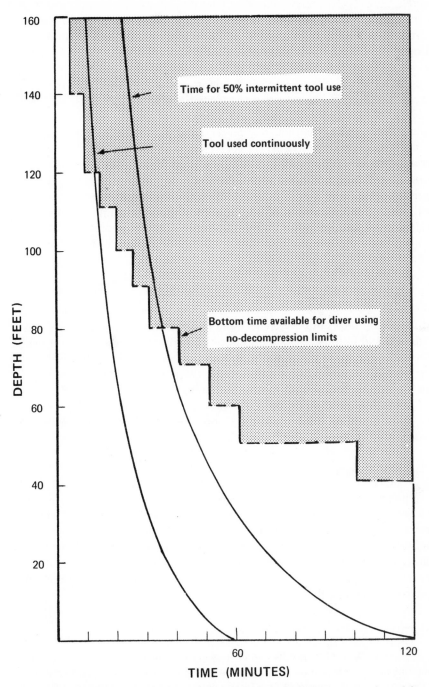

Figure IX-8. Available tool running time vs. depth for a 20-SCFM pneumatic tool for a cryogenic power module with a 50-liter dewar (Black 1971).

6. *Fuel Cells as Power Sources*

A fuel cell produces dc electrical power by electrochemical combination of hydrogen and oxygen. Its primary advantage is high energy density, due in part to the 1600 W-hr/lb energy density of hydrogen–oxygen reactants and in part to the high efficiency of the fuel cell process at a cell power density that results in low fixed weight. Other advantages relevant to a limited environment include long life and good reliability, self-starting capability, and instant response to load demand (which simplifies control and exhaust management). These advantages have aroused considerable interests in fuel cells as a power source for use under the sea.

Prospective uses for fuel cells include habitat power supplies for lighting, environmental conditioning, communications, and instrumentation; as propulsion and hotel power for vehicles; as power during construction or repair, for welding and using electric or hydraulic-driven tools; and power for unmanned, unattended applications (Sanderson and Landau 1970).

7. *Comparison of Various Power Sources*

A wide variety of power sources for undersea application are available today, or exist in varying stages of development and qualification. Parameters relating both to the mission and to the power supply must be coordinated and evaluated to judge the applicability of a given power system to a specific mission. Power systems considered below include several battery and fuel cell types, radioisotope thermoelectric and dynamic systems, and nuclear reactor thermoelectric and dynamic systems. Mission factors include logistic supply costs, "elevator charges," cost apportionment of pressure hulls, buoyancy management considerations, and thermal load charges or benefits.

a. Power System Parameters

While weight is an important consideration, volume and density are frequently more significant. Few power systems will sink unless rather heavily ballasted, an important consideration in all undersea missions. Form factor, both for surface and underwater movement, can be important. In some missions, byproduct heat may be an important factor, in initial investment or for support requirements. A summary of key parameters for a number of closed power plants is given in Table IX-7. Attention is confined to nonpropulsive applications.

b. Umbilical Power Supply

Umbilical power supply has been used successfully in a number of both nearshore and deep-sea missions. Estimated costs for nominal surface and bottom conditions are shown in Figure IX-9. It is interesting to note that cost dependence on power level is not strong. Cost of procuring and operating shore or surface terminus facilities must be reckoned separately. In the case of a barge terminus, apportionment of barge

Table IX-7
Power Plant Parameters[a]

Power plant type	Unit weight, lb/kW[b]	Unit volume, ft³/kW[b]	Unit cost, $/kW	Fuel consumption			Density, lb/ft³	Thermal efficiency, %	Nominal life, hr
				lb/kW-hr	ft³/kW-hr	$/kW-hr			
Battery: Leclanche laminated	c	c	c	20	0.23	6	87	90	~2,000
Battery: alkaline	c	c	c	21	0.18	21	117	90	~5,000
Battery: mercury	c	c	c	18	0.15	90	120	90	>5,000
Battery: silver/zinc	c	c	c	12.5	0.10	150	125	90	10,000
Fuel cell: Hydrox (pressure storage)	100	3	10,000	19	0.32	0.10	59[d]	60	3,000
Fuel cell: Hydrox (cryostorage)	100	3	10,000	0.94	0.032	0.10	29[d]	60	3,000
Fuel cell: zinc/O₂ (pressure storage)	450	9.4	10,000	7.0	0.096	15–20	73[d]	80	40,000
Fuel cell: zinc/O₂ (cryostorage)	450	9.4	10,000	3.1	0.032	15–20	97[d]	80	40,000
Radioisotope: thermoelectric	540–1,100	1.8–3.6	215,000–4,000,000	e	e	e	300	5–10	40,000
Radioisotope: dynamic	540–1,100	1.8–3.6	50,000–150,000	e	e	e	300	10–20	10,000–20,000
Reactor: thermoelectric	700–10,000	35–600	20,000–70,000	e	e	e	20–30	5–10	20,000–80,000
Reactor: dynamic	16–4,000	0.5–200	700–20,000	e	e	e	20–30	10–30	10,000–20,000

[a] From Wimmer (1970) by permission of the author. Copyright 1970 by Marine Technology Society, Washington, D. C.
[b] All power levels are kW electric.
[c] Not applicable. Batteries are not limited by power density considerations. See fuel consumption.
[d] Fuel only, packaged.
[e] Not applicable. Nuclear fuels are considered a part of the power plant.

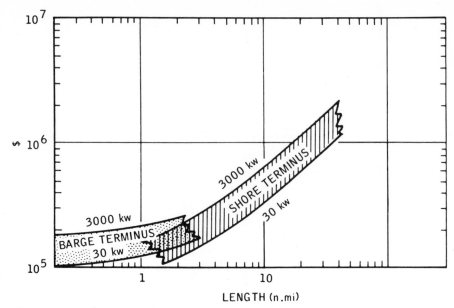

Figure IX-9. Undersea powerline costs. [From Wimmer (1970) by permission of the Marine Technology Society, Washington, D. C.]

costs between power supply and other functions (i.e., life support, mission logistics, safety, etc.) must also be considered.

c. Mission Parameters

Cost of pressure hulls was found to be approximately proportional to volume, with due allowances for fittings, etc. Figure IX-10 shows the variation of containment cost with design depth, and is an applicable cost apportionment to the power system. Also shown are hull weights, for purposes of ballasting calculations. While special cases may exist where it is convenient to balance off high-density items against other low-density items of the mission, it is generally recommended that neutral buoyancy for the power system as a whole be a specified design objective.

Charges must be apportioned to the cost of delivery or resupply of the power plant and its fuel (if any). The portion of this cost associated with the vertical delivery (surface to station) is termed the "elevator charge" and is based on buoyancy, Aluminaut rental rates, fleet submarine operating cost estimates, etc. The elevator charge is not strongly dependent on depth, and is a "round trip" charge, since it is assumed that waste products must be removed from continually resupplied missions, in a manner that maintains overall neutral buoyancy. In the case of low-density items appropriate ballast penalties must be assigned (Wimmer 1970).

The cost of delivering or resupplying at the surface point is termed the "logistic rate." This depends strongly on mission conditions and constraints, some of which are:

1. Distance from supply port (or air base).
2. Sea conditions, weather, etc. (average and extreme).
3. Prevailing local currents.

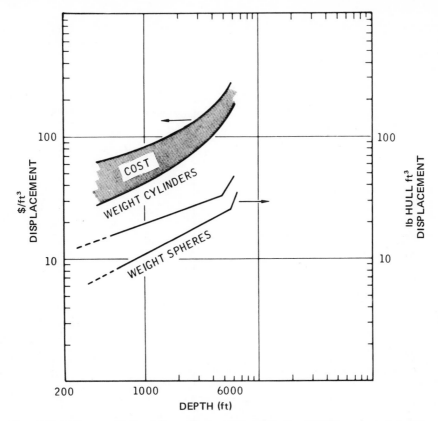

Figure IX-10. Pressure hull weights and costs. [From Wimmer (1970) by permission of the
Marine Technology Society, Washington, D. C.]

4. Type of surface tender.
5. Requirements for special handling procedures, facilities, etc.
6. Requirements for interim storage.
7. Criticality of delivery schedule.
8. Convert operations problems, enemy interference, etc.

Other factors may be found pertinent for specific missions. In 1970 a rate of $500/lb
was specified as "average" for a U. S. Navy world-wide surface activity, and rates as
high as several thousand dollars per pound have been experienced in difficult circumstances. Optimistic assumption of a logistics rate of no less than $10/lb was suggested
at that time, which included fair apportionment of handling, surface tender, manpower, and other costs for a near-shore installation of easy access.

An assessment of manpower costs may be necessary where the power plant
operation requires human attention. Cost estimates in 1970 ranged from $20 to $130
per deep-submergence man-hour, depending on circumstances. Surface manpower
should generally be accounted separately, where appropriate, and is much cheaper.

Temperature and heat balance requirements are potentially important power
plant selection factors in cases where process heat is available and needed. A simple
and generally valid way to treat this is to consider process heat "free" and available

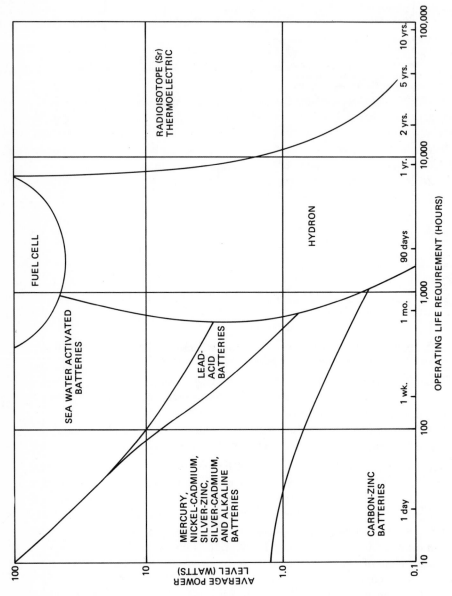

Figure IX-11. Hydron batteres fill the void in the center of the chart, encompassing power levels from roughly 0.1 to 30 W and operating life of 1 month to 1 yr. [From Ketler (1971) by permission of the Marine Technology Society, Washington, D. C.]

to offset electrical heating requirements. If cooling is required, then costs of insulation and refrigeration due to the power plant process heat load must be suitably apportioned.

Power profile requirements are also important. In addition to mission duration and total kW-hr requirement, it is also desirable for the power system designer to know the maximum, minimum, and average power requirements, and nature of the time variation. In some cases it is possible to realize overall savings of a factor of ten or more by astute use of storage cells and hybrid systems (Wimmer 1970).

d. General Selection Guides

It is evident that a power system can be properly evaluated only with full knowledge of the mission and application circumstances, requiring bilateral cooperation between mission planner and power supplier. Figure IX-11 is a comparison of power supplies with average power of each plotted against operating life requirement. Figure IX-12 shows operating requirements for some typical mission applications.

The following are qualitative evaluation guides:

1. Batteries are likely to be preferred as the prime power source only for the shortest missions, where their relatively low cost overcomes other considerations.

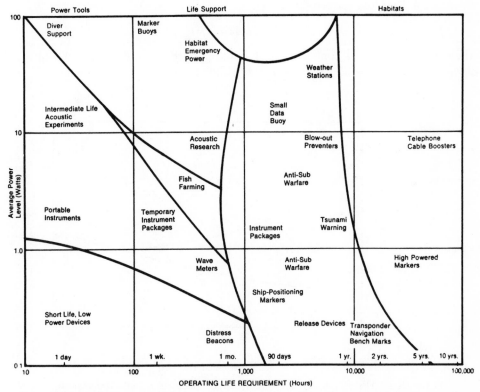

Figure IX-12. Operating requirements for typical mission applications. [From Ketler (1971) by permission of the Marine Technology Society, Washington, D. C.]

2. Fuel cells gain advantage over batteries with increasing mission duration, due to their higher energy density.

3. Nuclear power systems are preferred for intermediate- to long-duration missions where extremely high energy density is of utmost importance. Reactor systems are preferred over isotopes, except at low-power levels, due to the lower fuel and handling costs.

C. *Personal Equipment*

1. *Diving Suits*

One of the primary limitations to the effectiveness of a diver is his inability to sustain the large heat loss from his body as he dives deeper and deeper. Thus, divers require thermal protection to conserve body heat and the problems of thermal insulation increase with the depth and duration of dives. The relationship between thermal conductivity k and suit thickness L as depth increases is shown in Figure IX-13. Total heat loss per second is

$$Q = kAT/L$$

where k is the thermal conductivity, A is the surface area, T is the temperature difference across the suit, and L is the suit thickness.

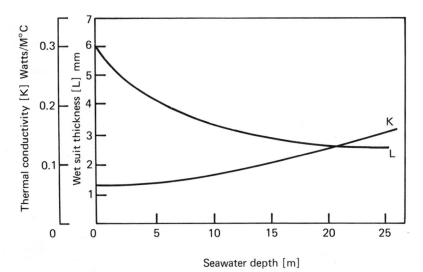

Figure IX-13. The relationship between thermal conductivity K and suit thickness L as depth increases. [From Common and Kettle (1972); reprinted from *Underwater Journal* by permission of IPC Science and Technology Press, Ltd.]

In the standard wet suit, gas is trapped in the closed cells of the foamed material and is in equilibrium with atmospheric pressure at sea level. As the diver goes deeper, the gas becomes progressively compressed as water pressure increases (Figure IX-14). At a depth of 90 m, for example, the gas in the cells will be contracted to one-tenth of its original volume; the thickness of the insulating barrier will be reduced; the thermal conductivity will be increased. Since heat loss per second Q, is proportional to k/L, there is a rapid rise in heat loss from a wet-suited diver with increasing depth. Similar problems arise with the dry suit but not to the same extent. In this case, the gas in the clothing is compressed, the fibers of the weave are flattened, and the insulation value is greatly diminished (Common and Kettle 1972).

> The wet suit has been modified to take a source of hot water pumped down from the surface. Although an improvement, this modification has disadvantages. Some flexibility is destroyed when water channels are made into the suit. A trailing umbilical cord with suitable valves is required, and a continuous source of hot water must be available under heat and pressure controlled conditions.
>
> Experiments with electrically heated suits have been conducted. . . . Shortcomings of this suit have been the proper distribution and control of the heat. A portable power pack is required on a free swimming diver. A trailing umbilical cord adds to the weight and mass required by a tethered diver. Like the water-heated suit, this type of suit loses some of its flexibility when the resistance wire is impregnated into the suit material.
>
> Isotope heat suits have been manufactured either by electrical type resistance or by liquid heat energy, but isotope cost is prohibitive. In addition, the problem of protecting the wearer from radiation has not been solved successfully. The care, storage and recharging of the isotope is another major consideration in this type of suit (Thompson 1969).

a. Neoprene Suits

1. Double-skin Neoprene. Three basic designs are available in wet-suit material: double-skin, single-skin, and nylon-lined. Double-skin, a layer of foam Neoprene sandwiched between two layers of smooth Neoprene rubber, is regarded as the standard. Foam Neoprene inside provides necessary insulation and outside smooth layers give the material strength.

2. Single-skin Neoprene. Single-skin Neoprene has a smooth layer on only one side. "This material is generally available in 1/8-inch thickness and offers tremendous stretchability. It is used in the fabrication of hoods and shorties, where comfort and flexibility are the primary requirements. Single-skin neoprene is seldom used for entire wet suits because it is more fragile than double-skin, and the 'open side' of the neoprene foam absorbs water like a sponge. It requires a longer drying time after use" (Lentz 1967b).

3. Nylon-lined Neoprene. Nylon-lined Neoprene, a flexible single-skin Neoprene to which fine-weave nylon fabric is bonded, is currently the most popular form of wet suit material. The nylon backing is cleverly woven to stretch all four ways. Liquid Neoprene bonds the two materials, forming a nylo-Neoprene skin for the "other side" of the formerly single-skin material. Nylon-lined wet suits are much more durable, noticeably in the longer life of the booties. Also the nylon backing serves as a rugged base for sewn seams.

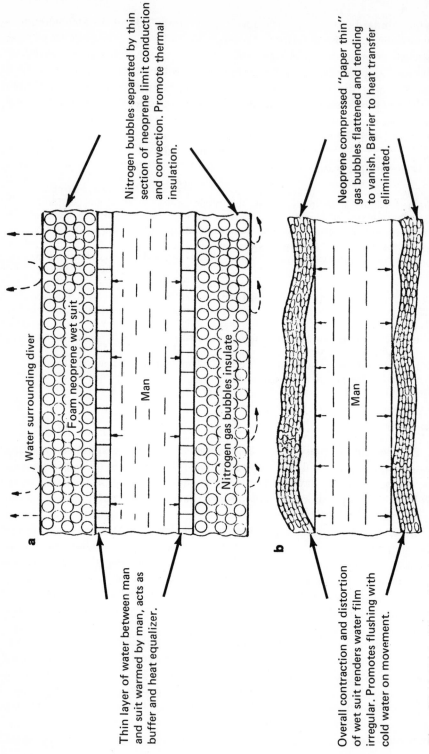

Figure IX-14. Value of thermal insulation of wet-suit material reduced by compression as depth increases. (a) Wet-suit in shallow water. (b) Wet-suit in deep water (> 55 m → 250 m). [From Common and Kettle (1972); reprinted from *Underwater Journal* by permission of IPC Science and Technology Press, Ltd.]

4. Suit thickness. Three commonly used thicknesses of Neoprene diving-suit material are 1/8, 3/16, and 1/4 in. Three-sixteenths inch is widely used for diving in water 60–70°F. This material is available in single- or double-skin design. One-eighth-inch material is split 1/4-in. Neoprene. It is single-skin, very flexible, and fragile, and is used for diving in waters 70°F and above. For water 60°F and less or when long-duration or deep diving is necessary, 1/4-in. Neoprene is usually selected. "For 'arctic' diving conditions, a multi-suit arrangement seems to work best. A 1/8-inch or 3/16-inch suit under a 1/4-inch suit is a warmer arrangement than a single thick suit because of added insulating surfaces. One suit over another also fits better because it is more flexible, and the better a suit fits, short of being tight, the warmer the diver" (Lentz 1967b).

5. Colored Neoprene. Wet suits dyed fashionable colors do not work well under actual use. Coloring compounds cut down on elasticity and add to the density of the rubber. A suit made entirely of colored Neoprene would not be as warm as the ordinary black suit and would tend to tear easily at the flex joints.

6. Embossed patterns (textured suits). A textured surface may be produced on Neoprene by embossing the normally smooth outside skin to afford a nonskid grip, especially for boots and gloves, where the extra grip is a safety advantage. Also, a textured surface tends to add a little flexibility to the suit.

7. Sewn seams. "Sewing and double-gluing of seams on nylon-lined suits has become a standard practice. This method not only makes a better bond at the neoprene butt-joint, but also ties the thin edges of nylon fabric. The edges of the wet suit are coated with liquid neoprene cement and allowed to dry thoroughly. Then a second layer of cement is applied and the edges are butted together and hard-pressed into a secure, bubble-free seam. Double-gluing gives a tougher, longer-lasting suit seam, and the sewed nylon makes the joint virtually untearable. In a sewed-seam suit, the stitching goes up the entire length of the seam" (Lentz 1967b).

8. Spinal pads. It is important to keep the sensitive spinal area of the back warm in cold water. Water can circulate in the pocket created by the indentation along the spine. Suit manufacturers have designed a simple Neoprene suit filler, called a spinal pad, to avoid chilling and water flow. Usually made of 1/4-in. foam Neoprene glued inside the wet suit jacket, the pad is about 3 in. wide and runs from the base of the skull to the tailbone.

9. Suspenders design. A wet suit with the waist extended to the high part of the wearer's chest is a good design for consistent use in extremely cold water. One model is the suspenders-type pants which have two Neoprene straps that come up over the shoulders and fix to the pants like modified suspenders. This design is excellent for cold water since the diver's torso area is wrapped in a double layer of Neoprene.

10. Zippers. Although convenient, a zipper constitutes a weak spot in the insulative barrier enveloping the diver's body. Cold water seeps in through zipper teeth, chilling the diver. Backup flaps can help keep cold-water seepage down to a minimum. It should be noted here that in the Swedish-designed Unisuit (briefly described below) a basic element is the 5-ft zipper which starts at the chest and passes down between the legs and up the back to the shoulder blades. While the designers have succeeded in producing a water-proof and pressure-proof zipper, it must still be handled with great care (Jennings 1971).

11. Booties and shoes. Footwear is important to the diver. Foot coverings must

fit snugly to be warm, yet must not cramp the feet since loss of circulation can also chill extremities. There are two distinct kinds of diving footwear, booties or socks for boat diving and hard-soled Neoprene shoes designed to protect a diver's feet when he walks over rocks, gravel beaches, or any other surface where punctures are a danger. The booties are Neoprene socks, comfortable and generally warmer than diving shoes. Neoprene shoes may require larger size fins. It is extremely important that the fins fit comfortably over the booties or shoes to prevent foot cramping (Lentz 1967*a*).

b. Noncompressible Wet Suits

Because of the loss of thermal insulation due to compression of the gas bubbles in Neoprene-foam suits, studies have investigated other forms of low-conductance materials, such as envelopes filled with jellies or oils loaded with hollow particles. Unlike solids, which are essentially dense and rigid, these materials would remain flexible. One of the more promising of these materials was developed by the Minnesota Mining and Manufacturing (3M) Company. The material is an elastic gel (elastomer and mineral oil mixed), which contains a high volume of hollow glass microbubbles.

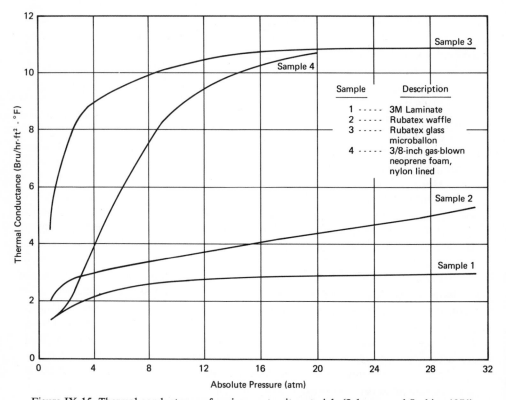

Figure IX-15. Thermal conductance of various wet-suit materials (Johnson and Jenkins 1971).

Since the 3M material cannot be bonded to itself or any other material, a post-fastening design laminates the three layers, which consists of a 1/8-in. thick inner suit of nylon-lined foamed Neoprene, a 1/3-in. layer of 3M material, and a 3/32-in. outer covering of foamed Neoprene. Figure IX-15 shows the thermal conductance of Neoprene foam and 3M material under pressure. The Naval Ship Research and Development Laboratory in Panama City tested and evaluated this material and recommended its use for deep-diving application (Johnson and Jenkins 1971).

c. Dry Suits

The principle of the dry suit is to maintain a layer of air between the diver's skin and the suit material. The greatest disadvantage of this is that the suits must be completely watertight. Also, as the depth increases, the air compresses unless some means of adding air is provided. In the standard deep-sea dress, the suit and helmet are one unit and the amount of air within the suit is varied by regulating the supply and exhaust helmet valves. Additional thermal protection is provided by various undergarments (Penzias and Goodman 1973).

d. Pressure-Compensated Constant-Volume Dry Suit

This modified dry suit has a constant supply of gas, either a separate gas supply or the diver's breathing gas supply, added to the suit to prevent collapse under pressure. The disadvantages of this type of suit are the same as those of a regular dry suit, and in addition there is about a 3-psi pressure difference between the head and the feet of the diver in a vertical position so that the bottom of the suit tends to collapse. It is available in several commercial models (Penzias and Goodman 1973).

e. Heating Systems for Divers' Suits

Various techniques of supplying heat from external sources are being explored in attempts to provide divers with thermal insulation. Several are described below.

1. Hot-water heat. The Naval Civil Engineering Laboratory (NCEL) has developed a prototype system which uses an oil-fired, closed-cycle, fresh-water steam generator and a heat exchanger to supply heated sea water to divers wearing open-cycle, hot-water-heated suits. Hot and cold sea water are blended to deliver the desired temperature. This system is designed for temperatures as low as 40°F at depths to 600 ft.

2. Chemical heat. The NCEL laboratory is also exploring the idea of utilizing the latent heat of crystallization. A material absorbs large quantities of heat while changing from a solid to a liquid and this stored heat is given up during recrystallization. Advantages of a suit heated by this method would be that a constant temperature can be maintained while crystallization is taking place, and also that no umbilical hose system would be required (USNCEL 1971).

3. Nuclear heat. Several investigations have been made on nuclear-heated diving suits. One model being developed utilizes thulium-170 and -171 as fuel (Groves 1967). Such isotope fuels, used in a heating system to produce 500 thermal watts at mission

start, can power a thermoelectric generator and pump to circulate 110°F water through-out the diver's undersuit. "The low radiation dose rate involved and the absence of a neutron dose are features of this system" (Groves 1967).

Another model of nuclear-heated suit was evaluated by the U. S. Naval Medical Research Institute. It derives its thermal energy from the decay of plutonium-234 and delivers 420 thermal watts to a heat-exchange module. It was found that the thermo-nuclear generator was ineffective for deep saturation diving, and had limitations imposed by the radiation hazard (Tauber *et al.* 1970).

f. Unisuit

This is a recent, improved version of the pressure-compensated, constant-volume dry suit, designed originally in Sweden and now manufactured in the United States. It fills the need for a suit that will protect a diver from cold in long-duration shallow dives. The suit, made of $\frac{1}{4}$-in. foam Neoprene coated with nylon on both sides, is worn over nylon fur underwear. The feet are an integral part of both the suit and the underwear. Air is let into the suit by pressing an intake valve, forming an insulating layer and also filling the cells of the material. The suit has been successfully tested under the most severe Arctic conditions. Air is let out of the suit by pressing the exhaust valve, thus making possible the easy regulation of buoyancy. The mobility of the wet suit is combined with the comfort of deep-diving dress. Major advantages are (1) warmth, (2) dryness, (3) buoyancy compensation, and (4) elimination of squeezing and chafing. With slight modification, the suit can be used with hard-hat, surface supply equipment (Covey 1972).

g. The Armored 1-atm Suit

Interest in pressurized diving suits lapsed after World War II, but increased activity in the offshore oil industry has revived it. The suit being developed by a British company, DHB Construction Ltd., is almost more like a chamber than a suit (Figure IX-16). It is made of magnesium alloy, weighs 1100 lb in air with diver, and is supplied with all-purpose handlike manipulators at the end of each arm enclosure, which can be changed for special purpose devices. Buoyancy is highly adjustable by rearrangement of the external weights. The life-support system carries 800 liters of oxygen and 12 lb of soda lime, which allows for work missions of 4 hr with an emergency reserve of from 8 to 16 hr. Oxygen partial pressure is monitored by a Beckman Minos instrument. Obvious advantages are the elimination of compression and decompression delays, greater bottom time, comfort, security, and ease of surface communication for the diver. Depth changes become unimportant. A dive can be immediately repeated. The suit is transportable by helicopter (Barton 1973).

2. *Breathing Apparatus*

Most of the limitations in diving are imposed by the large increase in pressure with increasing depth. This is especially true in solving the problem of how to supply

Figure IX-16. Armored, 1-atm diving suit (U. S. Navy photograph).

Table IX-8
Depth, Pressure, and Partial Pressures[a]

Depth, ft	Total pressure		Partial pressure, mm Hg	
	atm	mm Hg	Nitrogen	Oxygen
0 (surface)	1	760	600	160
33	2	1520	1200	320
66	3	2280	1800	480
99	4	3040	2400	640
132	5	3800	3000	800

[a] From Miles (1969) by permission of the author.

the diver with the proper oxygen partial pressure. Several techniques have been employed, using either air or mixed gases. Table IX-8 illustrates the effect of depth upon pressure and upon partial pressures of the breathing gases.

"Regardless of whether a breathing system provides air or mixed gas to the diver, the supply of breathing gas is delivered in one of two ways—by continuous flow or by demand. In a continuous-flow system a regulatable amount of gas continuously flows past the diver's face, completely independent of breathing rate. In a demand-supply system, gas flows only when the diver inhales" (Kenny 1972).

It is also possible to categorize life-support systems by the exhaust system. A system is called open circuit when the gas expired by the diver is vented into the water. "If the gas expired from the diver is cycled through a purifying system to remove the expired carbon dioxide and oxygen is added for reuse, the system is a closed-circuit system. If part of the gas is reused and part is vented, the system is classified as semiclosed circuit" (Kenny 1972).

a. Air-Breathing Systems

Air can be employed as the diving gas in either open-circuit, self-contained systems or in tethered, continuous-flow open circuits. The major problem with using air for the gas supply is the depth limitation imposed. The oxygen content of atmospheric air is approximately 20% or 0.2 atm partial pressure. (The depth-related physiological problems of oxygen toxicity and nitrogen narcosis are discussed in Chapter IV.) Air is the most economical and simplest diving gas and has been used successfully to depths of 165–220 ft.

b. Mixed-Gas Systems

It is technically possible but economically unfeasible to use synthetic-mix gases in standard air-breathing apparatus. Cost of the constituent gas makes it mandatory to recycle the mixture, either in a semiclosed- or closed-circuit system. In each case, either self-contained or umbilically supplied equipment can be used. The most common mixture today is helium–oxygen, but neon and hydrogen are being used experimentally as diluents. Advantages and disadvantages of each are discussed in Chapter IV.

c. Open-Circuit Underwater Breathing Apparatus

(i) Self-Contained. This system is based on demand breathing. The basic units are a storage cylinder to carry the high-pressure air supply, valves to reduce the high-pressure air to ambient water pressure, and a demand mechanism to deliver air to the diver with minimal respiratory effort. The pressure-reducing mechanisms and demand valving mechanisms are integrated into one unit called a regulator.

1. *Cylinders.* High-pressure storage cylinders are normally designed for pressures between 1800 and 3000 psi and are rated by the volume of standard atmospheric air they contain when pressurized. The cylinder most frequently used in civilian diving has a working pressure (the pressure to which a tank can be filled without causing excessive metal fatigue) of 2250 psi, and a 64.7 ft³ capacity. When charged to 2475 psi, or 10% over working pressure, the capacity is 71.2 ft³. This is indicated by a plus (+) symbol stamped near the neck. Other cylinder sizes available (at 10% over working pressure) are 26, 38, 50, 52.8, 75 and 100 ft³ (Somers 1972).

There are two basic valve mechanisms for the high-pressure valve, although the exact configuration will vary according to manufacturer. The straight-through "K" valve works on a simple "on/off" mechanism. The reserve-type "J" valve contains a spring-loaded shutoff that automatically closes when pressure within the storage cylinder reaches a preset minimum. This cuts off the air to the diver until he operates a lever to restore the air flow. In this manner, the diver is automatically warned of a low air supply in his cylinder (Kenny 1972).

2. *Regulators.* There are two major categories of regulator. One is the two-hose regulator, developed by Cousteau and Gagnon; the other is the more modern and widely used single-hose regulator (see Figure IX-17). Subcategories include single-stage, two-hose; two-stage, two-hose; and two-stage, single-hose.

In the two-hose regulator (see Figure IX-18), the entire demand mechanism is housed in one unit. There are two corrugated hoses, one supplying air from the air chamber to one side of the diver's mouthpiece and the other carrying expired air out the other side of the diver's mouthpiece to the water chamber where it is discharged. A flexible, watertight, airtight diaphragm separates the air and water chambers. As explained by Drayton (1972), "the regulator mechanism functions on the principle of pressure differential. Air inside the air chamber is equal to surrounding water pressure. When the diver inhales, he creates a pressure differential in the air chamber, the diaphragm is depressed and air flows to the diver. When the diver exhales, pressure on each side of the diaphragm equalizes and air flow is discontinued."

Two-hose regulators may be either one-or two-stage, stage referring to a reduction in pressure. In one-stage regulators there is one reduction from tank pressure to breathing pressure. In the two-stage regulator the first stage functions to reduce gas pressure to an intermediate level, normally around 100 psi, to provide for demand flow uninfluenced by cylinder pressure.

All single-hose regulators are two-stage. They differ from two-hose regulators in location of the demand mechanism, which is housed in the second stage. A flexible, intermediate-pressure hose separates the stages and acts as an intermediate chamber to transport air. Connected to the tank valve, the first-stage mechanism reduces the

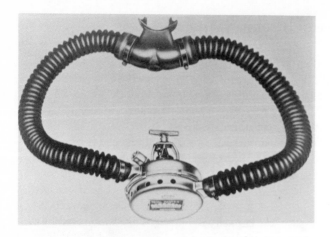

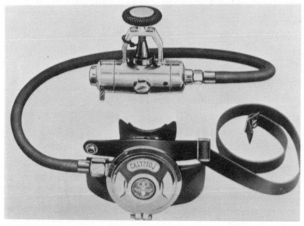

Figure IX-17. The regulator is the heart of the scuba system. The two-hose regulator (top) was the first dependable demand scuba regulator to be commercially available; improved versions are still popular in military diving activity. The single-hose regulator (bottom) was perfected in the 1950's and is in general use today. [From Kenny (1972) by permission of Gulf Publishing Company, Houston, Texas.]

cylinder pressure to an intermediate pressure (usually 85–140 psi). Air transported through the hose is breathed at ambient pressure from the second stage.

An advantage of the two-hose regulator is that bubbles are exhausted behind the diver where they cannot interfere with vision. Since the hoses contain air, the mouthpiece is lighter and more comfortable over prolonged use. Generally it is more difficult to inhale using a two-hose than a single-hose regulator, but it is easier to exhale. The U. S. Navy has conducted medical tests proving "that it is more fatiguing to exhale against a resistance than to inhale against a resistance. . . . A regulator that draws slightly hard on inhalation but exhales easily is much less fatiguing to the diver than one that draws more easily but exhales harder" (Drayton 1972).

Air is made available to the diver in any of several different ways, most commonly through a bite-type mouthpiece designed to be clenched in his teeth. On the two-hose regulator, the mouthpiece forms part of the hose system, and the entire regulator mechanism is built into a single utilized housing that attaches directly to the air storage system valve. When the diver inspires, the pressure reduction within the hose activates the mechanism in the regulator. In the single-hose configuration, the mouthpiece forms part of the demand system itself. Demand valving and diaphragm are contained in a small housing with the mouthpiece attached. A single interconnecting hose carrying air at intermediate pressure connects the demand mechanism, or second stage, to the pressure reducing first stage, which is attached to the tank valve (Kenny 1972).

There are many different manufacturers and models of regulators. Table IX-9 gives a small sampling comparing different makes and models (McKenney 1971).

(ii) Tethered. The simplest and most universal of all diving rigs is a continuous-flow, open-circuit, shallow-water diving system composed basically of a compressor, an umbilical hose, and a mask with a supply- and exhaust-valve system (Kenny 1972).

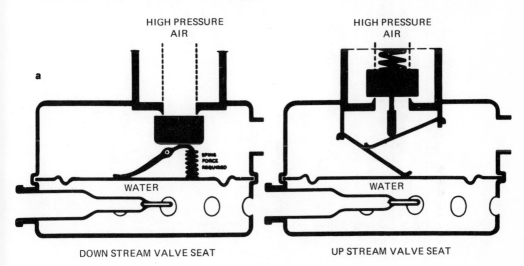

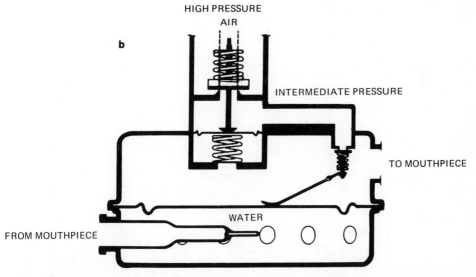

Figure IX-18. Cross section of common double-hose regulators: (a) One-stage regulator, including diagram of upstream and downstream valve seats; (b) Two-stage regulator. [From Somers (1972); photo courtesy of U. S. Divers Co., reprinted by permission.]

Table IX-9

A Comparison of Seven Models of Regulators[a]

Model																		
Dacor Olympic 400	Diaphragm balanced	Factory preset external adjust.	Downstream	External adjust. through hose connection	0–3000 psi	140 psi	30 cfm	20 cfm	1	No	2	11/16 in.	1 in.	25-1/16 in.	2 lb 5 oz	13 oz	Solid	Cycolac knurled knob	90
Healthways Scubair Sonic	Piston not balanced	Factory preset internal adjust.	Downstream	Internal adjustment	0–3000 psi	135 ± 5 psi	30 cfm @ 2250	24–26 cfm	1	No	1	11/16 in.	7/8 in.	24-3/4 in.	1 lb 11 oz	12 oz	Solid	Acetal-resin knurled knob	90
Seamless/Nemrod Downstream	Piston not balanced	Preset at factory internal adjust.	Downstream	Internal adjustment	0–3500 psi	135 ± psi	35 cfm	24–26 cfm	1	No	2	5/8 in.	7/8 in.	25 in.	13 oz	12 oz	Swivel	Metal T Bar	Any position
Sportways ATM 750	Piston balanced	External variable pressure adjust.	Downstream	Internal adjustment	0–3500 psi	130 psi	80 cfm	36 cfm	2	Yes	1	11/16 in.	27/32 in.	25 in.	2 lb 12 oz	11 oz	Solid	Cycolac wheel	90
U. S. Divers Calypso 111	Piston balanced	Preset at factory internal adjust.	Downstream	Internal adjustment	0–3000 psi	125–135 psi	120 cfm	106 cfm	1	No	1	1 in.	1 in.	25-3/8 in.	1 lb 11 oz	11 oz	Solid	Rubber and metal wheel	90
Voit MR-12	Diaphragm balanced	Factory preset external adjust.	Downstream	Internal adjustment	0–3000 psi	115–135 psi	30 cfm	20 cfm	2	Yes	1	7/8 in.	1-3/4 in.	25-1/4 in.	1 lb 3 oz	11 oz	Solid	Cycolac knurled knob	90
White Stag Sea Lung 5000	Piston balanced	Factory preset internal adjust.	Downstream	Internal adjust. through hose connection	0–3000 psi	135–140 psi	40 cfm+	26 cfm+	1	No	1	7/8 in.	1-5/8 in.	25-1/4 in.	1 lb 6 oz	13 oz	Solid	Metal T Bar	90

[a] From McKenney (1971) by permission of *Skin Diver*.

For prolonged and difficult work in shallow water, the mask component is replaced by helmet-diving gear.

The present helmet-diving apparatus is essentially the same as it was when it was first conceived in the 17th century. Improvements have been made in materials used, in air supply, pumps, and compressors. The helmet is no longer made of copper, but of polyester resin reinforced with glass fiber or spun brass, which makes better shaping possible and is less fatiguing to the diver. The helmets have also become smaller and are equipped with large, hinged visors, which are a great advantage over old ones, which severely restricted the field of vision. New helmets are now designed to be worn with any type of suit, which aids the mobility of the diver (Haux 1971).

Hoses for shallow-water diving are made of flexible reinforced rubber. A 1200-lb (minimum) tensile strength nylon lifeline and a communication telephone wire are taped to it with waterproof electrical tape at 36 in. intervals (Kenny 1972). "Continuous taping or single construction of umbilical with integrated lifeline and communication wire is not feasible with standard hose construction techniques because the hose lengthens when pressurized," usually 3–5 ft/100 ft of hose.

Most shallow-water diving systems supply air to divers with a two-stage air compressor driven by a small diesel engine. The engine and compressor are mounted on a receiver tank at least 60 ft^3 in volume. This tank maintains a steady air flow and reserves enough breathing gas for the diver to surface if the compressor should fail. The system maintains a pressure at least 50 psi above ambient water pressure of the diver's depth. A one-way check valve on the diver's mask prevents mask flooding or squeeze if the air hose is severed or the air flow is stopped (Kenny 1972). Various types of diving helmets and masks are listed in Table IX-10. Table IX-11 is a compilation of model designations, abbreviations, nicknames, manufacturers, and vendors of breathing apparatus, helmets, masks, and regulators.

d. Semiclosed-Circuit Breathing Apparatus

In a semiclosed-system the gas mixture is a function of flow. In this system, a premixed gas is provided to the diver. The exhaled gas is then partially vented to the water and the remainder passes through a canister containing a chemical compound that absorbs CO_2. The purified gas from the canister, or scrubber, is then added to the gas supply system and mixed with fresh gas. The sum of the oxygen content of the recirculated gas and the fresh gas must give a level of oxygen sufficient to prevent anoxia but low enough to avoid oxygen toxicity or hypoxia (Kenny 1972). A semi-closed-circuit typically contains the following elements (Riegel 1970):

Mouthpiece: This can be either a jaw-held mouthbit or an oral–nasal mask. The mouthpiece is the interface between the diver and the breathing apparatus.

Check Valves: These are commonly mounted at the mouthpiece, and they are arranged so that the diver will inhale from one hose and exhale into another. By providing for unidirectional flow, it is assured that the diver inhales only pure gas and that no significant amount of his CO_2-laden exhaled gas will be reinhaled.

Hoses: These connect the mouthpiece to the rest of the breathing apparatus with one delivering exhaled gas to the system and the other bringing pure gas to the diver. They are as flexible as possible to allow the diver to move his head freely.

Breathing Bags: These are flexible gas reservoirs that expand to receive exhaled gas

Table IX-10

Diving Helmets and Masks[a,b]

Manufacturer or vendor name	Model number or designation
Advanced Diving Equipment Co.	Series 2000 air helmet
	Series 3000 air–helium helmet
David Clark Co.	Model S-2002 air helmet
	Model S-5005 helium/mixed-gas helmet
Desco	Abalone diver's, Agar diver's, Sponge diver's: lightweight helmets
	Standard commercial diver's helmet
	U.S. Navy Std. diving helmet Mark V Mod 1
	U.S. Navy helium helmet
	Diving hat
	U.S. Navy Std. shallow-water diver's mask ("Jack Browne")
	Commercial free-flow mask
	Pool-cleaning mask
	Demand regulator scuba mask
Drager	Diving helmet (for surface tether or DM 40)
General Aquadyne	Air helmet
	Mixed-gas helmets
International Latex Corp.	Swim suit helmet (prototype)
Miller-Dunn	Shallow water helmet
Mine Safety Appliance	"Bugeye" standard mask
	Scuba elliptical faceplate mask
Morse Diving Equipment	Commercial diver's helmet
	Divecon integral canister helmet
	Shallow water helmet
Normalair	Facemask
Oceaneering International (formerly Caldive)	Rat hat
Piel	E. P. 100 helmet
Savoig Research and Development	Helmet
Scott Aviation	Constant-flow shallow-water diving mask
	Standard scuba mask
	Universal mask
	Scottoramic mask
	Air helmets
	Deep-diving injector helmet
	Seacrown
Siebe Gorman	Deep-sea air and helium helmets
TOA Diving Apparatus Co.	Kirby-Morgan air and mixed-gas helmets
	KMB-8 band mask
U.S. Divers, Commercial Diving Division	Kirby-Morgan NW-1, 2, 3, 4B, 4C "clamshell" helmets

[a] From Penzias and Goodman by permission of John Wiley and Sons.

[b] Miscellaneous notes: "Jap hat" manufacturer–vendor not identified. Desco "diving hat" couples to Lindbergh-Hammar "Pura 787" for recirculating CO_2 scrubbing capability for helium–oxygen diving. Examples of fiberglass fabrication: advanced diving equipment air and air–helium helmets; David Clark S-2002 and S-5005; General Aquadyne air and mixed-gas helmets; Savoie helmet; Caldive "rat hat"; U.S. Divers Kirby-Morgan "clamshell" helmets. Weight (in air) of various units: Advanced diving air helmet, 22 lb; David Clark S-5005 helmet, 10 lb; Desco lightweight helmets, 39 lb; Desco standard commercial helmet, 54 lb; U.S. Navy Mark V Mod 1 helmet, 68 lb; U.S. Navy helium helmet, 93 lb; Drager helmet, 41 lb; General Aquadyne air helmet, 25 lb. Helmet and mask selection criteria (based on following functions: protection; breathing apparatus interfacing; speech communications housing; diver visibility), include: effective dead space volume; sealing against overpressures; visual fields and distortion; buoyancy and couplings; swimming and mobility; fittings.

Table IX-11

Model Designations, Abbreviations, Nicknames, Manufacturers, and Vendors of Breathing Apparatus, Helmets, Masks, and Regulators[a]

Equipment	Manufacturer–Vendor	Equipment	Manufacturer–Vendor	Equipment	Manufacturer–Vendor
Abalone	Westinghouse	Avalon	Voit	Div-Air	Arpin
Abalone diver's helmet	Desco	Baltic	Drager	Divator Marine	AGA
Admiralty neck salvus	Siebe Gorman	Band Mask	U. S. Divers	Divator universal	AGA
Admiralty six-bolt helmet	Siebe Gorman	Blair	Old Dominion	Divecon integral canister helmet	Morse
Agar diver's helmet	Desco	Bocamat I, II, III	Drager	Diving hat	Desco
Airflo	Healthways	Browne lung	Desco	Diving lung	Dacor
Air helmet	Advanced Diving Equipment; General Aquadyne; Savoie; TOA; U. S. Divers	Calypso	U. S. Divers	DM 20	Drager
Air-lung	Northill	Calypso J	U. S. Divers	DM 40	Drager
Air-saver	Emerson	CCR-1000	Biomarine	Dolphin	Voit
Amphibian MK I, II, III, IV	Siebe Gorman	C.D.B.A.	Dunlop	Dual air	Sportsways
Aqua-div deluxe	U. S. Divers	Chariot	Siebe Gorman	Dual Oxymatic	AGA
Aqua-lung	U. S. Divers	Clamshell	U. S. Divers	Electrolung	Oceanic Equipment
Aqua-master	U. S. Divers	Clipper	Dacor	E.P. 100	Piel
Aqua-matic	U. S. Divers	Commercial diver's helmet	Desco; Morse	Explorer	Pirelli; Voit
Aquanaut	Spartan	Commercial free-flow mask	Desco	Fathom Master	White Stag
Aqua-sprint	—	Conshelf	U. S. Divers	FGG I, II, III (FGA)	Drager
Aquilon	Spirotechnique	Constant partial pressure apparatus	Old Dominion	Flatus II	Emerson
Arawak I, II, III	Westinghouse	Cristal	Spirotechnique	Forty Fathom	Voit
Atlantic	Drager; Submarine Products	Dart D1, D2, DR2	Dacor	Gas Helmet	Advanced Diving Equipment; General Aquadyne; U. S. Divers; TOA
Atlantis	Spartan	Davis	Siebe Gorman	Hydrolung	U. S. Divers
ATM-1000	Sportsways	DC-SS	Fenzy	Hydronaut	Sportsways
		Deep diving helmet	Siebe Gorman	Hydropack	Scott Aviation
		Deep star	U. S. Divers	Hydrotwin	Sportsways
		Delphin	Drager	Jap hat	—
		Detached cylinder apparatus	Siebe Gorman		

Table IX-11—*Cont.*

Equipment	Manufacturer–Vendor	Equipment	Manufacturer–Vendor	Equipment	Manufacturer–Vendor
Jet air	U. S. Divers	Mine recovery apparatus	Siebe Gorman	Pro standard	Rose Aviation
KSR-5	Westinghouse	Min-o-lung	Westinghouse	Rat hat	Oceaneering International
Lar II, III	Drager	Minor	Vespa	Reef diver	White Stag
LARU, T4	Emerson	Mistral	U. S. Divers	Reef diver deluxe	White Stag
LARU, USA/OSS	Ohio Chemical and Manufacturing	Model 57	Cressi	Rex	Old Dominion
Les	Old Dominion	Model A lung	Desco	Royal aqua-master	U. S. Divers
Little gem	Voit	Model B lung	Desco	Royal mistral	U. S. Divers
Lt. Lund II	Drager	Model C lung	Desco	S-701	Pirelli
Malibu diver	Sportsways	Marghile	—	S-901	Pirelli
MARK I	Demone; Desco; Scubapro	Nautilus	Bioengionics; Spartan	S-2002	David Clark
MARK II	Demone; M.S.A.; Scubapro	Navcon VI	U. S. Divers	S-5005	David Clark
MARK III	Scubapro	Navy unit	Sportsways	Scaphandre autonome	Spirotechnique
MARK III Mod 0, Mod 1	Scott Aviation	Navy V66	Voit	Scuba	Healthways
MARK IV	Scott Aviation	Nohl	—	Scubair	Healthways
MARK V	Emerson; Scubapro	Norseman	Aerotec	Scubair sonic 300	Healthways
MARK VI	Scott Aviation	Ocean diver	White Stag	Scuba star	Healthways
MARK VII	Scott Aviation	Olympic 100, 200, 400, 800	Dacor	Seacrown	Siebe Gorman
MARK VIII	Scott Aviation	Orca	Sportsways	Sealab III breathing apparatus (Fink rig)	Westinghouse
MARK IX	Scott Aviation; Hydrospace Developments	Oxygers	Fenzy	Sealung 1000	White Stag
MARK XI	Westinghouse	P68	Fenzy	Self-contained compressed air demand apparatus	Siebe Gorman
Master diver	Sportsways	PA 37/1600	Drager	S.G. 700	Siebe Gorman
Medi-nixe	Drager	PA 38/2800	Drager	Shallow water helmet	Miller-Dunn; Morse
Mercury MK 2	Siebe Gorman	PA 61/S	Drager	Skagerrak	Drager
Merlin MK VI (MGM I, II, III)	Siebe Gorman	Pacific	Drager	Sladen MK III	—
Mike 55	Old Dominion	Para min-o-lung	Westinghouse	SMI, II, III, IV	Drager
		PO68	Fenzy	SM III-5	Drager
		Polaris	Voit		
		Pool-cleaning mask	Desco		
		Poseidon	Spartan		
		Pro 57	Rose Aviation		

SM 5-1	Drager	Surface demand diving equipment	Siebe Gorman	U.S.N. helmet MK V Mod 1	Desco; Morse
Snark II, III	Seamless	S.W.B.A.	Dunlop	Varbell	—
Snark II servo	Seamless	Swimaster MR 12	Voit	Visionaire	Scubapro
Snark II super	Seamless	Titan II	Voit	Vista	Siebe Gorman
Sponge diver's helmet	Desco	Tricheco	Galeazzi	Waterlung	Sportsways
Sport diver	Sportways	Trident	Spartan	X-5	Mako Products
Sportsman	Pirelli	Trieste	Voit	800	Flowmatic
SS-1000	Sterling Electronics	Universal	Siebe Gorman		
Standard commercial diver's helmet	Desco	U. S. Army diving unit T4	Emerson		
Super pro	Rose Aviation	U.S.B.A.	Dunlop		
Super stag sea lung 5000	White Stag	U.S.N. helium helmet	Desco		

[a] From Penzias and Goodman (1973) by permission of John Wiley and Sons.

and contract to deliver inhaled gas. A rebreather system could not work without breathing bags, since the breathing apparatus would not be able to store and redeliver the tidal volume of the diver.

Canister: This is a container for a chemical absorbent for CO_2. The diver's exhaled gas is circulated through the canister before it is reinhaled. This removes the CO_2 and makes the exhaled gas suitable for rebreathing.

Gas Supply: As the diver breathes he consumes oxygen. A gas supply to the system from either backpack bottles or an umbilical hose provides a steady flow of mixed gas (oxygen plus a diluent gas such as helium) sufficient to replace the oxygen consumed under the severest work conditions. This gas flow is commonly called the "liter flow." The ratio of diluent gas to oxygen is chosen to keep pO_2, the partial pressure of oxygen, below a toxic limit.

A common method of regulating the gas flow is to use a sonic-flow orifice to meter gas to the rig. A pressure regulator located upstream from the orifice maintains orifice pressure at a constant level above that necessary to provide critical flow at the maximum depth. Thus, the mass flow rate of the supply gas is invariant.

Exhaust Valve: Since a mixed gas is supplied at a constant rate and since only oxygen is consumed, all diluent gas must escape from the system. An exhaust valve allows gas to vent when a certain system pressure is reached. Since gas is generally supplied at a rate that is far less than the diver's breathing rate, gas escapes only at the end of each exhalation and only in an amount about equal to what was supplied during that particular breathing cycle.

Bypass Valve: As the diver descends, increasing ambient pressure causes gradual collapse of the breathing bags. A bypass valve in the gas supply line usually is provided to allow the diver to keep the apparatus properly inflated as he descends.

A typical arrangement of the system is shown in Figure [IX-19]. It is not necessary to arrange the components precisely as shown. For example, it is not necessary to provide two breathing bags. One bag will do, but when two are provided as shown, flow of gas may proceed through the canister at a lower maximum rate, thus tending to reduce pressure drop and the pulmonary work needed to overcome it. The exhaust valve generally is located somewhere between the exhalation check valve and the canister. The vented gas then will contain some CO_2 and this choice of exhaust-valve location thus will reduce the amount of CO_2 that the canister must absorb. Also, since peak exhalation pressure normally occurs at the time that the valve is exhausting, breathing effort can be reduced by putting the exhaust valve as close as possible to the mouthpiece to reduce flow losses between mouthpiece and valve.

The gas supply may be brought into the system at any point, but it is obvious that, if gas is brought in between the mouthpiece and the exhaust valve, vented gas will be richer in O_2 than would be the case if only exhaled gas were vented. This would cause unnecessary waste of gas, so the gas usually is supplied to the system at a point downstream from the exhaust valve and upstream from the man.

A list of semiclosed-circuit breathing apparatus is given in Table IX-12. Discussion of the various models can be found in Kenny (1972) and Penzias and Goodman (1973).

e. Closed-Circuit Breathing Apparatus

There are two basic types of closed-circuit systems. In one the breathing mixture is pumped from a diving bell to a diver's helmet and back to the bell, where the carbon dioxide is scrubbed out and oxygen is added. Pressure in the diver's helmet must be equalized approximately with the ambient pressure of the diver. The second type is more complicated. It is a self-contained unit utilizing a gas mixture with only enough

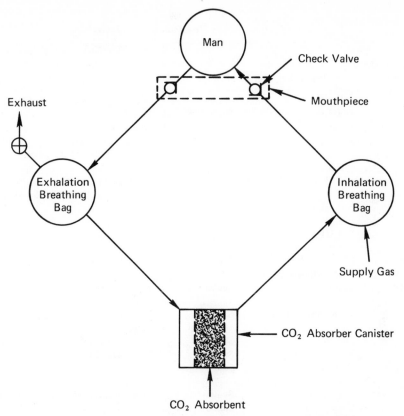

Figure IX-19. Schematic of typical semiclosed scuba. [From Riegel (1970) by permission of the Marine Technology Society, Washington, D. C.]

oxygen to supply the metabolic needs of the diver. To compensate for increasing ambient pressure as the diver descends, inert gas is added. This system can be used for long-duration dives (Parker, F. A., 1971).

The closed-circuit breathing apparatus is very similar to the semiclosed apparatus except that the closed system vents none of the gas to the water and controls the oxygen level in the gas supplied to the diver by measuring the oxygen partial pressure. Oxygen is automatically metered into the recirculating gas to maintain the proper level. This type of system is by far the most operationally economical, but has only recently become feasible because of the complex measuring and control electronics required (Kenny 1972). It has been calculated that the ratio of the cost of helium required to maintain a working diver in the water at 600 ft by open-circuit compared to semiclosed compared to closed is 1.00:0.13:0.004.

A comparison of diving duration capability of open-, semiclosed-, and closed-circuit breathing apparatus is shown in Figure IX-20. A list of various closed-circuit, mixed-gas underwater breathing apparatus and miscellaneous devices is given in Table IX-13.

Table IX-12
Semiclosed-Circuit Breathing Apparatus[a]

Manufacturer or vendor	Model designation	Primary mode	Miscellaneous comments: chronology, technology, availability
Drager	DM-40	Scuba	Air–oxygen unit for helmeted diver
	FGG I	Scuba/tethered	Developmental model
	FGG II	Scuba/tethered	Developmental model
	FGG III	Scuba/tethered	Tethered mode (SDC/PTC) uses FGZ III or FGB III mixed-gas supply
	SM I	Scuba	—
	SM II	Scuba	"Self-mixing" mixed-gas diving units developed for Royal Swedish Navy
	SM III	Scuba	—
	SM IV	Scuba	—
	SM III-5	Tethered	—
	SMS-4	Tethered	—
Dunlop	Clearance diver's breathing apparatus	Scuba	Pendulum-circuit rig
Emerson	Flatus II	Scuba	Modification of 1952 Lambertsen amphibious respiratory unit
	U. S. Navy Mark V	Scuba	Initial "fleet" semiclosed mixed-gas rig
Fenzy	DC-55	Scuba	Military model
	P68	Scuba	Commercial model
Hydrospace Developments	Mark IX	Scuba	—
Old Dominion Research and Development	Blair	Scuba	Interim-developmental design models
	Les	Scuba	
	Mike 55	Scuba	
Scott Aviation	U. S. Navy Mark VI	Scuba	Current "fleet" rig
	U. S. Navy Mark VIII	Tethered	Scuba emergency mode
	U. S. Navy Mark IX	Tethered	Scuba emergency mode
Siebe Gorman	Mine recovery apparatus	Scuba	World War II rig
	S.G. 700	Scuba/tethered	Incorporates closed-circuit capability
Westinghouse	Abalone Breadboard	Tethered	Scuba emergency mode
	U. S. Navy Mark XI interim fleet equipment	Tethered	Production model Abalone
	SEALAB III breathing apparatus ("Fink" rig)	Tethered	Scuba emergency mode

[a] From Penzias and Goodman (1973) by permission of John Wiley and Sons.

f. Cryogenic Breathing Apparatus

A modern development, not yet fully exploited, is the cryogenic breathing apparatus. "Technical developments within aerospace industries have now increased the availability of super-insulants and liquid gases and attempts are being made to take advantage of the possibilities of gas storage in the liquid phase for diving" (Common 1969). Storage and release of gas are controlled by varying the degree of

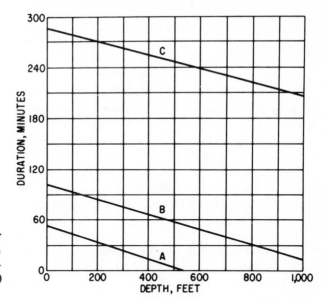

Figure IX-20. Comparison of diving duration capability of (A) open-circuit scuba, (B) semi-closed-circuit scuba, and (C) closed-circuit scuba.

insulation and vapor pressure. However, the loss of buoyancy imposed by storage in the liquid phase and the addition of complex heat exchange and insulation devices tend to offset the gain over gas-phase storage. Valve freezeup with altitude change is also a problem (Common 1969). Nevertheless, the combined use of chemical scrubbers and cryogenic processing can result in major weight and volume saving over backpack design (Fischel 1970).

Table IX-13

Closed-Circuit Mixed-Gas Underwater Breathing Apparatus and Miscellaneous Concepts and Devices for Increasing Underwater Endurance[a]

Oceanic Equipment Corp., Electrolung, closed-circuit mixed-gas scuba; automatic mixture control: electrochemical O_2 sensor (polarographic)

Biomarine Industries, Portable Closed-Cycle Breathing System, CCR-1000, closed-circuit mixed-gas scuba; automatic mixture control: electrochemical O_2 sensor

Westinghouse Electric Corp., Underseas Division, Krasberg Scubarig, KSR-5, closed-circuit mixed-gas scuba; automatic mixture control: electrochemical O_2 sensor (gold–cadmium fuel cell)

Sterling Electronics Inc., Ocean Technology Division, SS-1000, closed-circuit mixed-gas scuba; automatic mixture control: cryogenic system, O_2 liquid–vapor phase equilibrium regulated by rate of liquid nitrogen refrigerant boiloff

Westinghouse Electric Corp., Underseas Division, Arawak, tethered, closed-circuit compressor–depressor system

Mako Products, X-5, liquid-air (cryogenic) "cryo-lung" scuba

Ayres Gill-type Underwater Breathing Equipment and Methods for Reoxygenating Exhaled Breath (U. S. Patent 3,228,394)

Bodell Artificial Gill (U. S. Patent 3,333,583)

Strauss Gill-type Underwater Breathing Apparatus (U. S. Patent 3,318,306)

Lambertsen (prototype built by J. H. Emerson) Air Saver, partial-rebreather scuba

[a] From Penzias and Goodman (1973) by permission of John Wiley and Sons.

3. *Ancillary Equipment*

a. Masks

An important part of a diver's personal equipment is a mask or faceplate, regardless of whether he is snorkling, scuba diving, or diving with surface-supplied air. It provides a shield between his eyes and the sea, aiding his underwater vision. Today there are over 150 models of facemasks on the American market alone. Some important factors to consider in selecting a mask are the following.

(i) Visibility. Goggles are the smallest type of facemask but they greatly restrict a diver's visual field. Facemasks that provide a wide field of view improve stereoacuity. (See Chapter IV, discussion of vision; also see the discussion of corrective lenses, contact lenses, etc., which are designed for underwater use.)

(ii) Watertightness. Not all types of masks will fit every face and provide a watertight seal. The mask strap can be tightened to a certain point to provide a tighter seal, with a tradeoff of increased discomfort, but a proper seal is essential to maintaining the shield purpose of the mask. Also, if the mask covers a diver's nose, there is danger of inhaling water if the mask leaks badly.

(iii) Purge System. Every diver should be able to purge his mask manually. Basically this means turning on his side, breaking the seal of his mask, and exhaling air into the mask to force the water out and upward. This technique is difficult for some divers, and masks are now equipped with a purge valve. This is normally kept shut by the outside pressure of the water but, by exhaling into his mask, a diver can increase the pressure enough to open the valve and thus blow water out of his mask.

(iv) Displacement. As a diver goes to increased depths, his mask produces a "squeeze" on his face because the water pressure against his mask is greater than the air pressure inside it. To compensate for this, the diver exhales air into his mask to equalize the pressure. With goggles, however, there is no way to equalize the pressure and this limits the depth to which the diver can go. If the mask has a large displacement, more air must be used to equalize the pressure. This air is lost for rebreathing, a major factor to snorkel divers, but rarely important to scuba or surface-supplied divers. Some masks offer the low displacement of goggles and also have a rubber nose piece which fits down over the nose to allow the diver to equalize the pressure without using a great deal of air.

Aside from these factors, a mask is usually selected for personal factors, such as comfort, corrective lens requirement, and weight. Figure IX-21 shows the wide variety of styles available, and Table IX-14 is a tabulation of some of the masks available, indicating the various features of each.

In addition to masks used with regular scuba equipment, full-face diving masks, which use surface-supplied compressed air via an umbilical, are available. With this type of mask, an emergency air supply bottle is usually carried by the diver in the event that his umbilical becomes tangled or cut. A control valve allows the diver to switch to either gas supply. An example of this type of mask is the U. S. Navy Kirby-Morgan KMB-8 band mask. This is a lightweight mask which also has a communication system incorporated in it. The mask is usually worn with a Neoprene hood which secures the communication speakers over the diver's ears. The microphone is

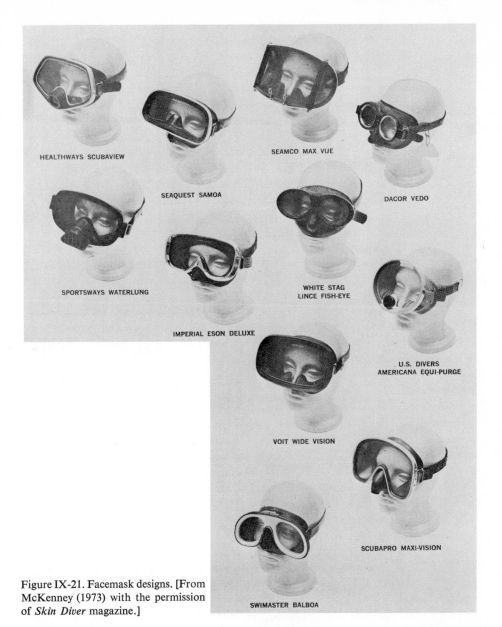

HEALTHWAYS SCUBAVIEW

SEAQUEST SAMOA

SEAMCO MAX VUE

DACOR VEDO

SPORTSWAYS WATERLUNG

IMPERIAL ESON DELUXE

WHITE STAG
LINCE FISH-EYE

U.S. DIVERS
AMERICANA EQUI-PURGE

VOIT WIDE VISION

SCUBAPRO MAXI-VISION

SWIMASTER BALBOA

Figure IX-21. Facemask designs. [From McKenney (1973) with the permission of *Skin Diver* magazine.]

located in the oral–nasal cavity of the mask (Reimers and Lebenson 1972). Figure IX-22 shows a variety of full-face diving masks. Table IX-10 gives available diving helmets and masks.

b. Fins

Fins increase a diver's speed and range by giving him added thrust and more efficient propulsion for each leg kick. There are basically two types of fins, the full

Table IX-14
Facemasks[a]

Mask Cat. No.	Basic shape	Purge size and desc.	Mask weight, oz	Retaining band	Band fastener	Strap fastener	Volume	Special features
AMF SWIMASTER								
2M35 Wide View	O	11/16 in. PP	12-1/2	SS	SL	SMB	SM	Heavy construction/all-Neoprene skirt
2M43 Polaris III	R	7/16 in. BN	16	SS	C	CB	L	Exterior clamps to actuate nose equalizers
2M44 Atlantide	R/RN	NP	14-1/2	CP	PF	CB	S	Double face seal/close fit for wide vision
2M45 Full View	R/RN	BO	14-1/2	SS	2R2FS	CB	ML	Double face seal/features two side windows
2M69 Baja	RO/RN	NP	10-1/2	P	PF/B	CB	SM	Double face seal/fits close for good vision
2M72 Balboa	G/RN	NP	8-1/2	WC	S	CB	S	Double face seal/mask fits extremely close
AMF VOIT								
B59BKS Wide Vision	R	NP	13	P	PF	CB	L	Double face seal/finger wells for equalization
B60BKS Competition	R/RN	5/16 in. BN	12-1/2	SS	PC	CB	ML	Double face seal/lens rim comes in colors
2M37 Nassau	O	3/4 in. PP	12-1/2	SS	SL	CB	ML	Comes with custom nose pad to be inserted
DACOR								
DM-16 Vedo	G/RN	NP	8	P/MER	PF	CB	S	Optical eyepieces available
DM-18 Tenue	O/RN	NP	11	SS	2B	CB	M	White lens liner provides attractive contrast
DM-27 Equinaut	RO/RN	NP	10-1/2	P	NB	CB	ML	Double face seal/large plate for wide view
DM-28 Dacor Professional	O	3/4 in. PP	11-1/2	SS	NB	CB	ML	Sturdy construction/large-diameter purge
HEALTHWAYS								
1271 Seaview w/p	O	9/16 in. PP	13-1/2	SS	SL	CB	ML	Yellow frame/finger wells for equalization
1272 Scubaview	TS	9/16 in. PP	15	SS	SL	CB	ML	Unusually large glass but small skirt opening
1275 Scubacat	G/RN	NP	7-1/2	P	B	CB	S	Strap clevis bolts eliminate strap slippage
1276 Scubacat w/p	G/RN	5/16 in. BN	7-1/2	P	B	CB	S	Positive lock on head strap
1280 Scubamaster	O/RN	NP	13-1/2	CP	2B	CB	ML	Larger glass area with small skirt opening
IMPERIAL D AND H								
1828 Marine Gold Deluxe	O	5/8 in. PP	10-1/2	SS	SL	CB	M	Accordian finger wells/heavy-duty strap
1830 Seaside Deluxe	R	5/8 in. PP	11-1/2	SS	SL	CB	M	Neoprene pressure pad/heavy-duty strap
1831 Eson Deluxe	O	NP	11-1/2	SS	SL	CB	M	Double face seal/soft skirt
SCUBAPRO								
219 Supervision	RO	NP	16	SS	4B	CB	L	Heavy-duty strap/side windows
220 Supervision w/p	RO	3/4 in. PP	16-1/2	SS	4B	CB	L	Heavy-duty strap/side windows
204 Maxi-Vison	RO/RN	NP	13-1/2	SS	2C	CB	ML	Double face seal/heavy-duty strap
205 Maxi-Vision w/p	RO/RN	7/16 in. BN	14	SS	2C	CB	ML	Double face seal/heavy-duty strap
228 Optical	RO/RN	NP	10	SS	2C	CB	S	Double face seal/...

Model	Basic shape	Purge valve description	Size	Retaining band	Band fastener	Strap fastener	Volume	Remarks
SEAMCO								
99-6500 Maxvue	R	3/8 in.	22	SS	2NB	CB	L	Largest front glass area of all masks/exterior
99-6501 Oporto	O	5/8 in. PP	12-1/2	SS	C	CB	ML	wire nose clamp to activate equalizer
99-6509 Denia	R/RN	3/8 in. BN/BO	20	SS	2NB	CB	L	Large glass area with side windows
99-6512 Coronado	O	3/8 in.	13	SS	C	CB	L	Double face seal/finger wells for equalization
99-6522 Toedo	R/RN	3/8 in.	13	SS	2C	CB	ML	Double face seal/close fit for wide vision
SEAQUEST								
DM50 Samoa	R/RN	5/16 in. SP	11-1/2	SS	SL	CB	ML	Double face seal/futuristic design/plastic frame
DM60 Super Marina	R/RN	5/16 in. SP	13	SS	SL	CB	ML	Double face seal/inner lens rim in colors
DM70 Atlantide	R/RN	NP	11	P	PF	CB	ML	Double face seal/fits close to eyes
SPORTSWAYS								
1626 Waterlung Mask	R	7/8 in. PP	13	P	2S	CB	M	All-neoprene body with heavy Neoprene skirt
1654 Scene-O-Rama w/p	R	7/8 in. HIG	16	SS	4B	MB	L	Large protective plastic cage over purge valve
1655 Scene-O-Rama	R	NP	15-1/2	SS	4B	MB	L	Features side windows for maximum vision
1656 Avalon w/p	R	7/8 in. HIG	12	P	2S	MB	M	Large protective plastic cage over purge valve
U. S. DIVERS								
5026 Wrap-Around	R/RN	3/8 in. BN/BO	14-1/2	SS	2R4N	PB	ML	Double face seal/side windows
5050 Pacifica	O	5/8 in. PP	16	SS	NB	CB	M	Double face seal/pressure pad for equalization
5055 Aqua-Naut	O/RN	3/8 in. BN/BO	15	SS	NB	CB	ML	Double face seal/heavy white rubber lens frame
5060 Naso	R/RN	NP	12	CP	PF	CB	M	Double face seal/chromed retaining band
5070 Falco	G/RN	NP	7	P	PF	SPB	S	Close fit to eyes/yellow plastic frame
5071 Fisheye	G/RN	NP	7-1/2	P	S	CB	S	Double face seal/small displacement/fits close
5206 Equi-Rama (silicone)	O	NP	11-1/2	SS	SL	CB	M	Silicone rubber skirt/accordian finger wells
5208 Equi-Purge (silicone)	O	1/2 in. HIG	11-1/2	SS	SL	CB	M	Accordian finger wells for equalization
5211 Americana Equi-Purge	O	1/2 in. HIG	11	SS	SL	CB	M	Accordian finger wells for equalization
WHITE STAG								
56610 Lince Fish-Eye	G/RN	NP	8	P	B	CB	S	Double face seal/fits close for good vision
52612 Super Stag	R/RN	NP	14	CP	2B	CB	M	Double face seal/chromed band, soft skirt
52660 Lince Fish-Eye w/p	G/RN	7/16 in. BN	8	P	B	CB	S	Double face seal/fits close for good vision
52665 UDT	O	9/16 in. PP	15-1/2	SS	NB	CB	ML	Rubber pressure pad for equalization
52662 Super Stag w/p	R/RN	7/16 in. BN	14	CP	2B	CB	ML	Double face seal/polished band for appearance
52716 Pinova	O/RN		9-1/2	P	2B	CB	SM	Double face seal/soft skirt

a From McKenney (1973) by permission of *Skin Diver*.

Legend: Basic shape: G, goggle type. O, oval. R, rectangular. RN, rubber nose. RO, rectangular/oval. TS, triangular/square.
Purge valve description: NP, no purge. PP, pinocchio type purge. BO, built in, but optional; opening has to be cut by diver if purge required. HIG, purge is inserted through round hole in glass. SP, mask comes with separate purge valve for diver installation. BN, purge is located at bottom of rubber nose. WC, wire cable.
Retaining band: SS, stainless steel. P, plastic. MER, metal eye rims. CP, chrome-plated. WC, wire cable.
Band fastener: PF, press fit. B, bolt(s). NB, nut and bolt. SL, snap lock. C, clamp(s). S, screw(s). PC, plastic catch. R, threaded rods. FS, female shank. SPB, single plastic buckle. PB, plastic buckle at each side of mask. MB, metal buckle at each side of mask. SMB, single metal buckle.
Strap fastener: CB, clevis bolt. MST, metal slip through both sides of mask. MSTS, metal slip through, one only on strap.
Volume: S, small. M, medium. L, large.

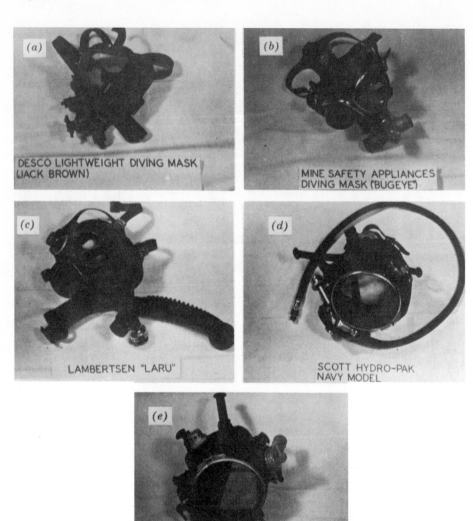

Figure IX-22. Full-face diving masks used with surface-supplied compressed-air or self-contained diving methods. [From Penzias and Goodman (1973) by permission of John Wiley and Sons, New York.]

foot-pocket fin and the open-heel fin. The open-heel fin is available in a fixed-heel-strap version and in an adjustable-heel-strap version. Varieties of these basic types include straight or curved blades, flexible or rigid blades, and blades equipped with slots, venturis, stabilizers, and vents. The large, rigid-bladed fins, or power fins, such as U. S. Divers Co. Aqua Lung Professional and Voit Super UDT, provide the diver with much more power, but also require very strong leg muscles. When using these fins, the kick stroke is shorter and slower, with emphasis placed on the down kick

(Lentz 1967*a*). Lighter fins with a slightly curved, flexible blade are generally used when speed and range are not important. With this type of fin, the kick stroke is wider and more rapid with less thrust than the power fins. Nearly as much force is used on the upstroke as on the downstroke kick. Examples of this type of fin are the Owen Churchill fin, Scuba Pro Jet-Fin, the Otarie model of U. S. Divers Co., and Voit Custom Duck Feet (Lentz 1967*a*).

(i) Fixed Open-Heel Fin. An open-heel fin was the first swim fin available and was used by the military during World War II. It has a simpler straight blade and about 60% of the thrust is achieved on the downstroke (Figure IX-23*a*).

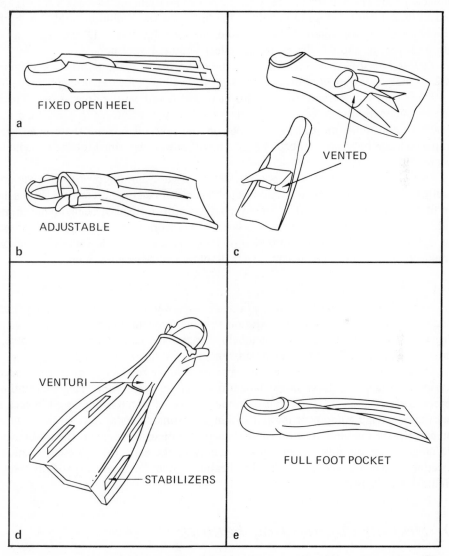

Figure IX-23. Swim-fin design. [From Parker (1971) by permission from *Skin Diver* magazine.]

(ii) Adjustable Open-Heel Fin. An adjustable fin (Figure IX-23b) is easier to put on, but, because one model is made to adjust to fit all foot sizes, the pocket for the foot is wide and long. This is a disadvantage because small feet tend to slide in the pocket from side to side. Nevertheless, it is a popular style, especially for the military, since only one size of fin must be stocked, and it can be easily adjusted to most standard foot sizes. Many of the adjustable open-heel fins are designed with additional features—venturis, stabilizers, or slots. Whether these features actually increase the propulsion efficiency is still under debate (Parker, G., 1971). A vent is essentially a hinged flap on the upper part of the blade which opens to allow backflow of water to enhance thrust. The flap then closes with the return kick (Figure IX-23c). Venturis are used to tunnel water from the middle of the fin to the tip. Stabilizers added to both sides of the blade help eliminate fin twist and sideward slip. Figure IX-23d shows a fin with both venturi and stabilizers, the Seamless Venturi Power fin.

(iii) Full Foot-Pocket Fin. Another type of fin fits over the entire foot (Figure IX-23e), making it both comfortable and safer for walking on coral or rough sea beds. The blade is usually curved, which helps increased thrust on the downstroke. There are two minor disadvantages to this style. One is that it is harder to carry than the open heel, and much more difficult to put on and take off because there is no heel strap which can be easily grabbed. Various modifications described for the adjustable fin are also available in this style.

(iv) Selection. The most important factors in choosing the proper fin style are comfort and application, with the additional criteria of personal preference. If fins do not fit well, they will only add to the diver's discomfort instead of being an aid to his underwater capability. Application is related to the work to be accomplished; speed fins differ from endurance-optimized fins. Personal preference is usually based on familiarity, build, leverage, and muscular development or conditioning.

c. Localization Equipment

In certain tasks, such as relocation of underwater installations, localization capability is needed. Acoustic localization systems frequently consist of an acoustic pinger on the installation, and a handheld directional sonar receiver for the diver. These systems are difficult for the already-encumbered diver to manage.

In order to avoid the necessity of carrying a sonar receiver, a system has been developed which simulates in-air binaural direction perception. The localization beacon consists of a supersonic emission, modulation of which is impressed in such a way that receiving equipment located within a reasonable distance from the beacon can perceive the direction from which the signal is emitted. Two sensors (hydrophones) are mounted on the diver's head in the vicinity of the ears; these present binaural directional signals to the diver. This localization system has been successfully tested (Bohman and Juels 1971).

D. *Human Engineering Factors*

The importance of having tools designed for underwater use lies in the fact that any tool is or may be more difficult to use underwater than on dry land. The adversity of the environment for tool use is discussed at the beginning of this chapter, in the section on hand-held tools. Tool categories are listed and the elements of good human engineering design are prescribed. This section extends that information and reviews certain selected performance studies.

1. *The Diver as a Resource*

A diver is placed in an underwater work site to do useful work. In spite of difficulties, the properly trained and equipped diver is a versatile and competent resource.

Comparing divers with submersible manipulator controllers in undersea tasks, Pesch *et al.* (1970) found that, in various manipulative tasks, divers outperformed the remote manipulator. A summary of their data is given in Table IX-15, comparing diver performance against four types of control for the remote manipulator; Tables IX-16 and IX-17 provide the diver–remote manipulator ratios. These data indicate the performance potential of the diver compared with other available means for manipulative tasks. It should be noted (from Table IX-17) that the diver and remote manipulator are nearly equally capable once the tool is positioned and ready to use. It is the other behavioral components—positioning, grasping, aligning the tool—that the diver performs much faster than the controller of a remote manipulator. It is, in brief, the versatile competence of the diver that constitutes his major capability.

Table IX-15

Comparison of Operator as Diver and Remote Manipulator Controller[a]

Task		Diver	Push button, fixed rate	Push button, variable rate	Joy stick, fixed rate	Joy stick, variable rate
Sample gathering		0.18	2.59	2.12	2.45	2.00
Hooking		0.15	2.43	2.36	2.99	2.79
Valve turning		0.09	0.95	0.83	0.84	0.83
Quick disconnect		0.21	5.63	6.13	5.10	6.59
Drilling	Task time	2.12	3.39	3.39	5.94	5.30
	Angle error	3.4°	4.0°	4.4°	4.8°	6.2°
	Qual. rtg.	2.7/3	2.2/3	2.2/3	2.5/3	1.9/3
Tapping	Task time	2.47	3.71	3.21	3.46	3.46
	Angle error	5.8°	3.6°	4.6°	3.6°	4.2°
	Tap depth	2.4	2.5	2.6	2.2	2.5
	Qual. rtg.	2.1/3	2.6/3	2.2/3	2.7/3	2.6/3
Threading		0.18	1.80	1.64	1.51	1.78
Unbolting		0.20	1.76	1.72	2.50	1.94
Behavioral elements						
Simple travel		0.04	0.49	0.40	0.45	0.35
Complex travel		0.05	0.59	0.46	0.67	0.63
Grasp		0.01	0.36	0.31	0.36	0.34
Alignment		0.02	0.36	0.41	0.52	0.44
Tool use		0.58	0.70	0.64	1.22	1.04

[a] From Pesch *et al.* (1970) by permission of General Dynamics, Groton, Connecticut. Except as noted, all figures are task time in minutes.

Table IX-16
Comparison of Operator as Diver and Remote Manipulator Controller: Data in Terms of Diver/Remote Manipulator Ratio[a]

| | Rate control | | | |
| | Discrete actuator switches, buttons | | Combined actuator joystick | |
	PBFR, fixed rate	PBVR, variable rate	JSFR, fixed rate	JSVR, variable rate
Sample collection	1/14.4	1/11.8	1/13.6	1/11.1
Valve manipulation	1/10.5	1/9.2	1/9.3	1/9.2
Rigging chain, hooks	1/16.2	1/15.7	1/19.9	1/18.6
Bolt removal with power tool	1/8.8	1/8.6	1/12.5	1/9.7
Tapping with power tool	1/1.5	1/1.3	1/1.4	1/1.4
Threading with power tool	1/10	1/9.1	1/8.4	1/9.9
Drilling with power tool	1/1.6	1/1.6	1/2.8	1/2.5
Connect–disconnect	1/26.8	1/29.2	1/24.3	1/31.4

[a] From Pesch et al. (1970) by permission of General Dynamics, Groton, Connecticut. Ratios expressed with respect to tasks.

Multiple behavioral components are involved in the completion of most underwater tasks. Andersen (1972a) reports on three tasks undertaken under shallow (baseline) conditions and at 200 ft, analyzing performance times into four components: preparation, work, rest/idle, and service–maintenance.

The tasks were: (a) load handling—moving a test stand rig by means of a gas blown buoyant Aqualift device; (b) abrasive saw cutting—abrasive saw cutting ½ in.

Table IX-17
Comparison of Operator as Diver and Remote Manipulator Controller: Data in Terms of Diver/Remote Manipulator Ratio[a]

| | Rate control | | | |
| | Discrete actuator switches, buttons | | Combined actuator joystick | |
	PBFR, fixed rate	PBVR, variable rate	JSFR, fixed rate	JSVR, variable rate
Simple travel	1/12.2	1/10	1/11.3	1/8.8
Complex travel	1/11.8	1/9.1	1/13.3	1/12.5
Simple grasp	1/36	1/31	1/36	1/34
Alignment	1/17.8	1/20.4	1/26	1/22
Tool use	1/1.2	1/1.1	1/2.1	1/1.8

[a] From Pesch et al. (1970) by permission of General Dynamics, Groton, Connecticut. Ratios expressed with respect to behavioral elements.

Table IX-18

Comparative Task Activity Times, Test-Stand Load-Handling
(Andersen 1972*a*)

Task activity classification	Baseline Point Mugu		Baseline Makai Range		200-ft Dive	
	Mean time, min	Percent	Mean time, min	Percent	Mean time, min	Percent
Preparation	5.92	43	1.50	22	13.00	46
Work	6.25	47	4.50	64	15.00	54
Rest/idle	Negligible		Negligible		Not observed	
Service/maintenance	1.38	10	1.00	14	No data	
Total mean time	13.55	100	7.00	100	28.00	100

by 6 in. steel plate and 1½-in.-square steel bar stock; and (c) combination task—eyebolt installation in a 1-in.-wide steel using impact tool and drill/tap method.

Data presented in Tables IX-18–IX-20 indicate the relative lengths of time spent in each work phase and the changes that occurred from baseline to 200-ft conditions. Such data comprise a working foundation for dive management, in demonstrating the effects of depth and in facilitating the improvement of tool design and tool usage procedures.

A third example of diver work capability is reported by Vaughan and Mavor (1972). In their report the authors state:

Six 4-hr., open-sea test trials were conducted with a wet submersible. The purpose of these trials was to assess the effects of long exposure to cold (16.5°C) water on man's ability to perform basic submersible control tasks.

The subjects were experienced submersible pilots who had a minimum of 20 hours training prior to the experimental trials. Skin and rectal temperatures were continuously recorded from both the pilot and rider of the submersible. A continuous record of vehicle depth and water temperature was also obtained. The pilot's task was to maintain a prescribed depth while performing a sequence of source changes for a 4-hr. period of submergence. Depth error variance was correlated with pilot care and skin temperature changes over time, and although pilot core temperature fell as much as 1.83°C, no degradation in depth-control performance was apparent.

Table IX-19

Comparative Task Activity Times, Hydraulic Abrasive Saw Cuts
(Andersen 1972*a*)

Task activity classification	Baseline Point Mugu		200-ft Dive	
	Mean time, min	Percent	Mean time, min	Percent
Preparation	1.62	4	3.00	12
Work: Cut ½-in. by 6-in. steel plate	4.74	41	13.75	53
Cut 1½-in.-square steel bar	4.75	41	5.75	22
Rest/idle	Negligible		Not observed	
Service/maintenance	1.62	14	3.25	13
Total mean time	11.54	100	25.75	100

Table IX-20

Comparison of Selected Tasks Performed Both in Shallow Water and at 200 ft during the Saturation Dive (Andersen 1972*a*)

| | Condition mean performance time, min | | | |
| | Baseline Point Mugu depth 40–50 ft | Saturation test Makai Range depth 200 ft | Time difference, min | Factor increase in performance time for saturation dive |
Task				
Test-stand load-handling from Aegir to work site	13.55	28.00	14.45	2.07
Test-stand load-handling from work site to Aegir deck	11.08	23.50	12.42	2.12
Combination task—eye-bolt installation in 1-in. mild steel using drill and tap method	8.28	12.01	3.73	1.45
Abrasive saw cutting $1\frac{1}{2}$-in.-square steel bar and $\frac{1}{2}$-in. by 6-in. steel plate	11.54	25.75	14.21	2.23

2. *Evaluation Procedure*

Andersen (1972*b*) recommends the following procedure:

Obtain data on the physiques of the intended user population. Although human beings vary widely in size and shape, their variability follows certain patterns. However, the limits of the desired percentile range must be determined by actual measurement of samples of the group to be accommodated, since the measurement values of divers may differ significantly from those groups for which measurements are available.

Select and measure a small group of representative divers. Select subjects representing both ends (around the 5th and 95th percentiles) of the height and weight distribution of the divers to be accommodated. This will give reasonable assurance that the range of other dimensions will be approximated. Ideally, divers should be measured in the detail desired for the operator group. For example, in order to determine a control handle size that will accommodate 90 percent of the expected user population, it would be more exact to determine such a design feature from hand measurements than from height and weight alone.

Outfit divers used as test subjects in the widest range of diving apparel and life support equipment that might be worn while operating the tool. All operational aspects of the tool should accommodate 90 percent of the anticipated user group in the full range of the diving apparel used when operating the tool.

Have the diver test subjects operate the tool under water under environmental conditions (depth, water temperature, and visibility) expected in an operational setting. The tool operators should perform all the motions normally required in an operational work setting. The test performance should last as long as an actual work task. Make sure the diver can operate the tool on all normal work surface orientations (deck, vertical and overhead).

Note the difficulties encountered by the test divers with respect to handling ease, personal comfort, quality of work output, vision, and safety hazards caused by body size and operator capability.

Table IX-21

Maximum Bare-Hand Torque (Mean Values,
in.-lb) (Andersen 1972a)

	5th percentile	95th percentile
Pronation	63	151
Supination	54	115

An anthropometric study of the type recommended above was conducted at the Navy Experimental Diving Unit (Beatty and Berghage 1972). Fifty-four anthropometric measurements were made on 100 navy divers, and two pulmonary function measurements were made on 41 of the divers. As a result of this study, it was concluded that obtaining diver anthropometrics is essential to comfort and to operational efficiency. It was recommended that the work be carried further by making measurements on fully equipped divers.

3. Biomechanical Capabilities

Data regarding arm and hand movements, with force emission capabilities and limitations, are necessary to tool design criteria. Some of the factors which influence a diver's physical effectiveness are his buoyancy, his dress, his position, and the direction of force emission. Tables IX-21–IX-24 demonstrate the effect of some of these factors on task performance. Much of the data in these tables were adapted from aerospace human engineering data, with an effort made to select data most relevant to the diver's requirements.

In determining the maximum amount of force required for a given task, the strength of the weakest operator should be the deciding factor.

The biomechanical capability of the diver can be augmented by the use of a restraint system when the tools used require great and/or sustained force (see Table IX-25). Andersen (1972b) recommends that the restraint system should be compatible with the following requirements.

Table IX-22

Sustained (4 sec) and Impulse (Instantaneous)
Hand Torque Values (Anderson 1972b)

		Torque, in.-lb	
	Variables	Sustained	Impulse
Mode	Underwater neutral-buoyancy simulation	212.5	369.3
Suits	Shirtsleeve (scuba)	218.3	334.7
	Advanced extravehicular suit	220.3	344.6
Restraints	Waist	140.9	298.7
	Shoes	231.4	347.0
	Waist and shoes	228.7	329.8
Tools	L-Handle	283.9	470.9
	T-Handle	148.2	194.8

Table IX-23

Mean Sustained Pushing Force for Restrained Divers
(Andersen 1972b)

Work surface	No. of arms used	Diver force (lb), duration of			Mean diver force, lb	
		6 min	3 min	1 min	Each arm	Both arms
Overhead	One	18.8	21.4	29.0	23.1	35.5
	Two	32.3	38.9	72.9	48.0	
Vertical	One	21.8	23.6	34.5	26.6	36.6
	Two	35.7	41.4	62.6	46.6	
Deck	One	17.5	18.4	35.4	23.8	39.6
	Two	42.2	44.0	80.4	55.5	
Mean	One	19.3	21.1	32.9	24.5	—
	Two	36.7	41.4	71.9	50.0	—
	Combined	28.0	31.2	52.4	—	37.2

A restraint system must be flexible enough to provide a diver with a wide degree of freedom to position his body with respect to his work surface. The system must provide the diver with a selection of restraint points that will fit the multiple or variable body positions that may be required.

The basic restraint equipment must be usable with all standard forms of diver apparel and life-support systems. These will include hard-hat, deep-dive dress and wet-suit scuba, self-contained life-support backpacks, and surface-supported air hose and mixed-gas systems.

The diver must be able to enter and exit a restraint device with and without gloved hands and without the assistance of support or tender divers.

The limited and sometimes zero-visibility conditions under which divers sometimes work require that the restraint system release and adjustment mechanisms be operable under "no-sight" conditions. In addition, all mechanisms must be operable with gloved hands.

A restraint system should be readily transportable between work sites by a diver. Any portion of the restraint system worn should not encumber the diver or affect his buoyancy under water.

Table IX-24

Sustained Pushing Force Range Based on
Individual Diver Mean Scores
(Andersen 1972b)

No. of arms used	Test range	Diver force (lb) for duration of		
		6 min	3 min	1 min
Two	High	50.0	61.3	102.5
	Low	13.0	22.7	28.3
One	High	32.3	33.3	54.0
	Low	8.8	12.0	20.8

Table IX-25

Mean Sustained and Impulse Forces (lb) as a Function of Restraint Systems[a]

Direction	None		Hand		Waist		Shoes		Restraints Hand and waist		Hand and shoes		Waist and shoes		Hand, waist, and shoes		Mean	
	S	I	S	I	S	I	S	I	S	I	S	I	S	I	S	I	S	I
Push	0	35	1	41	15	43	4	46	29	57	30	62	35	58	43	69	19.6	51.4
Pull	0	43	2	43	22	46	4	48	31	51	33	61	37	57	38	61	20.9	51.3
Up	0	19	5	21	10	23	17	28	14	23	18	31	17	28	17	29	12.3	25.3
Down	2	23	9	26	10	23	21	33	16	26	26	37	19	30	21	32	15.5	28.8
Right	0	18	10	22	12	22	9	22	15	25	16	28	14	23	16	27	11.5	23.4
Left	0	18	17	29	12	22	8	23	17	28	21	34	15	25	19	30	13.6	26.1
Mean	0.3	26.0	7.3	30.3	13.5	29.8	10.5	33.3	20.2	35.0	24.0	42.2	22.8	36.8	25.7	41.3	—	—

[a] From Norman (1969) by permission of *Human Factors*. S = sustained force (4-sec interval). I = impulse force (1-sec interval).

E. *Environmental Equipment*

With increased knowledge of the underwater environment, man has endeavored to accomplish more work and more detailed scientific research for longer periods of time. This has resulted in an ever-increasing demand and supply of support equipment, such as swimmer vehicles, manned and unmanned submersibles, and habitats which allow man to actually live on the sea floor. When a diver is supported by additional equipment, many of the burdens imposed by the underwater environment are alleviated. For example, the diver is no longer required to carry all the tools he might need. Long, heavy umbilical or gas-supply cylinders may be eliminated by a gas storage system carried on support equipment. A power source, though sometimes quite limited, may aid in lighting, supplying heat to the diver, and providing power for his tools. Communication is simpler with an on-the-site support subsystem than with the surface, and the diver can be provided a place for resting, protected from extreme cold and hostile marine life.

Obviously, the type of work being done, whether it is pipeline inspection, salvage, or experimental research, determines the type of support equipment desired and, many times, several different types of support equipment are used as subsystems to an entire complex system. Actual design of this support equipment has not been well coordinated with the main goal of developing a piece of equipment that will accomplish a particular task. Insufficient attention has been paid to diver safety in many instances. At the other extreme, some equipment has been designed around diver safety, with the result that the equipment is only marginally useful to the diver because his task requirements were not extensively considered. In the future, engineers, scientists, and divers will have to work together in order to develop the ultimate in underwater support equipment that not only will help the diver to perform the particular task necessary, but will also allow for the greatest diver safety and provide the diver with the most comfortable surroundings and support possible. Safety considerations as regards the various types of environmental conditions are discussed in Chapter VIII. Here we are principally concerned with design and function.

Although design and development of each type of support equipment have been extremely varied, there are basic underlying principles. An attempt is made in this section to explain these and to give specific examples of actual hardware in use.

The three basic types of working equipment for underwater tasks are diving systems, habitats, and submersibles. Where a platform or surface ship is available, and the task is localized, the diving system is the ideal choice. With the development of saturation-diving techniques, the habitat is becoming more frequently used as a base for scientific operations, since it permits a more continuous and flexible approach to the research project. When a large area is to be covered, the requirement is better met by the submersible, either manned or unmanned, lockout or 1 atm, depending on the nature of the task (Baume 1972).

1. *Diving Systems*

The term *diving system* is used very broadly to cover many different systems used for diving. In general, it implies the use of three basic components:

1. *A surface ship* or platform serves as a support center.
2. *A deck decompression chamber* (DDC), easily monitored, large and comfortable enough for several divers, is located on board the surface ship or platform.
3. *A transfer chamber,* which can be either a personnel transfer capsule (PTC) or a submersible decompression chamber (SDC), transports divers between surface and underwater work site and can be mated to the DDC.

These basic components can be used in a wide variety of different combinations, depending upon the task to be performed.

Diving systems may be divided into two general types. The first is the advanced diving system (ADS), which was developed commercially and is also used by the U. S. Navy. The basic unit is the SDC, a tethered pressure chamber which provides divers a warm, dry environment while they are transported between the surface ship or platform and the underwater work site. It also provides an underwater refuge for the diver and serves as a dry, warm decompression area. This greatly reduces the time a diver must spend in an uncomfortable, alien environment.

The second type of diving system is the deep-diving system (DDS) developed by the U. S. Navy. This type of system was designed to provide the capability of making observation dives, bounce dives, or long-term saturation dives. The basic unit of the DDS is the PTC. This capsule is a type of submersible decompression chamber, permitting dry decompression under water, and can also be used to transfer personnel under pressure. This provides the capability of using saturation-diving techniques. This capability differentiates it from an SDC. Although an SDC can be used for saturation-diving techniques, the chamber is usually too small to be comfortable for more than short-term decompressions. When used for saturation diving, the PTC is used to transfer personnel from the surface ship to the underwater work area while maintaining the pressure to which they are saturated. In this type of diving the DDC then becomes the surface habitat for the divers, in which they sleep and eat. The PTC can also be used as an observation chamber and in this case it is maintained at 1 atm of pressure.

a. Advanced Diving System

The ADS II was the first of its type developed. The SDC for this system has two chambers, which allows an observer to remain at 1 atm in the upper chamber, while the lower chamber can be pressurized to ambient underwater pressure. On descent both chambers remain at 1 atm and, once at the work site, the lower chamber is pressurized. One diver then exits from the bottom hatch using an umbilical from the SDC to provide breathing gas and communications. A second diver remains in the SDC to monitor the umbilical and is available for emergency assistance. This type of chamber is extremely useful in that an engineer can be present right at the work site to supervise the work and it also gives the working diver more flexibility and safety than he would have diving with a long, vertical umbilical from a surface support. When the work is completed, the diver reenters the lower chamber and decompression begins immediately. The chamber can then be removed from the water and decompression continues with the SDC aboard ship.

The ADS III utilizes the same type of SDC but also incorporates a deck decompression chamber, which can be mated to the lower chamber of the SDC. This allows for a second crew of divers to be transported to the work site while the first crew is decompressing in the DDC. The ADS III can also work with an underwater habitation, which can be used as a dry working area or a dry refuge at the work site. Figure IX-24 shows a diagrammatic sketch of the ADS III.

The ADS IV incorporates a single-chamber SDC, which can be mated to the deck compression chamber. It can be used both as a submersible decompression chamber or it can be lowered at 1 atm of pressure for observation tasks. The advantage of this system over the others is that it is of smaller size and weight and has a much less complicated support system.

b. Deep Diving Systems

The U. S. Navy used the ADS IV (now termed the SDS-450) for its deep diving operations up until the development of their own Mark I Deep Dive System. Although the SDS-450 has been used for deep saturation diving, the chamber is so small that it is unsuitable for long decompressions. The Mark I deep diving system was designed to meet the following specifications of the Navy (Milwee 1970):

1. A diving capability for work to depths of 850 ft.
2. A diving capability for observation dives at 1 atm pressure, bounce dives of short-term nonsaturation, and long-term saturation dives.
3. Air transportability for rapid deployment.
4. The capability to use a variety of surface support ships or platforms, the most frequently used of which is the ARS class salvage ship.

The actual diving system has a mode of operation similar to that described under advanced diving systems. However, two decompression chambers are available for use, so that two diving crews can be decompressed simultaneously, even if the decompression times are not the same. An entry lock connects the two DDC's and is the transfer area between the DDC and the PTC. The PTC will accommodate three men. Each DDC will accommodate two men comfortably but can accommodate three. The deck decompression chambers are equipped with sleeping bunks, shower, sanitation facilities, and a complete life-support system. The PTC has a special cable carrying power, communications, and instrumentation wiring. The gas supply for the PTC is surface supplied, but it also carries its own on-board emergency gas supply. Umbilicals used by divers for excursions from the PTC are located inside the chamber. A diagrammatic sketch of the Mark I Deep Dive System is shown in Figure IX-25 (Milwee 1970).

The Mark II Deep Dive System was designed to provide a depth capability of 1000 ft for eight divers to make either conventional dives or saturation dives for periods up to 14 days. It is quite similar in principle to the Mark I DDS but the Mark II system incorporates two PTC's, two main-control consoles (MCC), and two DDC's. The Mark II Deep Dive System is capable of operating in six modes, each with a different type of control requirement (Hall 1973):

1. Saturation diving.
2. Hydrostatic diving with the PTC maintained at 1 atm.

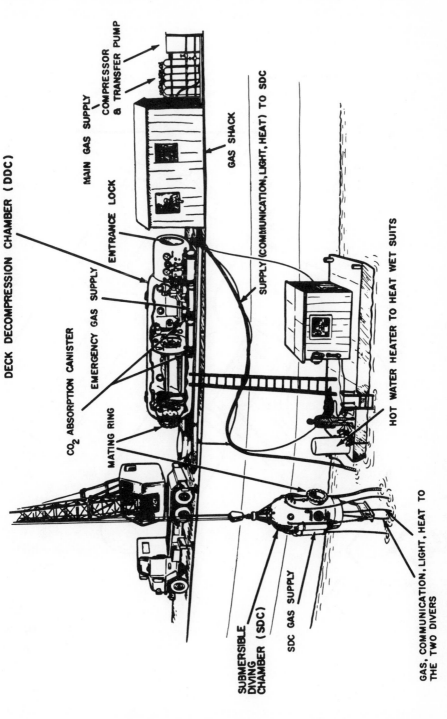

Figure IX-24. ADS III shown here includes deck decompression chamber with which the submersible chamber can mate through a special fitting. Divers can then decompress in safety and comfort ashore. [From Kowal (1967) by permission from *Sea Frontiers*.]

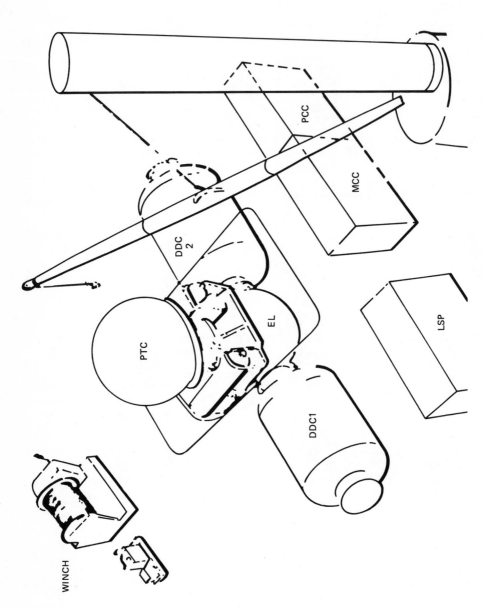

Figure IX-25. Mark I deep dive system. [From Milwee (1970) by permission of the Marine Technology Society, Washington, D. C.]

3. Air diving with the PTC pressurized from surface-supplied air hoses.
4. Routine diver support with the DDC being used as an emergency decompression chamber if a diving casualty should occur.
5. Submarine rescue in which each DDC could accommodate 34 men at 5 atm and with a mating adapter to allow mating to the Deep Submergence Rescue Vehicle (DSRV).
6. Decompression, whereby saturated divers are decompressed at 1 atm.

This diving system was used in support of the SEALAB Project.

c. The 1-atm Work System

A system recently developed by Lockheed for the Shell Oil Company consists of a chamber installed over a wellhead, marked by a buoy, and a service capsule capable of accommodating a crew of four, which contains the necessary operational equipment and control devices. The capsule mates to the chamber for transfer of personnel; both are maintained at atmospheric pressure. Air and power are supplied to the capsule by an umbilical from the surface ship. The system appears to be economical and efficient, due largely to the fact that work is done under atmospheric pressure in dry conditions, using ordinary tools and equipment, thus eliminating the need for highly specialized divers and supply installations (Anon. 1973).

2. Habitats

There are many types of habitats, but each basically is composed of the same subsystem and serves the same general purpose. A habitat may be defined as any system which permits men to live and work for extended periods in the sea at ambient pressure. The simplest habitats are those used in shallow water, where divers do not require extensive decompression. This eliminates a large part of the surface support usually required for decompression chambers and diver safety. "SUBLIMNOS" and "EDALHAB" are examples of this type of shallow-water habitat.

With increased depth, saturation diving techniques are normally employed. This type of diving is based on the fact that after being exposed to a pressurized inert gas for approximately 24 hr, a diver's tissues are essentially saturated with that gas at that pressure. The tissues have taken up almost all the inert gas that they will absorb at that pressure; therefore, the decompression time remains the same, no matter how long they remain at that pressure. Habitats can be used to provide living and working quarters for divers during the time they remain on the bottom; they then undergo only one decompression at the end of their sojourn.

Habitats have generally proven to be economically unfeasible for commercial operations and have been used almost exclusively for underwater research and experimental work (Tenny 1971).

The total habitat system, particularly for saturation diving, must be considered as a complex made up of a surface support ship or shore-based center, a sea-floor habitat/living area, and a transfer device such as a PTC to transport personnel and supplies between the habitat and the surface ship. A habitat contains its own

Table IX-26

Classification of Underwater Habitats[a]

Program or habitat name	Weight	Dimensions	Ballast	Operations	Crew	Depth, m	Duration	Atmosphere	Mode of supply Power	Mode of supply Gas	Comments
ADELAIDE (Australia)	—	—	—	1967-68	—	—	—	—	Pontoon, barge	—	—
AEGIR (Habitat II) (USA)	200 t	L = 2 × 4.6 m, D = 2.8 m, Ball D = 3 m, Total L = 15.2 m	—	Hawaii 1969	4-6	147	14 days	Variable	Ship	Autonomous	Can ascend and descend by completely internal control
AQUATAT I AQUATAT II (USA)	—	—	—	—	3-5	—	—	—	—	—	Six viewports 1.8 m² viewing area (shallow water)
BACCHUS BAH I (Germany)	—	L = 6 m, D = 2 m	—	Baltic Sea, Sept. 1968, Ost See, June 1970	2	10, 10	11 days, 14 days, 2-5 mo	Air	Ship	Ship	—
CARIBE I (Cuba)	—	L = 3 m, D = 15 m	—	1966	2	15	3 days	—	Ship	Partly autonomous	—
CHERNOMOR I (USSR)	—	L = 8 m, D = 3 m	—	Gelendehuk, Black Sea	4	5, 14	30 days	—	—	—	—
CHERNOMOR II (USSR)	70 t	—	—	Design	4	25, 35	4 weeks	—	—	—	—
EDALHAB (USA)	—	L = 3.7 m, D = 2.4 m	5.5 t	Alton Bay, New Hamps.	4	7.6	36 h	Air	Land	Land	—
GLAUCUS (UK) HEAVY DUTY	—	—	—	1965	4+	9	1 week	—	Surface	—	—
SEA BED VEHICLE (UK)	—	L = 12 m, W = 7 m, H = 4.2 m	—	Design		180	5 days	—		—	—
HEBROS I (Bulgaria)	—	L = 5.5 m, D = 2 m	—	Bay of Warna, July 1967	2	10	—	—	—	—	—
HEBROS II (Bulgaria)	—	L = 6.7 m, D = 2.5 m	—	Cape Maslenos, 1968	2	30	10 days	—	Surface	—	Effective volume of 30 m³; depth can be controlled autonomously by crew (about 64 t)
HELGOLAND (Germany)	75 t	L = 9.0 m, D = 2.5 m, H = 6 m	—	North Sea, July 1969	4	23	10 days	Air	Buoy	—	Capable of depths to 100 m; numerous occupations since 1966
HYDROLAB	—	L = 4.9 m, D = 2.4 m	40 t	July 1966 continuous to 1970	—	12	—	—	—	—	Transfer from submersible at either ambient or atmospheric pressure possible
ICHTHYANDER 66 (Idctiandr) (Ikhtiandr) (USSR)	600 kg	L = 22 m, W = 1.6 m, H = 2 m	—	Crimean Coast, Black Sea, August 1966	1-2	11	7 days	Air	Land	—	Single-chambered lab, 6.8 m³, four viewports
ICHTHYANDER 67	4.5 t	W = 8.6 m, H = 7.0 m	27 t	Crimean Coast, Black Sea, August 1967	2-5	12	2 weeks each	Air	Surface	—	Three chambers

ICHTHYANDER 68	—	—	Crimean Coast, Black Sea, Sept. 1968	Several crews	20	8 days total	—	Land	—	15 m³ displacement
ICHTHYANDER 69	—	—	Design 1968	5	20	—	—	—	—	—
KARNOLA (Czechoslovakia)	L = 5.6 m, D = 2.6 m	—	—	4	8–15	—	—	—	—	Volume = 30 m³, three chambers (converted railway tank car)
KITIJESCH (USSR)	D = 1.9 m, H = 4.6 m	—	Crimean Coast, Summer 1968		15	—	—	Shore	—	—
KOCKELBOCKEL (Netherlands)	—	9.5 t	Sloterplas 1867	2–4	15	Short period	Air	—	Autonomous	—
KRAKEN (UK)	L = 2.6 m, W = 1 m, H = 2.3 m	—	Firth of Lorne, Oban Argyle, Scotland	2	30	Several weeks	7% O₂, 93% N₂	—	—	Proposed; two compartments
MALTHER I (East Germany)	L = 4.2 m, D = 1.8 m, H = 3.5 m	14 t (1.4 m³ of iron) 11 mp	Malther Dam, Nov.–Dec. 1968	2	8	2 days	Air	Land	Autonomous	Volume = 10 m³
MAN-IN-SEA I	L = 3.2 m, D = 0.9 m	1.9 t	Villefranche, Mediterranean, Sept. 1962	1	61	1 day	3% O₂, 97% He	—	Ship	Aluminum cylinder
MAN-IN-SEA II (SPID) (USA)	L = 2.4 m, D = 1.2 m	—	Great Stirrup Cay, Bahamas, June 1964	2	142	2 days	4% O₂, 96% He	—	Ship	Flexible rubber tent (submersible, portable, inflatable dwelling)
MEDUSA I (Poland)	L = 2.2 m	3 t	Lake Klodno, July 1967	2	24	3 days	37% O₂, 63% N₂	—	Land	—
MEDUSA II (Poland)	L = 3.6 m, W = 2.2 m, H = 1.8 m, total H = 2.5	—	Baltic Sea, July 1968	3 / 3	30 / 26	14 days / 7 days	Air	—	Ship	Autonomous operation up to 50 hr
MINITAT (USA)	H = 3.5 m, D = 2.4 m	4.5 t	Design	2	50	—	Air	—	—	Four view ports
OKTOPUS (USSR)	—	—	Crimean Coast, Black Sea, July 1967	3	10	Several	Air	—	—	—
PERMON II (Czechoslovakia)	L = 2 m, W = 2 m	—	Split, Yugoslavia, Adriatic Sea, July 1966	2	3	Discont.	—	—	Autonomous	Displacement 5 m³; discontinued; decompression within habitat possible
PERMON III	—	1.5 t	Czech. Lake March 1967	2	10	4 days	—	Land	Autonomous	—
PRECONTINENT I (CONSHELF I) (France)	L = 5.2 m, D = 2.5 m	—	Marseille, Mediterranean, Sept. 1962	2	10	1 week	Air	—	Land	—
PRECONTINENT II (CONSHELF II) (France)	—	—	Shaab-Rumi Reef, Red Sea, July 1963	5 / 2	11 / 27	30 days / 1 week	Air / 5% O₂, 20% N₂, 75% He	Ship	—	—
PRECONTINENT III (CONSHELF III) (France)	L = 14 m, sphere D = 7.5 m, H = 8 m	70 t	France, Mediterranean, Sept 1965	6	100	22 days	1.9–2.3% O₂, 1% N₂, balance He	Surface	Autonomous	—

Table IX-26—Cont.

Program or habitat name	Weight	Dimensions	Ballast	Operations	Crew	Depth, m	Duration	Atmosphere	Mode of supply		Comments
									Power	Gas	
ROBIN II (Italy)	—	—	—	Genoa, Mediterranean, March 1969	1	17	7 days	—	—	—	Light, permeable plastic hull
ROBINSUB I (Italy)	—	$L = 2.5$ m, $W = 1.5$ m, $H = 2$ m	—	Ustica Is., Mediterranean, July 1968	1	10	—	Air	Land	—	Wire cage, plastic tent, volume 5 m³
ROMANIA LSI (Romania)	—	—	—	Bicaz Lake 1968	2	45	6 days	—	Ship (Capable of volume 14 m³ ship or land)	—	—
SADKO I (USSR)	—	Sphere $D = 3$ m	8.5 t	Caucasian Coast, Black Sea, Oct. 1966	2, 2	40, 25	6 hr, 1 month	—		—	—
SADKO II (USSR)	—	Twin spheres, $D = 3$ m	21 t	Caucasian Coast, Black Sea, Summer 1967	2	25 (50–60)	6 days	—	Land ship	Autonomous	Buoyancy 12 t
SADKO III (USSR)	—	—	—	Zukhumy Bay, Black Sea	6	25	6 days	—	—	—	—
SD-M 1 (UK)	—	$L = 2.7$ m, $W = 1.5$ m, $H = 2.1$ m	—	Malta, Mediterranean	—	9.1	—	Air	Autonomous	Resupply tanks from surface	Discontinued
SD-M/2 (UK)	—	—	—	Malta, Mediterranean	1–2	6.1	10 man-days	Air	Autonomous	Resupply tanks from surface	—
SEALAB I (USA)	—	$L = 12.2$ m, $D = 2.7$ m, $H = 4.5$ m	—	Argus Is., Bermudas, July 1964	4	59	11 days	4% O_2, 17% N_2, 79% He	Ship	—	Double chamber
SEALAB II (USA)	200 t	$L = 17.4$ m, $D = 3.7$ m	—	LaJolla, Calif.	28	60	10 days	4% O_2	Ship	Autonomous	Three teams, ten days each plus one man, 29 days
SEALAB III (USA)	—	—	—	Pacific Ocean	5–12	183	—	2% O_2, 6% N_2, 92% He	—	—	Suspended
SUBLIMNOS (Canada)	—	$H = 2.7$ m, $D = 2.4$ m	9 t	Little Dunks Bay, Tobermory, L. Ontario installed June 1969 to date	2–4	10	Up to 24 hr	Air	Land	—	Designed for "day-long" occupation—overnight accommodations feasible for short periods
TEKTITE I (USA)	—	Twin cylinders, $H = 5.5$ m, $D = 3.8$ m	79 t	Lameshur Bay, U. S. Virgin Is. 1969	4	12.7	59 days	—	Ship	—	—
TEKTITE II (USA)	—	Twin cylinders, $H = 5.5$ m, $D = 3.8$ m	79 t	Lameshur Bay, April–Nov. 1970	10–5	12.7	14–21 days	—	Land	—	—

a From Parrish et al. (1972) by permission of IPC Science and Technology Press.

life-support system, food preparation area, sanitation facilities, and waste disposal system. It can be mobile or stationary, but is seldom self-propelled and must be towed by the surface ship. Life-support gas can be supplied to the habitat either from its own high-pressure cylinders or from the surface ship via an umbilical. Likewise, power may be derived from a source at the habitat or from a power supply line from the surface ship (Garnett and Achurch 1969).

A tabulated list of design features of habitats is presented in Table IX-26. Since each habitat is a unique system, reference should be made to a report on a particular habitat for detailed system analysis.

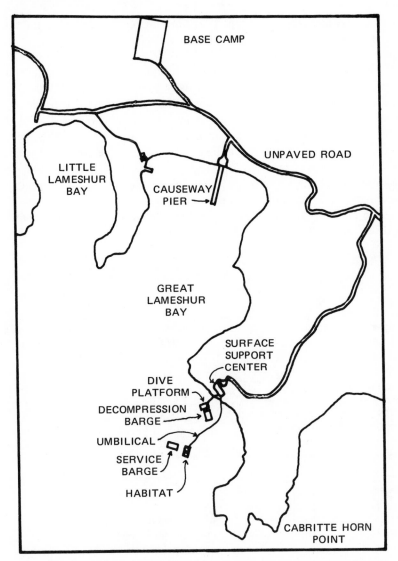

Figure IX-26. TEKTITE II operational site at St. John, U. S. Virgin Islands (Miller *et al.* 1971).

Although each habitat has different characteristics and design features, TEKTITE II is a good representative system. The operational facilities of TEKTITE II consisted of a base camp, a causeway pier, a shore-based surface-support center, a dive platform, a towable decompression barge, a service barge, and an underwater habitat. These are shown in Figure IX-26 at the operational site.

The base camp served as the principal accommodation facility for support personnel. It consisted of 13 tropical huts, which served various purposes, such as kitchen-dining area, office-dispensary, toilet and shower facilities. Power was supplied by electricity from the local power company with diesel generators as a backup system.

The causeway pier extended about 270 ft into the bay and served as the principal landing site for shuttle boats, supply vessels, and other small boats supporting the program.

The surface-support center contained the command van and the utilities trailer. Continuous monitoring of the habitat, visually and aurally, was done from the command van, as well as monitoring of the habitat atmosphere and trunkway water level. The command van also provided communications equipment. The utilities trailer (Figure IX-27) contained diesel generators to supply electrical power to the habitat, the decompression barge, the dive platform, and the command van. It also contained the air compressors used to charge the scuba equipment and to supply the makeup air to the habitat. Fresh water for the habitat was stored at the support center and was gravity-fed to the habitat via the umbilical bundle.

The dive platform served as an equipment storage area and a duty station for support divers. It also had a scuba-charging compressor and air bank.

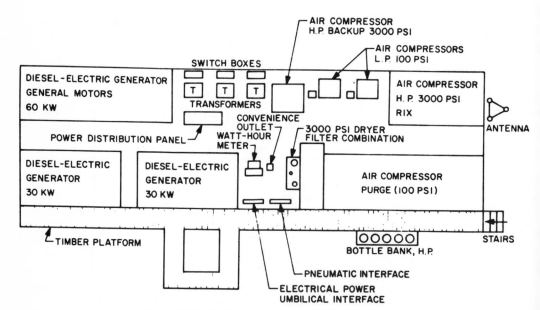

Figure IX-27. Utilities trailer layout (Miller *et al.* 1971).

The decompression barge was the primary decompression facility, mounted on a pontoon catamaran to make it towable. The decompression complex consisted of the decompression control console, a deck decompression chamber, and a personnel-transfer capsule. The decompression console contained gas-monitoring equipment, temperature probes, valves and gauges to control the pressure in the personnel-transfer chamber and the deck decompression chamber, as well as the master communications unit for the chambers. The deck decompression chamber was a double-lock chamber with a 5-ft entry lock and a 12-ft main chamber. Four bunks were available in the main chamber and a small pressure-flushing commode and fresh-water shower were located in the entry lock. The personnel-transfer capsule was a sphere, 66 in. in diameter, which could be mated to the main chamber of the DDC. A side double-door hatchway was used for mating and surface entry and a bottom double-door hatchway was used for diver lock-in and lock-out. An umbilical bundle contained compressed-air lines, communication cables, and gas-sampling lines. High-pressure bottles mounted on the exterior were available for on-board oxygen and air supply.

The service barge, 45 by 18 ft, served for equipment transport, work platform, and site evacuation. It was a staging site for transfer of material, equipment, and personnel to and from the habitat.

The underwater habitat was made up of two vertical cylinders, each divided into two compartments attached to a steel base. Diagrammatic layouts are shown in Figures IX-28 and IX-29. Two entry trunks were located at the base. One gave access to the open ingress–egress wetroom hatchway, while the other gave access to the normally closed emergency hatch located beneath the crew quarters.

3. *Swimmer Vehicles*

Many types of swimmer vehicles, towing-type or self-powered, are available to aid the free-swimming scuba diver. Towed vehicles, which depend on a surface ship for propulsion, are usually of simple design with some degree of directional control. Their use is limited by drag restrictions imposed on the diver and by reliance on the surface craft.

One of the simplest towed vehicles is the aquaplane—a board which, when tilted downward or sideways, provides a dynamic thrust to counter the corresponding pull on the towing cable. The addition of a broom handle seat and proper balancing of the towing points permit one-handed control of flight path. With this device, a diver may be towed at speeds of 2 or 3 knots by a rubber boat, the maximum speed being limited by the hydrodynamic forces that tend to tear off the diver's mask. There are several modifications of this basic design available, which incorporate improved controls and a face shield in order to increase maneuverability and pilot comfort. Aberdeen Marine Laboratory has developed the "MOBEL," which has excellent maneuverability, and the Russians are reported to have a complex towed submersible, the "ARLANT I BATHYPLANE," now being used in trawl net observations.

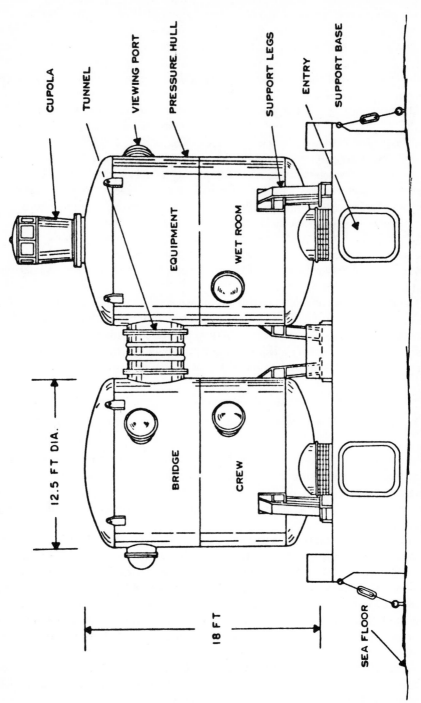

Figure IX-28. Side view of the TEKTITE II habitat (Miller *et al.* 1971).

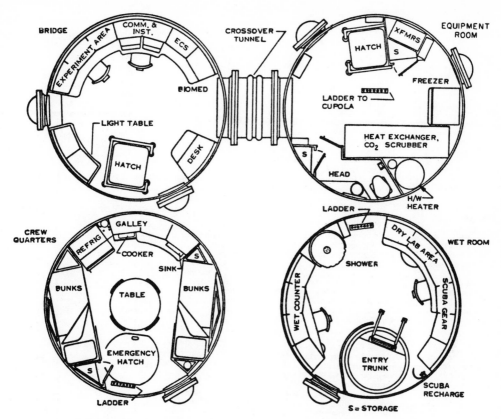

Figure IX-29. Plan views of the habitat compartments (Miller *et al.* 1971).

Even more helpful are self-powered craft. These generally provide some degree of shielding for the diver so that faster speeds may be attained. There is some confusion of terminology in this type of equipment, as indeed there is in the entire field of environmental and working equipment. Two terms frequently used are swimmer delivery vehicle (SDV) and diver propulsion vehicle (DPV). The larger vehicles are sometimes referred to as wet submersibles.

One example of a craft with its own electric propulsion unit is the "PEGASUS." Self-powered devices offer maneuverability to the diver free from a surface support vehicle and are equipped with comprehensive navigational equipment, but the costly batteries used in these vehicles make them too expensive for most research purposes (Woods and Lythgoe 1971). A complete coverage of the variable designs and configurations is beyond the scope of this chapter; however, the following three types are illustrative.

1. The *ECOR*-1 from East Coast Oceanic Research Inc. (Anon. 1972*a*) is a two-man unit, operated by a $1\frac{1}{2}$ hp electric motor, capable of maintaining an underwater speed of 4 knots for $2\frac{1}{2}$ hr with a depth limit of 300 ft. The hull is constructed of polyester fiberglass.

Table
Comparison of Wet Submersibles
(Black and

Description	General function	Diver support functions	Propulsion system	Maximum submerged speed, knots	Operational time at maximum speed
Buoyancy transport vehicle— experimental	Designed to prove concept of free-swimming vehicle that provides forklift or yard crane functions at an underwater site; i.e., transport and position relatively large payloads	1. Transport and position payloads (on bottom) 2. Power tools (limited capability)	Vertical, horizontal, surface, hover	1.3	1 hr
Buoyancy transport vehicle— (projected data) prototype	Underwater forklift/yard crane; move and position relatively large payloads	1. Transport and accurately position large (multi-thousand pound) payloads 2. Provide hydraulic power for tools	Vertical and horizontal, surface and submerged	1.5	Unlimited on umbilical, 1 hr on batteries
Construction assistance vehicle— experimental	Designed as an experimental diver support platform equipped with tools, power sources, and cargo area	1. Carry tools and cargo underwater 2. Power tools 3. Transport divers 4. Stable bottom platform	Vertical, horizontal, surface, hover, turn-in-place	2.5	4 hr
Construction assistance vehicle prototype	Extension of experimental model to include interchangeable work modules for drilling, excavating, cable installation, stabilization, etc.	Same as above plus: 5. Crawl on bottom 6. Translate through mild surf (3–4 ft) 7. Heavy work functions such as drilling, coring, excavating	Vertical, horizontal, surface, bottom crawl, dry land, surf zone	1–2	4 hr
Swimmer propulsion units	Designed for transportation of a single diver; commercially available units mostly designed for recreation	1. Transport divers 2. Visual survey or photography	Horizontal, surface	2–3	1–4 hr
Swimmer delivery vehicle; "Shark Hunter," commercial	Designed to transport two divers, tools, and limited cargo to and from underwater sites	1. Carry divers and tools to underwater work site 2. Survey bottom	Horizontal, surface	2–4	1–3 hr
Swimmer delivery vehicle	Several commercial models available; military models provide increased speed, endurance, cargo, and cost	—	—	—	—
Surface support platform	Ship or other moored platform with compressors, generators, etc., mounted on deck	1. Support working diver 2. Tools and lifting and positioning limited by sea state 3. Divers burdened by umbilicals	Surface and possibly a sled towed under water for survey	NA	NA

2. The *Farallon DPV MK-1* (Anon. 1971) combines a diver propulsion unit with a light source. It will operate to a depth of 300 ft while pulling one or more divers at a speed up to three times normal swimming speed. It is powered by two 12-V dc batteries.

3. The *CAV* or Construction Assistance Vehicle was developed for the Naval Civil Engineering Laboratory (Black and Elliott 1972). It can maintain a maximum speed of $2\frac{1}{2}$ knots for 4 hr with a depth limit of 120 ft. The electrohydraulically powered craft is capable of delivering up to 1500 lb of wet-weight cargo.

A comparison of some of the wet submersibles for supporting scuba divers is presented in Table IX-27.

X-27
or Supporting Scuba Divers
Elliott 1972)

Navigational capability	Maintainability	Payload capability	Tool power	Dry wt., lb	L × B × H, ft	Operating depth, ft	Comments
Visual	One technician full time (large part of effort is maintenance of surplus silver–zinc batteries)	1000 lb on cargo hook	6 gpm at 1800 psi (oil hydraulic)	1,800	8 × 6 × 6	850	Steering and depth control by propulsion; buoyancy control by dewatering sphere
Visual plus limited compass	Low man-hour requirements	3000 lb can be supplemented by modular buoyancy packages	10 gpm at 2000 psi	2,500	8 × 8 × 8	130	Steering and depth control by propulsion motors; automatic buoyancy or depth control system incorporated into design
Visual and limited compass	During operational period requires one skilled technician half time	1300 lb in 4 × 7-ft cargo bed or slung underneath	12 pgm at 1200 psi (oil hydraulic) 20 cfm pneumatic	18,630	26 × 9½ × 7½	130	Variable speed control; steering and depth control by propulsion; fabrication cost: $75,000
Visual compass; adaptable to future developments, such as transponders and pingers	Designed for compatibility with fleet, low maintenance: performed in the field by fleet personnel	2000 lb in cargo bed or lift	Total power to vehicle ~ 20–50 hp; power can be directed to diver tool (hydraulic)	10,000–15,000	—	Diver-limited	All specifications preliminary; derived for interpolation with performance of existing vehicle; both land and underwater final specifications will follow preliminary design
Usually direct visual	Relatively low maintenance because of simplicity of vehicle	None	None supplied	50–100	3 × 1 × 1	150	Primarily designed as a propulsion device; cost: $400 and up
Visual	Low maintenance because of simplicity of structure and components	Usually inside vehicle	None supplied	1,200–2,500	16 × 8 × 5	150–300	Step speed control; steering and depth controlled by planes and rudders; cost: $5,000–$500,000
—	—	100–300 lb; 5–10 ft³	—	—	—	—	
Standard surface ship techniques	Standard ship maintenance; performed by regular diver	Ship lift capacity	Limited only by size of umbilical diver can carry	NA	NA	NA	Support of diver limited by safety factors relating to sea state and umbilicals

4. Manned Submersibles

A submersible is sometimes defined as an underwater diving vehicle that requires surface support; it is thus differentiated from a submarine, which is autonomous and requires no surface support. However, some of the larger submersibles are virtually autonomous, so that a better distinction might be the somewhat arbitrary one of differentiation by function. A large submersible vessel designed for military use is generally termed a submarine, and does not come within the scope of this handbook. Most submersibles could be defined as dry, 1-atm vehicles in which the pilot and observers are protected from the outside pressure by a pressure hull or cabin. The shape of the hull can be spherical, ellipsoidal, cylindrical, etc. The spherical design is the best for withstanding high external pressures but the others are more hydro-dynamically efficient when a high degree of maneuverability is desired.

Submersibles are used for a wide variety of tasks, such as observation, measuring currents, exploring the sea floor, taking core samples, and conducting other types of oceanographic research. They are used in search missions for lost objects and in search and rescue missions for submarines.

Many parameters could be used to categorize submersibles but most can be termed either tethered to a surface ship or free, self-propelled.

a. Tethered Vehicles

Tethered submersibles can be subdivided into three general categories:

1. *Vertically tethered*—where the tether serves as a suspension cable which holds the vehicle at a desired depth.
2. *Self-propelled tethered*—where the tether supplies the power for propulsion, lighting, heating, and other equipment.
3. *Towed*—where the tether is used for towing as well as a supply line for communications, power for lighting, tools, etc.

The great advantage of a tethered submersible versus a free submersible is the long endurance which can be achieved by receiving power from the surface. However, there are the disadvantages of limited flexibility and mobility because of the umbilical and of dependence upon surface support.

b. Self-Propelled Vehicles

Self-propelled, free submersibles have a considerable mobility advantage over tethered submersibles in that they are self-sufficient, highly maneuverable, and free from the danger of entanglement with an umbilical. Most of the smaller ones are designed to carry only two men and are limited in range because of the small amount of power storage they have. The bigger submersibles are very useful for studies covering a large area, but are not as applicable to small work areas where high maneuverability is a requirement.

Most submersibles have some sort of special feature which makes them applicable for a particular task. These features are indicated in Table IX-28, which gives a compilation of some submersibles in existence today.

c. Lockout Capability

A recent advance in maneuverable submersibles is lockout capability whereby a submersible can provide a 1-atm environment for the crew and diving personnel up until the time a diver is actually needed, and then the diver compartment can be pressurized. This greatly reduces a diver's exposure time and either he can be decompressed in the submersible itself, or he can be transferred to a deck decompression chamber on the surface-support vessel. Handling of the submersible through the interface is more difficult than that of a submersible decompression chamber (SDC), since most submersibles are about 2–2½ times heavier than an SDC and generally require a swimmer for hookup and retrieval (Bailey *et al.* 1972). Nevertheless, the maneuverability and mobility of the submersible give it many advantages over the tethered SDC, which requires an umbilical from the surface and is limited to essentially vertical mobility. Current submersibles with lockout capability are limited in size and the power requirements impose severe limitations. To date, most are powered by lead–acid batteries, but nuclear power is a possibility for the future. Examples of submersibles with lockout capability are DEEP DIVER (PLC4) and SHELF DIVER (PLC4B), which were designed for economically feasible continental shelf applications (Stevens 1970).

d. Manipulator Capability

Prior to 1965, few submersibles had any degree of manipulator capability. "TRIESTE" and "ALVIN" used an electromechanical type of manipulator in which oil-filled and pressure-compensated dc motors powered the joints while additional pressure-compensated oil volume protected the other vital components. These manipulators, small and of relatively low capacity, were basically modifications of commercially available units primarily used in the nuclear energy industry (rated for loads in the 50-lb range).

In later modifications, the dc motors were replaced by hydraulic actuators. Like the electrohydraulic units, the motions are rate-controlled, and a portable miniaturized control station capable of being operated from any viewport contained the on–off and rate-control switches for various motions. The pump also must be selected carefully to function in a high-ambient-pressure environment. For example, in-line piston pumps which depend on cavitation of the pressure chamber to uncover an inlet port will not function properly because the power required to cavitate under the ambient backpressure is basically the same as that required to generate a system pressure equal to the ambient pressure.

Most manipulator-equipped submersibles since 1965 have used the same basic system, either electromechanical or electrohydraulic, with rate control. Outwardly, the configurations may vary; generally they are fitted to the various shaped hulls for both protection and optimum placement relative to the viewing system and visible "working" volume. The general appearance and vulnerability to damage have been cleaned up by concealing or eliminating exposed hydraulic lines and electrical conductors.

The control console can be either portable or fixed, depending on the vehicle-viewing system and the size of the submersible. Large walkaround submersibles such as "ALUMINAUT" use portable units to permit manipulator operation from any of a number of viewports.

Other vehicles use stationary optical systems; the operator remains seated in a central location and monitors the operation through periscopes and other visual aids at a fixed console.

Some typical units used on various submersibles are:

DOWB	rate-controlled, electromechanical
Deep Quest	two rate-controlled, electrohydraulic

Table XIII-28

Some Recent Manned Submersibles Currently (1972) in Service[a]

Name	Operating depth, m	Maximum length, m	Crew	Payload kg	Viewing ports	Speed,[a] knots	Life support,[e] hr	Personnel transfer	Diver lock-out	Manipulator	Drop wts.	Depth maint.	Depth change
Towed													
BK-9468-1 (1968) (France)	170	1.5[b]	2	?	c	?	?	?	Yes	No	No	Yes	Yes
CI-K9728-4 (1968)	335	1.7[b]	3	360	c	?	S	No	Yes	No	?	Yes	Yes
DI-K320 (1968)	335	1.7[b]	3	360	9	?	S	No	Yes	No	No	Yes	Yes
Guppy (1970)	305	3.6	2	135	3	4 max, 2 cr	8	No	No	No	No	Yes	Yes
Self-propelled tethered													
Reading & Bates Mark IV A Tractor (1969)	180	1.5[b]	2	?	8	0.5	18	No	Yes	No	No	Yes	Yes
Small, free, self-propelled													
Shinkai (1968) (Japan)	600	15.2	4	?	6	3.5 max, 1.5 cr	?	Yes	No	Yes	No	Yes	Yes
Ben Franklin (1968)	610	15.0	12	4500+	29	4 max, 1.1 cr	4320	Yes	No	No	No	Yes	Yes
Deep Quest (1967)	2430	12.0	4	3175	5	4 max, 2 cr	24	Yes	No	Yes	Yes	Yes	Yes
Aluminaut (1965)	2430	15.5	7	2720	4	3.8 max, 3 cr	36	Yes	No	Yes	No	Yes	Yes
Argyronete (1972) (France)	580	27.5	10	1815	2	6 surf, 4 div	?	Yes	Yes	?	No	Yes	Yes
DSRV-1 (1970)	1520	15.2	27	?	7	4.5 max, 3 cr	12	No	No	Yes	No	Yes	Yes
DSRV-2 (1972)	1520	15.2	27	?	7	4.5 max, 3 cr	12	No	No	Yes	No	Yes	Yes
Alvin (Redesigned 1972)	1800	6.7	2	545	?	6 max, 2 cr	10	Yes	Yes	Yes	No	Yes	Yes
Beaver Mk-IV (1968)	600	7.3	5	725+	11	5 max, 2 cr	12	Yes	Yes	Yes	No	Yes	Yes
Johnson Sea-Link (1971)	900	7.0	5	?	c	4	48	No	Yes	No	?	Yes	Yes

Deep Star 2000 (1969)	610	6.1	3	225	2	3 max, 0.5 cr	12	Yes	No	?	No	Yes	Yes
Deep Star 4000 (date?)	1220	5.5	3	225	2	3 max, 0.5 cr	12	No	No	Yes	Yes	Yes	Yes
DOWB (1967)	2000	5.2	3	545	c	2 max, 1.5 cr	6	No	?	Yes	No	Yes	Yes
Star III (1966)	610	7.5	2	455	5	4 max, 1 cr	20	Yes	No	Yes	No	Yes	Yes
Tiger Shark (1970) (Germany)	35	3.8	2	?	2	3.5	?	Yes	No	No	No	Yes	Yes
Turtle (1970)	2000	7.6	3	450	3	1.5 max, 1 cr	34	No	No	Yes	Yes	Yes	Yes
Nekton Alpha (date?)	305	4.6	2	90	23	3 max, 1.5 cr	5	Yes	No	Yes	No	Yes	Yes
PC3A (date?)	90	5.6	2	340	?	4 max, 3 cr	12	Yes	No	No	No	Yes	Yes
PC3B (date?)	180	6.7	2	160	14	4.5 max, 4 cr	24	Yes	No	No	No	Yes	Yes
Pisces I (date?) (Canada)	540	?	2	680	2	?	?	Yes	No	Yes	No	Yes	Yes
Pisces II (date?) (Canada)	900	?	3	?	2	?	6	Yes	No	Yes	No	Yes	Yes
Pisces IV (1970) (Canada)	610	?	3	?	?	?	18	Yes	Yes	Yes	No	Yes	Yes
Porpoise (1968)	45	3.2	1	109	c	3.5 max, 2 cr	6	Yes	No	No	No	Yes	Yes
Seacliff (1968)	2000	7.6	3	544	?	2.5 cr	?	?	No	No	?	Yes	Yes
Sea Graphics Snooper (1969)	305	3.5	2	68	12	2 max, 1 cr	5	Yes	No	No	No	No	No
Shelf Diver (1968)	245	7.0	4	820	26	4 max, 2 cr	?	Yes	Yes	Yes	No	Yes	Yes
Star II (1966)	365	5.0	2	225	6	3 max, 1 cr	?	Yes	No	Yes	No	Yes	Yes

[a] Condensed from Parrish *et al.* (1972) by permission of IPC Science and Technology Press.
[b] Spherical.
[c] 360° viewing.
[d] max = maximum; cr = cruising; surf = surfacing; div = diving.
[e] S = surface support.

Table IX-29

Some Characteristics of Underwater Manipulators[a]

Manufacturer	Underwater manipulator name or model number	Type	Assigned submersible	Design environmental conditions	Actuator power source	Manipulator-motion control devices	Terminal devices available
Programmed and Remote Systems	150w, 300w, 1000w through 7000w	General purpose; designed and built to specific order	Various	Max. depth 36,000 ft; Temp. range 50°–250°F; Pressure 18,000 psi; Immersion time and operating cycle, continuous	Oil-hydraulic, water hydraulic, electric	Switches, position controller	Parallel-jaw, hand, hook hand, tong hand, saw
Central Research Laboratories	CRL Model B—canal manipulator; CRL Model D—special underwater manipulator	Model B—standard, off-the-shelf, shallow-water, master-slave manipulator; Model D—standard; Both are general purpose	None	—	Manual, electric	Master-slave	Tong tips

General Electric	Manipulator arms for the ALUMINAUT submarine	One-of-a-kind general purpose	ALUMINAUT submarine Max. depth 15,000 ft Temp. range 34–120°F Pressure 7500 psi	Oil-hydraulic	Switches	
Litton Industries, Applied Science Div.	ASD Model-162 manipulator	General purpose, underwater environment, designed specifically for ALVIN vehicle, not a shelf item; manipulator now in use	ALVIN vehicle Max. depth 10,000 ft Temp. range 28–100°F Pressure 4400 psi Immersion time no limit Operating cycle continuous	Electric	Toggle switches	Three-fingered claw, clamshell bucket, hook
Westinghouse	OFRS Model No. 2. manipulator	General purpose	DEEPSTAR 4000 Max. depth 4000 ft	Oil-hydraulic	Micro-switches	Hydraulic-powered three-fingered claw

[a] From Johnson and Corliss (1971) by permission of John Wiley and Sons.

DSRV	rate-controlled, electrohydraulic
Pisces class	rate-controlled, electrohydraulic
Turtle and Sea Cliff	rate-controlled, electrohydraulic
Roughneck IV	rate-controlled, electrohydraulic

Characteristics of representative underwater manipulators are shown in Table IX-29.

e. Navigation

Because of the limits on electromagnetic propagation in water, the navigation system must be acoustic. Where the acoustic sources may be either transponders or beacons, acoustic navigation with respect to the bottom usually takes one of the three forms discussed below.

(i) Long-Baseline System. Utilizing multiple sources (transponders) on the bottom and a single receiver or interrogator (hydrophone) on the submersible, the long-baseline system determines position by measuring the arrival time of signals from the bottom-mounted sources. Three transponders, placed roughly in the shape of an equilateral triangle (Figure IX-30), make three-dimensional navigation possible. In deep water, the spacing can be as much as two miles; in shallow water, closer spacing is required.

(ii) Short-Baseline System. A short-baseline system utilizes a single transponder on the bottom and multiple hydrophones on the submersible. Position is determined by measuring the arrival time of the signal to each receiver. This system is useful for such tasks as tracking a towed vessel from the mother ship and maintaining position over a wellhead. Analysis shows that the short-baseline system is less accurate than the long-baseline system (Fain 1969, Cestone and St. George 1972).

(iii) Dead-Reckoning Navigation. For under-ice conditions, such as transporting oil from Arctic regions by submarine tanker, dead-reckoning navigation must be depended upon, since benchmarks and radio navigation fixes occur only at infrequent intervals. This type of navigation consists of the measurement of speed and direction. It has recently been greatly improved by the development of the Doppler sonar, which eliminates a major defect of other speed-measuring devices in that it is referenced to the bottom rather than to the surrounding water mass. The Doppler sonar (Figure IX-31) has one pair of beams fore and aft, and another to port and starboard. Forward and sidewise velocity can thus be measured, and the addition of a gyro or magnetic compass gives directional information. Navigation by Doppler sonar is a useful alternative to transponder navigation when accuracy requirements are moderate.

Bottom mapping can be accomplished with the help of side-scanning or forward-scanning high-resolution sonar; navigation depends on recognizable bottom features or on artificial benchmarks.

Visual navigation underwater is difficult for obvious reasons, but visual aids exist. Rolled-up, numbered plastic sheets which unroll on impact with the bottom have been used by TRIESTE. When dropped in a grid pattern, they have enabled TRIESTE to search bottom areas as large as 600 ft^2 (Cestone and St. George 1972).

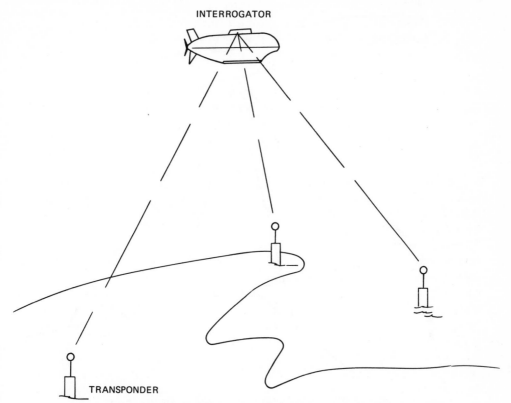

INTERROGATOR

TRANSPONDER

Figure IX-30. Acoustic transponder navigation (Cestone and St. George 1972).

5. *Unmanned Submersibles*

Although much of the interest of current underwater operation centers around manned submersibles, the unmanned submersible is still a valuable research and work tool. One of the greatest advantages of these systems versus manned systems is the cost, weight, and space savings achieved by eliminating human-safety and life-support systems.

Unmanned submersibles can be classified into three categories—towed, cable-controlled, and cableless vehicles. There is a wide range of vehicles in each category and only a few will be covered as a review of representative systems.

a. Towed Submersibles

An excellent example of the usefulness of an unmanned towed submersible is a vehicle developed by the University of California at Berkeley which is capable of operating at a constant depth with bottom contouring of 10–50 ft above the bottom. The vehicle is used to gather large amounts of data in a short time for the purpose of

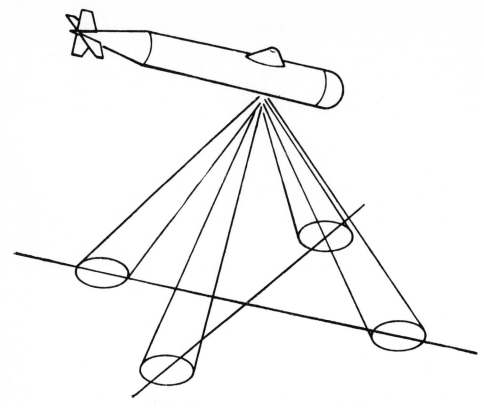

Figure IX-31. Doppler sonar (Cestone and St. George 1972).

measuring environmental parameters for pollution assessment in estuarine and coastal waters. It has an airplane configuration with fixed wings and movable rear elevators: The depth-control system, consisting of a pressure sensor and an echo sounder, permits operation either at a constant depth or at a constant distance from the bottom (Conti *et al.* 1971).

Another type of towed vehicle, being developed by the Continental Oil Company, is a submerged barge designed to transport crude oil in Arctic regions, to be towed at a controlled depth by an ice breaker. The adoption of this concept is awaiting completion and testing of a one-tenth scale model (Sudbury 1972).

b. Cable-Controlled Submersibles

Unmanned submersibles of the cable-controlled type possess the ability to move under the remote control of a surface operator, and the ability to carry out specific tasks via manipulators. In addition, those described here carry observational and sensing equipment such as echo-sounders, magnetometers, sonar, and cameras. Generally they are of open-frame design and operate at low speeds. They are not suitable for surveying extensive areas of the ocean bottom (Fenning 1972a).

(i) Military Worksubs. The Navy's Cable-operated Unmanned Recovery Vehicle (CURV) is used primarily to recover test torpedoes. Propulsion is supplied by three

10-hp motors powered by umbilical line from the surface ship. Two are used for horizontal movement and one for vertical movement. Passive sonar is used to guide CURV to within 300 ft of the target and active sonar is then used for determining the approach. Two closed-circuit underwater television cameras provide close-range vision to the operator on board the surface ship. One of the most spectacular achievements of the CURV was recovery of an H-bomb off the coast of Spain at a depth in excess of 2500 ft.

CURV II and CURV III resulted from design improvements and offer greater depth capability. CURV III has an operational design for 7000 ft of water (Fenning 1972a).

The Remote Unmanned Work System (RUWS) is a project of the Ocean Technology Department of the Naval Undersea Center, San Diego, California. It represents an updated version of the CURV with an operational depth of 20,000 ft. It is equipped with an advanced search sonar and an advanced manipulator. It is remotely controlled from the surface by a human operator via a head-coupled television whereby the manipulator duplicates the operator's movements. The manipulator is a clawlike device which can also operate power tools such as a cable cutter and a lift device (Booda 1973).

(ii) Industrial Worksubs. Designed specifically for work tasks such as drilling, inspection of installations, and operation of valves, the Sperry Worksub consists of a single large flotation sphere suspended under a symmetric triangular frame. It weighs 700 lb and has a forward speed of 3 knots and an operating depth of 2000 ft. It is equipped with an underwater television and a manipulator which is controlled by two operators on the surface ship who pilot the vehicle and control the manipulator via sonar information and the television system (Fenning 1972b).

Shell MOBOT is another unit designed for underwater engineering tasks. A drive shaft, monitored on television, can operate a wrench, cutter head, or other tools. Horizontal movement is controlled by two thruster units, but vertical movement is controlled by a winch on the oil rig for which MOBOT is designed. Variability in design of these submersibles is dictated by the specific tasks to be done.

There is room for further development of small, free-swimming, unmanned, surface-controlled submersibles which are hydraulically powered, wire-controlled, equipped with force-reflecting capability, and observable from the surface. Such vehicles could perform functions which cannot now be performed by the larger manned units using the simple rate-controlled manipulators. With the use of the force-feedback manipulator and miniaturization in size and structure, units of this type could evolve to become the real deep divers (Wood 1971).

c. Cableless Submersibles

The cableless, unmanned submersible is designed for oceanographic data collection or bottom observation along a preprogrammed course. Generally these submersibles are in the early experimental stage and are not yet economically competitive with surface-controlled vehicles for most underwater tasks. They would probably be most useful in collecting *midwater* oceanographic data, and in carrying out seabed search tasks in deep water (Fenning 1972*b*). Described below are several unmanned submersibles developed recently.

1. *SPURV*. The Self-Propelled Underwater Research Vehicle (SPURV) developed by the Applied Physics Laboratory, University of Washington, has an operating depth of 150–9000 ft and a range of 15,000 ft.

2. *Sea Drone*. Designed by Oceanographic Industries, Inc., for oceanographic payloads, the Sea Drone can operate to a depth of 20,000 ft at a speed of 7 knots for 6 hr.

3. *UARS*. The Unmanned Arctic Research Submersible (UARS) system was developed by the University of Washington as part of an ARPA–ONR sponsored program. An acoustic tracking system is used to keep track of the UARS, which has an operating depth of 1500 ft, is powered by silver–zinc batteries, and maintains a speed of 3.7 knots for 10 hr. The UARS is capable of a wide variety of under-ice operations, which include acoustical measurements, bathymetric surveying, and basic oceanographic research (Francois 1973).

6. *Buoyancy Devices*

In recovery or manipulation of objects underwater, a latent buoyancy device is an important accessory of a manned or unmanned work submersible, or of any kind of underwater work system. These devices usually consist of an inflatable bag and an actuating mechanism, contained in a compact, neutrally buoyant package. A chemically operated system can release nine times as much gas as a commercial gas cylinder of the same size. The three most commonly used types of devices may be illustrated by the three models from Ocean Recovery Systems, Inc. First, the more sophisticated, chemically operated type, such as the Model 100/1000A, is capable of 1000 lb of lift at 1000 ft, and can be equipped with accessories such as gripping tools and location aids that function upon surfacing. Second, a simpler chemically operated system, the modular salvage unit, weighs 2 lb in air and has a lift capability of 200 lb at 50 ft. It is used singly or in multiples, when economy is paramount. Third, Model CG-400/350A is operated by a compressed-gas cylinder, has a lift capability of 400 lb at 350 ft, and is actuated by hydrostatic pressure. Actuation can be preset for depths of from 30 to 350 ft. This device is designed primarily for attachment behind a towed object. In search and recovery, a buoyancy device is particularly important to a submersible, since lifting even a light load is a hazardous task for a submersible to attempt unaided. Buoyancy devices may be held, attached, and actuated by manipulator arms, although divers are also frequently used for this purpose. Table IX-30 shows the applications of these three types of devices (Baccaglini and Irgon 1972).

7. *Hyperbaric Facilities*

In order to study the effects of depth and increased pressure on man, hyperbaric chambers have been designed and developed. For the purpose of this handbook, a *hyperbaric chamber* is defined as a land-based, high-pressure chamber which will

Table IX-30

Offshore Applications of Submersibles with Buoyancy Devices[a]

Programmed object recovery	Model 1000/1000A
of a static load	
of a variable load	Model CG 400/350A
Load support functions	Modular salvage unit
to relieve excess weight or strain	
to lift, horizontally transport, and position objects	
to handle unwieldy loads	Model 1000/1000A
Emergency buoyancy	Model 1000/1000A
for submersibles	
for submerged equipment	Modular salvage unit and Model CG 400/350A
for floating vessels, barges, and platforms	Modular salvage unit
Marine salvage tool	Modular salvage unit
with inflatable bags for internal deballasting	
with inflatable or rigid pontoons for external lift	
with grappling devices to share the load	
Some auxiliary uses	Modular salvage unit
breakout force for embedded objects	
portable gas supply for airlift dredges	
portable power for pneumatic diver tools	
pressurized inert atmosphere	
messenger buoy and location aid	
portable heat for diver or habitat	
cathodic protection of submerged equipment	

[a] Adapted from Baccaglini and Irgon (1972) by permission of the Marine Technology Society, Washington, D. C.

accommodate men at varying pressures to simulate depths inside the chamber, while the external pressure remains at 1 atm. One or more of these chambers constitute land-based research centers in which man's reactions can be studied and carefully monitored by personnel who are not themselves exposed.

a. Design Factors and Certifications

A basic consideration in the design of any hyperbaric chamber is the fact that the chamber is subjected to the stress of internal pressure only, as opposed to external only, as seen in 1 atm submarines, or to both internal and external, as seen in the submersible decompression chamber. The two most commonly used shapes have been cylindrical and spherical, with various combinations of the two (Penzias and Goodman 1973). For example, the Navy Experimental Diving Unit employs two cylinders at right angles to each other.

In the design of any high-pressure chamber, safety factors are of the utmost importance; they are the primary criteria for certification of a chamber, whether it is an existing chamber or one being constructed (Richards 1973). (The U. S. Navy initiated its certification program in February 1968; the U. S. Coast Guard began certification procedures in 1970.)

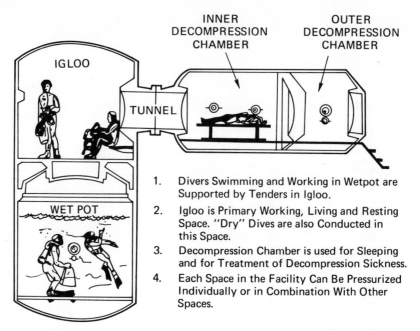

1. Divers Swimming and Working in Wetpot are Supported by Tenders in Igloo.
2. Igloo is Primary Working, Living and Resting Space. "Dry" Dives are also Conducted in this Space.
3. Decompression Chamber is used for Sleeping and for Treatment of Decompression Sickness.
4. Each Space in the Facility Can Be Pressurized Individually or in Combination With Other Spaces.

Figure IX-32. Typical experimental hyperbaric facility (Reimers and Hansen 1972).

b. Examples of Hyperbaric Chambers

(i) Navy Experimental Diving Unit. The facilities include two high-pressure chamber systems. Each system is composed of three basic units: a decompression chamber, connected to the outside of the unit by a safety chamber; an igloo, in which the diver prepares to descend; and a diving tank or "wet pot," which is 10 ft in height and approximately $9\frac{1}{2}$ ft in diameter, for simulation of deep-water excursions. The maximum working pressure of the complex is equivalent to 1000 ft of sea water (445 lb). A monitoring system is employed to measure total pressure, O_2 partial pressure, CO_2 partial pressure, temperature, and humidity. For dives deeper than 190 ft or for long-duration dives, a mixture of oxygen and helium is employed, with the oxygen content varying according to depth to prevent oxygen toxicity or anoxia. Figure IX-32 shows an artist's illustration of the facility (Reimers and Hansen 1972).

(ii) Navy Ocean-Pressure Simulation Facility. The mission of the facility is "simulation of ocean environments to a depth of 2250 ft to conduct research, develop tests, and evaluate systems involving man and/or machine with emphasis placed on the man–machine interface" (Mossbacher 1973). The high-pressure system consists of five dry chambers and one wet chamber, which is ellipsoidal, 15 ft in diameter, 15 ft high, and 30 ft long. It was designed to provide the capability to test most two- and three-man submersibles. The facility is located at the Naval Coastal Systems Laboratory at Panama City, Florida. A diagrammatic sketch of the high-pressure system is shown in Figure IX-33 and the specification are summarized in Table IX-31. Arrangement of the chambers and interconnecting lock permit use in several combinations. The dry chambers are designed as living and/or working space (Montgomery 1969).

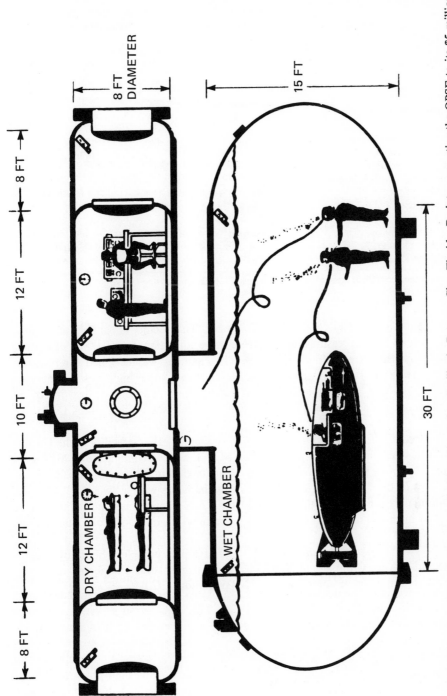

Figure IX-33. Diagram of the Navy's Ocean-Pressure Simulation Facility in Panama City, Florida. By interconnecting the OPSF to its $5 million hybrid computer complex, the Navy can simulate complete missions in real environments under laboratory conditions. [From Montgomery (1969) by permission of *Undersea Technology*.]

Table IX-31
Summary of Specifications for Navy Ocean-Pressure
Simulation Facility[a]

	Wet chamber	Dry chambers Main (2)	Locks (2)	Center section Lock	Trunk
Dimensions					
Diameter (ID), ft	15	8	8	8	8
Length (internal), ft	47	12	8	10	6½
Diameter: Access doors, in.	48	42	42	42 (2)	48
Hatches, ft	15	—	—	48 in.	—
Volume (internal), ft³	7000	600	300	520	330
Environmental control ranges					
Pressure, psig	0–1000		0–1000	0–1000	
Temperature, °F	29–110		29–110	29–110	
Salinity	As required		—	—	
Turbidity	As required		—	—	
Relative humidity, %	—		10–100	10–100	
Atmospheric gas control[b]	Yes		Yes	Yes	

[a] From Montgomery (1969) by permission of Compass Publications, Arlington, Virginia.
[b] Mixtures of oxygen, helium, and nitrogen available.

(iii) Duke University's Hyperbaric Chamber. This facility is capable of employing any desired gas at pressures ranging from those equivalent to 150,000 ft of altitude to 1000 ft of sea water (Linderoth 1973). The chamber was designed for research, including research on acoustics speech modifications, the high-pressure nervous syndrome, and CO_2 effects related to inert-gas narcosis. A schematic drawing of the floor plan and a side view of the diving chambers are shown in Figures IX-34 and IX-35.

(iv) Institute for Environmental Medicine. The hyperbaric facility at the University of Pennsylvania consists of six chambers in an L-shaped configuration. Maximum pressure capability is 90 psi with an equivalent maximum altitude of 150,000 ft. At the bottom of the "L" are three high-pressure chambers with a "wet pot" beneath. Maximum pressure is equivalent to 2000 fsw. Temperature and humidity can be controlled in the chambers and almost any combination of respiratory gases can be used (Covey 1971).

(v) Taylor Hyperbaric Complex. A high-pressure complex built by Taylor Diving and Salvage Company of New Orleans consists of three chambers with a "wet pot." Maximum pressure can simulate 2200 fsw (Anon. 1970).

c. Current Research Facilities

Combined facilities for clinical medicine, biomedical research, and diving research are listed below. Those with "wet pot" facilities are marked with an asterisk (Penzias and Goodman 1973):

* Duke University, Durham, N. C.
  Kantonsspital, Zurich, Switzerland

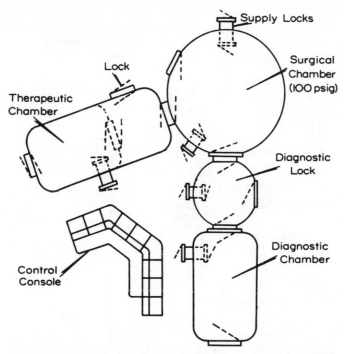

Figure IX-34. Plan view of the main chamber floor showing the arrangement of five of the chambers in an interconnected Vee formation. Scale, 1 in. = 10 ft. [From Linderoth (1973) by permission of the Marine Technology Society, Washington, D. C.]

* Karolinska Institut, Aviation & Nautical Medical Department
  Ohio State University, Columbus, Ohio
* State University of New York at Buffalo, Buffalo, N. Y.
* University of Pennsylvania, Philadelphia, Pa.

Current facilities for development of equipment and procedures as well as for diving research and training can be found at the following locations (Penzias and Goodman 1973):

* Compagnie Maritime D'Expertise, Hyperbaric Research Center, Marseille, France
* Deep Trials Unit—Royal Naval Physiological Laboratory, Alverstoke, U. K.
* Defense Research Establishment, Toronto, Ontario, Canada
* Diving Medical and Technical Centers, Royal Netherlands Navy, Den Helder, The Netherlands
  Diving Medical Research Laboratory, Japanese Maritime Self-Defense Force, Yokosuka, Japan
* Experimental Diving Unit, Panama City, Fla.
* Groupe d'Etudes et de Recherches Sous-Marines, Toulon, France
* Institute of Aviation Medicine (of D.V.L.R.) Bad Godesberg, West Germany

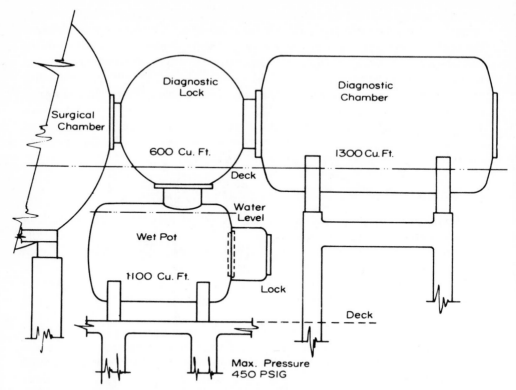

Figure IX-35. Side view of the diving chamber including the wet pot below the smaller sphere. Scale, $\frac{1}{4}$ in. = 1.0 ft. [From Linderoth (1973) by permission of the Marine Technology Society, Washington, D. C.]

* Institute of Navigation Medicine, Kiel-Kronshagen
  J. and J. Marine Diving Company, Pasadena, Texas
  Life Support Systems Division of U. S. Divers, Santa Ana, California
  Naval Medical Research Institute, Bethesda, Maryland
* Naval Research and Development Laboratory (formerly Mine Defense Laboratory), Panama City, Florida
* Office Francais de Recherches Sous Marines, Marseille, France
* Royal Norwegian Navy, Haakonsvern, Norway
* Rushcutter, H.M.A.S., Sydney, Australia
* Submarine Medical Research Laboratory, New London, Connecticut
* Taylor Diving and Salvage Company, New Orleans, Louisiana
* Westinghouse Ocean Research and Engineering Center, Annapolis, Maryland

8. *Environmental Control*

Control and maintenance of a livable environment are of the utmost importance in all the systems discussed above. Internal environments of deep-submergence research vehicles and fleet-type or nuclear submarines are controlled at 1 atm to allow operation at any desirable depth with the capability of rapid descent or ascent without the need for pressurization or decompression. Deep-submergence saturation-diving habitats and deep-diving systems control internal environments at pressures equivalent to the ambient underwater pressure. This allows divers to exit into the surrounding area for work tasks or research. In either case strict environmental conditioning and control systems are required. These systems must provide for maintaining the oxygen partial pressure and for removal of carbon dioxide and other contaminants; they must control temperature, humidity, and circulation and provide extravehicular support to divers if they are to make excursions from the unit.

a. Maintenance of Desired Partial Pressure of Oxygen

Oxygen is necessary to maintain the metabolic requirements of man. At sea level or normal atmospheric pressure, air consists of about 21% oxygen or 0.2 atm partial pressure of oxygen. Anoxia (low oxygen supply to tissues) occurs when the oxygen content falls below about 16% at normal atmospheric conditions. Pulmonary oxygen toxicity (see Chapter IV) can occur if the oxygen partial pressure is increased to 0.6 atm. Therefore, strict control of the oxygen partial pressure must be maintained in any diving system.

Determining the desirable partial pressure of oxygen depends to some extent on the type of mission. Nuclear submarines and other vehicles that maintain a 1-atm environment would maintain the normal 0.2 atm oxygen partial pressure. A high oxygen partial pressure around 1.2 atm is desirable for deep nonsaturation dives because it reduces the decompression time by decreasing the inert-gas partial pressure (Parker and Burt 1970). For normal diving operations and long-term saturation dives, pressures are generally maintained between 0.21 atm (160 mm Hg) and 0.33 atm (250 mm Hg). Oxygen partial pressure is maintained at the proper level as depth increases by increasing the amount of diluent gas. Nitrogen is generally used down to 150–200 ft, and helium or a helium–nitrogen mixture is used for greater depths. Figure IX-36 shows the decrease in the percentage of oxygen and the increase in the percentage of diluent gas, in this case helium, as depth, shown as atmospheric pressure, increases (Gussman *et al.* 1971).

For habitats and other surface-supported vehicles, the simplest means of maintaining oxygen partial pressure is by an air-supply system and venting of the vehicle. This type of system was used on the TEKTITE project, where compressed air was continually supplied by low-pressure air compressors located on the surface-support center. It was supplied to the habitat via an umbilical to maintain an oxygen partial pressure between 151 and 165 mm Hg. The flow rate was manually controlled, based on measured P_{O_2} (partial pressure of oxygen) levels and was approximately 16–24 SCF/hr. A continual outflow of the habitat atmosphere was provided through vents to the sea. This continuous inflow and outflow of air maintained the total atmospheric pressure of the habitat in equilibrium with the water-depth pressure. A Servomex O_2 sensor was used in the habitat to monitor the P_{O_2} and a Servomex A0150 O_2 analyzer was used on the surface. As backup equipment, an MSA O_2 meter was used in the habitat and a Beckman F3 O_2 analyzer was used on the surface. A schematic diagram

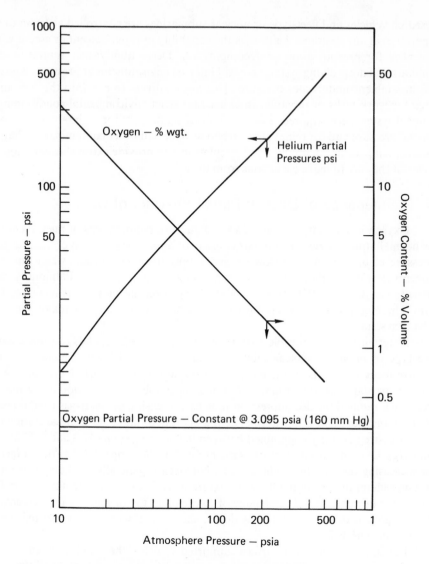

Figure IX-36. Oxygen–helium atmosphere pressure vs. composition. P_{O_2} = 160
mm Hg; T = 20°C (Gussman *et al.* 1971).

of the habitat gas-supply systems used on TEKTITE I is shown in Figure IX-37 (Pauli
and Cole 1970).

For deeper operations where surface support is not practical and for free, self-
propelled submersibles and vehicles, the atmosphere must be controlled using stored
gas. Two modes are in general use today—high-pressure gas cylinders and cryogenic-
gas storage. Adequate sensors must be used to monitor the P_{O_2}. The sensor should be
able to cover the range of P_{O_2} from 0.1 to 2.0 atm with 5% accuracy and a response
time of 10 sec (Reimers 1972). More than one sensor should be used in order to

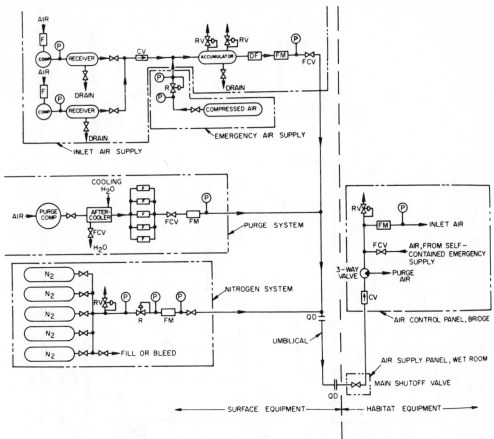

Figure IX-37. Habitat gas supply systems. F = filter, P = pressure gauge, CV = check valve, RV = relief valve, FCV = flow control valve, R = regulator, DF = desiccant filter, FM = flowmeter, QD = quick disconnect. (Pauli and Cole 1970.)

provide both a check system and a redundancy backup system in case of failure. Both galvanic and polarographic sensors are small and inexpensive (Parker and Burt 1970). A simple method of controlling a desired oxygen level is with a flow-limited solenoid valve. When the oxygen level falls below the set point by a fixed amount, the valve opens; when the level rises above the set point by a fixed amount, the valve closes. Audible and visible alarms, activated when the sensor reading deviates from the control point, should be incorporated in the system.

A schematic diagram of a self-contained oxygen supply system using high-pressure gas cylinders is shown in Figure IX-38, and one using cryogenic gas storage is shown in Figure IX-39. A third method of supplying oxygen is impractical except on large, nuclear-powered submarines because the power requirement is too high. This method utilizes water electrolysis to generate oxygen (Penzias and Goodman 1973).

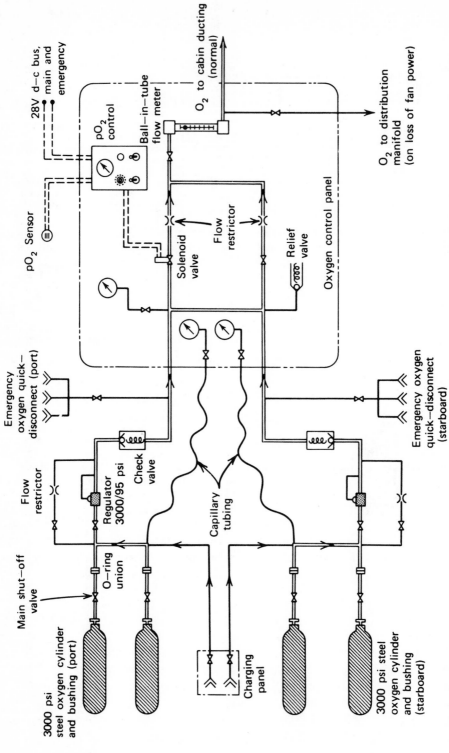

Figure IX-38. Metabolic oxygen system schematic for small submersible. [From Penzias and Goodman (1973) by permission of John Wiley and Sons, New York.]

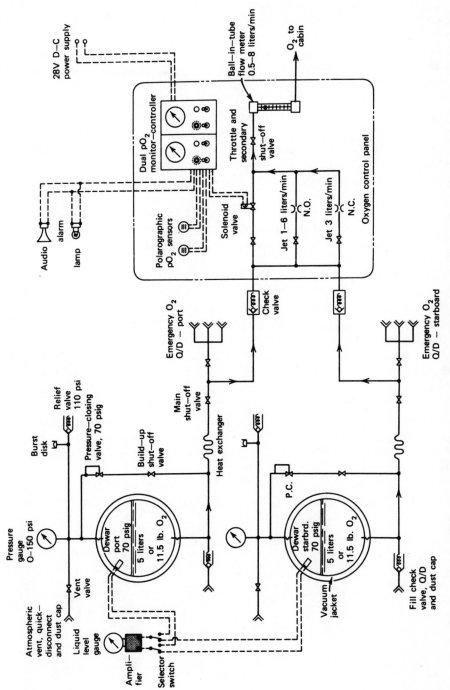

Figure IX-39. Cryogenic oxygen system. [From Penzias and Goodman (1973) by permission of John Wiley and Sons, New York.]

b. Removal of Carbon Dioxide and Trace Contaminants

Removal of CO_2 is one of the prime requirements of most environmental control systems. Rebreathing of expired air or gas is common in decompression chambers, submersibles, habitats, underwater vessels, and submarines. Therefore, exhaled carbon dioxide must be removed to avoid toxic effects. Carbon dioxide's physiological action is a function of its partial pressure, and toxicity depends on duration of exposure. The desirable upper limit is usually set at 0.5% by volume of surface equivalent, or 3.8 mm Hg. This decreases to approximately 0.014% at a depth of 1000 ft (Reimers 1972). A good control system is required to maintain the P_{CO_2} at this low level.

The usual method of removing CO_2 is by passing the air or gas through a canister containing a chemical absorber, the most common being Baralyme, Sodasorb (soda lime), and lithium hydroxide. Other methods and techniques have been investigated, but will not be covered in this section. These include the monoethanolamine (MEA) scrubbing system used on submarines, the use of molecular sieves, freezing out of CO_2, etc. Table IX-32 shows the characteristics of the three common absorbents.

Many factors influence the efficiency of a chemical absorption system. One of the more important of these is temperature. The rate of carbon dioxide absorption is considerably lower at 40°F than at 70°F. This means that, at 40°F, the useful life of a canister designed for 4 hr of use at 70°F may be reduced to about 2 hr (Reimers 1972). Also, maximum absorption can be obtained only when the relative humidity is above 70%. Chemical absorbers are deactivated when they become wet and lithium hydroxide and soda lime both form caustic solutions with water. Physical properties of the chemical absorbers are also factors, such as granular size, porosity, and moisture content. The properties of the absorbent bed also contributed to its efficiency. These include cross-sectional area, bed depth and volume, flow rate, and packing density with resulting pressure drop across the bed, and total pressure under which the bed operates. There is a great variety of absorbent bed configurations. One of the most important criteria is that a uniform distribution of flow prevents channelling of the gas and, therefore, inefficient use of all of the chemical absorbent. Flow diagrams of four typical canister configurations are shown in Figure IX-40. Chamber air is either forced through or drawn through by a blower or fan.

One of the most important considerations in a carbon dioxide removal system is to provide for an adequate monitoring system. This becomes extremely important at

Table IX-32

Characteristics of Three Carbon Dioxide Absorbents[a]

Absorbent	Density, ft³	Theoretical efficiency, lb CO₂/lb		Water generated, lb/lb CO₂	Theoretical heat of absorption, BTU/lb CO₂
		Weight basis	Volume basis		
Lithium hydroxide	28	0.92	25.6	0.41	875[b]
Baralyme	65.4	0.39	25.5	0.41	670[c]
Sodasorb	55.4	0.49	27.2	0.41	670[c]

[a] From Lower (1970) by permission of Marine Technology Society, Washington, D. C.
[b] Based on gaseous H_2O generation.
[c] Based on calcium hydroxide reaction only.

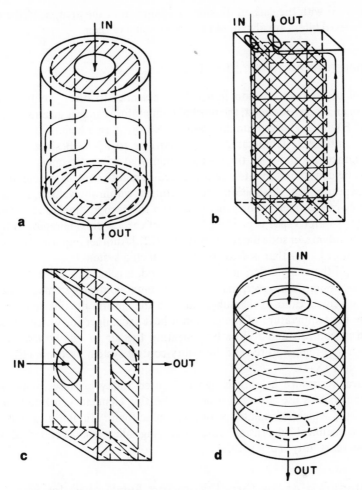

Figure IX-40. Channelling surface areas for four types of canisters.
(a) Annular flow (b) flat, transverse flow (c) flat, longitudinal flow
(d) solid cylinder. (Lower, 1970.)

increased depths because of the very small percentage of carbon dioxide that is permissible to avoid toxic effects. An adequate sensor should be capable of detecting carbon dioxide in the range of 0–7.6 mm Hg (0–1.0 surface equivalent by volume) with an accuracy of 5% and a 30-sec response time (Reimers 1972). To date there is not an adequate analyzer to meet all the needs of every diving system. Some systems merely employ a method to determine whether the CO_2 content has risen above a certain maximum value. TEKTITE II used a Lira CO_2 sensor in the habitat and a Varian gas chromatograph on the surface. As backups, a detector tube was used in the habitat and a Beckman 1R215 infrared sensor (Miller *et al.* 1971).

Except in the case of long-duration dives, as in nuclear submarines, the trace contaminants result chiefly from construction materials, lubricants, etc. These are usually removed with the carbon dioxide or vented from the area, as in the case of a habitat. Charcoal filters are also used to remove some of them.

c. Temperature and Humidity Control

Control of temperature and humidity is essential to diver comfort and work efficiency. In an air environment, the normal temperature range is generally considered to be 68–80°F with a relative humidity range of 40–70%. In a helium–oxygen environment, the comfortable temperature range narrows to 85–89°F since helium has a thermal conductivity about six times higher than that of air (Reimers 1972). Factors which must be considered in a temperature and humidity control system are ambient water temperature, heat dissipation and generation within the vehicle, humidity gain from the crew, and water vapor generated from the carbon dioxide removal system. [About 1 ft^3 of water vapor is produced per ft^3 of carbon dioxide removed by lithium hydroxide, Baralyme, or soda lime (Canty *et al.* 1972).] Another important consideration is the amount of power that is available to run such a system.

Because of the many different factors involved, it is hard to make generalizations about specific designs. However, many environmental-control systems circulate gas or air from the vehicle cabin or chamber through the carbon dioxide removal system, then reduce the temperature of the gas via a heat exchanger for water condensation and humidity control, and follow by reheating the air or gas to the appropriate temperature. The temperature- and humidity-control system used on TEKTITE I is shown in Figure IX-41.

Electrical resistance heating is generally used today because of its simplicity. Smaller submersibles use electrical resistance heating elements cemented on the outside of the hull, which is insulated with a syntactic foam. Humidity can also be controlled by a chemical adsorption unit employing silica gel, activated alumina, or molecular sieves (Canty *et al.* 1972).

d. Maintenance of Air Circulation and Extravehicular Support

Although they are important factors to be considered in an environmental-control system, maintenance of air circulation and extravehicular support are so dependent upon the overall design and function of a particular underwater vehicle or habitat that they are not described here. The most important factors to consider are the additional power that will be consumed and the additional breathing gas necessary to support excursion divers.

9. *Instrumentation*

In the last two decades the military, economic, and scientific demands to know more about the oceans has reached a high pitch. Electronics is playing a very large role in carrying man into the ocean, but electronic instrument designers need to appreciate the problems of working underwater.

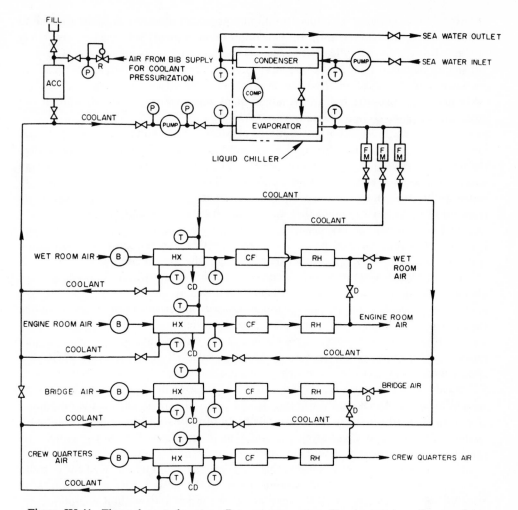

Figure IX-41. Thermal control system. P = pressure gauge, T = temperature, R = regulator, ACC = Accumulator, FM = flowmeter, B = blower, Hx = heat exchanger, CF = charcoal filter, RH = reheater, D = diameter, CD = condensate drain. (Pauli and Cole 1970.)

Since volume considerations are extremely important, miniaturization of instruments is vital. Every kilogram inside the hull means an overall increase in 2–3 kg because of the need for additional external buoyancy material. Instrument volume is important to the efficient use of space in a submersible, a factor critical to mission effectiveness. Instrument power consumption should be as low as possible since demands of propulsion leave very little for other purposes. Waste heat is not so often a problem as in other environments because of the excellent heat sink properties of water.

Some limitations of weight and volume can be negated by placing instrumentation in pressure-tight or pressure-balanced compartments external to the pressure hull;

however, this technique introduces additional problems of electrical cable glands and difficult access for maintenance. Instrumentation for submersibles must be absolutely reliable because of the difficulty of repair or maintenance tasks underwater. Carrying of spares is limited by the lack of space. The close proximity of instruments leads to acoustic and electrical interference between circuits. Electrical plugs and sockets external to the pressure hull can result in many pieces of equipment flooding if not carefully designed (Haigh 1971).

Some of the most important methods of ocean investigation are being developed in the field of acoustic technology. The extraordinary role of acoustics for oceanology is based on the unique properties of acoustic waves in water; no other form of energy can propagate in water to such long distances. This methodology is very effective in solving the problems of acoustic vision, telemetering, communication, and remote control of underwater systems that arise in underwater work and/or research.

Another significant advancement is the single-wire communication cable (especially the electrically isolated wire). When an acoustic telemetering or communication link is too expensive or too complicated in design, the thin isolated wire (with multichannel electronic system) should be used. Multiwire cables will find no place in the coming technology for reasons of reliability and weight.

New noncorrosive materials are the underwater materials of the future, including alloys with an aluminum base and plastics such as acrylic.

A review of the developing trends in instrument modification for physical and main hydrochemical measurements is shown in Table IX-33 (Mikhaltsev 1971). The last column characterizes the information processing techniques for research work. There are several reasons for the importance of these techniques. First, the observational methods of making measurements and physical experiments in the ocean and the high demand for the quantitative reliability of the data allows too little time for determining the final result from the collected information. Results from measurements obtained while an experiment proceeds make immediate corrections possible. This can sometimes show a considerable economy, by avoiding the need for a return visit to repeat the experiment. Second, there is the need for precise quantitative analysis of the characteristics of ocean processes, impossible without comprehensive computer facilities onboard. The need for the best and most explicit form for the final results of the analysis demands a prompt assessment by the scientist. Third, the economy of automatic information processing and the elimination of human labor from routine operations must be considered. One important example is reduction of the number of auxiliary personnel required on research vessels on long oceanic voyages.

The present state and trends in the development of techniques used in biological geological, and chemical oceanography are shown in Table IX-34. "The active role of the scientist-operator in gathering material, fulfilled by remote devices, will characterize the methods of work in this group of ocean sciences in the 1970s" (Mikhaltsev 1971). A topical outline of certain aspects of the hyperbaric chamber environment as related to instrumentation is given in Table IX-35. Not indicated, but of particular significance, are the many interactions involved. For example, minimum decompression time calls for breathing high-oxygen concentrations and this increases the fire hazard (Hamilton *et al.* 1970).

Table IX-33

Technical Means for Ocean Research in Hydrophysics [a]

	Methods widely used in the 1960's	Possible methods for use in the 1970's	Possible 1970's methods for information processing and analysis
Currents	Self-recording current meters from moored surface buoys; pinger-buoys of neutral buoyancy	Autonomous moored buoys (AMB) with radiotelemetering and self-contained memory; current–temperature–electroconductivity-meters, local optical and acoustical characteristics, acoustical noise and meteorology in the air–sea interaction layer	—
Temperature	Reversing mercury thermometers; thermographs with stylo- or photo-recording	Towed systems (TS) with temperature, electroconductivity, and other measuring gauges	Input of all the data in the information-memory and processing computer; processing in accordance with given programs and printing results in the form of drawings, charts, plans, etc.
Electroconductivity (salinity); fluctuations in turbulence, temperature, electroconductivity, sound velocity, and flow speed	Bathymetric samples with laboratory analysis	Sondes (S); temperature, salinity, depth, optical, and local acoustical characteristics; thermoanemometers, acoustical meters of velocity and flow; galvanic and electromagnetic conductivity meters (on AMB, TS, and S)	—
Fluctuations in atmospheric layer (see surface interaction zone), humidity, wind speed, and heat flow	—	Conductivity thermometers, thermoanemometers, optical meters of wind fluctuations	—
Surface state: waves and level	Quasistatic pressure gauges; string gauges	Acceleration meters, strain gauges on buoys; radar and stereophoto-grammeters	—
Less common characteristics: propagation velocity, damping and scattering for electromagnetic (radio and optic) and acoustical waves, radiation characteristics	—	Special instrumentation and probes for mass measurements on AMB and from ships	Computer processing to special programs

[a] From Mikhaltsev (1971) by permission of IPC Science and Technology Press.

Table IX-34

Technical Means for Ocean Research in Biology, Geology, and Chemistry[a]

Type of research	Methods widely used in the 1960's	Possible methods for use in the 1970's	Possible 1970's methods for information processing and analysis
Biology			
Ecology, biogeography, systematica: mass fauna collection (pelagic, bottom)	Plankton and ichtyo-nets for vertical movement; pelagic trawls, bottom trawls	Trawls and nets for multihorizontal catch with thermo- and bathystats; photo and introscopic tools for viewing in the waters; sonar with object-quantity-defining systems; sound–vision systems; bioluminescence counters; volume and surface	Special counting systems (television microscopes, counters, filter particle counters, microphotometers, etc.)
Productivity experiments	—	counters; isotopic counters; IR spectrophotometry; standard instrumentation for chemical and phys-	Automation of analysis
Biochemistry, biophysics, biotechnics (bionics)	—	ical experiments and for methods of treating live samples	—
Geology			
Bottom surface	Echosounders, deepwater photography	Narrow-beam echosounder with digital indicator and digit-code output; Towed deepwater side-bottom-sonar and photo-television systems	See Table IX-33
Surface layer sediments, rock material	Samplers and specimen corers, dredges, grab samplers; bottles for suspension gathering	Echosounders or determination of mechanical characteristics of surface layers, with adequate technique of sample gathering (piston- and vibrocorers, rotation dredges, drills, etc.) Deep drilling equipment; manned and unmanned submersibles with automatic manipulators; deep-water (towed or sonde) filtering techniques	Automation of optimum methods of physical-chemical analysis: plasma, X-ray structure, mass-spectrometer, radiation-energy, and other types of analysis

			See Table IX-33
Deep crust layers	Geophysical surveys using ship seismo-profiling systems with radio-buoys, towed hydrophone lines, towed sparkers, boomers, etc., and hydrophone lines; ship-towed proton, ferrosonde, or quantum magnetometer; ship gravity-meter; heat flow measurements with thermogradientometer	Towed systems of the full geophysical survey complex with digital output; Automatic self return-to-surface systems for complex measuring of: seismic waves, heat flow, near-bottom currents, magnetic variations, with built-in digital recording system	
Chemistry Geochemistry	Same as geology of surface layer and sediments		
Hydrochemistry	Bottles	Microprobes with automatic synchronous line analysis of all important chemical elements and structures; technical and chemical methods of monitoring all chemical elements of sea water	Standardizing of all chemical analysis of water for automatic printing and storage of information

[a] From Mikhaltsev (1971) by permission of IPC Science and Technology Press. The above information concerns only the main aspects of ocean research.

Table IX-35

Constraints of Instrumentation in Hyperbaric Chambers[a]

Condition	Remarks
Pressure	Protection is required for some equipment, or modification
	Special penetrations needed for electrical wiring, etc.
	Special techniques are required for external gas sampling, and gas analysis is complicated
	Pass-through "lunch" locks can be used for handling samples, etc.
	Subject, investigators, samples, and equipment are subject to "decompression sickness"
	Inadvertent loss of pressure or failure of lines or vessels may be catastrophic; consequences of failure should be a planning factor
	Wall tension in a tube, at a given pressure, is proportional to radius, so small tubes can safely hold high pressures; this permits topological manipulations of the actual pressure domain
Isolation	To contain an experimental high-pressure environment involves a wall of up to 5 in. of steel; huge, heavy hatches; only a few small thick windows; not nearly enough space and, inevitably, too few penetrations
	Information, things, and energy must be transmitted through the chamber wall; if pressure is high, transmission of power and mechanical movements may require special techniques; one is by means of a sealless magnetic drive
	A communication system is required, even if it is only banging with a leather (sparkless) mallet, or shouting
	When communication is impossible due to helium speech, the subjects in the chamber are in some ways as isolated as if in an orbiting spacecraft
	Some things will not fit into the lock or even into the chamber
	Mechanical isolation and decompression obligation may complicate rescue of sick and injured persons inside
Atmosphere	Oxygen causes a special fire hazard
	Helium may distort speech
	Helium permeates equipment
	Helium conducts heat readily, disturbing some equipment and measurements [e.g., tympanic (ear) temperatures are not usable in helium]; fans designed for air may not provide sufficient flow in helium, causing overheating
	Helium as a background gas may upset gas analyzers (e.g., infrared carbon dioxide analyzer and nitrogen analyzer)
	General toxicological problems of a captive atmosphere are present
Electricity	Main aspect of electrical safety is fire
	Shock is a special hazard because people inside are generally well grounded, crowded, and confined; in case of severe shock, rescue is difficult or impossible
	Sea water and perspiration increase shock hazard and electrical leakage
	Chamber itself is a superb electrostatic and electromagnetic shield, but only from interference outside the chamber
	Some instrumentation may shock subjects via internal defects and intentionally applied electrodes
Fire	Use of oxygen-enriched atmospheres or compressed air under pressure has high fire risk
	"Zone of no combustion" may eliminate risk
	People are "trapped" inside; in case of fire, extinguishing and rescue are difficult
	Very few materials will not burn in high-oxygen atmospheres
	Electric power is the most likely source of ignition

[a] From Hamilton *et al.* (1970) by permission of *Transactions of the New York Academy of Sciences.*

Terminology

The following definitions have been used in this chapter to describe various underwater operational equipment. Many times, different names are used for the same basic piece of equipment and assigned according to the function of the equipment. An example of this is a pressure chamber. This term implies an enclosed space which can withstand pressure. This can then be divided into three categories: (1) a chamber that can withstand internal pressure only; (2) a chamber that can withstand external pressure only; and (3) a chamber that can withstand both internal and external pressure. An example of the first is a hyperbaric chamber. This is designed to withstand internal pressure while the outside remains at 1 atm of pressure. An example of the second type is an underwater observation chamber, such as a bathysphere, which is designed to withstand the internal pressure imposed by the ambient underwater pressure while maintaining an internal pressure of 1 atm. An example of the third type is a submersible decompression chamber. This is designed to withstand the external pressure imposed by the ambient underwater pressure as well as being able to withstand internal pressurization. A pressure chamber is also named for its function. Thus, a hyperbaric chamber implies that it is used to simulate high pressures. A decompression chamber implies that it is used to gradually decrease the pressure to which a diver has been exposed. A recompression chamber implies that it is used to increase the pressure to return a diver to the exposed depth, such as when treating a case of the bends.

There are many combinations of basic terms which are used together to define equipment. Thus, a swimmer delivery vehicle is used to describe a transport device that will carry a swimmer for limited distances. A swimmer propulsion unit is used to define a small device that will aid in propelling a swimmer through the water. .

Many times several basic pieces of equipment are used together to form a total system. When one uses the term SDC/DDC (submersible decompression chamber/ deck decompression chamber), this implies a total diving system with no need to include mention of the support ship or platform and also implies that the system includes the capability of mating the SDC to the DDC.

New combinations of basic terms constantly appear in the literature. An example of this is the personnel transfer submersible, used to describe search and rescue vehicles such as the Deep Submergency Rescue Vehicle (DSRV). Since a submersible implies an underwater vehicle that requires surface support, the terminology is self-descriptive, even though sometimes confusing.

Therefore, as an aid to understand the many terms used when describing underwater operational equipment, or land-based equipment used in support of underwater diving projects, the following terms are presented with their conventional definitions or functional definitions.

Bathyscaph. A navigable submersible ship for deep-sea exploration having a spherical, watertight cabin attached to its underside; an example is TRIESTE.

Bathysphere. A tethered, strongly built diving sphere for deep-sea exploration; an example is CACHALOT.

Mesoscaph. The same as a bathyscaph except for depth of operation; an example is BEN FRANKLIN.

Diving system. A system composed of three basic subsystems which can be used in various combinations; the subsystems are (1) a surface support ship or platform; (2) a deck decompression chamber used to decompress divers on the surface; (3) a tethered capsule, which is used to transport divers from the surface ship to the underwater work site. The capsule is capable of being pressurized internally and can be used to transfer divers under pressure, whereby it is called a personnel transfer capsule, or it can be used to decompress divers, whereby it is called a submersible decompression chamber. An example of a complete diving system is the ADVANCED DIVING SYSTEM IV and the U. S. Navy DEEP DIVE SYSTEM MARK II.

Submersible. A dry, 1-atm vehicle where the operator and crew are protected from outside by a pressure hull and which requires surface support although it can operate autonomously for short periods of time; it may have diver lockout facility. An example is ALVIN.

Submarine. A propelled underwater vehicle, used primarily for military missions, which can operate autonomously for a long period of time.

Habitat. A life-support system of limited mobility, capable of providing functional living and working space in the underwater environment; highly dependent upon land- or sea-based support equipment; has an internal pressure equal to the pressure of the ambient underwater pressure; it has free access to enter and leave it through an open hatch in the bottom. An example is SEALAB.

Fixed bottom stations. Underwater work sites that are maintained at 1 atm of pressure; highly dependent upon land- or sea-based support equipment. An example is an underwater welding chamber.

Vessel. A hollow structure designed for carrying or transporting something underwater.

Vehicle. The same as a vessel.

Wet pot. One chamber of a hyperbaric facility capable of being filled with water and pressurized to simulate a given underwater depth.

Swimmer vehicle. Any one of a number of devices used to aid the swimmer in attaining swim speeds greater than he could accomplish using fins.

Hyperbaric facility. The entire group of systems and subsystems used to support a high-pressure chamber or chambers, used to simulate high pressures; may include a wet pot to simulate an actual underwater environment.

Lock out capability. A vehicle, usually maintained at a dry 1 atm pressure; has a chamber which can be pressurized to the ambient underwater pressure to allow egress of a diver or divers; if necessary, it can be used to decompress the divers back to 1 atm of pressure.

Decompression chamber. An enclosed space used to gradually decrease the pressure to which a diver is exposed from the ambient underwater pressure back to 1 atm.

Recompression chamber. An enclosed space used to rapidly increase the pressure to which a diver has been exposed to return him to the ambient underwater pressure; especially used when treating a diver with the bends.

References

ANDERSEN, B. G. Human engineering criteria for the design of diver-operated underwater tools. San Diego, Calif., Oceanautics, Inc., Rep. OI-TR-1-Sk (Jan. 1971).

ANDERSEN, B. G. Diver performance and human engineering tests. Salvage equipment evolution. Program U.S.N./Makai Range Aegir Habitat, Rep. OI-TR-72/1-Poi. La Jolla, Calif., Oceanautics, Inc. (1972a).

ANDERSEN, B. G. Human factors guide for the design of diver-operated hand and power tools. Landover, Md., Oceanautics, Inc., Rep. OI-TR-72/2-Sk. (July 1972b).

ANONYMOUS. Taylor hyperbaric complex simulates 2200-ft depths. Oceanol. Int. 5:15–16 (Aug. 1970).

ANONYMOUS. What's new for '71? Skin Diver 20:20 (Feb. 1971).

ANONYMOUS. Skin Diver 21:33 (Mar. 1972).

ANONYMOUS. Offshore-Bohrlochfassung mit Montagelift. Meerestechnik 4:41–42 (Apr. 1973).

BACCAGLINI, R., and J. IRGON. The use of command buoyancy systems with submersibles in the recovery or manipulation of subsea equipment. In: 1972 Offshore Technology Conference, May 1–3, Houston, Texas, Preprints, Vol. II, pp. 181–190. Published by the Conference (1972).

BAILEY, V. R., J. LACERDA, and J. F. MANUEL. Driver lockout and observation submersibles: A perspective of participation in offshore operations. In: 1972 Offshore Technology Conference, May 1–3, Houston, Texas. Preprints, Vol. I, pp. 548–556. Published by the Conference (1972).

BARTON, R. Armoured suit has 1000 ft capability. Offshore Serv. 6:18–12 (May 1973).

BAUME, A. D. Underwater habitats—a review. Meerestechnik 3:239–242 (Dec. 1972).

BAYLES, J. J. Salvage work projects—SeaLab III. U. S. Nav. Civ. Eng. Lab., Rep. NCEL-TR-684 (June 1970).

BEATTY, H. T., and T. E. BERGHAGE. Diver anthropometrics. U. S. Navy Exp. Diving Unit, Rep. NEDU-10-72 (June 1, 1972).

BLACK, S. A. Submersible diver tool power sources; electrohydraulic and cryogenic pneumatic. U. S. Nav. Civ. Eng. Lab., Rep. NCEL-TN-1174 (Aug. 1971).

BLACK, S. A., and F. B. BARRETT. Technical evaluation of diver-held power tools. U. S. Nav. Civ. Eng. Lab., Rep. NCEL-TR-729 (June 1971).

BLACK, S. A., and R. E. ELLIOTT. Construction assistance vehicle (CAV), the design, fabrication and technical evaluation of an experimental underwater vehicle. U. S. Nav. Civ. Eng. Lab., Rep. NCEL-TR-R-762 (Mar. 1972).

BOHMAN, C. E., and R. J. JUELS. An experimental underwater localizer (for scuba divers). In: Marine Technology Society 7th Annual Conference, Aug. 1971, Washington, D. C. Preprints, pp. 329–341. Washington, D. C., Marine Technology Society (1971).

BOODA, L. L. Navy has new unmanned deep ocean vehicle. UnderSea Technol. 14:11 (Feb. 1973).

BOWEN, H. M., and A. HALE. Study, feasibility of undersea salvage simulation. Darien, Conn., Dunlap Ass., NAVTRADEVCEN-69-C-0116-1 (May 1971).

BRIGGS, E. M. Fluids for deep sea applications. Presented at the American Society of Mechanical Engineers Meeting, September 1971, Houston, Texas, as paper #71-UnT-4 (1971).

CANTY, J. M., E. H. LANPHIER, and R. A. MORIN. Engineering evaluation: Study of environmental conditioning systems for high pressure research vessels. Buffalo, N. Y., State Univ. N. Y., Dept. Physiol., High Pressure Lab., Final Rep. on contract N00014-72-C-0125 (June 28, 1972).

CESTONE, J. A., and E. ST. GEORGE. Hydrospheric navigation. Navigation 19:199–208 (Fall 1972).

COMMON, R. P. Diving technology—Part II. Underwater Sci. Technol. J. 1:59–67 (Sept. 1969).

COMMON, R. P., and M. P. KETTLE. Diver suit heating. Underwater J. 4:20–29 (Feb. 1972).

CONTI, U., P. WILDE, and T. L. RICHARDS. Towed vehicle for constant depth and bottom contouring operations. In: 1971 Offshore Technology Conference, April 19–21, Houston, Texas. Preprints, Vol. II, pp. 385–392. Published by the Conference (1971).

COVEY, C. W. Measuring human tolerance to environmental extremes. UnderSea Technol. 12:13–18, (Mar. 1971).

COVEY, C. W. Unisuit takes the chill out of diving. *UnderSea Technol.* **13**:39–42 (Sept. 1972).

DRAYTON, J., JR. Selecting a regulator. *Skin Diver* **22**:68–71 (Nov. 1972).

EDWARDS, R. N. Underwater power plants for the working diver. In: *Equipment for the Working Diver. Symposium proceedings, February 1970, Columbus, Ohio*, pp. 301–321. Washington, D. C., Marine Technology Society (1970).

FAIN, G. Error analysis of several bottom referenced navigation systems for small submersibles. In: *The Decade Ahead, 1970–1980*, pp. 220–234. Washington, D. C., Marine Technology Society (1969).

FENNING, P. Unmanned submersibles. Part I, *Hydrospace* **5**:49–51 (Feb. 1972*a*).

FENNING, P. Unmanned submersibles. Part II. *Hydrospace* **5**:44–46 (Apr. 1972*b*).

FISCHEL, H. Closed-circuit cryogenic scuba. In: *Marine Technology 1970. Preprints*, Vol. I, pp. 139–150. Washington, D. C., Marine Technology Society (1970).

FRANCOIS, R. E. The unmanned Arctic research submersible system. *Mar. Technol. Soc. J.* **7**:46–48 (Jan./Feb. 1973).

GARNETT, M. A., and I. C. ACHURCH. A comparative study. *Underwater Sci. Technol. J.* **1**:68–74 (Sept. 1969).

GROVES, D. Atomic diving suit. *Sea Frontiers* **13**:304–307 (Sept./Oct. 1967).

GUSSMAN, R. A., A. M. SACCO, J. BEECKMANS, C. E. BILLINGS, and R. ABILOCK. Handbook of aerosol behavior in saturation diving environments. Waltham, Mass., Billings and Gussman, Rep. OC-108 (Oct. 1, 1971).

HACKMAN, D. J. Power tools underwater. *Oceanol. Int.* **5**:19–21 (April 1970).

HAIGH, K. R. Future electronic instrumentation for submersibles, habitats, and divers. *Radio Electron. Eng.* **41**:225–236 (May 1971).

HALL, D. A. Mark II diving system. In: Professional diving symposium, New Orleans, November 1972. *Mar. Technol. Soc. J.* **7**:10–12 (Mar./Apr. 1973).

HAMILTON, R. W., JR., T. D. LANGLEY, and V. A. DORR. Safe instrumentation for physiological research in the hyperbaric environment. *Trans. N. Y. Acad. Sci.* **32**:358–370 (Apr. 1970).

HAUX, G. What is the position of diving technics today? *Tauchtechnik Inf.* **5**:23–24 (Mar. 1971).

JENNINGS, C. Inflatable diving suit. *Skin Diver* **20**:40–42, 66–67 (Apr. 1971).

JOHNSEN, E. G., and W R. CORLISS. Human factors applications in teleoperator design and operation. New York, Wiley–Interscience (1971).

JOHNSON, R. K., and W. T. JENKINS. Evaluation of a noncompressible wet suit. U. S. Nav. Ship Res. Develop. Cent., Rep. NSRDL/PC 3475 (Sept. 1971).

KENNY, J. K. *Business of Diving*. Houston, Texas, Gulf Publishing (1972).

KETLER, A. E. The large seawater battery—a new submersible power source. *Mar. Technol. Soc. J.* **5**:52–54 (Nov./Dec. 1971).

KOWAL, J. P. Advanced diving systems. *Sea Frontiers* **13**:372–379 (Nov./Dec. 1967).

LENTZ, T. Fins, the swimming machines. *Skin Diver* **16**:18–20 (Apr. 1967*a*).

LENTZ, T. The wet suit story. Part I. *Skin Diver* **16**:20–23, 68–60 (Sept. 1967*b*).

LIFFICK, G. L., and S. A. BLACK. Power sources for underwater hydraulic tools. In: *The Working Diver 1972. Symposium Proceedings, February 1972, Columbus, Ohio*, pp. 215–238. Washington, D. C., The Marine Technology Society (1972).

LINDEROTH, L. S., JR. Duke University's hyperbaric chamber. In: Professional diving symposium, New Orleans, November 1972. *Mar. Technol. Soc. J.* **7**:53–57 (Mar./Apr. 1973).

LOWER, B. R. Removal of CO_2 from closed-circuit breathing apparatus. In: *Equipment for the Working Diver. Symposium Proceedings, February 1972, Columbus, Ohio*, pp. 271–282. Washington, D. C., Marine Technology Society (1970).

McKENNEY, J. Regulator roundup. *Skin Diver* **20**:11–15 (May 1971).

McKENNEY, J. Up to our snorkels in masks. *Skin Diver* **22**:18–21 (June 1973).

MIKHALTSEV, I. E. Main trends in the development of ocean research techniques. A Russian view. *Underwater J.* **3**:72–81 (Apr. 1971).

MILES, S. *Underwater Medicine*. Third edition. Philadelphia, J. B. Lippincott (1969).

MILLER, J. W., J. G. VANDERWALKER, and R. A. WALLER. *Tektite II. Scientists-in-the-Sea.* Washington, D. C., U. S. Department of the Interior (1971).

MILWEE, W. I., JR. Operational U. S. Navy diving systems. In: *Progress into the Sea. Transactions of the Symposium, October 1969, Washington, D. C.*, pp. 289–296. Washington, D. C., Marine Technology Society (1970).

MONTGOMERY, S. New hyperbaric facility will test man and machine. *UnderSea Technol.* **10**:23–29 (Dec. 1969).

MOSSBACHER, R. R. Navy's new Ocean Simulation Facility. In: Professional diving symposium, New Orleans, November 1972. *Mar. Technol. Soc. J.* **7**:58–62 (Mar./Apr. 1973).

NORMAN, D. G. Force application in simulated zero gravity. *Hum. Factors* **11**:489–506 (Oct. 1969).

PARKER, F. A. Mixed gas systems. *Oceanol. Int.* **6**:38–40 (May 1971).

PARKER, F. A., and J. A. BURT. An integrated life support system for habitats, bells, and submersibles. In: *Equipment for the Working Diver. Symposium Proceedings, February 1970, Columbus, Ohio*, pp. 245–260. Washington, D. C., Marine Technology Society (1970).

PARKER, G. Fins. *Skin Diver* **20**:10–11, 28–30 (Jan. 1971).

PARRISH, B. B., E. F. AKYUZ, J. ANDERSON, D. W. BROWN, W. HIGH, J. M. PERIS, and J. PICCARD. Submersibles and underwater habitats: A review. *Underwater J.* **4**:149–167 (Aug. 1972).

PAULI, D. C., and H. A. COLE, eds. Project Tektite I. Off. Nav. Res., Rep. DR-153 (Jan. 16, 1970).

PENZIAS, W., and M. W. GOODMAN. *Man Beneath the Sea.* New York, Wiley–Interscience (1973).

PESCH, A. J., R. H. HILL, and W. F. KLEPSER. Capabilities of operators as divers vs submersible manipulator controllers in undersea tasks. Groton, Conn., General Dynamics, Electric Boat Division, Rep. U-417-70-043 (1970).

PIERCE, B. F. Effects of wearing a full-pressure suit on manual dexterity and tool manipulation. *Hum. Factors* **5**:479–484 (Oct. 1963).

PORTER, R. B., and R. H. BANKS. Electrical-electronic equipment in high-pressure HeO environments. In: *1970 IEEE International Conference on Engineering in the Ocean Environments, Panama City, Fla., 21 September. Digest of Tutorial Papers*, pp. 6–11. New York, Lewis Winner, 1970.

REIMERS, S. D. Atmospheric control in the hyperbaric environment. U. S. Navy Exp. Diving Unit, Rep. NEDU-26-72 (1972).

REIMERS, S. D., and O. R. HANSEN. Environmental control for hyperbaric applications. U. S. Navy Exp. Diving Unit, Rep. NEDU 25-72 (1972).

REIMERS, S. D., and B. S. LEBENSON. KMB-8 band mask evaluation. Final report. U. S. Navy Exp. Diving Unit, Rep. NEDU-17-72 (Aug. 1972).

RICHARDS, E. L. Certifying hyperbaric chambers. In: Professional diving symposium, New Orleans, November 1972. *Mar. Technol. Soc. J.* **7**:28–31 (Mar./Apr. 1973).

RIEGEL, P. S. Liter flow and mix selection in semiclosed-circuit scuba. *Mar. Technol. Soc. J.* **4**:17–26 (Mar./Apr. 1970).

SANDERSON, R. A., and M. B. LANDAU. Development of fuel cells for commercial undersea power. In: *Marine Technology 1970. Preprints*, Vol. II, pp. 865–877. Washington, D. C., Marine Technology Society (1970).

SOMERS, L. H. Research diver's manual. Ann Arbor, Mich., Univ. Mich., Sea Grant Tech. Rep. 16, MICHU-SG-71-212 (Aug. 1972).

STEVENS, R. C., JR. The lock-out submersible. A new dimension for the working diver. In: *Equipment for the Working Diver. Symposium Proceedings, February 1970, Columbus, Ohio*, pp. 403–424. Washington, D. C., Marine Technology Society (1970).

SUDBURY, J. P. Controlled depth submerged barge for Arctic transport is feasible and economical. *Offshore* **32**:47–50 (Aug. 1972).

TAUBER, J. F., J. S. P. RAWLINGS, and K. R. BONDI. Evaluation of a diver's thermonuclear swimsuit heater system. U. S. Nav. Med. Res. Inst., M4306.07-1003, Rep. 3 (Feb. 18, 1970).

TENNY, J. B., JR. Future of research habitats. *Oceanol. Int.* **6**:23–24 (Aug. 1971).

THOMPSON, T. Diving equipment for professionals. *Oceanol. Int.* **4**:42–45 (Mar./Apr. 1969).

USNCEL (U. S. Naval Civil Engineering Laboratory). Heating system for divers' suits. In: *NCEL Ocean Engineering Program FY 1971*. Port Hueneme, Cal., published by the Laboratory (1971).

USNSRDL (U. S. Naval Ship Research and Development Laboratory). *Handbook of Fluids and Lubricants for Deep Ocean Applications.* Annapolis, Md., published by the Laboratory (Dec. 1969).

VAUGHAN, W. S., and A. S. MAVOR. Diver performance in controlling a wet submersible during four-hour exposures to cold water. *Hum. Factors* **14**:173–180 (Apr. 1972).

WIMMER, R. E. Comparative evaluation of power system types for undersea applications. In: *Marine Technology 1970. Preprints*, Vol. II, pp. 1389–1399. Washington, D. C., Marine Technology Society (1970).

WOOD, N. H. What's new in underwater manipulators? *Oceanol. Int.* **6**:30–32 (Nov. 1971).

WOODS, J. D., and J. N. LYTHGOE, eds. *Underwater Science. An Introduction to Experiments by Divers*, pp. 1–31. London, Oxford University Press (1971).

X

Underwater Communications

A. *Introduction*

A reliable communication capability is necessary in all but a few underwater operations in order to have efficient team coordination and to provide a prime safety factor for the diver. To solve the problems of underwater speech communications, a total systems approach is needed. In the past, the simplest and most dependable methods

Table X-1
Standard U. S. Navy Line-Pull Signals[a]

Tender to diver:	1 pull	Are you all right? (Or, when diver is descending: Stop.)
	2 pulls	Going down. (Or, during ascent: Go back down until I stop you.)
	3 pulls	Standby to come up.
	4 pulls	Come up.
	2-1 pulls	I understand. (Or: Answer the telephone.)
Diver to tender:	1 pull	I am all right.
	2 pulls	Give me slack. (Or: Lower me.)
	3 pulls	Take up my slack.
	4 pulls	Haul me up.
	2-1 pulls	I understand. (Or: Answer the telephone.)
	3-2 pulls	Give me more air.
	4-3 pulls	Give me less air.
Emergency from diver:	2-2-2 pulls	I am fouled and need assistance of another diver.
	3-3-3 pulls	I am fouled but can free myself.
	4-4-4 pulls	Haul me up immediately.
Search signals to diver:	7 pulls	Search signals will follow (at beginning of search sequence).
	1 pull	Stop and search where you are.
	2 pulls	Move away from tender if given slack. Move toward tender if line is tightened. (If using circling line: Move away from the weight.)
	3 pulls	Move to your right.
	4 pulls	Move to your left.
	7 pulls	End of search signals.

[a] From Penzias and Goodman (1973) by permission of John Wiley and Sons.

811

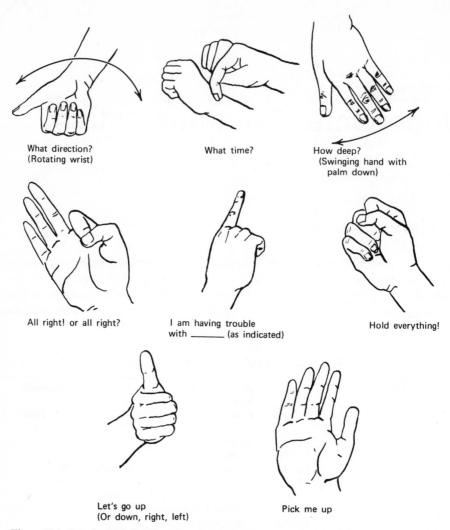

Figure X-1. Standard U. S. Navy diver hand signals. [From Penzias and Goodman (1973) by permission of John Wiley and Sons, New York.]

were tugs on a tether rope or hand signals (see Table X-1 and Figures X-1 and X-2). These *standardized-through-use* signals are still used by scuba divers and, as a backup system, by professional divers working for science, the military, or industry. However, for today's sophisticated underwater operations, an efficient, highly sensitive, hopefully simple, reliable, and flexible voice communication system is necessary. Such a system should be able to transmit the common language code without the diver having to learn any other.

It is certainly not enough to concentrate developmental efforts on one or even several of the components that make up the system. The best microphone will be useless if the transmitter portion of the system does not function adequately. Even if a very good microphone and efficient transmitter section are used, if the receiver and/or

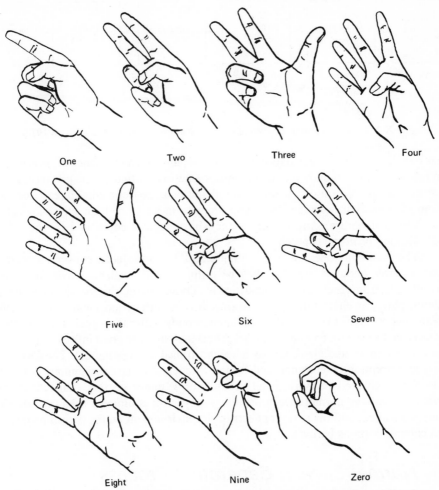

Figure X-2. Standard U. S. Navy diver numeral hand signals. [From Penzias and Goodman (1973) by permission of John Wiley and Sons, New York.]

the earphones (or loud speaker) is poor, the total system will not function efficiently. Since the divers and diving personnel must serve both a talking and listening function, they become actual components of the communication system in addition to using it merely as a tool. A highly simplified diagram of a total communication system is shown in Figure X-3.

The initial function shown in Figure X-3 is the speech generation process. This is accomplished by the talker. Hence, understanding the nature of normal speech production is prerequisite to comprehending the total communication *chain*. And since the talker (in this case the diver) is in an unusual environment, it is essential to understand the special characteristics of the second component in the diagram, the environment and its interaction effects upon the talker. Some factors affecting communications in the underwater environment include extraneous noise caused by exhaust breathing

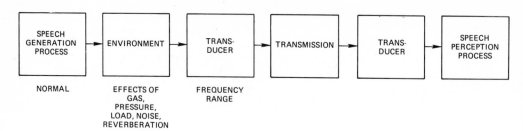

Figure X-3. General diagram of a voice communication system (Wathen-Dunn 1972).

gas bubbles, inhalation–exhalation valve *chatter*, unwanted resonance or reverberation in helmet or mask, mouthpiece configuration, increased atmospheric pressure, and the various gas mixtures used for breathing. Additional environmental interactions arise from differences in exertion levels and changes in stress of both physiological and psychological nature. Any or all of the above factors contribute to the adverse modification of normal speech.

The third component shown in Figure X-3, the transducer, converts speech into transmittable impulses, whether they be propagated through the environment electrically, electromagnetically, or acoustically. Transmission, the fourth component, is almost purely an engineering problem since it is essentially independent of the effects of the underwater environment on the human body. After the signal is transmitted, it is received by the fifth component, where the signals are converted into electrical signals and then into an audible sound by a second transducer (earphone). The final stage, speech perception, is accomplished by the listener. This last communication system function is covered in more detail in Chapter V under hearing. This chapter will briefly explain the normal speech process, the types of equipment used in underwater communication, and the various effects of the environment on the speech process and on communications hardware.

B. *Normal Speech Generation Process*

Acoustically, speech is a vibratory excitation of the molecules of the gas used for breathing (usually air). The alternating condensations and rarefactions of the molecules are propagated as sound waves. Speech sounds are generated by the air being forced from the lungs through the *vocal tract*. The actual noise-sound generator is the periodic opening and closing of the glottis controlled by the mass and tension of the vocal cords. This modified breath stream is further modified by the larynx, the pharynx, the oral cavity (including the various positionings of the tongue, teeth, lips, and jaw), and the nasal cavities. Different positioning of the *supraglottal* structures produces various modifications to the breath stream and results in various speech sounds and combinations of speech sounds. These supraglottal cavities do not act as a single resonator, since each area constitutes a different resonating chamber as each presents a different cross-sectional area to the acoustic energy. Obviously, there is considerable interaction among the various resonating cavities. Figure X-4 illustrates the various cross-sectional areas from the glottis to the lips, in a simplified form.

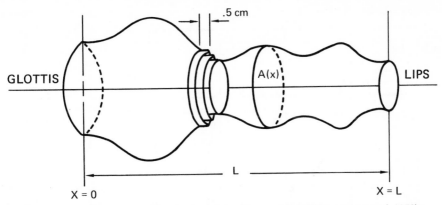

Figure X-4. Acoustic tube that approximates the vocal tract (Giordano *et al.* 1972).

As indicated above, the distribution of energy during the production of speech is determined by the shape of the vocal tract cavities, but additionally, can be influenced by the velocity of sound in the breathing gas mixture and in the external environment. (The influence of sound-velocity change upon speech production is discussed later under gas mixtures.) The sounds so produced can be classified rather arbitrarily into two general categories—vowels and consonants. Vowels are produced by the vibrating column of air forced from the lungs and set into vibration by the vocal cords. Other than changes in the shape of the resonating cavities, there is minimal blockage or constriction of the air flow, i.e., almost no pressure buildup. The consonants, however, are produced by some constriction or obstruction of the air flow in such a manner as to result in a definite pressure component.

Consonant sounds have received many subclassifications. The most fundamental is that all consonants are either *voiced* or *voiceless*. This means that in addition to the specific constriction or obstruction afforded by the tongue, teeth, lips, etc., the column of air is set into vibration by the action of the vocal cords (voiced sounds), or it is not, and only given constriction by the supraglottal structures (voiceless sounds). In other words, the same positioning of the articulators plus adding voice for the consonant sound /b/ will be heard as a /p/ if the voicing aspect is withheld.

Each of the two general voiced–voiceless consonant classifications can be further separated into two other subgroups, stops (sometimes labeled plosives) and continuants (of which fricatives is a further subclassification). Stop consonants are produced by completely obstructing the vocal tract at some point, allowing air pressure to build up behind the closure and then releasing it, such as when producing /b/ (voiced) and /p/ (voiceless). A continuant is produced by some specific set of constrictions in the vocal tract allowing increased pressure buildup behind the constriction but not interrupting the airflow. The voiced–voiceless duality is preserved in such consonant productions as /v/ and /f/ or as /z/ and /s/. Other continuant consonants, such as /m/, /l/, /n/, etc., do not have the voiced–voiceless dichotomy.

The vocal tract causes the vibrating air column flow to resonate at certain frequencies depending upon which speech sound is being produced. For vowel sounds in particular, there are several regions of acoustic energy maxima which specify the

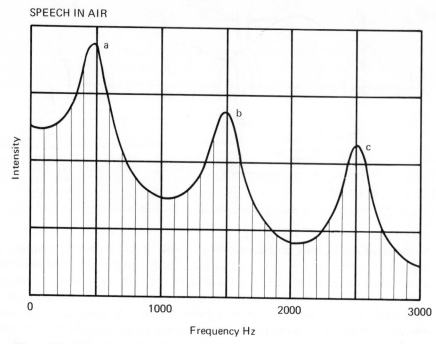

Figure X-5. Acoustic spectra showing vocal resonances for a neutral vowel (Gill 1972).

vowel and these maximum frequency regions caused by resonances of the vocal tract cavities are called formants. The neutral vowel /ʌ/ (the vowel sound used in the word b*u*t), when produced by a typical male speaker whose vocal tract is about 7 in. long when completely relaxed, has resonance regions centered at 500, 1500, and 2500 Hz, etc. (Gill 1972). An acoustic spectrum showing the formant resonant regions for the above neutral vowel is shown in Figure X-5. The voiced consonants also exhibit resonant formant regions, especially the continuants, but do not contribute as much to speech sound specificity as do the place and manner of articulation.

C. *Communication Systems*

1. *General*

Almost any voice communication system (communication *chain*, see Figure X-3) is composed of some version of the following elements or functions:

1. The human input—the voice signal.
2. Sensing and conversion device for transduction of the acoustic signal to electronic (electrical) signal.
3. Amplifier and its energy source.
4. Transmitting transducer matched to the transmitting medium.
5. Appropriate receiver and/or the listener's ear.

2. *Underwater Communication Systems*

A variety of underwater communication systems have been devised. These include considerations of (a) acoustic systems (direct projection and modulated carrier), (b) hard-line systems, (c) electromagnetic radiation, and (d) electric field potentials.

a. Acoustic Systems

(i) Direct Projection. The simplest transmission method is to project the voice directly into the water much like an underwater loudspeaker. The signals so produced can be heard by divers without the aid of any special auditory processing equipment or by a hydrophone placed in the water. The range is limited since the transducer size must be kept to a minimum and the signals are readily interfered with by any and all water-borne noise. Examples are:

> *Raytheon Yack-Yack:* Transduces voice signals at normal speech frequencies into the water using an especially built microphone and speaker system. A surface receiver is also available.
>
> *Bendix Watercom:* This system is similar electronically to Yack-Yack in that it has a self-contained power pack, amplifier, and projector, but uses a throat microphone rather than one placed within a mouthcup.

(ii) Modulated Carrier Frequency. This system type uses some form of a carrier frequency which may be modulated in several ways by the speech signal. A few of the modulation schemes include AM (amplitude modulation), FM (frequency modulation), or SSB (single sideband) suppressed carrier. The system consists of a microphone, power module, amplifier, modulator, and underwater transducer. To obtain the carrier frequency deviation necessary for intelligible voice communication, FM techniques generally must utilize a high ultrasonic frequency, which attenuates more rapidly in sea water than do lower frequencies. Currently, few FM systems are in use. AM systems have used frequencies as low as 10,000 Hz with some positive results. However, the disadvantage of AM is that at least half of the total energy output goes into the carrier frequency, where it is subsequently lost. Thus, the SSB method is the one generally used.

The U. S. Navy uses 8.0875 kHz for their standard SSB communicator. Commercial units using this frequency have reported speech transmission to a range of 900–1000 m or better, depending on the ocean environment (Martin and Adams 1970). For any of these systems, speech can be understood only by a diver or surface monitor having an appropriate receiver and demodulator. Several examples are:

> *Aquasonics QU-42:* A 42-kHz acoustic carrier frequency is transduced into the water after being modulated by the speech signals. This compound signal is picked up by a receiving coil, demodulated, and heard by the diver over earphones as normal speech.
>
> *Aquasonics 811:* This uses an SSB suppressed carrier frequency of 8.0875 kHz. The system consists of a microphone in a Nautilus mouthcup wired to

a transmitter worn on the diver's belt or body. This unit packages a solid-state modulator and amplifier, a power pack, and a projector. The receiving unit is a small demodulator and contact earphones (bone conduction) worn near the ear on the mask strap.

PQC-2: This consists of a transceiver, worn on the belt or body, whose system components include a power pack, a solid-state modulator/demodulator and amplifier circuit, a hydrophone/projector, and a five-position control (off, send, receive, homing, and CW). A headpiece consists of a headband and two contact microphone/receivers placed on the diver's temples.

Transmission carrier frequency is determined by considerations of acceptable size and of ocean transmission parameters. With the same power output, greater range is obtained by using lower frequencies. As mentioned earlier, however, as frequency decreases, ocean noise levels increase, necessitating an increase in the size of the transmitter. Usually a tradeoff region can be selected somewhere within the range of 8–40 kHz. The lower (8 kHz) is subject to considerable noise, while the higher (40 kHz) is fairly quiet; however, the transmission range for 8 kHz is five times as great as that for 40 kHz for the same power output (Martin and Adams 1970).

b. Hard-Line Systems

This type of system employs a closed system comparable to a telephone, including a microphone, an amplifier, a cable over which the signals are transmitted, and a receiver. Examples are:

Aquaphone: An inexpensive diver-to-surface system with a battery-operated telephone on the surface and a throat-microphone and an earphone on the diver.

British Buddy Line: A diver-to-diver system utilizing bone conduction (contact) transducers positioned on each diver's skull connected by a 10-ft length of wire which carries the signal between the transducers.

c. Electromagnetic Radiation Systems

This transmission method utilizes ultrasonic electromagnetic radiation, similar to radio frequencies and using the same modulation techniques. The advantage of this mode is that communication can be accomplished in or out of the water. Due to the extremely rapid deterioration of propagation of electromagnetic energy in sea water, the underwater range is limited to only 50–100 ft. One example is:

Sea Tel: This system employs a magnetic carrier with the modulating circuits and power module packaged in the diver's helmet. Earphones are mounted in the helmet flap and the microphone is positioned in a Bioengionics muzzle.

d. Electric Field Potential Systems

Another communication mode uses electric field potentials modulated by the voice signal. The range is determined by the amount of power applied into the water and the separation between the plates generating the electric field. The power output is limited by the sensation a diver can withstand (a mild shock is felt when transmitting). The separation of the field plates is limited generally by the diver's height. The range is limited to a few hundred feet.

(A number of underwater communications systems are compared by speech intelligibility scores later in this chapter under human factors considerations.)

3. *Special Microphone and Earphone Considerations*

There are four major difficulties in developing microphones and earphones for underwater use.

1. Both of these elements must be able to withstand the high pressures in the diving environment, increasing with depth, yet be sensitive enough to pick up the acoustic signals of speech generated by the diver (partially solved by the use of piezoelectric ceramic elements).

2. They must maintain watertight integrity while still retaining the necessary sensitivity and frequency response.

3. They should provide a wide enough frequency response (especially microphones) to allow speech formant frequency shifts to be picked up for processing wherever the diver is talking using breathing-gas mixtures in which the velocity of sound in that gas is increased significantly over that of air. (For further discussion see effects of gas mixtures, later in this chapter.)

4. Elements and systems must be adequately tested under environmental conditions.

Early underwater communication system fabrication relied heavily upon the modification of existing equipment, largely by coating or enveloping it with a potting compound. However, thick coatings deadened the response of the element, while thin coatings eventually leaked under increased or prolonged pressure. Later, piezoelectric elements were developed successfully for headphones, but few attempts for microphone usage were productive because of their limited sensitivity and/or frequency response. Prototype ceramic elements able to withstand pressures exceeding 500 ft-lb/in.^{2} with less than a 3-dB change in frequency response over the voice range were tested in SEALAB communication systems and resulted in reasonably good intelligibility. Reliable calibration of equipment under these conditions is difficult. Usually the extant methods have been developed along with the equipment to be tested (Hunter 1968).

D. *Factors Distorting Speech in the Underwater Environment*

One axiom basic to all voice communication systems is that any meaningful signal (speech) should reach the receiver at a level as close to 20 dB above the total noise level (unwanted sound) as is possible to achieve. Most investigators agree that a signal-to-noise ratio S/N in excess of 20 dB usually does not further enhance speech intelligibility test scores. Before adequate systems can be well designed, possible sources of environmental noise and factors which may otherwise distort the speech signal should be known, since good engineering practices, if applied, can minimize those effects.

1. *Noise Distortion Factors Due to Reverberation and Mouthpieces*

A number of sources of interference create problems in clear transmission of speech underwater (see Table X-2). Intake of breathing mixtures produces several

Table X-2

Causes of Transmission Losses for Sound in the Sea[a]

1. Absorption	In pure sea water, absorption increases exponentially with frequency, setting a practical limit to low-power transmission at about 0.5 MHz
2. Scattering	Most important in turbid water or in the presence of large concentrations of fish; serious losses occur at the deep scattering layer consisting of swarms of fish which rise to near the surface at night and sink below the depth limit of normal diving during the hours of daylight
3. Refraction	(a) In isothermal water, the speed of sound decreases with pressure, and hence with increasing depth, at a rate of approximately 1.8 m/sec per 100
	(b) In the thermocline, the speed of sound increases as the water becomes colder, bending the sound waves downward in the summer thermocline, located at about the depths commonly attained by divers (see Chapter IX); a temperature gradient of about 0.07°C/m counters the pressure effect; higher gradients cause increasingly sharp downward refraction
	Communication along nearly horizontal paths is most seriously affected, while vertical paths (e.g., between the surface and a diver directly below) are not
4. Multiple-path distortion	(a) Multiple reflections off the sea surface give the Lloyd's mirror effect, which leads to fading when the receiver is near the surface.
	(b) Multiple reflections from a rocky sea floor or in caves lead to signal distortion and fading; the loss of signal clarity may be overcome to some extent by exploiting binaural acuity (see discussion of hearing)

[a] From Woods and Lythgoe (1971) by permission of Oxford University Press.

noise sources. One noise source results from expansion of the gas under pressure to ambient pressure from a relatively small feeder line into a larger cavity, i.e., jet noise. A second source could arise from the configurations of the inhalation and/or exhalation valves set into vibration by the gas flow. These noises can be picked up by the speech microphone, thereby reducing the S/N being transmitted as well as masking the feedback of the diver's voice to his own ears (affecting his auditory self-monitoring mechanisms). Sound reverberations within the hard-hat helmet or full-face mask interfere with (distort) production of clear speech. Noise from the exhalation gas bubbles escaping into the water or other noisy equipment close to the diver's head and microphone creates interference in addition to that from biological and geological noise sources. Mouthpieces, facemasks, and helmets not only generate noise and reverberation problems, but to some extent invariably restrict free movement of the vocal mechanisms (Penzias and Goodman 1973). As a rough rule of thumb, the restrictions of speech musculature and reverberation effects upon speech production are less in a large mask (see Figure X-6).

Most mouthpiece units used for underwater communication are modifications of the standard flange-type found in scuba diving equipment. These usually incorporate a fairly large oral-muzzle enclosure which, for most individuals, permits nearly adequate movement of the speech articulators. The gas demand regulator works on the principle of supplying the gas to the diver at the proper pressure to equalize the breathing-gas pressure between the diver's lungs and the outside water pressure. To prevent water from entering the muzzle, a check valve ensures that a slightly higher gas pressure than ambient water pressure is used to expel gas from the muzzle. This condition forces the diver to speak against a somewhat elevated backpressure. During

Figure X-6. Various masks that permit a diver to speak under water. He encounters least difficulty in speaking and the resulting speech is clearest when the diver wears a mask enclosing a maximum volume of air. Speech distortion is the major difficulty hindering communication between divers. [From Woods and Lythgoe (1971) by permission of Oxford University Press, London.]

sustained conversation at 1 ATA the nominal pressures generated during speech range from $\frac{3}{4}$ to $1\frac{1}{2}$ cm H_2O. Studies have shown that the peak exhalation valve-release pressure is around 4 cm H_2O. In a closed scuba system where the gas is recirculated, the pressures may get as high as 6 cm H_2O. The effect of speaking against this increased pressure on intelligibility is shown in Figure X-7. There is a 4% decrease in intelligibility between zero and 1 cm H_2O; above 4 cm H_2O, there is about a 1% decrease for each 1 cm H_2O pressure increase. As can be seen in Figure X-7, the greatest decrease in intelligibility (averaging 20%) occurs upon merely donning the

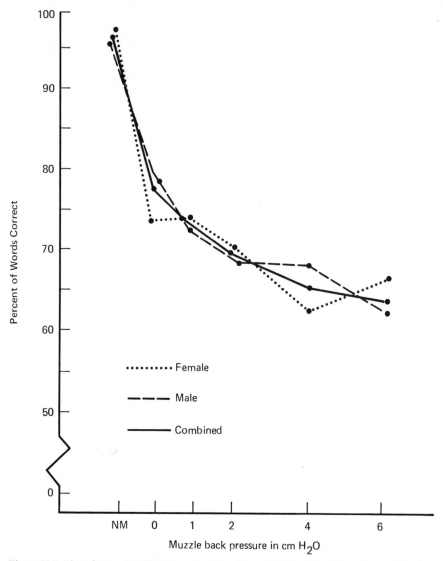

Figure X-7. Plot of mean intelligibility scores for six reading conditions for three male speakers, two female speakers, and all five speakers combined. NM: No oral muzzle was worn by the speakers. (Coleman and Krasik 1971.)

oral muzzle. This probably adds an additional acoustic chamber to the vocal tract, thus changing its resonant characteristics.

2. *Effects of Pressure upon Speech*

Considerable research has been done on the effect of increased pressure on man's speech. White's initial studies in 1955 (Tolhurst 1972) and later work by Hollien *et al.* (1971) indicated a steady decline in intelligibility scores with increases in depth. Also, speech took on a nasal quality. A general rule derived from the accumulated data is that speech intelligibility decreases approximately 4% for each increase of 1 atm. of pressure (Kenny 1971). Thus, by calculation, the intelligibility at 190 ft would decrease from the values at the surface (96–93%) to approximately 70%. Measured values are reported as 68.8% (Kenny 1971) and 63.2% (Giordiano *et al.* 1972). The discrepancies may be due to the measurements being made under simulated conditions in a dry atmosphere and/or instrumentation differences.

However real, the cause of pressure distortions and general decreased intelligibility due to pressure is not completely understood. Research has shown that there is a reduction in the consonant-to-vowel sound-pressure level ratio as pressure increases (Gill 1972). One possible explanation may be that the voice level rises to partially compensate for increased backpressure in the mask muzzle, effecting an almost shouting condition during which the consonant-to-vowel ratio reduces. In other words, the sound pressure level (SPL) of consonants is increased relative to the average SPL of vowels. There is not as significant a reduction in the intelligibility of nasals (such /m/, /n/, and /ŋ/ [ng as in *sing*]) as a function of pressure, but this could be due to the increased nasal quality of the voice increasing as pressure increases (Giordiano *et al.* 1972).

In an effort to understand the increases of the nasal quality of speech with increased depth, Fant and Sonesson (1967) studied vowel formant frequency shifts in speech produced at 1 and at 6 ATA. They found that although the fundamental frequency F_0 of the voice did not change, there was a nonlinear shift at the higher formants (Figure X-8). Figure X-9 shows the differences in predicted versus measured formant frequency transpositions under a pressure change of 50 m, equivalent depth.

Using speech synthesis techniques, Fant determined experimentally that the factor which gave a nasal quality to speech was the greater than normal increase in the shift of the first resonance region, or formant F_1, of the vocal tract (Lindqvist 1972). The lowest possible resonant frequency region of the vocal tract is the resonance of the combined vocal tract cavity air mass plus the compliance of the enclosed air volume. As the pressure of this enclosed volume of gas increases, the resonant frequency of F_1 also increases in proportion to the square root of the pressure increase. And since the walls of the vocal tract are not rigid, the lowest possible resonant frequency region is limited by the shunting effect of the walls. This has been measured for a closed-lip position (such as when saying /b/) and was found to be between 150 and 200 Hz at 1 ATA (Lindqvist 1971). The upward shift in formant frequency is more pronounced, the lower the frequency that is being considered. Hence, the nonlinear shift is more noticeable with the first formant F_1.

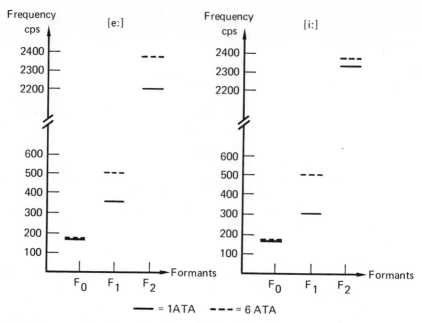

Figure X-8. Graphical representation of a speech spectrogram of the vowels /e/ and /i/ obtained in the decompression chamber at normal air pressure (1 ATA) and at increased air pressure (6 ATA). Note that the rise in the ambient air pressure caused no change of fundamental tone F_0 of the voice, whereas the frequencies of the first F_1 and second F_2 formants were increased. [From Fant and Sonnesson (1967) by permission of *Military Medicine*.]

3. *Effects of Gas Mixtures*

Gas mixtures used for life support at depths greater than approximately 100 ft tend to affect speech communication adversely. At the present time, a mixture of helium and oxygen typically is used to prevent the narcotic effects of nitrogen. This He–O_2 gas mixture changes the normal resonant frequencies of the vocal tract, shifting them upward, due to the increased velocity of sound in that gas mixture. Intelligibility decreases as the percentage of helium in the breathing-gas mixture increases, and deteriorates to almost complete unintelligibility in mixtures of 90% or more of helium (Hollien *et al.* 1971).

This higher-than-normal shift in the resonant frequency components during speech production makes the resultant speech appear to have a much faster rate. The actual fundamental frequency, or the periodicity of vocal fold actions, determining the pitch of the voice, remains unchanged in a helium–oxygen mixture compared to that in air (see Figure X-10). The upward shift in the frequency region of the formants gives the impression of a faster speaking voice, or Donald Duck effect, since the voice formant shift is approximately equal to some 2.93 times the increase in the velocity of sound in pure helium over that in air (Morrow 1971). Several recent engineering endeavors have resulted in limited solutions to what had been a major problem in modern deep-diving operation; one example is the ARL Helium Speech Converter (see this chapter, on helium unscramblers).

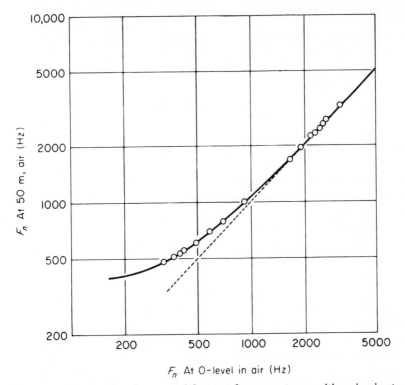

Figure X-9. Predicted and measured formant frequency transpositions in air at 50 m, equivalent depth (164 fsw). Average of four male subjects. The solid line indicates predicted values; circles show measured values; and the dashed line represents $k = 1$. [From Fant *et al*. (1971) by permission of Academic Press.]

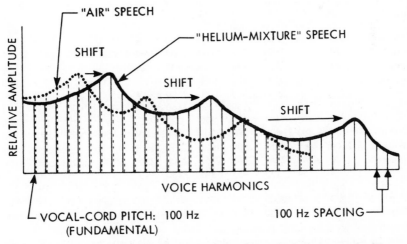

Figure X-10. Formant-shifting, the basic phenomenon in helium speech. [From Adams (1971) by permission from the Marine Technology Society, Washington, D. C.]

a. Summary of Speech He–O₂ Research

Numerous investigations have been conducted to provide quantitative data on the effects of He–O_2 on speech communication, since this gas mixture is used most frequently in deep diving.

Beil (1962) did a frequency analysis of vowels produced in a helium-rich environment. His study showed an upward shift of the formant frequency regions for the helium vowels when compared to vowels produced on air, although the ratios of frequency separation among the formants remained nearly constant. He attributed the slight elevation in fundamental frequency to the increased velocity of sound in the helium-filled vocal cavities. This formant shifting is illustrated in Table X-3.

Frequency information combined with temporal variations and the frequency–temporal translations between the consonants and vowels comprising syllables or words provide important acoustic cues which allow discriminations between speech sounds (phonemes). Phonemic analysis is another method of studying the factors of speech intelligibility. The upward shifts in speech formants in He–O_2 imply that variations from the normal would become more apparent and be revealed in an analysis of phonemic perception, even though the frequency separation ratios for specific vowels remain essentially the same in both air and helium. Timing aspects for other speech sounds, such as plosives or fricatives, also remain essentially the same for the two conditions. However, one major change that would be expected is a wider change, i.e., increased slope, of the frequency transition interval between vowels and consonants. Under these circumstances, those speech sounds whose identification acoustic

Table X-3

Formant and Fundamental Frequency Mean Values (in Hz) for Normal Vowels (N) and Helium Vowels (H) [a]

		/i/		/ɪ/		/æ/		/ɑ/		/ɔ/		/u/	
	Subj.	N	H	N	H	N	H	N	H	N	H	N	H
I	Fund.	148	148	133	133	111	114	132	135	128	128	112	114
	Form. 1	300	400	450	600	750	1050	800	1150	700	950	300	500
	Form. 2	2400	3300	2050	2850	1700	2400	1100	1550	950	1350	700	900
	Form. 3	2950	[b]	2550	3700	2300	3500	1650	2650	1900	2400	2450	3400
II	Fund.	127	134	128	136	120	129	121	125	122	152	150	158
	Form. 1	225	350	350	500	700	950	750	1000	550	700	250	400
	Form. 2	2800	3750	2500	3300	2200	2600	2050	1500	800	1050	900	1150
	Form. 3	3350	4700	2900	3900	2750	3150	2600	3350	2600	2950	2550	3000
III	Fund.	145	162	138	148	128	137	125	141	128	139	158	151
	Form. 1	225	400	400	550	700	900	700	800	550	625	500	450
	Form. 2	2100	2400	1950	2500	1800	2250	1000	1150	850	950	850	1000
	Form. 3	3700	4200	2750	3750	2350	2700	2100	2700	2800	2900	2100	[b]
IV	Fund.	144	154	145	153	141	150	130	144	135	147	142	153
	Form. 1	200	300	300	400	550	700	550	700	450	550	200	300
	Form. 2	2200	3150	2000	2800	1650	2000	900	1200	750	950	600	750
	Form. 3	2700	4000	2600	2800	2150	2550	2800	3650	2750	3900	3000	5000

[a] R. G. Beil's data on helium voice characteristics.
[b] Indicates the third formant was insufficiently defined to give a reliable frequency value.
International Phonetic Association symbols: /i/ as in "heed"; /ɪ/ as in "hid"; /æ/ as in "had"; /ɑ/ as in "cod"; /ɔ/ as in "saw"; /u/ as in "boot."

Table X-4

Ratios of Formant Frequencies of Speech in Air and in a Helium–Oxygen Mixture[a]

Section	Frequency Air	Frequency He–O$_2$	Ratio He–O$_2$/Air
1	2400	3850	1.60
	3400	5300	1.56
2	2350	4050	1.72
	3350	5300	1.58
	3900	5700	1.46
3	2250	3250	1.44
	3550	5250	1.48
4	2350	3350	1.43
	3800	5350	1.41
5	1700	2450	1.44
	2300	3650	1.59
	3750	5250	1.40
6	2100	3650	1.74
	3250	4950	1.52
	3950	5550	1.41
7	1600	2350	1.47
	3300	4950	1.50
	3750	5450	1.45
		Mean	1.51

[a] From Sergeant (1963) by permission of Aerospace Medical Association.

cues normally depend upon transition frequency-change rates should suffer intelligibility degradation in a helium-rich environment.

Sergeant (1963) did an acoustic and phonemic analysis of helium speech with the talkers using a mixture of 81% helium and 19% oxygen at normal atmospheric pressure. His results showed that the talker did not change his fundamental voice pitch (determined by the fundamental frequency of vocal-fold action), even though the perception of the total helium speech had the "typical" Donald Duck characteristic. The shift in the higher formant regions (F_1, F_2, F_3, etc.) changed by a mean value of 1.51 times, as shown in Table X-4. For the same ambient background noise level, mean intelligibility scores were 35% for He–O$_2$ mixtures compared to 69% for speech spoken in air. In a later study, Sergeant (1966) compared helium speech and normal speech by constructing a phonemic-confusion matrix using well-known phonemic classification categories (see Figure X-11). The consonants /l/ and /m/ were most often recognized correctly; /n/ was the second most correctly recognized sound; while /s/ was the consonant most frequently missed. He concluded that the formant frequency changes caused by breathing the He–O$_2$ mixtures interfered with the normal intelligibility rankings (on air) of specific consonants; however, the general patterns of discrimination according to phonemic classification were not radically affected.

An extensive series of recordings was made prior to and during the SEALAB II operation utilizing the crew members, who produced *standardized* samples of speech

	Voiceless fricatives		Voiceless plosives			Voiced plosives			Nasals		Semivowels		
	f	s	p	t	k	b	d	g	m	n	r	l	w
f	48	10	4	3	5	3	3	0	0	0	1	0	0
s	10	23	0	2	3	1	0	0	0	1	0	0	1
p	3	17	59	22	19	0	0	2	0	0	2	0	0
t	18	14	10	49	16	2	3	1	2	1	0	0	1
k	11	4	8	22	49	1	1	1	0	0	0	0	0
b	3	7	1	4	1	75	9	7	1	4	1	1	1
d	1	3	1	1	1	7	55	6	0	1	1	0	1
g	0	3	0	0	0	1	13	73	0	0	1	0	0
m	1	1	0	0	0	0	0	2	78	18	0	0	0
n	0	0	0	0	0	0	0	1	11	61	2	0	1
r	0	0	1	1	0	0	2	1	1	2	55	7	26
l	1	2	2	4	0	2	3	2	5	4	27	90	14
w	0	1	0	0	0	0	0	0	0	3	7	0	48
y	0	0	0	1	0	0	3	0	0	1	0	0	0

Figure X-11. Stimulus–response matrix for 13 consonants spoken in a mixture of 80% helium and 20% oxygen. Entries have been converted to percent response to a particular phoneme (Sergeant 1966).

talking on air and He–O$_2$. MacLean (1966) and his co-workers made acoustic analyses of both types of speech and drew the following conclusions:

1. The usual quality of the He–O$_2$ speech was due to the upward shifts of the characteristic speech formants.
2. These formant shifts are nonlinear, the first (F_1) being greater than the higher ones (F_2, F_3, etc.).
3. Energy associated with fricative sounds (/s/, /f/, etc.) was found to shift upward in frequency.
4. Fundamental frequency shifts, determining the pitch of the voice, usually were not significant.
5. The general quality of the speech sounded more natural after the crew had spent several days in the helium-rich environment.

Changes in formant frequencies and the ratio of formant frequency changes for one crew member are given in Table X-5.

By means of acoustic spectral analysis, the distorting effects upon speech in a pressurized helium-rich environment at five depths of submersion were studied by Brubaker and Wurst (1967). Their results indicated that all of the voice factors, the vocal fundamental, the vowel formants, and the frequency-separation ratios between the consonant-vowel combinations, were elevated under these conditions.

Studies of speech intelligibility as a function of helium–nitrogen–oxygen mixtures and increased ambient pressure were conducted under pressure-chamber conditions

Table X-5

Changes in Formant Frequency and Ratio of Changes to
Formant Frequency for One Crew Member
(MacLean 1966)

Vowel sound	Formant			Ratio to talkers' normal sound		
	F_1	F_2	F_3	F_1	F_2	F_3
Mid-front: RED (e)						
Normal voice	550	1700	2450			
Early SEALAB	1000	1900	3250	1.8	1.12	1.33
Late SEALAB	900	1800	2950	1.6	1.06	1.205
Low-front: AND (æ)						
Normal voice	600	1800	2500			
Early SEALAB	1100	2100	3600	1.8	1.17	1.44
Late SEALAB	950	1900	3300	1.59	1.06	1.32
High-black: BLUE (u)						
Normal voice	400	1200	2300			
Early SEALAB	950	2200	3150	2.4	1.8	1.37
Late SEALAB	800	2100	2850	2.0	1.75	1.24
Low-back: DOWN (aU)						
Normal voice	700	1400	2350			
Early SEALAB	1450	2800	4300	2.07	2.00	1.83
Late SEALAB	1100	2300	3700	1.57	1.64	1.57
	Average early			2.03	1.53	1.49 ⟨1.68⟩
	Average late			1.70	1.38	1.33 ⟨1.47⟩

ᵃ From MacLean (1966) by permission of the *Journal of the Acoustical Society of America*.

simulating 200 and 450 ft by Hollien *et al.* (1968a). Their results, shown in Table X-6, indicated that at 200 ft (helium–nitrogen–oxygen mixture 79/17/4%) the speech intelligibility scores were less than 50%, while at 450 ft. (helium–nitrogen–oxygen mixture of 90/8/2%) the scores dropped to less than 20%.

In a further study, Hollien *et al.* (1971) attempted to establish firm baseline data concerning gas mixture, pressure, and intelligibility relationships. Recordings of standardized speech materials were made in helium–nitrogen–oxygen mixtures (79.3/6.5/4.2; 90.2/7.8/2.0; and 92.4/6.0/1.6) at pressure-chamber simulated depths of 200, 450, and 600 ft, respectively. From this study they drew the following conclusions:

1. Speech intelligibility scores are depressed markedly and monotonically as depth is increased. This corresponds to the functional relationships between a joint rise in pressure and the helium content of the gas mixture.
2. They tentatively postulate that intelligibility is decreased by approximately 50% for each 200 ft of depth.
3. There is considerable variability in speaker intelligibility among different divers under helium-rich environments.
4. For a particular talker, high intelligibility scores at the surface do not necessarily correspond to high intelligibility at depth.

Table X-6
Speech Intelligibility under Pressure in
Helium–Nitrogen–Oxygen[a]
(Hollien *et al.* 1968*a*)

Depth	Percent intelligibility
0[b]	92.3
200[c]	49.7
450[d]	19.5

[a] All recordings were made during SEALAB III training at EDU. $N = 10$ listeners.
[b] Thirty-seven diver/talkers.
[c] Twenty-eight diver/talkers.
[d] Twenty-one diver/talkers.

b. Helium Unscramblers

Since speech becomes increasingly unintelligible under increased pressures and the breathing-gas mixtures required for deep diving, there have been numerous attempts to devise systems that will restore helium-generated speech to intelligible speech and hopefully to a more "natural" sounding quality. A fairly complete history of this development by the Navy has been outlined by Tolhurst (1972).

Processes for unscrambling helium speech generally fall into the categories of (1) heterodyning, (2) vocodering, and (3) time-domain or time-sampling processing. In heterodyning, the signal is filtered into several frequency bands, which are then shifted downward (for helium speech) to approximate their placement in normal speech. Because only a limited number of filtered bands is practical, the harmonic structure of the voice frequencies is not preserved, thus interjecting another major source of distortion.

Vocoder techniques also employ discrete frequency-band filtering in which the energy in a particular band is allowed to modulate the output of a predetermined, preset lower band. This process is relatively inflexible unless automatically tunable band-pass filters can be utilized, i.e., formant vocoder tracking. Theoretically this latter type of processing should result in highly intelligible and natural-sounding speech. Such processing has been done using large and powerful computers, but no operational instruments are available at present.

Time-sampling or time-domain processing forms the basis for the most practical current systems, mainly because the electronic circuitry is available and amenable to microminiaturization. This type of processing relies on the fact that there are physical limitations to the number of changes a voice can generate in any given second. Formant frequency regions can be sampled repeatedly—about 400 or more times per second—and little information content is lost between samplings. The *dead time* between samplings is discarded and the samples are usually read into a storage unit. They are then retrieved (read out) at a slower rate, much like lengthening each sample, to the amount required (predictable by knowing the percent of helium in the breathing mixture) to bring the formant frequencies into the proper range of normal speech.

SPEECH IN OXY-HELIUM AT 1500 FT.

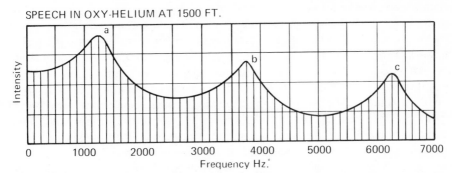

Figure X-12. Acoustic spectra showing vocal resonances for a neutral vowel (Gill 1972).

One such method was developed into an instrument by the Admiralty Research Laboratory. This scheme places time sections of speech into a temporary memory store and, later, reads them out at a lower rate. Voiced sounds are sampled at the rate of the most intense part of each vocal-fold period and the rest rejected. During unvoiced sound production, samples (sections) are taken more frequently, the period triggered by an electronic clock circuit. In March 1970, an experimental model of this device was used successfully during a 1500-ft saturation experimental dive at the Royal Naval Physiological Laboratory, Alverstoke (Gill 1972). It is now produced commercially as Model ARL Helium Speech Converter, type 023. It was used during a U. S. Navy experimental dive to 1600 ft in the summer of 1973 (Undersea Medical Society 1973).

Figures X-10 and X-12 show the speech changes resulting from He–O$_2$ mixtures. Figure X-13 shows the principle of the operation of the ARL unscrambler. One of the prime reasons that this unit works as well as it does is the insistence of the developers on using a wideband microphone to pick up the helium-shifted speech signals. Without such a broadband initial transducer the shifted speech frequencies would be filtered out

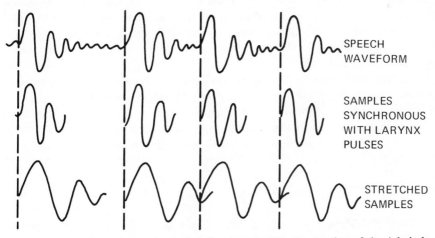

Figure X-13. Waveform diagrams to describe the principle of operation of the Admiralty Research Laboratory Processor (Gill 1972).

when a narrowband microphone was used and would be lost for processing in any unscrambler system, The signals from the microphone are filtered through a 12-kHz low-pass filter to eliminate spurious transients and then are sent through a shaping network. From there the signals are sent to a laryngeal-pulse detector and then into an analog-to-digital convertor. The shaping network can provide up to 20 dB of amplification to the upper part of the speech spectrum to enhance the level of the unvoiced sounds. The initial 2.5-msec sections (time samples) of each vocal-fold period are then encoded at a 30-kilosamples/sec rate into eight-bit unit pulse-code modulators and in this form are read into the shift register memory storage. The shift registers not occupied with being read in are sequentially read out with the readout signals interleaved and decoded. The signals then pass into a digital-to-analog conver-tor, the resultant reconstructed speech is passed through a 4-kHz low-pass filter, which tends to average the signals and remove unwanted products of the system. Time-expansion ratios can be adjusted as required from 1:1 to 1:3 (Gill 1972).

E. *Human Factors Considerations in Underwater Communication*

As mentioned previously in this chapter, any communication system is bonded on either end by the human factor. Consideration of these human factors is virtually important to any communication system, especially those for use underwater. Most of the early attempts to solve these problems did stem from a rather thorough knowl-edge of electronic circuitry and of the principles of underwater sound propagation. Some of the early system developments, as well as some of those designed more recently, exhibit little understanding of the total underwater speech communication process, however. Many of the systems failed to meet user expectations and needs, not because signals could not be transmitted from point to point underwater, but because inade-quate consideration was given to the capabilities and limitations of the diver and his interactions with his life-support and operational-work equipment. This lack of consideration is reflected in the intelligibility scores measured on some early communi-cation systems of less than 5%.

The human factors to be taken into account in total system design are far-ranging. They include (1) psychological considerations, which encompass talker variability, measures of total system intelligibility, speech feedback, and training; (2) physiological factors of diver safety and effects of physical exertion; (3) com-munication system packaging; and (4) total design factors.

1. *Psychological Factors*

Discussion of the psychological factors that may influence communication effi-ciency will be limited in this chapter to but a few. For a discussion of the complex role of psychological factors in deep diving refer to Chapters V and VI.

Hollien *et al.* (1971) found a considerable variability among divers in the efficiency with which they were able to initiate speech at depth and attributed this observation to two characteristics:

1. Some individuals expend considerably more effort than do others in an attempt to compensate for the degradation of speech in underwater environments.
2. The relative success or failure of divers to manipulate their articulatory mechanisms in an attempt to improve their speech intelligibility ultimately affects all of their subsequent attempts to communicate in this environment.

a. Adaptation to Voice Changes

The recordings, mentioned earlier, made during the SEALAB II operation indicated to MacLean (1966) that over a period of several days at depth speakers had a rather pronounced voice-quality change. He suggested that the psychophysiological adaptations of the speaker that produced a more natural-sounding speech were either intuitive or learned. This speaker adaptation is in agreement with other observations of the experimentally or clinically modified sidetone of talkers with impaired vocal cord function or other vocal tract anomalies that require compensation or adaptation to improve voice quality.

According to Adams (1971), a psychological feedback occurs so that individuals attempt to compensate for the unnatural sound of their own voices. After speaking for several days under $He-O_2$ mixture conditions, the diver is able to arrive at some modification of his speech production toward more normal-sounding words.

Brubaker and Wurst (1968) found that in a high-pressure helium environment, which may both modify the diver's auditory threshold and degrade his vocal output, a diver tends to speak louder as one factor is an effort to regain intelligibility and restore his speech to what he perceived as normal. This increased effort results in a pitch elevation. However, a diver's speech does improve with time in an $He-O_2$ mixture, which might represent a feedback adaptive learning effect. If such an improvement can be attributed to the articulatory-compensatory efforts on the part of the diver, then an appropriately instrumented modification of the way a diver hears himself could result in better intelligibility in an $He-O_2$ environment.

A fact which has been known for many years is that whenever an individual must use a new system (including communication systems) his operations will be more efficient if he has had some prior actual-conditions training. This is especially true in deep diving, where considerable indoctrination is needed for the diver to compensate as well as he can to the distortions arising from speaking against increased breathing pressures, different gas mixtures, and noise.

b. General Intelligibility Interference Factors

Many of the difficulties encountered in engineering effective communication systems are not in the inherent mechanics or electronics of the systems, but in the message construction of the talker's input and in factors influencing the divers' ability to produce intelligible speech, such as ambient pressure, immersion in water,

Table X-7

Intelligibility Scores[a]

Communication system	Percent intelligibility
Aquasonics A	52.3
Aquasonics B	46.4
British Buddy Line	35.9
Yack-Yack B	33.2
Yack-Yack A	22.9
Aquaphone A	15.4
Bendix watercom	16.4
MAS	9.7
Aquaphone B	6.5
PQC-B	2.3
PQC-A	0.9

[a] From Martin and Adams (1970) by permission of the Marine Technology Society, Washington, D. C.

isolation of the nasal from the oral cavity by certain facemasks, constriction of the articulators by the head harness or mouthcup, and backpressure resulting from the action of the breathing-gas regulator (Hollien *et al.* 1968*b*). Further, other factors to be considered are: (1) limitations imposed by the breathing equipment flow and noise, (2) the effects upon hearing when water fills the ear canal, and (3) the acoustic characteristics of the particular mouthpiece or facemask.

The design of mouthpieces and facemasks has been largely by trial and error. Special cavity-type mouthpieces and full-face masks have been designed to free the diver's speech articulators for easier word formulation. However, very few measurements of the acoustic parameters have been published. Acoustical and perceptual measurements on proprietary waterproof microphones indicate that if good intelligibility is to be achieved, the microphone and the mouthpiece cavity must be considered as a functional unit (Morrow 1971).

Intelligibility scores of a number of communication system evaluations are given in Table X-7.

2. Physiological Considerations

a. Basic Factors

Certain basic physiological considerations needed in the human engineering design and fabrication of underwater communication equipment should include the following:

1. For safety reasons, minimize or eliminate any external cables.
2. Develop equipment which will not increase the difficulty of performing emergency operations.

3. Design mouthpieces or facemasks which will lessen jaw fatigue as well as providing additional life protection in extreme-exposure diving.
4. Provide easily accessible and dependable operating switches and other equipment simplifications to help compensate for possible loss of diver's mental acuity due to high nitrogen partial pressures (narcosis) or cold exposures.

b. Effects of Exertion

Performing tasks underwater has a definite effect upon voice communication. Initial studies of the effects of exercise upon speech at atmospheric pressure have shown an alteration in both the intensity and the rate of speech (Otis and Clark 1968). Both factors affect speech intelligibility, but no attempts were made to assess these effects separately in these early studies.

Later studies by Murry *et al.* (1972) did determine speech intelligibility under varying exercise loads and at increased ambient pressures. Their results agreed with other reports (see earlier discussions of the effects of pressure) that there was a definite decline in intelligibility scores as ambient pressure increased. Also, as shown in Figure X-14, there was a decrease in intelligibility at the onset of exercise, labeled in the figure as a 50-W power load, but there were no further decreases as the amount of exercise increased to 100- and 150-W power loads. Figure X-15 graphically illustrates intelligibility as a function of depth. At the surface there was a 6.75% loss in intelligibility from a zero workload to the highest exercise condition. As can be seen, the largest decrease took place from zero to the 50-W load, with essentially no further drop as the exercise loads increased to 150-W; the latter level is somewhat below the level of exhaustion for adult males. As ambient pressure increased to an equivalent 150-ft depth, there was a 15% loss in intelligibility accompanying the onset of exercise. These results suggest that the drop in intelligibility scores with the onset of exercise is due to the physical movements alone and that once a diver adjusts to the diving task, increased physical activities do not affect further his ability to communicate (Murry *et al.* 1972).

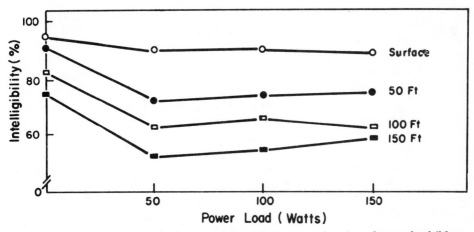

Figure X-14. Mean speech intelligibility scores for all talkers as a function of power load (Murry *et al.* 1972).

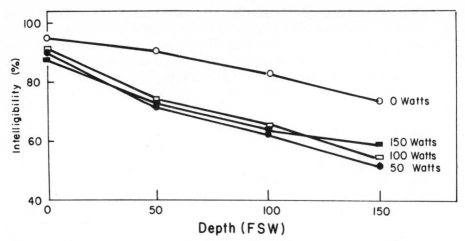

Figure X-15. Mean speech intelligibility scores for all talkers as a function of depth (Murry *et al.* 1972).

3. *Packaging of Equipment*

Human engineering factors influencing design packaging of underwater communications equipment are: size, weight, wearability, operational complexity, number of cables, range requirements, mode of communication, and communication system battery life.

Largely through trial and error, underwater communications equipment has evolved. During this evolution, considerable data have accumulated through actual use and handling of diver underwater communicators. These empirical data have resulted in better packaging configurations for design and manufacture of this equipment. Recent work at the University of Florida has provided much of the quantitative information necessary to allow divers to produce intelligible speech while wearing scuba equipment (Martin and Adams 1970).

Different packaging techniques for communicators have included a single cylinder clamped on the diver's tank with various output terminations, a complete helmet package with no external wires, and small separate units connected by short wires and to the output terminations. The main advantage of the single cylinder is that it can be readily attached to the diver's tank(s) and is usually well out of the way of his arms and legs, which allows good diver maneuverability. However, the necessity of having one or more cables to connect the transducer, microphone, earphones, and controls presents a potential hazard to the diver if he becomes entangled (Martin and Adams 1970).

The complete-helmet package has the advantage of no external cables or connectors since the helmet houses all electronics (including power supply), plus the microphone–earphone combination, and the transducer. The main problem is that the helmet may become top-heavy and cumbersome to the diver, changing his center of gravity underwater.

A third communicator-packaging approach is to use separate transmitter and receiver units. The small receiver package mounts near the diver's ear, attached to the facemask strap. The transmitter attaches to the belt and has two interconnecting cables going to the microphone and to the transmit switch. This presents the same cable hazard as with the single-cylinder package (Martin and Adams 1970).

The positioning of the transducer is important also. The best transmission radiation pattern is obtained when the transducer is mounted on the diver's head. If the communicator package mounts on the diver's belt, care must be taken to position the transducer element as far away as possible from pockets of air (such as the diver's lungs or diving tanks) since entrapped air acts as a baffle, changing the radiation pattern of the signals through the water.

4. Design Problem Considerations and Recommendations

An underwater communications equipment designer must utilize areas of knowledge not usually relevant to engineering problems. Thus, the diver's psychological attitude, his physiological reactions to stressful and restrictive situations, and the compatability of the communication equipment with his basic life-support system are design factors. Many of these have been discussed previously in this chapter but are listed here as summary. An outline of certain problems inherent in underwater speech communication given by Martin and Adams (1970) is as follows:

1. The human input, as a generator of language and as a talker–listener in an unusual underwater environment.
2. The facemask or enclosure into which the diver speaks.
3. The microphone and its positioning.
4. Adequate electronic circuitry for transmitting and receiving.
5. Mode of propagation of the acoustic (or other) energy.
6. The headset, earphones, or loudspeaker, and their position in relation to the diver's ear.

From the diver's standpoint, the design factors to be considered in any two-way, nonwired, underwater communications system are:

1. The size and shape of the equipment should be as small as possible but are somewhat dictated by the mode of propagation.
2. Communication equipment must be worn on the diver.
3. The controls must be simple and easy to operate.
4. The equipment must approximate neutral buoyancy so that no adjustments in weights are required.
5. The equipment must not hinder the diver's movements when maneuvering.
6. The equipment must be acceptable to the diver.
7. The equipment must provide adequate intelligibility.

The emphasis in the above factors is upon intelligibility, diver mobility, and speech propagation methods. Table X-8 gives a ranking of various underwater communication systems, evaluated as to their intelligibility score rankings.

Table X-8

Intelligibility Levels, in Rank Order, of the Diver
Communication Systems Evaluated[a]

		Percent intelligibility	
Communication system	Number of diver/talkers	Raw scores[b]	Corrected for talker's differences
Aquasonics 811 (double-hose regulator: Nautilus muzzle)	7	56.5	53.8
Aquasonics UO-42 (A) (double-hose regulator; Nautilus muzzle)	12	52.3	52.3
British Buddy Line (double hose regulator; Nautilus muzzle; microphone placed on forehead)	4	40.2	35.9
Raytheon Yack-Yack (B) (double-hose regulator; Nautilus muzzle)	5	37.2	33.2
Raytheon Yack-Yack (A) (Single-hose regulator; Raytheon muzzle)	12	22.9	22.9
Sea-Tel (prototype Canadian mfg.) (double-hose regulator; Nautilus muzzle)	8	22.8	23.3
ERUS-2-3A (French mfg.) (double-hose regulator; U. S. Diver's prototype full-face mask)	9	18.9	18.9
PQC-2 (military) (double-hose regulator; Nautilus muzzle)	6	17.2	16.6
Aquaphone (A) (inexpensive model) (single-hose regulator, Raytheon muzzle)	12	15.4	15.4
Bendix Watercom (not in production) (double-hose regulator, Bencom muzzle)	7	12.3	16.4
MAS (prototype) (not in production) (single-hose regulator, modified Desco mask)	5	10.9	9.7
Aquaphone (B) (inexpensive model) (single-hose regulator, standard mouthpiece)	12	6.5	6.5
PQC-la (B) (military) (double-hose regulator; Nautilus muzzle)	9	2.6	2.3
PQC-la (A) (military) (double-hose regulator, Scott mask)	9	1.0	0.9

[a] From Hollien et al. (1970) by permission of Compass Publications, Arlington, Va.
[b] All scores are based on at least least ten listeners per talker.

F. *Summary*

Present-day underwater operation requires a speech communication system that can be carried by the diver without hampering mobility or safety; can provide easy two-way intelligible speech; is reliable, flexible, efficient, and simple to operate; has a reasonable signal propagation range; and allows the diver to talk on any breathing-gas mixture best suited for his life support. Whenever a sure and fail-safe diver navigation system is developed (a modification of the speech communication system may be the key), then the task capabilities of the diver(s) will be greatly enhanced and diver safety will be expanded markedly.

Terminology

Bandwidth. When frequency is plotted against a linear scale any interval along the scale measures the difference $f_b - f_a$ between the frequency of the upper boundary of the band represented and the frequency of the lower boundary. This difference is known as the bandwidth of the frequency band. Bands having equal widths are represented on a linear frequency scale by intervals of equal length.

Consonant. One of a class of speech sounds produced by constriction or closure at one or more points in the breath channel; any letter of the English alphabet except /a/, /e/, /i/, /o/, /u/.

Continuant. A consonant or combination of consonants that may be continued or prolonged without alteration during one emission of breath; sound is produced by frictional passage of expired breath against a narrowing at some point in the vocal tract; synonymous to fricative; examples are /f/, /th/.

Decibel. A unit for expressing the ratio of two amounts of acoustic signal power equal to ten times the common logarithm of this ratio.

Electroacoustic transducers. Hydrophones, together with microphones, telephone receivers, and loudspeakers, are known generically as electroacoustic transducers, which are transducers for changing sound or audio waves to electrical impulses or vice versa.

Formant. A characteristic component of the quality of a speech sound; any of several resonance bands used to determine the phonetic quality of a vowel; any of several areas of acoustic energy maxima that specify vowels.

Frequency. The frequency of a periodic function is the reciprocal of the period, usually expressed in hertz (cycles per second) or in revolutions per minute.

Fricative. See Continuant.

Fundamental frequency. The lowest natural frequency of an oscillating system.

Hertz. A unit of frequency equal to one cycle per second; abbreviated Hz.

Hydrophone. An electroacoustic transducer that responds to water-borne sound waves and delivers essentially equivalent electric waves. The conversion from sound energy and electrical energy is achieved through the use of either the piezoelectric or magnetostrictive effect. The varying potential generated across an piezoelectric material when it is subjected to a varying mechanical force can be coupled to an amplifier as an electrical signal with the same frequency characteristics as those of the mechanical vibration that excited the material.

Kilohertz. One thousand hertz; abbreviated kHz.

Microphone. An electroacoustic transducer that responds to sound waves and delivers essentially equivalent electric waves.

Modulation. Variation in the value of some parameter characterizing a periodic oscillation. The best known of these are amplitude modulation (AM) and frequency modulation (FM). Another type uses a single-sideband suppressed carrier (SSB).

AM and FM radio receivers are common radio receiver devices for detecting these modulations.

Plosive. Speech sound made by impounding the air stream for a moment, then suddenly releasing it.

Phoneme. Smallest unit of speech that distinguishes one utterance from another.

Piezoelectricity. Property exhibited by some asymmetric crystalline materials which, when subjected to strain in suitable directions, develop electrical polarization proportional to the strain.

Sound pressures (underwater). Expressed in decibels or in dynes per square centimeter.

Sound velocity. Rate of travel at which sound energy moves through a medium, usually expressed in feet per second. The velocity of sound in sea water is a function of temperature, salinity, and the changes in pressure associated with changes in depth. An increase in any of these factors tends to increase the velocity. Temperature has the greatest influence, with pressure and salinity exercising less influence. (For detailed discussion of sound velocity in sea water see Chapter I.) At 70°F and normal salinity of 34 parts per thousand by weight, the sound velocity is 4935 ft/sec at the ocean surface (speed of sound in air is 1090 ft/sec).

Stop. A pause or breaking off of speech.

Transducer. Any device for converting energy from one form to another (electrical, mechanical, or acoustical).

Ultrasonic. Above audible frequency range, i.e., above 20,000 Hz.

Velocity of sound. See Sound velocity.

Vowel. One of a class of speech sounds in the articulation of which the oral part of the breath channel is not blocked and is not constricted enough to cause audible friction; in the English alphabet /a/, /e/, /i/, /o/, /u/, and sometimes /y/.

Wavelength of sound. Ratio of speed to frequency.

References

ADAMS, R. L. Can you communicate? Underwater? In: *1971 Offshore Technology Conference, Houston, Texas, April 1971, Preprints*, Vol. II, p. 747–752. Published by the Conference (1971).

BEIL, R. G. Frequency analysis of vowels produced in a helium-rich atmosphere, *J. Acoust. Soc. Amer.* 34:347–349 (1962).

BRUBAKER, R. S., and J. W. WURST. The correction of helium speech distortion. State College, Pa., H. R. B. Singer, Inc., Final Report (1967).

BRUBAKER, R. S., and J. W. WURST. Spectrographic analysis of divers' speech during decompression. *J. Acoust. Soc. Amer.* 43(4):798–802 (1968).

COLEMAN, R. F., and W. R. KRASIK. Oral muzzle pressure effects in underwater communication. Gainesville, Fla., Univ. Fla., Commun. Sci. Lab. Rep. CSL/ONR #19 (Jan. 19, 1971).

FANT, G., and B. SONNESSON. Diver's speech in compressed air atmospheres. *Milit. Med.* 132:434–436 (1967).

FANT, G. M., J. LINDQVIST, B. SONNESON, and H. HOLLIEN. Speech distortion at high pressures. In: Lambertsen, C. J., ed. *Underwater Physiology. Proceedings of the Fourth Symposium on Underwater Physiology*, pp. 293–299. New York, Academic Press (1971).

GILL, J. S. The Admiralty Research Laboratory Processor for Helium Speech. In: Sergeant, R. L., and T. Murry, eds. *Processing Helium Speech: Proceedings of a Navy-Sponsored Workshop, Aug. 1971*, pp. 34–38. U. S. Nav. Submar. Med. Res. Lab., Rep. NSMRL 708 (May 22, 1972).

Giordano, T., H. B. Rothman, and H. Hollien. Helium speech unscramblers—A critical review of the state of the art. Gainesville, Fla., Univ. Fla., Commun. Sci. Lab., Rep. CSL/ONR 45 (Aug. 1, 1972).

Hollien, H., J. F. Brandt, and J. Malone. Abstracts of four studies in underwater communication. Gainesville, Fla., Univ. Fla., Commun. Sci. Lab., Rep. CSL/ONR 20 (Mar. 1, 1968a.)

Hollien, H., R. F. Coleman, C. L. Thompson, and K. Hunter. Intelligibility of diver communication systems, Gainesville, Fla., Univ. Fla., Commun. Sci. Lab., Rep. CSL/ONR 11 (Dec. 15, 1968b).

Hollien, H., R. F. Coleman, C. L. Thompson, and K. Hunter. Evaluation of diver communications systems under controlled conditions. In: *Undersea Technology Handbook*, Chapter 11. Arlington, Va., Compass Publications (1970).

Hollien, H., C. L. Thompson, and B. Cannon. Speech intelligibility as a function of ambient pressure and HeO₂ atmosphere. Gainesville, Fla., Univ. Fla., Commun. Sci. Lab., Rep. 18, CSL/ONR (Feb. 15, 1971) (AD722-371).

Hunter, E. K. Problems of diver communication. *IEEE Trans. Audio Electroacoust.* AU-16:118–120 (Mar. 1968).

Kenny, J. E. Diver communications. *Ocean Ind.* 6:36–37 (Feb. 1971).

Lindqvist, J. Present and future work in underwater communications at the speech transmission Laboratory in Stockholm. In: Sergeant, R. L., and T. Murry, eds. *Processing Helium Speech: Proceedings of a Navy Sponsored Workshop, Aug. 1971*, pp. 22–25. U. S. Nav. Submar. Med. Res. Lab., Rep. NSMRL 708 (May 22, 1972).

MacLean, D. J. Analysis of speech in a helium–oxygen mixture under pressure. *J. Acoust. Soc. Amer.* 40(3):625–627 (1966).

Martin, W. R., and R. L. Adams. Physiologic factors in the design of underwater communication. *Mar. Technol. Soc. J.* 4:55–58 (May–June 1970).

Morrow, C. T. Speech in deep-submergence atmospheres. *J. Acoust. Soc. Amer.* 50(3, Part 1):715–728 (1971).

Murry, T., E. J. Nelson, and E. W. Swenson. Speech intelligibility during exercise at normal and increased atmospheric pressures. U. S. Nav. Submar. Med. Cent., Rep. NSMRL 700 (Feb. 9, 1972).

Otis, A. B., and R. G. Clark. Ventilatory implications of phonation and phonatory implications of ventilation. *Ann. N.Y. Acad. Sci.* 155:122–128 (1968).

Penzias, W., and M. W. Goodman. *Man Beneath the Sea.* New York, Wiley–Interscience (1973).

Sergeant, R. L. Speech during respiration of mixtures of helium and oxygen. *Aerosp. Med.* 34:826–829 (Sept. 1963).

Sergeant, R. L. The effect of frequency passband upon the intelligibility of helium speech in noise. U. S. Nav. Submar. Med. Cent. Rep. SMRL 480 (Aug. 1966).

Tolhurst, G. C. History of helium speech processing instrumentation. In: Sergeant, R. L., and T. Murry, eds. *Processing Helium Speech: Proceedings of a Navy Sponsored Workshop, August 1971*, pp. 4–10. U. S. Nav. Submar. Med. Res. Lab., Rep. NSMRL 70B (May 22, 1972).

Undersea Medical Society, U. S. Navy Supervisor of Diving dive—1600 ft. *Pressure Med. Physiol.* 2(Spec. Insert. 2):1–4 (July–Aug. 1973).

Wathen-Dunn, W. Workshop summary and comments. In: Sergeant, R. L., and T. Murry, eds. *Processing Helium Speech: Proceedings of a Navy Sponsored Workshop, August 1971*, pp. 59–63. U. S. Nav. Submar. Med. Res. Lab., Rep. NSMRL 708 (May 22, 1972).

Woods, J. D., and J. N. Lythgoe. Apparatus and methods for the diving scientist. In: Woods, J. D., and J. N. Lythgoe, eds. *Underwater Science*, pp. 1–31. London, Oxford University Press (1971).

Selection and Training of Divers

The evidence is incontrovertible that if one wishes a task to be completed efficiently and safely, one must carefully select the individual who is to perform the task according to standard techniques and then see that there is proper motivation and adequate training. In underwater activity there is usually no problem of motivation, for most of the individuals are highly motivated volunteers, but there is a problem of selection and training. For selection, it is largely a matter of selecting out, i.e., not allowing those with disqualifying defects to come into underwater activity.

Few, if any, countries have control over either selection or training, except for their military personnel. Reliance must be placed on educating the public and this is a difficult task. Even though information programs have been rather successful at promoting training in the U. S., where most scuba enthusiasts are proud of their "certificate," in many cases the training has been rather inadequate. No attempt, however, is made at selection by any of the training groups. This matter is left to the individual and his personal physician, neither of whom, in most cases, has any idea what the problems are. Education is the only answer.

A. *Selection*

Although much is known about selection, further research is necessary to establish definitive guidelines. This opinion was expressed by Captain Bornmann, Medical Corps, U. S. Navy (Bornmann 1967) when he expressed the hope that perhaps we can "develop tests to eliminate from deep diving the individuals who would be bends-prone"—overage, overweight, physically unfit, and the diver who has already had the bends several times. Research is going on and will continue to lead to development of better tests and screening methods. Here we present what is generally accepted by the professional community as having at least face validity in selecting for underwater activity.

1. Objectives of Selection

It cannot be stressed too strongly that the primary objective of selection is protection of the individual. Psychological and physiological idiosyncrasies characteristic of individuals must be carefully evaluated in light of possible harmful responses to the psychological and physiological stresses that are part of the undersea activity. Also, Penzias and Goodman (1973) point out that "adherence to promulgated standards enhances input uniformity of personnel, thereby assisting in definition of meaningful operational and safety routines by means of controlled intra-group variability. Data accumulation, for future statistical appraisal, is a subsidiary, but no less important objective of a comprehensive program. Future evaluation of the validity of the certification standards must be based on such information."

The objectives of selection (or criteria for disqualification) are set forth by King (1970) in three broad groups: any pathological condition that might lead to entrapment of compressed air, any condition that increases susceptibility to compressed air illness, any condition with symptoms that might be confused with those of severe compressed air illness.

It must be pointed out, however, that in most selective physical examinations, standards for acceptance or disqualification cannot be shown to be related in a statistically significant way to the production of compressed air illness, to the likelihood of accidents, or to work proficiency. Some of the standards, for example, are based on prima facie evidence that a person likely to lose consciousness by reason of uncontrolled diabetes or epilepsy should never be allowed to dive; or that a person with skin disease or active venereal disease should restrict diving, at least where clothing and apparatus must be shared. The consensus of opinion for each element will be discussed in this chapter under the topic, physical and medical examination.

2. Medical History

The examining physician must see that the history is particularly complete in four areas (Merer 1968): neuropsychical, respiratory, cardiovascular, and otorhinolaryngological. In taking the history it is well to keep in mind the various diseases and conditions considered to be disqualifying or contraindicated. Merer lists the following conditions as permanent contraindications for diving: "Chronic necrotic otitis; Ocena (a disease of the nose with an offensive discharge); chronic tubular catarrh with obstruction; old (unhealed) perforation of the tympanic membrane; chronic allergic or infectious conditions of the nasal sinuses with hyperplastic mucosa polyps, etc. It is necessary to be extremely cautious regarding old skull injuries with attendant unconsciousness, or otorhinolaryngological operations."

The Manual of the Medical Department of the U. S. Navy indicates "that a history of the following diseases shall be disqualifying: tuberculosis, asthma, chronic pulmonary disease; chronic or recurrent sinusitis, otitis media, otitis externa, chronic or recurrent orthopedic pathology; chronic or recurrent gastrointestinal disorder; chronic alcoholism; and history of syphilis, unless adequately treated and symptom free; and emotional, temperamental and intellectual condition shall be normal."

A most extensive list of contraindications for diving activity was compiled by the Belgian Medical Commission (Mortier and Wellens 1971) as follows:

Cardiovascular contraindications:
  Anatomical alteration of the coronary network
  Sequela of myocardial damage (rheumatic or other) with serious modification of
      function as determined by auscultation, ECG, or circulatory response;
  Serious arrythmias
  Cardiomyopathies
  Arteriosclerosis with myocardial or coronary involvement
  Arteriosclerosis with serious alteration of the peripheral or cerebral arterial
      circulation
  Systolic–diastolic hypertension of Grade III and IV.
  In general all cardiovascular affections likely to lead to shock or fainting.
Oto-rhino-laryngological contraindications:
  Serious vestibular alteration of any origin
  Serious cochlear alteration of any origin
  Serious alteration of the middle or external ear of any origin
  Acute sinusitis
Pulmonary contraindications:
  Tuberculosis in all of its forms
  Serious respiratory insufficiency
  Acute pathology of the bronchial tree
Neurological contraindications
  Conditions seriously affecting neuromuscular coordination
  Uncontrolled epilepsy
  Recurrent migraine
  Sequela of cerebral trauma with altered EEG
  Tetany
Urological and gynecological contraindications:
  All acute kidney, ureter, vesicular and gynecological infections
Psychiatric contraindications:
  All ailments leading to serious instability
  All ailments leading to alteration of the intellect
Hormone and metabolic contraindications:
  All serious ailments of the endocrine glands and particularly uncompensated
      diabetes
Infections that are contraindicated:
  All active infectious diseases and particularly involvement of the liver
Serious fatigue and periods of convalescence
Hypersensitivity to cold, allergy to cold, or to water, or to organisms present in the
    underwater environment
Pharmacological contraindications:
  All drugs affecting performance
  Alcohol

The Committee on Diver Training and Performance Standards of the U. K. Society for Underwater Technology (Society for Underwater Technology 1972), in the section on "Medical examination of commercial divers," lists the following conditions as disqualifying a man from diving.

Chronic catarrh of upper air passages, in particular recurrent sinus infection
Perforated eardrum, chronic otitis media, or mastoid operation
Inability to "clear the ears"

Any chronic lung disease, past or present; bronchial asthma, history of pneumo-
thorax
Heart disease; essential hypertension
Peptic ulcer
Hernia
Epilepsy; severe head injury; cranial injury; disease of the central nervous system
Severe hearing or visual defects
Excessive obesity
Diabetes
Psychiatric disorder
Gross abnormalities of the renal tract

The British CIRIA Underwater Engineering Group (CIRIA 1972) put together a questionnaire to be completed by the diver candidate which is quite comprehensive (Figure XI-1) and which precedes the medical examination by the Medical Officer. They also provided a Memorandum for Medical Officers (Figure XI-2), which lists 13 disqualifying conditions.

MEDICAL EXAMINATION OF COMMERCIAL DIVERS
Section 1. To be completed by the candidate

Full name _____ Date of birth _____
Address _____
1. How long have you been diving _____ Max. depth _____ Max. duration at that depth _____
2. Where were you trained as a diver _____
_____
3. Are you being treated by a doctor at present _____
4. Are you having injections, medicines or tablets _____
5. Are you subject to colds _____
6. Are you ever seasick _____
7. Have you ever had any of the following :

Tuberculosis _____	Pneumonia _____	Bronchitis _____
Asthma _____	Hay fever _____	Spitting of blood _____
Nose bleeding _____	Sinusitis _____	Ear discharge _____
Rheumatic fever _____	Heart disease _____	High blood pressure _____
Digestive troubles _____	Gastric or duodenal ulcers _____	
Nervous breakdown _____	Fits _____	Recurrent headaches _____
Typhoid _____	Dysentery _____	Malaria _____

Other fever _____
Serious injuries _____ Concussion _____ Operation _____
Anaesthetic _____
Decompression sickness ("Bends," "Chokes," or "Staggers," etc.) _____
Damaged ear drums _____ Other diving "accident" _____
8. Is there epilepsy, mental disease, tuberculosis, or asthma in your family _____
9. Give your previous occupations and duration _____
10. Give dates of last inoculations :

Typhoid _____ Tetanus _____ B.C.G. _____ Smallpox _____
Polio _____ Cholera _____ Typhus _____ Yellow fever _____
11. Has your chest been X-rayed _____ When _____
12. Have your bones and joints been X-rayed _____ When _____

I declare the above answers to be true

Signed _____ Date _____

Figure XI-1. Questionnaire to be completed by a commercial diving candidate before the medical examination (CIRIA 1972).

MEMORANDUM FOR MEDICAL OFFICERS

The certification of a man to be fit to dive implies that he is both physically and psychologically healthy. A diver may have to undergo considerable physical strain under adverse conditions and upon his reaction to an emergency may depend not only his own life but also the lives of others. During the examination he should be carefully observed for psychological defects.

A certificate of fitness should not be given when a man is undergoing treatment for any acute condition.

The following conditions disqualify a man for diving :
1. Chronic catarrh of upper air passages, in particular recurrent sinus infection
2. Perforated eardrum, chronic otitis media or mastoid operation
3. Inability to "clear the ears"
4. Any chronic lung disease, past or present, bronchial asthma, history of pneumothorax
5. Heart disease, essential hypertension
6. Peptic ulcer
7. Hernia
8. Epilepsy, severe head injury, cranial surgery, disease of the central nervous system
9. Severe hearing or visual defects
10. Excessive obesity
11. Diabetes
12. Psychiatric disorder
13. Gross abnormalities of the renal tract

Exercise tolerance test
There are tests of different severity but a candidate must not fail to pass the following simple test: After stepping up onto a chair five times in 15 sec the pulse should return to the preexercise rate in 45 sec.

Blood pressure
The systolic blood pressure should not exceed 140 mm/Hg. The diastolic blood pressure should not exceed 90 mm/Hg.

Skinfold test
With the left arm hanging relaxed the skin in the mid triceps area should give a "Harpenden caliper" reading below 15 mm.

Joint radiography
Good definition of trabecular structure of bone is important.

Shoulders—AP projection. Positioning should be such that there is a clear joint space and that the acromion does not overlap the humeral head.

Hips—AP projection. Feet at 90° to table top. Center over the head of the femur. The gonads should be protected.

Knees—AP and lateral projections. Center at the upper border of the patella. The field should include the distal third of the femur and the proximal third of the tibia.

Age
The minimum age for a commercial diver is 18 years.

Figure XI-2. Guidelines to medical officers for certifying commercial divers (CIRIA 1972).

Cigarette smoking, although not reported by any of the authors to be disqualifying, was studied by the Navy Neuropsychiatric Research Unit in connection with the Physical Fitness Scale (PFS). The subjects were 241 Navy enlisted volunteers for Underwater Demolition Team training. Both sports interests and smoking habits were correlated with the PFS and with each other. Interest in sports was positively correlated with PFS scores. Smoking correlated negatively with both the PFS and with sports interest. The results suggest that habits, such as smoking, can be used to specify some of the behavior associated with physical fitness (Biersner *et al.* 1972).

Chronic alcoholism is a disqualifying condition in all of the lists but King (1970) put it most picturesquely: "Chronic alcoholism also disqualifies a man from working in compressed air. One of the workers' terms for neurological bends is 'the staggers.' When 'the staggers' are 80 proof in origin, diagnosis and treatment become inordinately complex."

An insurance official of Lloyd's of London considers medical, mental, and dental checkups every six months absolutely necessary, and further states: "Care should be taken to avoid absolutely the use of alcoholics or psychiatric patients in diving" (Dawson 1972).

3. Physical and Medical Examination

An essential consideration in examining candidates is that "fitness for diving implies fitness for the performance of heavy physical exertion . . . and limitations which prohibit the performance of hard physical work must be considered, as practically absolute disqualifying characteristics" (Penzias and Goodman 1973). There are many specific factors and conditions to be considered and they will be taken up in the order given in the U. S. Navy Diving Manual (1970). In each case the first part of the paragraph reports the Navy standard and the additional material is taken from other reports as indicated.

a. Age

"Candidates beyond the age of 30 years shall not be considered for initial training in diving, the most favorable age being from 20 to 30. All divers upon reaching the age of 40 shall be examined in accordance with sub-article 15-30 (U. S. Navy 1970). Miles (1966) takes a more lenient view: "Professional bodies do not usually accept men for training of over 35 years but many divers continue active work until very much older and a few still carry on diving in their eighties. Age is not really of great importance and provided there is a good standard of fitness, it should not debar the enthusiast." It must be remembered that diving involves heavy physical exertion and strenuous activity. King (1970) says: "The question of age and its effect on being able to work safely under compressed air is a point of much debate. Miners and other manual workers are discouraged if they are over 45; if they are over 55 they no longer qualify."

b. Weight

"Diving candidates should be rugged individuals without tendency toward obesity. Fat absorbs about five times the volume of nitrogen as does lean tissue. Because of the low circulatory rate of fatty tissue, nitrogen is eliminated very slowly, thus increasing the incidence of bends. It is considered in general that candidates should present no greater than 10% variation from standard age–height–weight tables . . . Consideration will be given, however, to applicants whose overweight is considered to be due to heavy bone and muscular structure" (U. S. Navy 1970).

King (1970) says that obese workers are more prone to bends than workers of a leaner build. He goes on to say:

> In an attempt to determine obesity, the circumference of six muscles (both biceps, forearms, and calves) and the circumferences of six bony structures (both wrists, knees and ankles) are incorporated into a formula with the circumference of the shoulders and the height to determine an anthropometric, or ideal body weight. The applicant's anthropometric weight is subtracted from his actual weight; the result is the number of pounds overweight [Table XI-1]. Since certain body builds tend to skew these calculations, a clinical estimate is also made of the patient's obesity. If the clinical estimate is greater than the calculated obesity, the clinical weight is then used as the more accurate figure. If found overweight, the applicant is disqualified until the excess weight is lost.

The overweight consideration according to Meckelnburg (1972) "is more that the person needs to be in good physical condition than merely his total body mass. It should be assumed that if good physical condition is achieved then the weight will take care of itself."

Among the indices which should be employed with judgment, Penzias and Goodman (1973) list weight, which should be considered "in terms of the individual candidate, and the capacity in which he will be expected to function." Matthys (1970) considers candidates over- or underweight by more than 10% to be undesirable. Behnke (1963) made a plea for allowing the older man to dive under saturation conditions: " Prolonged residence in compressed air without the need for daily decompressions, makes possible the employment of the older man who, although corpulant, has the proved qualifications of competency, judgment, and stability." Wise (1963) did a retrospective study of men who had experienced decompression sickness and those who had never had any trouble to determine if there was a relationship between constitutional factors (age, height, weight, and body build) and decompression sickness. He found that, "None of the obtained mean differences between the divers incurring bends and those who did not were of sufficient magnitude to distinguish between the groups. Therefore, no selection criteria could be established." He went on to say: "The findings suggest that the role of adipose tissue in the etiology of decompression sickness is not as great as has been thought." This section is well summarized by Penzias and Goodman (1973): "Physiological criteria of decompression sickness susceptibility have not, to date, been incorporated into selection criteria . . . neither age, nor height, nor body weight, nor type of physique can be shown, with statistical validity, to be useful for selection predictors."

Although the Manual of the Medical Department of the Navy does not consider the problem of bone and joint involvement, many contemporary writers stress the need for long bone and major joint survey by radiology, "since any indication of aseptic bone necrosis or any other bone lesion that might facilitate its development, is disqualifying" (King 1970). Osteoarthritis of the shoulder and hip are disqualifying according to Godefroid (1965). The French take this problem quite seriously and Chauderon (1971) says that: "Hips and shoulders are x-rayed at the time of hiring, again 15 days later, and then after every three months." The British apparently are more conservative: Miles (1966) suggests that "it might be advisable therefore to

Table XI-1

Anthropometric (B') Weight as a Monitor of Individual Overweight: Relative Overweight = Body Weight − B' Weight[a]

B' cm	Factor	B' cm	Factor	B' cm	Factor	h cm	Factor	h cm	Factor	h in.	Factor	h in.	Factor
220	15.71	200	12.99	180	10.52	200	8.14	180	7.56	79	17.95	69	16.33
219	15.57	199	12.86	179	10.40	199	8.11	179	7.53	78.5	17.87	68.5	16.25
218	15.43	198	12.73	178	10.29	198	8.09	178	7.50	78	17.79	68	16.17
217	15.29	197	12.60	177	10.17	197	8.06	177	7.47	77.5	17.71	67.5	16.09
216	15.15	196	12.47	176	10.06	196	8.03	176	7.44	77	17.63	67	16.00
215	15.01	195	12.34	175	9.94	195	8.00	175	7.41	76.5	17.55	66.5	15.91
214	14.87	194	12.22	174	9.83	194	7.97	174	7.39	76	17.47	66	15.83
213	14.73	193	12.09	173	9.72	193	7.94	173	7.36	75.5	17.39	65.5	15.75
212	14.59	192	11.97	172	9.60	192	7.91	172	7.33	75	17.31	65	15.66
211	14.45	191	11.84	171	9.49	191	7.88	171	7.30	74.5	17.23	64.5	15.58
210	14.32	190	11.72	170	9.38	190	7.86	170	7.27	74	17.15	64	15.49
209	14.18	189	11.60	169	9.27	189	7.83	169	7.24	73.5	17.07	63.5	15.41
208	14.05	188	11.47	168	9.16	188	7.80	168	7.21	73	16.99	63	15.32
207	13.91	187	11.35	167	9.05	187	7.77	167	7.18	72.5	16.91	62.5	15.24
206	13.78	186	11.23	166	8.95	186	7.74	166	7.15	72	16.82	62	15.15
205	13.64	185	11.11	165	8.84	185	7.71	165	7.12	71.5	16.74	61.5	15.07
204	13.51	184	10.99	164	8.73	184	7.68	164	7.09	71	16.66	61	14.98
203	13.38	183	10.87	163	8.63	183	7.65	163	7.06	70.5	16.58	60.5	14.90
202	13.25	182	10.75	162	8.52	182	7.62	162	7.03	70	16.50	60	14.81
201	13.12	181	10.64	161	8.42	181	7.59	161	6.99	69.5	16.42	59.5	14.73

[a] From Behnke (unpublished) by permission of the author.

Conversion factors: B' weight = B' factor × stature (h) factor.

B' weight is the ponderal equivalent of extremity girths.

Equations (I) B' weight (kg) = $(B'/55.5)^2 \times h^{0.7}$ (stature, cm).

(II) B' weight (lb) = $(B'/55.5)^2 \times h^{0.7}$ (stature, in.) × 0.843.

Stature h: Measure with headboard against a wall chart, head in the Frankfort plane. (I) Measure to the nearest cm. (II) Measure to the nearest half inch.

B' Girths: Measure to the nearest mm with flexible tape, the following girths of the right and left upper and lower extremities: *Examinee sitting*: biceps, full flexion; forearm, maximal girth, arm extended; wrist: minimal (circumcarpal) girth. *Examinee standing on elevated platform*: knee, with patella relaxed, measure at mid-patellar level; calf, maximal girth of the leg; ankle, minimal girth of the leg.

B' = sum of the 12 girths/2.

consider periodic radiological examination of hips, knees, and shoulders, say every three to five years."

That the problem is important is shown by the work of Werts and Shilling (1972), who found 218 scientific articles in a world-wide survey dealing with what is now known as dysbaric osteonecrosis, and pointing out the rather alarming incidence of the condition, and its permanent and disabling character.

c. Vision

The Navy requires a "minimum of 20/30 vision bilateral, corrected to 20/20" and that the ophthalmoscopic examination shall be normal. Most of the others writing on selection do not even mention vision. A few say that those who must wear glasses to see are disqualified for "technical reasons" (Matthys 1970).

d. Color Vision

The Navy requires normal color vision. Others either ignore it or say color-blindness is not important.

e. Teeth

The Navy requires a complete dental examination, but does not disqualify unless bridges, dentures, or malocclusion "interfere with the effective use of self-contained underwater breathing apparatus (scuba)." This latter requirement is also stressed by Miles (1966), "If the swimmer uses a mouthpiece it is essential that, with or without dentures, he is able to make a good seal with it." "Cavities and abscesses in the teeth can entrap air and cause severe pain" (King 1970). The teeth and mouth should be cleaned regularly before hyperbaric activity.

f. Ears

The Navy says "acute or chronic disease of the auditory canal, tympanic membrane, middle or internal ear shall be disqualifying. Perforation or marked scarring and/or thickness of the drum shall be disqualifying. The eustachian tubes must be freely patent for equalization of pressure changes. Hearing of each ear shall be normal." There is universal agreement on the disqualification of individuals with various forms of pathology of the ear. Individuals with perforated eardrums must not be allowed to dive because of "the possibility of having cold ocean or quarry water enter directly into the middle ear and cause extreme vertigo" (Meckelnburg 1972). "Ears must be very carefully checked. A middle ear exposed to pressure variations by an unhealed perforation in the tympanum will cause loss of balance. Chronic ear or sinus infections disqualify because patent tubes in the ENT system are essential" (King 1970). "It is essential that the diver should be able to clear both eustachian tubes. Though frequently a temporary condition, repeated failure of either tympanic membrane to give a positive Valsalva reaction is an indication for stopping diving" (Miles 1966). As noted in Figure XI-2, "perforated eardrum, chronic otitis media or mastoid operation" and "inability to 'clear the ears' " are disqualifying conditions (CIRIA 1972).

"Ear drum scars which arch over during Valsalva maneuvers and chronic infections cannot be tolerated in divers" (Matthys 1970).

Although in most cases hearing defects do not disqualify a worker, it is important to obtain a preemployment audiogram in order to evaluate possible hearing loss or potential complaints of hearing loss (King 1970). Not mentioned in the Navy criteria but mentioned by others is that "disturbances in orientation ability and labyrinthine sense preclude diving" (Matthys 1970). Fifty-four healthy male subjects were examined using the Coriolis acceleration test, and were classified as tolerant or intolerant to vestibular stimuli (Galle *et al.* 1971). It is concluded that the test is quite detailed and "can be used in the occupational screening of a wide range of personnel."

g. Nose and Throat

"Obstruction to breathing or chronic hypertrophic or atrophic rhinitis shall disqualify. Septal deviation is not disqualifying in the presence of adequate ventilation. Chronically diseased tonsils shall be disqualifying pending tonsillectomy. Presence or history of chronic or recurrent sinusitis is cause for rejection" (U. S. Navy 1970). "Chronic catarrh of upper air passages, in particular recurrent sinus infections" (CIRIA 1972) are disqualifying conditions. Acute and chronic sinus problems are considered by Meckelnburg (1972) as "relative contraindications" for skin and scuba diving, for such conditions " do respond with time and the proper treatment." All individuals with "chronic rhinopharyngitis, . . . must be eliminated" (Godefroid 1965). As in most physical conditions, it is a matter of judgment related to what the applicant wants to do. Surely there is a difference between sport diving and arduous tunnel work.

h. Respiratory System

"The lungs shall be normal as determined by physical examination and 14- by 17-inch chest X-ray" (U. S. Navy 1970). "It is important that there should be no evidence of any chronic respiratory disease such as tuberculosis, catarrh, sinusitis, asthma or emphysema" (Miles 1966). Conditions disqualifying a man for diving are: "Any chronic lung disease, past or present, bronchial asthma, history of pneumo-thorax" (CIRIA 1972). "X-rays of the thorax are examined for calcification in the lungs, pleurisy, and bronchial obstructions, all of which are grounds for disqualifi-cation" (Matthys 1970). Once again for the sports diver, Meckelnburg (1972) takes a more lenient position: "Asthma and emphysema—The severity of the condition dictates whether or not the individual can participate in this type of sporting event. Mild to moderate degrees of either of these conditions probably do not constitute adequate reason for eliminating the individual whereas severe degrees of either condition would mean that this type of activity would be contraindicated."

i. Cardiovascular System

"The cardiovascular system shall be without significant abnormality in all respects as determined by physical examination and tests. The blood pressure shall not exceed 145 mm systolic, or 90 mm diastolic. In cases of apparent hypertension, repeated daily blood-pressure determinations should be made before final decision" (U. S. Navy 1970). CIRIA (1972) recommends that the systolic blood pressure should not exceed 140 mm Hg and the diastolic blood pressure should not exceed 90 mm Hg. "It should be kept in mind that valuable indication of undesirable excitable tempera-ment is often revealed by vasomotor manifestations. Persistent tachycardia and arrhythmia except of sinus type, evidence of arteriosclerosis (an ophthalmoscopic examination of the retinal vessels should be included in the examination), varicose veins, marked or symptomatic hemorrhoids shall be disqualifying" (U. S. Navy 1970). Miles (1966) says: "There must be no gross impairment of exercise tolerance or evidence of cardiovascular disease." He gives the limits of resting blood pressure as systolic 140 mm Hg, but allows a diastolic of 100 mm Hg. "Obviously mild to moder-ate elevations of the blood pressure probably would have no effect whatsoever on the individual while he is diving. To date there is no experimental evidence that pressure externally applied influences the intravascular pressures. A severe elevation of blood pressure with concomitant possibilities of causing imminent catastrophies in different organs of the body would contraindicate skin or scuba activities" (Meckelnburg 1972).

j. Gastrointestinal System

Candidates subject to gastrointestinal disease are disqualified by the U. S. Navy (1970). Peptic ulcer is also a disqualifying condition (CIRIA 1972).

k. Genitourinary System

The Navy disqualifies divers for the following conditions:

Chronic or recurrent genitourinary disease or complaints (normal urinalysis required).
Active veneral disease or repeated venereal infection.
History of clinical or serological evidence of active or latent syphilis within the past 5 years, or of cardiovascular or central nervous system involvement at any time. An applicant who has had syphilis more than 5 years before must have negative blood and spinal fluid serology.

Matthys (1970) says that varicoceles and hydroceles preclude diving and that a urinalysis, including sediment, belongs in every routine examination, but he believes that kidney and urinary infections are a temporary impediment to a diver.

l. Skin

"There shall be no active, acute, or chronic disease of the skin on the basis of infectiveness and or offensiveness in close working conditions and interchange of diving apparel" (U. S. Navy 1970).

m. Pregnancy

Although the Navy is silent on this condition, no deleterious effects of increased pressure upon the developing fetus have been reported, and the woman diver should make her own decision.

4. Requirements in Other Countries

Although in no way binding for any U. S. activity, we believe it worthwhile to look at some of the requirements for selection of divers as practiced in other countries. For example, in the French Navy, there are five classes of divers (Joly 1972):

1. "Ordinary" divers, who dive occasionally in open-circuit compressed-air apparatus to 30–40 m.
2. Mine-clearing divers, who go to 60–70 m on open-circuit compressed-air apparatus and occasionally use semiclosed high-oxygen mixtures at lesser depths, and who carry out tasks in cold, turbid water.
3. Combat swimmers, who are qualified to carry out numerous tasks and who operate on closed-circuit pure-oxygen equipment at depths shallow enough to preclude the dangers of oxygen toxicity.
4. "Hard-hat" divers, who are experienced divers with additional qualifications enabling them to work long hours under difficult conditions, to depths of 60 m, on air.
5. Intervention divers, selected from the most experienced mine-clearing divers, who go to depths greater than 200 m, on experimental and test dives, on binary or tertiary helium-breathing mixtures.

The medical examination given to all candidates is briefly described as follows: The respiratory system, the cardiovascular system, and the nervous system must show no abnormalities; other examinations include: hematological and urinological; dental (removable dentures are contraindicating); vision (requirements are not high); oto-laryngological (these are of the utmost importance, and any chronic disorders of the passages are definite contraindications, while acute conditions are temporary contraindications). Three essential tests are (1) audiogram, (2) vestibular function, (3) tube permeability. For the depth divers, psychological tests are important. Life history questionnaires are valuable. Personality tests to determine social attitudes and personal interviews directed especially toward motivation are important. Stability, maturity, judgment, flexibility of intelligence, and sharpness of memory and attention, plus a high degree of sociability are necessary qualifications. Motivation is fundamentally important; it must stem from a genuine ambition accompanied by a healthy self-criticism and not from a neurotic wish to be a hero.

Some interesting observations concerning examination and selection of personnel in Monaco are worth reporting (Aquadro 1965):

> Routine accessory studies should include dental examination, including dental radiographs; chest radiographs (lungs, heart and vascular system, etc); and analysis of blood and urine. Special lung, heart, vascular and other function studies may be required in questionable cases in which abnormalities are suspected or for participants who may be subjected to unusual trials or great effort. Tests of ability to accommodate to pressure changes and of oxygen tolerance within a recompression chamber are important supplements to the medical examination.

It is pointed out that both psychological and physiological considerations are interrelated and in many cases cannot be clearly separated for the purposes of selection. Although it is not necessary to be a superman, it is essential to have satisfactory functioning of the respiratory, cardiovascular, nervous, skeletal, and other body systems. A thorough physical examination is suggested, in addition to a past medical history, including serious illnesses, surgical procedures, and serious injuries.

5. *Psychiatric Examination*

Frequently the psychiatrist and the clinical psychologist are part of the selection team. The U. S. Navy (1970) assumes the presence of a psychiatrist when they speak of "temperament" and for selection say:

> The special nature of diving duties requires a careful appraisal of the candidate's emotional, temperamental, and intellectual fitness. Past or recurrent symptoms of neuropsychiatric disorder or organic disease of the nervous system shall be disqualifying. No individual with a history of any form of epilepsy, head injury with sequelae, or personality disorder shall be accepted. Neurotic trends, emotional immaturity or instability and asocial traits, if of sufficient degree to militate against satisfactory adjustment, shall be disqualifying. Stammering or other speech impediment is disqualifying. Intelligence must be at least normal.

No tests are suggested, thus leaving all of the above to the intuitive judgment of the examining physician. However, in evaluating the potential for performance of an individual it is not enough to measure any single attribute. An attempt must be made to measure the *entire man*, and certainly to do this one must take into consideration the most important factors of all—social, emotional and ideological makeup, and above all motivation.

In the memorandum for medical officers (Figure XI-2) CIRIA stresses the need for the diver to be "psychologically healthy" and states: "During the examination he should be carefully observed for psychological defects." Epilepsy, severe head injury, cranial surgery, disease of the central nervous system, and psychiatric disorders are listed as conditions for which the candidate should be disqualified.

"Persons affected by sensibility disturbances, crippling of the peripheral nerves, or with the after effect of poliomyelitis, meningitis, encephalitis, severe skull traumata, or tendency to lose consciousness are not suited for diving . . . Neurotic tendencies, emotional and intellectual immaturity, as well as professional and social shortcomings must be detected at an early stage and the subject prevented from diving" (Matthys 1970).

Neurological contraindications listed by Patte (1973) are: "affections seriously altering neuromuscular coordination; uncontrolled epilepsy; severe migraine headache; sequela of cerebral concussion with alteration of the E.E.G. and tetany."

Probably the most important single item in the selection battery is that the candidate must be a volunteer. "Self selection or natural selection presumably functions as an initial portion of the selection battery: without sufficient motivation training never begins in the first place" (Penzias and Goodman 1973). "Both professionals and amateurs are primarily volunteers for underwater work or recreation and must have some special reason for wishing to do so. Even with the professional, money is rarely the primary motivating factor. More often it is a genuine love of water and a sense of adventure. In some cases there is a desire for individual achievement and in others a wish to belong to an exclusive group. Quite a few take up diving as a means to an end as with archeologists, photographers, cave explorers and fishermen who simply wish to extend their sphere of interest and activity" (Miles 1966). As pointed out by Wise (1963), although psychological fitness has perhaps less face validity for diving, it has been given sufficient weight for the formulation of the universal rule that all divers must be volunteers.

Although the general problem of motivation is considered in Chapter VI, it is worthwhile here to mention that motivation plays a most important part in adaptation to the underwater environment. Siffre (1963) stated it eloquently. "Man's faculty of adaptation to conditions absolutely different from any known on the earth's surface is more and more being demonstrated, and found to be almost unlimited . . . This power . . . is closely linked with the individual's motivation . . . It is impossible to exaggerate how greatly a dynamic spirit can extend the physiological limits set for human endurance."

The need for selection tests that will indicate motivation was set forth by Joly (1972) for the French navy. "For the depth divers, psychological tests are important. Life history questionnaires are valuable. Personality tests to determine social attitudes, and personal interviews directed especially toward motivation are important. Stability, maturity, judgment, flexibility of intelligence, and sharpness of memory and attention, plus a high degree of sociability, are necessary qualifications. Motivation is fundamentally important; it must stem from a genuine ambition accompanied by a healthy self-criticism, and not from a neurotic wish to be a hero."

It is important to select persons who will be able to stand up under periods of stress (see Chapter VI) and persons with evidence of past emotional decompensation during periods of stress should be considered as unacceptable. In the absence of any such history, stability is difficult to predict, and thus, it is difficult to select for ability to withstand stress, but Aquadro (1965) states that tendencies which militate against satisfactory adjustment are below-average intelligence, tendencies to claustrophobia, unhealthy motivations, history of past personal ineffectiveness, difficulties in interpersonal relations, and lack of adaptability.

Claustrophobia is a condition requiring careful evaluation. It would seem that anyone with a morbid dread of being shut up in a confined space would not want to be a candidate for diving, submarine duty, or living in a habitat. But a certain number of borderline cases do apply and some start training. "An individual with true claustrophobia would be adversely affected by having a mask applied over the face. Most individuals with a mild degree of claustrophobia can overcome it with proper training. However, the individual who cannot overcome this fear obviously cannot dive" (Meckelnburg 1972).

Lanphier and Dwyer (1954) noted that, "aside from claustrophobia as a distinct entity, the sense of confinement caused by absence of light, lack of communication, and awareness of solitude can certainly elicit strong reactions in susceptible men. Unfortunately, few of the fundamental facts on the psychiatric aspects of selection of men are actually known." The best test is the confinement of the pressure chamber. In examining men for the submarine service, the men who became emotionally disturbed, and asked to be let out of the pressure chamber were usually found not qualified for submarine duty.

Closely related to claustrophobia are a number of anxiety states mentioned by

Edmonds and Thomas (1972), but there is no way to examine for these reactions that are brought on by stress. For example, they mention what they call the "over-reaction anxiety state" and say: "Owing to some misfortune—e.g., entry of water into the face mask—the subject panics unnecessarily and is likely to behave irrationally. This may result in ascent to the surface without taking adequate precautions, frantic grabs for air supplies, lack of concern for other personnel, etc. This is more likely in those divers with a normal or above normal neuroticism gradient."

The electroencephalogram (EEG) has been used by many different groups to study such environmental effects as nitrogen narcosis and oxygen intoxication, but only the French appear to have used the EEG as a selection device. Early in 1954 during a study of nitrogen narcosis (Cabarrou 1963) the idea was generated that . . . "it seemed possible to establish a relationship between the EEG of a man at rest under atmospheric pressure and the aptitude of this subject for diving." A study was conducted in which all candidates had a pretraining EEG and a forecast was made as to their potentials as divers. Marks were then given in the training course for underwater performance. "Of the 154 pupils completing the course in 8 two-month courses, the forecast was accurate in 88 and wrong for 66. However, in each of the 8 courses, the three pupils with the lowest marks had been given an unfavorable forecast." Although stressing the need for further refinement of the work, Cabarrou concluded that the EEG is valuable in selecting divers.

Concerning contributions of the EEG to the medical examination of divers, Palem and Palem (1971) indicated that the first contribution is the detection of epilepsy, and the second is that it can detect other conditions which make the subject unfit for diving. For example (although they admit that identifying a normal EEG is a difficult and complicated problem), they say, "There are types of recordings that indicate possible unfitness: the flat and rapid pattern indicates nervous anxiety; slow posterior waves might indicate a reckless nature."

Even more positive is Sala Matas (1971), who declares, "Systematic EEG examinations have confirmed, decisively and definitely, the great importance of the EEG in the medical examination of divers . . . The results obtained justify the struggle through the years to require the EEG in medical examination of divers. It has made possible one more step in the prevention of diving accidents." At present in the U. S. we have no experience with the EEG as a selection tool for divers.

Unfortunately the psychiatric examination must be subjective rather than objective and there are those who do not believe that for the purpose of selection of divers it is worth the time of a psychiatrist.

6. Psychological Tests

During the past 35 years well over 100 tests have been devised and tried on divers. These tests were designed to measure human performance (see Chapter VI) in the underwater environment, and covered all human abilities from single-finger dexterity to the most complicated of mental functions; that is, most tests focused on psychomotor and cognitive aspects of human behavior in the hyperbaric and underwater situation. Sensory performance has also been extensively studied. But in almost every case the tests have been designed to measure man's ability to function in a hostile environment, and have not been used for purposes of selection.

Miller (1967) asks and then answers the question of why we want to measure performance:

> Men always have been measuring each other's performance, be it memory, intellectual capacity, motor skills, athletic prowess, or business skill. In the underwater world we are interested in performance measurement for several reasons:
>
> (1) to determine whether man should be included as a free swimmer;
> (2) to increase the probability of survival;
> (3) to better estimate the probability of man performing successfully;
> (4) to establish safety practices and medical limitations;
> (5) to properly select and plan specific tasks;
> (6) to decide on equipment requirements;
> (7) to assist in the selection of performance aids in the form of equipment, drugs, or training;
> (8) scientific curiosity.
>
> There are, of course, numerous additional reasons. It is obvious that in each case we are concerned with determining how well man can do "something," either that which we are specifically measuring or some task in which man plays a role.

Perhaps the most important, and certainly the most unique psychological test for divers is the U. S. Navy's SINDBAD (System for Investigation of Diver Behavior at Depth). This is described in Chapter VI, but since one of the aims is the "development of selection criteria " (Reilly and Cameron 1968), it is worth brief mention here. It is an integrated and partially automated system for the study of human performance in the underwater environment, and can measure human mental and perceptual motor functions at ambient pressures up to 444 lb/in^2. (equivalent to a depth of 1000 ft). The system permits remote administration and scoring of 26 specific tests ranging from simple reaction time to complex manual tracking, and from monitoring a simple display to solving difficult mental arithmetic and symbolic problems. "As a formal test battery and general research tool, the system is expected to have extensive application in the areas of (1) specification of human underwater performance capabilities, (2) delineation of factors of the diving environment which affect performance, and (3) development of diver selection criteria."

The most extensive psychological testing program to date was conducted in connection with TEKTITE II, which put a total of 53 aquanauts, largely scientists, on the ocean floor at a depth of 50 ft. Some of the aquanauts spent as long as 60 days in the underwater habitat, located near the Virgin Islands (Miller et al. 1971). The aquanauts were formed into teams and psychological testing and inventorying of each individual were completed before the team entered the habitat. They were then carefully observed in the "socially confined and stressful environment" (Gilluly 1970). Thus the behavioral scientists were afforded a unique opportunity to observe men in a stressful situation for long periods of time.

The Life History Questionnaire (LHQ) was the most important instrument to emerge. The impetus for the development of the LHQ was to "understand and explain differences among Tektite aquanauts in their ability to work effectively under water, to get along with fellow team mates and to adjust generally to this stressful, isolated and confining environment" (Helmreich *et al.* 1971). The life history question-naire is designed to be wholly objective, requiring no interpretations by the subject, and it could be inaccurate only if the subject deliberately falsified the answers.

The most productive aquanauts, generally speaking, were from a stable, small-town environment and were first-borns. But the most important single item was that they had worked as children. Evidently they learned good work habits that carried over into later life. In presenting conclusions, Helmreich *et al.* (1971) say that it was proven that both male and female aquanauts could adapt sucessfully to the habitat environment, and could perform even more effectively than under normal conditions. But from the standpoint of selection his conclusion of importance is that "the LHQ has proven to be a most effective predictive instrument."

Scuba divers may not require exactly the same psychological characteristics as habitat divers. Using a sample of 71 male diver trainees from scuba diving courses, a study was made (Deppe 1971) of 29 behavioral elements, positively and negatively defined. Instructors reported on their own students, and these data were converted to standard scores, which were "subjected to iterative principal factor analysis followed by varimax rotation to simple structure." The four factors which emerged were diving skill, intellectual ability, task orientation, and emotional maturity. It was noted that social compatibility is apparently unimportant; in fact, it is probable that the success-ful diver would be the type who would shun social activity. Underwater performance and invulnerability to stress were grouped under one factor.

Although social adjustment, compatibility, group orientation, gregariousness, etc., are not important in qualifying candidates for scuba diving, these traits are of paramount importance in selecting a team for long periods of confinement in a habitat. "The compatibility and continued effectiveness of the workers as a member of a group or crew is often as important as his individual capabilities. Cohesion of the group, particularly for prolonged periods in restricted quarters, is of considerable importance in effective team performance and successful accomplishment of its mission" (Aquadro 1965). Man as part of a system and man as part of a group are important but tests to evaluate his contribution are probably different from tests to select man as a separate entity.

In TEKTITE II the overall performance was markedly influenced by the inter-action of the team members. According to Helmreich (1971), "The implications of these findings are that in selecting professionally mixed teams for isolated environ-ments, not only the professional qualifications but also the breadth of the candidate's interests should be considered. The individual who is not only skilled in his own profession but eager to acquire knowledge about other lines of endeavor should be the most effective type of group member."

In two studies of diver personality, one (Ross 1968) failed to demonstrate any significant difference between student divers and a matched control group, and the other (Nichols n.d.) showed a significant difference between divers and the general population "on nine of the first-order factors, generally in the direction to be expected from participation in sport." But neither of these authors nor the many others who have worked in this field were sufficiently confident of their results to suggest that the tests be used to select divers.

The characteristic of personal autonomy was studied in 147 scuba diver trainees at the University of California at Los Angeles by Weltman and Egstrom (1969) with the use of the Pensacola Z scale. They hoped that "observed relationships between psychological characteristics and diving success might be applied to trainee selection and to optimal development of training methods." But although "diver trainees described themselves as individualistically as did groups of astronauts and Antarctic scientists, their Z-scale did not differ significantly from other campus groups. Further, the Z-scale did not significantly differentiate between successful trainees and dropouts in the course."

It cannot be too strongly stated that almost all validation of tests is based on success in the training phase and may have little or no relationship to final success as a diver. There should be some carryover, but the proof is lacking. For example, a study by Wise (1963) found that scores in mechanical and arithmetic tests bore the closest relationship to the final grade of the trainee. "A combined cut-off score of 80 on these tests would eliminate the largest number of potential failures and concurrently reject very few potential successful candidates." This is a worthwhile effort, for it saves time and disappointment on the part of the trainee and time and thus money on the part of the Navy. But these tests may not predict sucess as a diver in the actual working situation.

The study of personality is a complex field and many factors must be considered. Are divers being selected for general duty, or for a particular task? One hears talk of test performance versus work performance; and evaluating diver performance versus performance of divers. It is particularly important in studying performance to take into consideration the total environment, the complexity of the task, the type of life-support equipment, and the tools being used. Test evaluation from one piece of research to another may not be possible because the variables have not been stated or even considered.

7. Training Course as Selection Device

Regardless of the care exercised or the type of tests administered, it is quite impossible to predict which applicants will make successful divers with any high degree of accuracy. It is, of course, possible to eliminate the obvious misfits—physically, psychiatrically, and psychologically—but of those who pass all the tests, a number drop out of training prior to completion. In this group, the reason is often psychological. They find they do not like the underwater work—it is work and not the glamorous recreation they had thought. A latent claustrophobia may become evident. As they learn of the obvious dangers in underwater work, they decide it is not for them.

The French naval research psychologist, Dr. Emile-Jean Caille (Mensh 1970) argues that personality characteristics are indisputably "the most indispensable," and reports that about 30% of candidates fail for psychological reasons. The items of review are education, parental relations, intellectual level, psychosomatic and psychopathological history, social adaptation, reactions to frustration and other stresses, and motivation. But in spite of the careful physical and psychological examination and the elimination of many candidates, 44% fail the course for free diving—47% of these for ineptitude, and 27% for personality or psychophysiological reasons. However, another French researcher (Cabarrou 1963) found only nine dropouts when he studied 163 trainees in eight two-month diving training courses in the French Navy.

A British scientist (Flemming 1972) reports findings more in line with those of Caille:

> ... even after a full Royal Navy medical examination and further selection tests, about 35% of a diving training course usually drop out. This failure rate is a great waste of time and money. Unfortunately the numerous characteristics which make up the personality of the diver do not seem to differ from average in any easily detectable way which would enable one to select divers more accurately, and the standard method is still fundamentally to start training and let the unhappy ones drop out over a period of weeks or months.

According to Deppe (1971):

> Without going into the complexities of motivation ... for the purpose of this study it may be defined as a drive towards a particular goal. For motivation to affect success in the diving course it must persist for the duration of the course. Thus it is unlikely that the person who shows bursts of essentially short-term enthusiasm will succeed or the person who has unrealistic expectations regarding the nature of the course or the satisfaction it provides.

Precourse interest in diving is an important element of motivation.

Achievement and survival under the sea depend on mental reeducation and development of new patterns of behavior. This goes on during the training period and some do not make the adjustment.

Thus, regardless of the apparent loss of time and money, the training course must be considered as the final selection procedure.

8. Pressure Test

The *U. S. Navy Diving Manual* (U. S. Navy 1970), under the heading "Ability to equalize pressure," sets forth the following requirement: "All candidates shall be subjected in a recompression chamber to a pressure of 50 psi to determine their ability to clear their ears effectively and otherwise to withstand the effects of pressure. Due consideration must be given to the presence of an upper respiratory infection which temporarily may impair the ability to equalize pressure because of congestion of the eustachian tube." What they do not mention is that the pressure test is an excellent test for claustrophobia and many a candidate for submarine duty has been eliminated at this level.

The Transit Compressed Air Medical Center in San Francisco (King 1970) requires a "pressurization test" to 14.7 psi, and after three or four failures to equalize pressure in the ears the candidate is "disqualified for work in compressed air." Matthys (1970) considers, "Pressure chamber tests are useful in measuring the ability of various skull orifices, especially the middle ear to equalize pressure."

Most advocates of the pressure test as a selection device are interested in ability to "clear the ears," but Matthys (1970) advocates breathing air in a pressure chamber at a depth of 50 m to detect signs of "depth intoxication" and thus to eliminate those unsuited for diving.

9. *Arctic, Antarctic, or Under-the-Ice Diving*

Special selection criteria are required for picking the men who are going on polar diving expeditions. As pointed out by Jenkins (1973), a major problem in assembling a unit for under-the-ice work is the "lack of personnel with adequate training in cold weather operation and maintenance." Of crucial importance is physical fitness and an active exercise program. Age is probably a factor in that persons below the age of 17 and those over 40 seem to be more susceptible to cold injury. Race is an important factor, with the Nordics being more successful in both the physiological and the psychological adjustment to the environment. "The Negro, with all environmental conditions equalized, is approximately six times more vulnerable to cold injury than Caucasians, and his injury is usually more severe." Attention must be paid to emotional stability and ability to work with a small group in a confined space, for there is a great deal more psychological stress under arctic conditions than in the temperate zone. Once again, the volunteer is important, for: "Obtaining a man's commitment to a mission in the arctic is probably the major task of cold weather indoctrination."

B. *U. S. Navy Divers*

Although probably not applicable to civilian selection and training, the U. S. Navy diver designations, qualification for various types of training, and Navy schools and their curricula are of sufficient interest to be briefly reviewed.

1. *Diver Designation*

a. Officers

According to the Bureau of Naval Personnel Manual, diving officers are divided into four classifications according to their specialization or degree of qualification as follows:

Description	NOBC
1. Ship Salvage Operations Officer	9375
2. Diving Officer (General)	9312
3. Deep Sea (He–O_2) Diving Officer	9313
4. Ship Salvage Diving Officer	9314

In addition to the above there are four Naval Officer Billet Classifications (NOBC), which include a degree of diving qualification collateral to the basic classification. They are:

Description	NOBC
1. Explosive Ordnance Disposal Officer	9230
2. Underwater Demolition Team Officer	9294
3. Sea Officer	9293
4. Submarine Medical Officer	0090

b. Enlisted Divers

The Bureau of Naval Personnel Manual divides qualified enlisted divers into the following Navy Enlisted Classification (NEC) Codes:

Description	NEC
1. Master Saturation Diver	5346
2. Master Diver	5341
3. Deep Dive Systems Diver	5311
4. Diver First Class	5442
5. Medical Deep Sea Diving Technician	8493
6. Salvage Diver	(Mobilization only)
7. Diver Second Class	5343
8. Scuba Diver	5345

In addition to the above eight Navy Enlisted Classifications, other NEC codes that require a degree of diving qualification include:

Description	NEC
1. Underwater Demolition Team Swimmer	5321
2. Underwater Demolition Team Swimmer/ Explosive Ordnance Disposal Technician	5322
3. Combatant Swimmer (Seal Team)	5326
4. Combatant Swimmer (Seal Team Explosive/ Ordnance Disposal Technician)	5327
5. Explosive Ordnance Disposal Technician	5332
6. Underwater Photographer	8136
7. Hospital Corpsman, Special Operations/ Technician	8492

2. *Availability of Training*

a. Qualifying Criteria

Within the Navy, diver training is available to the following groups:

1. Officers: (a) Line officers, Ensign through Lieutenant; (b) Medical Officers, selected by BUMED; (c) Engineering Duty Officers, selected by BUPERS.

2. Enlisted: (a) Diver First Class—Job Classification, BM, GM, TM, DC, SF, MR, MM, EN, CE, EO, EM, BU, SW, UT, IC, and in paygrade E-4 or above. (b) Diver Second Class—same as above, plus job classifications MN, EM, EA, in paygrade E-3 above and SN/FN who desire to pursue advancement to one of the aforementioned ratings. (c) SCUBA Diver—same as above, plus job classifications PH and HM in paygrades E-4 and above who are in training for underwater photographer or Hospital Corpsman Special Operations Technician. (d) Medical Deep Sea Diving Technician—HM in paygrade E-5 or above.

b. Selection Procedure

From this defined population, personnel are selected using the following procedures:

1. Must be a volunteer and psychologically adapted to diving duty.
2. Must be recommended by the Commanding Officer.
3. Be physically qualified in accordance with Article 15-30 Manual of the Medical Department.
4. Be certified a Swimmer First Class in accordance with BUPERS Manual, Article D-2502.
5. Be interviewed by a designated Diving Officer to ascertain aptitude and motivation.
6. Satisfactorily complete a test dive in a deep sea diving suit.
7. Have a combined ARI plus MECH score of 105 (can be waived).

3. *Naval Schools of Diving*

Listed below are the major centers for diver training in the U. S. Navy, and with each facility are the various types of courses provided.

a. Naval School of Diving and Salvage, Washington Navy Yard, Washington, D. C. 20390

Course title	Length, weeks	Skill identifier
Ship Salvage Diving Officer	16	9314
He–O_2 Diving Officer	23	9313
He–O_2 Cross Training	7	9313
ARS/ATF—Prospective XO and CO	3	None
Engineering Duty Officers	8	9314
Medical Department Diving Officer	10	None
Recognition and Treatment Of Diving Casualties	1	None
Master Diving Qualification	5	5341
Diver Second Class	11	5343
Medical Deep Sea Diving Technician	23	8493

b. Naval School, Divers Second Class, Service School Command Annex, U. S. Naval Station, San Diego, California

Course title	Length, weeks	Skill identifier
Diver Second Class	10	5343
Scuba Diver	9	5345

c. Naval Amphibious School, Coronado, San Diego, California 92155

Course title	Length, weeks	Skill identifier
Basic Underwater Demolition	—	9294
Seal/BUBS/Training	20	5321

d. Naval School, Explosive Ordnance Disposal, Naval Ordnance Station, Indian Head, Maryland 20640

Course title	Length, weeks	Skill identifier
Navy UDT/SEAL EOD	5	None
Navy Explosive Ordnance Disposal Basic	30	5332
Scuba Diver	4	5345

e. Naval School, Underwater Swimmers, Naval Station, Key West, Florida

Course title	Length, weeks	Skill identifier
Diver Second Class	10	5343
Scuba Diver	4	5345
EOD Scuba Diver	8	5332
Special Operations Technician	8	8492
UDT Scuba Diver	6	5321

f. Naval Submarine School, Box 700, Groton, Connecticut 06340

Course title	Length, weeks	Skill identifier
Scuba Diver	4	5345

g. Naval Submarine Training Center Pacific (Pearl Harbor) FPO San Francisco, California

Course title	Length, weeks	Skill identifier
Diver Second Class	10	5343
Scuba Diver	4	5345

h. Harbor Clearance Unit Number Two, Naval Station, Norfolk, Virginia

Course title	Length, weeks	Skill identifier
Diver Second Class	10	5343
Scuba Diver	4	5345

4. Course Unit Contents

The course material for the major instructional units in the Navy diving program are listed in Table XI-2, and are from the Naval School of Diving and Salvage in Washington, D. C. Because each of the diving schools operates relatively independently, the material may vary slightly from unit outlines listed. All class periods represent 55 min of actual instruction. Practical periods vary in length. The units required for the various courses are given in Table XI-3.

C. Training—Civilian Sector

Until quite recently diver training for the civilian segment of the community was quite inadequate. "As late as 1958, for example, there was no acceptable national

Table XI-2
Outline of Courses at the Washington, D. C., Naval School of Diving and Salvage

Unit	Topic	Class	Periods practical	Total
1. Diving orientation	1. Introduction and orientation to the diving school	1	0	1
	2. Physical examination and oxygen tolerance test	0	3	3
	3. Swimming qualification test	0	2	2
	4. Deep-sea diving outfits; function and nomenclature	3	0	3
	5. Dressing the diver	1	2	3
	6. Deep sea diving techniques and procedures	1	0	1
	7. Tending the diver and communications	1	1	2
	8. Familiarization diving with standard deep-sea diving equipment	1	17	18
	9. Review and examination	1	1	2
	Total	9	26	35
2. Diving physics and physiology	1. Principles of diving physics	2	0	2
	2. Formula application; pressure computation and air supply	1	1	2
	3. Air decompression tables and decompression procedures	2	3	5
	4. Recompression chamber and associated equipment	1	0	1
	5. Trainee participation in recompression chamber runs	0	2	2
	6. Review and examination	1	1	2
	Total	7	7	14
3. Underwater basic	1. Underwater work	0	35	35
4. Scuba	1. Scuba indoctrination	3	0	3
	2. Scuba maintenance	1	4	5
	3. Charging scuba	1	1	2
	4. Open-circuit scuba swimming	2	22	24
	5. Open-water swimming with scuba	0	28	28
	6. Scuba descent	1	2	3
	7. Scuba accessories	1	0	1
	8. Review and examination	2	2	4
	Total	11	59	70
5. Underwater tools	1. Application of underwater tools	1	34	35
6. Helmet and dress repair	1. Dress repair	1	4	5
	2. Belt and shoe repair	0	4	4
	3. Maintenance of live line and air hose	0	4	4
	4. Helium–oxygen helmet repair	0	5	5
	5. Air helmet repairs	0	8	8
	6. Shallow-water mask repairs	0	6	6
	Total	1	31	32
7. Lightweight diving	1. Introduction to lightweight diving	2	0	2
	2. Lightweight diving procedures and techniques	1	20	21

Table XI-2—*Cont.*

Unit	Topic	Class	Periods practical	Total
7. Lightweight diving (*Cont.*)	3. Underwater work using lightweight diving equipment	0	12	12
	Total	3	32	35
8. Medical aspects of diving	1. Divers' disease and injuries	1	0	1
	2. Treatment of divers' diseases and injuries	4	5	9
	3. Anatomy and physiology related to diving	1	0	1
	4. Primary and secondary effects of pressure	1	0	1
	5. First-aid for divers	1	0	1
	6. Gas hazards	1	0	1
	7. Effects of underwater blasts on the human body	1	0	1
	8. Pressure runs in recompression chamber, 200 ft	0	3	3
	9. Review and examination	1	2	3
	Total	11	10	21
9. Deep air diving	1. Deep air diving (pressure tank)	0	35	35
10. Submarine rescue chamber	1. History and development of the submarine rescue chamber	1	0	1
	2. Nomenclature and construction	2	0	2
	3. Preparation for operation	0	4	4
	4. Operation of chamber	0	28	28
	Total	3	32	35
11. Underwater cutting, welding	1. Introduction to underwater cutting	1	0	1
	2. Oxy-arc cutting techniques	2	0	2
	3. Oxy-arc cutting	0	14	14
	4. Introduction to metablic arc	0	13	13
	5. Practical application of metablic arc cutting	0	13	13
	6. Introduction to underwater welding	1	0	1
	7. Underwater welding equipment	1	0	1
	Total	5	40	45
12. Underwater advance	1. Underwater work (advanced)	0	28	28
	2. Qualifying 200-ft dive	0	7	7
	Total	0	35	35
13. Underwater demolition	1. Introduction to underwater demolition	1	0	1
	2. Characteristics of underwater explosives	2	0	2
	3. Firing systems and equipment	2	0	2
	4. Application of explosives	2	19	21
	5. Review and examination	1	1	2
	Total	8	20	28
14. Helium–oxygen diving orientation	1. Introduction to helium–oxygen diving	4	0	4
	2. Helium–oxygen diving techniques	3	0	3
	3. Helium–oxygen rack and control panel	1	6	7
	4. Helium–oxygen decompression tables	3	15	18
	5. Review and examination	1	2	3
	Total	12	23	35
15. Helium–oxygen mixing and analyzing	1. Methods of splitting and mixing	1	0	1
	2. Formulas for computing desired percentages of helium–oxygen mixtures	2	0	2

Table XI-2—*Cont.*

Unit	Topic	Class	Periods practical	Total
15. Helium–oxygen mixing and analyzing (*Cont.*)	3. Practical application of helium–oxygen splitting and mixing	2	0	2
	4. Oxygen transfer pump	1	6	7
	5. Analyzing helium and oxygen mixtures with the model "C" Beckman oxygen analyzer	0	4	4
	6. Review and examination	1	1	2
	Total	7	11	18
16. Surface decompression procedures	1. Surface decompression procedures for air and helium–oxygen diving	3	14	17
17. Helium–oxygen diving (pressure tank)	1. Helium–oxygen diving	0	49	49
18. Open-sea diving using air and he-lium–oxygen	1. Preparations for diving from shipboard	0	6	6
	2. Open-sea diving using air	0	8	8
	3. Open-sea diving using helium–oxygen	0	56	56
	Total	0	70	70
19. Salvage seamanship	1. Salvage vessels	2	0	2
	2. Deep-sea towing	2	0	2
	3. Mooring	2	0	2
	4. Highline gear	2	0	2
	5. Beach gear	3	10	13
	Total	11	10	21
20. Salvage theory	1. Types and methods of ship salvage	2	0	2
	2. Legal aspects of salvage	4	0	4
	3. Dewatering methods	2	0	2
	4. Preparations for salvage operations	4	0	4
	5. Review and examination	1	1	2
	Total	13	1	14
21. Salvage machinery	1. Fundamentals of salvage machinery	2	0	2
	2. Operation of salvage machinery	0	5	5
	Total	2	5	7
22. Ship salvage project	1. Organization of ship salvage plan	2	0	2
	2. Preparation for raising a vessel	0	26	26
	3. Raising a sunken vessel	0	21	21
	4. Removal and/or restoration of ship salvage project	0	7	7
	Total	2	54	56
23. Administrative procedures				
24. Advance gas mixing				
25. Diving accidents	1. Decompression sickness	3	0	3
	2. Air embolism	3	0	3
	3. Drowning	2	0	2
	4. Oxygen poisoning	4	0	4
	5. Shallow-water blackout	1	0	1
	6. Marine hazards	1	0	1
	Total	14	0	14

Table XI-2—*Cont.*

Unit	Topic	Class	Periods practical	Total
26. Clerical and property accounting procedures	1. Reports and forms peculiar to diving	3	0	3
	2. Reports and forms used on small ships	2	0	2
	3. Fiscal and supply procedures	3	0	3
	4. The health record	2	0	2
	5. Selection of personnel	2	2	4
	Total	12	2	14
27. Respiratory physiology	1. Preparatory respiratory anatomy	4	0	4
	2. Pulmonary functions and lung volumes	5	0	5
	3. Respiratory functions	3	0	3
	4. Oxygen consumption in underwater swimmers	2	0	2
	Total	14	0	14
28. Mixed-gas decompression tables				
29. Treatment of diving accidents and diseases	1. Derivation and history of treatment tables	1	0	1
	2. Use and dependability of the Navy treatment tables	1	0	1
	3. Mechanical resuscitators	1	1	2
	4. External cardiac massage	1	0	1
	5. Simulated treatment of diving accidents utilizing the recompression chamber	0	12	12
	6. Case reports—selected cases of decompression sickness	4	0	4
	7. Case reports—selected cases of air embolism	4	0	4
	8. Review and examination	2	1	3
	Total	14	14	28
30. Advance medicine				
31. Salvage computations				

standard syllabus for sport diving" (Kenny 1972). Most training courses to qualify sports divers had no entrance requirements (many do not even now), and in many cases the entire course consisted of as little as 12 hr, with no training in the ocean or in deep water. Now, however, for each type of diving activity, there is a discrete form of training, and there has been an explosive growth of all types of training to meet the demands for commercial, scientific, and sport divers.

1. Background

A recent survey completed by the National Oceanographic and Atmospheric Agency (NOAA 1974) indicates that there probably are 1,890,000 individuals in the United States who possess diving skills (see Table XI-4). Considering various factors, it was determined that the actual total "is not believed to be less than 1.5 million nor more than 2.1 million." This summary of the national diving population gives an estimate of the upper and lower limits in addition to a "most likely" population.

Table XI-3
Units Required for Courses

Course	Weeks	Units
Salvage Officers	16	1-2-8-4-4-7-3-12-5-11-13-31-19-20/21-22-22
He–O$_2$ Officers	23[a]	1-2-8-4-4-7-3-12-5-11-13-31-19-20/21-22-22 14-15/16-9/17-18-18-10
He–O$_2$ Cross Trainee	7	14-15/16-9/17-9/17-18-18-10
ARS/ATF-PXO/PCO	3	1-2-4-7-8-13-19-20/21-22-30-31
EDO	6	1-2-8-5-13-11-30-3-7-8-19-20/21-22-31
Medical Officers	8	1-7-3-2-8-19-20/21-10-30-14-15/16-9/17-18
Medical Officers	1	Special
Masters	5	Varies
Diver First Class	17	2-5-8-6-11-13-19-20/21-22-22-14-15/16-9/17 9/17-18-18-10
Diver Second Class	12	1-2-3-12-6-8-19-4-7-11-21-5
HM's	23	1-2-8-4-4-7-3-12-5-13-19-20/21-22-22-14-15/16- 9/17-9/17-18-18-10-25/26-27/29
Police Scuba	4	2-8-4-4

[a] First 16 weeks same as Salvage Officers.

The first three categories shown in the table are additive, while the fourth category can overlap any or all of the other three. It should be noted that in the second and third categories diving is primarily a skill used by the individual, rather than a vocation.

Another piece of evidence that there is need for a careful and thorough development of both selection and training criteria is the amount of money spent on diving. In Table XI-5 it will be noted that the estimated annual expenditure for civilian diving purposes is $316,100,000 (NOAA 1974). And this does not include government spending for the Man-in-the-Sea program, which was reported to be $43,400,000 in 1970.

For the analyses presented in Tables XI-4 and XI-5 and for general consideration, civilian divers and categories of diving are defined as follows (NOAA 1974):

Diver—Any individual in direct or clothing contact with the water who uses equipment which supplies a breathing-gas mixture from a compressed source that enables the individual to breathe under water at ambient pressure.

Commercial Diver—A diver who engages in diving in connection with non-training (of divers) employment with a "for profit" firm. This group may be further divided into "Full-time Payroll Employment" and "Free-lance/Part-time Employment."

Scientific/Educational Institutional Diver—A diver who engages in diving in connection with research, studies, or employment, at an institution of higher education, associate degree (two year) awarding institution or research consortium: this group is further divided into "College and Universities," and "Associate Degree Institutions and Research Consortia." These were then subdivided into the type of individual diving, i.e. "professional staff" or "student."

Governmental Diver—A diver who engages in diving in connection with direct employment by a federal, state, or local government agency or a police or fire department.

Recreational Diver—A diver who engages in diving solely for personal enjoyment at the time of the activity (without financial compensation).

"In past years, except for a small group of highly trained specialists who followed oceanographic careers in marine laboratories or universities (usually on government-

Table XI-4

Estimated U. S. Civil Diving Population (NOAA 1974)

Diver category	Estimated population		
	Minimum	Most likely	Maximum
Commercial:			
Full-time payroll employment	1,450	1,530	1,750
Free lance/part-time	650	775	1,000
Total	2,100	2,305	2,750
Scientific/educational/institutional			
Colleges and universities:			
Professional staff	540	500	600
Students	1,300	1,360	1,425
Support divers	60	65	75
Total	1,900	2,020	2,100
Associate degree institutions and			
research consortia:			
Staff	30	45	50
Students	230	250	280
Support divers	15	25	30
Total	275	320	360
Nonmilitary governmental:			
Federal agencies	590	600	650
State and local	300	335	500
Police departments	6,500	8,060	9,700
Fire departments	2,500	5,000	7,500
Total	9,890	13,995	18,350
Recreational:			
Individuals with diving skills	1,500,000	1,890,000	2,300,000
Individuals practicing in the sport	375,000	474,000	600,000

sponsored projects), engineers, and technicians working on ocean projects came from non-ocean oriented disciplines or trades " (Stephan 1969). This is as it should be, for many believe that commercial diving should take a leaf from the U. S., British, and Italian navies, for example, and first train engineers, electricians, etc., then train them to be divers.

In the past decade scientific advances, technological development, and industrial efforts to meet increasing demands for fuel or food supply have stimulated the development of a great variety of ocean-related training programs, including on-the-job training, high school work, technical school work both in trade and military service schools, and junior college, college, and continuing education programs.

Graduates of these programs can look forward to a broad range of challenging career opportunities: Veterans of the military, Coast Guard, Coast and Geodetic Survey, Merchant Marine, and fishing fleet services often have already acquired skills of great value to prospective employers or which may have qualified them to seek academic advancement.

Civilian training in marine sciences can include anything from programs on swimming, water safety, and scuba diving to navigation or nautical engineering to

Table XI-5

Estimated Annual Expenditures
for Civil Diving by Purpose (NOAA 1974)

Purpose	Expenditures, in millions of dollars
Marine science and engineering (Government and Academic):	
Direct diving cost	2.6
Equipment and replacements	0.3
Industrial applications:	
Diving services	63.1
Diver equipment replacement	2.0
Training	0.9
Recreational:	
Basic training equipment	13.0
Advanced equipment and replacements	41.6
Personnel expenditures	186.7
Public safety:	
Direct diving cost	3.3
Equipment and replacement	2.6
Total	316.1

underwater research. Training may be for sport or hobby, or for a highly specialized scientific career as a marine scientist or aquanaut. The problem for an individual is to locate a suitable program or curriculum.

2. Recreational Diver Training

The most active and widely publicized form of diver training is in recreational or sport diving. A survey in 1972 indicated that 86.9% of those responding had been trained by one or more of five major organizations, the YMCA, the National Association of Underwater Instructors (NAUI), the Professional Association of Diving Instructors (PADI), the National Association of Skin Diving Schools (NASDS), or Los Angeles County (LA Co.) (NOAA 1974). A total of 1,095,000 divers were trained by these organizations by 1972 (Table XI-6).

Table XI-6

Recreational Divers
Trained in the United States (NOAA 1974)

Organization	1970 & before	1971	1972	Total	Percent of total
YMCA	172,000	12,000	42,000	226,000	20.7
NAUI	212,000	54,000	67,000	333,000	30.4
PADI	47,000	37,000	52,000	136,000	12.4
NASDS	133,000	52,000	55,000	240,000	21.9
LA Co.	140,000	10,000	10,000	160,000	14.6
Total	704,000	165,000	226,000	1,095,000	100

These organizations (particularly the YMCA and NAUI) have established minimum standards of course content, student performance, and instructor qualification. The average sport diver training course consists of 36 hr of lecture and pool work, including three ocean dives. "At conclusion of the course, the neophyte diver should be capable of diving without endangering himself or his diving partner. He should be capable of intelligently evaluating the safety of diving situations; he should understand diving physiology well enough to conduct himself safely underwater; he should be familiar enough with the ocean to understand potential hazards; and he should be sufficiently adept at handling his equipment so that it presents no problems to him" (Kenny 1972).

The National YMCA offers five courses in skin and scuba diving: a skin diving course for beginner divers ages 10 and up; a course for ages 15 and up, covering theory, skills, and techniques for safe use of scuba equipment; an advanced scuba diver course; and courses for training scuba leaders and instructors. The YMCA scuba program, although not at every YMCA, is broadly available and offers recognized certification. It serves marine education institutions as well as local sportsmen. Further information is available from the National YMCA Scuba Headquarters, 1611 Candler Building, Atlanta, Georgia 30303 (YMCA 1973).

The evidence seems to indicate that of the over one million individuals who have had scuba training and who have received C-cards very few have ever had any open-sea training, and this is a " national scandal of the 70s " (Tzimoulis 1971). Although most of the recognized organizations require at least 2 hr of open-sea diving, very few candidates actually receive the training and there is no check on the laxness of the local instructors. There is no question but that this lack in training has led to many quite unnecessary deaths.

The more progressive scuba schools are offering three separate open-water training experiences to their basic scuba course curriculum (Tzimoulis 1971).

> For example, UCLA scuba students are required to make three full-day ocean outings. The first is a skin diving orientation trip, utilizing nothing more than mask/ fins/snorkel, plus wet suit, weight belt and safety accessories. It is a beach dive, requiring a surf entry and the use of a float. The second open water trip is a scuba diving orientation, designed to familiarize the student with his new (and totally different) environment. It is not a testing session, but instead a learning experience. The third open water trip involves the final check-out, where the student must demonstrate all of the skills learned in the pool. By now, the student is familiar with his new environment and comfortable enough to go through the series of designated proficiency tests. It is this gradual, step-by-step progression from pool to open water which provides the safest, most efficient transition possible.

As pointed out above, much of the scuba training consists of lectures and demonstrations with experience in a swimming pool and a final dive in a rock quarry or nearby lake. This means that the instructors talk about pressure phenomena but have no way of exposing the trainee to actual pressure experience. In a San Francisco dive shop (Leckie 1968) a pressure chamber used routinely to demonstrate pressure effects has proven to be so successful that it is called the "ultimate teaching aid."

The "buddy system"—never diving without a partner to help in time of trouble— has been called into question because of the number of double deaths that have been reported. But the general consensus is that it is not the system that is at fault but inadequate training in such techniques as buddy breathing, underwater rescue, or underwater first-aid. Once again it is the training that is at fault.

One final thought is that "good physical condition is a prime factor in preventing serious injury or death to sport divers" (Bove 1969). The evidence seems to point to fatigue and exhaustion as leading causes of accidents. Thus keeping fit is an important aspect of all diving, whether scuba or other types of underwater activity. Unfortunately, the sport diver usually comes into training with an inadequate physical examination that has not considered physical condition.

The requirements for scuba training in the USSR are as follows (Bulenkov 1969):

> Instruction for swimming under water with the apparatus may begin after the trainees have passed examinations on the physiology of diving operations, construction and maintenance of breathing apparatus and knowledge of signals. For instruction the swimmers are divided into divers and safety men. The safety men control the movements of their comrades and maintain communications with them by means of a signal line.
>
> During instruction, they are trained in the following sequence:
>
> 1. Having been inclosed in the apparatus on the shore, breathing is tried when standing in the water up to the chest. Lying on the bottom for 2–3 minutes observe the correctness of breathing. While maintaining a slow rate of breathing stand on the knees, lie on the back, and on the right and left sides.
>
> 2. Swim 100 meters at a depth of 1.5–2 meters.
>
> 3. Standing in water up to the chest, inhale while not inclosed in the apparatus, sit down under water, put the mouthpiece in the mouth, turn over on the left side and with a sharp breath discharge the water from the breathing tubes. Do this exercise under water 6–8 times.
>
> 4. Stand in the water up to the chest. Put on the mask that is included in the apparatus. Dive into the water. Swim for a period of 3–5 minutes.
>
> 5. Stand in the water up to the chest. Put on the apparatus, place the face in the water, fill the mask with water while standing. Throw the head backward so that the glass in the mask is turned upward, and by exhaling with the nose, discharge the water from the mask.
>
> 6. Once more put water in the mask, stand up, press in the upper part of the mask with the fingers. Evacuate the water by exhaling through the nose.
>
> 7. Do the same under water.
>
> 8. Swim for 100–200 meters, while underway take water into the mask and discharge it, and also take the mouthpiece out of the mouth and once more put it in.
>
> 9. At a depth of two meters stand on the knees, remove the shoulder and belt straps, take the apparatus by the tanks and, passing it over the head, place it in front. By holding the apparatus with one hand, remove the mouthpiece from the mouth and push it under the automatic breathing apparatus. While exhaling air swim to the surface.
>
> 10. Having inhaled at the surface, dive to the apparatus with the purpose of putting the face in the automatic breathing apparatus. Taking the apparatus in one hand lie on the chest on the bottom, tilt the apparatus to the left and, making a sharp breath, discharge the water from the breathing tubes of the apparatus. Putting the arms under the shoulder straps, put on the tanks, put the apparatus over the head, tighten the belt and shoulder straps and come to the surface.
>
> 11. Standing on the bottom with the apparatus breathe through the breathing tube. Switch to breathing in the apparatus. Repeat shifting to the breathing tube.
>
> 12. Swim 100–200 meters in the apparatus, while from time to time swimming to the surface and shifting to breathing through the tube. Then submerge while shifting to the apparatus.
>
> 13. Swim in the apparatus 200–300 meters while changing depth from the surface

to 5–6 meters and becoming completely accustomed to equalizing pressure in the middle ear and in the space under the mask.

14. One of the scuba divers descends to a depth of 2 meters with his apparatus. The other, in a mask and wearing fins, dives to the first, takes the mouthpiece, exhales and inhales and returns the mouthpiece to the first. Repeat these acts several times.

15. A diver with the apparatus swims on the left side, the other, without apparatus and on the right side, faces the first. Both breathe from one apparatus in turn.

16. Place the apparatus on the bottom along with the mask and fins. Dive, turn on the apparatus and put it on. Find the mask, put it on and discharge the water from it. Put on the fins. Swim 100 meters and take off all equipment and swim to the surface.

3. *Commercial Diver Training*

"Commercial divers are workers who dive—not divers who work" (Mikalow 1972). "Every diver is proud of the depths to which he has gone. In commercial diving, however, it is not how far down you go that counts, but what you do when you get there." A commercial diver must be able to do many different jobs that would be handled by different specialists if the work were not under water. Thus he is looked upon as a jack-of-all-trades. A high-school education is a must and a college education is a help, as is a knowledge of mechanical drafting, blueprint reading, pipe fitting, carpentry, mechanics, metal working, and welding! The industry asks three things of a diver: Does he know how to use all of the different kinds of equipment? Can he handle many different kinds of commercial jobs? Can he work in the most adverse conditions? Obviously very special training is necessary for this type of diving activity.

Although not necessarily applicable in the U. S., the British requirements for training a commercial diver are of sufficient interest to be quoted in their entirety (Society for Underwater Technology 1972):

Syllabus for Basic Training Course

The syllabus outlined below is intended to be the minimum requirement for a competent air diver. The intention is to train a diver so that the physical part of the diving comes naturally and easily to him, leaving his mind free to concentrate on the work in hand. On successful completion of the basic course the trainee will be able to descend and ascend safely and efficiently to a depth of 50 m or 150 ft in open water using a range of specified equipment and techniques.

Conditions under which the trainees must be competent to operate
  Cold water
  Black water (nil visibility)
  Obstructed water (e.g. underwater structures, wreckage, etc)
  Tidal water
  Open sea
  Strong currents
Use and maintenance of self-contained underwater breathing apparatus (SCUBA)
  operating on air.
Use and maintenance of surface demand equipment with bale out.
Use and maintenance of standard equipment.
Use and maintenance of drysuits and wetsuits.
Use of lifejackets.
Conduct of diving operations.

Emergency surfacing procedure.
Rigging diving boats and bottom lines.
Use and maintenance of high-pressure and low-pressure compressors.
Use and maintenance of decompression chamber.
Knowledge of hazardous diving conditions.
Knowledge of the following aspects of diving:
  Adaptation to the water environment: vision, touch, hearing, smell and taste, weightlessness and disorientation, cold, psychological problems, voice changes with pressure, seasickness, electrocution.
  The body: metabolism and respiration, heart and circulation, lungs, blood and gas transport.
  Physics: pressure air/water units, gas laws (Boyle's, Dalton's), solubility, partial pressure (Henry's), specific gravity, buoyancy, viscosity and density of breathing gases.
  Effects of exposure to pressure on gas uptake—oxygen, CO_2, nitrogen, and other inert gas on ears, sinuses, teeth and bowels; lungs and "squeeze" barotrauma in free ascent.
  Decompression: principles of safe decompression and the construction of decompression tables.
  Decompression sickness (type I, type II): bone necrosis and "burst lung," danger of flying after diving, treatment of decompression sickness.
  Reversed ear.
  Nitrogen narcosis.
  Oxygen poisoning.
  First aid including: artificial respiration and cardiac massage; use of resuscitators.
  Hypoxia: "dilution hypoxia," CO_2 accumulation in the set.
  Underwater blast injuries.

4. *College, University, and Institutional Programs*

a. Degree Programs in the Marine Sciences and Related Fields

Persons seeking information about college or university training will certainly find helpful a pamphlet by the Oceanographer of the Navy, *University Curricula in the Marine Science and Related Fields* (write to Superintendent of Documents, Government Printing Officer, Washington D. C., for pamphlet #43, 1971, $1.75). This pamphlet describes programs in different schools, colleges, and universities, listing courses and faculty and describing facilities and degree requirements.

There may be confusion about the terms "marine science" and "oceanography." According to most institutions, oceanography is a multidisciplinary science, whereas marine science refers to the marine branch of a particular scientific discipline. "Related fields" include academic programs for training ships' officers and maritime engineers, naval architects, marine lawyers, marine (science and engineering) technicians, and applied fisheries scientists and technicians (Oceanographer of the Navy 1971).

Academic degree programs are offered in six fields, Marine Science, Ocean Engineering, Fisheries, Marine Law, Maritime Officers, and Marine Technicians, but many universities award only graduate degrees in some fields. According to the

Oceanographer of the Navy the general requirements for admission to graduate schools are as follows.

Oceanography and Marine Science:

1. A baccalaureate degree from an accredited college or university with a major in biology, chemistry, engineering, geology, mathematics, oceanography, or physics.
2. A cumulative undergraduate grade average of approximately B (or 3.00 on a 4.00 scale).
3. Mathematics through differential and integral calculus.
4. One year of chemistry, with laboratory.
5. One year of physics, with laboratory.
6. One semester of geology, with laboratory.
7. One semester of biology, with laboratory.
8. A broad background in the humanities.

Minimum requirements may be modified or waived for admission to a specific program. Satisfactory completion of additional courses, such as the following, is regarded as highly desirable for admission to study in particular disciplines: advanced calculus, differential equations, thermodynamics, organic chemistry, historical and structural geology, and vertebrate and invertebrate zoology. For admission to a specific discipline within the marine sciences the student should have an undergraduate major in the same or a closely related discipline, with appropriate minors. For instance, for admission to study in physical oceanography a student should have had a major in physics and a minor in mathematics or vice versa.

Ocean Engineering. For students applying to graduate ocean engineering programs, requirements are not as clearly specified as those for the marine sciences. Broad general requirements can be identified as:

1. A baccalaureate degree in engineering or science from an accredited college or university.
2. A cumulative undergraduate grade average of approximately B (or 3.00 on a 4.00 scale).
3. Mathematics through differential equations.

Satisfactory completion of additional courses in mathematics, especially in advanced calculus, probability, and statistics, in physical and life sciences, and in economics is highly desirable.

Other requirements commonly include the taking of the Graduate Record Examinations and submission of recommendations from major professors. Deadlines for submission of applications and the schedules for fees vary greatly from institution to institution (Oceanographer of the Navy 1971).

For information regarding financial aid for education in the marine sciences, the following booklets may prove helpful:

"Federal Benefits for Veterans and Dependents"
 (VA Fact sheet 1S-1)—Free of charge
Veterans Benefits Office
Veterans Administration
Washington, D. C. 20420

"More Education, More Opportunity"—Free of charge
U. S. Office of Education
Division of Student Financial Aid
Washington, D. C. 20202

"Need a Lift"—50¢ per copy (prepaid)
American Legion, Department S
P. O. Box 1055
Indianapolis, Indiana 46206

"Opportunities in Oceanography"—
  $1.25 per copy
Smithsonian Press
Smithsonian Institution
Washington, D. C. 20560

"Scholarships for American Indian Youth"—
  Free of charge
Bureau of Indian Affairs
Department of the Interior
Washington, D. C. 20240

"Financing a College Science Education"—
  15¢ per copy
Superintendent of Documents
U. S. Government Printing Office
Washington, D. C. 20402

(i) Academic Fields. Academic programs related to undersea operations are available in six degree fields as follows.

Marine Sciences. Only a few institutions award a B.S. in oceanography or marine sciences, although many supplement their regular degree curriculum with an inter-disciplinary program which includes undergraduate courses in aquatic sciences. Most of the larger universities located near the sea offer extensive graduate programs.

Ocean Engineering. Many universities offer special work and a few award an undergraduate degree in ocean engineering; however, most programs are graduate school programs.

Fisheries. Although most Fisheries programs are graduate degree programs, some colleges and universities offer a B.S. in Wildlife and Fisheries and/or a B.S. in Food Science and Technology.

Marine Law. Several law schools offer a L.L.M. and/or J.D. in Ocean Law.

Maritime Officers. This program is primarily an undergraduate program, designed to prepare students for the U. S. Coast Guard license examination and, sometimes, for commission as an ensign in the Naval Reserve. Students may seek a B.S. in Nautical Science or a B.S. in Marine Engineering.

Marine Technicians. A two-year course in marine technology may result in an Associate in Arts degree in Ocean Engineering or an Associate in Science degree in Marine Technology, Commercial Fishing Technology, Oceanographic Technology, or Applied Science. This program is widely available at junior and community colleges in Florida, California, and other states where large numbers of organizations are involved in marine activities.

According to the American Association of Junior Colleges, projections indicated that 15,000 oceanographic technicians would be needed within 20 years and that 38,000 technicians would be needed to support marine technology by 1974 (Chan 1968). An Associate degree in this field should qualify a student to work in such areas as marine oil and mining, oceanography, scientific research, fisheries technologies, hardware technologies, and aquaculture.

A general training program includes:

1. Mathematics (including some algebra and trigonometry).
2. Broad training in biology, chemistry, and physics.
3. Basic training with special materials, skills, and methods (including collecting and processing data; operation, maintenance, and repair of equipment; drafting and graphics; seamanship; and safety procedures).
4. Communication skills related to the technology of marine science (Chan 1968).

Figure XI-3 is an example of a course of study for certification as a marine scientist.

MARINE TECHNICIAN CURRICULUM

Certificated Major in Marine Technology Associate of Science (A.S.) Degree

1st Semester	Units	2nd Semester	Units
Geology 15 (Gen. Oceanography)	3	Biology 10 (General)	3
†Math 53 (Technical Math)	3	Computer Science 50	3
Electronics 61 (Intro.)	4	Electronics 62 (Intro.)	4
††Engineering 51 AB (Drafting)	2	American Studies 1A (History)	3
English or Communications	3	Electives	2
Physical Education	½	Physical Education	½
	15½		15½

SUMMER—Marine Experience
College of Marin's
    Bolinas Marine Station—Course work
U.S. Bureau of Mines at Tiburon—
    Ship experience and work experience

3rd Semester	Units	4th Semester	Units
Chemistry 11 (General Intro.)	4	MMT 59 (Machine Processes)	3
Physics 55 (General Instru.)	4	Physics 50 (Marine Instrumentation)	2
Geology 1A (Intro to Physical Geol.)	4	American Studies 1B (Government)	3
English or Communications	3	Electives	7
Physical Education	½	Physical Education	½
	15½		15½

For Biological Technician option, please contact the Director for approval.

†Eligibility for higher math, Math B, may satisfy requirement
††May be satisfied by one year of high school drafting

Suggested Electives:
Biology 20A—Marine Biology
Biology 21A, 21B—Marine Ecology
Chemistry 71—Chem. Instrumentations
Electronics 60A—Fabrications
Geology 56—Mineralogy
Art 48—Photography

Figure XI-3. A plan for a course of study to train marine technologists (prepared by College of Marin, Kentfield, California).

For information about marine technician programs in various states, write to: The American Association of Junior Colleges, One Dupont Circle, N.W., Washington, D. C., 20036.

(ii) Scientists-in-the-Sea Program. A special program for training postdoctoral and graduate scientists in underwater skills is the Scientists-in-the Sea (SITS) program, sponsored by the National Sea Grant Program, the Manned Undersea Science and Technology Program of the National Oceanic and Atmospheric Administration (NOAA), the State University System of Florida, and the Naval Coastal Systems Laboratory (Dunham 1972).

The purpose of the SITS program is to train skilled scientists in advanced diving technology, scientific methods adapted for use to divers, and diver management and safety (Dunham and Haythorn 1972). Training includes navigation and search, communications, semiclosed- and open-circuit scuba, advanced umbilical diving techniques, planning and implementation of diving programs, exposure to saturation diving principles and methods from oceanography, biological sciences, geology, engineering science, behavioral sciences, diving medicine, and archeology. Successful completion of the program results in State University System Institute of Oceanography (SUSIO) certification in academic scientific diving and 12 quarter-hours of graduate credit (Dunham 1972). For further information concerning the SITS Program write to the following:

Dr. R. Smith, Principal Investigator
State Univ. System Institute of
  Oceanography
830, First Street, South
St. Petersburg, Florida 33701

Dr. R. M. Dunham, SITS II Coordinator
Psychology Department
Florida State University
Tallahassee, Florida 32306

Stringent qualifications criteria governing selection of applicants assures a high level of program safety and academic quality as well as a high rate of success for those who are accepted into the SITS program. To make initial application it is necessary to (1) be in a relevant graduate training program or have suitable career interests, (2) submit academic transcript, (3) pass an FAA third-class flight physical or Navy diving physical, (4) have scuba diving certification, (5) have three recommendations from academic advisors or scientific colleagues, and (6) offer warranty concerning physical fitness (criteria included with application). Applicants should be between the ages of 21 and 39.

If the application is accepted, program acceptance depends on subsequent qualification in the Navy diver physical examination, tests for diving and swimming skills, respiratory physiology factors, review, and psychological factors review. SITS program and swimming requirements—based on *BUPERS Manual* (NAVPERS 1579A), *BUMED Manual* (NAVMED P-117), and the *Navy Diving Manual* (NAVSHIPS 250–538)—are described in the application form appendices (Figure XI-4).

b. Nondegree Programs in Marine Sciences and Related Fields

A variety of special training programs relating to marine science and technology are offered by schools, communities, universities, institutions, and organizations. Typical of these programs are the Yale University Scuba Diving Program, the National

PHYSICAL AND SWIMMING REQUIREMENTS FOR SITS CANDIDATES

Prior to participation in the SITS program, all candidates must meet the physical and swimming requirements for Navy divers as follows:

Swimming Requirements (First Class Swimmers Test)
1. Enter water feet first from a minimum height of 10 ft and remain afloat for 10 min.
2. Swim 220 yards using any stroke or strokes desired.
3. Approach a person of approximately his own size while in the water, demonstrate one break or release, get him in a carry position, and tow him 25 yards.
4. Enter water feet first and immediately swim under water for 25 yards. Swimmer is to break the surface twice for breathing during this distance at intervals of approximately 25 ft.
5. Remove trousers in water and inflate for support.

Scuba Diving Test (Similar to a Certification Test)
   Skills: Ditch and Don (includes mask clearing and free ascent)
          Buddy breathing
          Band mask (test for claustrophobia)

Physical Fitness (based in part on minimal male standards)

Pushups	15
Situps	25
Mile run	10 min maximum time
Consecutive knee bends	25

Physical Examination
Diving physical examinations shall be performed by a Submarine Medical Officer who has been trained in this area of medicine. The salient features of the examination are as follows:

1. *History of disease.* Any of the following shall be disqualifying:
   (a) Tuberculosis, asthma, chronic pulmonary disease.
   (b) Chronic or recurrent orthopedic pathology.
   (c) Chronic or recurrent sinusitis, otitis media, otitis externa.
   (d) Chronic or recurrent gastrointestinal disorder.
   (e) Chronic alcoholism.
2. *Age.* Be at least 21 years of age and not older than 39 (waiver of age limits is sometimes possible).
3. *Weight.* No greater than 10% variation from standard age–height–weight tables.
4. *Color-vision.* Normal color perception is required of all candidates.
5. *Teeth.* A complete dental examination shall be conducted by a dental officer.
6. *Ears.* Acute or chronic disease of the auditory canal, tympanic membrane, middle or internal ear shall be disqualifying. Perforation or marked scarring and/or thickening of the drum shall be disqualifying. The eustachian tubes must be freely patent for equalization of pressure changes. Hearing of each ear shall be normal.
7. *Nose and throat.* Obstruction to breathing or chronic hypertrophic or atrophic rhinitis shall disqualify. Septal deviation is not disqualifying in the presence of adequate ventilation. Chronically diseased tonsils shall be disqualifying pending tonsillectomy. Presence or history of chronic or recurrent sinusitis is cause for rejection.
8. *Respiratory system.* The lungs shall be normal as determined by physical examination and 14 by 17 in. chest X-ray.
9. *Cardiovascular system.* The cardiovascular system shall be without significant abnormality in all respects as determined by physical examination and tests. The blood pressure shall not exceed 145 mm systolic or 99 mm diastolic. Persistent tachycardia and arrhythmia,

(Cont.)

Figure XI-4. Requirements for acceptance in SITS program (adapted from the SUSIO Scientists-in-the-Sea application form).

except for sinus type, evidence of arteriosclerosis, varicose veins, marked or symptomatic hemorrhoids shall be disqualifying.

10. *Gastrointestinal system.* Candidates subject to gastrointestinal disease shall be disqualified.
11. *Genitourinary system.* The following shall be disqualifying:
    (a) Chronic or recurrent genitourinary disease or complaints (normal urinalysis required).
    (b) Active veneral disease or repeated veneral infection.
    (c) History of clinical or serological evidence of active or latent syphillis within the past 5 yr, or of cardiovascular or central nervous system involvement at any time. An applicant who has had syphillis more than 5 yr before must have negative blood and spinal fluid serology.
    (d) Skin. There shall be no active acute or chronic disease of the skin which is contagious or offensive in close working conditions or in interchange of diving apparel.

Recompression Chamber Pressure Test

All candidates shall be subjected in a recompression chamber to a pressure of 50 psi to determine their ability to clear their ears effectively and otherwise to withstand the effects of pressure.

Oxygen Tolerance Test

Each candidate must demonstrate his ability to tolerate breathing pure oxygen at a simulated depth of 60 ft for 30 min at rest. The purpose of this test is to eliminate from the program those individuals who are susceptible to oxygen toxicity to a degree that may be hazardous to themselves and others.

Figure XI-4—*Cont.*

YMCA Scuba Diving Program, the Corning Community College Underwater Physiology Course, the Scripps Institute of Oceanography training classes, and the University of Texas Aquamedics Program. Special programs are usually created to meet local demands and developed to fulfill enlarging needs for trained personnel.

(i) Yale University Scuba Diving Program. The goal of the Yale diving program is to train competent, nonprofessional divers, providing basic knowledge and skills for safe open-water activity using scuba equipment. The program is divided into three parts.

1. Medical screening—a stringent examination, especially to detect scarring of the lungs or wheezing, or difficulty in clearing the ears.

2. Swimming qualifications—students who are not throughly proficient in swimming skills are expected to complete qualifying swimming courses before scuba training.

3. Scuba instruction—this training includes required lectures on water safety and use of equipment and, after thorough training in the pool, the course ends with at least one closely controlled open-water dive to gain competence experience (and thus safety) (Sawyer 1972).

(ii) Corning Community College Underwater Physiology Course. In an effort to stimulate interest among the large community of undergraduate students, the Marine Technology Society and the Committee of Man's Underwater Activities have sponsored a pilot program at Corning Community College in Corning, New York. Designed to "introduce students to the principles of underwater physiology as a basis for understanding human body reactions and adaptability in water," the four-week program "provides students the opportunity to explore his capability, limitation and compatibility in the hydrospace for living, useful work and safe recreational activities" (Gee 1972). The credit course includes lectures, laboratory experiments, field work at Seneca Lake, and scuba training (Gee 1971). The scuba training is conducted by professional instructors employing the NAUI or the YMCA standard

MBI BASIC DIVE COURSE FOR AQUAMEDICS
(30 hr minimum)

Resultant certification in National Association of Underwater Instructors and UTMB qualification.
All equipment is provided except one's suit and text.
You are encouraged to purchase some basic equipment of your own well into the course.
Fees for the course consist of text purchase, possible wet-suit rentals, and Houston facility fees.
Requalification dives are scheduled during the fall and spring for those divers whose qualification
has lapsed due to infrequent diving.

MBI BASIC DIVE COURSE
PART I—WATERMANSHIP PREREQUISITE

A. Swimming Skills; less swim aids   2 hr
1. 200-yard continuous swim
2. 20-yard underwater (u/w) swim
3. 10 min tread water
4. 10 min float
5. Surface dive and recover object
6. Tow an inert swimmer 40 yards

B. Equipment Skills   2 hr
1. "Clearing" a mask
2. Use of snorkel
3. Entries (two types)
4. 200-yard continuous swim using mask, fins, snorkel
5. 20-yard u/w swim
6. Surface dives (two types)
7. Introduction to buoyancy compensator/safety vest

PART II—INTRODUCTION TO SCUBA

A. Lecture Material

	Chapters	Hours
1. Basic scuba equipment and its use	1,5	1
2. Diving physics and physiological effect	2,4	2
3. Medical aspects, including first aid	3,7	2
4. Mechanics of scuba	5	1
5. Diving safety	6,7,9	1
6. Environmental and marine life	8	1
7. Decompression tables	Appendix	2
8. Review	Entire book	2
9. Exam		1

B. Pool Sessions

1. Scuba familiarization: surface use, shallow water, mask and mouthpiece clearing, swims u/w	5,6	2
2. Pressure equilization: deep-water drills, snorkel use, buoyancy control	3,4	1
3. Swims, entries	6	
4. Buddy breathing, ditch/don		2
5. Review; various pool skills and exercises		2
6. 40-ft dive, tank, pool environment only		1
7. Open-water checkout, bay environment		1
8. Open-water dive, ocean environment		1

(Cont.)

Figure XI-5. Prerequisites and course requirements for the University of Texas Aquamedics Program.

PART III—HYPERBARIC FACILITIES FAMILIARIZATION

	Hours
A. Field Trips (Required)	
1. NASA/MSC treatment facilities for hyper- and hypobaric conditions	1
2. Commercial and/or industrial recompression facility for treatment and research	1
B. Practical Work (Required)	
1. Chamber dive to 165 ft as "diver," and then operator	1
2. Chamber dive to 30 ft, use oxygen	1
3. Shipboard system familiarization	1
C. Recommended:	
1. Chamber operator course for physicians	
2. NAUI diving physician course	
3. Visits to operating treatment and research facilities	
4. Several open water dives each year	

Text: *The New Science of Skin and Scuba Diving*, third revised edition, 1970
References: *U. S. Navy Diving Manual*
                 NASA Underwater Swimmer Course Manual

Figure XI-5—*Cont.*

training program, and the physiology part of the program is given by college staff and guest lecturers. Appropriate topics are emphasized, such as "(1) body temperature changes and heat loss, shivering and hypothermia in cold water, and the design of thermal protection; (2) the function of various breathing media in diving and their effects on body tissues; (3) the effects of deep diving and cold stress on metabolic, respiratory, and cardiovascular functions; and (4) man's acclimatization to water as a result of prolonged exposure to the medium" (Gee 1971).

(iii) Scripps Institution of Oceanography (SIO). In addition to holding training classes for scuba divers and instructors, the SIO conducts classes "designed to blend the science and technology of underwater work." Some of the subjects which have been covered in such classes are the use of underwater communication systems, cameras, closed-circuit television, and hand-held sonar as well as methods of collecting scientific data (Stewart 1972).

(iv) University of Texas Aquamedics Program. The Aquamedics, a team of physicians skilled in dealing with medical problems related to the water, serve primarily as a medical support team for undersea activities conducted by the Marine Biomedical Institute of University of Texas Medical Branch (UTMB) (*Galveston Daily News* 1972). In addition they provide rescue and emergency medical care for hazardous water-sport activities (for example, the Galveston powerboat races).

The Aquamedics began as a volunteer group of local citizens, many of whom had no medical training. Organized by Dr. Charles Moore of the UTMB Department of Surgery in 1968, the Aquamedics became by 1970 an adjunct activity of UTMB's emergency care facility, where the medical director assured leadership and began reorganizing the group into a highly skilled team of medical professionals.

Classes are offered to medical personnel who are interested in joining a formal diving course, sanctioned by the National Association of Underwater Instructors. This class trains medical students, faculty members, and house staff for certification as scuba divers. Those who complete training become members of the Aquamedics. An outline of prerequisites and courses required in the Aquamedics program is given in Figure XI-5.

D. *Training Pays Off*

1. *Safety*

Newspaper reports and the scientific literature both indicate that lack of training and experience is a major cause of fatal accidents in both skin and scuba diving. In one study (Webster 1966) analysis of national figures showed that only 17 of the 86 fatalities that year were experienced divers, while 69 did not have proper training or experience. Other studies have indicated the same situation.

A study of 21 deaths associated with scuba diving in Michigan during the years 1956–1965 indicated that only 6 of the 21 had formal diving training with greater than 2 yr of experience. Of the remaining 15 inexperienced divers, 3 were receiving instruction at the time of their accident and another 3 died on their first attempt at diving. Of the 21 deaths, 18 occurred in water not over 25 ft deep (Denney and Read 1965).

Miles (1967) reports that of 200 accidents investigated, 150 would probably not have occurred if accepted practice had been observed. Predisposing factors were inadequate safety precautions, inadequate training, and hazardous diving. All of these are related to faulty training and lack of proper experience. The medical profession and the diving community must warn the public of the dangers.

Panic, followed by a loss of control, is the most common cause of accidents and of death for the diver. It has been estimated that for the diver who has had the usual scuba training at least 10–12 open-sea dives are needed before he is comfortable. It is believed (Egstrom and Bachrach 1971) that some, at least, of this necessary self-reliance can be achieved during the training program, if tasks are programmed from simple to complex and if they are repeated enough times so that the response becomes automatic. Learning should become established under conditions as closely related as possible to the conditions under which the response will be required. For example, working under water while holding the breath is a first task. When the diver becomes confident of his ability under this condition, he is asked to use his breath to inflate a safety vest. He cannot believe it is possible, but he will have enough air left for a safe return to the surface.

2. *Productivity*

All of the studies indicate that training and experience make for a marked increase in work productivity under water. Weltman *et al.* (1969) studied work performance improvement of diver teams on a complex assembly task and found that novices were about 60% slower than the experienced teams in the tank. When the experiment was moved to the ocean, the novices' times were increased by another 26%. The experts' times did not change. According to Bowen and Hale (1970), *activity analysis* revealed why the experienced divers were doing better. The most significant finding was that the advantage of the experienced divers was not in working faster but in spending less time doing nonproductive activities, such as checking instructions, communications, in idle observations, etc. They were at ease in this environment and thus could do their best work.

Visual orientation in experienced and novice divers is discussed by Ross (1971). Typically experienced divers are better than novices at maintaining their orientation to the vertical and/or horizontal, they judge size and distance of objects better, and they suffer less error in judgment due to face mask distortion (see discussion of vision, Chapter V).

Although only two examples have been given, the literature makes the conclusion inescapable that productivity is increased with training and experience in diving in the open water.

3. Pleasure

Although personal satisfaction is perhaps not so important as safety and productivity, it is true that many novices give up because they find no real pleasure in diving. The problem usually is that they are not sufficiently at home in the water to accomplish complex tasks. A hobby such as underwater photography requires that a diver be so trained and experienced that underwater activity is like second nature, and all attention can be on taking the picture. It is the same with other hobbies—proficiency makes them fun.

References

ANONYMOUS. Aquamedics are taking a new dimension in problem solving, *The Galveston Daily News*, Section B, p. 1 (Nov. 5, 1972).

AQUADRO, C. F. Examination and selection of personnel for work in underwater environment. *J. Occup. Med.* 7:619–625 (Dec. 1965).

BEHNKE, A. R. Problems of living on the sea bed. In: Eaton, B., ed. *The Undersea Challenge. Proceedings of the Second World Congress of Undersea Activities*, pp. 63–68. London, The British Sub-Aqua Club (1963).

BEHNKE, A. R. Unpublished tables. Furnished by the author.

BIERSNER, R. J., E. K. E. GUNDERSON, and R. H. RAHE. Relationships of sports interests and smoking to physical fitness. *J. Sports Med.* 12:124–127 (June 1972).

BORNMANN, R. C. Decompression after saturation diving. In: Lambertsen, C. J., ed. *Underwater Physiology. Proceedings of the Third Symposium on Underwater Physiology, 23–25 March 1966, Washington, D. C.*, pp. 109–121. Baltimore, Md., Williams and Wilkins (1967).

BOVE, A. A. Physical fitness of sport dives. *J. Occup. Med.* 2(5):281–284 (1969).

BOWEN, H. M., and A. HALE. Study, feasibility of undersea salvage simulation. Technical report: NAVTRADEVCEN 69-C-0116-1. Dunlap and Associates, Inc. (1970).

BULENKOV, S. Ye, *et al. Manual of Scuba Diving.* USSR (11 April 1969) JPRS 47828 translation.

CABARROU, P. Selecting divers by brain impulses. In: Eaton, B., ed. *The Undersea Challenge. Proceedings of the Second World Congress of Undersea Activities*, pp. 68–70. London, The British Sub-Aqua Club (1963).

CHAN, G. L. A study to determine the demand for marine technicians in the State of California. Kentfield, Calif., College of Marin (1967).

CHAN, G. L. The education and training of marine technicians. Washington, D. C., American Association of Junior Colleges (1968).

CHAUDERON, J. Recent Moroccan legislation for protection of underwater sportsmen and compressed air workers. In: L'Huillier, J.-R., ed. Medecine de plongee. *Gaz. Hop.* 35:1023–1024 (Dec. 20, 1971).

CIRIA Underwater Engineering Group. The principles of safe diving practice. Report UR2, London, U. K., published by the Group (March 1972).

DAWSON, J. Safety in the sixth continent. In: Professional diving safety. Second annual symposium. New Orleans, November 1971. *Mar. Tech. Soc. J.* 6:28–31 (Jan/Feb. 1972).

DENNEY, M. K., and C. READ. Scuba-diving deaths in Michigan. *J.A.M.A.* **192**:120–122 (1965).

DEPPE, A. H. A study of psychological factors relevant to the selection of scuba divers. Ph. D. Thesis, Faculty of Arts, University of South Africa (Jan. 1971).

DUNHAM, R. M. Scientists in the Sea II (pamphlet). Tallahassee, Florida, Florida State Univ. (1972).

DUNHAM, R. M., and W. W. HAYTHORN. Selection and training in Scientist-in-the Sea Training Program. Presented at the annual convention of the American Psychological Association, Hawaii (1972).

EDMONDS, C., and R. L. THOMAS. Medical aspects of diving—Part 1. *Med. J. Aust.* **2**:1199–1201 (Nov. 18, 1972).

EGSTROM, G. H., and A. J. BACHRACH. Diver panic. *Skin Diver* **20**:35–38 (1971).

FLEMMING, N. C. Diver training and education. In: *Diving Applications in Marine Sciences, Research Seminar. Proceedings of a seminar, National Institute of Oceanography, Godalming, Surrey, U. K. December 1971*, pp. 136–167. Published by the Institute (1972).

GALLE, R. R., B. V. USTYUSHKIN, L. N. GAVRILOVA, and E. I. KHELEMSKIY. Evaluating vestibular tolerance. *Space Biol. Med.* **5**:99–107 (June 17, 1971) (JPRS 53, 388).

GEE, G. K. Man in a water environment. *Mar. Technol. Soc. J.* **5**:49–50 (Mar/Apr. 1971)

GEE, G. K. Educational programs for studying man's reactions in water. In: *Eighth Annual Conference of the Marine Technology Society, Sept. 11–13, 1972. Preprints*, pp. 173–201. Washington, D. C., Marine Technology Society (1972).

GILLULY, R. H. Tektite: Unique observations of men under stress. *Sci. News* **98**:400–401 (Nov. 21, 1970).

GODEFROID, A. Physio-pathology of submarine diving. Accidents in which underwater workers are exposed. *Arch. Mal. Prof. Med. Trav. Secur. Soc.* **26**(3):149–158 (1965).

HELMREICH, R. Human reactions to psychological stress. The TEKTITE II human behavior program. Social Psychological Lab. Univ. of Texas (1971) (AD 721364).

HELMREICH, R., R. BAKEMAN, and R. RADLOFF. The life history questionnaire: prediction of performance in Navy diver training. Austin, Tex., Univ. of Texas, Social Psychol. Lab., Rep. TR-18 (Sept. 1971) (AD 733444).

JENKINS, W. T. A summary of diving techniques used in polar regions. U. S. Nav. Coastal Syst. Lab., Prelim. Rep. under ONR Res. Proj. RF 51-523-101 (July 1973).

JOLY, R. Les problems d'aptitude a la plongee en scaphandre dans la Marine Nationale. *Med. Sport* **46**(3):164–169 (1972).

KENNY, J. E. *Business of Diving*. Gulf Publishing Co., Houston, Texas (1972).

KING, W. One clinic's criteria for employment in a compressed air environment. *J. Occup. Med.* **12**:113–116 (Apr. 1970).

LANPHIER, E., and J. DWYER, JR. Selection and training for the use of self-contained underwater breathing apparatus. U. S. Navy Exp. Diving Unit, Spec. Rep. 3-54 (1954).

LECKIE, B. The ultimate teaching aid. *Skin Diver* **17**:38–39, 60–61 (1968).

MATTHYS, H. Die medizinische Tauchtauglichkeitasabklarung. *Schweiz. Med. Wschr.* **100**:459–461 (1970).

MECKELNBURG, R. L. To dive or not to dive. *Delaware Med. J.* **44**:91–92 (Mar. 1972).

MENSH, I. N. La plongee et les plongeurs. *Eur. Sci. Notes* **24**(1):1–2 (Jan. 31, 1970).

MERER, P. Physiological aptitude to diving. *Rev. Physiol. Subaquatique Med. Hyperbare.* **1**:94–102 (Oct/Dec. 1968).

MIKALOW, A. How to select a commercial deep diving school. *Skin Diver* **21**:66–67 (Feb. 1972).

MILES, S. *Underwater medicine*. Second edition. London, Staples Press (1966).

MILES, S. Medical hazards of diving. In: *The Effects of Abnormal Physical Conditions of Work. Proceedings of a meeting of the British Occupational Hygiene Society, the Ergonomics Research Society and the Society of Occupational Medicine. 5–6 January 1967, London*, pp. 111–128. Edinburgh, E. and S. Livingston (1967).

MILLER, J. W. Psychophysiological aspects of deep saturation exposures in the sea. In: Lambertsen, C. J. ed. *Underwater Physiology. Proceedings of the Third Symposium on Underwater Physiology, 23–25 March 1966, Washington, D. C.*, pp. 122–127. Baltimore, Williams and Wilkins (1967).

MILLER, J. W., J. G. VAN DERWALKER, and R. A. WALLER. Tektite 2: Scientists-in-the-sea, Washington, D. C., U. S. Department of Interior (August 1971).

MORTIER, J.-P., and W. WELLENS. Recommendations concerning bathing and swimming. In: L'Huillier, J.-R., ed. Medecine de plongee. *Gaz. Hop.* **35**:1044 (Dec. 20, 1971).

NICHOLS, A. K. The personality of divers (and other sportsmen). In: Lythgoe, J. N., and E. A. Drew, eds. *Underwater Association Report 1969*, pp. 62–66. Guildford, U. K., Guildford Science and Technology Publications (n.d.).

NOAA. An analysis of the civil diving population of the United States. Unpublished (1974).

OCEANOGRAPHER OF THE NAVY. University Curricula in the Marine Sciences and Related Fields. Washington, D. C., U. S. Government Printing Office (1971) (Pamphlet #43).

PALEM, R. M., and L. PALEM. Que peut apporter l'EEG a l'examen medical des plongeurs? In: L'Huillier, J.-R., ed. Medecine de plongee. *Gaz. Hop.* **35**:1050–1052 (Dec. 20, 1971).

PATTE, M. Medecine de plongee sous-marine. *Brux. Med.* **53**:237–266 (Apr. 1973).

PENZIAS, W., and M. W. GOODMAN. *Man Beneath the Sea. A Review of Underwater Ocean Engineering.* New York, Wiley Interscience (1973).

REILLY, R. E., and B. J. CAMERON. An integrated measurement system for the study of human performance in the underwater environment. Falls Church, Virginia, BioTechnology, Inc., Rep. on Contract N00014-67-C-00410 (Dec. 1968).

ROSS, H. E. Personality of student divers. In: Lythgoe, J. N., and J. D. Woods, eds. *Underwater Association Report 1968*, pp. 59–62. Published by the Association (1968).

ROSS, H. E. Spatial perception under water. In: Woods, J. D., and J. N. Lythgoe, eds. *Underwater Science: An Introduction to Experiments by Divers.* pp 69–101. London, Oxford University Press (1971).

SALA MATAS, J. Importance de l'electroencephalogramme dans l'examen medical des plongeurs. In: L'Huillier, J.-R., ed. Medecine de plongee. *Gaz. Hop.* **35**:1056–1057 (Dec. 20, 1971).

SAWYER, R. N. Some aspects of scuba in college health. *J. Amer. Coll. Health Assoc.* **20**:323–327 (June 1972).

SIFFRE, M. *Hors du temps.* Rene Julliard, Paris (1963).

SOCIETY FOR UNDERWATER TECHNOLOGY. Report by the committee on diver training and performance standards. Guildford, U. K., IPC Science and Technology Press (1972).

STEPHAN, C. R. Education to prepare for man's progress into the sea. In: *Progress into the Sea, Transactions of the Symposium, 20–22 October 1969*, pp. 27–34, Washington, D. C., Marine Technology Society (1970).

STEWART, J. R. Scuba training program. In: *Institute of Marine Resources*, pp. 21. La Jolla, Calif., University of California (1972).

SUSIO. Scientists-in-the-Sea application form, St. Petersburg, Florida, State University System, Institute of Oceanography (1972).

TZIMOULIS, P. J. Scandal of the 70's. *Skin Diver* **20**:3–4 (Dec. 1971).

U. S. NAVY DEPARTMENT, *U. S. Navy Diving Manual*, Appendix F, Section F-B, 1970. Washington, D. C., the U. S. Navy Department (March 1970) (NAVSHIPS 0994-001-9010).

WEBSTER, D. P. Skin and scuba diving fatalities in the United States. Public Health Reports 81, Vol. 8 (August 1966).

WELTMAN, G., and G. H. EGSTROM. Personal autonomy of scuba diver trainees. *Res. Quart. Amer. Ass. Health Phys. Educ.* **40**:613–618 (Oct. 1969).

WELTMAN, G., G. H. EGSTROM, R. A. CHRISTIANSON, and T. P. CROOKS. Underwater work measurement techniques: 1968 studies. Report No. 69-19, BioTechnology Laboratory, Univ. of California, Los Angeles (1969).

WERTS, M., and C. W. SHILLING. Dysbaric Osteonecrosis. An annotated bibliography with preliminary analysis. Washington, D. C., The George Washington University Medical Center, Biol. Sci. Commun. Proj., Rep. GW-BSCP-72-10P (July 1972).

WISE, D. A. Constitutional factors in decompression sickness. U. S. Navy Exp. Diving Unit, Res. Rep. 2-63 (Apr. 26, 1962).

WISE, D. A. Aptitude selection standards for the U. S. Navy's first class diving course. Research Rep. 3-63, U. S. Navy Experimental Diving Unit, Washington, D. C. (1963).

YMCA. National YMCA Scuba diving program (pamphlet). Atlanta, Georgia, National YMCA (1973).

Appendix

Table A-1 [a]

Six units have been adopted to serve as the base for the International System:

Length . . . . meter
Mass . . . . kilogram
Time . . . . second
Electric current . . . . ampere
Thermodynamic temperature . . . . degree Kelvin
Light intensity . . . . candela

Some of the other more frequently used units of the SI and their symbols and, where applicable, their derivations are listed below.

Quantity	Unit	Symbol	Derivation
	Supplementary units		
Plane angle	radian	rad	
Solid angle	steradian	sr	
	Derived units		
Area	square meter	m^2	
Volume	cubic meter	m^3	
Frequency	hertz	Hz	(s^{-1})
Density	kilogram per cubic meter	kg/m^3	
Velocity	meter per second	m/s	
Angular velocity	radian per second	rad/s	
Acceleration	meter per second squared	m/s^2	
Angular acceleration	radian per second squared	rad/s^2	
Force	newton	N	$(kg \cdot m/s^2)$
Pressure	newton per square meter	N/m^2	
Kinematic viscosity	square meter per second	m^2/s	
Dynamic viscosity	newton-second per square meter	$N \cdot s/m^2$	
Work, energy, quantity of heat	joule	J	$(N \cdot m)$
Power	watt	W	(J/s)
Electric charge	coulomb	C	$(A \cdot s)$
Voltage, potential difference, electromotive force	volt	V	(W/A)
Electric field strength	volt per meter	V/m	
Electric resistance	ohm	Ω	(V/A)
Electric capacitance	farad	F	$(A \cdot s/V)$
Magnetic flux	weber	Wb	$(V \cdot s)$
Inductance	henry	H	$(V \cdot s/A)$
Magnetic flux density	tesla	T	(Wb/m^2)
Magnetic field strength	ampere per meter	A/m	
Magnetomotive force	ampere	A	
Flux of light	lumen	lm	$(cd \cdot sr)$
Luminance	candela per square meter	cd/m^2	
Illumination	lux	lx	(lm/m^2)

Table A-1^a—*Cont.*

Prefixes

The following prefixes, in combination with the basic unit names, provide the multiples and submultiples in the International System. For example, the unit name "meter," with the prefix "kilo" added, produces "kilometer," meaning "1000 meters."

Multiples and submultiples	Prefixes	Symbols	Pronunciations
10^{12}	tera	T	tĕr'à
10^9	giga	G	jĭ'gà
10^6	mega	M	mĕg'à
10^3	kilo	k	kĭl'ŏ
10^2	hecto	h	hĕk'tŏ
10	deka	da	dĕk'à
10^{-1}	deci	d	dĕs'ĭ
10^{-2}	centi	c	sĕn'tĭ
10^{-3}	milli	m	mĭl'ĭ
10^{-6}	micro	μ	mĭ'krŏ
10^{-9}	nano	n	năn'ŏ
10^{-12}	pico	p	pē'cŏ
10^{-15}	femto	f	fĕm'tŏ
10^{-18}	atto	a	ăt'tŏ

^{a} From CHISHOLM, L. J. Units of weight and measure, National Bureau of Standards Misc. Publ. 286. Department of Commerce, U.S. G.P.O. (May 1967).

Table A-2

Some Units and Their Symbols [a]

Unit	Symbol	Unit	Symbol	Unit	Symbol
acre	acre	fathom	fath	millimeter	mm
are	a	foot	ft	minim	minim
barrel	bbl	furlong	furlong	ounce	oz
board foot	fbm	gallon	gal	ounce, avoirdupois	oz avdp
bushel	bu	grain	grain	ounce, liquid	liq oz
carat	c	gram	g	ounce, troy	oz tr
Celsius, degree	°C	hectare	ha	peck	peck
centare	ca	hectogram	hg	pennyweight	dwt
centigram	cg	hectoliter	hl	pint, liquid	liq pt
centiliter	cl	hectometer	hm	pound	lb
centimeter	cm	hogshead	hhd	pound, avoirdupois	lb avdp
chain	ch	hundredweight	cwt	pound, troy	lb tr
cubic centimeter	cm³	inch	in	quart, liquid	liq qt
cubic decimeter	dm³	International		rod	rod
cubic dekameter	dam³	Nautical Mile	INM	second	s
cubic foot	ft³	Kelvin, degree	°K	square centimeter	cm²
cubic hectometer	hm³	kilogram	kg	square decimeter	dm²
cubic inch	in³	kiloliter	kl	square dekameter	dam²
cubic kilometer	km³	kilometer	km	square foot	ft²
cubic meter	m³	link	link	square hectometer	hm²
cubic mile	mi³	liquid	liq	square inch	in²
cubic millimeter	mm³	liter	liter	square kilometer	km²
cubic yard	yd³	meter	m	square meter	m²
decigram	dg	microgram	μg	square mile	mi²
deciliter	dl	microinch	μin	square millimeter	mm²
decimeter	dm	microliter	μl	square yard	yd²
dekagram	dag	micron	μm	stere	stere
dekaliter	dal	mile	mi	ton, long	long ton
dekameter	dam	milligram	mg	ton, metric	t
dram, avoirdupois	dr avdp	milliliter	ml	ton, short	short ton
				yard	yd

[a] From Chisholm (1967). (See footnote to Table A-1.)

Table A-3
Units of Pressure Conversion Table

From \ To	Atm abs	ATA	Bar	dyn/cm²
			Multiply by	
Atm abs		1.03323	1.01325	1.01325×10^6
ATA	0.967841		0.980665	9.80665×10^5
Bar	1.01325	0.980665		1×10^6
dyn/cm²	9.86923×10^{-7}	1.01972×10^{-6}	1×10^{-6}	
ft sea water	0.030238	0.031243	0.0306391	3.06391×10^4
ft pure water	0.029499	0.030480	0.0298898	2.98898×10^4
kg-f/cm²	0.967841	1.0	0.980665	9.80665
kg-f/m²	9.67841×10^{-5}	1×10^{-4}	9.80665×10^{-5}	98.0665
Meter sea water	0.099206	0.102503	0.100522	1.00522×10^5
Meter pure water	0.096784	0.10	0.0980665	9.80665×10^4
mm Hg	1.31579×10^{-3}	1.35951×10^{-3}	1.33323×10^{-3}	1.33323×10^3
Pa	9.86923×10^{-6}	1.01972×10^{-5}	1×10^{-5}	10.0
PSI	0.0680457	0.0703087	0.0689474	6.89474×10^4

From \ To	ft sea water	ft pure water	kg-f/cm²	kg-f/m²
			Multiply by	
Atm abs	33.071	33.8995	1.03323	1.03323×10^4
ATA	32.007	32.808	1.00	10,000
Bar	3.2638	33.4562	1.01972	1.01972×10^4
dyn/cm²	3.26380×10^{-5}	3.34562×10^{-5}	1.01972×10^{-6}	1.01972×10^{-2}
ft sea water		1.0250	3.12427×10^{-2}	312.427
ft pure water	0.975610		0.03048	304.8
kg-f/cm²	32.007	32.808		1×10^4
kg-f/m²	3.2007×10^{-3}	3.2808×10^{-3}	1×10^{-4}	
Meter sea water	3.28085	3.36305	0.102503	1025.03
Meter pure water	3.2007	3.2808	0.10	1000
mm Hg	4.35145×10^{-2}	4.46046×10^{-2}	1.35951×10^{-2}	13.5951
Pa	3.26380×10^{-4}	3.34562×10^{-4}	1.01972	0.101972
PSI	2.25031	2.30672	7.03068×10^2	703.068

From \ To	Meter sea water	Meter pure water	mm Hg	Pa	PSI
			Multiply by		
Atm abs	10.080	10.3323	760.0	1.01325×10^5	14.6960
ATA	9.7559	10.00	735.56	9.80665×10^4	14.223
Bar	9.94810	10.1972	750.06	1×10^5	14.5038
dyn/cm²	9.94810×10^{-6}	1.01972×10^{-5}	7.50058×10^{-4}	0.10	1.45038×10^{-5}
ft sea water	0.30480	0.312427	22.9809	3.06391×10^3	0.444377
ft pure water	0.297350	0.3048	22.4192	2.98898×10^3	0.433517
kg-f/cm²	9.7559	10.0	735.557	9.80665×10^4	14.223
kg-f/m²	9.7559×10^{-4}	1×10^{-3}	0.073557	9.80665	1.4223×10^{-3}
Meter sea water		1.02503	75.3968	1.00522×10^4	1.45793
Meter pure water	0.97559		73.556	9.80665×10^3	1.4223
mm Hg	1.32632×10^{-2}	1.35951×10^{-2}		1.33323×10^{-2}	
Pa	9.94810×10^{-5}	1.01972×10^{-4}	7.50058×10^{-3}		1.45038×10^{-4}
PSI	0.685896	0.703068	51.7147	6.89474×10^3	

Table A-4

Length—International Nautical Miles and Kilometers [a]

Basic relation: International Nautical Mile = 1.852 Kilometers

Int. nautical miles	Kilometers	Int. nautical miles	Kilometers	Kilometers	Int. nautical miles	Kilometers	Int. nautical miles
0		50	92.600	0		50	26.9978
1	1.852	1	94.452	1	0.5400	1	27.5378
2	3.704	2	96.304	2	1.0799	2	28.0778
3	5.556	3	98.156	3	1.6199	3	28.6177
4	7.408	4	100.008	4	2.1598	4	29.1577
5	9.260	5	101.860	5	2.6998	5	29.6976
6	11.112	6	103.712	6	3.2397	6	30.2376
7	12.964	7	105.564	7	3.7797	7	30.7775
8	14.816	8	107.416	8	4.3197	8	31.3175
9	16.668	9	109.268	9	4.8596	9	31.8575
10	18.520	60	111.120	10	5.3996	60	32.3974
1	20.372	1	112.972	1	5.9395	1	32.9374
2	22.224	2	114.824	2	6.4795	2	33.4773
3	24.076	3	116.676	3	7.0194	3	34.0173
4	25.928	4	118.528	4	7.5594	4	34.5572
5	27.780	5	120.380	5	8.0994	5	35.0972
6	29.632	6	122.232	6	8.6393	6	35.6371
7	31.484	7	124.084	7	9.1793	7	36.1771
8	33.336	8	125.936	8	9.7192	8	36.7171
9	35.188	9	127.788	9	10.2592	9	37.2570
20	37.040	70	129.640	20	10.7991	70	37.7970
1	38.892	1	131.492	1	11.3391	1	38.3369
2	40.744	2	133.344	2	11.8790	2	38.8769
3	42.596	3	135.196	3	12.4190	3	39.4168
4	44.448	4	137.048	4	12.9590	4	39.9568
5	46.300	5	138.900	5	13.4989	5	40.4968
6	48.152	6	140.752	6	14.0389	6	41.0367
7	50.004	7	142.604	7	14.5788	7	41.5767
8	51.856	8	144.456	8	15.1188	8	42.1166
9	53.708	9	146.308	9	15.6587	9	42.6566
30	55.560	80	148.160	30	16.1987	80	43.1965
1	57.412	1	150.012	1	16.7387	1	43.7365
2	59.264	2	151.864	2	17.2786	2	44.2765
3	61.116	3	153.716	3	17.8186	3	44.8164
4	62.968	4	155.568	4	18.3585	4	45.3564
5	64.820	5	157.420	5	18.8985	5	45.8963
6	66.672	6	159.272	6	19.4384	6	46.4363
7	68.524	7	161.124	7	19.9784	7	46.9762
8	70.376	8	162.976	8	20.5184	8	47.5162
9	72.228	9	164.828	9	21.0583	9	48.0562
40	74.080	90	166.680	40	21.5983	90	48.5961
1	75.932	1	168.532	1	22.1382	1	49.1361
2	77.784	2	170.384	2	22.6782	2	49.6760
3	79.636	3	172.236	3	23.2181	3	50.2160
4	81.488	4	174.088	4	23.7581	4	50.7559
5	83.340	5	175.940	5	24.2981	5	51.2959
6	85.192	6	177.792	6	24.8380	6	51.8359
7	87.044	7	179.644	7	25.3780	7	52.3758
8	88.896	8	181.496	8	25.9179	8	52.9158
9	90.748	9	183.348	9	26.4579	9	53.4557
		100	185.200			100	53.9957

[a] From Chisholm (1967). (See footnote to Table A-1.)

Table A-5
Conversion Factors Relating to Thermophysiology [a]

Temperature

	1 C or K	1.80 F or R
	C = 273 = K	F = 459 = R
Thus:	0 C	32 F
	10	50
	20	68
	30	86
	40	104

Power

1 kilogram calorie/hr = 3.97 BTU/hr
= 1.16 watts

Work

1 kilogram calorie = 3.97 BTU
= 1.16 watt-hr

Insulation

1 CLO = $0.18°C$ per kcal/m²·hr⁻¹
= $0.88°F$ per BTU/ft²·hr⁻¹
= that amount of insulation which will transfer 5.56 kcal/m²·hr⁻¹ per °C
= 1.14 BTU/ft²·hr⁻¹ per °F

[a] Adapted from BECKMAN, E. L. Thermal protective suits for underwater swimmers. *Milit. Med.* **132**:195–209 (Mar. 1967). Used with permission of the author.

Table A-6
Units of Length—Conversion Factors [a,b]

To convert from centimeters	
To	Multiply by
Inches	0.393 700 8
Feet	0.032 808 40
Yards	0.010 936 13
Meters	**0.01**

To convert from meters	
To	Multiply by
Inches	39.370 08
Feet	3.280 840
Yards	1.093 613
Miles	0.000 621 37
Millimeters	**1 000**
Centimeters..........	**100**
Kilometers	**0.001**

To convert from inches	
To	Multiply by
Feet	0.083 333 33
Yards	0.027 777 78
Centimeters	**2.54**
Meters	**0.025 4**

To convert from feet	
To	Multiply by
Inches	**12**
Yards	0.333 333 3
Miles...................	0.000 189 39
Centimeters	**30.48**
Meters	0.304 8
Kilometers	0.000 304 8

In using conversion factors, it is possible to perform division as well as the multiplication process shown here. Division may be particularly advantageous where more than the significant figures published here are required. Division may be performed in lieu of multiplication by using the reciprocal of any indicated multiplier as divisor. For example, to convert from centimeters to inches by division, refer to the table headed "To convert from inches" and use the factor listed at "centimeters" (2.54) as divisor.

To convert from yards	
To	Multiply by
Inches	**36**
Feet	**3**
Miles...................	0.000 568 18
Centimeters	**91.44**
Meters	0.914 4

To convert from miles	
To	Multiply by
Inches	**63 360**
Feet	**5 280**
Yards	**1 760**
Centimeters	**160 934.4**
Meters	1 609.344
Kilometers	1.609 344

[a] From Chisholm (1967). (See footnote to Table A-1.)
[b] All boldface figures are exact; the others generally are given to seven significant figures.

Table A-7
Units of Capacity or Volume, Liquid Measure—Conversion Factors [a]

To convert from milliliters	
To	Multiply by
Minims................	16.230 73
Liquid ounces	0.033 814 02
Gills	0.008 453 5
Liquid pints............	0.002 113 4
Liquid quarts...........	0.001 056 7
Gallons................	0.000 264 17
Cubic inches	0.061 023 74
Liters	**0.001**

To convert from liters	
To	Multiply by
Liquid ounces........	33.814 02
Gills.................	8.453 506
Liquid pints	2.113 376
Liquid quarts	1.056 688
Gallons	0.264 172 05
Cubic inches	61.023 74
Cubic feet	0.035 314 67
Milliliters	**1 000**
Cubic meters	**0.001**
Cubic yards	0.001 307 95

To convert from cubic meters	
To	Multiply by
Gallons	264.172 05
Cubic inches	61 023.74
Cubic feet	35.314 67
Liters	**1 000**
Cubic yards	1.307 950 6

To convert from liquid quarts	
To	Multiply by
Minims	**15 360**
Liquid ounces.......	**32**
Gills...............	**8**
Liquid pints	**2**
Gallons	**0.25**
Cubic inches	**57.75**
Cubic feet	0.033 420 14
Milliliters	**946.352 946**
Liters	**0.946 352 946**

To convert from liquid ounces	
To	Multiply by
Minims...............	**480**
Gills	**0.25**
Liquid pints...........	**0.062 5**
Liquid quarts..........	**0.031 25**
Gallons...............	**0.007 812 5**
Cubic inches	1.804 687 5
Cubic feet.............	0.001 044 38
Milliliters	29.573 53
Liters	0.029 573 53

To convert from cubic feet	
To	Multiply by
Liquid ounces	957.506 5
Gills	239.376 6
Liquid pints	59.844 16
Liquid quarts	29.922 08
Gallons..........	7.480 519
Cubic inches	**1 728**
Liters	**28.316 846 592**
Cubic meters	**0.028 316 846 592**
Cubic yards	0.037 037 04

To convert from cubic inches	
To	Multiply by
Minims	265.974 0
Liquid ounces......	0.554 112 6
Gills..............	0.138 528 1
Liquid pints	0.034 632 03
Liquid quarts	0.017 316 02
Gallons	0.004 329 0
Cubic feet	0.000 578 7
Milliliters	**16.387 064**
Liters	**0.016 387 064**
Cubic meters	**0.000 016 387 064**
Cubic yards	0.000 021 43

To convert from cubic yards	
To	Multiply by
Gallons..........	201.974 0
Cubic inches	**46 656**
Cubic feet	**27**
Liters	**764.554 857 984**
Cubic meters	**0.764 554 857 984**

[a] From Chisholm (1967). (See footnote to Table A-1.)

Index

899